FOREWORD

It is generally accepted that photosynthesis, whereby solar is transformed into geobiochemical energy, is one of the most important processes in the biosphere. When we realize that the human population, most of which is starving, will double in 20-30 years, and that photosynthesis is in practice the only source (direct or indirect) of food not only for men, but for all living beings, its importance becomes even plainer.

This fact was recognized from the outset of the International Biological Programme, when photosynthesis was selected as one of the principal problems to be concentrated on. The significance of the IBP in initiating a broader-based approach and worldwide collaboration need not be emphasized here; it is merely necessary to consider the rapid increase in the number of papers on photosynthesis since its inception.

It is of fundamental importance for scientists working on photosynthesis to get information on what has been done in recent years in this discipline. Thus as complete as possible a bibliography is badly needed. The complexity of this problem requires also that the progress of ecology, chemistry and physics, as far as they concern photosynthesis, be followed. Such a bibliography, covering at least the last ten years, will be indispensable for final synthesis of the results of the IBP photosynthetic programme.

The "Photosynthesis Bibliography", prepared by Z. ŠESTÁK and J. ČATSKÝ, fills a gap in this information and will certainly be appreciated by all scientists working in photosynthesis.

Professor Dr. EMIL HADAČ
Chairman of the
Czechoslovak National Committee
of IBP

Photosynthesis Bibliography

volume 1 1966/1970

References no. 1 - 5620 / AAS - MYR

Editors Z. Šesták & J. Čatský

Dr. W. Junk b.v. - Publishers - The Hague 1974

ISBN-13: 978-90-6193-039-6   e-ISBN-13: 978-94-010-2329-0
DOI: 10.1007/978-94-010-2329-0

© Dr. W. Junk b.v., Publishers, The Hague 1974
Printed in the Netherlands by Offsetdrukkerij Pasmans, The Hague

PREFACE

The need for a specialized source-book of references to the voluminous and continuously increasing literature in all fields of photosynthesis research has recently been intensified by the discontinuation of two of the most important photosynthesis bibliographies:

The Russian bibliography prepared by N.V. ARTSIKHOVSKAYA: Fotosintez. Ukazatel' Otechestvennoĭ i Inostrannoĭ Literatury (Izd. Moskovskogo Universiteta, Moskva, 1961, 1968, 1970), covers in two volumes (in five issues plus a cumulative subject index) papers in all fields of photosynthesis published all over the world from 1951 to 1962. The references in this bibliography contain all bibliographic data and are reliable but, unfortunately, the individual volumes appeared with a considerable delay.

The American bibliography prepared in the Charles F. Kettering Research Laboratory at Yellow Springs, Ohio and published in mimeographed form and later printed in the journal "Photochemistry and Photobiology" (Parts I to XXIII in Volumes 3 to 13), has been focussed mainly on papers of biochemical and biophysical character. Although this bibliography has large gaps in the East European and Asian literature and the references are often incomplete, the staff members of the Laboratory, headed by L.P. VERNON, contributed significantly to the extension of photosynthetic information from 1963 to 1971.

At present, two bibliographies with different scopes are published:

The "Bibliography: Photosynthesis and Related Subjects" prepared at the University of Tokyo, Japan, is issued monthly and considered rather as a current information source suitable mainly for preparing card-indexes.

The bibliography published regularly in the journal "Photosynthetica" presents only references of review articles and methodological papers.

This evident lack of a comprehensive, well indexed guide to current literature dealing with the physiology, ecology, chemistry and physics of photosynthesis has encouraged us to launch this volume. It covers the years 1966 to 1970, which have formed the basis of present research. Subsequent annual volumes are being prepared and will appear in the year following the year they cover.

The bibliography includes papers in all fields of photosynthesis research - from studies of model biochemical and biophysical systems of the photosynthesis mechanism to primary production studied by so-called growth analysis. In addition to papers devoted entirely to photosynthesis, papers on other topics are included if they contain data on photosynthetic activity, photorespiration, chloroplast structure, chlorophyll and carotenoid synthesis and destruction, etc., or if they contain valuable methodological information (measurement of selected environmental factors, leaf area, etc.). In many branches it has been very difficult to define the limits of interest for photosynthesis researchers. This problem has arisen e.g. in topics dealing with the transport of gases, where - in addition to the papers on CO_2 transfer - some papers on water vapour transfer are included, these being of general application. On the other hand, many papers dealing with the anatomy and physiology of stomata have been omitted.

To maximize the value of the bibliography the references are arranged alphabetically by authors' names, and each volume is provided with three indexes. The author index to this volume contains all names of authors and editors over a five-year period; subsequent annual volumes - Vol. 2 to Vol. 5 - will include only the names of co-authors and editors.

The subject index covers only primary items chosen according to their interest for photosynthesis researchers. In this volume its preparation was based mainly on the paper titles.

In the index of plant names, only important crop plants and selected plant types and groups are indexed.

Cumulative indexes will accompany every fifth volume, i.e. Vols. 1, 6, 11, etc.

We have tried to cover fully the relevant papers which have appeared in the most important scientific periodicals and books. Articles published in local journals, mimeographed booklets, abstracts of theses and of symposia contributions, etc., were chosen mostly from reprints and lists of publications received direct from the authors. Material which reached us after the deadline for this volume will be included in the following volume and marked with an asterisk.

Since some 2000 relevant papers are currently published every year and included in this bibliography, and since the majority of citations have been checked with the originals, collecting and preparing for publication such a large amount of material would have been impossible without the collaboration of the authors of the relevant publications. The courtesy of those authors who have already supplied us with reprints and lists of their publications is highly appreciated.

We acknowledge with thanks the cooperation of our colleagues from the Department of Physiology of Photosynthesis and Water Relations of the Institute of Experimental Botany of the Czechoslovak Academy of Sciences in Prague, especially Mrs. DRAHOMÍRA TĚŽKÁ who helped in preparing the card material, retyping the manuscript and preparing the author index, Dr. INGRID TICHÁ who helped with completing and checking the references and reading the proofs, and Mr. PETR ZÁZVORKA who supplied us with rare periodicals. In addition, the librarian of the Institute, Mrs.ALENA ŠTĚTINOVÁ, helped us with checking the references.

Dr. Z. ŠESTÁK and Dr. J. ČATSKÝ
Institute of Experimental Botany
Czechoslovak Academy of Sciences
Flemingovo nám. 2

160 00 PRAHA 6

Czechoslovakia

INSTRUCTIONS FOR USE

All references are arranged alphabetically according to authors' names and year of publication. They are numbered and these numbers are used in the indexes. In case of a book title, the number is preceded by B.

The references contain the original unabbreviated title of the paper (book). English, French and German titles are cited in the original language. Titles in other languages are supplemented by an English translation (sometimes using the title of the respective English abstract or a shortened title, "deadweight" words being omitted). Titles of Japanese, Chinese, etc., papers are given in English translation only. The journals' names are abbreviated mainly according to the "Style Manual for Biological Journals" (Second Edition, Amer. Institute of Biological Sciences, Washington, D.C., 1964), e.g.:

Abhandlungen	central	imperial	Quarterly
Abstract	chemical	inorganic	Radiation
Abteilung	Chemistry	Institute	Radiobiology
Academy	chinese	international	Report
Acta	Chromatography	Investigation	Research
Africa	Communication	italian	Review
agricultural	comparative	Jahrbuch	royal
Agriculture	Comptes rendus	japanese	russian
Agronomy	Conference	Japan	russkiĭ
Akademie	Congress	jewish	Science
Algology	Contribution	Journal	Series
allgemeine	Cytochemistry	Klasse	Society
american	Cytology	Laboratory	sovetskiĭ
America	czechoslovak	Magazin	soviet
analytical	Department	Mathematics	special
Anatomy	Deutschland	miscellaneous	SSSR
angewandte	Disease	molecular	Station
Annals	Dissertation	Monograph	Supplement
annual	Doklady	moskovskiĭ	Survey
anorganisch	Ecology	Mycology	Symposium
applied	Embryology	national	Technology
Arbeit	Encyclopedia	natural	Transactions
Archiv	Engineer	Naturforschung	Travail(-aux)
Atmosphere	Enzymology	neerlandicus	tropical
atomic	experimental	Netherlands	ukrainian
Australia	Experiment	New Zealand	UK
Beiheft	Faculty	nuclear	UN
Belgique	Federation	Optics	US, USA
Bericht	Forestry	organic	USSR
biochemical	Forschung	original	University
Biochemistry	France	Pathology	Virology
biokhimicheskiĭ	Gazette	Pflanzen-	Virusforschung
Biokhimiya	general	Philosophy	Volume
biological	genetical	physical	Weekblad
Biology	Genetics	Physics	Wissenschaft
biophysical	Gesellschaft	physiological	Wochenschrift
Biophysics	Giornale	Physiology	Zeitschrift
Bodenkunde	helveticus	Phytopathology	Zeitung
bolgarskiĭ	Hereditas	Plant(-arum)	Zentralblatt
botanical	Histochemistry	Proceedings	Zhurnal
Botany	Histology	Publication	
british	Horticulture	Publishers	
Canada	Husbandry	quantitative	

The numbers at the end of each reference for a journal article denote: volume (issue): first page - last page, year of publication. The number of the issue is given only for journals where each issue is paginated separately.

Book titles are cited according to the title page, not to the book jacket or cover (if the names of the editors are not given on the title page, they are not cited in the reference). The publishing house, place and year of publication are included.

Parentheses at the end of the reference give bibliographic details and explanations of the contents not given in the original. The following abbreviations are used most often:

ab	abstract	Latv.	Latvian
Belorus.	Belorussian	Lithu.	Lithuanian
Bulg.	Bulgarian	Norweg.	Norwegian
Car	carotenoids	PC	paper chromatography
CC	column chromatography	PhAR	photosynthetically active radiation
Chin.	Chinese		
Chl	chlorophyll	Pol.	Polish
Croat.	Croatian	Ps	photosynthesis
E	English	R	Russian
F	French	Rum.	Rumanian
G	German	Span.	Spanish
GC	gas chromatography	Swed.	Swedish
Hung.	Hungarian	TLC	thin-layer chromatography
IRGA	infra-red gas analyser	Tr	transpiration
Ital.	Italian	Ukr.	Ukrainian
Jap.	Japanese	Uz.	Uzbeg

The transliteration of Cyrillic characters is in accordance with the BSI-ASA/SC-Z39 draft table, i.e.:

Translit.	Cyrill.	Translit.	Cyrill.
a	а	r	р
b	б	s	с
ch	ч	sh	ш
d	д	shch	щ
e	в	t	т
f	ф	ts	ц
g	г	u	у
i	и	v	в
ī	й	y	ы
k	к	ya	я
kh	х	yu	ю
l	л	z	з
m	м	zh	ж
n	н	"	ъ
o	о	'	ь
p	п		

Several exceptions apply for Ukrainian and Belorussian:

	Translit.		Cyrill.
Ukrainian:	y	=	и
	i	=	i
	ī	=	ī
Belorussian:	ū	=	у

Authors' names are given in the spelling used in the original paper. If this spelling does not correspond to the original spelling used by the author (e.g. Russian papers of English authors), one spelling is referred to the other in the Author index.

Printers' errors in the original papers are marked by underlining the respective words (letters).

1 - AASE, J.K., KEMPER, W.D., DANIELSON, R.E.: Response of corn to white and black ground covers. - Agron. J. *60*: 234-236, 1968. [Chl.]

2 - AASEN, A.J., EIMHJELLEN, K.E., LIAAEN JENSEN, S.: An extreme source of β-carotene. - Acta chem. scand. *23*: 2544-2545, 1969. [Isolation from *Dunaliella salina.*]

3 - AASEN, A.J., LIAAEN JENSEN, S.: Bacterial carotenoids XXI. Isolation and synthesis of 3,4,3',4'-tetrahydro-spirilloxanthin. - Acta chem. scand. *21*: 371-377, 1967.

4 - AASEN, A.J., LIAAEN JENSEN, S.: Bacterial carotenoids XXIII. The carotenoids of *Thiorhodaceae*. 6. Total synthesis of okenone and related compounds. - Acta chem. scand. *21*: 970-982, 1967.

5 - AASEN, A.J., LIAAEN JENSEN, S.: Bacterial carotenoids XXIV. Carotenoids of *Thiorhodaceae*. 7. Cross-conjugated carotenals. - Acta chem. scand. *21*: 2185-2204, 1967.

6 - ABAEVA, S.S., STESNYAGINA, T.Ya., KHODZHAEV, D.K.: Znachenie nekotorykh mikroelementov v prisposoblenii khlopchatnika k neblagopriyatnym faktoram sredy. [Significance of some microelements in adaptation of cotton to unfavourable environmental conditions.] - Trudy bot. Inst. Im. V.L. Komarova Akad. Nauk SSSR, Ser. 4 - eksp. Bot. *19* (Ekol.-fiziol. Osoben. introduts. Rast.): 140-155, 1967. [Ps, chl; in R.]

7 - ABDEL-WAHAB, M.F., HASSAN, H.M., SOBHY, C.M.: Study of the dependence of photosynthetic yields on the water and mineral supply. Part 3: Effect of mineral deficiency on photosynthesis using double labelling technique. - Isotopenpraxis *4*: 112-117, 1968.

8 - ABDULLAEV, Kh.A., TAGEEVA, S.V., KAS'YANENKO, A.G., USMANOV, P.D., NASYROV, Yu.S.: Biokhimicheskiĭ mutant *Arabidopsis thaliana* (L.) HEYNH., vosstanavlivayushchiĭ ul'trastrukturu khloroplastov pod deĭstviem ekzogennogo leĭtsina. [Biochemical mutant of *Arabidopsis thaliana* (L.) HEYNH. reducing chloroplast ultrastructure under the influence of exogenous leucine.] - Dokl. Akad. Nauk Tadzh. SSR *13* (6): 47-50, 1970. [In R.]

9 - ABDULLAEVA, S.K., GILLER, Yu.E., SAPOZHNIKOV, D.I.: O zakonomernostyakh vklyucheniya feofitina v iskusstvennyĭ pigment-belkovyĭ kompleks. [Regularities of incorporation of pheophytin in an artificial pigment-protein complex.] - Dokl. Akad. Nauk Tadzh. SSR *13* (12): 61-64, 1970. [In R.]

10 - ABDULLAEVA, S.K., KASYMOVA, R.F.: O zakonomernostyakh svyazyvaniya khlorofilla v iskusstvennom pigment-belkovom komplekse. [Regularities of chlorophyll bounds in an artificial pigment-protein complex.] - In: Tezisy Dokladov Molodykh Uchenykh na Vtoroĭ Respublikanskoĭ Nauchnoĭ Konferentsii, Posvyashchennoĭ 50-Letiyu VLKSM. P. 166. Dushanbe 1968. [In R.]

11 - ABDULLAEVA, T.M.: Vliyanie fiziologicheski aktivnykh veshchestv na korneobrazovanie u listovykh cherenkov i soderzhanie v nikh pigmentov. [Effect of physiologically active substances on root formation in leaf petioles and their pigment content.] - Bot. Zh. *52*: 999-1003, 1967. [In R.]

12 - ABDULLAEVA, T.M., YUSUFOV, A.G.: Soderzhanie pigmentov plastid i nukleinovykh kislot v ukorenennykh list'yakh. [Amount of plastid pigments and nucleic acids in rooted leaves.] - In: Trudy bot. Inst. Im. V.L. Komarova Akad. Nauk SSSR, Ser. 4 - eksp. Bot. *19* (Ekol.-fiziol.Osoben. introduts. Rast.): 169-173, 1967. [In R.]

13 - ABDURAKHMANOVA, Z.N.: Metabolizm ugleroda C^{14} v svyazi s obezvozhivaniem lista khlopchatnika. [Metabolism of carbon ^{14}C in connection with the dehydration of cotton plant leaf.] - In: Issledovaniya po Fotosintezu. Pp. 59-66. Akad. Nauk Tadzh. SSR, Dushanbe 1967. [In R, ab: Tadzh.]

14 - ABDURAKHMANOVA, Z.N., BELAN, N.F., KHODZHAEVA, R., NASYROV, Yu.S.: Fotosintez i assimilyatsiya CO$_2$ u khlorofil'nykh mutantov *Arabidopsis thaliana*. [Photosynthesis and CO$_2$ assimilation in chlorophyll mutants of *Arabidopsis thaliana*.] - In: Tezisy Dokladov Vtorogo Vsesoyuznogo Biokhimicheskogo S"ezda, Sektsiya "Problemy Fotosinteza". Pp. 24-25. FAN Uzb. SSR, Tashkent 1969. [In R.]

15 - ABDURAKHMANOVA, Z.N., BELAN, N.F., LEBEDEVA, G.P.: Assimilyatsiya uglekisloty i
     usloviya ee osushchestvleniya. [Carbon dioxide assimilation and conditions of
     its realization.] - In: Trudy I. Konferentsii Biokhimikov Respublik Srednei
     Azii i Kazakhstana. Pp. 94-97. FAN, Tashkent 1967. [In R.]

16 - ABDURAKHMANOVA, Z.N., KHODZHAEVA, R., BELAN, N.F., NASYROV, Yu.S.: Regulator-
     naya rol' sootnosheniya putei assimilyatsii CO_2 v chasy maksimuma i depressii
     fotosinteza. [A regulatory role of the ratio of CO_2 assimilation pathways
     during maximum and depressed photosynthesis.] - In: Fotosintez i Ispol'zovanie
     Energii Solnechnoi Radiatsii. P. 37. Dushanbe 1967. [In R.]

17 - ABELES, F.B., HOLM, R.E., GAHAGAN, H.E.: Abscission: the role of aging. - Plant
     Physiol. 42: 1351-1356, 1967. [Chl.]

18 - ABILOV, Z.K.: Issledovanie formirovaniya razlichnykh form khlorofilla "a" v
     khode zeleneniya rastenii metodami izmereniya spektrov nizkotemperaturnoi
     lyuminestsentsii i pogloshcheniya. [Study of the formation of different forms
     of chlorophyll a during greening of plants by methods of measuring spectra of
     low temperature luminescence and absorbance.]-In: Materialy nauch.-teoret.
     Konf. molod. Uchen., Ser. biol. Nauk. Pp. 7-9. Baku 1970. [In R.]

19 - ABILOV, Z.K., GASANOV, R.A., KURBANOVA, I.M.: Issledovanie spektrov fluorest-
     sentsii nativnykh, ekstragirovannykh i rekonstruirovannykh khloroplastov pri
     20 °C i 77 °K. [Studies of fluorescence spectra of natural, extracted and re-
     constructed chloroplasts at 20 °C and 77 °K.] - In: Materialy yubil. nauch.
     Konf. AzNIIZ, posvyashch. 100-let. so Dnya Rozhd. V.I. Lenina i 50-let. Ustan.
     sov. Vlasti v Azerbaidzhane. Pp. 132-133. Baku 1969. [In R.]

20 - ABILOV, Z.K., GASANOV, R.A., LITVIN, F.F.: Universal'naya ustanovka dlya izme-
     reniya lyuminestsentnykh kharakteristik fotosinteziruyushchikh organizmov.
     [Universal device for measuring luminescence characteristics of photosynthesi-
     zing organisms.] - Materialy I. zakavk. Konf. po Fiziol. Rast. Pp. 39-41.
     Izdat. Akad. Nauk Azerb. SSR, Baku 1967. [In R.]

21 - ABOU KHALED, A.: Optical properties of leaves in relation to their energy-
     balance, photosynthesis and water use efficiency. - Diss. Abstr. 28: 410-B,
     1967.

22 - ABOU-KHALED, A., HAGAN, R.M., DAVENPORT, D.C.: Effects of kaolinite as a re-
     flective antitranspirant on leaf temperature, transpiration, photosynthesis,
     and water-use efficiency. - Water Resources Res. 6: 280-289, 1970.

23 - ABRAHAMSEN, M., MAYER, A.M.: Photosynthetic and dark fixation of $^{14}CO_2$ in de-
     tached soybean cotyledons. - Physiol. Plant. 20: 1-5, 1967.

24 - von ABRAMS, G.J., PRATT, H.K.: Interaction of naphthalene-acetic acid and kine-
     tin in the senescence of detached leaves. - Plant Physiol. 41: 1525 -1530, 1966.
     [Chl.]

25 - von ABRAMS, G.J., PRATT, H.K.: The effect of kinetin and naphthalenacetic acid
     upon localized accumulation as related to senescence in detached leaves. -
     Planta 76: 306-308, 1967. [Chl.]

26 - von ABRAMS, G.J., PRATT, H.K.: Effect of the kinetin-naphthaleneacetic acid
     interaction upon total RNA and protein in senescing detached leaves. - Plant
     Physiol. 43: 1271-1278, 1968. [Chl.]

27 - ABUTALYBOV, M.G., RAKHMANOVA, S.A., ALIEV, D.A.: [Effect of mineral elements on
     the photosynthetic rate in Solanum melongena L.] - Dokl. Akad. Nauk Azerb. SSR
     23 (7): 72-75, 1967. [In Azerb., ab: R.]

28 - ABUTALYBOV, M.G., RAKHMANOVA, S.A., ALIEVA, D.A.: [Effect of trace elements on
     photosynthetic rate in eggplant.] - Dokl. Akad. Nauk Azerb. SSR 23 (9): 63-66,
     1967. [In Azerb., ab: R.]

29 - ACOCK, B., THORNLEY, J.H.M., WARREN WILSON, J.: Spatial variation of light in
     the canopy. - In: Prediction and Measurement of Photosynthetic Productivity.
     Pp. 91-102. PUDOC, Wageningen 1970.

30 - ADACHI, T., KATAYAMA, Y.: [Induction of callus tissue and its chlorophyll
     formation in leaf tissue cultures of Sansevieria.] - Bull. Fac. Agr. Univ.

Miyazaki *16:* 68-76, 1969. [In Jap., ab: E.]

31 - ADAMCOVÁ-BÍNOVÁ, J.: PP/IBP - Initial level experiments in South Bohemia. - Annu. Rep. algol. Lab.Třeboň *1967:* 161-170, 1968. [Growth analysis.]

32 - ADAMS, J.A., BAIRD, I.E.: Chlorophyll a and zooplankton standing crop. - Ann. Biol. *23:* 92-93, 1968.

33 - ADAMS, M.S.: Adaptations of *Aplectrum hyemale* to the environment: Effects of preconditioning temperature on net photosynthesis. - Bull. Torrey bot. Club *97:* 219-224, 1970.

34 - ADAMS, M.S., STRAIN, B.R.: Photosynthesis in stems and leaves of *Cercidium floridum*: spring and summer diurnal field response and relation to temperature. - Oecol. Plant. *3:* 285-297, 1968.

35 - ADAMS, M.S., STRAIN, B.R.: Seasonal photosynthetic rates in stems of *Cercidium floridum* BENTH. - Photosynthetica *3:* 55-62, 1969.

36 - ADAMS, M.S., STRAIN, B.R., TING, I.P.: Photosynthesis in chlorophyllous stem tissue and leaves of *Cercidium floridum*: Accumulation of ^{14}C from $^{14}CO_2$. - Plant Physiol. *42:* 1797-1799, 1967.

37 - ADEDIPE, N.O., ORMROD, D.P., MAURER, A.R.: The response of pea plants to low concentrations of Cycocel, Phosphon and B-nine. - J. amer. hort. Soc. Sci. *94:* 321-323, 1969. [Chl.]

38 - ADEÏSHVILI, N.I.: Vliyanie intensivnosti osveshcheniya i vozrasta lista na soderzhanie plastidnykh pigmentov v list'yakh chaĭnogo rasteniya. [Effect of illuminance and leaf age on the content of plastid pigments in tea leaves.] - Subtrop. Kul'tury *1966* (4): 78-84, 1966. [In R.]

39 - ADEÏSHVILI, N.I., DUMBADZE, V.Z.: Nakoplenie plastidnykh pigmentov v list'yakh tsitrusovykh rasteniĭ v svyazi s ikh morozostoĭkost'yu. [Accumulation of plastid pigments in Citrus leaves in relation to their frost resistance.] - Subtrop. Kul'tury *1970* (3): 48-60, 1970. [In R.]

40 - ADIJUWANA, H., SOERIANEGARA, I.: Fotosintesa dan produksi lateks pada tiga klon karet *(Hevea brasiliensis)*. [Photosyntesis and latex production of three *Hevea brasiliensis* clones.] - Commun. Agricult. (Bogor) *3* (3): 40-60, 1970. Menara Perkebunan (Bogor) *39* (5-6): 77-86, 1970. [In Indonesian, ab: E.]

41 - ADLER, K.: Spezifische Rolle der Carotinoidabsorption bei der photosynthetischen Sauerstoffentwicklung. - Planta *75:* 220-227, 1967.

42 - ADLER, K.: Blaugrün-Effekt und Emerson-Effekt in der photosynthetischen Sauerstoffentwicklung. - Studia biophys. *5:* 225-232, 1967.

43 - ADLER, K.: Eine Anordnung zur Messung kleiner Sauerstoffaustauschraten photosynthetisierender Algen bei geringen Lichtintensitäten. - Kulturpflanze *15:* 151-160, 1967.

44 - ADLER, K.: Der Emerson-Enhancement-Effekt während der Induktionsperiode der Photosynthese. - Photosynthetica *2:* 17-23, 1968.

45 - AEROV, I.L.: Nekotorye osobennosti fotosinteticheskogo apparata v svyazi s rostom i razvitiem soi na raznykh fotoperiodakh. [Some characteristics of the photosynthetic apparatus in relation to growth and development of soybean in different photoperiods.] - In: Puti Povysheniya Intensivnosti i Produktivnosti Fotosinteza. Pp. 144-151. Nauk. Dumka, Kiev 1966. [In R.]

46 - AEROV, I.L.: Opticheskie svoĭstva verkhneĭ i nizhneĭ storony list'ev. [Optical properties of upper and lower leaf sides.] - Fiziol. Biokhim. kul't. Rast. *1:* 191-196, 1969. [In R, ab: E.]

47 - AEROV, I.L., GULYAEV, B.I., MANUIL'SKIĬ, V.D.: Spektral'nye svetovye krivye fotosinteza list'ev svekly v svyazi s soderzhaniem pigmentov. [Spectral light curves of sugarbeet leaves in relation to pigment contents.] - Dokl. Akad. Nauk SSSR *187:* 1194-1197, 1969. [In R.]

48 - AEROV, Ï.L., LIKHOLAT, D.A.: Odnochasne vyznachennya vmistu pigmentiv khloroplastiv ta mitsnosti zv'yazku ïkh z bilkovo-lipoĭdnym kompleksom v lystkakh

roslyn. [Simultaneous determination of the content of chloroplast pigments and the strength of their bond with the protein-lipoid complex in plant leaves.] - Dopovidi Akad. Nauk URSR *1966* (12): 1599-1602, 1966. [In Ukr., ab: R, E.]

49 - AEROV, I.L., LIKHOLAT, D.A.: Izmeneniya pigmentnoĭ sistemy u raznykh po vozrastu i raspolozheniyu list'ev yabloni v techenie vegetatsii. [Changes in pigment system in apple leaves of different age and insertion level during vegetation.] - In: Puti Povysheniya Intensivnosti i Produktivnosti Fotosinteza. Vol. 2. Pp. 176-184. Naukova Dumka, Kiev 1967. [In R.]

50 - AEROV, Ĭ.L., LIKHOLAT, D.A.: Zalezhnist' mizh poglynannyam promenevoĭ energiĭ lystkamy roslyn i vmistom u nykh khlorofylu. [Relationship of absorption of radiant energy by plant leaves and their chlorophyll content.] - Dopovidi Akad. Nauk URSR B *1968* (8): 753-756, 1968. [In Ukr., ab: E, R.]

51 - AEROV, I.L., LIKHOLAT, D.A.: Pigmenty i opticheskie svoĭstva osennikh list'ev drevesnykh rasteniĭ. [Pigments and optical properties of autumn leaves of woody plants.] - In: Puti Povysheniya Intensivnosti i Produktivnosti Fotosinteza. Vol. 3. Pp. 176-182. Naukova Dumka, Kiev 1969. [In R.]

52 - AEROV, I.L., LIKHOLAT, D.A.: Izmeneniya opticheskikh svoĭstv list'ev rasteniĭ v zavisimosti ot soderzhaniya pigmentov khloroplastov. [Changes of optical properties of leaves in relation to the content of chloroplast pigments.] - Fiziol. Biokhim. kul't. Rast. *2*: 318-323, 1970. [In R, ab: E.]

53 - AEROV, Ĭ.L., MANUĬL'SKIĬ, V.D.: Vyvchennya stanu khlorofilu v lystkakh za dopomogoyu spektrofotometrii. [Spectrophotometric studies of chlorophyll state in leaves.] - Dopovidi Akad. Nauk URSR *29* B: 740-743, 1967. [In Ukr., ab: E,R.]

54 - AFANAS'EVA, T.A.: Osobennosti sutochnogo khoda fotosinteza teplichnykh ogurtsov v zone belykh nocheĭ. [Peculiarities of the daily course of photosynthesis in greenhouse cucumbers in the zone of white nights.] - In: Informatsionnyĭ Byulleten'.Vol. 6. Sibir. Inst. Fiziol. i Biokhim. Rast. SO Akad. Nauk SSSR. Pp. 69-70. Irkutsk 1970. [In R.]

55 - AFANAS'EVA, T.A.: Sutochnyĭ khod fotosinteza u teplichnykh ogurtsov v usloviyakh Zapolyar'ya. [Daily course of photosynthesis in greenhouse cucumbers behind the Polar circle.] - Fiziol. Rast. *17*: 259-264, 1970. [In R, ab: E.]

56 - AGAVERDIEV, A.Sh., TARUSOV, B.N.: Zavisimost' intensivnosti sverkhslabogo izlucheniya zelenykh list'ev ot stepeni nakopleniya pervichnykh produktov fotosinteza. [Dependence of intensity of ultra-weak radiation of green leaves on accumulation of primary products of photosynthesis.] - Biofizika *14*: 754-756, 1969. [In R, ab: E.]

57 - AGHION, J.: Propriétés spectroscopiques des complexes chlorophylliens d'*Euglena gracilis*. - Physiol. vég. *4*: 67-73, 1966.

58 - AGHION, J.: Action du méthanol sur le spectre d'absorption des complexes pigmentés extraits des feuilles d'Épinard. Mise en évidence de deux complexes chlorophylliens. - Physiol. vég. *4*: 389-399, 1966.

59 - AGHION, J., AGHION, C.: Étude de l'agrégation in vitro des complexes chlorophylliens extraits de feuilles d'épinards. - Rev. can. Biol. *27*: 1-7, 1968.

60 - AGHION, J., BOURRET, R.L.: Agrégation des complexes de pigments chlorophylliens extraits de plantes vertes. II. Effets des variations de concentrations de divers solvants. - Physiol. vég. *7*: 297-303, 1969.

61 - AGHION, J., BROYDE, S.B., BRODY, S.S.: Surface reactions of chlorophyll *a* monolayers at a water-air interface. Photochemistry and complex formation. - Biochemistry *8*: 3120-3126, 1969.

62 - AGHION, J., BROYDE, S.B., BRODY, S.S.: Quelques propriétés photochimiques de monocouches de chlorophylle *a* situées à une interface eau-air. - Bull. Soc. roy. bot. Belg. *103*: 107-114, 1970.

63 - AGHION, J., CREVIER, L.: Effect of basic solvents on the absorption spectrum of *Rhodospirillum rubrum* pigment complexes. - Physiol. Plant. *21*: 1045-1053, 1968.

64 - AGZAMOV, A., KHODZHAEV, A.S.: O pigmentnom sostave semyan saksaula i ikh soder-
zhanii. [Pigment composition and content in *Haloxylon* seeds.] - In: Materialy
XV Nauchnoi Konferentsii Molodykh Uchenykh i Aspirantov po Fiziologii Rastenii.
Pp. 129-135. Inst. eksp. Biol. Rast. Akad. Nauk Uzb. SSR, Tashkent 1967. [In R.]

65 - AHMADJIAN, V., GANNUTZ, T.P., FRISHMAN, S.: Photosynthesis and respiration of
Antarctic lichens. - Antarctic J. U.S. 2: 100-101, 1967.

66 - AHMED, A.M.M., RIES, E.: The pattern of $^{14}CO_2$ fixation in different phases of
the life cycle and under different wavelengths in *Chlorella pyrenoidosa*. - In:
METZNER, H. (ed.): Progress in Photosynthesis Research. Vol. III. Pp. 1662-1668.
Tübingen 1969.

67 - AIGA, I., SASA, T.: Studies on chlorophyllase of *Chlorella protothecoides*. II.
Formation of atypical chlorophyllide a. - Plant Cell Physiol. 11: 161-165, 1970.

68 - AIHARA, M.S., YAMAMOTO, H.Y.: Occurrence of antheraxanthin in two *Rhodophyceae*
Acanthophora spicifera and *Gracilaria lichenoides*. - Phytochemistry 7: 497-499,
1968.

69 - AITZETMÜLLER, K., STRAIN, H.H., SVEC, W.A., GRANDOLFO, M., KATZ, J.J.: Loroxan-
thin, a unique xanthophyll from *Scenedesmus obliquus* and *Chlorella vulgaris*. -
Phytochemistry 8: 1761-1770, 1969.

70 - AITZETMÜLLER, K., SVEC, W.A., KATZ, J.J., STRAIN, H.H.: Structure and chemical
identity of diadinoxanthin and the principal xanthophyll of *Euglena*. - Chem.
Commun. *1968*: 32-33, 1968.

71 - AÏVAZYAN, S.A.: Sovmestnoe vliyanie streptomitsina i radiatsii na kolichestven-
noe soderzhanie khlorofilla v prorostkakh pshenitsy. [Combined effect of streptomycin and radiation on the chlorophyll level in wheat shoots.] - Biol. Zh.
Armenii 23 (2): 103-104, 1970. [In R.]

72 - AÏVAZYAN, S.A., BABAYAN, V.O.: Sovmestnoe deistvie streptomicina i radiacii na
zelenenie prorostkov pshenitsy. [Joint effect of streptomycin and radiation on
the greenness of wheat shoots.] - Biol. Zh. Armenii 21 (2): 72-75, 1968. [Chl;
in R, ab: Armen.]

73 - AÏVAZYAN, S.A., BABAYAN, V.O., ABRAMYAN, A.G.: Vliyanie streptomytsina na pro-
rostki pshenitsy. [Effect of streptomycin on wheat seedlings.] - Biol. Zh. Armenii 19 (10): 28-33, 1966. [Chl; in R, ab: Armen.]

74 - AKAIKE, S., YAMADA, S., YASUMATSU, N.: [Changes in pigments and color during
air-curing of Shirodaruma tobacco.] - Bull. Hatana Tobacco exp. Sta. [Hatano
Tabako Shikensho Hokoku] 57: 71-79, 1966. [In Jap.]

75 - AKAZAWA, T.: The structure and function of fraction-I protein. Regulatory as-
pects of photosynthetic CO_2-fixation in chloroplasts. - In: REINHOLD, L.,
LIWSCHITZ, Y. (ed.): Progress in Phytochemistry. Vol. 2. Pp. 107-141. Intersci.
Publ., London-New York-Sydney-Toronto 1970.

76 - AKAZAWA, T., SATO, K., SUGIYAMA, T.: Structure and function of chloroplast proteins. VIII. Some properties of ribulose-1,5-diphosphate carboxylase of *Athiorhodaceae* in comparison with those of plant enzyme. - Arch. Biochem. Biophys.
132: 255-261, 1969.

77 - AKAZAWA, T., SATO, K., SUGIYAMA, T.: Ribulose-1,5-diphosphate carboxylase of
Chromatium strain D. - Plant Cell Physiol. 11: 39-46, 1970.

78 - AKAZAWA, T., SUGIYAMA, T. KATAOKA, H.: Further studies on ribulose-1,5-diphos-
phate carboxylase from *Rhodopseudomonas spheroides* and *Rhodospirillum rubrum*. -
Plant Cell Physiol. 11: 541-550, 1970.

79 - AKAZAWA, T., SUGIYAMA, T., NAKAYAMA, N., ODA, T.: Structure and function of
chloroplast proteins. VI. Further studies on the PCMB-treatment of spinach leaf
RuDP carboxylase. - Arch. Biochem. Biophys. 128: 646-653, 1968.

80 - AKININA, D.K.: Zavisimost' svetovogo nasyshcheniya dvukh massovykh vidov dinoflagellat ot ryada faktorov. [The dependence of light saturation of two mass
species of dinoflagellates on many factors.] - Okeanologiya 6: 861-868, 1966.
[In R.]

81 - AKININA, D.K.: O zavisimosti fotosinteza *Prorocentrum micans* i *Gymnodinium kowa-lewskii* ot intensivnosti solnechnoT radiatsii. [Dependence of photosynthesis of *Prorocentrum micans* and *Gymnodinium kowalewskii* on sun irradiance.] - Fiziol. Rast. *13*: 226-230, 1966. [In R, ab: E.]

82 - AKININA, D.K.: Nekotorye dannye o vliyanii mineral'nogo fosfora na fotosintez dinoflagellat. [Some data on the effect of inorganic phosphorus on photosynthe-sis of dinoflagellates.] - In: Biologiya i Raspredelenie Planktona Yuzhnykh MoreT. Pp. 35-40. Nauka, Moskva 1967. [In R.]

83 - AKININA, D.K.: Osedanie i fototaksis dvukh massovykh vidov dinoflagellat v svya-zi s ikh fotosinteticheskoT aktivnost'yu. [Sedimentation and phototaxis of two mass species of dinoflagellates in relation to their photosynthetic activity.] - In: Voprosy Biookeanografii. Pp. 95-100. Naukova Dumka, Kiev 1967. [In R.]

84 - AKININA, D.K., BURLAKOVA, Z.P.: Metod massovoT peresadki kletok planktonnyh vo-dorosleT iz odnoT sredy v druguyu. [A method for mass transfer of cells of plankton algae from one medium to another.] - Fiziol. Rast. *13*: 1094-1096, 1966. [In R, ab: E.]

85 - AKITA, S., MIYASAKA, A.: Studies on the differences of photosynthesis among species. II. Effect of oxygen-free air on photosynthesis. - Proc. Crop Sci. Soc. Jap. *38*: 525-534, 1969.

86 - AKITA, S., MIYASAKA, A., MURATA, Y.: Studies on the differences of photosynthe-sis among species. I. Differences in the response of photosynthesis among spe-cies in normal oxygen concentration as influenced by some environmental factors. - Proc. Crop Sci. Soc. Jap. *38*: 507-524, 1969.

87 - AKITA, S., MURATA, Y., MIYASAKA, A.: [On light-photosynthesis curves of rice leaves.] - Proc. Crop Sci. Soc. Jap. *37*: 680-684, 1968. [In Jap., ab: E.]

88 - AKOYUNOGLOU, G.: The effect of age on the phytochrome-mediated chlorophyll for-mation in dark-grown bean leaves. - Physiol. Plant. *23*: 29-37, 1970.

89 - AKOYUNOGLOU, G., ARGYROUDI-AKOYUNOGLOU, J.H.: Effects of intermittent and con-tinuous light on the chlorophyll formation in etiolated plants at various ages. - Physiol. Plant. *22*: 288-295, 1969.

90 - AKOYUNOGLOU, G., ARGYROUDI-AKOYUNOGLOU, J.H.: Mechanism of the first carboxyla-tion reaction in photosynthesis. - In: METZNER, H. (ed.): Progress in Photosyn-thesis Research. Vol. III. Pp. 1519-1528. Tübingen 1969.

91 - AKOYUNOGLOU, G., ARGYROUDI-AKOYUNOGLOU, J.H., GUIALI, A., DASSIOU, C.: On the relationship between ribulose diphosphate carboxylase and protochlorophyllide holochrome of *Phaseolus vulgaris* leaves. - Plant Physiol. *45*: 443-446, 1970.

92 - AKOYUNOGLOU, G., ARGYROUDI-AKOYUNOGLOU, J.-H., METHENITOU, H.: Studies on the active site of the enzyme ribulose-diphosphate carboxylase. - Biochim. Biophys. Acta *132*: 481-491, 1967.

93 - AKOYUNOGLOU, G., ARGYROUDI-AKOYUNOGLOU, J.H., MICHEL-WOLWERTZ, M.R., SIRONVAL, C.: Effect of intermittent and continuous light on chlorophyll formation in etiolated plants. - Physiol. Plant. *19*: 1101-1104, 1966.

94 - AKOYUNOGLOU, G., ARGYROUDI-AKOYUNOGLOU, J.H., MICHEL-WOLWERTZ, M.R., SIRONVAL, C.: Chlorophyll *a* as a precursor for chlorophyll *b*. Synthesis in barley leaves. - Chimika Chronika *32 A*: 5-8, 1967.

95 - AKOYUNOGLOU, G., ARGYROUDI-AKOYUNOGLOU, J.-H., MICHEL-WOLWERTZ, M.-R., SIRONVAL, C.: L'origine de la chlorophylle *b*. - In: SIRONVAL, C. (ed.): Le Chloroplaste, Croissance et Vieillissement. Pp. 91-98, Masson et Co., Paris 1967.

96 - AKOYUNOGLOU, G.A., SIEGELMAN, H.W.: Protochlorophyllide resynthesis in dark-grown bean leaves. - Plant Physiol. *43*: 66-68, 1968.

97 - AKSENOV, S.I., VERKHOTUROV, V.N., KAUROV, B.S., RUBIN, A.B.: DvukhluchevoT spektrofotometr dlya registratsii raznostnykh spektrov v fotosinteziruyushchikh ob"ektakh. [Double-beam spectrophotometer for registration of difference spec-tra in photosynthesizing bacteria.] - Nauch. Dokl. vyssh. Shkoly, biol. Nauki *11* (8): 123-128, 1968. [In R.]

98 - AKULOVA, E.A.: Vydelenie khloroplastov pri issledovanii fotovosstanovleniya
     NADF. [Isolation of chloroplasts for the study of NADP photoreduction.] - In:
     KIRICHENKO, E.B. (ed.): Metody Vydeleniya Khloroplastov. Pp. 39-51. Pushchino-
     na-Oke 1970. [In R, ab: E.]

99 - AKULOVA, E.A., MALKINA, I.S., KHAZANOV, V.S., TSEL'NIKER, Yu.L., SHISHOV, D.M.:
     O metodike izucheniya svetovogo rezhima v lesu. [Methods of studying illumina-
     tion in a forest.] - Bot. Zh. 51: 681-686, 1966. [In R.]

100 - AKULOVA, E.A., MUKHIN, E.N.: Ob uchastii plastotsianina iz list'ev *Pisum sati-
      vum* L. v fotovosstanovlenii NADF fragmentami khloroplastov. [Participation of
      plastocyanin from pea leaves in the photoreduction of NADP by chloroplast frag-
      ments.] - Dokl. Akad. Nauk SSSR 180: 734-737, 1968. [In R.]

101 - AKULOVA, E.A., MUKHIN, E.N.: O prirodnom ingibitore fotosinteticheskikh reaktsiĭ.
      [Natural inhibitor of photosynthetic reactions.] - Dokl. Akad. Nauk SSSR 185:
      702-704, 1969. [In R.]

102 - AKULOVICH, N.K., GODNEV, T.N., ORLOVSKAYA, K.I.: Osobennosti spektral'nykh izme-
      neniĭ protokhlorofill-(id)-golokhroma etiolirovannykh list'ev v protsesse ego
      formirovaniya. [Features of spectral transformation of protochlorophyll(ide)
      holochrome of etiolated leaves during its formation.] - Dokl. Akad. Nauk SSSR
      191: 1406-1409, 1970. [In R.]

103 - AKULOVICH, N.K., ORLOVSKAYA, K.İ.: Spektral'naya kharakteristika protokhloro-
      fill(id) golokhroma v protsesse ego obrazovaniya i prevrashcheniya v khlorofill
      u raznykh tipov etiolirovannykh rasteniĭ. [Spectral characteristics of proto-
      chlorophyll(ide) holochrome during its formation and transformation into chlo-
      rophyll in various types of etiolated plants.] - In: Metabolizm i Stroenie Fo-
      tosinteticheskogo Apparata. Pp. 37-52. Nauka i Tekhnika, Minsk 1970. [In R.]

104 - ALAM, A.U., COUCH, J.R., CREGER, C.R.: The carotenoids of the marigold, *Tagetes
      erecta.* - Can. J. Bot. 46: 1539-1541, 1968.

105 - ALAMUTI, N., LÄUGER, P.: Fluorescence of thin chlorophyll membranes in aqueous
      phase. - Biochim. Biophys. Acta 211: 362-364, 1970.

106 - AL'BASSAM, Kh.: Biokhimicheskie izmeneniya, sopryazhennye s mutatsiyami poteri
      antotsianinovoĭ okraski tsvetka u nekotorykh rasteniĭ. [Biochemical alterations
      connected with mutations of anthocyanine loss in flowers of some plants.] -
      Dokl. Akad. Nauk SSSR 168: 1405-1408, 1966. [Chl; in R.]

107 - ALBERDA, T.: Dry matter production and light interception of crop surfaces. IV.
      Maximum herbage production as compared with predicted values. - Neth. J. agr.
      Sci. 16: 142-153, 1968.

108 - ALBERDA, T.: The effect of low temperature on dry matter production, chloro-
      phyll concentration and photosynthesis of maize plants of different ages. -
      Acta bot. neerl. 18: 39-49, 1969.

109 - ALBERDA, T.: The influence of carbohydrate reserves on respiration, photosyn-
      thesis, and dry matter production of intact plants. - In: Proc. XI Int. Grass-
      land Congr. Pp. 517-522. Univ. Queensland Press, 1970.

110 - ALBERDA, T., SIBMA, L.: Dry-matter production and light interception of crop
      surfaces. 3. Actual herbage production in different years as compared with po-
      tential values. - J. brit. Grassland Soc. 23: 206-215, 1968.

111 - ALBRECHT, E.: Zur Physiologie der Hill-Reaktion. - Studia biophys. 5: 143-148,
      1967.

112 - ALCAIDE, A., MUNICIO, A.M.: Inhibition and uncoupling of photophosphorylation.
      - Biochim. biophys. Acta 131: 195-197, 1967.

113 - ALCAIDE, A., MUNICIO, A.M., RIBERA, A., STAMM, M.D.: Influence of structural
      characteristics on inhibition of the Hill reaction. - An. Real Soc. Espan. Fis.
      Quim. Ser. B 62: 1931-1402, 1966.

114 - ALCOCK, M.B., LOVETT, J.V.: The electronic measurement of the yield of growing
      pasture. I. A statistical assessment. - J. agr. Sci. 68: 27-38, 1967.

115 - **ALDERFER, R.G.**: Environmental regulation of photosynthesis in plant canopies.
      - Plant Physiol. *46* (Suppl.): 8, 1970.

116 - **ALEKSANDROV, E.E., SIYANOVA, N.S.**: Posledetstvie povyshennykh temperatur na
      sveto-indutsirovannoe sokrashchenie khloroplastov, izolirovannykh iz list'ev
      gorokha. [After-effect of increased temperature on light-induced shrinkage of
      chloroplasts isolated from pea leaves.] - In: Funktsional'nye Osobennosti Khlo-
      roplastov. Pp. 35-43. Kazan'. Univ., Kazan' 1969. [In R.]

117 - **ALEKSANDROVA, I.V., IVANOV, E.A.**: Izuchenie totosinteza khlorelly pri neizmen-
      nykh usloviyakh rosta. [Study of photosynthesis of *Chlorella* grown in constant
      conditions.] - In: Problemy Kosmicheskoi Biologii. Vol. 7. Pp. 470-474. Nauka,
      Moskva 1967. [In R.]

118 - **ALEKSANYAN, R.V.**: Vliyanie udobrenii na produktivnost' i kachestvo kukuruzy.
      [Effect of nutrients on productivity and quality of maize.] - Biol. Zh. Armen.
      21 (6): 84-88, 1968. [Ps; in R, ab: Armenian.]

119 - **ALEKSEEV, V.A.**: O propuskanii solnechnoi radiatsii pologom drevostoev. [Pene-
      tration of solar radiation through a forest canopy.] - In: Svetovoi Rezhim,
      Fotosintez i Produktivnost' Lesa. Pp. 15-35. Nauka, Moskva 1967. [In R.]

120 - **ALEKSEEV, V.A.**: Fotosinteticheski aktivnaya radiatsiya (F.A.R.) i prirost fito-
      massy drevostoev nekotorykh tipov biogeotsenozov. [Photosynthetically active
      radiation (PhAR) and increment of phytomass in standing timber of certain types
      of biogeocoenoses.] - Dokl. Akad. Nauk SSSR *175*: 954-957, 1967. [In R.]

121 - **ALEKSEEVA, S.A.**: Fungitsidnoe detstvie kolloidnoi sery protiv parshi i muchnis-
      toi rosy yabloni i ee vliyanie na fiziologicheskie protsessy. [Fungicide effect
      of colloid serum against scap and powdery mildew and its effect on physiologi-
      cal processes.] - Sb. nauch. Rabot Aspir. Kabardin-Balkar Univ. *2*: 286-288,
      1968. [In R.]

122 - **ALIEV, D.A.**: Produktivnost' fotosinteza i radiatsionnyi rezhim posevov ovosh-
      chnykh kul'tur v svyazi s kornevym pitaniem. [Productivity of photosynthesis
      and radiation regime of vegetable stands as related to root nutrition.] - In:
      Materialy I. zakavk. Konf. po Fiziol. Rast. Pp 41. Izdat. Akad. Nauk Azerb. SSR,
      Baku 1967.

123 - **ALIEV, D.A.**: Produktivnost' fotosinteza i radiatsionnyi rezhim posevov bakla-
      zhana v svyazi s kornevym pitaniem. [Productivity of photosynthesis and radia-
      tion regime of aubergine stands in relation to root nutrition.] - Izv. Akad.
      Nauk Azerb. SSR, Ser. biol Nauk *1968* (1): 13-24, 1968. [In R, ab: Azerb.]

124 - **ALLAWAY, W.G., MANSFIELD, T.A.**: Stomatal responses to changes in carbon dioxide
      concentration in leaves treated with 3-(4-chlorophenyl)-1,1-dimethylurea. -
      New Phytol. *66*: 57-63, 1967.

125 - **ALLAWAY, W.G., MANSFIELD, T.A.**: Automated system for following stomatal beha-
      vior of plants in growth cabinets. - Can. J. Bot. *47*: 1995-1998, 1969.

126 - **ALLAWAY, W.G., MANSFIELD, T.A.**: Experiments and observations on the aftereffect
      of wilting on stomata of *Rumex sanguineus*. - Can. J. Bot. *48*: 513-521, 1970.
      [CO_2 compensation point.]

127 - **ALLEN, C.F., FRANKE, H., HIRAYAMA, O.**: Identification of a plastoquinone and
      two naphthoquinones in *Anacystis nidulans* by NMR and mass spectroscopy. - Bio-
      chem. biophys. Res. Commun. *26*: 562-568, 1967.

128 - **ALLEN, C.F., HIRAYAMA, O., GOOD, P.**: Lipid composition of photosynthetic sys-
      tems. - In: GOODWIN, T.W. (ed.): Biochemistry of Chloroplasts. Vol. I. Pp. 195-
      200. Academic Press, London-New York 1966.

129 - **ALLEN, M.B.**: Distribution of the chlorophylls. - In: VERNON, L.P., SEELY, G.R.
      (ed.): The Chlorophylls. Pp. 511-519. Academic Press, New York-London 1966.

130 - **ALLEN, M.B.**: Nannoplankton and the carbon cycle in tropical waters. - Stud.
      trop. Oceanogr. Inst. mar. Sci. Univ. Miami *5*: 273-279, 1967.

131 - **ALLEN, M.B.**: Structure, physiology, and biochemistry of the *Chrysophyceae*. -
      Annu. Rev. Microbiol. *23*: 29-46, 1969. [Ps, chl.]

132 - **ALLEN, M.M.:** Photosynthetic membrane system in *Anacystis nidulans*. - J. Bacteriol. *96*: 836-841, 1968.

133 - **ALLEN, M.M., SMITH, A.J.:** Nitrogen chlorosis in blue-green algae. - Arch. Mikrobiol. *69*: 114-120, 1969.

134 - **ALLEN, W.A., GAYLE, T.V., RICHARDSON, A.J.:** Plant-canopy irradiance specified by the Duntley equations. - J. opt. Soc. Amer. *60*: 372-376, 1970.

135 - **ALLERUP, S.:** Fotosyntesen, kilden til dannelse af organisk stof. [Photosynthesis, the source for creation of organic matter.] - In: Dansk Natur - Dansk Skole. Årsskrift 1970. Pp. 3-19. Copenhagen 1970. [In Danish.]

136 - **ALLISON, J.C.S.:** Some crop physiological aspects of breeding for yield in maize. - In: Proc. 3 Cong. South African genet. Soc. 1966. Pp. 1-4. Minerva, Pretoria 1968. [Ps.]

137 - **ALLISON, J.C.S.:** Effect of plant population on the production and distribution of dry matter in maize. - Ann. appl. Biol. *63*: 135-144, 1969.

138 - **ALLISON, J.C.S.:** Net assimilation rates of Rhodesian and foreign maize (*Zea mays* L.) varieties. - Rhod. J. agr. Res. *8*: 79-80, 1970.

139 - **ALLISON, J.C.S., EDDOWES, M.:** Climate and optimum plant density for maize. - Nature *220*: 1343-1344, 1968.

140 - **ALLISON, J.C.S., WATSON, D.J.:** The production and distribution of dry matter in maize after flowering. - Ann. Bot. *30*: 365-381, 1966. [Growth analysis.]

141 - **ALLISON, J.C.S., WEINMANN, H.:** Effect of absence of developing grain on carbohydrate content and senescence of maize leaves. - Plant Physiol. *46*: 435-436, 1970. [Photosynthates and age of leaves.]

142 - **AL'PEROVICH, L.I., BABAEV, T.B., POPOVA, S.I.:** O vliyanii mezhmolekulyarnogo vzaimodeĭstviya na opticheskie svoĭstva khlorofilla *a* i β-karotina. [Effect of intermolecular relationships on optical properties of chlorophyll *a* and β-carotene.] - In: Teplovoe Dvizhenie Molekul i Mezhmolekul. Vzaimodeĭstvie v Zhidkostyakh i Rastvorakh. Pp. 206-210. Samarkand 1969.

143 - **AL-SHAHINE, F.O.:** Photosynthesis, respiration and dry matter production of Scots pine (*Pinus silvestris* L.) seedlings originating from Poland (Nowy Targ) and Turkey (Eskishaher). - Acta Soc. Bot. Pol. *38*: 355-369, 1969.

144 - **ALVIM, R., ALVIM, P. De T.:** Efeito da densidade de plantio no aproveitamento da energia luminosa pelo milho (*Zea mays*) e pelo feijão (*Phaseolus vulgaris*), em culturas exclusivas e consorciadas. [Effect of plant density on the use of luminous energy in *Zea mays* and *Phaseolus vulgaris* in monocultures and mixed stands.] - Turrialba *19*: 389-393, 1969. [In Port.]

145 - **AMESZ, J., FORK, D.C.:** Quenching of chlorophyll fluorescence by quinones in algae and chloroplasts. - Biochim. biophys. Acta *143*: 97-107, 1967.

146 - **AMESZ, J., FORK, D.C.:** The function of P700 and cytochrome *f* in the photosynthetic reaction center of system 1 in red algae. - Photochem. Photobiol. *6*: 903-912, 1967.

147 - **AMESZ, J., FORK, D.C.:** Role of P700 and cytochrome *f* in the reaction center of photosystem 1. - Carnegie Inst. Year Book *66*: 149-155, 1968.

148 - **AMESZ, J., FORK, D.C.:** Quenching by quinones of chlorophyll fluorescence in vivo. - Carnegie Inst. Year Book *66*: 165-171, 1968.

149 - **AMESZ, J., FORK, D.C., NOOTEBOOM, W.:** Function of the reaction center of system 1. - Studia biophys. *5*: 175-181, 1967.

150 - **AMESZ, J., NOOTEBOOM, W., SPAARGAREN, D.H.:** Quenching of chlorophyll fluorescence by quinones in vivo and spill-over of excitation energy from photosystem II to I. - In: METZNER, H. (ed.): Progress in Photosynthesis Research. Vol. II. Pp. 1064-1072. Tübingen 1969.

151 - **AMESZ, J., van den BOS, P., DIRKS, M.P.:** Oxidation-reduction potentials of photosynthetic intermediates. - Biochim. biophys. Acta *197*: 324-327, 1970.

152 - **AMESZ, J., VREDENBERG, W.J.:** Near-infrared action spectra of fluorescence, cy-
      tochrome oxidation and shift in carotenoid absorption in purple bacteria. - Bio-
      chim. biophys. Acta *126*: 254-261, 1966.

153 - **AMESZ, J., VREDENBERG, W.J.:** Absorbancy changes of photosynthetic pigments in
      various purple bacteria. - In: THOMAS, J.B., GOEDHEER, J.C. (ed.): Currents in
      Photosynthesis. Pp. 75-83. Donker, Rotterdam 1966.

154 - **AMESZ, J., VREDENBERG, W.J.:** Reaction kinetics of photosynthetic intermediates
      in intact algae. - In: GOODWIN, T.W. (ed.): Biochemistry of Chloroplasts. Vol.
      II. Pp. 593-600. Academic Press, London-New York 1967.

155 - **AMIRDZHANOV, A.G.:** Struktura vinogradnogo kusta kak faktor ego fotosintetiches-
      koǐ produktivnosti. [Structure of vine shrub as a factor of its photosynthetic
      productivity.] - In: Vazhneǐshie Problemy Fotosinteza v Rastenievodstve. Pp.
      254-262. Kolos, Moskva 1970. [In R.]

156 - **AMMANN, E.C.B., FRASER-SMITH, A.C.:** Gas exchange of algae. IV. Reliability of
      Chlorella pyrenoidosa. - Appl. Microbiol. *16*: 669-672, 1968.

157 - **AMMANN, E.C.B., LYNCH, V.H.:** Purine metabolism by unicellular algae. III. The
      photochemical degradation of uric acid by chlorophyll. - Biochim biophys. Acta
      120: 181-182, 1966.

158 - **AMMANN, E.C.B., LYNCH, V.H.:** Gas exchange of algae. II. Effects of oxygen, he-
      lium, and argon on the photosynthesis of *Chlorella pyrenoidosa.* - Appl. Micro-
      biol. *14*: 552-557, 1966.

159 - **AMMANN, E.C.B., LYNCH, V.H.:** Gas exchange of algae. III. Relation between the
      concentration of carbon dioxide in the nutrient medium and the oxygen production
      of *Chlorella pyrenoidosa.* - Appl. Microbiol. *15*: 487-491, 1967. [Algae culture
      apparatus with pCO_2 electrodes.]

160 - **AMMANN, E.C.B., REED, L.L.:** Microbiological life support system: Photosynthesis
      versus chemosynthesis. - In: Chemical Engineering in Medicine and Biology. Pp.
      541-571. Plenum Press, New York 1967.

161 - **AMMERAAL, R.N., VENNESLAND, B.:** Effect of CO_2 and HCN on the quinone Hill reac-
      tion with *Anabaena variabilis.* - Arch. Biochem. Biophys. *117*: 429-436, 1966.

162 - **AMMON, R., FRIEDRICH, G.:** Das Verhalten von Fermenten in der *Euglena gracilis.*
      1. - Acta biol. med. german. *19*: 669-672, 1967. [Chl.]

163 - **AMSTER, R.L.:** A spectroscopic investigation of aggregations in chlorophyll so-
      lutions. - Photochem. Photobiol. *9*: 331-338, 1969.

164 - **AMSTER, R.L., PORTER, G.:** Solvate and dimer equilibria in solutions of chloro-
      phyll. - Proc. roy. Soc. (London) A *296*: 38-44, 1967.

165 - **ANDERSEN, W.R., WILDNER, G.F., CRIDDLE, R.S.:** Ribulose diphosphate carboxylase.
      III. Altered forms of ribulose diphosphate carboxylase from mutant tomato
      plants. - Arch. Biochem. Biophys. *137*: 84-90, 1970.

166 - **ANDERSON, G.C.:** Subsurface chlorophyll maximum in the northeast Pacific Ocean.
      - Limnol. Oceanogr. *14*: 386-391, 1969.

167 - **ANDERSON, J.L., SCHAELLING, J.P.:** Effect of pyrazon on bean chloroplast ultra-
      structure. - Weed Sci. *18*: 455-459, 1970.

168 - **ANDERSON, J.M.:** Derivative absorbance and fluorescence spectra of chloroplast
      fractions. - Carnegie Inst. Year Book *65*: 479-481, 1967.

169 - **ANDERSON, J.M., BOARDMAN, N.K.:** Fractionation of the photochemical systems of
      photosynthesis. I. Chlorophyll contents and photochemical activities of partic-
      les isolated from spinach chloroplasts. - Biochim. biophys. Acta *112*: 403-421,
      1966.

170 - **ANDERSON, J.M., FORK, D.C., AMESZ, J.:** P700 and cytochrome F in particles ob-
      tained by digitonin fragmentation of spinach chloroplasts. - Biochem. biophys.
      Res. Commun. *23*: 874-879, 1966.

171 - **ANDERSON, J.M., FORK, D.C., AMESZ, J.:** P700 and cytochrome *f* in particles ob-

tained by digitonin fragmentation of spinach chloroplasts. - Carnegie Inst.
Year Book *65:* 481-483, 1967.

172 - ANDERSON, J.M., PYLIOTIS, N.A.: Studies with manganese-deficient spinach chloroplasts. - Biochim. biophys. Acta *189:* 280-293, 1969.

173 - ANDERSON, J.M., THORNE, S.W.: The fluorescence properties of manganese-deficient
spinach chloroplasts. - Biochim. biophys. Acta *162:* 122-134, 1968.

174 - ANDERSON, J.M., VERNON, L.P.: Digitonin incubation of spinach chloroplasts in
Tris (hydroxymethyl) methylglycine solutions of varying ionic strengths. - Biochim. biophys. Acta *143:* 363-376, 1967.

175 - ANDERSON, J.W., ROWAN, K.S.: The effect of 6-furfurylaminopurine on senescence
in tobacco-leaf tissue after harvest. - Biochem. J. *98:* 401-404, 1966. [Chl.]

176 - ANDERSON, L., FULLER, R.C.: The rapid appearance of glycolate during photosynthesis in *Rhodospirillum rubrum.* - Biochim. Biophys. Acta *131:* 198-201, 1967.

177 - ANDERSON, L., FULLER, R.C.: Photosynthesis in *Rhodospirillum rubrum.* I. Autotrophic carbon dioxide fixation. - Plant Physiol. *42:* 487-490, 1967.

178 - ANDERSON, L., FULLER, R.C.: Photosynthesis in *Rhodospirillum rubrum.* II. Photo-
heterotrophic carbon dioxide fixation. - Plant Physiol. *42:* 491-496, 1967.

179 - ANDERSON, L., FULLER, R.C.: Photosynthesis in *Rhodospirillum rubrum.* III. Metabolic control of reductive pentose phosphate and tricarboxylic acid cycle enzymes. - Plant Physiol. *42:* 497-502, 1967.

180 - ANDERSON, L., WORTHEN, L.E., FULLER, R.C.: The role of ribose-5-P isomerase in
regulation to the Calvin cycle in *Rhodospirillum rubrum.* - In: SHIBATA, K.,
TAKAMIYA, A., JAGENDORF, A.T., FULLER, R.C. (ed.): Comparative Biochemistry and
Biophysics of Photosynthesis. Pp. 379-386. Univ. Tokyo Press, Tokyo, Univ. Park
Press, State College, Pa. 1968.

181 - ANDERSON, L.E., FULLER, R.C.: Energy metabolism and control of an enzyme of the
Calvin cycle. - In: METZNER, H. (ed.): Progress in Photosynthesis Research.
Vol. III. Pp. 1618-1623. Tübingen 1969.

182 - ANDERSON, L.E., FULLER, R.C.: Photosynthesis in *Rhodospirillum rubrum.* IV. Isolation and characterization of ribulose-1,5-diphosphate carboxylase. - J. biol.
Chem. *244:* 3105-3109, 1969.

183 - ANDERSON, L.E., PRICE, G.B., FULLER, R.C.: Molecular diversity of the ribulose-
1,5-diphosphate carboxylase from photosynthetic microorganisms. - Science *161:*
482-484, 1968.

184 - ANDERSON, M.C.: Some problems of simple characterization of the light climate
in plant communities. - In: BAINBRIDGE, R., EVANS, G.C., RACKHAM, O. (ed.):
Light as an Ecological Factor. Pp. 77-90. Blackwell sci. Publ., Oxford 1966.

185 - ANDERSON, M.C.: Stand structure and light penetration. II. A theoretical analysis. - J. appl. Ecol. *3:* 41-54, 1966.

186 - ANDERSON, M.C.: Photon flux, chlorophyll content, and photosynthesis under
natural conditions. - Ecology *48:* 1050-1053, 1967.

187 - ANDERSON, M.C.: A comparison of two theories of radiation in crops. - Agr. Meteorol. *6:* 399-405, 1969.

188 - ANDERSON, M.C.: Radiation climate, crop architecture and photosynthesis. - In:
Prediction and Measurement of Photosynthetic Productivity. Pp. 71-78. PUDOC,
Wageningen 1970. [Monogram of extinction profiles of diffuse radiation.]

189 - ANDERSON, M.C., DENMEAD, O.T.: Short wave radiation on inclined surfaces in
model plant communities. - Agr. J. *61:* 867-872, 1969.

190 - ANDERSON, M.M., Mc CARTY, R.E.: The effects of plastocyanin on photophosphorylation. - Biochim. biophys. Acta *189:* 193-206, 1969.

191 - ANDERSON, R.L., FLACCUS, E.: Seasonal chlorophyll change in a leafy liverwort.
- J. Minn. Acad. Sci. *35:* 40-41, 1968.

192 - ANDRÉ, M.: Un monochromateur solaire à usage biologique. - In: THOMAS, J.B.,
GOEDHEER, J.C. (ed.): Currents in Photosynthesis. Pp. 305-312. Donker, Rotterdam 1966.

193 - ANDRÉ, M., GUÉRIN DE MONTGAREUIL, P., SEIMANDI, N.: Effet du rouge lointain
sur des capacités à produire de l'oxygène photosynthétique. - Compt. rend.
Acad. Sci. Paris D 265: 540-543, 1967.

194 - ANDREENKO, S.S.: Fotosintez rasteniĭ kukuruzy. [Photosynthesis of maize plants.]
- In: RUBIN, B.A. (ed.): Fiziologiya Sel'skokhozyaĭstvennykh Rasteniĭ. Tom 5,
Fiziologiya Kukuruza i Risa. Pp. 112-133. Izdat. Mosk. Univ., Moskva 1969. [In
R.]

195 - ANDREENKO, S.S., KERECHKI, B.: O vliyanii ponizhennoĭ temperatury v zone korneĭ
na nekotorye fiziologicheskie protsessy v rasteniyakh kukuruzy. [The influence
of decreased temperature in the root zone on some physiological processes in
maize.] - Dokl. Akad. Nauk SSSR 168: 947-950, 1966. [Chl, Ps.]

196 - ANDREEVA, T.F.: Fiziologiya fotosinteza. [Physiology of photosynthesis.] - In:
OPARIN, A.I. (ed.): Fiziologiya Sel'skokhoyaĭstvennykh Rasteniĭ. Vol. 2. Pp.
267-308. Izdat. Mosk. Univ., Moskva 1967. [In R.]

197 - ANDREEVA, T.F., AVDEEVA, T.A.: Vliyanie azotnogo pitaniya na fotosinteticheskuyu aktivnost' i biosintez belka. [Influence of nitrogen nutrition on photosynthetic activity and protein biosynthesis.] - In: Mineral'nye Elementy i Mekhanizm Fotosinteza. Pp. 128-136. Kishinev 1969. [In R.]

198 - ANDREEVA, T.F., AVDEEVA, T.A.: Belok "fraktsii 1" i fotosinteticheskaya aktivnost' list'ev. [Protein of "fraction 1" and photosynthetic activity of leaves.]
- Fiziol. Rast. 17: 225-233, 1970. [In R.]

199 - ANDREEVA, T.F., PERSANOV, V.M.: Vliyanie prodolzhitel'nosti fosfornogo golodaniya na intensivnost' fotosinteza i rost list'ev v svyazi s produktivnost'yu kon-
skikh bobov. [Effect of duration of phosphorus deficiency on photosynthetic
rate and leaf growth with regard to productivity of horse bean.] - Fiziol. Rast.
17: 478-484, 1970. [In R.]

200 - ANDREW, R.H.: A technique for measuring root volume in vivo. - Crop Sci. 6:
384-386, 1966.

201 - ANDREWS, T.J., HATCH, M.D.: Properties and mechanism of action of pyruvate,
phosphate dikinase from leaves. - Biochem. J. 114: 117-125, 1969.

202 - ANDRIANOV, V.K., KURELLA, G.A., LITVIN, F.F.: O vzaimosvyazi potentsiala pokoya
i fotosinteza. [Relationship between potential of dormancy and photosynthesis.]
- Trudy mosk. Obshch. Ispyt. Prirody, Otd. biol. 28: 200-206, 1968. [In R.]

203 - ANDRIANOV, V.K., KURELLA, G.A., LITVIN, F.F.: Vliyanie ingibitorov dykhaniya i
fotosinteza na potentsial pokoya kletok Nitella i ego fotoindutsirovannye izmeneniya. [Effect of inhibitors of respiration and photosynthesis on the resting potential of Nitella cells and its photoinduced changes.] - Tsitologiya 11:
1014-1020, 1969.[In R, ab: E.]

204 - ANGOT, M.: Rapports entre la concentration en chlorophylle a, de taux d'assimilation du carbone et la valeur de l'énergie lumineuse en eau tropical littorale.
- Cah. ORSTOM, Ser. Oceanogr. 5: 39-45, 1967.

205 - ANGOT, M., GERARD, R.: Hydrologie et phytoplancton de l'eau de surface en avril
1965 à Nosy-Be. - Cah. ORSTOM, Ser. Oceanogr. 4: 95-136, 1966.

206 - ANISIMOV, A.A.: K voprosu o putyakh vozdeĭstviya azotnogo pitaniya na peredvizh-
henie assimilyatov u kartofelya. [Modes of action of nitrogen nutrition on
translocation of assimilates in potato.] - Fiziol. Rast. 15: 13-18, 1968. [In
R, ab: E.]

207 - ANISTRATOVA, N.A., FEDIN, V.M., KOROLENOK, L.N., BALUEVA, G.R.: Ob infrakrasnom
spektre khlorelly. [Infra-red spectrum of Chlorella] - Iz. sib. Otd. Akad. Nauk
SSSR 1968 (15); Ser. biol.-med. Nauk 3: 69-71, 1968. [In R.]

208 - ANSLOW, R.C.: The production of dry matter by swards of perennial ryegrass,
differing in average age of foliage. - J. brit. Grassland Soc. 23: 195-201, 1968.

209 - **ANTONENKO, M.K.**: Kolichestvo khloroplastov v kletkakh i fotokhimicheskaya aktivnost' khlorofilla list'ev sakharnoĭ svekly urozhaĭnogo i sakharistogo napra-
vleniĭ. [Amount of chloroplasts in cells and photochemical activity of chlorophyll of sugar beet leaves in respect to crop and sugar content.] - In: Puti
Povysheniya Intensivnosti i Produktivnosti Fotosinteza. Pp. 152-159. Kiev 1966.
[In R.]

210 - **ANUCHIN, N.P.**: Novyĭ sposob ustanovleniya tekushchego prirosta drevesnykh stvo-
lev po ob'emu. [A new method of the evaluation of the current growth of tree
trunks by volume.] - Vest. sel'skokhoz. Nauki *1969* (6): 70-73, 1969. [In R, ab:
E, G, F.]

211 - **AOKI, S., HASE, E.**: Ribonucleic acids appearing during the process of chloroplast regeneration in the "glucose-bleached" cells of *Chlorella protothecoides*.
- Plant Cell Physiol. *8*: 181-195, 1967.

212 - **D'AOUST, M.J., TAYLER, R.S.**: The interaction between nitrogen and water in the
growth of grass swards: II. Leaf area index and net assimilation rate. - J.agr.
Sci. *72*: 437-443, 1969.

213 - **APEL, P.**: Rhythmische Änderungen der Stomataapertur und der CO_2-Aufnahme bei
Keimblättern von Gerste. - Ber. deut. bot. Ges. *79*: 279-288, 1966.

214 - **APEL, P.**: Die Bedeutung der Grannen für die Kornentwicklung III. Photosyntheseintensität der Ähren verschiedener Gersten- und Weizensorten. - Kulturpflanze
14: 163-169, 1966.

215 - **APEL, P.**: Herstellung CO_2-haltiger Gasgemische für Photosynthesemessungen mit
dem URAS. - Flora A *157*: 330-333, 1966.

216 - **APEL, P.**: Über rhythmisch verlaufende Änderungen in der CO_2-Aufnahme von Blättern. - Ber. deut. bot. Ges. *80*: 3-9, 1967.

217 - **APEL, P.**: Photosynthesemessungen an Chlorophyllmutanten von Gerste (Lichtkurven,
"Lichtatmung", Starklichtempfindlichkeit). - Studia biophys. *5*: 105-110, 1967.

218 - **APEL, P.**: Potentielle Photosyntheseintensität von Gerstensorten des Gaterslebe-
ner Sortiments. - Kulturpflanze *15*: 161-174, 1967.

219 - **APEL, P.**: Zum systematischen Aspekt des Sauerstoffeinflusses auf die apparente
CO_2-Aufnahme höherer Pflanzen. - Kulturpflanze *17*: 191-204, 1969.

220 - **APEL, P., LEHMANN, C.O.**: Photosyntheseintensität von Winterweizen-Hybriden (F_1)
und ihren Eltern. - Züchter *37*: 377-378, 1967.

221 - **APEL, P., LEHMANN, C.O.**: Variabilität und Sortenspezifität der Photosynthesera-
te bei Sommergerste. - Photosynthetica *3*: 255-262, 1969.

222 - **APPELQVIST, L.-Å., BOYNTON, J.E., HENNINGSEN, K.W., STUMPF, P.K., WETTSTEIN,
D. von**: Lipid biosynthesis in chlosophyll mutants in barley. - J. Lipid Res. *9*:
513-524, 1968.

223 - **APPELQVIST, L.-Å., BOYNTON, J.E., STUMPF, P.K., WETTSTEIN, D. von**: Lipid biosynthesis in relation to chloroplast development in barley. - J. Lipid Res. *9*:
425-436, 1968.

224 - **APPELQVIST, L.A., STUMPF, P.K., WETTSTEIN, D. von**: Lipid synthesis and ultrastructure of isolated barley chloroplasts. - Plant Physiol. *43*: 163-187, 1968.

225 - **APPLEMAN, D., FULCO, A.J., SHUGARMAN, P.M.**: Correlation of α-linolenate of
photosynthetic O_2 production in *Chlorella*. - Plant Physiol. *41*: 136-142, 1966.

226 - **ARAKAWA, N.**: Studies on cover-culture of tobacco. On carbon-dioxide environment
in cover-culture. - Proc. Crop Sci. Soc. Jap. *37*: 150-155, 1968. [Ps.]

227 - **ARCHER, L.J.**: Regulatory mechanisms of pigmentation and plant tissue culture.
- Brotéria, Cienc. natur. *35*: 179-203, 1966. [Chl, car.]

228 - **ARCHER, M.C. Jr.**: Hill reaction rates of chloroplasts from virus-infected tobacco (*Nicotiana tabacum* L.) - Diss. Abstr. Int. B *31*: 995-B, 1970.

229 - **ARCHIBALD, J.L., WALKER, D.M., SHAW, K.B., MARKOVAC, A., MacDONALD, S.F.**: The
synthesis of porphyrins derived from *Chlorobium* chlorophylls. - Can. J. Chem.

44: 345-362, 1966.

230 - ARGLEBE, C., HALL, T.C.: Extraction and comparison of bean cytoplasmic and chloroplast ribosomes. - Plant Cell Physiol. *10*: 171-182, 1969.

231 - ARGYROUDI-AKOYUNOGLOU, J., AKOYUNOGLOU, G.: On the protection of ribulose diphosphate carboxylase from iodoacetamide inhibition. - Biochem. biophys. Res. Commun. *32*: 15-22, 1968.

232 - ARGYROUDI-AKOYUNOGLOU, J.H., AKOYUNOGLOU, G.: Mechanism of action of carboxy-dismutase. - Nature *213*: 287-288, 1967.

233 - ARGYROUDI-AKOYUNOGLOU, J.H., AKOYUNOGLOU, G.: Photoinduced changes in the chlorophyll *a* to chlorophyll *b* ratio in young bean plants. - Plant Physiol. *46*: 247-249, 1970.

234 - ARIAS, E., MORALES, E.: Ecología del puerto de Barcelona y desarrollo de adherencias orgánicas sobre placas sumergidas durante los años 1964 a 1966. [Ecology of the harbor of Barcelona, and fouling, during the years 1964-1966.] - Invest. Pesq. *33*: 179-200, 1969.[Chl, Car; in Span., ab: E.]

235 - ARLOŬSKAYA, K.I., AKULOVICH, N.K., RASKIN, V.I., GODNEŬ, C.M.: Peratvarenne protakhlarafilavaga pigmentu ŭ khlarafilavy va ŭmovakh chastkovaga razburennya formy 650. [The conversion of protochlorophyll to chlorophyll under conditions of the partial destruction of the 650 nm form.] - Vestsi Akad. Navuk Belarus. SSR, Ser. biyal. Navuk *1970* (3): 45-49, 135, 1970. [In Beloruss., ab: R.]

236 - ARMSTRONG, F.A.J., STEARNS, C.R., STRICKLAND, J.D.H.: The measurement of upwelling and subsequent biological processes by means of the Technicon Autoanalyzer R and associated equipment. - Deep-Sea Res. *14*: 381-389, 1967.

237 - ARMY, T.J., GREER, F.A.: Photosynthesis and crop production systems. - In: SAN PIETRO, A., GREER, F.A., ARMY, T.J. (ed.): Harvesting the Sun. Pp. 321-332. Academic Press, New York-London 1967.

238 - ARN, H., GROB, E.C., SIGNER, R.: Die Auftrennung der Chlorophylle durch Gegenstromextraktion. - Helv. chim. Acta *49*: 851-854, 1966.

239 - ARNDT, F., KÖTTER, C.: Zur Selektivität von Phenmedipham als Nachauflaufherbizid in Betarüben. - Weed Res. *8*: 259-271, 1968. [Ps.]

240 - ARNOLD, W.: Light reaction in green plant photosynthesis. A method of study. - Science *154*: 1046-1049, 1966.

241 - ARNOLD, W., AZZI, J.R.: Chlorophyll energy levels and electron flow in photosynthesis. - Proc. nat. Acad. Sci. U.S.A. *61*: 29-35, 1968.

242 - ARNON, D.I.: The photosynthetic energy conversion process in isolated chloroplasts. - Experientia *22*: 273-287, 1966.

243 - ARNON, D.I.: On the energy conversion process in illuminated chloroplasts. - In: THOMAS, J.B., GOEDHEER, J.C. (ed.): Currents in Photosynthesis. Pp. 465-477. Donker, Rotterdam 1966.

244 - ARNON, D.I.: Photosynthetic activity of isolated chloroplasts. - Physiol. Rev. *47*: 317-358, 1967.

245 - ARNON, D.I.: Photosynthetic phosphorylation: facts and concepts. - In: GOODWIN, T.W. (ed.): Biochemistry of Chloroplasts. Vol. II. Pp. 461-503. Academic Press, London-New York 1967.

246 - ARNON, D.I.: Electron transport and phosphorylation in photosynthesis by chloroplasts. - In: SINGER, T.P. (ed.): Biological Oxidations. Pp. 123-170. Intersci. Publ., New York 1968.

247 - ARNON, D.I.: Role of ferredoxin in photosynthesis. - Naturwissenschaften *56*: 295-305, 1969.

248 - ARNON, D.I.: Role of ferredoxin in photosynthesis. - In: METZNER, H. (ed.): Progress in Photosynthesis Research. Vol. III. Pp. 1444-1473. Tübingen 1969.

249 - ARNON, D.I., CHAIN, R.K., McSWAIN, B.D., TSUJIMOTO, H.Y., KNAFF, D.B.: Evidence from chloroplast fragments for three photosynthetic light reactions. - Proc.

nat. Acad. Sci. U.S.A. *67*: 1404-1409, 1970.

250 - ARNON, D.I., TSUJIMOTO, H.Y., McSWAIN, B.D.: Ferredoxin and photosynthetic phosphorylation. - An. Edafol. Agrobiol. *26*: 259-272, 1967.

251 - ARNON, D.I., TSUJIMOTO, H.Y., McSWAIN, B.D.: Ferredoxin and photosynthetic phosphorylation. - Nature *214*: 562-566, 1967.

252 - ARNON, D.I., TSUJIMOTO, H.Y., McSWAIN, B.D., CHAIN, R.K.: Separation of two photochemical systems of photosynthesis by fractionation of chloroplasts. - In: SHIBATA, K., TAKAMIYA, A., JAGENDORF, A.T., FULLER, R.C. (ed.): Comparative Biochemistry and Biophysics of Photosynthesis. Pp. 113-132. Univ. Tokyo Press, Tokyo, and Univ. Park Press, State College, Pa. 1968.

253 - ARNTZEN, C.J.: Membrane differentiation in greening *Zea mays* plastids. - Plant Physiol. *46*(Suppl.): 22, 1970.

254 - ARNTZEN, C.J.: Binary membrane structure in chloroplast lamellae. - Diss. Abstr. int. B *31*: 1687-B, 1970.

255 - ARNTZEN, C.J., DILLEY, R.A., CRANE, F.L.: A comparison of chloroplast membrane surfaces visualized by freeze-etch and negative-staining techniques; and ultrastructural characterization of membrane fractions obtained from digitonintreated spinach chloroplasts. - J. Cell Biol. *43*: 16-31, 1969.

256 - ARONOFF, S.: The chlorophylls - an introductory survey. - In: VERNON, L.P., SEELY, G.R. (ed.): The Chlorophylls. Pp. 3-20. Academic Press, New York-London 1966.

257 - ARONOFF, S., ELLSWORTH, R.K.: The biogenesis of chlorophyll a. - Photosynthetica *2*: 288-297, 1968.

258 - ARONOFF, S., KIRK, P.: Deaggregation of chlorophyll a by xanthophylls. - Nature *213*: 722, 1967.

259 - ARONOFF, S., NISSEN, M.: Spectroscopy of concentrated solutions of the chlorophylls. - Fed. Proc. *25*: 736, 1966.

260 - ARPIN, N., FIASSON, J.-L., LEBRETON, P.: Méthodes modernes d'analyse structurale des caroténoïdes. - Prod. Probl. pharm. *24*: 630-644, 1969; *25*: 21-34, 1969.

261 - ARPIN, N., LIAAEN-JENSEN, S.: Carotenoids of higher plants - II. Rubixanthin and gazaniaxanthin. - Phytochemistry *8*: 185-193, 1969.

262 - ARTAMKINA, I.Yu.: Metod vydeleniya kisloroda iz rastitel'nykh pigmentov dlya mass-spektral'nogo analiza. [A method for separation of oxygen from plant pigments for mass spectrometer analysis.] - Fiziol. Rast. *17*: 193-197, 1970. [In R, ab: E.]

263 - ARTAMONOV, V.I.: O sinteze i razrushenii khlorofilla v rasteniyakh pod vliyaniem gibberellina i vitamina B$_2$. [Synthesis and destruction of chlorophyll in plants affected by gibberellin and vitamin B$_2$.] - Fiziol. Rast. *13*: 424-428, 1966. [In R.]

264 - ARTAMONOV, V.I.: Ob izmenenii sostoyaniya khlorofilla v rasteniyakh, obrabotannykh gibberellinom i riboflavinom. [Changes in chlorophyll state in gibberellin- and riboflavine-affected plants.] - Nauch. Dokl. vyssh. Shkoly, biol. Nauki *11* (5): 110-113, 1968. [In R.]

265 - ARTEM'EV, P.N., VECHER, A.S.: Razdelenie lipoproteidnogo kompleksa izolirovannykh khloroplastov. [Separation of the lipoproteid complex from isolated chloroplasts.] - In: Fiziologo-biokhimicheskie issledovaniya rasteniī. Pp. 18-22. Nauka i Tekhnika, Minsk 1970. [In R.]

266 - ARUGA, Y.: Ecological studies of photosynthesis and matter production of phytoplankton III. Relationship between chlorophyll amount in water and primary productivity. - Bot. Mag. (Tokyo) *79*: 20-27, 1966.

267 - ARUGA, Y., YOKOHAMA, Y., NAKANISHI, M.: Primary productivity studies in February-March in the northwestern Pacific off Japan. - J. oceanogr. Soc. Jap. *24*: 265-280, 1968.

268 - ASADA, K., DEURA, R., KASAI, Z.: Effect of sulfate ions on photophosphorylation by spinach chloroplasts. - Plant Cell Physiol. *9*: 143-146, 1968.

269 - ASAFOV, G.B.: K voprosu o metodike opredeleniya produktov fotosinteza v perekhodnyĭ period. [On the methods of determination of photosynthates during the transient period.] - Dokl. Timiryaz. sel.-khoz. Akad. *144*: 169-175, 1968. [In R.]

270 - ASAHI, T., MASAKI, S.: Purification and properties of a chloroplast-protein containing photoreducible disulfide bond. - J. Biochem. *60*: 90-92, 1966.

271 - ASANUMA, K., NAKA, J., TAMARI, K.: Effects of topping on the growth, the translocation and accumulation of carbohydrates in corn plants. - Proc. Crop Sci. Soc. Jap. *36*: 488, 1967.

272 - ASHKINAZI, M.S., DOLIDZE, I.A.: Sensibilizirovannoe khlorofillom fotookislenie oksi- i aminokislot. [Oxi- and amino acid photooxidation sensibilized with chlorophyll.] - Biokhimiya *32*: 1000-1003, 1967. [In R.]

273 - ASHKINAZI, M.S., DOLIDZE, I.A., EGOROVA, V.A.: Sensibilizirovannoe proizvodnymi khlorofilla fotookislenie tirozina. [Tyrosine photooxidation sensibilized by chlorophyll derivatives.] - Biofizika *12*: 427-432, 1967. [In R.]

274 - ASHKINAZI, M.S., GLIKMAN, T.S., ZAVGORODNYAYA, L.N.: Fotokhimicheskie okislitel'no-vosstanovitel'nye prevrashcheniya feofitinov margantsa. [Photochemical oxidation-reduction transformations of manganese pheophytins.] - Dokl. Akad. Nauk SSSR *170*: 1195-1197, 1966. [In R.]

275 - ASHOUR, N.I., EL-FOULY, M.M.: Effect of (2-chloroethyl)trimethyl ammonium chloride (CCC) on the photosynthetic pigments of cotton leaves. - Acta bot. Acad. Sci. hung. *15*: 211-216, 1969.

276 - ASHOUR, N.I., EL-FOULY, M.M., ABDALLA, F.E.: Effect of CCC on photosynthetic apparatus and P^{32} uptake and distribution in cotton plants affected with bicarbonate-induced chlorosis. - Flora A *160*: 533-537, 1969.

277 - ASHTON, F.M., BISALPUTRA, T., RISLEY, E.B.: Effect of atrazine on *Chlorella vulgaris*. - Amer. J. Bot. *53*: 217-219, 1966.

278 - ASLYNG, H.C.: Weather, water balance and plant production at Copenhagen 1955-1964. - Kong. veterinaer Landbohøjskole Arsskr. *1966*: 1-21, 1966.

279 - ASPANDIYAROVA, M.B.: Intensivnost' dykhaniya i produktivnost' sortov yarovykh pshenits, vozdelyaemykh na raznykh vysotakh. [Respiration rate and productivity of cultivars of spring wheat cultivated at various altitudes.] - Tr. Inst. Bot. Akad. Nauk Kazakh. SSR (Alma-Ata) *25* (Fiziologicheskie Protsessy i Produktivnost' Yarovoĭ Pshenitsy): 78-82, 1968. [In R.]

280 - ASROROV, K.A.: Fotosinteticheskaya deyatel'nost' posevov kukuruzy i sorgo v usloviyakh Gissarskoĭ doliny Tadzhikistana. [Photosynthetic activity of maize and sorghum crops in Gissar valley of Tadzhikistan.] - In: Tezisy Dokladov Konferentsii Molodykh Uchennykh. P. 41. Dushanbe 1966. [In R.]

281 - ASROROV, K.A.: Koefitsienty ispol'zovaniya sveta posevami khlopchatnika v zavisimosti ot ob"ema plodorodnogo sloya pochvy. [Coefficients of light utilization by cotton stands as related to volume of fertile soil layer.] - In: Fotosintez i Ispol'zovanie Energii Solnechnoĭ Radiatsii. Pp. 5-6. Dushanbe 1967.

282 - ASROROV, K.A.: Fotosinteticheskaya deyatel'nost' rasteniĭ v protsesse formirovaniya urozhaya. [Photosynthetic activity of plants during yield formation.] - Izv. Akad. Nauk Tadzh., Otd. biol. Nauk *1969* [2 (35)]: 76-82, 1969. [In R.]

283 - ASTON, A.R., MILLINGTON, R.J. PETERS, D.B.: Radiation exchange in controlling leaf temperature. - Agron. J. *61*: 797-801, 1969.

284 - ATANASIU, L.: Quantity variation of chlorophyll in leaves of some *Coniferae* and autumn cereal plants, during the winter. - Rev. roum. Biol., Sér. bot. *13*: 15-18, 1968.

285 - ATANASIU, L.: Fotosinteze in decursul iernii. [Photosynthesis during winter.] - Natura, Ser. biol. (Bucureşti) *21* (6): 14-25, 1969. [In Rum.]

286 - ATANASIU, L.: Photosynthesis and respiration of some lichens during winter. -
      Rev. roum. Biol., Sér. bot. *14*: 165-168, 1969.

287 - ATEN, W.C., BÜCHEL, K.H.: Hemmstoffe der Photosynthese. VIII. Der Einfluss von
      Substituenten auf die Ionisationskonstante von Benzimidazolen, Benztriazolen,
      Indazolen und Indolen. - Z. Naturforsch. *25 b*: 961-965, 1970.

288 - ATHERTON, N.M., GARBETT, K., GILLARD, R.D., MASON, R., MAYHEW, S.J., PEEL, J.L.,
      STANGROOM, J.E.: Spectroscopic investigation of rubredoxin and ferredoxin. -
      Nature *212*: 590-593, 1966.

289 - ATKIN, E.K., WAIN, R.L.: Studies on plant growth-regulating substances. XXIV.
      Factors influencing kinetin-induced phosphorus mobilization in detached radish
      cotyledonary leaves. - Ann. appl. Biol. *60*: 321-331, 1967.

290 - ATTIWILL, P.M., OVINGTON, J.D.: Determination of forest biomass. - Forest Sci.
      14: 13-15, 1968.

291 - AUBERT, B.: Étude de la résistance à la diffusion gazeuse au niveau de l'épi-
      derme foliaire du bananier *(Musa acuminata* COLLA cv. *sinensis)* et de l'ananas
      (Ananas comosus (L.) MERR.) en conditions naturelles. - Fruits *25*: 495-507,
      577-580, 1970.

292 - AUBERT, B.: Mesures au poromètre de la résistance à la diffusion gazeuse de
      l'avocat avant et après cueillette. - Fruits *25*: 717-723, 1970.

293 - AUBERT, B., ČATSKÝ, J.: The onset of photosynthetic CO_2 influx in banana leaf
      segments as related to stomatal diffusion resistance at different air humidi-
      ties. - Photosynthetica *4*: 254-256, 1970.

294 - AUBERT, B., de PARCEVAUX, S.: Résistance à la diffusion gazeuse au niveau de
      l'épiderme foliaire de quelques plantes fruitières tropicales et subtropicales.
      - Fruits *24*: 177-190, 1969.

295 - AUBERT,B.M., ČATSKÝ, J.: Photosynthesis and transpiration of banana leaf samp-
      les in a controlled environment. - Trop. Ecol. *10*: 256-269, 1969.

296 - AUFHAMMER, W.: Ertragsbildung der Wintergerste. - Umschau Fortschr. Naturwiss.,
      Med., Tech. *67*: 420, 1967.

297 - van AUKEN, O.W.: Photosynthetic rates, respiratory rates, and intracellular os-
      motic pressures of a halophylic *Chlamydomonas*. - Plant Physiol. *46* (Suppl.):
      11, 1970.

298 - van AUKEN, O.W., McNULTY, I.B.: A preliminary investigation of the pigment com-
      position of a halophilic *Chlamydomonas*. - Proc. Utah Acad. Sci. Arts Lett. *45*:
      313-314, 1968.

299 - AUSTIN, R.B., LONGDEN, P.C.: A rapid method for the measurement of rates of
      photosynthesis using $^{14}CO_2$. - Ann. Bot. *31*: 245-253, 1967.

300 - AVAKIMOVA, L.T.: Vremennoĭ khod vidimogo fotosinteza lista fasoli v zavisimosti
      ot dlitel'nosti vozdeĭstviya mannita. [Time course of apparent photosynthesis
      of French bean leaves in relation to the length of mannitol application.] -
      Dokl. mosk. sel'.-khoz. Akad. K.A. Timiryazeva *160*: 142-147, 1970. [In R.]

301 - AVDONIN, N.S., ARENS, I.P.: Vliyanie molibdena na biokhimicheskie protsessy v
      rasteniyakh i na kachestvo rastitel'noĭ produktsii. [Effect of molybdenum on
      biochemical processes in plants and on quality of plant production.] - Agrokhi-
      miya *1966* (3): 70-79, 1966. [Ps, Chl, Car.]

302 - AVERY, D.J.: The supply of air to leaves in assimilation chambers. - J.exp.Bot.
      17: 655-677, 1966.

303 - AVERY, D.J.: The temperatures of leaves in assimilation chambers, and in the
      open. - J. exp. Bot. *18*: 379-396, 1967.

304 - AVERY, D.J., BRIGGS, J.B.: The aetiology and development of damage in young
      fruit trees infested with fruit tree red spider mite, *Panonychus ulmi* (KOCH).
      - Ann. appl. Biol. *61*: 277-288, 1968.[Growth analysis.]

305 - AVRATOVŠČUKOVÁ, N.: Differences in photosynthetic rate of leaf disks in five
      tobacco varieties. - Photosynthetica *2*: 149-160, 1968.

306 - AVRON, M.: Mechanism of photoinduced electron transport in isolated chloroplasts.
      - Curr. Topics Bioenerg. *2*: 1-22, 1967.

307 - AVRON, M.: Photochemistry and biochemistry in photosynthesis. - In: SWANSON, C.P.
      (ed.): An Introduction to Photobiology. Pp. 143-156. Prentice-Hall, Inc., Engle-
      wood Cliffs, N.J. 1969.

308 - AVRON, M., BEN-HAYYIM, G.: Interaction between two photochemical systems in
      photoreactions of isolated chloroplasts. - In: METZNER, H. (ed): Progress in
      Photosynthesis Research. Vol. III. Pp. 1185 -1196. Tübingen 1969.

309 - AVRON, M., CHANCE, B.: The relation of light-induced oxidation reduction changes
      in cytochrome *F* of isolated chloroplasts to photophosphorylation. - In: THOMAS,
      J.B., GOEDHEER, J.C. (ed.): Currents in Photosynthesis. Pp. 455-464. Donker,
      Rotterdam 1966.

310 - AVRON, M., CHANCE, B.: Relation of phosphorylation to electron transport in
      isolated chloroplasts. - In: Energy Conversion by the Photosynthetic Apparatus.
      Brookhaven Symp. Biol. *19*: 149-160, 1967.

311 - AVRON, M., NEUMANN, J.: Photophosphorylation in chloroplasts. - Annu. Rev.
      Plant Physiol. *19*: 137-166, 1968.

312 - AWA, T.: [On the carotenoids and organic acid of *Basella rubra* LINN.] - Res.
      Bull. Fac. Education, Oita Univ., nat. Sci. [Oita Daigaku Kyoikugakubu Kenkyu
      Kiyo, Shizenkagaku] *3* (4): 53-62, 1969. [In Jap., ab: E.]

313 - AWASTHI, P.: Behaviour of catalase in some members of *Chlorophyceae* and *Cyano-
      phyceae*. - Rev. algol. *8*: 307-311, 1967. [Chl.]

314 - AYLOR, D.E., KRIKORIAN, A.D.: Transient heat and mass transfer from a leaf un-
      dergoing stomatal closure in a low pressure environment. - Plant Physiol. *46*:
      557-563, 1970.

315 - AZIMURATOVA, R.Zh., BUSHUEVA, T.M., MOKHOVA, L.S.: O fotosinteze i aktivnosti
      izolirovannykh khloroplastov u gorokha pri nedostatke kal'tsiya. [Photosynthesis
      and activity of isolated chloroplasts in Ca-deficient pea plants.] - Vestn. le-
      ningrad. Univ. *22* (3 Biol. - 1): 124-130, 1967. [In R.]

316 - van BAALEN, C.: The effects of ultraviolet irradiation on a coccoid blue-green
      alga: survival, photosynthesis, and photoreactivation. - Plant Physiol. *43*:
      1689-1695, 1968.

317 - BABAEVA, T.N.: O zavisimosti fotosinteza mutantov *Arabidopsis thaliana* ot kont-
      sentratsii uglekisloty. [Dependence of photosynthesis of mutants of *Arabidopsis
      thaliana* on carbon dioxide concentration.] - In: Fotosintez i Ispol'zovanie
      Energii Solnechnoi Radiatsii. Pp. 37-38. Dushanbe 1967.

318 - BABAEVA, T.N.: Manometricheskie dannye o fotosinteza khlopchatnika. [Manometric
      data on cotton photosynthesis.] -Izv. Akad. Nauk Tadzh. SSR, Otd. biol. Nauk
      1969 [2 (35)]: 72-75, 1969. [In R, ab: Tadzh.]

319 - BABAEVA, T.N., NASYROV, Yu.S., TOLIBEKOV, D.T., BOBODZHANOV, V.A.: Deĭstvie ki-
      netina na intensivnost' fotosinteza pigmentnykh mutantov gorokha. [Effect of
      kinetin on photosynthetic rate of pigment mutants of pea.] - Dokl. Akad. Nauk
      Tadzh. SSR *12* (11): 55-57, 1969. [In R, ab: Tadzh.]

320 - BABAEVA, T.N., NIKITINA, A.N., BOBODZHANOV, V.A.: Fotosintez mutantov gorokha i
      deĭstvie kinetina na intensivnost' protsessa. [Photosynthesis of pea mutants and
      effect of kinetin on the rate of the process.] - In: Tezisy Dokladov Vtorogo
      Vsesoyuznogo Biokhimicheskogo S"ezda. Sektsiya "Problemy Fotosinteza". P. 50.
      FAN. Uzb. SSR, Tashkent 1969. [In R.]

321 - BABALOLA, O., BOERSMA, L., YOUNGBERG, C.T.: Photosynthesis and transpiration of
      Monterey pine seedlings as a function of soil water suction and soil tempera-
      ture. - Plant Physiol. *43*: 515-521, 1968.

322 - BABAYAN, G.B.: K metodike kolichestvennogo ucheta massy kornei na gornykh lu-
      gakh. [Method for the quantitative determination of the root-mass in alpine
      meadows.] - Biol. Zh. Armenii *23* (2): 10-16, 1970. [In R, ab: Armen.]

323 - **BABENKO, D.K.**: Opredelenie optimal'noĬ gustoty stoyaniya sosny obyknovennoĬ
       starshe pyatiletnego vozrasta na nizhnedneprovskikh peskakh. [Determination of
       optimum stand density of pine trees more than five years old on sands of Low
       Dnepr.] - Lesovod. Agrolesomelior. (respub. mezhvedom. temat. nauch. Sb.) *14*:
       30-38, 1968. [Chl; in R.]

324 - **BABICH, A.A., MAKAROV, O.V.**: AnaliticheskiĬ metod izucheniya protsessa formiro-
       vaniya listovoĬ poverkhnosti soi. [The analytical method of the study of the
       process of the formation of leaf surface in soybean.] - Vest. sel'skokhoz.
       Nauki (Moskva) *1969* (1): 97-102, 1969. [In R, ab: E,F,G.]

325 - **BABITSKIĬ, A.F.**: Fotofosforilirovanie v khloroplastakh kukuruzy. [Photophosphor-
       ylation in maize chloroplasts.] - Sb. Tr. Aspirantov molodykh nauch. Sotrud.
       vsesoyuz. nauch.-issled. Inst. Rastenievod. *10* (14): 192-199, 1969. [In R.]

326 - **BABUSHKIN, L.N.**: O fotosinteze u tomatov pri oroshenii. [Photosynthesis of ir-
       rigated tomato.] - Tr. mold. nauch.-issled. Inst. orosh. Zemled. Ovoshchevod.
       8: 5-18, 1968. [In R.]

327 - **BABUSHKIN, L.N., SHAKHOV, A.A.**: O fotoimpul'snoĬ stimulyatsii energeticheskikh
       protsessov i urozhaĬnosti u rasteniĬ. [Light-impulse stimulation of energetic
       processes and yield of plants.] - In: Svetoimpul'snoe Obluchenie RasteniĬ (Tr.
       Lab. evolyuts. i ekol. Fiziol. B.A. Kellera 6). Pp. 85-92. Nauka, Moskva 1967.

328 - **BACCARINI, A.**: Autotrophic incorporation of $C^{14}O_2$ in *Cuscuta australis* in rela-
       tion to its parasitism. - Experientia *22*: 46-47, 1966.

329 - **BACCARINI, A.**: Fissazione di $C^{14}O_2$ in colture di tessuti vegetali "in vitro".
       [$^{14}CO_2$ fixation in plant tissue cultures "in vitro".] - G. bot. ital. *101*:
       161-166, 1967. [In Ital.]

330 - **BACCARINI, A., MELANDRI, B.A.**: Studies on *Orobanche hederae* physiology: Pigments
       and CO_2 fixation. - Physiol. Plant. *20*: 245-250, 1967.

331 - **BACCARINI, A., MELANDRI, B.A.**: Relationship between increased NADP-linked gly-
       ceraldehyde-3-phosphate dehydrogenase activity and protein synthesis during the
       greening of etiolated pea seedlings. - Physiol. Plant. *23*: 444-451, 1970.

332 - **BACCARINI-MELANDRI, A., GEST, H., SAN PIETRO, A.**: A coupling factor in bacterial
       photophosphorylation. - J. biol. Chem. *245*: 1224-1226, 1970.

333 - **BACHMANN, M.D., ROBERTSON, D.S., BOWEN, C.C.**: Developmental changes in the fine
       structure of chloroplasts in etiolated leaves of *Zea mays*. - Amer. J. Bot. *53*:
       603, 1966.

334 - **BACHMANN, M.D., ROBERTSON, D.S., BOWEN, C.C.**: Thylakoid anomalies in relation
       to grana structure in pigment-deficient mutants of *Zea mays*. - J. Ultrastruct.
       Res. *28*: 435-451, 1969.

335 - **BACHMANN, M.D., ROBERTSON, D.S., BOWEN, C.C., ANDERSON, I.C.**: Chloroplast de-
       velopment in pigment deficient mutants of maize. I. Structural anomalies in
       plastids of allelic mutants of the w_3 locus. - J. Ultrastruct. Res. *21*: 41-60,
       1967.

336 - **BACHOFEN, R.**: Photooxidation of manganese by chloroplasts. - In: Energy Conver-
       sion by the Photosynthetic Apparatus. Brookhaven Symp. Biol. *19*: 478-484, 1967.

337 - **BACHOFEN, R.**: Effect of macrotetralide antibiotics on photosynthetic reactions
       in chloroplasts.-Verh. schweiz. naturforsch. Ges. *150*: 154-155, 1970.

338 - **BACHOFEN, R., ARNON, D.I.**: Crystalline ferredoxin from the photosynthetic bac-
       terium *Chromatium*. - Biochim. biophys. Acta *120*: 259-265, 1966.

339 - **BACHOFEN, R., SPECHT-JÜRGENSEN, I.**: Durch Säure-Base-Übergang induzierte ATP-
       P_a-Austauschreaktion in Chloroplasten. - Z. Naturforsch. *22 b*: 1051-1054, 1967.

340 - **BACHOFEN, R., SPECHT-JÜRGENSEN, I.**: Die Wirkung von Dinactin auf die cyclische
       Photophosphorylierung, die lichtinduzierte ATP-P_a-Austauschreaktion und den
       lichtinduzierten Protonentransport in Chloroplasten. - Planta *90*: 66-79, 1970.

341 - **BACK, H.L.**: An evaluation of an electronic instrument for pasture yield estima-
       tion. Part 1. General relationships. - J. brit. Grassland Soc. *23*: 216-222,
       1968.

342 - BACON, M.F.: Artifacts from chromatography of chlorophylls. - Biochem. J. *101*: 34-36, 1966.

343 - BACON, M.F., HOLDEN, M.: Changes in chlorophylls resulting from various chemical and physical treatments of leaves and leaf extracts. - Phytochemistry *6*: 193-210, 1967.

344 - BACON, M.F., HOLDEN, M.: Chlorophyllase of sugar-beet leaves. - Phytochemistry *9*: 115-125, 1970.

345 - BADANOVA, K.A., LEVINA, V.V.: O vliyanii gibberellina i retardanta CCC na zasukhoustoĭchivost' yachmenya. [On the effect of gibberellin and CCC on resistance of barley to water stress.] - Fiziol. Rast. *17*: 568-574, 1970. [Chl, in R., ab: E.]

346 - BAGAUTDINOVA, R.I.: Regulyatsiya fotosinteza rostovymi protsessami (na primere dvukh sortov soi). [Regulation of photosynthesis by growth processes (shown on two cultivars of soybean).] - Uchen. Zap. ural'. gos. Univ. Ser. biol. *58*: 95-104, 1967. [In R.]

347 - BAGAUTDINOVA, R.I.: Vliyanie ekzogennykh metabolitov na sintez uglevodov i organicheskikh kislot v list'yakh kartofelya. [Effect of exogenous metabolites on synthesis of carbohydrates and organic acids in potato leaves.] - Uchen. Zap. ural'. gos. Univ. Ser. biol. *113*: 89-97, 1970. [Ps, in R.]

348 - BAIG, S.R., YENTSCH, C.S.: A photographic means of obtaining monochromatic spectra of marine algae. - Appl. Opt. *8*: 2566-2568, 1969.

349 - BAILEY, J.L., KREUTZ, W.: Characterization of pigment-protein complexes related to photosystems I and II. - In: METZNER, H. (ed): Progress in Photosynthesis Research. Vol. I. Pp. 149-158. Tübingen 1969.

350 - BAILEY, W.A., KLUETER, H.H., KRIZEK, D.T., STUART, N:W.: CO_2 systems for growing plants. - Trans. ASAE *13*: 263-268, 1970. [CO_2 measurements.]

351 - BAILISS, K.W.: Infection of cucumber cotyledons by cucumber mosaic virus and the participation of chlorophyllase in the development of chlorotic lesions. - Ann. Bot. *34*: 647-655, 1970.

352 - BAIN, J.M.: A crystalline inclusion in the chloroplasts of the outer hypodermal cells of the banana fruit. - Aust. J. biol. Sci. *21*: 421-427, 1968.

353 - BAIRD, I.E., WETZEL, R.G.: A method for the determination of zero thickness activity of ^{14}C labeled benthic diatoms in sand. - Limnol. Oceanogr. *13*: 379-382, 1968.

354 - BAJAJ, Y.P.S., McALLAN, J.W.: Chlorophyll formation in excised potato roots under various light treatments. - Amer. J. Bot. *55*: 708, 1968.

355 - BAJAJ, Y.P.S., McALLEN, J.W.: Effect of various light treatments on chlorophyll formation in excised potato roots. - Physiol. Plant. *22*: 25-28, 1969.

356 - BAKANIDZE, M.Sh.: Vliyanie razlichnoĭ intensivnosti osveshcheniya na sostoyanie zelenykh plastid i nakoplenie plasticheskikh veshchestv v list'yakh chaya. [Effect of various illuminances on the state of chloroplasts and accumulation of plastic substances in tea leaves.] - Subtrop. Kul'tury *1966* (1): 104-109, 1966. [In R.]

357 - BAKARDJIEVA, N., IVANOVA, Y.: Effect of gibberellic acid and certain trace elements on the pigment extractability from hemp leaves. - Dokl. bolg. Akad. Nauk *21*: 1299-1302, 1968.

358 - BAKARDJIEVA, N., JORDANOV, N.: ESR studies on photoinduced changes in Mn^{2+} content in plant tissues and isolated chloroplasts. - Dokl. bolg. Akad. Nauk *20*: 719-722, 1967.

359 - BAKER, A.L., BROOK, A.J., KLEMER, A.R.: Some photosynthetic characteristics of a naturally occurring population of *Oscillatoria agardhii* GOMONT. - Limnol. Oceanogr. *14*: 327-333, 1969.

360 - BAKER, D.N.: Microclimate in the field. - Trans. ASAE *9*: 77-81, 84, 1966. [Ps.]

361 - BAKER, D.N., HESKETH, J.D.: Respiration and the carbon balance in cotton *(Gossypium hirsutum,* L.). - Beltwide Cotton Prod. Res. Conf. 1969.

362 - BAKER, D.N., MEYER, R.E.: Influence of stand geometry on light interception and net photosynthesis in cotton. - Crop Sci. *6*: 15-19, 1966.

363 - BAKER, D.N., MYHRE, D.L.: Effects of leaf shape and boundary layer thickness on photosynthesis in cotton *(Gossypium hirsutum)*. - Physiol. Plant. *22*: 1043-1049, 1969.

364 - BALAGUROVA, N.I.: Izmeneniya khloroplastov v list'yakh kartofelya pri delstvii na rasteniya zamorozkov. [Chloroplast changes in potato leaves of frost affected plants.] - Tsitologiya *10*: 95-101, 1968.

365 - BALASUBRAMANIAM, S., WILLIS, A.J.: Stomatal movements and rates of gasous exchange in excised leaves of *Vicia faba*. - New Phytol. *68*: 663-674, 1969.

366 - BALATINECZ, J.J., FORWARD, D.F., BIDWELL, R.G.S.: Distribution of photoassimilated $C^{14}O_2$ in young Jack pine seedlings. - Can. J. Bot. *44*: 362-364, 1966.

367 - BALAUR, N.S.: Struktura i funktsiya khloroplastov kukuruzy pri osveshchenii zhelto-zelenym svetom. [Structure and function of maize chloroplasts illuminated by yellow-green light.] - In: Fotosinteticheskaya Deyatel'nost' i Vliyanie na nee Mineral'nogo Pitaniya. Pp. 133-141. Kishinev 1970. [In R.]

368 - BALAUR, N.S., SHAKHOV, A.A.: Vliyanie infrakrasnogo izlucheniya na formirovanie membrannol sistemy khloroplastov kukuruzy. [Effect of infra-red radiation on formation of membrane system in maize chloroplasts.] - In: Khloroplasty i Mitokhondrii. Pp. 199-208. Nauka, Moskva 1969. [In R.]

369 - BALDRY, C.W., BUCKE, C., COOMBS, J.: Light/phosphoenolpyruvate dependent carbon dioxide fixation by isolated sugar cane chloroplasts. - Biochem. biophys. Res. Commun. *37*: 828-832, 1969.

370 - BALDRY, C.W., BUCKE, C., COOMBS, J.: Effects of some phenoloxidase inhibitors on chloroplasts and carboxylating enzymes of sugar cane and spinach. - Planta *94*: 124-133, 1970.

371 - BALDRY, C.W., BUCKE, C., COOMBS, J., GROSS, D.: Phenols, phenoloxidase, and photosynthetic activity of chloroplasts isolated from sugar cane and spinach. - Planta *94*: 107-123, 1970.

372 - BALDRY, C.W., BUCKE, C., WALKER, D.A.: Incorporation of inorganic phosphate into sugar phosphates during carbon dioxide fixation by illuminated chloroplasts. - Nature *210*: 793-796, 1966.

373 - BALDRY, C.W., BUCKE, C., WALKER, D.A.: Temperature and photosynthesis. I. Some effects of temperature on carbon dioxide fixation by isolated chloroplasts. - Biochim. biophys. Acta *126*: 207-213, 1966.

374 - BALDRY, C.W., COCKBURN, W., WALKER, D.A.: Inhibition, by sulphate, of the oxygen evolution associated with photosynthetic carbon assimilation. - Biochim. biophys. Acta *153*: 476-483, 1968.

375 - BALDRY, C.W., COOMBS, J., GROSS, D.: Isolation and separation of chloroplasts from sugar cane. - Z. Pflanzenphysiol. *60*: 78-81, 1968.

376 - BALDRY, C.W., WALKER, D.A., BUCKE, C.: Calvin-cycle intermediates in relation to induction phenomena in photosynthetic carbon dioxide fixation by isolated chloroplasts. - Biochem. J. *101*: 642-646, 1966.

377 - BALDWIN, B.C.: On the mode of action of bipyridylium herbicides in photosynthesis. - In: METZNER, H. (ed.): Progress in Photosynthesis Research. Vol. III. Pp. 1737-1741. Tübingen 1969.

378 - BALDWIN, B.C., CLARKE, C.B., WILSON, I.F.: Paraquat in chloroplasts. - Biochim. biophys. Acta *162*: 614-617, 1968.

379 - BALDY, C., JONARD, P.: Contribution à l'étude de rendement photosynthétique du blé. - Compt. rend. Acad. Sci. Paris, Ser. D *268*: 1296-1299, 1969.

380 - BALEGH, M.S.E.D.A.: Photosynthetic studies on the bean plant. - Diss. Abstr. int., Ser. B *30*: 3045-B, 1970.

381 - BALEGH, S.E., BIDDULPH, O.: The photosynthetic action spectrum of the bean plant. - Plant Physiol. *46*: 1-5, 1970.

382 - BÁLINT, J., KERECSÉNYI, L., SÁRKÁNY, D., HORVÁTH, M.: Change of total chlorophyll in vegetable plants. - Acta agron. Acad. Sci. hung. *17*: 437-440, 1968.

383 - BALLANTINE, J.E.M., FORDE, B.J.: The effect of light intensity and temperature
on plant growth and chloroplast ultrastructure in soybean. - Amer. J. Bot. *57*:
1150-1159, 1970.

384 - BALLESTER, A.: Fiziologicheskii krugooborot razlichnykh pigmentov fitoplanktona i ekologicheskoe znachenie sostava pigmentov. [Physiological cycle of different phytoplankton pigments and ecological significance of pigment composition.]
- In: 2-T Mezhdunarodnyi Okeanograficheskii Kongress 1966. Tezisy Dokladov. P.
14. Nauka, Moskva 1966.

385 - BALLESTER, A.: Crítica de los métodos espectrofotométrico y cromatográfico en
el estudio de los pigmentos del plancton. [Criticism of spectrophotometric and
chromatographic methods of studying plankton pigments.] - Invest. Pesquera *30*:
613-630, 1966. [In Span., ab: E.]

386 - BALLESTER, A., PLANA, A.: Estudio cromatográfico y autorradiográfico de los pigmentos fotosintetizadores en Algas marinas. [A chromatographic and autoradiogra-
phic study of photosynthetizing pigments in marine algae.] - Proc. int. Seaweed
Symp. *6*: 427-433, 1969. [In Span., ab: E.]

387 - BALLSCHMITER, K.: Recent investigations on chlorophylls. - Angew. Chem. (int.
Ed.) *8*: 617, 1969.

388 - BALLSCHMITER, K., COTTON, T.M., STRAIN, H.H., KATZ, J.J.: Chlorophyll-water interactions. Hydration, dehydration and hydrates of chlorophyll. - Biochim. biophys. Acta *180*: 347-359, 1969.

389 - BALLSCHMITER, K., KATZ, J.J.: Long wavelength forms of chlorophyll. - Nature
220: 1231-1233, 1968.

390 - BALLSCHMITER, K., KATZ, J.J.: An infrared study of chlorophyll-chlorophyll and
chlorophyll-water interactions. - J. amer. chem. Soc. *91*: 2661-2677, 1969.

391 - BALLSCHMITER, K., TRUESDELL, K., KATZ, J.J.: Aggregation of chlorophyll in non-
polar solvents from molecular weight measurements. - Biochim. biophys. Acta
184: 604-613, 1969.

392 - BALNY, C., BRODY, S.S., HUI BON HOA, G.: Absorption and fluorescence spectra of
chlorophyll-a in polar solvents as a function of temperature. - Photochem.
Photobiol. *9*: 445-454, 1969.

393 - BALTSCHEFFSKY, H.: Energy coupling at different coupling sites in photophosphorylation. - In: GOODWIN, T.W. (ed.): Biochemistry of Chloroplasts. Vol. II.
Pp. 581-591. Academic Press, London-New York 1967.

394 - BALTSCHEFFSKY, H.: Biochemistry of electron transport and photophosphorylation.
- In: METZNER, H. (ed.): Progress in Photosynthesis Research. Vol. III. Pp.
1420-1424. Tübingen 1969.

395 - BALTSCHEFFSKY, H., BALTSCHEFFSKY, M., von STEDINGK, L.-V.: Inorganic pyrophosphate, bacterial photophosphorylation, and evolution of biological energy
transformation. - In: METZNER, H. (ed.): Progress in Photosynthesis Research.
Vol. III, Pp. 1313-1318. Tübingen 1969.

396 - BALTSCHEFFSKY, H., von STEDINGK, L.V.: Bacterial photophosphorylation in the
absence of added nucleotide. A second intermediate stage of energy transfer in
light-induced formation of ATP. - Biochem. biophys. Res. Commun. *22*: 722-728,
1966.

397 - BALTSCHEFFSKY, H., von STEDINGK, L.-V.: Energy transfer from two coupling sites
in bacterial photophosphorylation. - In: THOMAS, J.B., GOEDHEER, J.C. (ed.):
Currents in Photosynthesis. Pp. 253-262. Donker, Rotterdam 1966.

398 - BALTSCHEFFSKY, H., von STEDINGK, L.-V., HELDT, H.-W., KLINGENBERG, M.: Inorganic
pyrophosphate: formation in bacterial photophosphorylation. - Science *153*:
1120-1122, 1966.

399 - BALTSCHEFFSKY, M.: Inorganic pyrophosphate and ATP as energy donors in chromatophores from *Rhodospirillum rubrum*. - Nature *216*: 241-243, 1967.

400 - BALTSCHEFFSKY, M.: Inorganic pyrophosphate as an energy donor in photosynthetic
    and respiratory electron transport phosphorylation system. - Biochem. biophys.
    Res. Commun. 28: 270-276, 1967.

401 - BALTSCHEFFSKY, M.: Inorganic pyrophosphate as energy donor in photosynthetic
    and respiratory structures. - In: JÄRNEFELT, J. (ed.): Regulatory Functions of
    Biological Membranes. Pp. 277-286. Elsevier Publishing Co., Amsterdam-London-
    New York 1968.

402 - BALTSCHEFFSKY, M.: Energy conversion-linked changes of carotenoid absorbance
    in *Rhodospirillum rubrum* chromatophores. - Arch. Biochem. Biophys. 130: 646-652,
    1969.

403 - BALTSCHEFFSKY, M.: Reversed energy conversion reactions of bacterial photophos-
    phorylation. - Arch. Biochem. Biophys. 133: 46-53, 1969.

404 - BALTSCHEFFSKY, M.: Reversed energy transfer in *Rhodospirillum rubrum* chromato-
    phores. - In: METZNER, H. (ed.): Progress in Photosynthesis Research. Vol. III.
    Pp. 1306-1312. Tübingen 1969.

405 - BALTSCHEFFSKY, M., BALTSCHEFFSKY, H., von STEDINGK, L.-V.: Light-induced energy
    conversion and the inorganic pyrophosphatase reaction in chromatophores from
    Rhodospirillum rubrum. - In: Energy Conversion by the Photosynthetic Apparatus,
    Brookhaven Symp. Biol. 19: 246-257, 1967.

406 - BAMBERG, S., SCHWARZ, W., TRANQUILLINI, W.: Influence of daylength on the pho-
    tosynthetic capacity of stone pine (*Pinus cembra* L.). - Ecology 48: 264-269,
    1967.

407 - BAMBERGER, E., PARK, R.B.: The effect of hydrolytic enzymes on the photosynthe-
    tic efficiency and morphology of chloroplasts. - Israel J. Chem. 4: 81 p, 1966.

408 - BAMBERGER, E.S., PARK, R.B.: Effect of hydrolytic enzymes on the photosynthetic
    efficiency and morphology of chloroplasts. - Plant Physiol. 41: 1591-1600, 1966.

409 - BAMJI, M.S., JAGENDORF, A.T.: Amino acid incorporation by wheat chloroplasts. -
    Plant Physiol. 41: 764-770, 1966. [Chl.]

410 - BAMJI, M.S., KRINSKY, N.I.: The carotenoid pigments of a radiation-resistant
    Micrococcus species. - Biochim. biophys. Acta 115: 276-284, 1966.

411 - BANERJEE, A.K., TOLLIN, G.: Reversible light-induced single electron transfer
    reactions between chlorophyll and hydroquinone in solution. - Photochem. Photo-
    biol. 5: 315-322, 1966.

412 - BANERJI, D., LALORAYA, M.M.: Chlorophyll formation in isolated pumpkin cotyle-
    dons in the presence of kinetin and chloramphenicol. - Plant Cell Physiol. 8:
    263-268, 1967.

413 - BANNISTER, T.T.: Effect of carbonyl cyanide p-trifluoromethoxyphenylhydrazone
    on chlorophyll fluorescence and photosynthesis. - Biochim. biophys. Acta 143:
    275-278, 1967.

414 - BANNISTER, T.T., RICE, G.: Parallel time courses of oxygen evolution and chlo-
    rophyll fluorescence. - Biochim. biophys. Acta 162: 555-580, 1968.

415 - BANNISTER, W.H., CHOUSSY, M., VUILLAUME, M., BARBIER, M.: The prosthetic group
    of the green chromoprotein of *Patella* ova. - Experientia 26: 1211, 1970. [Chl.]

416 - BANSE, K., ANDERSON, G.C.: Computations of chlorophyll concentrations from
    spectrophotometric readings. - Limnol. Oceanogr. 12: 696-697, 1967.

417 - BANTHORPE, D.V., WIRZ-JUSTICE, A.: Terpene biosynthesis. Part I. Preliminary
    tracer studies on terpenoids and chlorophyll of *Tanacetum vulgare* L. - J. chem.
    Soc. (London) C 1969: 541-549, 1969.

418 - BARA, M., GALSTON, A.W.: Experimental modification of pigment content and pho-
    totropic sensitivity in excised *Avena* coleoptiles. - Physiol. Plant. 21: 109-
    118, 1968.

419 - BARABAL'CHUK, K.A.: Reaktsiya termolabil'nykh i termostabil'nykh funktsiĭ ras-
    titel'noĭ kletki na deĭstvie teplovoĭ zakalki. [Response of thermolabile and
    thermostabile functions of plant cells to the action of heat hardening.] - Tsi-

tologiya *11*: 1021-1032, 1969. [Chloroplast; in R, ab: E.]

420 - BARABAL'CHUK, K.A.: Izmenenie intensivnosti fluorestsentsii khlorofilla v list'-yakh tradeskantsii posle deTstviya nagreva. [Heating-induced change of intensity of chlorophyll fluorescence in *Tradescantia* leaves.] - Tsitologiya *12*: 1009-1019, 1970. [In R, ab: E.]

421 - BAR-AKIVA, A., LAVON, R.: Visible symptoms and some metabolic patterns in micro-nutrient-deficient Eureka lemon leaves. - Israel J. agr. Res. *17*: 7-16, 1967. [Chl.]

422 - BAR-AKIVA, A., LAVON, R.: Peroxidase activity as an indicator of the iron requirement of citrus plants. - Israel J. agr. Res. *18*: 145-153, 1968. [Chl.]

423 - BARANAŬ, A.A., SHLYK, A.A.: Ab kharaktary zdabyvannya khlarafilu z kletak eugleny malapalyarnym rastvaral'nikam. [Chlorophyll extraction from *Euglena* cells with solvents of a low polarity.] - Vestsi Akad. Nauk Belorus. SSR, Ser. biyal. Nauk *1967* (2): 41-51, 1967. [In Belorus.]

424 - BARANINA, I.I., MAKHARINETS, S.N.: O vozmozhnosti zimnego dykhaniya i produktiv-nogo fotosinteza u ozimoT pshenitsy. [On the possibility of winter respiration and productive photosynthesis in winter wheat.] - In: Materialy IV. Konferent-sii Molodykh Uchennykh Moldavii, 1964. Sekts. Fiziologii, Biokhimii i Genetiki RasteniT. Pp. 7-10. Kishinev 1966. [In R.]

425 - BARASHKOVA, E.A.: Fotosintez kustistykh lishaTnikov *Cladonia alpestris* (L.) RABH. i *Cladonia rangiferina* (L.) WEB. na TaTmyre. [Photosynthesis of fructi-cose lichens *Cladonia alpestris* (L.) RABH. and *Cladonia rangierina* (L.) WEB. in the TaTmyr peninsula.] - Bot. Zh. *55*: 284-292, 1970. [In R.]

426 - BARBER, J.: Biophysical aspects of photosynthesis. A report of the British Bio-physical Society Meeting held in London on 17 and 18 December 1969. - FEBS Lett. *6*: 289-294, 1970.

427 - BARBER, J., KRAAN, G.P.B.: Salt-induced light emission from chloroplasts. - Biochim. biophys. Acta *197*: 49-59, 1970.

428 - BARBER, R.T., WHITE, A.W., SIEGELMAN, H.W.: Evidence for a cryptomonad symbiont in the ciliate, *Cyclotrichium meunieri*. - J. Phycol. *5*: 86-88, 1969. [Chl, car.]

429 - BARBIERI, G., DELOSME, R., JOLIOT, P.: Comparaison entre l'émission d'oxygène et l'émission de luminescence à la suite d'une série d'éclairs saturants. - Photochem. Photobiol. *12*: 197-206, 1970.

430 - BARBOUR, M.: Early growth in annual and perennial ryegrass. - Agron. J. *59*: 204-205, 1967. [Ps, Chl.]

431 - BARD, S.A., GORDON, M.P.: Studies on spinach chloroplast and nuclear DNA using large-scale tissue preparations. - Plant Physiol. *44*: 377-384, 1969.

432 - BARJONA DE FREITAS, A.G.: Enrichissement en sucres et accroissement du volume des baies: Mécanisme, facteurs. Rôle du feuillage pour le rendement et la qua-lité du raisin. Productivité du feuillage: Rapport portugais. - Bull. OIV Office int. Vigne Vin. *41*: 862-874, 1968.

433 - BARKER, H.: Methods of measuring leaf surface area of some conifers. - Can. Forestry Dept. Publ. *1219*: 1-5, 1968.

434 - BARNES, D.K., PEARCE, R.B., CARLSON, G.E., HART, R.H., HANSON, C.H.: Specific leaf weight differences in alfalfa associated with variety and plant age. - Crop Sci. *9*: 421-423, 1969.

435 - BARNES, D.L., LYND, J.Q.: Factors in paraquat-induced chlorosis with *Phaseolus* foliar tissue. - Agron. J. *59*: 364-366, 1967.

436 - BAROOVA, S.R., HORVÁTH, I., SZÁSZ, K.: Dry weight and carbohydrate changes in tomato seedlings germinated in dark and light. - Acta biol. Szeged *16* (3-4): 73-78, 1970.

437 - BARR, R., CRANE, F.L.: Comparative studies on plastoquinones. III. Distribution of plastoquinones in higher plants. - Plant Physiol. *42*: 1255-1263, 1967. [Chl.]

438 - BARR, R., CRANE, F.L.: Comparative studies on plastoquinones. V. Changes in li-
       pophilic chloroplast quinones during development. - Plant Physiol. *45*: 53-55,
       1970. [Chl.]

439 - BARR, R., HENNINGER, M.D., CRANE, F.L.: Comparative studies on plastoquinone.
       II. Analysis for plastoquinones A, B, C and D. - Plant Physiol. *42*: 1246-1254,
       1967. [Chl, car.]

440 - BARR, R., MAGREE, L., CRANE, F.L.: Quinone distribution in horse-chestnut chlo-
       roplasts, globules and lamellae. - Amer. J. Bot. *54*: 365-374, 1967. [Chl.]

441 - BARRENTINE, J.L., WARREN, G.F.: Selective action of terbacil on peppermint and
       ivyleaf morningglory. - Weed Sci. *18*: 373-377, 1970. [Ps.]

442 - BARRS, H.D.: Effect of cyclic variations in gas exchange under constant environ-
       mental conditions on the ratio of transpiration to net photosynthesis. - Physiol.
       Plant. *21*: 918-929, 1968.

443 - BARRS, H.D., KLEPPER, B.: Cyclic variations in plant properties under constant
       environmental conditions. - Physiol. Plant. *21*: 711-730, 1968. [Stomata]

444 - BARTA, A.L., HODGES, H.F.: Characterization of photosynthesis in cold hardening
       winter wheat. - Crop Sci. *10*: 535-538, 1970.

445 - BARTELS, P.G., HOSHAW, R.W.: Cylindrical structures in the chloroplasts of *Siro-
       gonium melanosporum*. - Planta *82*: 293-298, 1968.

446 - BARTELS, P.G., HYDE, A.: Chloroplast development in 4-chloro-5-(dimethylamino)-
       2-(α,α,α-trifluoro-m-tolyl)-3(2H)-pyridazinone(Sandoz 6706)-treated wheat seed-
       lings. A pigment, ultrastructural, and ultracentrifugal study. - Plant Physiol.
       45: 807-810, 1970.

447 - BARTELS, P.G., WEIER, T.E.: Particle arrangements in proplastids of *Triticum
       vulgare* L. seedlings. - J. Cell Biol. *33*: 243-253, 1967.

448 - BARTELS, P.G., WEIER, T.E.: The effect of 3-amino-1,2,4-triazole on the ultra-
       structure of plastids of *Triticum vulgare* seedlings. - Amer. J. Bot. *56*: 1-7,
       1969.

449 - BARTHOVÁ, J., LEBLOVÁ, S., KOŠTÍŘ, J.: The influence of light, darkness and
       changes in CO_2 and O_2 concentration in the atmosphere on growth and gas exchan-
       ge in pea *(Pisum sativum)*. - Biol. Plant. *9*: 173-181, 1967.

450 - BARTLETT, L., KLYNE, W., MOSE, W.P., SCOPES, P.M., GALASKO, G., MALLAMS, A.K.,
       WEEDON, B.C.L., SZABOLCS, J., TÓTH, G.: Optical rotatory dispersion of carote-
       noids. - J. chem. Soc. Sect. C *1969* (18): 2527-2544, 1969.

451 - BARTON, R.: Fine structure of mesophyll cells in senescing leaves of *Phaseolus*.
       - Planta *71*: 314-325, 1966.

452 - BARTON, R.: The production and behaviour of phytoferritin particles during se-
       nescence of *Phaseolus* leaves. - Planta *94:* 73-77, 1970. [Chloroplast.]

453 - BARTOŠ, J.: Productivity of photosynthesis of cultures of algae and leaves of
       field products under laboratory conditions. - Acta Univ. Carolinae - Biol.
       1966 (Suppl. 1/2): 121, 1966.

454 - BARTOŠOVÁ, E., KONÍČEK, J.: The use of combustion calorimetry in the study of
       the metabolism of algae. - Photosynthetica *1*: 13-17, 1967.

455 - BARUA, D.N.: Light as a factor in metabolism of the tea plant (*Camellia sinen-
       sis* L.). - In: LUCKWILL, L.C., CUTTING, C.V. (ed.): Physiology of Tree Crops.
       Pp. 307-322. Academic Press, New York-London 1970. [Ps.]

456 - BARUA, R.K., BARUA, A.B.: Oxidation of zeaxanthin. Isolation and properties of
       3-hydroxyertinene. - Biochem. J. *101*: 250-255, 1966.

457 - BARUFFINI, A., BORGNA, P., CALDERARA, G., MAZZA, M.: Attività fitotossica di
       p-alchiltioanilidi. [The phytotoxicity of *p*-alkylthioanilides.] - Farmaco, Ed.
       sci. *25*: 427-441, 1970. [Inhibition of Hill reaction; in Ital., ab: E.]

458 - BARUFFINI, A., BORGNA, P., CALDERARA, G. MAZZA, M., GIALDI, F.: *p*-alchiltioani-
       lidi inibenti la reazione di Hill. [*p*-alkylthioanilidine inhibitors of the Hill

reaction.] - Farmaco, Ed. sci. *25*: 10-35, 1970. [In Ital., ab: E.]

459 - BARUFFINI, A., BORGNA, P., PAGANI, G.: Acidi ossanilici e derivati. [Oxanilic
acid derivatives.] - Farmaco, Ed. sci. *22*: 717-734, 1967. [Inhibitors of Hill
reaction; in Ital., ab: E.]

460 - BARUFFINI, A., BORGNA, P., PAGANI, G., GIALDI, F.: Anilidi inibenti la fotosintesi clorofilliana ad attività fitotossica selettica. [Selective phytotoxic
anilides which inhibit chlorophyll photosynthesis.] - Farmaco, Ed. sci. *22*: 895-
916, 1967. [In Ital., ab: E.]

461 - BARUFFINI, A., BORGNA, P., PAGANI, G., PONCI, R.: Anilidi inibenti la reazione
di Hill. [Anilides which inhibit the Hill reaction.] - Farmaco, Ed. sci. *22*:
590-611, 1967. [In Ital., ab: E.]

462 - BARUFFINI, A., BORGNA, P., PAGANI, F., PONCI, R.: Attività fitotossica di anilidi inibenti la reazione di Hill. [Phytotoxic effect of Hill reaction inhibiting analides.] - Farmaco, Ed. sci. *22*: 612-626, 1967. [In Ital., ab: E.]

463 - BASIEV, K.Kh.: O soderzhanii khlorofilla v list'yakh geterozisnykh mezhlineĭnykh
gibridov kukuruzy i ikh roditel'skikh samoopylennykh liniĭ. [Chlorophyll content
in leaves of heterotic interline hybrids of maize and their parental selfpolli-
nated lines.] - Tr. Gorsk. sel'.-khoz. Inst. *27*: 137-142, 1967. [In R.]

464 - BASLAVSKAYA, S.S., BAMBUROVA, L.S.: K kharakteristike assimilyatsionnoĭ deyatel'-
nosti i sostava kletok *Scenedesmus quadricauda* raznogo vozrasta. [Characteris-
tics of photosynthetic activity and cell composition of *Scenedesmus quadricauda*
of different age.] - Vest. mosk. Univ. Biol. Pochvoved. *1966* (5): 27-34, 1966.
[In R.]

465 - BASLAVSKAYA, S.S., KULIKOVA, R.F., KURKOVA, E.B.: Deĭstvie azotnogo golodaniya i
azotnykh podkormok na fotosintez. Soderzhanie khlorofilla i rost kul'tury protokokkovykh vodorosleĭ. [Effect of nitrogen deficiency and nitrogen nutrition on
photosynthesis. Chlorophyll content and growth of culture of protococcous algae.]
- Byul. mosk. Obshch. Ispyt. Prirody, Otd. biol.*71* (2): 107-117, 1966. [In R.]

466 - BASSHAM, J.A.: Photosynthesis. - In: SCOTT, A.F. (ed.): Survey of Progress in
Chemistry. Vol. 3. Pp. 1-54. Academic Press, New York-London 1966.

467 - BASSHAM, J.A.: The Calvin photosynthetic cycle in algae and leaves. - In:
COOMBS, J. (ed.): Photosynthesis in Sugar Cane. Pp. 1-14. Tate and Lyle Ltd.,
London 1969.

468 - BASSHAM, J.A., EL-BALDRY, A.M., KIRK, M.R., OTTENHEYM, H.C.J., SPRINGER-LEDERER,
H.: Photosynthesis of isolated chloroplasts. V. Effect of fixation rate and
metabolite transport from the chloroplast caused by added fructose-1,6-diphos-
phatase. - Biochim. biophys. Acta *223*: 261-274, 1970.

469 - BASSHAM, J.A., JENSEN, R.G.: Photosynthesis of carbon compounds. - In: SAN
PIETRO, A., GREER, F.A., ARMY, T.J. (ed.): Harvesting the Sun. Photosynthesis
in Plant Life. Pp. 79-110. Academic Press, New York-London 1967.

470 - BASSHAM, J.A., KIRK, M.: Dynamic metabolic regulation of the photosynthetic carbon reduction cycle. - In: SHIBATA, K., TAKAMIYA, A., JAGENDORF, A.T., FULLER,
R.C. (ed.): Comparative Biochemistry and Biophysics of Photosynthesis. Pp.
365-378. Univ. Tokyo Press, Tokyo; Univ. Park Press, State College, Pa. 1968.

471 - BASSHAM, J.A., KIRK, M., JENSEN, R.G.: Photosynthesis by isolated chloroplasts.
I. Diffusion of labeled photosynthetic intermediates between isolated chloroplasts and suspending medium. - Biochim. biophys. Acta *153*: 211-218, 1968.

472 - BASSHAM, J.A., KRAUSE, G.H.: Free energy changes and metabolic regulation in
steady-state photosynthetic carbon reduction. - Biochim. biophys. Acta *189*:
207-221, 1969.

473 - BASSHAM, J.A., SHARP, P., MORRIS, I.: The effect of Mg^{2+} concentration on the
pH optimum and Michaelis constants of the spinach chloroplast ribulosediphosphate carboxylase (carboxydismutase.) - Biochim. biophys. Acta *153*: 898-900,
1968.

474 - BASU, P.K.: A rapid method for the determination of the leaf area of *Oxalis cor-niculata* L. and *Tephrosia purpurea* (LINN.) PERS. - Ann. Bot. *33*: 77-82, 1969.

475 - BATALOVA, A.G.: VIiyanie podsushivaniya list'ev na osushchestvlenie svetovoĭ reaktsii prevrashcheniya ksantofillov. [Effect of drying of leaves on light-induced xanthophyll transformation.] - In: Tezisy Dokladov Molodykh Uchenykh na Vtoroĭ Respublikanskoĭ Konferentsii, Posvyashchennoĭ 50-letiyu VLKSM. P. 162. Dushanbe 1968.

476 - BATE, G.C., d'AOUST, A., CANVIN, D.T.: Calibration of infra-red CO_2 gas analyzers. - Plant Physiol. *44*: 1122-1126, 1969.

477 - BATTIN, G.A.W.: Nomograms for chlorophyll determinations. - J. mar. biol. Ass. U.K. *47*: 407-414, 1967.

478 - BAUER, H., HUTER, M., LARCHER, W.: Der Einfluss und die Nachwirkung von Hitze-und Kältestress auf den CO_2 Gaswechsel von Tanne und Ahorn. - Ber. deut. bot. Ges. *82*: 65-70, 1969.

479 - BAUM, S.J., ELLSWORTH, R.K.: The chromatographic separation of magnesium pro-toporphyrin IX dimethyl esters from zinc protoporphyrin IX dimethyl esters. - J. Chromatogr. *47*: 503-505, 1970.

480 - BAUMGARTNER, A.: Ecological significance of the vertical energy distribution in plant stands. - In: ECKARDT, F.E. (ed.): Functioning of Terrestrial Ecosys-tems at the Primary Production Level. Pp. 367-374. UNESCO, Paris 1968.

481 - BAUMGARTNER, A.: Meteorological approach to the exchange of CO_2 between the atmosphere and vegetation, particularly forest stands. - Photosynthetica *3*: 127-149, 1969.

482 - van BAVEL, C.H.M.: Physiological limitations of productivity. - Proc. S. Conf. Forest Tree Impr. *10*: 2-5, 1969.

483 - BAVRINA, T.V.: VIiyanie dlitel'noĭ temnoty na pigmentnyĭ apparat neĭtral'nykh, dlinnodnevnykh i korotkodnevnykh vidov. [Effect of prolonged darkness on the pigment apparatus of neutral, long-day and short-day species.] - Dokl. Akad. Nauk SSSR *167*: 464-467, 1966. [In R.]

484 - BAVRINA, T.V.: VIiyanie dliny dnya na khlorofill-belkovo-lipoidnyĭ kompleks rasteniĭ. [Effect of daylength on the chlorophyll-protein-lipoid complex in plants.] - Fiziol. Rast. *13*: 578-584, 1966. [In R, ab: E.]

485 - BAVRINA, T.V., AKSENOVA, N.P., KONSTANTINOVA, T.N.: K voprosu ob uchastii fotosinteza v fotoperiodizme. [On the participation of photosynthesis in photo-periodism.] - Fiziol. Rast. *16*: 381-391, 1969. [In R, ab: E.]

486 - BAXTER, J.H.: Absorption of chlorophyll phytol in normal man and in patients with Refsum's disease. - J. Lipid Res. *9*: 636-641, 1968. [Pheophytin-^{14}C prepa-ration and analysis.]

487 - BAXTER, J.H., STEINBERG, D.: Absorption of phytol from dietary chlorophyll in the rat. - J. Lipid Res. *8*: 615-620, 1967. [TLC of ^{14}C-pheophytins.]

488 - BAYER, E., DIETER, J.: Untersuchungen zur Struktur des Pflanzenferredoxins. - Hoppe-Seyler's Z. physiol. Chem. *351*: 537-543, 1970.

489 - BAYER, E., HAGENMAIER, H.: Structure and reconstitution of plant ferredoxin. - In: METZNER, H. (ed.): Progress in Photosynthesis Research. Vol. III. Pp. 1427-1432. Tübingen 1969.

490 - BAYER, E., JOSEF, D., KRAUSS, P., HAGENMAIER, H., RÖDER, A., TREBST, A.: Abbau und Resynthese des Aktivzentrums von Pflanzenferredoxin. - Biochim. biophys. Acta *143*: 435-437, 1967.

491 - BAYFIELD, R.F., BARRETT, J.D., FALK, R.H.: Determination of lipids in biologic-al materials by paper chromatography. - J. Chromatogr. *28*: 363-370, 1967. [Chl, Car.]

492 - BAYFIELD, R.F., FALK, R.H., BARRETT, J.D.: The separation and determination of α-tocopherol and carotenoids in serum or plasma by paper chromatography. - J. Chromatogr. *36*: 54-62, 1968.

493 - BAZANOVA, T.B., AKOPOVA, K.M.: Vliyanie naftenovykh kislot na nekotorye storony obmena i urozhaT tonkovoloknistogo khlopchatnika v razlichnykh usloviyakh pitaniya. [Effect of naphthoic acids on some aspects of metabolism and yield of fine-thread cotton under various nutrition.]-Izv. Akad. Nauk Turkm. SSR, Ser. biol. Nauk 5: 53-58, 1966. [Ps, Chl; in R.]

494 - BAZHANOVA, N.V., GEVORKYAN, A.G.: O vliyanii dlinnovolnovoT ul'trafioletovoT radiatsii na sootnoshenie i prevrashchenie plastidnykh pigmentov u vysokogornykh rasteniT. [Effect of long-wave UV radiation on the ratio and transformation of plastid pigments in alpine plants.] - Dokl. Akad. Nauk Arm. SSR 50: 111-117, 1970. [In R.]

495 - BAZHANOVA, N.V., GEVORKYAN, A.G., OGANESYAN, D.A.: Reaktsii vzaimoprevrashcheniya ksantofillov u vysokogornykh rasteniT. [Transformation reactions in xanthophylls of alpine plants.] - Biol. Zh. Armenii 22: 58-64, 1969. [In R.]

496 - BAZHANOVA, N.V., PODIN', V.S., SAPOZHNIKOV, D.I.: Zavisimost' svetovoT reaktsii prevrashcheniya ksantofillov v list'yakh gerani ot temperaturnykh usloviT. [Dependence of light reaction of xanthophyll transformation of geranium leaves on temperature.] - In: Issledovaniya po Fotosinteza. Pp. 19-26. Akad. Nauk Tadzh. SSR, Dushanbe 1967. [In R.]

497 - BAZZAZ, M., MOHANTY, P., GOVINDJEE: Photosynthetic study on two necrotic corn mutants. - Plant Physiol. 46 (Suppl.): 41, 1970.

498 - BEALE, S.I.: The regulation of chlorophyll biosynthesis in *Chlorella*. - Diss. Abstr. int. B 31: 1756-B-1757-B, 1970.

499 - BEALE, S.I.: The biosynthesis of δ-aminolevulinic acid in *Chlorella*. - Plant Physiol. 45: 504-506, 1970. [Chl.]

500 - BEBIASHVILI, Sh.L., ZHGENTI, T.G., NAMICHEÏSHVILI, O.M.: MatematicheskiT raschet plotnosti poseva sel'skokhozyaTstvennykh kul'tur. [Mathematical approach to the density of crop stands.] - Vestn. sel'.-khoz. Nauki (Moskva) 1969: 105-109, 1969. [In R, ab: E, F, G.]

501 - BECACOS-KONTOS, T., SVANSSON, A.: Relation between primary production and irradiance. - Mar. Biol. 2: 140-144, 1969.

502 - BECK, E., SELLMAIR, J., KANDLER, O.: Biosynthese der Hamamelose. I. Die intramolekulare ^{14}C-Verteilung in Hamamelose nach Assimilation von $^{14}CO_2$ und ^{14}C-positionsmarkierter Glucose durch Blätter von *Primula clusiana* TAUSCH. - Z. Pflanzenphysiol. 58: 434-451, 1968.

503 - BECKER, J.-D., DÖHLER, G., EGLE, K.: Die Wirkung monochromatischen Lichts auf die extrazelluläre Glykolsäure-Ausscheidung bei der Photosynthese von *Chlorella*. - Z. Pflanzenphysiol. 58: 212-221, 1968.

504 - BECKER, W.: Ein sehr empfindliches Manometer für Gasstoffwechseluntersuchungen mit dem Cartesianischen-Taucher-Mikrorespirometer. - Z. Biol. 116: 114-120, 1968.

505 - BEDELL, G., GOVINDJEE: Quantum yield of oxygen evolution and the Emerson enhancement effect in deuterated *Chlorella*.-Science 152: 1383-1385, 1966.

506 - BEDENKO, V.P.: Ispol'zovanie energii solnechnoT radiatsii na fotosintez v posevakh yarovoT pshenitsy, vozdelyvaemykh v gornykh usloviyakh. [Utilization of solar radiation for photosynthesis in stands of spring wheat grown in mountains.] - In: Aktinometriya i Optika Atmosfery. Pp. 370-375. Valgus, Tallin 1968. [In R.]

507 - BEDENKO, V.P., FEDYUSHIN, A.A.: O nekotorykh resul'tatakh izmereniya pryamoT solnechnoT fotosinteticheski aktivnoT radiatsii (FAR). [Some results of measuring direct solar photosynthetically active radiation.] In: Lesnaya Selektsiya, Semenovodstvo i Introduktsiya v Kazakhstane. Pp. 121-122. Alma-Ata 1969. [In R.]

508 - BEDENKO, V.P., VOÏNOVSKAYA, K.K., MYTS, A.M., NAZARENKO, S.D., USHAROVA, G.P.: Fotosinteticheskaya deyatel'nost' yarovykh pshenits, vozdelyvaemykh v predgormoT i srednegornoT zonakh ZailiTskogo Alatau. [Photosynthetic activity of spring wheat cultivars grown in low and middle zones of ZailiTsk Alatau.] - In: Trudy I. Konferentsii Biokhimikov Respublik SredneT Azii i Kazakhstana. Pp. 135-138. FAN, Tashkent 1967. [In R.]

509 - BEDENKO, V.P., VOĬNOVSKAYA, K.K., MYTS, A.M., NAZARENKO, S.D., USHAROVA, G.P.: Formirovanie urozhaya i fotosinteticheskaya deyatel'nost' yarovykh pshenits v svyazi s ikh vozdelyvaniem i reproduktsieĭ v gorakh. [Yield formation and photosynthetic activity of spring wheats in relation to their agrotechnics and reproduction in mountains.] - Tr. Inst. Bot. Akad. Nauk Kazakh. SSR (Alma-Ata) 25 (Fiziologicheskie Protsessy i Produktivnost' Yarovoĭ Pshenitsy): 47-60, 1968. [In R.]

510 - BEDNAR, T.W., SMITH, D.C.: Studies in the physiology of lichens. VI. Preliminary studies of photosynthesis and carbohydrate metabolism of the lichen *Xanthoria aureola*. - New Phytol. 65: 211-220, 1966.

511 - BEEVERS, H.: Metabolic sinks. - In: EASTIN, J.D., HASKINS, F.A., SULLIVAN, C.Y., van BAVEL, C.H.M. (ed.): Physiological Aspects of Crop Yield. Pp. 169-184. Amer. Soc. Agron. & Crop Sci. Soc. Amer., Madison, Wisc. 1969. [Ps.]

512 - BEEVERS, L.: Effect of gibberellic acid on the senescence of leaf discs of nasturtium *(Tropaeolum majus)*. - Plant Physiol. 41: 1074-1076, 1966. [Chl.]

513 - BEEVERS, L., GUERNSEY, F.S.: Interaction of growth regulators in the senescence of nasturtium leaf disks. - Nature 214: 941-942, 1967. [Chl.]

514 - BEEVERS, L., LOVEYS, B., PEARSON, J.A., WAREING, P.F.: Phytochrome and hormonal control of expansion and greening of etiolated wheat leaves. - Planta 90: 286-294, 1970.

515 - BEGG, J.E., JARVIS, P.G.: Photosynthesis in Townsville lucerne *(Stylosanthes humilis* H.B.K.). - Agr. Meteorol. 5: 91-109, 1968.

516 - BEGG, J.E., LAKE, J.V.: Carbon dioxide measurement: a continuous conductimetric method. - Agr. Meteorol. 5: 283-290, 1968.

517 - BEGROV, V.V.: Fotosinteticheskiĭ koeffitsient kak funktsiya sostava biomassy rasteniĭ. [Photosynthetic coefficient as a function of the composition of plant biomass.] - Zh. obshch. Biol. 31: 99-105, 1970. [In R, ab: E.]

518 - BEHNKE, H.D.: Cell structures in relation to translocation phenomena in plants. - In: BEEMSTER, A.B.R., DIJKSTRA, J. (ed.): Viruses of Plants. Pp. 28-43. North-Holland Publ.Co., Amsterdam 1966.

519 - BEHNKE, H.-D.: Die Siebröhren-Plastiden der Monocotyledonen. Vergleichende Untersuchungen über Feinbau und Verbreitung eines charakteristischen Plastidentyps. - Planta 84: 174-184, 1969.

520 - BEHNKE, H.-D.: Über Siebröhren-Plastiden und Plastidenfilamente der *Caryophyllales*. Untersuchungen zum Feinbau und zur Verbreitung eines weiteren spezifischen Plastidentyps. - Planta 89: 275-283, 1969.

521 - BEIDERBECK, R., NITSCHE, E.: Der Chlorophyllgehalt von Tumoren an bleichem und grünen Blattgewebe von *Kalanchoe fedtschenkoi*. - Planta 92: 57-63, 1970.

522 - BEĬSOVA, M.P., SEMENOV, A.D., LOPATINA, L.N.: K lyuminestsentnomu opredeleniyu khlorofilla i feofitina v fitoplanktone. [On the luminescent determination of chlorophyll and pheophytin in phytoplankton.] - Gidrodkhim. Materialy 51: 203-210, 1969. [In R.]

523 - BEKINA, R.M.: Glavnye metodicheskie podkhody k issledovaniyu fotofosforilirovaniya. [Main methodological approaches to studying photophosphorylation.] - In: KIRICHENKO, E.B. (ed.): Metody Issledovaniya Fotofosforilirovaniya. Pp. 37-49. Pushchino-na-Oke 1970. [In R.]

524 - BEKINA, R.M., KRASNOVSKIĬ, A.A.: Khranenie izolirovannykh khloroplastov bez izmeneniya aktivnosti fotofosforilirovaniya. [Storage of isolated chloroplasts without changes in the activity of photophosphorylation.] - Biokhimiya 33: 178-181, 1968. [In R, ab: E.]

525 - BEKINA, R.M., KRASNOVSKIĬ, A.A.: Konservatsiya khloroplastov s sokhraneniem ikh sposobnosti k fotofosforilirovaniyu. [Conservation of chloroplasts with preserving their photophosphorylation activity.] - In: KIRICHENKO, E.B. (ed.): Metody Vydeleniya Khloroplastov. Pp. 32-38. Pushchino-na-Oke 1970. [In R, ab: E.]

526 - BEKINA, R.M., KRASNOVSKIŸ, A.A.: Fotofosforiliruyuschchaya aktivnost' pri raz-
      lichnykh metodakh razrusheniya khloroplastov. [Effect of various methods of
      chloroplast destruction on the photophosphorylating activity.] - Biokhimiya 35:
      132-139, 1970. [In R, ab: E.]

527 - BELAN, N.F., ABDURAKHMANOVA, Z.N.: Razdelenie produktov fotosinteza metodom
      khromatografii v tonkikh sloyakh. [Separation of photosynthates by thin-layer
      chromatography.] - Dokl. Akad. Nauk Tadzh. SSR 12 (10): 61-64, 1969. [In R, ab:
      Tadzhik.]

528 - BELIKOV, I.F., SEMKIN, B.I.: Raspredelenie mechenykh assimilyatov u loby i mor-
      kovi v pervyĭ god ikh zhizni. [Distribution of labelled assimilates in radish
      and carrot in the first year of life.] - Sel'skokhoz. Biol. 3: 918-920, 1968.
      [In R.]

529 - BELIKOV, I.F., SEMKIN, B.I.: O raspredelenii mechenykh assimilyatov u redisa.
      [Distribution of labelled assimilates in radish.] - Dokl. VASKHNIL 1968 (6):
      19-21, 1968. [In R.]

530 - BELIKOV, P.S., ASAFOV, G.B.: Vliyanie skorosti nagreva i okhlazhdeniya vozdukha
      na vremennoĭ khod fotosinteza. [Effect of the rates of heating and cooling the
      air on the time course of photosynthesis.] - Izv. Timiryaz. sel'.-khoz. Akad.
      1966 (6): 3-12, 1966. [In R.]

531 - BELIKOV, P.S., ASAFOV, G.B.: Ustanovka dlya izucheniya statsionarnykh i perek-
      hodnykh odnykh sostoyaniĭ fotosinteza. [An apparatus for measuring rates of
      steady-state and transient photosynthesis.] - Izv. Timiryaz. sel'.-khoz. Akad.
      1968 (2): 24-34, 1968. [In R, ab: E.]

532 - BELIKOV, P.S., AVAKIMOVA, L.G.: Opyt ispol'zovaniya aeroponiki dlya izucheniya
      vremennogo khoda fotosinteza. [Use of aeroponics for studying time course of
      photosynthesis.] - Izv. Timiryaz. sel'.-khoz. Akad. 1966 (1): 32-41, 1966.
      [In R.]

533 - BELIKOV, P.S., MALOFEEV, V.M.: Vremennoj khod intensivnosti fotosinteza i flyu-
      orestsentsii u otchlenennogo lista. [Time course of photosynthetic rate and
      fluorescence in a detached leaf.] - Izv. Timiryaz. sel'.-khoz. Akad. 1968 (1):
      30-39, 1968. [In R.]

534 - BELIKOV, P.S., MALOFEEV, V.M.: Bystrye izmeneniya nekotorykh opticheskikh
      svoĭstv u otrezannogo lista fasoli. [Rapid changes of some optical properties
      in a cut bean leaf.] - Izv. Timiryaz. sel'.-khoz. Akad. 1968 (3): 3-12, 1968.
      [In R, ab: E.]

535 - BELIKOV, P.S., MOTORINA, M.V.: Zavisimost' fotosinteza ot vnutrennykh i vnesh-
      nikh usloviĭ. (Itogi i perspektivy issledovaniĭ). [Control of photosynthesis
      by internal and external factors. (Results and perspectives of research.)] -
      Dokl. TSKhA (Moskva) 139: 273-286, 1968. [In R.]

536 - BELL, D.E.: Kinetics of $^{14}CO_2$ uptake in relation to photorespiration and nitrite
      reduction. - Diss. Abstr. int. B 31: 1757-B, 1970.

537 - BELL, L.N.: O fiziologicheskom znachenii zelenogo tsveta fotosinteziruyushchikh
      rasteniĭ. [Physiological significance of the green colour of plants.] - Fiziol.
      Rast. 13: 7-14, 1966. [In R, ab: E.]

538 - BELL, L.N.: O vremennom khode pokazaniĭ termoelektricheskikh detektorov izluche-
      niya i termoelektricheskikh fotokalorimetrov. [The time course of the response
      of thermoelectric radiation detectors and of thermoelectric photocalorimeters.]
      - In: Issledovaniya Radiatsionnogo Rezhima Atmosfery. Pp. 133-140. Akad. Nauk
      Est. SSR, Tartu 1967. [In R.]

539 - BELL, L.N., BUKINA, G.S.: O rezhimakh raboty amperometricheskikh yacheek, ispol'-
      zuemykh dlya izucheniya obmena kisloroda v biologicheskikh ob"ektakh. [On the
      operation of amperometric chambers employed in oxygen metabolism studies.] -
      Biofizika 12: 1043-1049, 1967. [In R.]

540 - BELL, L.N., FEDENKO, E.P., MIL'GRAM, V.D.: Optimal'nyĭ energeticheskiĭ vykhod
      fotosinteza. [Optimum energetic yield of photosynthesis.] - Dokl. Akad. Nauk
      SSSR 180: 1480-1483, 1968.

541 - BELL, L.N., LIN'KOVA, E.A., SLOBODSKAYA, G.A., SPEKTOROV, K.S., FEDENKO, E.P.,
BUKINA, G.S.: Fotoenergetika sinkhronnoĭ kul'tury khlorelly. [Photoenergetics
of a synchronous *Chlorella* culture.] - Fiziol. Rast. *14*: 866-871, 1967. [In R,
ab: E.]

542 - BELL, L.N., SHUVALOVA, N.P.: Zapasanie energii sinego sveta, nesopryazhennoe s
vydeleniem kisloroda. [Blue radiant energy storage uncoupled from oxygen evolution.] - Fiziol. Rast. *17*: 1019-1027, 1970. [In R, ab: E.]

543 - BELL, L.N., SHUVALOVA, N.P., MIRONOVA (BUKINA), G.S., NICHIPOROVICH, A.A.:
Spektr energeticheskogo vykhoda fotosinteza. Anomaliya v korotkovolnovoĭ oblasti vidimogo spektra. [Spectrum of energy yield of photosynthesis. An anomaly
in the short-wave range of visible spectrum.] - Dokl. Akad. Nauk SSSR *182*:
1439-1442, 1968.

544 - BELL, P.R., FREY-WYSSLING, A., MÜHLETHALER, K.: Evidence for the discontinuity
of plastids in the sexual reproduction of a plant. - J. Ultrastruct. Res. *15*:
108-121, 1966.

545 - BELLMANN, K., MEINL, G., RAEUBER, A., PFEFFER, C., WINKEL, A.: Mehrjährige Untersuchungen über Stoffbildung und Entwicklung des Maises. 3. Mitt.: Die phy-
siologischen Grundlagen der Ertragsbildung und das Modell des Maiswachstums
und seine Variabilität. - Züchter *37*: 324-341, 1967.

546 - BELLMANN, K., RAEUBER, A., PFEFFER, C., WINKEL, A., MEINL, G.: Mehrjährige Un-
tersuchungen über Stoffbildung und Entwicklung des Maises. 1. Mitteilung. Der
Wachstumsverlauf und seine Abhängigkeit von einigen Witterungsfaktoren. - Z.
Pflanzenzücht. *58*: 105-121, 1967.

547 - BELLOTTI, A., COGHI, E., BARUFFINI, A., PAGANI, G., BORGNA, P.: Attivitá ini-
bente la reazione di Hill e fitotossicitá di p.alchilanilidi. [Inhibition of
the Hill reaction and phytotoxicity of p-alkylanilides.] - Farmaco, Ed. sci.
23: 591-619, 1968. [In Ital., ab: E.]

548 - BELOSLYUDOVA, L.F., POLIMBETOVA, F.A., BEDENKO, V.P., KISELEVA, L.I., NAZAREN-
KO, S.D.: Fiziologicheskie protsessy i produktivnost' yarovoĭ pshenitsy pri
proizrastanii na razlichnykh vysotakh. [Physiological processes and productivity of spring wheat grown at different altitudes.] - Trudy Akad. Nauk SSSR, bot.
Inst. V.L. Komarova, Ser. 4 - eksp. Bot. *19*: 63-80, 1967. [In R.]

549 - BELOZEROVA, L.S.: Vliyanie malonata na fotosintez, dykhanie i prevrashchenie
organicheskikh kislot u sukkulentov na svetu. [Effect of malonate on photosynthesis, respiration and transformation of organic acids in succulents on light.]
- Vestn. leningrad. Univ. *22* (3 - Biol.-1): 119-123, 1967. [In R.]

550 - BELYAEVA, O.B.: Issledovanie promezhutochnykh fotokhimicheskikh stadiĭ v protsesse biosinteza khlorofilla. [Studies of transient photochemical steps in the
process of chlorophyll biosynthesis.] - Nauch. Dokl. vyssh. Shkoly, biol. Nauki
10: 147-148, 1967. [In R.]

551 - BELYAKOVA, Z.P.: Izmenenie pigmentov plastid v ontogeneze list'ev tabaka. [Chan-
ge in leaf pigments during ontogenesis of tobacco.] - Tabak *27* (2): 41-43, 1966.
[In R.]

552 - BELYAKOVA, Z.P.: Izmenenie pigmentov plastid pri tomlenii list'ev tabaka. [Chan-
ges in pigments in plastids during tobacco leaf sweating.] - Izv. vyssh. ucheb.
Zaved., pishch. Tekhnol. *1967*: 59-61, 1967. [Chl, Car.; in R.]

553 - BELYANIN, V.N., KOVROV, B.G.: K matematicheskoĭ modeli biosinteza v svetolimi-
tirovannoĭ kul'ture mikrovodorosleĭ. [Mathematical model of biosynthesis in a
light-limited culture of microalgae.] - Dokl. Akad. Nauk SSSR *179*: 1463-1466,
1968. [Ps, Chl; in R.]

554 - BELYANIN, V.N., SID'KO, F.Ya., EROSHIN, N.S.: Zavisimost' maksimal'noĭ produktivnosti kul'tury mikrovodorosleĭ ot obluchennosti. [Dependence of maximum productivity of a culture of microalgae on irradiance.] - In: Nepreryvnoe Upravlya-
aemoe Kul'tivirovanie Mikroorganizmov. Pp. 78-82. Nauka, Moskva 1967. [In R.]

555 - BELYANIN, V.N., SID'KO, F.Ya., EROSHIN, N.S., GEVEL', L.M.: Vliyanie svetovogo
rezhima na otnositel'noe soderzhanie pigmentov v biomasse i produktivnost' vo-

doroslei. [Effect of light regime on the relative content of pigments in biomass and productivity of algae.] - In: Nepreryvnoe Upravlyaemoe Kul'tivirovanie
Mikroorganizmov. Pp. 89-95. Nauka, Moskva 1967. [In R.]

556 - BELYANIN, V.N., TERSKOV, I.A., SID'KO, F.Ya.: Rost i produktivnost' mikrovodoros-
lei pri osveshchenii ikh svetom razlichnogo spektral'nogo sostava. [Growth and
productivity of microalgae under radiation of different spectral composition]. -
In: Upravlyaemyi Biosintez. Pp. 158-165. Nauka, Moskva 1966. [Chl; in R.]

557 - BELYANIN, V.N., TERSKOV, I.A., SID'KO, F.Ya., EROSHIN, N.S.: Effektivnost' ispol'zovaniya luchistoi energii v usloviyakh plotnostatnogo kul'tivirovaniya mi-
krovodoroslei. [Effectivity of radiant energy utilization in a flat static culture of microalgae.] - In: Nepreryvnoe Upravlyaemoe Kul'tivirovanie Mikroorganizmov. Pp. 82-86. Nauka, Moskva 1967. [In R.]

558 - BELYUSTINA, L.N., KOKINA, G.A.: Kachestvennoe issledovanie sistemy uravnenii fo-
tosinteza. [Qualitative study of the system of equations of photosynthesis.] -
In: FRANK, G.M. (ed.): Kolebatel'nye Protsessy v Biol. i Khim. Sist. Pp. 67-80.
Nauka, Moskva 1967. [In R.]

559 - BENDALL, D.S.: Oxidation-reduction potentials of cytochromes in chloroplasts
from higher plants. - Biochem. J. *109*: 46P-47P, 1968.

560 - BENDALL, D.S., HILL, R.: Haem-proteins in photosynthesis. - Annu. Rev. Plant
Physiol. *19*: 167-186, 1968.

561 - BENEDICT, C.R., KOHEL, R.J.: Repression of chloroplast pigment synthesis in virescent cotton mutants. - Plant Physiol. *43* (Suppl.): S-7, 1968.

562 - BENEDICT, C.R., KOHEL, R.J.: Characteristics of a virescent cotton mutant. -
Plant Physiol. *43*: 1611-1616, 1968. [Ps, Chl.]

563 - BENEDICT, C.R., KOHEL, R.J.: The synthesis of ribulose-1,5-diphosphate carboxylase and chlorophyll in virescent cotton leaves. - Plant Physiol. *44*: 621-622,
1969.

564 - BENEDICT, C.R., KOHEL, R.J.: Photosynthetic rate of a virescent cotton mutant
lacking chloroplast grana. - Plant Physiol. *45*: 519-521, 1970.

565 - BENEDICT, W.G.: Changes in levels of some chemical constituents of soybean
leaves following leaf spot infection. - Can. J. Plant Sci. *46*: 553-560, 1966.
[Chl.]

566 - BENGTSSON, E.L., HYLMÖ, S.: The effect of light on blonding and chlorophyll
content of peas. - Acta Agr. scand. *19*: 49-53, 1969.

567 - BEN-HAYYIM, G., AVRON, M.: Enhancement in isolated chloroplasts. - Israel J.
chem. *4* (1a): 73p, 1966.

568 - BEN-HAYYIM, G., AVRON, M.: Cytochrome *b* of isolated chloroplasts. - Europe. J.
Biochem. *14*: 205-213, 1970.

569 - BEN-HAYYIM, G., AVRON, M.: Involvement of photosystem two in non-oxygen evolving
non-cyclic, and in cyclic electron flow processes in chloroplasts. - Europe. J.
Biochem. *15*: 155-160, 1970.

570 - BEN-HAYYIM, G., AVRON, M.: Mn^{2+} as electron donor in isolated chloroplasts. -
Biochim. biophys. Acta *205*: 86-94, 1970.

571 - BEN-HAYYIM, G., GROMET-ELHANAN, Z., AVRON, M.: A specific and sensitive method
for the determination of NADPH. - Anal. Biochem. *28*: 6-12, 1969.

572 - BEN-HAYYIM, G., HOCHMAN, A., AVRON, M.: Phosphoadenosine diphosphate ribose, a
specific inhibitor of nicotinamide adenine dinucleotide phosphate enzymes. - J.
biol. Chem. *242*: 2837-2839, 1967.

573 - BEN'KOVSKAYA, L.A.: Nekotorye fiziologo-biokhimicheskie izmeneniya v list'yakh
i pobegakh persika, porazhennykh gribom *Clasterosporium carpophilum* (LEV.)
ADERH. [Some physiological and biochemical changes in leaves and branches of
peach tree infected by the fungus *Clasterosporium carpophilum* (LEV.) ADERH.] -
Mikol. Fitopatol. *4*: 256-258, 1970. [Chl; in R.]

574 - BENNETT, D.: Analysis of gas mixtures by gas chromatography. - J. Chromatogr.

26: 482-484, 1967. [Also O$_2$ and CO$_2$.]

575 - BENNOUN, P.: Réoxydation du quencher de fluorescence "Q" en présence de 3-(3,4-dichlorophényl)-1,1-diméthylurée. - Biochim. biophys. Acta *216*: 357-363, 1970.

576 - BENNOUN, P., JOLIOT, A.: Etude de la photooxydation de l'hydroxylamine par les chloroplastes d'épinards. - Biochim. biophys. Acta *189*: 85-94, 1969.

577 - BENNOUN, P., LEVINE, R.P.: Detecting mutants that have impaired photosynthesis by their increased level of fluorescence. - Plant Physiol. *42*: 1284-1287, 1967.

578 - BENNUN, A., RACKER, E.: Partial resolution of the enzymes catalyzing photophosphorylation. IV. Interaction of coupling factor 1 from chloroplasts with components of the chloroplast membrane. - J. biol. Chem. *244*: 1325-1331, 1969.

579 - BEN-SHAUL, Y., MARKUS, Y.: Effects of chloramphenicol on growth, size distribution, chlorophyll synthesis and ultrastructure of *Euglena gracilis*. - J. Cell Sci. *4*: 627-644, 1969.

580 - BEN-SHAUL, Y., OPHIR, I.: Effects of streptomycin on plastids in dividing *Euglena*. - Planta *91*: 195-203, 1970.

581 - BEN-SHAUL, Y., OPHIR, I.: Structural and developmental aspects of cycloheximide effects on the chloroplasts of *Euglena gracilis*. - Can. J. Bot. *48*: 929-934, 1970.

582 - BENYUSH, V.A., KOTOVA, E.N., MITROFANOV, A.I.: Kontrol' za fotosintezom pri kul'tivirovanii khlorelly. [Photosynthesis control during *Chlorella* cultivation.] - In: Upravlyaemyĭ Biosintez. Pp. 330-335, Nauka, Moskva 1966. [In R.]

583 - BERDYKULOV, Kh.A.: Izuchenie fotosinteza *Chlorella pyrenoidosa* CHICK. v. massovoĭ kul'ture. [Study of photosynthesis in *Chlorella pyrenoidosa* CHICK. in a mass culture.] - In: Materialy po Fiziologii i Ekologii Rasteniĭ Sredneĭ Azii. Pp. 128-132. FAN, Tashkent 1966. [In R.]

584 - BERDYKULOV, Kh.A.: Vliyanie kontsentratsii uglekislogo gaza na fotosintez *Chlorella pyrenoidosa* CHICK. [Effect of CO$_2$ concentration on photosynthesis in *Chlorella pyrenoidosa* CHICK.] - Uzb.biol.Zh. *12* (2): 18-20, 1968. [In R, ab:E,Uzb.]

585 - BERDYKULOV, Kh.A.: Vliyanie temperatury na fotosintez *Chlorella pyrenoidosa* CHICK. pri massovom kul'tivirovanii v otkrytykh basseĭnakh. [Effect of temperature on photosynthesis in *Chlorella pyrenoidosa* CHICK. in mass culture in open reservoirs.] - Uzb. biol. Zh. *14* (1): 20-22, 1970. [In R, ab: E, Uzb.]

586 - BEREZHNAYA, Z.G.: Fotosinteticheskaya deyatel'nost' skorospelykh sortov belogo lyupina. [Photosynthetic activity in early cultivars of white lupine.] - Dokl. TSKhA (Moskva) *131*: 85-91, 1967.

587 - BEREZIN, B.D., DROBYSHEVA, A.N.: Fotokhimicheskaya ustoĭchivost' metalloanalogov khlorofilla. [Photochemical stability of metal analogues of chlorophyll.] - Zh. fiz. Khim. *42*: 2092-2096, 1968. [In R.]

588 - BEREZOVSKIĬ, Yu.V.: O potentsial'noĭ intensivnosti fotosinteza gorokha razlichnogo ekologicheskogo proiskhozhdeniya. [Potential photosynthetic rate in pea of different ecological origin.] - Dokl. vsesoyuz. Akad. sel'skokhoz. Nauk *1967* (3): 10-12, 1967. [In R.]

589 - BEREZOVSKIĬ, Yu.V.: Vliyanie ekologicheskikh usloviĭ na intensivnost' fotosinteza i produktivnost' gorokha i phenitsy. [Effect of ecological conditions on photosynthetic rate and productivity of pea and wheat.] - In: Mineral'nye Elementy i Mekhanizm Fotosinteza. Pp. 235-241. Kishinev 1969. [In R.]

590 - BERG, H., KRAMARCZYK, H.: Das Chlorophyll-Radikal, generiert durch polarographische Reduktion. - Biochim. biophys. Acta *131*: 141-146, 1967.

591 - BERGER, C.: Aktivitätsänderungen einiger Enzyme synchronisierter *Chlorella*-Zellen im Licht-Dunkel-Wechsel. I. Veränderungen während des Entwicklungszyklus. - Flora A *157*: 211-232, 1966.

592 - BERGER, C., BERGMANN, L.: Farblicht und Plastidendifferenzierung im Speichergewebe von *Solanum tuberosum* L. - Z. Pflanzenphysiol. *56*: 439-445, 1967.

593 - BERGER, C., FEIERABEND, J.: Plastidenentwicklung und Bildung von Photosynthese-Enzymen in etiolierten Roggenkeimlingen. - Physiol. vég. 5: 109-122, 1967.

594 - BERGER, C., PIRSON, A.: Aktivitätsänderungen einiger Enzyme synchronisierter Chlorella-Zellen im Licht-Dunkel-Wechsel. II. Veränderungen bei gehemmtem Wachstum. - Flora A 158: 164-180, 1967.

595 - BERGERON, J.A., OLSON, J.M.: Low-temperature fluorescence emission and excitation spectra for Anacystis nidulans. - Biochim. Biophys. Acta 131: 401-404, 1967.

596 - BERGFELD, R.: Chloroplastenausbildung und Morphogenese der Gametophyten von Dryopteris filix-mas (L.) SCHOTT nach Applikation von Chloramphenicol und Actidion (Cycloheximid). - Planta 81: 274-279, 1968.

597 - BERGFELD, R.: Die Feinstruktur und Entwicklung der Plastiden einiger Süsswasserrotalgen. - Cytobiologie 1: 411-419, 1970.

598 - BERGFELD, R.: Feinstruktur der Chloroplasten in den Gametophytenzellen von Dryopteris filix-mas (L.) SCHOTT nach Einwirkung hellroter und blauer Strahlung. - Z. Pflanzenphysiol. 63: 55-64, 1970.

599 - BERGMAN, I.: Rapid-response atmospheric oxygen monitor based on fluorescence quenching. - Nature 218: 396, 1968.

600 - BERGMAN, I.: Metallized membrane electrode: Atmospheric oxygen monitoring and other applications. - Nature 218: 266, 1968.

601 - BERGMANN, H., LERCH, G., MÜNTZ, K.: Über die physiologische Wirkung von Sonne und Schatten auf Coffea arabica L. "caturra" unter Freilandbedingungen in Kuba: II. Stoffwechsel von Jungpflanzen. - Z. Pflanzenphysiol. 63: 444-460, 1970. [Ps, Chl.]

602 - BERGMANN, L.: Wachstum grüner Suspensionskulturen von Nicotiana tabacum Var. "Samsun" mit CO_2 als Kohlenstoffquelle. - Planta 74: 243-249, 1967.

603 - BERGMANN, L.: Photosynthesis and growth of suspension cultures of Nicotiana tabacum with carbon dioxide as carbon source. - In: Les Cultures de Tissue de Plantes. Pp. 213-221. Édit. CNRS, Paris, 1968.

604 - BERGMANN, L., BALZ, A.: Der Einfluss von Farblicht auf Wachstum und Zusammensetzung pflanzlicher Gewebekulturen. I. Mitteilung. Nicotiana tabacum var. "Samsun". - Planta 70: 285-303, 1966. [Ps, Chl.]

605 - BERGMANN, L., BERGER, C.: Farblicht und Plastidendifferenzierung in Zellkulturen von Nicotiana tabacum var. "Samsun". - Planta 69: 58-69, 1966.

606 - BERIDZE, T.G., ODINTSOVA, M.S., CHERKASHINA, N.A., SISAKYAN, N.M.: Ob uchastii nukleinovykh kislot v protsesse biosinteza khlorofilla. [Participation of nucleic acids in chlorophyll biosynthesis.] - Dokl. Akad. Nauk SSSR 166: 1454-1457, 1966. [In R.]

607 - BERIDZE, T.G., ODINTSOVA, M.S., CHERKASHINA, N.A., SISAKYAN, N.M.: Vliyanie ingibitorov sinteza DNK na obrazovanie khlorofilla. [Effect of inhibitors on DNA synthesis on chlorophyll formation.] - Dokl. Akad. Nauk SSSR 169: 466-468, 1966. [In R.]

608 - BERIDZE, T.G., ODINTSOVA, M.S., CHERKASHINA, N.A., SISSAKIAN, N.M.: The effect of nucleic acid synthesis inhibitors on the chlorophyll formation by etiolated bean leaves. - Biochim. biophys. Res. Commun. 23: 683-690, 1966.

609 - BERKALOFF, C.: Observations sur l'organisation infrastructurale d'une Volvocale. - Compt. rend. Acad. Sci. Paris, Sér. D 263: 1232-1234, 1966. [Chloroplast]

610 - BERKALOFF, C.: Modifications ultrastructurale du plaste et de divers autres organites cellulaires au cours du développement et de l'enkystement du Protosiphon botryoïdes (Chlorophycées). - J. Microscop. 6: 839-852, 1967.

611 - BERKALOFF, C.: Essai d'isolement des globules pigmentés extraplastidaux de l'algue Protosiphon botryoïdes. - Compt. rend. Acad. Sci. Paris, Sér. D 271: 1518-1521, 1970.

612 - BERKOVÁ, E., DOUCHA, J.: Chlorophyll synthesis and photosynthesis in synchronous

cultures of *Scenedesmus quadricauda*. - Annu. Rep. algol. Lab. Třeboň *1969*: 141-150, 1970.

613 - BERNARD, E.A.: Théorie des échanges gazeux et énergétiques entre la végétation et l'air. - In: ECKARDT, F.E. (ed.): Functioning of Terrestrial Ecosystems at the Primary Production Level. Pp. 67-83. UNESCO, Paris 1968.

614 - BERNATH, P., SWISHER, H.E.: Rapid quantitative method for determining carotenoids of the California sweet orange. - Food Technol. *23* (6): 107-110, 1969.

615 - BERNER, L.: Les pigments des algues. - Bull. Centr. étud. Rech. sci. Biarritz *8*: 77-83, 1970.

616 - BERNS, D.S.: Protein aggregation in phycocyanin - osmotic pressure studies. - Biochem. biophys. Res. Commun. *38*: 65-73, 1970.

617 - BERNS, D.S., MACCOLL, R., LEE, J.J.: The aggregation properties of C-phycocyanin. - Biochem. J. *119*: 14P-15P, 1970.

618 - BERRY, D.R., SMITH, H.: The inhibition by high concentrations of (2-chloroethyl)-trimethylammonium chloride (CCC) of chlorophyll and protein synthesis in excised barley leaf sections. - Planta *91*: 80-86, 1970.

619 - BERRY, J.A., DOWNTON, W.J.S., TREGUNNA, E.B.: The photosynthetic carbon metabolism of *Zea mays* and *Gomphrena globosa*: the location of the CO_2 fixation and the carboxyl transfer reactions. - Can. J. Bot. *48*: 777-786, 1970; *48*: 1499, 1970 - erratum.

620 - BERRY, R.E., RANEY, L.W.: A recording photometer for biological studies. - Ecology *49*: 161-162, 1968.

621 - BERSHTEĬN, B.I., OKANENKO, A.S.: Vliyanie nedostatka kaliya na fotosintez, dykhanie i fosfornyĭ obmen v ontogeneze sakharnoĭ svekly. [Effect of potassium deficiency on photosynthesis, respiration and phosphorus metabolism during ontogenesis of sugar beet.] - Fiziol. Rast. *13*: 629-639, 1966. [In R, ab: E.]

622 - BERSHTEĬN, B.I., REĬNGARD, T.A.: Osobennosti opredeleniya fotofosforilirovaniya i reaktsii Khilla v izolirovannykh khloroplastakh. [Peculiarities of determination of photophosphorylation activity and Hill reaction in isolated chloroplasts.] - In: Puti Povysheniya Intensivnosti i Produktivnosti Fotosinteza. Vol. 3. Pp. 141-150. Naukova Dumka, Kiev 1969. [In R.]

623 - BERSHTEĬN, B.I., VOLKOVA, N.V., OSTROVSKAYA, L.K., OKANENKO, A.S., REĬNGARD, T.A., YASNIKOV, A.A.: O mekhanizme obrazovaniya ATF pri fotosinteze. [Mechanism of ATP formation in photosynthesis.] - In: Fiziologiya i Biokhimiya Zdorovogo i Bol'nogo Rasteniya. Pp. 232-240. Izdat. mosk. Univ., Moskva 1970. [In R.]

624 - BERSHTEĬN, B.I., VOLKOVA, N.V., VOLOVIK, O.I., IVANISHCHEVA, S.Yu., OKANENKO, A.S., OSTROVSKAYA, L.K., PETRENKO, S.G., POLISHCHUK, A.I., PSHENICHNAYA, A.K., REĬNGARD, T.A., SEMENYUK, I.I., YASNIKOV, A.A.: Obrazovanie ATF pri fotofosforilirovanii i mekhanizm razobshcheniya aminami fosforilirovaniya i transporta elektronov. [ATP formation in photophosphorylation and the mechanism of disturbance of photophosphorylation and electron transport by amines.] - Fiziol. Biokhim. kul't. Rast. *1*: 21-26, 1969. [In R, ab: E.]

625 - BERSON, G.Z.: Vliyanie svetooblucheniya rassady ogurtsov i tomatov na ikh fotosintez v induktsionnyĭ i posleinduktsionnyĭ period. [Effect of illumination of cucumber and tomato seedlings on their photosynthesis during the induction and post-induction period.]- In: Informatsionnyĭ Byulleten' 6. Sibir. Inst. Fiziol. i Biokhim. Rast. SO Akad. Nauk SSSR. Pp. 71-72. Irkutsk 1970. [In R.]

626 - BERTAGNOLLI, B.L., NADAKAVUKAREN, M.J.: An ultratructural study of pyrenoids from *Chlorella pyrenoidosa*. - J. Cell Sci. *7*: 623-630, 1970.

627 - BERTHIER, J.: Organogenèse foliaire et axillaire de l'*Aulacomnium palustre* (HEDW.) SCHWAEGER. Régularisation de la caulogenèse par la photosynthèse et l'action locale des substances de type Kinine. - Compt. rend. Acad. Sci. Paris, Sér. D *270*:2174-2177, 1970.

628 - BERTSCH, A.: Über den CO_2-Gaswechsel einiger Flechten nach Wasserdampfaufnahme. - Planta *68*: 157-166, 1966.

629 - BERTSCH, A.: CO_2-Gaswechsel und Wasserhaushalt der aerophilen Grünalge *Apatococcus lobatus*. - Planta *70*: 46-72, 1966.

630 - BERTSCH, A.: CO_2-Gaswechsel, Wasserpotential und Sättigungsdefizit bei der Antrocknung epidermisfreier Blattscheiben von *Valerianella*. - Naturwissenschaften *54*: 204, 1967.

631 - BERTSCH, A.: Der Diffusionswiderstand der Spaltöffnungen; ein Vergleich des CO_2-Gaswechsels von Blättern mit und ohne Epidermis. - Planta *87*: 102-109, 1969.

632 - BERTSCH, A., DOMES, W.: CO_2-Gaswechsel amphistomatischer Blätter. 1. der Einfluss unterschiedlicher Stomaverteilung der beiden Blattepidermen auf den CO_2-Transport. - Planta *85*: 183-193, 1969.

633 - BERTSCH, A., DOMES, W.: CO_2-Gaswechsel amphistomatischer Blätte·. 3. Das unterschiedliche Zeitverhalten der Stomata beider Blattseiten in Abhängigkeit von der vorausgegangenen Dunkelzeit. - Planta *89*: 47-55, 1969.

634 - BERTSCH, W.: Milisecond delayed light emission from photoreaction II: two models for the reaction center. - In: METZNER, H. (ed.): Progress in Photosynthesis Research, Vol. II. Pp. 996-1005. Tübingen 1969.

635 - BERTSCH, W., AZZI, J.R., DAVIDSON, J.B.: Delayed light studies on photosynthetic energy conversion, I. Identification of the oxygen-evolving photoreaction as the delayed light emitter in mutants of *Scenedesmus obliquus*. - Biochim. biophys. Acta *143*: 129-143, 1967.

636 - BERTSCH, W., WEST, J., HILL, R.: Delayed light studies on photosynthetic energy conversion. II. Effect of electron acceptors and phosphorylation cofactors on the millisecond emission from chloroplasts. - Biochim. biophys. Acta *172*: 525-538, 1969.

637 - BERÜTER, J., TEMPERLI, A.T.: Influence of prometryne and ioxynil on photosynthesis and nucleic acid metabolism in plants. - Experientia *26*: 600-601, 1970.

638 - BERYASHVILI, T.V., CHIGVINADZE, T.D.: Produkty assimilyatsii radioaktivnoǐ uglekisloty v grozd'yakh vinogradnoǐ lozy. [Products of the assimilation of radioactive carbon dioxide by grapes.] - Soobshch. Akad. Nauk Gruz. SSR *49*: 601-606, 1968. [In Georgian, ab: R.]

639 - BERZBORN, R.: 2. Mitteilung über lösliche und unlösliche Chloroplasten-Antigene. Nachweis der Ferredoxin-NADP-Reduktase in der Oberfläche des Chloroplasten-Lamellarsystems mit Hilfe spezifischer Antikörper. - Z. Naturforsch. *23b*: 1096-1104, 1968.

640 - BERZBORN, R., MENKE, W., TREBST, A., PISTORIUS, E.: Über die Hemmung photosynthetischer Reaktionen isolierter Chloroplasten durch Chloroplasten-Antikörper. - Z. Naturforsch. *21b*: 1057-1059, 1966.

641 - BERZBORN, R.J.: Untersuchungen über die Oberflächenstruktur des Thylakoidsystems der Chloroplasten mit Hilfe von Antikörpern gegen die Ferredoxin-NADP-Reduktase. - Z. Naturforsch. *24b*: 436-446, 1969.

642 - BERZBORN, R.J.: Demonstration of ferredoxin-NADP$^+$ reductase on the surface of the lamellar system of chloroplasts by antibodies. - In: METZNER, H. (ed.): Progress in Photosynthesis Research. Vol. I. Pp. 106-114. Tübingen 1969.

643 - BERZINYA-BERZITE, R.V.: Izmenenie soderzhaniya pigmentov v list'yakh rastenǐ pod vliyaniem okislitelěǐ i vosstanovitelěǐ. [Changes in pigment content of plant leaves induced by oxidants and reductants.] - Uch. Zap. latv. Univ. *109*: 50-56, 1968 (1969). [In R.]

644 - BEUCHAT, L.R., LECHOWICH, R.V., SCHANDERL, S.H., CO, D.Y.C., McFEETERS, R.F.: Inhibition of bacterial growth by chlorophyllide *a*. - Mich. State Univ., agr. exp. Sta., quart. Bull. *48*: 411-416, 1966.

645 - BEUGELING, T.: Photochemical activities of $K_3Fe(CN)_6$-treated chromatophores from *Rhodospirillum rubrum*. - Biochim. biophys. Acta *153*: 143-153, 1968.

646 - BEUGELING, T.: Primary light-induced reactions in chemically bleached chromatophores from *Rhodospirillum rubrum*. - In: METZNER, H. (ed.): Progress in Photosynthesis Research. Vol. II. Pp. 1101-1109. Tübingen 1969.

647 - **BEUGELING, T., DUYSENS, L.N.M.**: P890 and cytochrome C422 in *Chromatium*. - In: THOMAS, J.B., GOEDHEER, J.C. (ed.): Currents in Photosynthesis. Pp. 49-56. Donker, Rotterdam 1966.

648 - **BEZUGLOV, V.K.**: Vliyanie usloviĭ mineral'nogo pitaniya na izmenenie okislitel'-no-vosstanovitel'nogo potentsiala suspenziĭ khloroplastov i gomogenatov list'ev pri osveshchenii. [Effect of mineral nutrition on changes of oxido-reduction potential of chloroplast suspensions and leaf homogenates during illumination.] - In: Funktsional'nye Osobennosti Khloroplastov. Pp. 47-50. Kazan. Univ., Kazan' 1969. [In R.]

649 - **BEZUGLOV, V.K.**: Deĭstvie bikarbonata na fotokhimicheskuyu aktivnost' izolirovan-nykh khloroplastov. [Effect of bicarbonate on photochemical activity of isolated chloroplasts.] - In: Funktsional'nye Osobennosti Khloroplastov. Pp. 51-55. Kazan. Univ., Kazan' 1969. [In R.]

650 - **BEZUGLOV, V.K.**: Zavisimost' svetoindutsiruemoĭ bioelektricheskoĭ reaktsii list'-ev ot uslovĭ mineral'nogo pitaniya rasteniĭ. [Dependence of light-induced bio-electric reaction of leaves on mineral nutrition of plants.] - In: Mineral'nye Elementy i Mekhanizm Fotosinteza. Pp. 102-107. Kishinev 1969. [In R.]

651 - **BEZUGLOV, V.K., BAKIROVA, I.A., SALEEVA, Z.A.**: Vliyanie ekzogennogo ATF na izmenenie okislitel'no-vosstanovitel'nogo potentsiala suspenziĭ khloroplastov pri osveshchenii. [Effect of exogenous ATP on changes of oxido-reduction potential of chloroplast suspensions during illumination.] - In: Trudy 3-eĭ Gorodskoĭ Nauchnoĭ Konferentsii Molodykh Uchenykh (Biol. Ser.). Pp. 72-74. Kazan' 1967. [In R.]

652 - **BEZUIDENHOUT, S.J.P.K.**: Die invloed van 'n toenemende watertekort op die gaswis-selingsprocesse by koringplante. [Influence of an increased water deficit on gas exchange processes in wheat.] - Tydskr. Natuurwetensk. *9*: 144-151, 1969. [In Afrikaans.]

653 - **BEZVERKHNAYA, T.M.**: Vzainosvyaz' mezhdu urozhaem i produktivnost'yu raboty list'-ev plodonosnykh i besplodnykh pobegov u vinogradnogo rasteniya. [Relationship of yield and productivity of leaves of bearing and barren shoots in vine.] - Dokl. TSKhA (Moskva) *143*: 69-75, 1968. [In R.]

654 - **BHAMBOTA, J.R., KANWAR, J.S.**: Effect of different salt concentrations on sweet-orange *(Citrus sinensis* L.) OSBECK. - Indian J. agr. Sci. *40*: 485-494, 1970. [Chl.]

655 - **BHAN, W.M., PANDE, H.K.**: Measurement of leaf area of rice. - Agron J. *58*: 454, 1966.

656 - **BIBLINA, B.I., KIRILLOVA, E.N.**: Vliyanie mineral'nogo pitaniya na soderzhanie khlorofilla i askorbinovoĭ kisloty v list'yakh yabloni. [Effect of mineral nu-trition on contents of chlorophyll and ascorbic acid in apple leaves.] - In: Funktsional'nye Rasstroĭstva i Zabolevaniya Kul'turnykh Rasteniĭ. Pp. 90-104. Kartya Moldovenyaske, Kishinev 1968. [In R.]

657 - **BIDDULPH, O.**: Mechanisms of translocation of plant metabolites. - In: EASTIN, J.D., HASKINS, F.A., SULLIVAN, C.Y., van BAVEL, C.H.M. (ed.): Physiological Aspects of Crop Yield. Pp. 143-167. Amer. Soc. Agron. & Crop Sci. Soc. Amer., Madison, Wisc. 1969.

658 - **BIDWELL, R.G.S.**: Photosynthesis and metabolism in marine algae. VII. Products of photosynthesis in fronds of *Fucus vesiculosus* and their use in respiration. - Can. J. Bot. *45*: 1557-1565, 1967.

659 - **BIDWELL, R.G.S., LEVIN, W.B., SHEPHARD, D.C.**: Photosynthesis, photorespiration and respiration of chloroplasts from *Acetabularia mediterrania*. - Plant Physiol. *44*: 946-954, 1969.

660 - **BIDWELL, R.G.S., LEVIN, W.B., SHEPHARD, D.C.**: Intermediates of photosynthesis in *Acetabularia mediterranie* chloroplasts. - Plant Physiol. *45*: 70-75, 1970.

661 - **BIDWELL, R.G.S., TURNER, W.B.**: Effect of growth regulators on CO_2 assimilation in leaves, and its correlation with the bud break response in photosynthesis. - Plant Physiol. *41*: 267-270, 1966.

662 – BIEDERMANN, M., DREWS, G.: Trennung der Thylakoidbausteine einiger *Athiorhoda-ceae* durch Gelelektrophorese. – Arch. Mikrobiol. *61*:48-58, 1968.

663 – BIEHL, B.: Membranfragmente mit unterschiedlichem Verhältnis von Chlorophyll *a* und Chlorophyll *b* nach Ultraschall-Fragmentierung von Spinat-Chloroplasten. – Z. Naturforsch. *21b*: 501-502, 1966.

664 – BIEHL, B.: Elektrophoretische Trennversuche und fragmentierten Membranen isolierter Chloroplasten. – Ber. deut. bot. Ges. *81*: 307-310, 1968/9.

665 – BIEHL, B.: Versuche zur elektrophoretischen Trennung fragmentierter Chloroplastenmembranen. – Z. Pflanzenphysiol. *60*:98-113, 1969.

666 – BIEHL, B.: Verteilung membranassoziierter Proteine in Fraktionen ultraschallfragmentierter Thylakoide aus Spinatchloroplasten. – Ber. deut. bot. Ges. *83*: 465-470, 1970.

667 – BIELORAI, H., MENDEL, K.: The simultaneous measurement of apparent photosynthesis and transpiration of citrus seedlings at different soil moisture levels. – J. amer. Soc. hort. Sci. *94*: 201-204, 1969.

668 – BIERHUIZEN, J.F., NUNES, M.A., PLOEGMAN, C.: Studies on productivity of coffee. II – Effect of soil moisture on photosynthesis and transpiration of *Coffea arabica*. – Acta bot. neerl. *18*: 367-374, 1969.

669 – BIGGINS, J.: Preparation of metabolically active protoplasts from the blue-green alga, *Phormidium luridum*. – Plant Physiol. *42*: 1442-1446, 1967. [Ps.]

670 – BIGGINS, J.: Photosynthetic reactions by lysed protoplasts and particle preparations from the blue-green alga, *Phormidium luridum*. – Plant Physiol. *42*: 1447-1456, 1967.

671 – BIGGINS, J., PARK, R.B.: CO_2 assimilation by etiolated *Hordeum vulgare* seedlings during the onset of photosynthesis. – Plant Physiol. *41*: 115-118, 1966.

672 – BIGGS, W., EDISON, A.R., EASTIN, J.D., BROWN, K.W., MARANVILLE, J.W., CLEGG, M.D.: Photosynthesis light sensor and meter. – Ecology *52*: 125-131, 1970.

673 – BILLINGS, W.D., CLEBSCH, E.E.C., MOONEY, H.A.: Photosynthesis and respiration rates of Rocky Mountain alpine plants under field conditions. – Amer. Midland Natur. *75*: 34-44, 1966.

674 – BILLINGS, W.D., GODFREY, P.J.: Photosynthetic utilization of internal carbon dioxide by hollow-stemmed plants. – Science *158*: 121-123, 1967.

675 – BILLINGS, W.D., MOONEY, H.A.: The ecology of arctic and alpine plants. – Biol. Rev. *43*: 481-529, 1968. [Also Ps, Chl.]

676 – BILLINGTON, R.W., HEYES, J.K.: Effects of inhibitors on light-stimulated synthesis in radish hypocotyl. – Nature *227*: 858-860, 1970. [Chl.]

677 – BILLMEYER, F.W. Jr., CARLON, H.R.: Estimation of refractive indices by a spectrophotometric method. – Appl. Optics *9*: 501-502, 1970. [For plastics used in Ps measurements.]

678 – BILLOT, J.: Recherches sur les pigments des feuilles d'*Acalypha*: Décoloration des anthocyanes et altération des chlorophylles. – Physiol. vég. *5*: 341-355, 1967.

679 – BILLOT, J.: Les pigments anthocyaniques, les chlorophylles et les carotenoïdes des bractées florales et des feuilles de bougainvillée. – Ann. Univ. Madagascar, Ser. Sci. nat. Math. *1967* (5): 41-51, 1967.

680 – BÍNOVÁ, J., PŘIBÁŇ, K.: PP/IBP initial level experiments in South Bohemia. – Annu. Rep. algol. Lab. Třeboň *1968*: 180-186, 1969. [Growth analysis.]

681 – BÍNOVÁ, J., PŘIBÁŇ, K., ONDOK, P.: PP/IBP initial level experiments in South Bohemia. – Annu. Rep. algol. Lab. Třeboň *1969*: 202-209, 1970. [Growth analysis.]

B682 – Bioenergetika i Biologicheskaya Spektrofotometriya. [Bioenergetics and Biological Spectrophotometry.] – Nauka, Moskva *1967*. [In R.]

683 - BIRECKA, H.: Contribution of carbon assimilated before and after heading to the accumulation of organic compounds in the grain of some cereals. - Acta Agr. scand. *16* (Suppl. 16): 177,1966.

684 - BIRECKA, H.: Effect of (2-chloroethyl)trimethylammonium chloride (CCC) on photosynthesis and photosynthate distribution in oats and wheat after heading. - Bull. Acad. pol. Sci., Sér. Sci. biol. *14*: 261-267, 1966.

685 - BIRECKA, H.: Influence of 2-chloroethyl trimethylammonium (CCC) on photosynthetic activity and assimilate distribution in wheat. - In: Proc. Symp. on Isotopes in Plant Nutrition and Physiology. Pp. 189-199. Int. At. Energy Agency, Vienna 1967.

686 - BIRECKA, H.: Translocation and redistribution of ^{14}C-assimilates in cereal plants deprived of the ear. I. Spring wheat. - Bull. Acad. pol. Sci., Sér. Sci. biol. *16*: 455-460, 1968.

687 - BIRECKA, H., DAKIĆ-WŁODKOWSKA, L.: Photosynthetic activity and productivity before and after ear emergence in spring wheat. - Acta Soc. Bot. Pol. *35*: 637-662, 1966.

688 - BIRECKA, H., SKIBA, T.: Translocation and redistribution of ^{14}C-assimilates in cereal plants deprived of ear. II. Oat and barley. - Bull. Acad. pol. Sci., Sér. Sci. biol. *16*: 595-601, 1968.

689 - BIRECKA, H., SKIBA, T., KOZŁOWSKA, Z.: Translocation and redistribution of ^{14}C-assimilates in cereal plants deprived of the ear. III. Assimilate distribution in root and respiration of culm in wheat plants. - Bull. Acad. pol. Sci., Sér. Sci. biol. *17*: 121-127, 1969.

690 - BIRECKA, H., SKUPIŃSKA, J., BERNSTEIN, I.: Photosynthetic activity and productivity before and after ear emergence in spring barley. - Acta Soc. Bot. Pol. *36*: 387-409, 1967.

691 - BIRECKA, H., SZCZYPA, E., KOZŁOWSKA, Z.: Transplantation and redistribution of ^{14}C-assimilates in cereal plants deprived of ear. IV. Aftereffect of ear removal on photosynthesis in wheat and oat. - Bull. Acad. pol. Sci., Sér. Sci. biol. *17*: 257-263, 1969.

692 - BIRECKA, H., WOJCIESKA, U.: Photosynthetic activity and productivity before and after heading in oats. - Acta Soc. Bot. Pol. *37*: 77-100, 1968.

693 - BIRECKA, H., WOJCIESKA, U., GŁAŻEWSKI, S.: Ear contribution to photosynthetic activity in winter cereals. I. Winter wheat. - Bull. Acad. pol. Sci., Sér. Sci. biol. *16*: 191-196, 1968.

694 - BIRECKA, H., WOJCIESKA, U., ZINKIEWICZ, E.: Ear contribution to photosynthetic activity in winter cereals. Part. II. Winter rye. - Bull. Acad. pol. Sci., Sér. Sci. biol. *16*: 257-260, 1968.

695 - BIRECKA, H., ŻEBROWSKI, Z.: Influence of (2-chloroethyl)trimethylammoniumchloride (CCC) on photosynthetic activity and frost resistance of tomato plants. - Bull. Acad. pol. Sci., Sér. Sci. biol. *14*: 367-373, 1966.

696 - BIRECKA, H., ZINKIEWICZ, E.: Photosynthesis and ^{14}C-assimilate translocation in winter rye after heading. - Bull. Acad. pol. Sci., Sér. Sci. biol. *16*: 323-330, 1968.

697 - BIRTH, G.S., McVEY, G.R.: Measuring the color of growing turf with a reflectance spectrophotometer. - Agr. J. *60*: 640-643, 1968.

698 - BISALPUTRA, T., BISALPUTRA, A.A.: The ultrastructure of chloroplast of a brown alga *Sphacelaria* sp. I. Plastid DNA configuration - the chloroplast genophore. - J. Ultrastruct. Res. *29*: 151-170, 1969.

699 - BISALPUTRA, T., BISALPUTRA, A.A.: The ultrastructure of chloroplast of a brown alga *Sphacelaria* sp. III. The replication and segregation of chloroplast genophore. - J. Ultrastruct. Res. *32*: 417-429, 1970.

700 - BISALPUTRA, T., BURTON, H.: The ultrastructure of chloroplast of a brown alga

Sphacelaria sp. II. Association between the chloroplast DNA and the photosynthetic lamellae. - J. Ultrastruct. Res. *29*: 224-235, 1969.

701 - BISALPUTRA, T., DOWNTON, W.J.S., TREGUNNA, E.B.: The distribution and ultrastructure of chloroplasts in leaves differing in photosynthetic carbon metabolism. I. Wheat, *Sorghum,* and *Aristida (Gramineae).* - Can. J. Bot. *47*: 15-21,
1969.

702 - BISHOP, D.G., SMILLIE, R.M.: The effect of chloramphenicol and cychloheximide
on lipid synthesis during chloroplast development in *Euglena gracilis.* - Arch.
Biochem. Biophys. *139*: 179-189, 1970.

703 - BISHOP, J.W.: Effects of zooplankton on photosynthesis by algae in lakes. -
Water Resources Res. Cent. Bull. *40*: 43-53, 1970.

704 - BISHOP, N.I.: Partial reactions of photosynthesis and photoreduction. - Annu.
Rev. Plant Physiol. *17*: 185-208, 1966.

705 - BISHOP, N.I.: Light-induced changes in the carotenoid complement of wild and
mutant strains of *Scenedesmus obliquus.* - Plant Physiol. *42* (Suppl.): S 34,
1967.

706 - BISHOP, N.I.: Comparison of the action spectra and quantum requirements for
photosynthesis and photoreduction of *Scenedesmus.* - Photochem. Photobiol. *6*:
621-628, 1967.

707 - BISHOP, N.I.: Oxygen metabolism of photosynthetic organisms. - In: SAN PIETRO,
A., GREER, F.A., ARMY, T.J. (ed.): Harvesting the Sun. Pp. 255-262. Academic
Press, New York - London 1967.

708 - BISHOP, N.I.: Photochemical characteristics of algal mutants lacking carotenoids. - Plant Physiol. *44* (Suppl.): 13, 1969.

709 - BISHOP, P.M., WHITTINGHAM, C.P.: The photosynthesis of tomato plants in a carbon dioxide enriched atmosphere. - Photosynthetica *2*: 31-38, 1968.

710 - BISHOP, R.S., DARROW, D.K., DODGE, E.J., SCHREIBER, R.W.: Isolated chloroplasts
lamellar systems: prolonged survival in association with aerobic bacteria. -
Exp. Cell Res. *50*: 646-649, 1968.

711 - BISHOP, R.S., PERRY, M.J., SCHREIBER, R.W.: Lamellar lipoprotein of *Nicotiana*
chloroplasts: in vivo incorporation of carbon dioxide, acetate, and leucine. -
Can. J. Bot. *47*: 667-673, 1969.

712 - BITTMANNOVÁ-NOGOVÁ, J.: Opakovaná působení chloru na prothallia *Dryopteris filix-
mas* (L.) SCHOTT. [Repeated chlorine effects on prothallia of *Dryopteris filix-
mas* (L.) SCHOTT.] - Acta Univ. Palack. Olomouc, Fac. nat. Biol. *9*: 15-24, 1967.
[Chl; in Czech.]

713 - BJÖRKMAN, O.: Comparative studies of photosynthesis and respiration in ecological races. - Brittonia *18*: 214-224, 1966.

714 - BJÖRKMAN, O.: The effect of oxygen concentration on photosynthesis in higher
plants. - Physiol. Plant. *19*: 618-633, 1966.

715 - BJÖRKMAN, O.: Photosynthetic inhibition by oxygen in higher plants. - Carnegie
Inst. Year Book *65*: 446-454, 1967.

716 - BJÖRKMAN, O.: Carboxydismutase activity in relation to light-saturated rate of
photosynthesis in plants from exposed and shaded habitats. - Carnegie Inst.
Year Book *65*: 454-459, 1967.

717 - BJÖRKMAN, O.: Carboxydismutase activity in shade-adapted and sun-adapted species
of higher plants. - Physiol. Plant. *21*: 1-10, 1968.

718 - BJÖRKMAN, O.: Further studies on differentiation of photosynthetic properties
in sun and shade ecotypes of *Solidago virgaurea.* - Physiol. Plant. *21*: 84-99,
1968.

719 - BJÖRKMAN, O.: Further studies of the effect of oxygen concentration on photosynthetic CO_2 uptake in higher plants. - Carnegie Inst. Year Book *66*: 220-228,
1968.

720 - BJÖRKMAN, O.: Carboxydismutase activity in shade-adapted and sun-adapted species of higher plants. - Carnegie Inst. Year Book 67: 487-488, 1969.

721 - BJÖRKMAN, O.: Characteristics of the photosynthetic apparatus as revealed by
laboratory measurements. - In: Prediction and Measurements of Photosynthetic
Productivity. Pp. 267-281. PUDOC, Wageningen 1970.

722 - BJÖRKMAN, O., BJÖRKMAN, M.: Carboxydismutase activity in sun and shade ecotypes
of *Solidago*. - Carnegie Inst. Year Book 66: 216-220, 1968.

723 - BJÖRKMAN, O., GAUHL, E.: Effect of temperature and oxygen concentration on
photosynthesis in *Marchantia polymorpha*. - Carnegie Inst. Year Book 67: 479-
482, 1969.

724 - BJÖRKMAN, O., GAUHL, E.: Carboxydismutase activity in plants with and without
β-carboxylation photosynthesis. - Planta 88: 197-203, 1969.

725 - BJÖRKMAN, O., GAUHL, E.: Application of a new O_2 sensing device to measurements
of higher plant photosynthesis. - Carnegie Inst. Year Book 68: 636-640, 1970.

726 - BJÖRKMAN, O., GAUHL, E.: Use of the zirconium oxide ceramic cell for measurements of photosynthetic oxygen evolution by intact leaves. - Photosynthetica 4:
123-128, 1970.

727 - BJÖRKMAN, O., GAUHL, E., HIESEY, W.M., NICHOLSON, F., NOBS, M.A.: Growth of
Mimulus, Marchantia, and *Zea* under different oxygen and carbon dioxide levels.
- Carnegie Inst. Year Book 67: 477-478, 1969.

728 - BJÖRKMAN, O., GAUHL, E., NOBS, M.A.: Comparative studies of *Atriplex* species
with and without β-carboxylation photosynthesis and their first-generation hybrid. - Carnegie Inst. Year Book 68: 620-633, 1970.

729 - BJÖRKMAN, O., HIESEY, W.M., NOBS, M., NICHOLSON, F., HART, R.W.: Effect of oxygen concentration on dry matter production in higher plants. - Carnegie Inst.
Year Book 66: 228-233, 1968.

730 - BJÖRKMAN, O., HOLMGREN, P.: Photosynthetic adaptation to light intensity in
plants native to shaded and exposed habitats. - Physiol. Plant. 19: 854-859,
1966.

731 - BJÖRKMAN, O., NOBS, M.A., HIESEY, W.M.: Growth, photosynthetic, and biochemical
responses of contrasting *Mimulus* clones to light intensity and temperature. -
Carnegie Inst. Year Book 68: 614-620, 1970.

732 - BJÖRN, L.O.: Chloroplasts in roots. - In: SIRONVAL, C. (ed.): Le Chloroplaste,
Croissance et Vieillissement. Pp. 313-322. Masson, Paris 1967.

733 - BJÖRN, L.O.: Some effects of light on excised wheat roots with special reference to peroxide metabolism. - Physiol. Plant 20: 149-170, 1967. [Chl.]

734 - BJÖRN, L.O.: The light requirement for different steps in the development of
chloroplasts in excised wheat roots. - Physiol. Plant. 20: 483-499, 1967.

735 - BJÖRN, L.O.: The effect of blue and red light on NADP-linked glyceraldehyde-
phosphate dehydrogenases in excised roots. - Physiol. Plant. 20: 519-527, 1967.
[Chl.]

736 - BJÖRN, L.O.: The effect of light on the development of plastids in plant roots.
- In: Book of Abstracts. European Photobiology Symposium "Photochemistry and
Photobiology in Plant Physiology". Pp. 3-4. Hvar, Yugoslavia 1967.

737 - BJÖRN, L.O.: Studies on the phototransformation and fluorescence of protochlorophyll holochrome in vitro. - In: METZNER, H. (ed.): Progress in Photosynthesis
Research, Vol. II. Pp. 618-629. Tübingen 1969.

738 - BJÖRN, L.O.: Action spectra for transformation and fluorescence of protochlorophyll holochrome from bean leaves. - Physiol. Plant. 22: 1-17, 1969.

739 - BJÖRN, L.O.: Effects of N-methylphenazonium methosulfate and pyocyanine on delayed light emission in *Chlorella* cells and spinach chloroplasts. - Carnegie
Inst. Year Book 68: 603-607, 1970.

740 - BJÖRN, L.O., ODHELIUS, I.: Chlorophyll formation in excised roots of cucumber and pea. - Physiol. Plant. *19*: 60-62, 1966.

741 - BLACK, C.C.: Evidence supporting the possibility of two sites of ATP production in green plant photosynthesis. - Plant Physiol. *42* (Suppl.): S-33-S-34, 1967.

742 - BLACK, C.C.: Partial resolution of the photosynthetic oxygen evolution system of spinach chloroplasts. - Plant Physiol. *43*(Suppl.): S 13, 1968.

743 - BLACK, C.C., CHEN, T.M., BROWN, R.H.: Biochemical basis for plant competition. - Weed Sci. *17*: 338-344, 1969. [Ps.]

744 - BLACK, C.C., LABER, L.J.: On the relationship of photophosphorylation to other biochemical activities of isolated spinach chloroplasts. - Fed. Proc. *27*: 344, 1968.

745 - BLACK, C.C., SAN PIETRO, A.: Vitamin B_6 activity in photosynthetic reactions. - Arch. Biochem. Biophys. *128*: 482-487, 1968.

746 - BLACK, C.C. Jr.: Chloroplast reactions with dipyridyl salts. - Biochim. Biophys. Acta *120*: 332-340, 1966.

747 - BLACK, C.C. Jr.: Evidence supporting the theory of two sites of photophosphorylation in green plants. - Biochem. biophys. Res. Commun. *28*: 985-990, 1967.

748 - BLACK, C.C. Jr.: Photosynthetic phosphorylation and associated reactions in the presence of a new group of uncouplers: Salicylanilides. - Biochim. biophys. Acta *162*: 294-296, 1968.

749 - BLACK, C.C. Jr., MAYNE, B.C.: P_{700} activity and chlorophyll content of plants with different photosynthetic carbon dioxide fixation cycles. - Plant Physiol. *45*: 738-741, 1970.

750 - BLACK, C.C. Jr., MYERS, L.: Some biochemical aspects·of the mechanisms of herbicidal activity. - Weeds *14*: 331-338, 1966.

751 - BLACK, J.N.: The utilization of solar energy by forests. - Forestry *1966* (Suppl.) 98-109, 1966.

752 - BLACK, L.L.: Studies on anaerobic metabolism and photosynthesis in *Albugo candida* infected radish cotyledons. - Diss. Abstr. *26*: 4949-4950, 1966.

753 - BLACK, L.L., GORDON, D.T., WILLIAMS, P.H.: Carbon dioxide exchange by radish tissue infected with *Albugo candida* measured with an infrared CO_2 analyzer. - Phytopathology *58*: 173-178, 1968.

754 - BLACK, M., SCHOOLAR, I.: Photosynthesis: mapping new pathways. - New Sci. *47*: 233-235, 1970.

755 - BLACKADAR, A.K., HOCEVAR, A.: A theory of the canopy wind profile. - Biometeorology *4* (11): 55-56, 1970.

756 - BLACKMAN, G.E.: The application of the concepts of growth analysis to the assessment of productivity. - In: ECKARDT, F.E. (ed.): Functioning of Terrestrial Ecosystems at the Primary Production Level. Pp. 243-259. UNESCO, Paris 1968.

757 - BLACKWELL, M.J., BLACKBURN, M.R.: Crop environment data aquisition. - In: WADSWORTH, R.M., CHAPAS, L.C., RUTTER, A.J., SOLOMON, M.E., WARREN WILSON, J., (ed.): The Measurement of Environmental Factors in Terrestrial Ecology. Pp. 213-224. Blackwell sci. Publ., Oxford-Edinburgh 1968.

758 - BLACKWELL, S.J., LAETSCH, W.M., HYDE, B.B.: Development of chloroplast fine structure in aspen tissue culture. - Amer. J. Bot. *56*: 457-463, 1969.

759 - BLAGONRAVOVA, L.N.: Vplyv khlororganichnykh otrutokhimikativ na aktyvnist' chlorofilazy deyakykh roslyn. [Effect of chlororganic herbicides on chlorophyllase activity of some plants.] - Ukr. bot. Zh. *26* (4): 103-106, 1969. [In Ukr., ab: E, R.]

760 - BLAIN, J.A.: Carotene-bleaching activity in plant tissue extracts. - J. Sci. Food Agr. *21*: 35-38, 1970.

761 - BLEASDALE, J.K.A.: Plant growth and crop yield. - Ann. appl. Biol. *57*: 173-182, 1966.

762 - BLISS, L.C.: Plant productivity in alpine microenvironments on Mt. Washington, New Hampshire. - Ecol. Monogr. *36*: 125-155, 1966.[Ps, Chl.]

763 - BLIXT, S.: Linkage studies in *Pisum*. XI. Linkage relations of the gene *chi 16* determining chlorophyll-deficiency *Chlorotica*. - Agri Hort. Gen. *26*: 107-110, 1968.

764 - BLIXT, S.: Linkage relations in *Pisum* XIV. The chlorophyll deficiency genes *chi 6, chi 9* and *chi 17*, determining *chlorotica*-type, and *ch 4* and *ch 5*, determining *chlorina*. - Agri Hort. Gen. *27*: 36-52, 1969.

765 - BLOKHIN, V.G., USTENKO, G.P., FETISOV, I.M.: Svyaz' fotosinteticheskoĭ deyatel'-nosti i urozhaya s pogloshchayushcheĭ sposobnost'yu kornevoĭ sistemy u gorokha. [Relationships of photosynthetic activity and yield with absorbing capacity of pea roots.] - Agrokhimiya *1966* (12): 54-67, 1966. [In R.]

766 - BLORE, T.W.D., WORMER, T.M.: The measurement of the leaf area of *Coffea arabica* L., pruned on the single-stem system. - Turrialba *16*: 86-87, 1966.

767 - BLUM, J.J., BÉGIN-HEICK, N.: Metabolic changes during phosphate deprivation in *Euglena* in air and in oxygen. - Biochem. J. *105*: 821-829, 1967. [Car.]

768 - BLUMBERG, W.E., PEISACH, J.: The optical and magnetic properties of copper in *Chenopodium album* plastocyanin. - Biochim. biophys. Acta *126*: 269-273, 1966.

769 - BLUMENTHAL-GOLDSCHMIDT, S., POLJAKOFF-MAYBER, A.: Comparison of the osmotic volume changes in chloroplasts isolated from different plants. - Plant Cell Physiol. *7*: 357-362, 1966.

770 - BOAG, J.W.: Techniques of flash photolysis. - Photochem. Photobiol. *8*: 565-577, 1968. [Chl.]

771 - BOARDMAN, N.K.: Protochlorophyll. - In: VERNON, L.P., SEELY, G.R. (ed.): The Chlorophylls. Pp. 437-479. Academic Press, New York-London 1966.

772 - BOARDMAN, N.K.: Chloroplast structure and development. - In: SAN PIETRO, A., GREER, F.A., ARMY, T.J. (ed.): Harvesting the Sun. Pp. 211-230. Academic Press, New York-London 1967.

773 - BOARDMAN, N.K.: The photochemical systems of photosynthesis. - Adv. Enzymol. *30*: 1-79, 1968.

774 - BOARDMAN, N.K.: Cytochromes of developing and mutant chloroplasts. - In: SHIBA-TA, K., TAKAMIYA, A., JAGENDORF, A.T., FULLER, R.C. (ed.): Comparative Biochem-istry and Biophysics of Photosynthesis. Pp. 206-213. Univ. Tokyo Press, Tokyo; Univ. Park Press, State College, Pa. 1968.

775 - BOARDMAN, N.K.: The photochemical systems of photosynthesis. - Aust. J. Sci. *32* (2): 36-45, 1969.

776 - BOARDMAN, N.K.: Physical separation of the photosynthetic photochemical systems. - Annu. Rev. Plant Physiol. *21*: 115-140, 1970.

777 - BOARDMAN, N.K., ANDERSON, J.M.: Fractionation of the photochemical systems of photosynthesis. II. Cytochrome and carotenoid contents of particles isolated from spinach chloroplasts. - Biochim. biophys. Acta *143*: 187-203, 1967.

778 - BOARDMAN, N.K., HIGHKIN, H.R.: Studies on a barley mutant lacking chlorophyll *b*. I. Photochemical activity of isolated chloroplasts. - Biochim. biophys. Acta *126*: 189-199, 1966.

779 - BOARDMAN, N.K., THORNE, S.W.: Studies on a barley mutant lacking chlorophyll *b*. II. Fluorescence properties of isolated chloroplasts. - Biochim. biophys. Acta *153*: 448-458, 1968.

780 - BOARDMAN, N.K., THORNE, S.W.: Fluorescence properties of fragments from soni-cated spinach chloroplasts. - Biochim. biophys. Acta *189*: 294-297, 1969.

781 - BOARDMAN, N.K., THORNE, S.W., ANDERSON, J.M.: Fluorescence properties of par-ticles obtained by digitonin fragmentation of spinach chloroplasts. - Proc. nat.

Acad. Sci. U.S.A. *56*: 586-593, 1966.

782 - BOASSON, R., LAETSCH, W.M.: Chlorophyll synthesis in tobacco callus: interaction of sugar and kinetin. - Experientia *23*: 968, 1967.

783 - BOASSON, R., LAETSCH, W.M.: Chloroplast replication and growth in tobacco. - Science *166*: 749-751, 1969.

784 - BODEA, C., TĂMAŞ, V., NEAMŢU, G.: Partialsynthesen von Dehydro-Carotinen. IV. Dehydro-Carotine aus Kryptoxanthin und Kryptoxanthin-Palmitat. - Rev. roum. Chim. *11*: 739-743, 1966.

785 - BODIN, K., NAUWERCK, A.: Produktionsbiologische Studien über die Moosvegetation eines klaren Gebirgsees. - Schweiz. Z. Hydrol. *30*: 318-352, 1968. [Ps.]

786 - BOE, A.A., SALUNKHE, D.K.: Ripening tomatoes: $C^{14}O_2$ uptake by green tomato fruit. - Experientia *23*: 779, 1967.

787 - BOGACHEVA, I.I.: Dvizhenie khloroplastov i ego energeticheskie osnovy. [Chloroplast movement and its energetics.] - In: Khloroplasty i Mitokhondrii. Pp. 227-247. Nauka, Moskva 1969. [In R.]

788 - BOGACHEVA, I.I., GOLUBKOVA, B.M., KISLYAKOVA, T.E.: Struktura i funktsii khloroplastov v ontogeneze kartofelya. [Structure and function of chloroplasts during potato ontogenesis.] - In: Ontogenez Vysshikh Rasteniĭ. Pp. 94-100. Izdat. Akad. Nauk Arm. SSR, Erevan 1970. [In R.]

789 - BOGDANOVIĆ, M.: The rate of biochemical conversion of protochlorophyll to chlorophyll in black pine dark grown seedlings. - Zemljište Biljka *17*: 287-293, 1968.

790 - BOGDASHEVSKAYA, O.V., RUNOVA, Yu.N.: Vliyanie γ-oblucheniya semyan rzhi na dinamiku nakopleniya khlorofilla i karotina. [Effect of γ-irradiation of rye seeds on the dynamics of accumulation of chlorophyll and carotene.] - Radiobiologiya *6*: 288-291, 1966. [In R.]

791 - BÖGER, P.: Photophosphorylierung mit Chloroplasten aus *Bumilleriopsis filiformis* VISCHER. - Z. Pflanzenphysiol. *61*: 85-97, 1969.

792 - BÖGER, P.: Ferredoxin-katalysierte Reaktionen im zellfreien System der Alge *Bumilleriopsis filiformis* VISCHER. - Z. Pflanzenphysiol. *61*: 447-461, 1969.

793 - BÖGER, P.: Ferredoxin aus *Bumilleriopsis filiformis* VISCHER. - Planta *92*: 105-128, 1970.

794 - BÖGER, P.: Wechselwirkung zwischen Ferredoxin und der Ferredoxin-NADP-Reduktase. - Ber. deut. bot. Ges. *83*: 471-472, 1970.

795 - BÖGER, P., BLACK, C.C., SAN PIETRO, A.: Photosynthetic reactions with pyridine nucleotide analogs. II. 3-pyridinealdehyde-diphosphopyridine nucleotide and 3-pyridinealdehyde-deamino-diphosphopyridine nucleotide. - Arch. Biochem. Biophys. *115*: 35-43, 1966.

796 - BÖGER, P., BLACK, C.C., SAN PIETRO, A.: Photosynthetic reactions with pyridine nucleotide analogs. III. N-Methylpyridinium iodides. - Biochemistry *6*: 80-88, 1967.

797 - BÖGER, P., BLACK, C.C., SAN PIETRO, A.: Photosynthetic reactions with pyridine nucleotide analogs. I. Isonicotinic acid hydrazide - NAD. - In: GOODWIN, T.W. (ed.): Biochemistry of Chloroplasts. Vol. II. Pp. 565-579, Academic Press, London-New York 1967.

798 - BÖGER, P., SAN PIETRO, A.: Ferredoxin and cytochrome *f* in *Euglena gracilis*. - Z. Pflanzenphysiol. *58*: 70-75, 1967.

799 - BÖGER, P., SAN PIETRO, A.: Oxidation of reduced triphosphopyridine nucleotide by pyridinium salts. - Arch. Biochem. Biophys. *120*: 379-383, 1967.

800 - BOGIN, E., WALLACE, A.: CO_2 fixation in preparations from Tunisian sweet lemon and Eureka lemon fruits. - Proc. amer. Soc. hort. Sci. *88*: 298-307, 1966.

801 - BOGORAD, L.: The biosynthesis of chlorophylls. - In: VERNON, L.P., SEELY, G.R. (ed.): The Chlorophylls. Pp. 481-510. Academic Press, New York-London 1966.

802 - **BOGORAD, L.**: Photosynthesis. - In: Plant Biology Today: Advances and Challenges. Pp. 27-56. Wadsworth Publ. Co. Inc., Belmont, Calif. 1966.

803 - **BOGORAD, L.**: Aspects of chloroplast assembly. - In: VOGEL, H.J., LAMPEN, J.O., BRYSON, V. (ed.): Organizational Biosynthesis. Pp. 395-418, 435-437. Academic Press, London-New York 1967.

804 - **BOGORAD, L.**: Chloroplast structure and development. - In: SAN PIETRO, A., GREER, F.A., ARMY, T.J. (ed.): Harvesting the Sun. Pp. 191-210. Academic Press, New York-London 1967.

805 - **BOGORAD, L.**: Control mechanisms in plastid development. - Developm. Biol. (Suppl. 1): 1-31, 1967.

806 - **BOGORAD, L.**: The organization and development of chloroplasts. - In: ALLEN, J.M. (ed.): Molecular Organization and Biological Function. Pp. 134-185. Harper & Row Publ., New York-Evanston-London 1967.

807 - **BOGORAD, L.**: Biosynthesis and morphogenesis in plastids. - In: GOODWIN, T.W. (ed.): Biochemistry of Chloroplasts. Vol. II. Pp. 615-631. Academic Press, London and New York 1967.

808 - **BOGORAD, L.**: Control mechanisms in plastid development. - In: LOCKE, M. (ed.): Control Mechanisms in Developmental Processes. Pp. 1-31. Academic Press, New-York-London 1967.

809 - **BOGORAD, L., LABER, L., GASSMANN, M.**: Aspects of chloroplast development: transitory pigment-protein complexes and protochlorophyllide regeneration. - In: SHIBATA, K., TAKAMIYA, A., JAGENDORF, A.T., FULLER, R.C. (ed.): Comparative Biochemistry and Biophysics of Photosynthesis. Pp. 299-312. Univ. Tokyo Press, Tokyo; Univ. Park Press, State College, Pa. 1968.

810 - **BOGORAD, L., TROXLER, R.F.**: The biogenesis of heme, chlorophylls, and bile pigments. - In: BERNFELD, P. (ed.): Biogenesis of Natural Compounds. 2nd Ed. Pp. 247-313. Pergamon Press, Oxford-London-Edinburgh-New York-Toronto-Sydney-Paris-Braunschweig 1967.

811 - **BÖHME, H., TREBST, A.**: On the properties of ascorbate photooxidation in isolated chloroplasts. Evidence for two ATP sites in noncyclic photophosphorylation. - Biochim. biophys. Acta *180*: 137-148, 1969.

812 - **BOÏCHENKO, E.A.**: Uchastie metallov v evolyutsii okislitel'no-vosstanovitel'nykh protsessov rasteniĭ. [Participation of metals in evolution of oxido-reduction processes in plants.] - Izv. Akad. Nauk SSSR, Ser. biol. *1968*: 24-33, 1968. [In R, ab: E.]

813 - **BOISSYA, C.L.**: Effect of kinetin and sucrose on the preservation of chlorophylls A and B in detached *Phaseolus vulgaris* leaves. - J. Assam Sci. Soc. *13*: 35-40, 1970.

814 - **BOJE, R.**: Chlorophyll. - Limnologica *4*: 397-401, 1966.

815 - **BOLDOR, O., ATANASIU, L.**: Variation of the chlorophyll quantity in higher plants, during the day. - Physiol. Plant. Rom. *1970*: 23-28, 1970.

816 - **BOLHÅR-NORDENKAMPF, H.**: Die Wirkung von Atrazin auf den pflanzlichen Gasstoffwechsel. - Biochem. Physiol. Pflanzen *161*: 342-357, 1970.

817 - **BOLIN, B.**: The carbon cycle.- Sci. Amer. *223* (3): 124-132, 1970.

818 - **BOLL, M.**: Oxidation of reduced nicotinamide-adenine-dinucleotide in *Rhodospirillum rubrum*. III. Properties of a NADH dehydrogenase solubilized from electron transport particles. - Arch. Mikrobiol. *69*: 301-313, 1969.

819 - **BOLL, M.**: Action of sodium dodecyl sulfate on electron transport enzymes of *Rhodospirillum rubrum*. - Experientia *26*: 956-957, 1970.

820 - **BOLL, M.**: Studies with Triton X-100 treated electron transport particles from *Rhodospirillum rubrum*. - Arch. Mikrobiol. *71*: 1-8, 1970.

821 - **BOLL, M.**: Effect of proteolytic and lipolytic enzymes on the electron transport particle fraction of *Rhodospirillum rubrum*. I. Proteases. - Z. Naturforsch. *25b*: 1448-1450, 1970.

822 – BOLLE, J., de MAHEAS, M.-R.: Rôles et interrelations des chlorophylles *a* et *b*.
      Réactivité des deux chlorophylles en présence de composés phosphorés. – Compt.
      rend. Acad. Sci. Paris, Sér. D *263*: 434-435, 1966.

823 – BOLLE-JONES, E.W.: Variations of chlorophyll and soluble sugar in oil palm
      leaves in relation to position, time of day and yield. – Oléagineaux *28*: 505-
      512, 1968.

824 – BOLLI, M., CIRI, S., FRENGUELLI, G.: Metodo gascromatografico per la determina-
      zione della CO_2 emessa ed assorbita dalle piante superiori. [Gas chromatographic-
      al method for determination of CO_2 evolved and absorbed by higher plants.] –
      Agrochimica *13*: 386-392, 1969. [In Ital., ab: E, G, F, Span.]

825 – BOLLIGER, F.R., KÖNIG, A.: Vitamine einschliesslich Carotinoide, Chlorophylle
      und biologisch aktive Chinone. – In: STAHL, E. (ed.): Dünnschicht-Chromatogra-
      phie. 2nd Ed. Pp. 253-302. Springer-Verlag, Berlin-Heidelberg-New York 1967.
      [TLC.]

826 – BOLTON, J.R., CLAYTON, R.K., REED, D.W.: An identification of the radical
      giving rise to the light-induced electron spin resonance signal in photosynthe-
      tic bacteria. – Photochem. Photobiol. *9*: 209-218, 1969.

827 – BOLTON, J.R., COST, K., FRENKEL, A.W.: A kinetic study of the production of
      light-induced ESR·signals in *Rhodospirillum rubrum* chromatophores. – Arch. Bio-
      chem. Biophys. *126*: 383-387, 1968.

828 – BOMSEL, J.-L.: Étude in vivo de la photophosphorylation cyclique dans les feuil-
      les de Blé. – Compt. rend. Acad. Sci. Paris, Sér.D *262*: 1706-1709, 1966.

829 – BONAVENTURA, C., MYERS, J.: Fluorescence and oxygen evolution from *Chlorella
      pyrenoidosa*. – Biochim. biophys. Acta *189*: 366-383, 1969.

830 – BOND, C.P.: The isolation of leaf components. III. The separation, identifica-
      tion and estimation of carotenoids and some quinones. – J. Sci. Food Agr. *18*:
      161-163, 1967.

831 – BONEY, A.D., WHITE, E.B.: Phycoerythrins from some *Acrochaetium* species. –
      Nature *218*: 1068-1069, 1968.

832 – BONHOMME, R.: Microclimat lumineux dans une culture de patate douce; incidence
      sur la photosynthèse. – In: Caribean Food Crop Soc., Martinique, Proc. 7th annu.
      Meeting. Pp. 279-293. Martinique-Guadeloupe. 1969.

833 – BONHOMME, R.: Spectral distribution of radiation from sun and sky, and energy
      available to plants. – Biometeorology *4*: 56, 1970.

834 – BONHOMME, R.: Surface relative des taches de soleil dans la végétation. – In:
      Techniques d'Étude des Facteurs Physiques de la Biosphère. Pp. 99-104. INRA,
      Paris 1970.

835 – BONHOMME, R.: Application de la technique des photographies hémisphériques in
      situ à la mesure de l'indice foliaire. – In: Techniques d'Étude des Facteurs
      Physiques de la Biosphère. Pp. 501-505. INRA, Paris 1970.

836 – BONHOMME, R., CHARTIER, P., GANS, F.: Mesure de la composition spectrale du
      rayonnement diffus. – Météorologie *1*: 295-305, 1967.

837 – BONHOMME, R., VARLET GRANCHER, C.: Comparaison d'un spectrophotomètre à disposi-
      tif optique et d'un spectroradiomètre. Composition spectrale du rayonnement
      global ou diffus. – In: Techniques d'Étude des Facteurs de la Biosphère. Pp.
      89-97. INRA, Paris 1970.

838 – BONN, B., FORSYTH, F.R., HALL, I.V.: A comparison of the rates of apparent
      photosynthesis of the cranberry and the common lowbush blueberry. – Nat. can.
      96: 799-804, 1969.

839 – BONNEMAIN, J.-L.: Sur le transport des produits de la photosynthèse chez la
      Tomate lors de la fructification. – Compt. rend. Acad. Sci. Paris, Sér. D *262*:
      366-369, 1966.

840 – BONNEMAIN, J.-L.: Sur les modalités de la distribution des assimilats chez la
      Tomate et sur ses mécanismes. – Compt. rend. Acad. Sci. Paris, Sér. D *262*: 1106-

1109, 1966.

841 - BONNEMAIN, J.-L.: Transport du ^{14}C assimilé à partir des feuilles de Tomate en voie de croissance et vers celles-ci. - Compt. rend. Acad. Sci. Paris, Série D *269*: 1660-1663, 1969.

842 - BONNER, B.A.: A short-lived intermediate form in the in vivo conversion of protochlorophyllide 650 to chlorophyllide 684. - Plant Physiol. *44*: 739-747, 1969.

843 - BONNETT, R., MALLAMS, A.K., SPARK, A.A., TEE, J.L., WEEDON, B.C.L., McCORMICK, A.: Carotenoids and related compounds. Part XX. Structure and reactions of fucoxanthin. - J. chem. Soc., Sect. C - org. Chem. *1969*: 429-454, 1969.

844 - BOOTH, J.R., MENNAGHAN, G.F.: The kinetics of a photosynthetic gas exchanger with laminar flow during low intensity illumination. - In: Chemical Engineering in Medicine and Biology. Pp. 477-493. Plenum Press, New York 1967.

845 - BOOTH, J.R. Jr.: The kinetics of a photosynthetic gas exchanger with laminar flow during low intensity illumination. - Diss. Abstr. *27*: 1130-B, 1966.

846 - BOOTS, M.R.: Conformational aspects of carbamates in the inhibition of the Hill reaction. - J. med. Chem. *12*: 426-428, 1969.

847 - BORCHERT, M.T., WESSELS, J.S.C.: Combined preparation of ferredoxin, ferredoxin-NADP$^+$ reductase and plastocyanin from spinach leaves. - Biochim. biophys. Acta *197*: 78-83, 1970.

848 - BORCHGREVINK, N.C., CHARLEY, H.: Color of cooked carrots related to carotene content: Determinations by chromatographic and spectrophotometric analyses. - J. amer. diet. Ass. *49*: 116-121, 1966.

849 - BORG, D.C., FAJER, J., FELTON, R.H., DOLPHIN, D.: The π-cation radical of chlorophyll *a*. - Proc. nat. Acad. Sci. U.S.A. *67*: 813-820, 1970.

850 - BORISENKO, T.T.: Fotosintez u poliploidnoĭ sakharnoĭ svekly pri razlichnoĭ vliyazhnosti pochvy. [Photosynthesis in polyploid sugar beet with different soil moisture.] - Fiziol. Biokhim. kul't. Rast. 2: 634-638, 1970. [In R, ab: E.]

851 - BORISOV, A.Yu.: O prirode fotosinteticheskikh edinits u fotosinteziruyushchikh organizmov. [On the nature of photosynthetic units in photosynthesizing organisms.] - Nach. Dokl. vyssh. Shkoly, biol. Nauki 10 (1): 149, 1967. [In R.]

852 - BORISOV, A.Yu.: Fotookislenie tsitokhromov kak kriteriĭ dlya razlicheniya fotosinteziruyushchikh organizmov s odnoĭ i dvumya fotokhimicheskimi pigmentnymi sistemami. [Photooxidation of cytochromes as a criterion of difference between photosynthesizing organisms with one and two photochemical pigment systems.] - Nauch. Dokl. vyssh. Shkoly. Biol. Nauki 10 (1): 150, 1967. [In R.]

853 - BORISOV, A.Yu.: Teoriya rezonansnoĭ migratsii energii v pigmentnykh kompleksakh fotosinteziruyushchikh organizmov. [Theory of energy resonance migration in pigment complexes of photosynthesizing organisms.] - Biofizika *12*: 630-636, 1967. [In R.]

854 - BORISOV, A.Yu.: Priroda fotosinteticheskikh edinits u fotosinteziruyushchikh organizmov. [The nature of photosynthetic units in photosynthesizing organisms.] - Dokl. Akad. Nauk SSSR *173*: 208-211, 1967. [In R.]

855 - BORISOV, A.Yu.: Svyaz' lyuminestsentnykh kharakteristik khlorofilla s elektronnym transportom pri fotosinteze. [Relation between the luminescence characteristics of chlorophyll and the electron transport in photosynthesis.] - Izv. Akad. Nauk SSSR, Ser. fiz. *32*: 1511-1516, 1968. [In R.]

856 - BORISOV, A.Yu.: Energetika fotosinteza (ot kvanta sveta do khimicheskikh reaktsiĭ). [Energetics of photosynthesis (from light quantum to chemical reactions).] - Uspekhi sovrem. Biol. *68*: 210-231, 1969. [In R.]

857 - BORISOV, A.Yu.: Termodinamicheskie soobrazheniya protiv odnokvantovogo elektronnogo transporta pri fotosinteze u rasteniĭ. [Thermodynamic objections to one-quantum electron transport in plant photosynthesis.] - Mol. Biol. (Moskva) *3*: 343-348, 1969. [In R, ab: E.]

858 - BORISOV, A.Yu., GODIK, V.I.: Fluorescence lifetime of bacteriochlorophyll and reaction center photooxidation in a photosynthetic bacterium. - Biochim. biophys. Acta *223*: 441-443, 1970.

859 - BORISOV, A.Yu., GODIK, V.I., CHIBISOV, A.K.: O tipakh migratsii energii pri bakterial'nom fotosinteze. [On the types of energy transfer in bacterial photosynthesis.] - Mol. Biol. (Moskva) *4*: 500-508, 1970. [In R, ab: E.]

860 - BORISOV, A.Yu., IL'INA, M.D.: Absorbtsionnye i fluorestsentnye svoĭstva khloroplastov gorokha i ikh fragmentov. [Absorption and fluorescence properties of pea chloroplasts and their fragments.] - Mol. Biol. (Moskva) *3*: 391-405, 1969. [In R, ab: E.]

861 - BORISOV, A.Yu., IVANOVSKIĬ, R.N.: Teoreticheskoe rassmotrenie funktsionirovaniya tsitokhromnykh perenoschikov zaryada pri fotosinteze v primenenii k issledovaniyu purpurnykh bakteriĭ. [Theory of electron transport via cytochromes in photosynthesis and its application to the study of purple bacteria.] - Mol. Biol. (Moskva) *4*: 642-654, 1970. [In R, ab: E.]

862 - BORISOV, A.Yu., IVANOVSKIĬ, R.N., SAMUILOV, V.D.: Differentsial'naya spektrofotometriya fotosinteziruyushchikh ob"ektov. [Difference spectrophotometry of photosynthesizing objects.] - Biofizika *14*: 676-683, 1969. [In R, ab: E.]

863 - BORISOV, A.Yu., KONDRAT'EVA, E.N., SAMUILOV, V.D., SKULACHEV, V.P.: Sootnoshenie fotovosstanovleniya NAD i fosforilirovaniya v khromatoforakh *Rhodospirillum rubrum*. [Relation between photoreduction of NAD and phosphorylation in chromatophores of *Rhodospirillum rubrum*.] - Mol. Biol. (Moskva) *4*: 795-807, 1970. [In R, ab: E.]

864 - BORKOWSKI, J.D., JOHNSON, M.J.: Long-lived steam-sterilizable membrane probes for dissolved oxygen measurement. - Biotechnol. Bioeng. *9*: 635-639, 1967.

865 - BOROJEVIĆ, S., ĆUPINA, T.: Istraživanja komponenti prinosa zrna kod različitih genotipova pšenice. [Studies of components of grain yield of different genotypes of *Triticum vulgare*.] - Savrem. Poljoprivreda *17*: 3-26, 1969. [Chl; in Croatian, ab: E.]

866 - BORRISS, H.: Kohlenstoff-Assimilation und diurnaler Säurerhythmus epiphytischer Orchideen. - Orchidee *18*: 396-406, 1967.

867 - BORRISS, H., KÖHLER, K.-H.: Die Bestimmung des Gesamtchlorophyllgehaltes von Laubblättern mit Hilfe des Spektralkolorimeters SPEKOL. - Jenaer Rundschau *13*: 232-236, 1968.

868 - BORSDORF, W.: Über die Beziehungen zwischen Assimilationsintensität und Ertrag bei Jungpflanzen einiger Pappelklone. - Züchter *37*: 300-306, 1967.

869 - BÖRTITZ, S., FUCHS, S., WEISE, G.: Gasstoffwechselphysiologisches und biochemisches Verhalten bei *Prunus laurocerassus* L., *Ledum palustre* L. und *Ilex agnifalum* L. bei Frostbelastung unter Freilandbedingungen. - Biol. Zentralbl. *86*: 67-77, 1967. [Ps.]

870 - BORZENKOVA, R.A.: Regulyatsiya fotosinteza v sutochnom tsikle. [Control of photosynthesis in a diurnal cycle.] - Uchen. Zap. ural'.gos.Univ., Ser. biol. *58*: 77-88, 1967. [In R.]

871 - BOSCHETTI, A., GROB, E.C.: Isolation and characterization of chloroplasts from *Chlamydomonas*. - In: METZNER, H. (ed.): Progress in Photosynthesis Research. Vol. I. Pp. 245-249. Tübingen 1969.

872 - BOSE, S., CRESPI, H.L., BLAKE, M.I., KATZ, J.J.: Comparative studies of photosynthetic processes in ordinary and fully deuterated algae. - Plant Cell Physiol. *8*: 545-555, 1967.

873 - BOSIAN, G.: Zum Gaswechselproblem: Beweisführung zur gravierenden Bedeutung der Temperatur für Transpiration und Respiration. - Ber. deut. bot. Ges. *79*: 385-400, 1966. [Method.]

874 - BOSIAN, G.: Die Bedeutung der Stomata, der Luftfeuchte und der Temperatur für den CO_2- und Wasserdampfgaswechsel der Pflanzen. - Photosynthetica *2*: 105-125, 1968.

875 - BOSIAN, G.: Relationships between stomatal aperture, temperature, illumination, relative humidity and assimilation determined in the field by means of controlled-environment plant chanbers. - In: ECKARDT, F.E. (ed.): Functioning of Terrestrial Ecosystems at the Primary Production Level. Pp. 321-328. UNESCO, Paris 1968.

876 - BOSKHARDT, Kh., ARNON, D.I.: Vliyanie plastokhinona i proizvodnykh vitamina K na fotosinteticheskoe fosforilirovanie i perenos elektronov. [Effect of plastoquinone and derivatives of vitamin K on photosynthetic phosphorylation and electron transport.] - In: Funktsional'naya Biokhimiya Kletochnykh Struktur. Pp. 97-103. Nauka, Moskva 1970. [In R.]

877 - BOSSHARD-HEER, E., BACHOFEN, R.: Die Wirkung von Dinactin auf Volumenänderungen ganzer Chloroplasten im Licht und im Dunkeln. - Planta 91: 204-211, 1970.

878 - BOTHE, H.: Ferredoxin als Kofaktor der cyclischen Photophosphorylierung in einem zellfreien System aus der Blaualge Anacystis nidulans. - Z. Naturforsch. 24b: 1574-1582, 1969.

879 - BOTHE, H.: The role of phytoflavin in photosynthetic reactions. - In: METZNER, H. (ed.): Progress in Photosynthesis Research. Vol. III. Pp. 1483-1491. Tübingen 1969.

880 - BOTHE, H., BERZBORN, R.J.: Wirkung von Antikörpern gegen die Ferredoxin-NADP-Reduktase aus Spinat auf photosynthetische Reaktionen in einem zellfreien System aus der Blaualge Anacystis nidulans. - Z. Naturforsch. 25b: 529-534, 1970.

881 - BOTKIN, D.B.: Prediction of net photosynthesis of trees from light intensity and temperature. - Ecology 50: 854-858, 1969.

882 - BOTKIN, D.B., MALONE, C.R.: Efficiency of net primary production based on light intercepted during the growing season. - Ecology 49: 438-444, 1968.

883 - BOTKIN, D.B., WOODWELL, G.M., TEMPEL, N.: Forest productivity estimated from carbon dioxide uptake. - Ecology 51: 1057-1060, 1970.

884 - BOTTOMLEY, W.: Some effects of Triton X-100 on pea ethioplasts. - Plant Physiol. 46: 437-441, 1970.

885 - BOTTRILL, D.E., HAWKER, J.S.: Chlorophylls and their derivatives during drying of sultana grapes. - J. Sci. Food Agr. 21: 193-196, 1970.

886 - BOTTRILL, D.E., POSSINGHAM, J.V.: Isolation procedures affecting the retention of water-soluble nitrogen by spinach chloroplasts in aqueous media. - Biochim. biophys. Acta 189: 74-79, 1969.

887 - BOTTRILL, D.E., POSSINGHAM, J.V.: The effect of mineral deficiency and leaf age on the nitrogen and chlorophyll content of spinach chloroplasts. - Biochim. Biophys. Acta 189: 80-84, 1969.

888 - BOTTRILL, D.E., POSSINGHAM, J.V., KRIEDEMANN, P.E.: The effect of nutrient deficiencies on photosynthesis and respiration in spinach. - Plant Soil 32: 424-438, 1970.

889 - BOUCHER, L.J., CRESPI, H.L., KATZ, J.J.: Optical rotatory dispersion of phycocyanin. - Biochemistry 5: 3796-3802, 1966.

890 - BOUCHER, L.J., KATZ, J.J.: The infrared spectra of metalloporphyrins (4000 - 160 cm^{-1}). - J. amer. chem. Soc. 89: 1340-1345, 1967.

891 - BOUCHER, L.J., KATZ, J.J.: Aggregation of metallochlorophylls. - J. amer. chem. Soc. 89: 4703-4708, 1967.

892 - BOUCHER, L.J., STRAIN, H.M., KATZ, J.J.: The far-infrared spectra of monomeric and aggregated chlorophylls a and b. - J. amer. chem. Soc. 88: 1341-1346, 1966.

893 - BOUCHET, R.-J., ROBELIN, M.: Évapotranspiration potentielle et réelle. Domaine d'utilisation - Portée pratique. - B.T.I. 238 (L3 - Agro - 321): 215-223, 1969. [Ps.]

894 - BOUMA, D.: Growth changes of subterranean clover during recovery from phosphorus and sulfur stresses. - Aust. J. biol. Sci. 20: 51-66, 1967.

895 - BOUMA, D.: Effects of nitrogen nutrition on leaf expansion and photosynthesis of *Trifolium subterraneum* L. I. Comparison between different levels of nitrogen supply. -.Ann. Bot. *34*: 1131-1142, 1970.

896 - BOUMA, D.: Effects of nitrogen nutrition on leaf expansion and photosynthesis of *Trifolium subterraneum* L. II. Comparison between nodulated plants and plants supplied with combined nitrogen. - Ann. Bot. *34*: 1143-1153, 1970.

897 - BOUNIAS, M.: Une microméthode d'analyse quantitative par chromatographie en couche mince des sucres, amino acides, pigments et co-facteurs éthéro-solubles de plants d'*Arabidopsis thaliana* (L.) HEYNH. - Chim. anal. *51*: 76-82, 1969.

898 - BOURDU, R.: Quelques progrès récents dans la connaissance des systèmes membranaires de la cellule. - Inf. sci. *1966* (4): 139-149, 1966.[Chloroplast.]

899 - BOURDU, R.: Sur la structure et les activités des chloroplastes d'orge isolés "intacts". - Stud. biophys. *5*: 71-76, 1967.

900 - BOURDU, R.: Les écosystèmes. Première partie. Généralités - définitions - limites et composition. - Inf. sci. *1967* (4): 143-149, 1967.

901 - BOURDU, R.: Les écosystèmes. Deuxième partie. Les cycles énergétiques et biochimiques. - Inf. sci. *1968* (1): 35-42, 1968.

902 - BOURDU, R., CHAMPIGNY, M.-L., LEFORT, M., MASLOW, M., RÉMY, R., MOYSE, A.: Hétérogénéité structurale et fonctionelle de l'appareil photosynthétique mise en évidence par des variations de nutrition azotée. - In: SINRONVAL, C. (ed.): Le Chloroplaste, Croissance et Vieillissement. Pp. 298-305. Masson, Paris 1967.

903 - BOURDU, R., LEFORT, M.: Structure fine, observée en cryodécapage, des lamelles photosynthétiques des Cyanophycées endosymbiotiques: *Glaucocystis nostochinearum* ITZIGS, et *Cyanophora paradoxa* KORSCHIKOFF. - Compt. rend. Acad. Sci. Paris, Sér. D *265*: 37-40, 1967.

904 - BOURDU, R., MATHIEU, Y., MIGINIAC-MASLOW, M., RÉMY, R., MOYSE, A.: Structure granaire, réduction du NADP et photophosphorylation des chloroplastes isolés de feuilles d'orge. - Planta *80*: 191-210, 1968.

905 - BOURQUE, D.P.: Correlation of physiological, ultrastructural, and macromolecular aspects of light-induced chloroplast development. - Diss. Abstr. int. B *31*: 3142-B, 1970.

906 - BOURQUE, D.P., NAYLOR, A.W.: Hill reaction activity in chloroplasts and cellular RNA distribution in greening leaves of *Canavalia ensiformis* (L) D.C. - Plant Physiol. *41* (Suppl.): IX, 1966.

907 - BOURRET, L.R., AGHION, J.: Agrégation des complexes de pigments chlorophylliens extraits de plantes vertes. I. - Effects des variations de température, entre 12,5°C et 60°C. - Physiol. vég. *6*: 259-267, 1968.

908 - BOUTARD, J.: Effets de la lumière et de l'alimentation en nitrate sur les variations de l'activité nitrate réductase de plantules d'Orge. - Physiol. vég. *4*: 105-123, 1966. [Chl.]

909 - BOUWKAMP, J.C., HONMA, S.: Physiological differences between a green and a tan dry podded line of snap bean. - HortScience *5*: 171-173, 1970.

910 - BOVARNICK, J.G., CHANG, S.-W., SCHIFF, J.A.: Streptomucin (SM) inhibition of the development of photosynthetic competence in illuminated dark grown non-dividing cells of *Euglena*. - Plant Physiol. *43* (Suppl.): S 6, 1968.

911 - BOVARNICK, J.G., ZELDIN, M.H., SCHIFF, J.A.: Differential effects of actinomycin D on cell division and light-induced chloroplast development in *Euglena*. - Developm. Biol. *19*: 321-340, 1969.

912 - BOVÉ, J.M., BOVÉ, C.: La rôle de la phosphorylation photosynthétique dans la synthèse du RNA viral. - Compt. rend. Séances Soc. Biol. *161*: 542-549, 1967.

913 - BOVEY, R.W., MILLER, F.R.: Phytotoxicity of paraquat on white and green hibiscus, sorghum and alpinia leaves. - Weed Res. *8*: 128-135, 1968. [Chl.]

914 - BOWERS, P.G., PORTER, G.: Quantum yields of triplet formation in solutions of chlorophyll. - Proc. roy. Soc. (London) A *296*: 435-441, 1967.

915 - **BOWERS, R.C.**: The effect of girdling on photosynthesis, respiration, carbohydrate levels, and nitrogen status of three year old Lisbon lemon seedlings. - Diss. Abstr. *27*: 2494-B, 1967.

916 - **BOWERS, S.A., HAYDEN, C.W.**: A simple portable reflectometer for field use. - Agron. J. *59*: 490-492, 1967. [Reflectance from leaves.]

917 - **BOWES, G., OGREN, W.L.**: The effect of light intensity and atmosphere on ribulose diphosphate carboxylase activity. - Plant Physiol. *46* (Suppl.): 7, 1970.

918 - **BOWES, G.W.**: Carbonic anhydrase in marine algae. - Plant Physiol. *44*: 726-732, 1969.

919 - **BOWMAN, G.E.**: The measurement of carbon dioxide concentration in the atmosphere. - In: WADSWORTH, R.M., CHAPAS, L.C., RUTTER, A.J., SOLOMON, M.E., WARREN WILSON, J. (ed.): The Measurement of Environmental Factors in Terrestrial Ecology. Pp. 131-139. Blackwell sci. Publ., Oxford-Edinburgh 1968.

920 - **BOWMAN, G.E.**: The control of carbon dioxide concentration in plant enclosures. - In: ECKARDT, F.E. (ed.): Functioning of Terrestrial Ecosystems at the Primary Production Level. Pp. 335-343. UNESCO, Paris 1968.

921 - **BOYD, C.E.**: Production, mineral accumulation and pigment concentrations in *Typha latifolia* and *Scirpus americanus*. - Ecology *51*: 285-290, 1970.

922 - **BOYER, J.S.**: Leaf enlargement and metabolic rates in corn, soybean, and sunflower at various leaf water potentials. - Plant Physiol. *46*: 233-235, 1970. [Ps.]

923 - **BOYER, J.S.**: Differing sensitivity of photosynthesis to low leaf water potentials in corn and soybean. - Plant Physiol. *46*: 236-239, 1970.

924 - **BOYER, J.S., BOWEN, B.L.**: Inhibition of oxygen evolution in chloroplasts isolated from leaves with low water potentials. - Plant Physiol. *45*: 612-615, 1970.

925 - **BOYER, Y.**: Déficit hydrique et intensité des échanges gazeux au niveau des frondes de *Platycerium stemaria* (BEAUV.) DESV. - Compt. rend. Acad. Sci. Paris, Sér. D *265*: 1377-1380, 1967.

926 - **BOYNTON, J.E.**: Chlorophyll-deficient mutants in tomato requiring vitamin B_1. I. Genetics and physiology. - Hereditas *56*: 171-199, 1966.

927 - **BOYNTON, J.E.**: Chlorophyll-deficient mutants in tomato requiring vitamin B_1. - II. Abnormalities in chloroplast ultrastructure. - Hereditas *56*: 238-254, 1966.

928 - **BOYNTON, J.E.**: Chlorophyll-deficient mutants of the tomato requiring vitamin B_1. - I. Genetics and physiology. II. Abnormalities in chloroplast ultrastructure. - Diss. Abstr. *27*: 2593-B-2594-B, 1967.

929 - **BOYNTON, J.E., HENNINGSEN, K.W.**: The physiology and chloroplast structure of mutants at loci controlling chlorophyll synthesis in barley. - Stud. biophys. *5*: 85-88, 1967.

930 - **BRACH, E.J., MACK, A.R.**: A radiant energy meter and integrator for plant growth studies. - Can. J. Bot. *45*: 2081-2085, 1967.

931 - **BRADBEER, J.W.**: Photosynthetic-carbon-cycle enzyme activities in leaves of a chlorophyll-less *Phaseolus vulgaris* plant. - Biochem. J. *114*: 11 P, 1969.

932 - **BRADBEER, J.W.**: The activities of the photosynthetic carbon cycle enzymes of greening bean leaves. - New Phytol. *68:* 233-245, 1969.

933 - **BRADBEER, J.W.**: Plastid development in primary leaves of *Phaseolus vulgaris*. An initial lag phase in light-induced chloroplast development. - New Phytol. *69*: 635-637, 1970.

934 - **BRADBEER, J.W., CLIJSTERS, H., GYLDENHOLM, A.O., EDGE, H.J.W.**: Plastid development in primary leaves of *Phaseolus vulgaris*. The effects of brief flashes of light on dark-grown plants. - J. exp. Bot. *21*: 525-533, 1970.

935 - **BRADBEER, J.W., GYLDENHOLM, A.O., WALLIS, M.E., WHATLEY, F.R.**: Studies on the biochemistry of chloroplast development. - In: METZNER, H. (ed.): Progress in Photosynthesis Research. Vol. I. Pp. 272-279. Tübingen 1969.

936 - BRADLEY, G.A., RHODES, B.B.: Carotenes, xanthophylls, and color in carrot varieties and lines as affected by growing temperatures. - J. amer. Soc. hort. Sci.
94: 63-65, 1969.

937 - BRAÏON, A.V.: Sezonnye izmeneniya sobstvennoĭ fluorestsentsii rastitel'nykh
tkaneĭ. [Seasonal changes of the appropriate fluorescence of plant tissues.] -
Uch. Zap. tartusk. Inst. *185*: 129-133, 1966. [In R.]

938 - BRAÏON, A.V.: Vliyanie dlitel'nykh ottepeleĭ na sostoyanie i nekotorye svoĭstva plastidnogo apparata zimuyushchikh rasteniĭ. [Effect of long warm periods
on the composition and some properties of plastid apparatus of hibernating
plants.] - In: Puti Povysheniya Intensivnosti i Produktivnosti Fotosinteza.
Vol. 3. Pp. 203-207. Naukova Dumka, Kiev 1969. [In R.]

939 - BRAND, J.J., KROGMANN, D.W., CRANE, F.L.: A lipid requirement for photosystem
I in photosynthesis. - Plant Physiol. *46* (Suppl.): 39, 1970.

940 - BRÄNDLE, R., ERISMANN, K.H.: Photosynthese-abhängige Sulfidaufnahme grüner
Pflanzen. - Naturwissenschaften *55*: 41, 1968.

941 - BRANDON, P.C.: Thiocyanato-indoles as energy-transfer inhibitors in photophosphorylation. - Arch. Biochem. Biophys. *138*: 566-573, 1970.

942 - BRANDT, A.B.: Kharakteristika opticheskikh pokazateleĭ suspenzii khlorelly kak
rasseivayushcheĭ sredy. [Characteristics of optical properties of a *Chlorella*
suspension as a scattering medium.] - In: Bioenergetika i Biologicheskaya Spektrofotometriya. Pp. 224-231. Nauka, Moskva 1967. [In R.]

943 - BRANDT, A.B., AKOPOV, E.I.: Avtomaticheskiĭ fotometricheskiĭ izmeritel' plotnosti odnokletochnykh vodorosleĭ. [Automatic photometer for determining density of
unicellular algae.] - Mikrobiologiya *35*: 369-373, 1966. [In R, ab: E.]

944 - BRANDT, A.B., SHARIPOV, K.A.: Nelineĭnye opticheskie yavleniya v list'yakh,
obuslovlennye fototaksicheskoĭ reaktsieĭ khloroplastov. [Non-linear optical
phenomena in leaves, caused by phototactic reaction of chloroplasts.] - Biofizika *14*: 91-97, 1969, [In R, ab: E.]

945 - BRANDT, A.B., SHARIPOV, K.A.: K voprosu o fototaksise khloroplastov. [On the
problem of chloroplast phototaxis.] - Fiziol. Rast. *16*: 557-558, 1969. [In R.]

B946 - BRANDT, A.B., TAGEEVA, S.V.: Opticheskie Parametry Rastitel'nykh Organizmov.
[Optical Parameters of Plant Organisms.] - Nauka, Moskva 1967.

947 - BRANTON, D.: Structural units of chloroplast membranes. - In: SIRONVAL, C.: Le
Chloroplaste, Croissance et Vieillissement. Pp. 48-54. Masson, Paris 1967.

948 - BRANTON, D.: Quantasomes: the power units of photosynthesis? - New Scientist
(London) *40* (617): 19-21, 1968.

949 - BRANTON, D.: Structure of the photosynthetic apparatus. - In: GIESE, A.C. (ed.):
Photophysiology. Vol. III. Pp. 197-224. Academic Press, New York-London 1968.

950 - BRANTON, D., PARK, R.B.: Subunits in chloroplast lamellae. - J. Ultrastruct.
Res. *19*: 283-303, 1967.

951 - BRAUNITZER, G., BAUER, G.: Über die Anzahl der Proteinkomponenten im Lamellar-
system der Chloroplasten. - Naturwissenschaften *54*: 70-71, 1967.

952 - BRAUNITZER, G., BAUER-STÄB, G.: Analytical characterisation and preparative
isolation of the proteins of chloroplast lamellae. - In: METZNER, H. (ed.):
Progress in Photosynthesis Research. Vol. I. Pp. 179-185. Tübingen 1969.

953 - BRAVDO, B.-A.: Decrease of net photosynthesis caused by respiration. - Plant
Physiol. *43*: 479-483, 1968.

954 - BRAZHENAS, G.R., KANOPKAÏTE, S.I.: Biosintez kobalaminov. 2. Vliyanie khloro-
fillov *a*, *b* i katalazy na biosintez kobalaminov. [Cobalamines biosynthesis. 2.
The effect of chlorophyll *a*, *b* and catalase of cobalamine biosynthesis.] - Tr.
Akad. Nauk Lit. SSR, Ser. C *1970*: 157-164, 1970. [In R.]

955 - BREHM, K.: Kationenaustausch bei Hochmoorsphagnen: Die Wirkung von an den Aus-
tauscher gebundenen Kationen in Kulturversuchen. - Beitr. Biol. Pflanzen *47*:
91-116, 1970. [Chl.]

956 - BRENNAN, R.D., de WIT, C.T., WILLIAMS, W.A., QUATTRIN, E.V.: The utility of a digital simulation language for ecological modeling. - Oecologia 4: 113-132, 1970.

957 - BRETON, J., MATHIS, P.: Mise en évidence de l'état triplet de la chlorophylle dans les lamelles chloroplastiques. - Compt. rend. Acad. Sci. Paris, Sér. D 271: 1094-1096, 1970.

958 - BREYHAN, T., HEILINGER, F.: Analytische Untersuchungen der Aminosäuren und der Chlorophyllbildung in Kartoffelkeimen. - Phytochemistry 5: 811-814, 1966.

959 - BRIANTAIS, J.-M.: Échanges d'oxygène induits par la lumière dans les fragments de chloroplastes. - Compt. rend. Acad. Sci. Paris, Sér. D 263: 1899-1902, 1966.

960 - BRIANTAIS, J.M.: Isolement et activité de particules chloroplastiques correspondant aux deux systèmes photochimiques. - Photochem. Photobiol. 5: 135-142, 1966.

961 - BRIANTAIS, J.M.: Rétablissement du lien entre deux structures chloroplastiques isolées par action du Triton X-100. - Biochim. biophys. Acta 143: 650-653, 1967.

962 - BRIANTAIS, J.-M.: Spectroscopie de la chlorophylle dans les chloroplastes entiers et des fragments chloroplastiques. - Photochem. Photobiol. 6: 155-162, 1967.

963 - BRIANTAIS, J.-M.: Isolement de structure liées aux deux systèmes photochimiques des chloroplastes. - Bull. Soc. franç. Physiol. vég. 14: 227-244, 1968.

964 - BRIANTAIS, J.-M.: Extraction différentielle des deux systèmes photochimiques des chloroplastes. - Compt. rend. Acad. Sci. Paris, Sér. D 267: 2207-2210, 1968.

965 - BRIANTAIS, J.-M.: Re-establishment of a link between two particles isolated from chloroplasts by Triton X-100. - In: METZNER, H. (ed.): Progress in Photosynthesis Research, Vol. 1. Pp. 174-178. Tübingen 1969.

966 - BRIANTAIS, J.-M.: Séparation physique et arrangement mutuel des deux systèmes photochimiques des chloroplastes. - Physiol. vég. 7: 135-180, 1969.

967 - BRIK, P.L.: Elektronnomikroskopicheskoe i biokhimicheskoe izuchenie khlorofil'-nykh mutatsiĭ kukuruznykh rasteniĭ. [Electron microscopic and biochemical study of chlorophyll mutants of maize.] - Izv. Akad. Nauk Mold. SSR, Ser. biol. khim. Nauk 1969 (5): 69-70, 1969. [In R.]

968 - BRIL, C., HOBBELEN, J.F., van MILTENBURG, J.C., SCHOUWSTRA, Y., THOMAS, J.B.: Action of hydrolytic enzymes on spinach chloroplast fragments. - Acta bot. neerl. 18: 339-342, 1969.

969 - BRIL, C., van der HORST, D.J., POORT, S.R., THOMAS, J.B.: Fractionation of spinach chloroplasts with sodium deoxycholate. - Biochim. biophys. Acta 172: 345-348, 1969.

970 - BRILLER, S., GROMET-ELHANAN, Z.: Ammonium chloride and valinomycin or nonactin as inhibitors of photophosphorylation in chromatophores and subchloroplast particles. - Israel J. Chem. 7: 144 P, 1969.

971 - BRILLER, S., GROMET-ELHANAN, Z.: Effect of ammonium salts, amines and antibiotics on proton uptake and photophosphorylation in Rhodospirillum rubrum chromatophores. - Biochim. biophys. Acta 205: 263-272, 1970.

972 - BRIN, G.P.: Pryamoe i fotosensibilizirovannoe vosstanovlenie metilviologena; fotoreaktivatsiya vosstanovlennykh piridinnukleotidov. [Direct and photosensibilized reduction of methylviologen; photoreactivation of reduced pyridinenucleotides.] - In: YAKOVLEV, V.A. (ed.): Mekhanizmy Dykhaniya, Fotosinteza i Fiksatsii Azota. Pp. 302-308. Nauka, Moskva 1967. [In R.]

973 - BRIN, G.P., KRASNOVSKIĬ, A.A.: Obratiomoe ingibirovanie reaktsii Khilla. [Reversible inhibition of Hill reaction.] - In: Fiziologiya i Biokhimiya Zdorovogo i Bol'nogo Rasteniya. Pp. 198-207. Izdat. mosk. Univ., Moskva 1970. [In R.]

974 - BRIN, G.P., LUGANSKAYA, A.N., KRASNOVSKIĬ, A.A.: Fotovosstanovlenie metilviologena, sensibilizirovannoe khlorofillom i ego analogami; ispol'zovanie tsisteina i tiomocheviny v kachestve donorov elektrona. [Photoreduction of methyl viologen, sensibilized by chlorophyll and its analogues; utilization of cystein and thiourea as electron donors.] - Dokl. Akad. Nauk SSSR 174: 221-224, 1967. [In R.]

975 - BRINGMAN, M., RODSKJER, N.: A small thermoelectric pyranometer for measurements of solar radiation in field crops. - Arch. Meteorol. Geophys. Bioklimatol., Ser. B *16*: 418-433, 1968.

976 - BRINTZINGER, H., PALMER, G., SANDS, R.H.: On the ligand field of iron in ferredoxin from spinach chloroplasts and related nonheme enzymes. - Proc. nat. Acad. Sci. U.S.A. *55*: 397-404, 1966.

977 - BRITTON, G., GOODWIN, T.W.: The occurrence of phytoene 1,2-oxide and related carotenoids in tomatoes. - Phytochemistry *8*: 2257-2258, 1969.

978 - BRIX, H.: An analysis of dry matter production in Douglas-fir seedlings in relation to temperature and light intensity. - Can. J. Bot. *45*: 2063-2072, 1967.

979 - BRIX, H.: Influence of light intensity at different temperatures on rate of respiration of Douglas-fir seedlings. - Plant Physiol. *43*: 389-393, 1968. [Photorespiration.]

980 - BRIX, H.: Effect of temperature on dry matter production of Douglas-fir seedlings during bud dormancy. - Can. J. Bot. *47*: 1143-1146, 1969.

981 - BRIX, H., EBELL, L.F.: Effects of nitrogen fertilization on growth, leaf area, and photosynthesis rate in Douglas-fir. - Forest Sci. *15*: 189-196, 1969.

982 - BROADBENT, D., RADLEY, M.E.: Some effects of 1-amino-2-nitrocyclopentane-1-carboxylic acid on flowering plants. - Ann. Bot. *30*: 763-777, 1966. [Chl.]

983 - BROCK, T.D.: Relationship between standing crop and primary productivity along a hot spring thermal gradient. - Ecology *48*: 566-571, 1967.

984 - BROCK, T.D.: Micro-organisms adapted to high temperature. - Nature *214*: 882-885, 1967. [Ps.]

985 - BROCK, T.D.: Taxonomic confusion concerning certain filamentous blue-green algae. - J. Phycol. *4*: 178-179, 1968. [Chl.]

986 - BROCK, T.D.: Vertical zonation in hot spring algal mats. - Phycologia *8*: 201-205, 1969.

987 - BROCK, T.D.: High temperature systems. - Annu. Rev. Ecol. Systematics *1*: 191-220, 1970. [Ps, Chl.]

988 - BROCK, T.D.: Photosynthesis by algal epiphytes of *Utricularia* in Everglades National Park. - Bull. mar. Sci. *20*: 952-956, 1970.

989 - BROCK, T.D., BROCK, M.L.: The measurement of chlorophyll, primary productivity, photophosphorylation, and macromolecules in benthic algal mats. - Limnol. Oceanogr. *12*: 600-605, 1967.

990 - BROCK, T.D., BROCK, M.L.: Effect of light intensity on photosynthesis by thermal algae adapted to natural and reduced sunlight. - Limnol. Oceanogr. *14*: 334-341, 1969.

991 - BROCK, T.D., BROCK, M.L.: The fate in nature of photosynthetically assimilated ^{14}C in a blue-green alga. - Limnol. Oceanogr. *14*: 604-607, 1969.

992 - BROCKMANN, H. Jr.: Zur absoluten Konfiguration des Bacteriochlorophylls. - Angew. Chem. *80*: 234, 1968.

993 - BROCKMANN, H. Jr.: Zur absoluten Konfiguration des Chlorophylls. - Angew. Chem. *80*: 233-234, 1968.

994 - BROCKMANN, H. Jr., BLIESENER, K.-M., INHOFFEN, H.H.: Zur weiteren Kenntnis des Chlorophylls und des Hämins, XX. Formyl- und Acetyl-substituierte Deuteroporphyrine. - J. Liebigs Ann. Chem. *718*: 148-161, 1968.

995 - BROCKMANN, H. Jr., KLEBER, I.: Zur absoluten Konfiguration des Bacteriochlorophylls *a*. - Angew. Chem. *81*: 626-627, 1969.

996 - BROCKMANN, H. Jr., KLEBER, I.: Bacteriochlorophyll *b*. - Tetrahedron Lett. *1970* *(25)*: 2195-2198, 1970.

997 - BRODY, M.: Chlorophyll studies. - In: BUETOW, D.W. (ed.): The Biology of *Eugle-*

na. Vol. 2. Pp. 215-283. Academic Press, New York-London 1968.

998 - BRODY, M., BRODY, S.S.: The existence of chlorophyll aggregates at room temperature in vivo and in vitro. - Biochim. biophys. Acta *112*: 54-57, 1966.

999 - BRODY, M., BRODY, S.S., DÖRING, G.: Effects of *Ricinus* leaf extracts on light induced changes in absorption of chloroplasts associated with System I and System II. - Z. Naturforsch. *25b*: 862-865, 1970.

1000 - BRODY, M., BROYDE, S.B., YEH, C.C., BRODY, S.S.: Chlorophyll-sensitized oxidation-reduction reactions of hemin in pyridine. - Biochemistry *7*: 3007-3015, 1968.

1001 - BRODY, M., NATHANSON, B., COHEN, W.S.: Enhancement of emission from chloroplasts at 698 nm by a naturally-occurring factor. - Biochim. Biophys. Acta *172*: 340-342, 1969.

1002 - BRODY, S.S.: Low-temperature fluorescence excitation spectra for long-wavelength emission as a function of greening in *Euglena gracilis* and chlorophyll *a* concentration in vitro: A mathematical model to describe both systems. - Biophys. J. *8*: 210-230, 1968.

1003 - BRODY, S.S.: Fluorescence changes accompanying protochlorophyll to chlorophyll conversion in *Euglena gracilis* during the first four hours of illumination. - Photosynthetica *3*: 279-284, 1969.

1004 - BRODY, S.S.: The effects of linolenic acid and extracts of *Ricinus* leaf on System I and System II. - Z. Naturforsch. *25 b*: 855-859, 1970.

1005 - BRODY, S.S.: Low temperature reactions in chloroplasts. - Z. Naturforsch. *25 b*: 860-862, 1970.

1006 - BRODY, S.S., BRODY, M., DÖRING, G.: Effects of linolenic acid on system II and system I-associated light induced changes in absorption of chloroplasts. - Z. Naturforsch. *25 b*: 367-372, 1970.

1007 - BRODY, S.S., BROYDE, S.B.: Low temperature absorption spectra of chlorophyll *a* in polar and nonpolar solvents. - Biophys. J. *8*: 1511-1533, 1968.

1008 - BRODY, S.S., ZIEGELMAIR, C.A., SAMUELS, A., BRODY, M.: Effect of method of preparation on the states of chlorophyll in *Euglena* chloroplast fragments as determined by fluorescence spectroscopy. - Plant Physiol. *41*: 1709-1714, 1966.

1009 - BROERMAN, B.F.S., GATHERUM, G.E., GORDON, J.C.: A controlled-environment chamber for measurement of gas-exchange of tree seedlings. - Forest Sci. *13*: 207-209, 1967.

1010 - BROERMAN, F.S., GATHERUM, G.E.: Relationship of nitrogen and light intensity to growth, photosynthesis and respiration of green ash seedlings. - Iowa State J. Sci. *42*: 137-148, 1967.

1011 - BRONCHART, R.: Structure de la membrane des thylakoides. - In: SIRONVAL, C. (ed.) Le Chloroplaste, Croissance et Vieillissement. Pp. 55-59. Masson, Paris 1967.

1012 - BRONCHART, R.: Effets de la durée des jours sur la structure du chloroplaste d'épinard. - In: SIRONVAL, C. (ed.): Le Chloroplaste, Croissance et Vieillissement. Pp. 325-327. Masson, Paris 1967.

1013 - BRONZAFT, B.D., BERENSHTEÏN, O.P., KESEL'MAN, A.P., ASAUL, Z.I., MOSHKOVA, N.O.: Sklotrubnyĭ aparat dlya masovoĭ kul'tury khlorely. [Glass-pipe apparatus for mass culture of *Chlorella.*] - Ukr. bot. Zh. *25 (4)*: 87-92, 1968. [In Ukr., ab: E, R.]

1014 - BROOKHOUSE, J.K.: Processing the output of paper tape recording equipment. - In: WADSWORTH, R.M., CHAPAS, L.C., RUTTER, A.J., SOLOMON, M.E., WARREN WILSON, J. (ed.): The Measurement of Environmental Factors in Terrestrial Ecology. Pp. 243-254. Blackwell sci. Publ., Oxford-Edinburgh 1968.

1015 - BROOKMAN, J.S.G., OWEN, T.R.: A description and working recommendation for using a modified galvanic cell oxygen electrode. - Biotechnol. Bioeng. *10*: 693-697, 1968.

1016 - BROOKS, K., CRIDDLE, R.S.: Enzymes of the carbon cycle of photosynthesis. I.
       Isolation and properties of spinach chloroplast aldolase. - Arch. Biochem. Bio-
       phys. *117*: 650-659, 1966.

1017 - BROOKS, M.M.: Atomic energy levels explaining photosynthesis. - Adv. Frontiers
       Plant Sci. *16*: 51-70, 1966.

1018 - BROOKS, M.M.: Photosynthesis without chlorophyll. - J. gen. Physiol. *50*: 2508,
       1967.

1019 - BROOKS, M.M.: Molecular oxygen and glucose of photosynthesis without chlorophyll.
       Fed. Proc. *27*: 508, 1968.

1020 - BROOKS, M.M.: Further interpretations of photosynthesis. - Adv. Frontiers Plant
       Sci. *23*: 55-63, 1969.

1021 - BROUGHAM, R.W.: Aspects of light utilization, leaf development and senescence
       and grazing on grass legume balance and productivity of pastures. - Proc. N.Z.
       ecol. Soc. *13*: 58-65, 1966.

1022 - BROUNSHTEÏN, A.M., KOZYREV, B.P., LEBEDEVA, K.D., SIVKOV, S.I.: Nekotorye rezul'-
       taty issledovaniya spektral'nykh kharakteristik chernykh i belykh pokrytiĭ, pri-
       menyaemykh dlya priemnykh poverkhnosteĭ aktinometricheskikh priborov. [Some re-
       sults of studying spectral characteristics of black and white surfaces used as
       acceptor surfaces of actinometric devices.] - Trudy G.G.O. (Leningrad) *1968
       (213)*: 3-12, 1968. [In R.]

1023 - BROUNSHTEÏN, A.M., LEBEDEVA, K.D., SIVKOV, S.I.: Vliyanie spektral'nykh kharak-
       teristik priemnykh poverkhnosteĭ priborov na tochnost' izmereniya radiatsionnykh
       potokov. [Effect of spectral characteristics of acceptor surfaces of apparatuses
       on accuracy of radiant fluxes determinations.] - In: Aktinometriya i Optika At-
       mosfery. Pp. 218-226. Valgus, Tallin 1968. [In R, ab: E.]

1024 - BROWN, A., Mac FAYDEN, A.: Soil carbon dioxide output and small-scale vegetation
       pattern in a *Calluna* heath. - Oikos *20*: 8-15, 1969. [Chl.]

1025 - BROWN, A.P.: Volume contraction of isolated pea chloroplasts promoted by biva-
       lent cations. - Biochem. J. *102*: 791-800, 1967.

1026 - BROWN, A.P., HARVEY, M.J.: Photophosphorylation by isolated chloroplasts. -
       Plant Physiol. *43* (Suppl.): S 20, 1968.

1027 - BROWN, D.L., TREGUNNA, E.B.: Inhibition of respiration during photosynthesis by
       some algae. - Can. J. Bot. *45*: 1135-1143, 1967.

1028 - BROWN, D.L., WEIER, T.E.: Chloroplast development and ultrastructure in the
       freshwater red alga *Batrachospermum*. - J. Phycol. *4*: 199-206, 1968.

1029 - BROWN, J.C.: Iron chlorosis in soybeans as related to the genotype of rootstock:
       5. Differential distribution of photosynthetic C14 as affected by phosphate and
       iron. - Soil Sci. *105*: 159-165, 1968.

1030 - BROWN, J.S.: The fluorescence emission spectra of chlorophyll *a* forms from
       Euglena. - Biochim. biophys. Acta *120*: 305-307, 1966.

1031 - BROWN, J.S.: Fluorescence emission from the forms of chlorophyll *a*. - Carnegie
       Inst. Year Book *65*: 483-487, 1967.

1032 - BROWN, J.S.: Fluorometric evidence for the participation of chlorophyll *a*-695
       in System 2 of photosynthesis. - Biochim. biophys. Acta *143*: 391-398, 1967.

1033 - BROWN, J.S.: Fluorescence emission spectra of a diatom. - In: Abstracts Volume
       - 7th International Congress of Biochemistry. H-69. Tokyo 1967.

1034 - BROWN, J.S.: Chlorophyll absorption and fluorescence in *Ochromonas danica*. -
       Biochim. biophys. Acta *153*: 901-902, 1968.

1035 - BROWN, J.S.: Chlorophyll fluorescence in algae and chloroplasts. - Carnegie Inst.
       Year Book *66*: 192-196, 1968.

1036 - BROWN, J.S.: Absorption and fluorescence spectra of *Ochromonas danica*. - Carne-
       gie Inst. Year Book *66*: 196-197, 1968.

1037 - BROWN, J.S.: Factors affecting absorption and fluorescence spectra of natural chlorophyll complexes. - Carnegie Inst. Year Book *67*: 528-534, 1969.

1038 - BROWN, J.S.: Absorption and fluorescence of chlorophyll *a* in particle fractions from different plants. - Biophys. J. *9*: 1542-1552, 1969.

1039 - BROWN, J.S.: Studies on fractions of chlorophyll complexes from a variety of plants. - Carnegie Inst. Year Book *68*: 566-570, 1970.

1040 - BROWN, J.S.: Absorption and fluorescence of chlorophyll *a* in vivo. - Carnegie Inst. Year Book *68*: 570-572, 1970.

1041 - BROWN, J.S., MICHEL-WOLWERTZ, M.R.: Chlorophyll fluorescence near 720 mμ in *Euglena* extracts. - Biochim. biophys. Acta *153*: 288-290, 1968.

1042 - BROWN, J.S., PRAGER, L.: Absorption and fluorescence of fractions from several plants. - Carnegie Inst. Year Book *67*: 516-520, 1969.

1043 - BROWN, K.W.: Experimental considerations for the measurement of photosynthetic rates by means of carbon dioxide exchange in leaf chambers. - Univ. Nebraska Coll. Agr. Home Econ., Agr. exp. Sta., Progress Rep. *66*: 1-40, 1968.

1044 - BROWN, K.W.: A model of the photosynthesizing leaf. - Physiol. Plant. *22*: 620-637, 1969.

1045 - BROWN, K.W., ROSENBERG, N.J.: Errors in sampling and infrared analysis of CO_2 in air and their influence in determination of net photosynthetic rate. - Agron. J. *60*: 309-311, 1968.

1046 - BROWN, K.W., ROSENBERG, N.J.: Computer program for plotting time-dependent meteorological data. - Agr. Meteorol. *6*: 463-464, 1969.

1047 - BROWN, K.W., ROSENBERG, N.J.: Effect of windbreaks and soil water potential on stomatal diffusion resistance and photosynthetic rate of sugar beets *(Beta vulgaris)*. - Agron. J. *62*: 4-8, 1970.

1048 - BROWN, K.W., ROSENBERG, N.J.: Influence of leaf age, illumination, and upper and lower surface differences on stomatal resistance of sugar beet *(Beta vulgaris)* leaves. - Agron. J. *62*: 20-24, 1970.

1049 - BROWN, K.W., ROSENBERG, N.J.: Concentration of CO_2 in the air above a sugar beet field. - Month. Weather Rev. *98*: 75-82, 1970.

1050 - BROWN, K.W., ROSENBERG, N.J., DORAISWAMY, P.C.: Shading inverted pyranometers and measurements of radiation reflected from an alfalfa crop. - Water Resources Res. *6*: 1782-1786, 1970.

1051 - BROWN, R.H., COOPER, R.B., BLASER, R.E.: Effects of leaf age on efficiency. - Crop Sci. *6*: 206-209, 1966.

1052 - BROWN, R.H., LYNN, W.S.: P/2e⁻ ratios approaching 4 in isolated chloroplasts. - Fed. Proc. *25*: 225, 1966.

1053 - BROWN, S., PREBBLE, J.: The subcellular distribution of carotenoids in cauliflower bud tissue. - Biochem. J. *100*: 54 P, 1966.

1054 - BROWN, S.R.: Absorption coefficients of chlorophyll derivatives. - J. Fish. Res. Board Can. *25*: 523-540, 1968.

1055 - BROWN, S.R.: Bacterial carotenoids from freshwater sediments. - Limnol. Oceanogr. *13*: 233-241, 1968.

1056 - BROWN, T.E., RICHARDSON, F.L.: The effect of growth environment on the physiology of algae: Light intensity. - J. Phycol. *4*: 38-54, 1968.

1057 - BROWN, T.E., RICHARDSON, F.L., VAUGHN, M.L.: Development of red pigmentation in *Chlorococcum wimmeri (Chlorophyta: Chlorococcales)*. - Phycologia *6*: 167-184, 1967. [TLC of Chl, Car.]

1058 - BROWNELL, P.F., NICHOLAS, D.J.D.: Some effects of sodium on nitrate assimilation and N_2 fixation in *Anabaena cylindrica*. - Plant Physiol. *42*: 915-921, 1967.

1059 - BROYDE, S.B., BRODY, S.S.: Spectral studies of a chlorophyll pigment with fluorescence maximum at 698 mμ. - Biophys. J. *6*: 353-366, 1966.

1060 - BROYDE, S.B., BRODY, S.S.: Emission spectra of chlorophyll-a in polar and non-polar solvents. - J. chem. Phys. 46: 3334-3340, 1967.

1061 - BROYDE, S.B., BRODY, S.S., BRODY, M.: Absorption difference spectroscopy of chlorophyll a in ethanol solution. - Biochim. biophys. Acta 153: 183-187, 1968.

1062 - BRUCE, D.L., DUFF, D.C., ANTIA, N.J.: The identification of two antibacterial products of the marine planctonic alga Isochrysis galbana. - J. gen. Microbiol. 48: 293-298, 1967. [PC of chlorophyll derivatives.]

1063 - BRUIN, W.J., NELSON, E.B., TOLBERT, N.E.: Glycolate pathway in green algae. - Plant Physiol. 46: 386-391, 1970.

1064 - BRUINSMA, J.: Effect of plant growth regulators on virus production and chlorophyll content of potato leaf discs. - Naturwissenschaften 53: 281, 1966.

1065 - BRUINSMA, J.: Analysis of growth, development and yield in a spacing experiment with winter rye (Secale cereale L.). - Neth. J. agr. Sci. 14: 198-214, 1966. [Chl.]

1066 - BRUMM, P., RÜPPEL, H.: Periodische Temperatursprung-Anregung. Untersuchungen an Komplexbildungs-Reaktionen. - Z. Naturforsch. 22 b: 980-982, 1967. [Ps.]

1067 - BRUN, W.A., COOPER, R.L.: Effects of light intensity and carbon dioxide concentration on photosynthetic rate of soybean. - Crop Sci. 7: 451-454, 1967.

1068 - BRUNE, D., SAN PIETRO, A.: Chlorophyllin a-catalyzed photoreduction of viologen dyes (Krasnovsky reaction). - Arch. Biochem. Biophys. 141: 371-373, 1970.

1069 - BRUNNHÖFER, H., SCHAUB, H., EGLE, K.: Die Beziehungen zwischen den Veränderungen der Malat- und Stärkekonzentrationen und dem CO_2- und O_2-Gaswechsel bei Bryophyllum daigremontianum. - Z. Pflanzenphysiol. 60: 12-18, 1968.

1070 - BRUNNHÖFER, H., SCHAUB, H., EGLE, K.: Der Verlauf des CO_2- und O_2-Gaswechsels bei Bryophyllum daigremontianum in Abhängigkeit von der Temperatur. - Z. Pflanzenphysiol. 59: 285-292, 1968.

1071 - BRYAN, G.W.: An experimental design for using the biological spectrograph. - Plant Physiol. 44 (Suppl.): 17, 1969. [Chl.]

1072 - BRYAN, G.W., ZADYLAK, A.H., EHRET, C.F.: Photoinduction of plastids and of chlorophyll in a Chlorella mutant. - J. Cell Sci. 2: 513-528, 1967.

1073 - BRYANT, F.D., SEIBER, B.A., LATIMER, P.: Absolute optical cross sections of cells and chloroplasts. - Arch. Biochem. Biophys. 135: 97-108, 1969. [Spectrophotometer for measuring in vivo absorbance without the scattered radiation.]

1074 - BUBICZ, M.: Changes in the carotenoids content in fruits of Berberis vulgaris L. and Berberis Thunbergii atropurpurea in the vegetation period. - Bull. Acad. pol. Sci., Sér. Sci. biol. 14: 285-291, 1966.

1075 - BUBICZ, M.: Comparison of changes in the carotenoid content in leaves and fruit of Berberis vulgaris during the vegetation period. - Bull. Acad. pol. Sci., Sér. Sci. biol. 17: 271-276, 1969.

1076 - BUCH, A., MALESZEWSKI, S.J.: Studies on air passage capacity of maize leaves at varying light intensities, temperatures and fertilization levels in field conditions. - Biol. Plant. 11: 270-276, 1969. [Viscous porometer.]

1077 - BUCHANAN, B.B.: The chemistry and function of ferredoxin. - Structure Bonding 1: 109-148, 1966.

1078 - BUCHANAN, B.B.: Role of ferredoxin in the synthesis of α-ketobutyrate from propionyl coenzyme A and carbon dioxide by enzymes from photosynthetic and non-photosynthetic bacteria. - J. biol. Chem. 244: 4218-4223, 1969.

1079 - BUCHANAN, B.B., ARNON, D.I.: Ferredoxin in plant and bacterial photosynthesis. - In: PEETERS, H. (ed.): Protides of the Biological Fluids. Pp. 143-158. Elsevier Publ. Co., Amsterdam-London-New York 1967.

1080 - BUCHANAN, B.B., ARNON, D.I.: Ferredoxins: Chemistry and function in photosynthesis, nitrogen fixation, and fermentative metabolism. - Adv. Enzymol. 33: 119-176, 1970.

1081 – BUCHANAN, B.B., BACHOFEN, R.: Ferredoxin-dependent reduction of nicotinamide-adenine dinucleotides with hydrogen gas by subcellular preparations from the photosynthetic bacterium, *Chromatium*. – Biochim. biophys. Acta *162*: 607-610, 1968.

1082 – BUCHANAN, B.B., EVANS, M.C.W.: The synthesis of phosphoenolpyruvate from pyruvate and ATP by extracts of photosynthetic bacteria. – Biochem. biophys. Res. Commun. *22*: 484-487, 1966.

1083 – BUCHANAN, B.B., EVANS, M.C.W.: Photoreduction of ferredoxin and its use in NAD(P)$^+$ reduction by a subcellular preparation from the photosynthetic bacterium, *Chlorobium thiosulfatophilum*. – Biochim. biophys. Acta *180*: 123-129, 1969.

1084 – BUCHANAN, B.B., EVANS, M.C.W., ARNON, D.I.: Ferredoxin-dependent carbon assimilation in *Rhodospirillum rubrum*. – Arch. Mikrobiol. *59*: 32-40, 1967.

1085 – BUCHANAN, B.B., KALBERER, P.P., ARNON, D.I.: Ferredoxin-activated fructose diphosphatase in isolated chloroplasts. – Biochem. biophys. Res. Commun. *29*: 74-79, 1967.

1086 – BUCHANAN, B.B., MATSUBARA, H., EVANS, M.C.W.: Ferredoxin from the photosynthetic bacterium, *Chlorobium thiosulfatophilum:* A link to ferredoxins from nonphotosynthetic bacteria. – Biochim. biophys. Acta *189*: 46-53, 1969.

1087 – BÜCHEL, K.H.: Hemmstoffe der Photosynthese. V. Herbizide Trifluormethyl-benzimidazole. – Z. Naturforsch. *25 b*: 934-944, 1970.

1088 – BÜCHEL, K.H.: Hemmstoffe der Photosynthese. VI. Synthesen von elektronegativ substituierten Benzimidazolen. – Z. Naturforsch. *25 b*: 945-953, 1970.

1089 – BÜCHEL, K.H., DRABER, W.: Structure-activity relationship of photosynthesis-inhibition by NH-acidic π-excessive heteroaromatica. – In: METZNER, H. (ed.): Progress in Photosynthesis Research. Vol. III. Pp. 1777-1788. Tübingen 1969.

1090 – BÜCHEL, K.H., DRABER, W., TREBST, A., PISTORIUS, E.: Zur Hemmung photosynthetischer Reaktionen in isolierten Chloroplasten durch Herbizide des Benzimidazol-Typs und deren Struktur-Aktivitäts-Beziehung unter Berücksichtigung des Verteilungskoeffizienten und des p_{K_A}-Wertes. – Z. Naturforsch. *21 b*: 243-254, 1966.

1091 – BÜCHEL, K.H., RÖCHLING, H., BAEDELT, H., GERHARDT, B., TREBST, A.: Hemmung der Photosynthese in *Anacystis* durch Alkylbenzimidazole. – Z. Naturforsch. *22 b*: 535-537, 1967.

1092 – BUCHWALD, H.-E., RÜPPEL, H.: Suppression of disturbing light signals in rapid flash kinetic measurements. – Nature *220*: 57-58, 1968.

1093 – BUCKE, C.: On the interconversion of α-tocopherol and α-tocopherolquinone in broad bean chloroplasts. – In: METZNER, H. (ed.): Progress in Photosynthesis Research. Vol. I. Pp. 325-331. Tübingen 1969.

1094 – BUCKE, C.: The distribution and properties of alkaline inorganic pyrophosphatase from higher plants. – Phytochemistry *9*: 1303-1309, 1970.

1095 – BUCKE, C., BALDRY, C.W., WALKER, D.A.: Photosynthetic carbon dioxide fixation by isolated chloroplasts in Good's buffers. – Phytochemistry *6*: 495-497, 1967.

1096 – BUCKE, C., HALLAWAY, M.: The distribution of plastoquinone C and the seasonal variation in its level in young leaves of *Vicia faba* L. – In: GOODWIN, T.W. (ed.): Biochemistry of Chloroplasts. Vol. I. Pp. 153-157. Academic Press, London and New York 1966.

1097 – BUCKE, C., LEECH, R.M., HALLAWAY, M., MORTON, R.A.: The taxonomic distribution of plastoquinone and tocopherolquinone and their intracellular distribution in leaves of *Vicia faba* L. – Biochim. biophys. Acta *122*: 19-34, 1966.

1098 – BUCKE, C., WALKER, D.A., BALDRY, C.W.: Some effects of sugars and sugar phosphates on carbon dioxide fixation by isolated chloroplasts. – Biochem. J. *101*: 636-641, 1966.

1099 – BUCKLE, K.A., EDWARDS, R.A.: Chlorophyll degradation products from processed pea purée. – Phytochemistry *8*: 1901-1906, 1969.

1100 - BUCKLE, K.A., EDWARDS, R.A.: Chlorophyll degradation and lipid oxidation in frozen unblanched peas. - J. Sci. Food Agr. *21*: 307-312, 1970.

1101 - BUCKMAN, R.E.: Estimation of cubic volume of shrubs (*Corylus* spp.). - Ecology *47*: 858-860, 1966.

1102 - BUDAGOVSKIĬ, A.I., ROSS, Yu.K.: Osnovy kolichestvennoĭ teorii fotosintetiches-koĭ deyatel-nosti posevov. [Principles of quantitative theory of photosynthetic activity of canopies.] - In: Fotosinteziruyushchie Sistemy Vysokoĭ Produktiv-nosti. Pp. 51-58. Nauka, Moskva 1966. [In R.]

1103 - BUDAGOVSKIĬ, A.I., ROSS, Yu.K., TOOMING, Kh.G.: Vertikal'noe raspredelenie po-tokov dlinnovolnovoĭ radiatsii i radiatsionnogo balansa v rastitel'nom pokrove. [Vertical distribution of fluxes of long-wave radiation, and of radiation ba-lance in plant cover.] - In: Aktinometriya i Optika Atmosfery. Pp. 299-307. Val-gus, Tallin 1968. [In R.]

1104 - BUDD, T.W., TJOSTEM, J.L., DUYSEN, M.E.: Ultrastructure of *Chlorella pyrenoido-sa* as affected by environmental changes. - Amer. J. Bot. *56*: 540-545, 1969.

1105 - BUDYKO, M.I.: Solar radiation and the use of it by plants. - In: Agroclimatolo-gical Methods. Pp. 39-53. UNESCO, Paris 1968.

1106 - BUDZIKIEWICZ, H., BRZEZINKA, H., JOHANNES, B.: Zur Photosynthese grüner Pflanzen, 2. Mitt.: Massenspektroskopische Untersuchungen an Carotinoiden. - Monatsch. Chem. *101*: 579-609, 1970.

1107 - BUDZIKIEWICZ, H., DREWES, S.E.: Zur weiteren Kenntnis des Chlorophylls und des Hämins, XXII. "M + 2" peaks. Hydrierung von Doppelbindungen im Massenspektrome-ter. - J. Liebigs Ann. Chem. *716*: 222-223, 1968.

1108 - BUDZIKIEWICZ, H., ECKAU, H.: Zur Photosynthese grüner Pflanzen. IV: Versuche mit 2-Acetylamino-1.3.4.-thiadiazolsulfonamid (Diamox) zur Hemmung eines O-Aus-tausches zwischen CO_2 und H_2O. - Z. Naturforsch. *25b*: 610-612, 1970.

1109 - BUDZIKIEWICZ, H., ECKAU, H., INHOFFEN, H.H.: Zur Photosynthese grüner Pflanzen. I. Versuche mit $H_2^{18}O$ und $K_2C^{18}O_3$ und *Chlorella pyrenoidosa* CHICK. - Z. Natur-forsch. *24 b*: 1147-1152, 1969.

1110 - BUDZIKIEWICZ, H., ECKAU, H., INHOFFEN, H.H.: Zur Photosynthese grüner Pflanzen. III: Untersuchungen der chemischen Vorgänge beim METZNERschen Photosynthese-Modell. - Z. Naturforsch. *25 b*: 525-528, 1970.

1111 - BUDZIKIEWICZ, H., INHOFFEN, H.H.: Experiments on the process of photosynthesis using ^{18}O labeled substrates. - In: METZNER, H. (ed.): Progress in Photosynthe-sis Research. Vol. II. Pp. 1009-1012. Tübingen 1969.

1112 - BUDZIKIEWICZ, H., TARAZ, K.: Zur weiteren Kenntnis des Chlorophylls und des Hämins. XXXI. Nachweis von Vinylgruppen in Porphin-Derivaten durch Addition von Diazoessigester an die Kupfer-Komplexe. - J. Liebigs Ann. Chem. *737*: 128-131, 1970.

1113 - BUETOW, D.E.: Acetate repression of chlorophyll synthesis in *Euglena gracilis*. - Nature *213*: 1127-1128, 1967.

1114 - BUETOW, D.E., MEGO, J.L.: Hydroxyurea inhibition of greening, macromolecule syn-thesis and cell division in *Euglena gracilis*. - Biochim. biophys. Acta *134*: 395-401, 1967.

1115 - BULL, T.A.: Photosynthetic efficiencies and photorespiration in Calvin cycle and C_4-dicarboxylic acid plants. - Crop Sci. *9*: 726-729, 1969.

1116 - BULLEY, N.R., NELSON, C.D.: Action spectra of photosynthesis and photorespira-tion in radish leaves. - Plant Physiol. *43* (Suppl.): S 20, 1968.

1117 - BULLEY, N.R., NELSON, C.D., TREGUNNA, E.B.: Photosynthesis: Action spectra for leaves in normal and low oxygen. - Plant Physiol. *44*: 678-684, 1969.

1118 - BULLEY, N.R., TREGUNNA, E.B.: Photosynthesis and photorespiration rates at the CO_2 compensation point. - Can. J. Bot. *48*: 1271-1276, 1970.

1119 - BULLEY, N.R., TREGUNNA, E.B.: Sensitivity of the infrared CO_2 gas analyzer to $^{14}CO_2$. - Can. J. Bot. *48*: 1292-1294, 1970.

1120 - BÜLTEMANN, V., RÜPPEL, H., VATER, J., WITT, H.T.: Photosynthetic water cleavage reaction. - Abh. deut. Akad. Wiss. Berlin, Kl. Med. *1966*: 321-324, 1966.

1121 - BUNT, J.S.: Some characteristics of microalgae isolated from Antarctic sea ice. - Antarctic Res. Ser. *11*: 1-14, 1967.

1122 - BUNT, J.S.: The CO_2 compensation point, Hill activity and photorespiration. - Biochem. biophys. Res. Commun. *35*: 748-753, 1969.

1123 - BUNT, J.S.: Observations on photoheterotrophy in a marine diatom. - J. Phycol. *5*: 37-42, 1969.

1124 - BUNT, J.S.: Some observations on the carbon dioxide burst in *Chlorella* and *Chlamydomonas*. - Plant Physiol. *45*: 139-142, 1970.

1125 - BUNT, J.S., van OWENS, O.H., HOCH, G.: Exploratory studies on the physiology and ecology of a psychrophilic marine diatom. _ J. Phycol. *2*: 96-100, 1966. [Ps.]

1126 - BURENIN, V.I., TRETYAKOVA, A.A.: Osobennosti assimilyatsionnogo apparata tetraploidnoT svekly. [The peculiarities of assimilation apparatus of tetraploid sugar beet.] - Sel'skokhoz. Biol. *4*: 280-282, 1969. [In R.]

1127 - BUREŠ, F.: Investigations on yield formation in fodder sugar beet in fodder crop rotation. - In: Productivity of Terrestrial Ecosystems. Production Processes. Czech.nat. Comm. IBP, PT-PP Report No. 1. Pp. 28-29. Praha 1970.

1128 - BURGER, G.: Kupfer als Antagonist des Magnesiums bei der Chlorophyllbildung. - Z. Pflanzenphysiol. *56*: 207-208, 1967.

1129 - BURIAN, K.: Die photosynthetische Aktivität eines *Phragmites-communis*-Bestandes am Neusiedler See. - Sitzungsber. österr. Akad. Wiss., math.-naturw. Kl., Abt. I *178 (1-4)*: 43-62, 1969.

1130 - BURIAN, K.: Die Photosynthese einiger Glaushauspflanzen bei Beleuchtung durch Fluoreszenzröhren mit unterschiedlichen Emissionsspektren. - Österr. bot. Z. *117*: 64-86, 1969.

1131 - BURIKOV, E.A.: Osobennosti radiatsionnogo rezhima rastitel'nogo pokrova v zavisimosti ot ego struktury. [The peculiarities of the radiation regime of the plant cover depending on its structure.] - In: Aktinometriya i Optika Atmosfery. Pp. 289-293. Valgus, Tallin 1968. [In R.]

1132 - BURKHOLDER, P.R., BURKHOLDER, L.M., ALMODOVAR, L.R.: Carbon assimilation of marine flagellate blooms in neritic waters of southern Puerto Rico. - Bull. mar. Sci. *17*: 1-15, 1967.

1133 - BURNS, E.R.: Influence of amitrole, dichlormate, and pyriclor on plastid pigment development in wheat. - Diss. Abstr. int. B *31*: 1689-B, 1970.

1134 - BURR, A., MAUZERALL, D.: The oxygen luminometer. An apparatus to determine small amounts of oxygen, and application to photosynthesis. - Biochim. biophys. Acta *153*: 614-624, 1968.

1135 - BURR, F.A.: Phylogenetic transitions in the chloroplasts of the *Anthocerotales*. I. The number and ultrastructure of the mature plastids. - Amer. J. Bot. *57*: 97-110, 1970.

1136 - BURR, G.O.: A note on the efficiency of sugarcane. - Taiwan Sugar exp. Sta. annu. Rep. *1968-1969*: 8-11, 1969. [Ps.]

1137 - BURRIS, J.S.: Carbohydrate metabolism in orchardgrass (*Dactylis glomerata* L.) as affected by light, nitrogen, CO_2 and age. - Diss. Abstr. *28*: 3968B-3969B, 1968.

1138 - BURROWS, F.J.: The diffusive conductivity of sugar beet and potato leaves. - Agr. Meteorol. *6*: 211-226, 1969.

1139 - BURTON, G.W., WILKINSON, W.S., CARTER, R.L.: Effect of nitrogen, phosphorus and potassium levels and clipping frequency on the forage yield and protein, carotene, and xanthophyll content of coastal bermudagrass. - Agron. J. *61*: 60-63, 1969.

1140 - BUSSER, J.H.: Light measurements. - BioScience *18*: 511-512, 1968.

1141 - BUTLER, W.L.: Spectral characteristics of chlorophyll in green plants. - In:
VERNON, L.P., SEELY, G.R. (ed.): The Chlorophylls. Pp. 343-379. Academic Press,
New York, London 1966.

1142 - BUTLER, W.L.: Fluorescence yield in photosynthetic systems and its relation to
electron transport. - Curr. Topics Bioenerg. *1*: 49-73, 1966.

1143 - BUTLER, W.L., BRIGGS, W.R.: The relation between structure and pigments during
the first stages of proplastid greening. - Biochim. biophys. Acta *112*: 45-53,
1966.

1144 - BUTLER, W.L., HOPKINS, D.W.: Higher derivative analysis of complex absorption
spectra. - Photochem. Photobiol. *12*: 439-450, 1970. [Algae and chloroplasts.]

1145 - BUTLER, W.L., HOPKINS, D.W.: An analysis of fourth derivative spectra. - Photochem. Photobiol. *12*: 451-456, 1970. [Chloroplasts.]

1146 - BUTT, A.M.: Vegetative growth, morphogenesis and carbohydrate content of the
onion plant as a function of light and temperature under field- and controlled
conditions. - Meded. Landbouwhogesch. Wageningen *68 (10)*: 1-211, 1968.

1147 - BUTTERFASS, T.: Das Muster aus zellspezifischen Chloroplastenzahlen im Blatt
und seine Ursachen. - Naturwiss. Rundschau *21*: 466-469, 1968.

1148 - BUTTERFASS, T.: Die Plastidenverteilung bei der Mitose der Schliesszellenmutter-
zellen von haploidem Schwedenklee (*Trifolium hybridum* L.) - Planta *84*: 230-234,
1969.

1149 - BUTTERY, B.R.: Effects of variation in leaf area index on growth of maize and
soybeans. - Crop Sci. *10*: 9-13, 1970.

1150 - BUTTROSE, M.S.: Some effects of light intensity and temperature on dry weight
and shoot growth of grape-vine. - Ann. Bot. *32*: 753-765, 1968.

1151 - BUZANOV, I.F., USTIMENKO-BAKUMOVSKIĬ, A.V., OSTROUSHKO, A.I.: Fiziologicheskie
i tekhnologicheskie osobennosti poliploidnykh sortov sakharnoĭ svekly. [Physio-
logical and technological peculiarities of polyploid varieties of sugar beet.]
- Sel'skokhoz. Biol. *1*: 654-665, 1966. [Ps; in R.]

1152 - BUZAS, M.A.: Foraminiferal species densities and environmental variables in an
estuary. - Limnol. Oceanogr. *14*: 411-422, 1969. [Chl.]

1153 - BYKOV, O.D.: K voprosu ob izmerenii i velichine izotopnogo effekta pri fotosinteze. [Measurement and value of the isotopic effect in photosynthesis.] - Dokl.
Akad. Nauk SSSR *183*: 1449-1451, 1968. [In R.]

1154 - BYKOV, O.D.: Opredelenie konstanty skorosti fotosinteticheskoĭ assimilyatsii
$C^{14}O_2$. [Determination of the rate constant of photosynthetic $^{14}CO_2$ assimilation.]
- In: Metody Kompleksnogo Izucheniya Fotosinteza. Tr. VNII Rastenievod. *40*
(Suppl.): 5-16, 1969. [In R.]

1155 - BYKOV, O.D.: Perevod edinits izmereniĭ pri raschete intensivnosti fotosinteza
po dannym radiometricheskogo metoda. [Calculation of photosynthetic rate measured by radiometric method.] In: Metody Kompleksnogo Izucheniya Fotosinteza. Tr.
VNII Rastenievod. *40* (Suppl.): 46-54, 1969. [In R.]

1156 - BYKOV, O.D.: Ob intensivnosti dykhatel'nogo gazoobmena pri fotosinteze. [On the
rate of respiratory gas exchange in photosynthesis.] - Byul. vsesoyuz. Ord.
Lenina Inst. Rastenievod. I.I. Vavilova *14*: 19-24, 1969. [In R.]

1157 - BYKOV, O.D.: The method of calculation of isotopic effect in photosynthesis and
its value. - Photosynthetica *4*: 195-201, 1970.

1158 - BYKOV, O.D., KOSHKIN, V.A.: Radiometricheskoe opredelenie potentsial'noĭ intensivnosti fotosinteza. [Radiometric determination of potential photosynthetic
rate.] - In: Metody Kompleksnogo Izucheniya Fotosinteza. Tr. VNII Rastenievod.
40 (Suppl.): 17-31, 1969. [In R.]

1159 - BYKOV, O.D., KOSHKIN, V.A.: Prigotovlenie i ispol'zovanie preparatov so sloem
polnogo pogloshcheniya dlya radiometricheskogo izmereniya intensivnosti fotosinteza. [Preparation and utilization of samples with a layer of full absorption

for radiometric determination of photosynthetic rate.] - In: Metody Kompleksnogo Izucheniya Fotosinteza. Tr. VNII Rastenievod. *40* (Suppl.): 32-45, 1969. [In R.]

1160 - BYRN, M., LINDSAY SMITH, J.R., CALVIN, M.: The role of bacteriochlorophyll in photosynthetic hydrogen transfer. - J. amer. chem. Soc. *88*: 3177-3178, 1966.

1161 - BYRN, M.P.: The role of chlorophyll as a chemical intermediate in photosynthesis. - Diss. Abstr. *27*: 3816B-3817B, 1967.

1162 - BYRNE, G.F., ROSE, C.W., SLATYER, R.O.: An aspirated diffusion porometer. - Agr. Meteorol. *7*: 39-44, 1970.

1163 - BYRNE, G.F., ROSE, C.W., TORSELL, W.R.: Resistive element solarimeter. - Agr. Meteorol. *6*: 453-455, 1969.

1164 - BYSTROVA, M.I., KRASNOVSKIĬ, A.A.: Izuchenie agregirovannykh form khlorofilla i ego analogov v svyazi so strukturnymi osobennostyami molekul pigmentov. [Study of aggregated forms of chlorophyll and its analogues in relation to structural peculiarities of pigment molecules.] - Mol. Biol. (Moskva) *1*: 362-372, 1967. [In R.]

1165 - BYSTROVA, M.I., KRASNOVSKIĬ, A.A.: Sravnitel'noe issledovanie lyuminestsentsii agregirovannykh form khlorofilla i ego analogov v tverdykh plenkakh. [Comparative study of luminescence of aggregated forms of chlorophyll and its analogues in solid films.] - Mol. Biol. (Moskva) *2*: 847-858, 1968. [In R.]

1166 - BYSTROVA, M.I., UMRIKHINA, A.V., KRASNOVSKIĬ, A.A.: Fotovosstanovlenie protokhlorofilla i protofeofitina. [Photoreduction of protochlorophyll and protopheophytin.] - Biokhimiya *31*: 83-92, 1966. [In R, ab: E.]

1167 - BYTEVA, I.M., GURINOVICH, G.P.: Kinetika reaktsii fotovosstanovleniya khlorofilla *"a"* fenilgidrazinom. [Reaction kinetics of photoreduction of chlorophyll *"a"* by phenylhydrazine.] - Biofizika *12*: 782-787, 1967. [In R.]

1168 - BYTEVA, I.M., PETSOL'D, O.M., PODDUBNAYA, V.M.: O fotovosstanovlenii khlorofilla i feofitina *a* v prisutstvii kisloroda. [Photoreduction of chlorophyll and phaeophytin *a* in the presence of oxygen.] - Biofizika *15*: 977-982, 1970. [In R, ab: E.]

1169 - BYTEVA, I.M., SARZHEVSKAYA, M.V.: Kvantovyĭ vykhod reaktsii fotovosstanovleniya khlorofila i rodstvennykh soedineniĭ. [Quantum yield of photoreduction of chlorophyll and related compounds.] - Biofizika *14*: 441-446, 1969. [In R, ab: E.]

1170 - CABALLERO, A., COSSINS, E.A.: Studies of intermediary metabolism in radish cotyledons. Turnover of photosynthetic products in $^{14}CO_2$ pulse-chase experiments. - Can. J. Bot. *48*: 1191-1198, 1970.

1171 - CAIN, J.C.: A portable economical instrument for measuring light and temperature intensity-time integrals. - HortScience *4*: 123-125, 1969.

1172 - CALABRESE, G., FELICINI, G.P.: Ricerche sui pigmenti della alghe rosse. I: Analisi qualitativa e quantitativa dei pigmenti di *Pterocladia capillacea* (GMEL.) BORN. et THUR. coltivata in presenza di alcune fonti minerali di azoto. [Study of pigments of red algae. I. Qualitative and quantitative analysis of pigments of *Pterocladia capillacea* (GMEL.) BORN. et THUR. cultivated in the presence of different nitrogen sources.] - Giornale bot. ital. *104*: 81-89, 1970. [In Ital., ab: E.]

1173 - CALDWELL, M.M.: Plant gas exchange at high wind speeds. - Plant Physiol. *46*: 535-537, 1970.

1174 - CALDWELL, M.M.: The effect of wind on stomatal aperture, photosynthesis, and transpiration of *Rhododendron ferrugineum* L. and *Pinus cembra* L.- Centralbl. gesamte Forstwesen *87*: 193-201, 1970.

1175 - CALLEJA, G.B., REYNOLDS, G.T.: A quantitative slide test for ATP. - Experientia *26*: 221-222, 1970.

1176 - CALVIN, M.: Perspectives. - In: METZNER, H. (ed.): Progress in Photosynthesis Research. Vol. 1. Pp. 5-7. Tübingen 1969. [Ps.]

1177 - CAMERON, R.J.: Translocation of carbon-14-labelled assimilates in shoots of *Pinus radiata*. - J. exp. Bot. *21*: 943-950, 1970.

1178 - CAMERON, R.J.: Light intensity and the growth of *Eucalyptus* seedlings. I. Ontogenetic variation in *E. fastigata*. - Aust. J. Bot. *18*: 29-43, 1970. [Ps.]

1179 - CAMERON, R.J.: Light intensity and the growth of *Eucalyptus* seedlings. II. The effect of cuticular waxes on light absorption in leaves of *Eucalyptus* species. - Aust. J. Bot. *18*: 275-284, 1970. [Ps.]

1180 - CAMPBELL, R.E., VIETS, F.G. Jr.: Yield and sugar production by sugar beets as affected by leaf area variations induced by stand density and nitrogen fertilization. - Agron. J. *59*: 349-354, 1967.

1181 - CAMPBELL, R.K., REDISKE, J.H.: Genetic variability of photosynthetic efficiency and dry-matter accumulation in seedling Douglas-fir. - Silvae Genet. *15*: 65-72, 1966.

1182 - CAMPO, del, F.F., RAMÍREZ, J.M., ARNON, D.I.: Stoichiometry of photosynthetic phosphorylation. - J. biol. Chem. *243*: 2805-2809, 1968.

1183 - CAMPO, del, F.F., RAMIREZ, J.M., PANEQUE, A., LOSADA, M.: Ferredoxin and the dark and light reduction of dinitrophenol. - Biochem. biophys. Res. Commun. *22*: 547-553, 1966.

1184 - CANNELL, R.Q.: Net assimilation rate in barley, oats and wheat. - J. agr. Sci. *68*: 157-164, 1967.

1185 - CANNELL, R.Q., BRUN, W.A., MOSS, D.N.: A search for high net photosynthetic rate among soybean genotypes. - Crop Sci. *9*: 840-841, 1969.

1186 - CANVIN, D.T.: Photorespiration. - In: METZNER, H. (ed.): Progress in Photosynthesis Research. Vol. I. Pp. 512-513. Tübingen 1969.

1187 - CAPERON, J.: Population growth in micro-organisms limited by food supply. - Ecology *48*: 715-722, 1967. [Chl.]

1188 - CAPIEL, M.: Estimating solar radiation intensity from other meteorological data. - J. agr. Univ. Puerto Rico *54*: 377-389, 1970.

1189 - CARELL, E.F.: Studies on chloroplast development and replication in *Euglena*. I. Vitamin B_{12} and chloroplast replication. - J. Cell Biol. *41*: 431-440, 1969.

1190 - CARELL, E.F., KAHN, J.S.: Purification and properties of a Mg^{2+}-dependent ATPase from chloroplasts of *Euglena gracilis*. - Biochim. biophys. Acta *131*: 571-579, 1967.

1191 - CAREY, F.G., TEAL, J.M.: Responses of oxygen electrodes to variables in construction, assembly, and use. - J. appl. Physiol. *20*: 1074-1077, 1965.

1192 - CARLETON, A.E., FOOTE, W.H.: Heterosis for grain yield and leaf area and their components in two x six-rowed barley crosses. - Crop Sci. *8*: 554-557, 1968.

1193 - CARLEY, H.E., WATSON, R.D.: A new gravimetric method for estimating root-surface areas. - Soil. Sci. *102*: 289-291, 1966.

1194 - CARLSON, L.W.: Effects of Vitavax on chlorophyll content, photosynthesis and respiration of barley leaves. - Can. J. Plant Sci. *50*: 627-630, 1970.

1195 - CARLUCCI, A.F., SILBERNAGEL, S.B.: Effect of vitamin concentrations on growth and development of vitamin-requiring algae. - J. Phycol. *5*: 64-67, 1969. [Ps, Chl.]

1196 - CARMELI, C.: Properties of ATPase in chloroplasts. - Biochim. biophys. Acta *189*: 256-266, 1969.

1197 - CARMELI, C.: Proton translocation induced by ATPase activity in chloroplasts. - FEBS Lett. *7*: 297-300, 1970.

1198 - CARMELI, C., AVRON, M.: A light-triggered ATP-Pi exchange in chloroplasts. - Israel J. Chem. *4*: 76 p, 1966.

1199 - CARMELI, C., AVRON, M.: A light-triggered ATP-Pi exchange activity in chloroplasts. - Biochem. biophys. Res. Commun. *24*: 923-928, 1966.

1200 - CARMELI, C., AVRON, M.: A light-triggered adenosine triphosphate-phosphate exchange reaction in chloroplasts. - Europe. J. Biochem. 2: 318-326, 1967.

1201 - CARMELI, C., AVRON, M.: Uncouplers as stimulators and inhibitors of ATPase activity in chloroplasts. - In: METZNER, H. (ed.): Progress in Photosynthesis Research. Vol. III. Pp. 1169-1175. Tübingen 1969.

1202 - CARMELI, C., LIFSCHITZ, Y.: The requirements of adenosine diphosphate for light-triggered ATP-ase and ATP-Pi exchange reactions in chloroplasts. - FEBS Lett. 5: 227-230, 1969.

1203 - CARMER, S.G.: Number of replications for precise estimation of optimum plant density and maximum corn yield. - Agron. J. 62: 357-359, 1970.

1204 - CARPENTER, W.J., NAUTIYAL, J.P.: Light intensity and air movement effects on leaf temperatures and growth of shade-requiring greenhouse crops. - J. amer. Soc. hort. Sci. 94: 212-214, 1969. [Chl.]

1205 - CARR, D.J., PATE, J.S.: Ageing in the whole plant. - In: Aspects in the Biology of Ageing. (Symp. Soc. exp. Biol. Vol. 21). Pp. 559-599. Univ. Press, Cambridge 1967. [Ps, translocation.]

1206 - CARR, J.L.: The primary productivity and physiology of Ceratophyllum demersum. II. Micro primary productivity, pH, and the P/R ratio. - Aust. J. mar. Freshwater Res. 20: 127-142, 1969.

1207 - CARR, N.G.: Growth of phototrophic bacteria and blue green algae. - In: NORRIS, J.R., RIBBONS, D.W. (ed.): Methods in Microbiology. Vol. 33. Pp. 53-77. Academic Press, London-New York 1969.

1208 - CARR, N.G.: Production and measurement of photosynthetically useable light. - In: NORRIS, J.R., RIBBONS, D.W. (ed.): Methods in Microbiology. Vol. 22. Pp. 205-212. Academic Press, New York-London 1970.

1209 - CARR, N.G., CRAIG, I.W.: The relationship between bacteria, blue-green algae and chloroplasts. - In: HARBORNE, J.B. (ed.): Phytochemical Phylogeny. Pp. 119-143. Academic Press, London-New York 1970.

1210 - CARR, N.G., HALLAWAY, M.: Quinones of some blue-green algae. - In: GOODWIN, T.W. (ed.): Biochemistry of Chloroplasts. Vol. I. Pp. 159-163. Academic Press, London-New York 1966.

1211 - CARR, N.G., HOOD, W., PEARCE, J.: Control and intermediary metabolism in blue-green algae. - In: METZNER, H. (ed.): Progress in Photosynthesis Research. Vol. III. Pp. 1565-1569. Tübingen 1969. [Ps.]

1212 - CARR, N.G., SANDHU, G.R.: Endogenous metabolism of polyphosphates in two photosynthetic micro-organisms. - Biochem. J. 99: 29 P-30P, 1966.

1213 - CARRIER, J.-M.: Oxidation-reduction potential of plastoquinones. - In: GOODWIN, T.W. (ed.): Biochemistry of Chloroplasts. Vol. II. Pp. 551-557, Academic Press, London-New York 1967.

1214 - CARTWRIGHT, P.M., PAPENFUS, H.D.: The contribution of leaves of different ages to the total assimilation of a crop canopy: Methods for measuring the daily net photosynthesis of individual leaves and some data from an experimental tobacco crop. - In: Phytochemistry and Photobiology in Plant Physiology. Europe. Photobiol. Sump. Hvar. Book of Abstracts. Pp. 5-7. Hvar 1967.

1215 - CASE, G.D., PARSON, W.W., THORNBER, J.P.: Photooxidation of cytochromes in reaction center preparations from Chromatium and Rhodopseudomonas viridis. - Biochim. biophys. Acta 223: 122-128, 1970.

1216 - CASSELTON, P.J.: Chemo-organotrophic growth of xanthophycean algae. - New Phytol. 65: 134-140, 1966. [Chl.]

1217 - CASWELL, A.H.: Kinetic studies on photosynthetic phosphorylation. - In: GOODWIN, T.W. (ed.): Biochemistry of Chloroplasts. Vol. II. Pp. 601-608. Academic Press, London-New York 1967.

1218 - CASWELL, A.H., PRESSMAN, B.C.: Electrometric monitoring of ferricyanide reduction in respiration and photosynthesis. - Anal. Biochem. *32*: 396-401, 1969.

1219 - CATALINA, L.: Identificación de pigmentos carotenoides en tallos de *Orobanche crenata*, FORSK. [Identification of carotenoid pigments of *Orobanche crenata* FORSK. shoots.] - Anal. Edafol. Agrobiol. *26*: 1363-1368, 1967. [In Span., ab: E.]

1220 - CATHEY, H.M.: Interference of chloroplast development in foliage and flowers with substituted pyridazinones. - HortScience *5* (4 Sec. 2): 345, 1970.

1221 - ČATSKÝ, J.: Méthodes et techniques de mesure de la concentration en anhydride carbonique dans l'air. - In: Techniques d'Étude des Facteurs Physiques de la Biosphère. Pp. 181-188. INRA, Paris 1970.

1222 - ČATSKÝ, J., CHARTIER, M., CHARTIER, P.: Mesure de la concentration en anhydride carbonique dans l'air par absorption des rayonnements infrarouges. - In: Techniques d'Étude des Facteurs Physiques de la Biosphère. Pp. 189-199. INRA, Paris 1970.

1223 - ČATSKÝ, J., NOVÁKOVÁ, J., ŠESTÁK, Z.: Daily carbon dioxide balance and its changes with age in a fodder cabbage plant grown in controlled conditions. - Photosynthetica *1*: 215-218, 1967.

1224 - ČATSKÝ, J., ŠESTÁK, Z.: Suitable indicators and an altered empirical equation for calculating the CO_2 concentration in colorimetric determinations of photosynthetic rate. - Biol. Plant. *8*: 60-72, 1966.

1225 - CAVAUDAN, P., MARCHAND, C., POUSSEL, H.: Action des hautes pressions sur la photosynthèse. - In: Compt. rend. 90e Congr. nat. Soc. savant. Nice 1965. Vol. II. Pp. 349-353. Paris 1966.

1226 - CAYLAND, R., MASSENGALE, M.A.: Use of area weight relationship to estimate leaf area in alfalfa. - Crop. Sci. *7*: 181-182, 1967.

1227 - ČECH, M.: Removal of the chloroplast fragments from plant extracts by the octanol flotation method. - Nature *212*: 1609-1610, 1966.

1228 - CEDERSTRAND, C.N., GOVINDJEE: Some properties of spinach chloroplast fractions obtained by digitonin solubilization. - Biochim. biophys. Acta *120*: 177-180, 1966.

1229 - CEDERSTRAND, C.N., RABINOWITCH, E., GOVINDJEE: Absorption and fluorescence spectra of spinach chloroplast fractions obtained by solvent extraction. - Biochim. biophys. Acta *120*: 247-258, 1966.

1230 - CEDERSTRAND, C.N., RABINOWITCH, E., GOVINDJEE: Analysis of the red absorption band of chlorophyll *a* in vivo. - Biochim. biophys. Acta *126*: 1-12, 1966.

1231 - CELLARIUS, R.A.: A model for the chloroplast: A study of the photochemical and spectral properties of pheophytin *a* adsorbed to the surface of small particles. - Diss. Abstr. B *27*: 3478-B, 1967.

1232 - CELLARIUS, R.A.: The incorporation of bacteriochlorophyll into photosynthetic membranes of nonsulfur, purple bacteria. - In: METZNER, H. (ed.): Progress in Photosynthesis Research. Vol. II. Pp. 655-661. Tübingen 1969.

1233 - CELLARIUS, R.A., MAUZERALL, D.: A model for the photosynthetic unit. Photochemical and spectral studies on pheophytin *a* adsorbed onto small particles. - Biochim. biophys. Acta *112*: 235-255, 1966.

1234 - CELLARIUS, R.A., PETERS, G.A.: Fluorescence studies on photosynthetic pigment development in *Rhodopseudomonas spheroides*. - Photochem. Photobiol. *7*: 325-330, 1968.

1235 - CELLARIUS, R.A., PETERS, G.A.: Photosynthetic membrane development in *Rhodopseudomonas spheroides*: Incorporation of bacteriochlorophyll and development of energy transfer and photochemical activity. - Biochim. biophys. Acta *189*: 234-244, 1969.

1236 - CERNUSCA, A.: Der Einsatz automatischer Datenerfassungssysteme für klimaökologische Untersuchungen im Rahmen der Produktivitätsforschung. - Photosynthetica *2*: 238-244, 1968.

1237 - CERNUSCA, A., MOSER, W.: Die automatische Registrierung produktionsanalytischer Messdaten bei Freilandversuchen auf Lochstreifen. - Photosynthetica *3*: 21-27, 1969.

1238 - ČERNÝ, V., VRKOČ, F., KŘIŠŤAN, F., STRNAD, P.: Yield formation in crop rotations in ecologically different regions of Bohemia, as affected by cultivation practices. - In: Productivity of Terrestrial Ecosystems. Production Processes. Czech. nat. Comm. IBP, PT-PP Report No. 1. Pp. 225-226, Praha 1970.

1239 - ČERVENKA, K.: Příspěvek ke studiu vlivu intenzity světla a vodního deficitu na asimilaci CO_2 jabloní. [Effects of irradiance and water deficit on CO_2 assimilation in apple trees.] - Věd. Práce ovoc. výzk. Ústavu ovoc. (Holovousy) *3*: 79-88, 1967. [In Czech.]

1240 - ČERVENKA, K.: Použití infračerveného analyzátoru plynů IREX ke sledování dynamiky asimilace CO_2 jabloní. [The application of the infra-red gas analyzer IREX to the examination of the dynamics of CO_2 assimilation in apple-trees.] - Rostl. Výr. *16*: 201-206, 1970. [In Czech, ab: E,R.]

1241 - CHACKERIAN, C. Jr., EGGERS, D.F. Jr.: The infrared spectrum of $^{12}C^{18}O_2$. - J. mol. Spectr. *27*: 59-71, 1968.

1242 - CHAÏLAKHYAN, M.Kh., AKSENOVA, N.P., BAVRINA, T.V., KONSTANTINOVA, T.N.: Fotoperiodizm rastenii i okislitel'noe i fotosinteticheskoe fosforilirovanie. [Photoperiodism of a plant and oxidative and photosynthetic phosphorylation.] - Zh. obshch. Biol. *30*: 515-527, 1969. [In R.]

1243 - CHAÏKA, M.T., PRUDNIKOVA, I.V., SAVCHENKO, G.E., PARAMONOVA, T.K.: Fraktsionirovanie pigmentnogo fonda list'ev kukuruzy raznogo vozrasta. [Fractionation of pigments of maize leaves of various age.] - In: Metabolizm i Stroenie Fotosinteticheskogo Apparata. Pp. 144-151. Nauka i Tekhnika, Minsk 1970. [In R.]

1244 - CHAIN, R.K., TSUJIMOTO, H.Y., McSWAIN, B.D., ARNON, D.I.: The stoichiometry of noncyclic photophosphorylation with plastocyanin as an added electron acceptor. - Plant Physiol. *43* (Suppl.): S 13, 1968.

1245 - CHAMPIGNY, M.L., BISMUTH, E.: Effet de l'ATP et de Mg^{2+} sur les activités ribulose-diphosphate carboxylase et ribose-5-phosphate isomérase et kinase d'une préparation enzymatique de chloroplastes. - Can. J. Bot.*48*: 1227-1233, 1970.

1246 - CHAMPIGNY, M.-L., GIBBS, M.: Photosynthesis by isolated chloroplasts in the presence of antimycin *A* and ascorbic acid. - In: METZNER, H. (ed.): Progress in Photosynthesis Research. Vol. III. Pp. 1534-1537. Tübingen 1969.

1247 - CHAMPIGNY, M.-L., MOYSE, A.: Influence de la carence en azote sur les activités de carboxylation et de transamination des chloroplastes isolés des feuilles de *Bryophyllum*. - Z. Pflanzenphysiol. *57*: 280-297, 1967.

1248 - CHAN, A.S.K., ELLSWORTH, R.K., PERKINS, H.J., SNOW, S.E.: Purification of dihydroporphyrins for specific activity determination by thin-layer chromatography. - J. Chromatogr. *47*: 295-299, 1970. [TLC of Chl.]

1249 - CHAN, H.W.-S., BASSHAM, J.A.: Metabolism of ^{14}C-labeled glycolic acid by isolated spinach chloroplasts. - Biochim. biophys. Acta *141*: 426-429, 1967.

1250 - CHANCE, B., CROFTS, A.R., NISHIMURA, M., PRICE, B.: Fast membrane H^+ binding in the light-activated state of *Chromatium* chromatophores. - Europe. J. Biochem. *13*: 364-374, 1970.

1251 - CHANCE, B., HIYAMI, T., NISHIMURA, M.: Determination of carotenes by thin-layer chromatography. - Anal. Biochem. *29*: 339-342, 1969.

1252 - CHANCE, B., KIHARA, T., DEVAULT, D., HILDRETH, W., NISHIMURA, M., HIYAMA, T.: Temperature-insensitive electron transfer in photosynthetic systems. - In: METZNER, H. (ed.): Progress in Photosynthesis Research. Vol. III. Pp. 1321-1364. Tübingen 1969.

1253 - CHANCE, B., McCRAY, J.A., BUNKENBURG, J.: Fast spectrophotometric measurement of H^+ changes in *Chromatium* chromatophores activated by a liquid dye laser. - Nature *225*: 705-708, 1970.

1254 - CHANCE, B., NISHIMURA, M., AVRON, M., BALTSCHEFFSKY, M.: Light-induced intrave-sicular pH changes in *Rhodospirillum rubrum* chromatophores. - Arch. Biochem. Biophys. *117*: 158-166, 1966.

1255 - CHANCE, B., DE VAULT, D., HILDRETH, W.W., PARSON, W.W., NISHIMURA, M.: Early chemical events in photosynthesis: Kinetics of oxidation of cytochromes of types *c* or *f* in cells, chloroplasts and chromatophores. - In: Energy Conver-sion by the Photosynthetic Apparatus. Brookhaven Symp. Biol. *19*: 115-131, 1967.

1256 - CHANDLER, M.T.: Induction transients in photosynthetic oxygen evolution. - Diss. Abstr. int. B *31*: 1758-B-1759-B, 1970.

1257 - CHANDLER, M.T., VIDAVER, W.E.: Photosynthetic oxygen induction transients in the alga *Ulva lactuca* L. - Phycologia *9*: 133-142, 1970.

1258 - CHANEY, S.G., BOYER, P.D.: Lack of detection of intermediates in the path of phosphate oxygen to water in photophosphorylation. - J. biol. Chem. *244*: 5773-5776, 1969.

1259 - CHANG, I.C., KAHN, J.S.: Isolation of a possible coupling factor for photophos-phorylation from chloroplasts of *Euglena gracilis*. - Arch. Biochem. Biophys. *117*: 282-288, 1966.

1260 - CHANG, I.C., KAHN, J.S.: Factors affecting the rate of photophosphorylation by isolated chloroplasts of *Euglena gracilis*. - J. Protozool. *17*: 556-564, 1970.

1261 - CHANG, J.-H.: The agricultural potential of the humid tropics. - Geogr. Rev. *58*: 333-361, 1968. [Ps.]

1262 - CHANG, S.B., VEDVICK, T.S.: Structural specificities of plastoquinones in pho-tophosphorylation with *Euglena* chloroplasts. - Fed. Proc. *27*: 825, 1968.

1263 - CHANG, S.B., VEDVICK, T.S.: A study of plastoquinones in photochemical reac-tions in the chloroplasts of *Euglena gracilis* strain Z. - Plant Physiol. *43*: 1661-1665, 1968.

1264 - CHANG, W.-H., TOLBERT, N.E.: Excretion of glycolate, mesotartrate and isoci-trate lactone by synchronized cultures of *Akistrodesmus braunii*. - Plant Phy-siol. *46*: 377-385, 1970. [Ps.]

1265 - CHANG-CHI, C.: Studies on the profiles of concentration of carbon dioxide in a sugarcane field at Tainan, Taiwan. - Taiwan Sugar exp. Sta. annu. Rep. *1968-1969*: 1-7, 1969.

1266 - CHANISVILI, Sh.Sh.: K issledovaniyu ritma peredvizheniya plasticheskikh vesh-chestv i fotosinteza yagod v vinogradnoĭ loze. [Rhythm of transport of plastic substances and photosynthesis in vine berries.] - Vestn. gruz. bot. Obshchest-va Akad. Nauk Gruz. SSR *2*: 50-57, 1967. [In R.]

1267 - CHAPLIN, J.F.: Inheritance and possible use of pale yellow character in tobac-co. - Crop. Sci. *9*: 169-172, 1969. [Chl.]

1268 - CHAPMAN, D., CHERRY, R.J., MORRISON, A.: Spectroscopic and electrical studies of all-trans β-carotene crystals. - Proc. roy. Soc. A *301*: 173-193, 1967.

1269 - CHAPMAN, D., FAST, P.G.: Studies of chlorophyll-lipid-water systems. - Science *160*: 188-189, 1968.

1270 - CHAPMAN, D.J., COLE, W.J., SIEGELMAN, H.W.: Chromophores of allophycocyanin and *R*-phycocyanin. - Biochem. J. *105*: 903-905, 1967.

1271 - CHAPMAN, D.J., COLE, W.J., SIEGELMAN, H.W.: The structure of phycoerythrobilin. - J. amer. chem. Soc. *89*: 5976-5977, 1967.

1272 - CHAPMAN, D.J., COLE, W.J., SIEGELMAN, H.W.: Phylogenetic implications of phyco-cyanobilin and *C*-phycocyanin. - Amer. J. Bot. *55*: 314-316, 1968.

1273 - CHAPMAN, D.J., COLE, W.J., SIEGELMAN, H.W.: Cleavage of phycocyanobilin from *c*-phycocyanin. - Biochim. biophys. Acta *153*: 692-698, 1968.

1274 - CHAPMAN, D.J., HAXO, F.T.: Chloroplast pigments of *Chloromonadophyceae*. - J. Phycol. *2*: 89-91, 1966.

1275 – CHAPMAN, G., RAE, A.C.: Excretion of photosynthate by a benthic diatom. – Mar. Biol. *3*: 341-351, 1969.

1276 – CHARLES-EDWARDS, D.A., CHARLES-EDWARDS, J.: An analysis of the temperature response curves of CO_2 exchange in the leaves of two temperate and one tropical grass species. – Planta *94*: 140-151, 1970.

1277 – CHARLTON, J.M., TREHARNE, K.J., GOODWIN, T.W.: Incorporation of [2-^{14}C] mevalonic acid into phytoene by isolated chloroplasts. – Biochem. J. *105*: 205-212, 1967.

1278 – CHARTIER, P.: Étude théorique de la photosynthèse globale de la feuille. – Compt. rend. Acad. Sci. Paris, Sér. D *263*: 44-47, 1966.

1279 – CHARTIER, P.: Étude théorique de l'assimilation brute de la feuille. – Ann. Physiol. vég. *8*: 167-195, 1966.

1280 – CHARTIER, P.: Étude du microclimat lumineux dans la végétation. – Ann. agron. *17*: 571-602, 1966. [Ps.]

1281 – CHARTIER, P.: Lumière, eau et production de matière sèche du couvert végétal. – Ann. agron. *18*: 301-331, 1967.

1282 – CHARTIER, P.: Modèle biophysique des échanges photosynthétiques de l'unité de surface de feuille. – Stud. biophys. *11*: 47-55, 1968.

1283 – CHARTIER, P.: Assimilation nette d'une culture couvrante. I. Détermination de l'assimilation nette de la culture à partir d'une analyse théorique. – Ann. Physiol. vég. *11*: 123-159, 1969.

1284 – CHARTIER, P.: Assimilation nette d'une culture couvrante. II. La réponse de l'unité de surface de feuille. – Ann. Physiol. vég. *11*: 221-263, 1969.

1285 – CHARTIER, P.: A model of CO_2 assimilation in the leaf. – In: Prediction and Measurement of Photosynthetic Productivity. Pp. 307-315. PUDOC, Wageningen 1970.

1286 – CHARTIER, P., CHARTIER, M., ČATSKÝ, J.: Resistances for carbon dioxide diffusion and for carboxylation as factors in bean leaf photosynthesis. – Photosynthetica *4*: 48-57, 1970.

1287 – CHARTIER, P., GANS, F.: Mesure de la composition spectrale du rayonnement diffus. – Z. angew. Meteorol. *5*: 324-325, 1968.

1288 – CHASOVNIKOVA, L.V.: Spektral'nye svoĭstva monosloev khlorofilla ne granitse faz zhidkost'-gaz, zhidkost'-zhidkost'. [Spectral properties of chlorophyll monolayers in the interphases liquid-gas and liquid-liquid.] – Nauch. Dokl. vyssh. Shkoly, biol. Nauki *10* (1): 148-149, 1967. [In R.]

1289 – CHASOVNIKOVA, L.V., NEKRASOV, L.I., KOBOZEV, N.I.: O spektral'nykh svoĭstvakh monomolekulyarnykh plenok khlorofilla na poverkhnosti vody. [Spectral properties of monomolecular films of chlorophyll on water surface.] – Zh. fiz. Khim. *40*: 1141-1144, 1966. [In R.]

1290 – CHASOVNIKOVA, L.V., NEKRASOV, L.I., KOBOZEV, N.I.: Vliyanie fosfolipida kefalina na spektral'nye svoĭstva monomolekulyarnykh sloev khlorofilla a. [Effect of the phospholipid cephalin on spectral properties of monomolecular layers of chlorophyll a.] – Zh. fiz. Khim. *40*: 1655-1657, 1966. [In R.]

1291 – CHASTON, I., WALKER, P.J.: A probe photometer. – Oikos *19*: 146-148, 1968.

1292 – CHAUDHARY, A.H., FRENKEL, A.W.: Effects of adenosine 3',5'-cyclic monophosphoric acid on certain light-induced reactions and on ATPase activity of isolated chromatophores from *Rhodospirillum rubrum*. – Biochem. biophys. Res. Commun. *39*: 238-246, 1970.

1293 – CHAYANOVA, S.S.: Aktivnost' pogloshcheniya kisloroda etioplastami, khloroplastami i plastidami, differentsiruyushchimisya v khloroplasty, pri zelenenii etiolirovannykh list'ev. [Rate of oxygen absorption by etioplasts, chloroplasts and plastids differentiating into chloroplasts during greening of etiolated leaves.] – Fiziol. Rast. *17*: 1182-1186, 1970. [In R.]

1294 – CHAYANOVA, S.S., KURSANOV, A.L.: Sravnitel'noe izuchenie belkovykh kompleksov etioplastov i formiruyushchikhsya iz nikh khloroplastov. [Comparative study of protein complexes of etioplasts and chloroplasts formed from them.] - Fiziol. Rast. *17*: 485-490, 1970. [In R.]

1295 - CHEBOTAR', A.A.: Razvitie i ul'trastruktura plastid zlakovykh. [Development and ultrastructure of plastids in cereals.] - In: Khloroplasty i Mitokhondrii. Pp. 122-145. Nauka, Moskva 1969. [In R.]

1296 - CHEN, C.H.: Assimilation of $^{14}CO_2$ by *Medicago sativa* leaves in relation to sulfur nutrition. - Diss. Abstr. B *29*: 867-B, 1968.

1297 - CHEN, J.E., JONES, R.F.: Multiple forms of phosphoenolpyruvate carboxylase from *Chlamydomonas reinhardtii*. - Biochim. Biophys. Acta *214*: 318-325, 1970.

1298 - CHEN, L.H., HUANG, B.K., SPLINTER, W.E.: Developing a physical-chemical model for a plant growth system . - Trans. ASAE *12*: 698-702, 1969.

1299 - CHEN, P.K., VENKETESWARAN, S.: Chlorophyllous nature and growth characteristics of teratoma and habituated tissue cultures of tobacco. - Physiol. Plant. *21*: 262-269, 1968.

1300 - CHEN, S., McMAHON, D., BOGORAD, L.: Early effects of illumination on the activity of some photosynthetic enzymes. - Plant Physiol. *42*: 1-5, 1967.

1301 - CHEN, T.M., BROWN, R.H., BLACK, C.C. Jr.: Photosynthetic activity of chloroplasts isolated from bermudagrass (*Cynodon dactylon* L.), a species with a high photosynthetic capacity. - Plant Physiol. *44*: 649-654, 1969.

1302 - CHEN, T.M., BROWN, R.H., BLACK, C.C. Jr.: Biochemical activity of chloroplasts isolated from species with high photosynthetic capacity. - Plant Physiol. *44* (Suppl.): 11, 1969.

1303 - CHEN, T.M., BROWN, R.H., BLACK, C.C.: CO_2 compensation concentration, rate of photosynthesis, and carbonic anhydrase activity of plants. - Weed Sci. *18*: 399-403, 1970.

1304 - CHENIAE, G.M.: Photosystem II and O_2 evolution. - Annu. Rev. Plant Physiol. *21*: 467-498, 1970.

1305 - CHENIAE, G.M., MARTIN, I.F.: Studies on the function of manganese in photosynthesis. - In: Energy Conversion by the Photosynthetic Apparatus. Brookhaven Symp. Biol. *19*: 406-417, 1967.

1306 - CHENIAE, G.M., MARTIN, I.F.: Photoreactivation of manganese catalyst in photosynthetic oxygen evolution. - Biochem. biophys. Res. Commun. *28*: 89-95, 1967.

1307 - CHENIAE, G.M., MARTIN, I.F.: Site of manganese function in photosynthesis. - Biochim. biophys. Acta *153*: 819-837, 1968.

1308 - CHENIAE, G.M., MARTIN, I.F.: Requirement of two pools of Mn within photosystem II and their function in O_2 evolution. - Plant Physiol. *43* (Suppl.): S 13, 1968.

1309 - CHENIAE, G.M., MARTIN, I.F.: Photoreactivation of manganese catalyst in photosynthetic oxygen evolution. - Plant Physiol. *44*: 351-360, 1969.

1310 - CHENIAE, G.M., MARTIN, I.F.: Sites of function of manganese within photosystem II. Roles in O_2 evolution and system II. - Biochim. biophys. Acta *197*: 219-239, 1970.

1311 - CHEPELEV, V.U.: K voprosu o filosofskom znachenii problemy fotosinteza rastenii. [Philosophical meaning of the problem of photosynthesis in plants.] - Tr. khar'-kov. sel'.-khoz. Inst. *48*: 146-153, 1966. [In R.]

1312 - CHERNAVINA, I.A.: O nekotorykh zakonomernostyakh biosinteza khlorofilla u vysshikh rastenii. [Some features of chlorophyll biosynthesis in higher plants.] - Sel'.-khoz. Biol. *2*: 522-528, 1967. [In R, ab: E.]

1313 - CHERNAVINA, I.A.: O nekotorykh zakonomernostyakh biosinteza khlorofilla u vysshikh rastenii. [On some regularities of chlorophyll biosynthesis in higher plants.] - In: Vazhneīshie Problemy Fotosinteza v Rastenievodstve. Pp. 97-107. Kolos, Moskva 1970. [In R.]

1314 - **CHERNAVINA, I.A., KARTASHOVA, E.R.**: Ob uchastii sistem med'proteinovykh fermentov v biosinteze khlorofilla. [On the participation of the system of copper-protein enzymes in chlorophyll biosynthesis.] - Sel'.-khoz. Biol. 2: 358-364, 1967. [In R.]

1315 - **CHERNAVINA, I.A., KRENDELEVA, T.E.**: O mekhanizme uchastiya zhelezosoderzhashchikh fermentov okislitel'nogo obmena v biosinteze khlorofilla. [Mechanism of participation of Fe-containing enzymes of the oxidizing metabolism in chlorophyll biosynthesis.] - Nauch. Dokl. vyssh. Shkoly. biol. Nauki 10 (1): 152-153, 1967. [In R.]

1316 - **CHERNAVINA, I.A., KRENDELEVA, T.E., GRACHEVA, N.K.**: O mekhanizme uchastiya zheleza v protsesse obrazovaniya khlorofilla. [On the mechanism of iron participation in chlorophyll synthesis.] - Vestn. mosk. Univ., Biol. Pochvoved. 1968 (4): 48-55, 1968. [In R.]

1317 - **CHERNAVINA, I.A., KRENDELEVA, T.E., SVERDLOVA, P.S.**: Vliyanie zheleza i margantsa na energeticheskiĭ obmen rasteniĭ s narushennym sintezom khlorofilla. [Effect of iron and manganese on energy transformation in plants with inhibited chlorophyll synthesis.] - Fiziol. Rast. 15: 1008-1014, 1968. [In R, ab: E.]

1318 - **CHERNAVINA, I.A., KUKARSKIKH, G.P.**: Fotosinteticheskoe fosforilirovanie u rasteniĭ s narushennym sintezom khlorofilla. [Photosynthetic phosphorylation in plants showing disturbances of chlorophyll synthesis.] - Dokl. Akad. Nauk SSSR 176: 959-962, 1967. [In R.]

1319 - **CHERNAVINA, I.A., LEKHOTSKI, E.**: Osobennosti okislitel'nogo dekarboksilirovaniya piruvata u vysshikh rasteniĭ. [Features of oxidative decarboxylation of pyruvate in higher plants.] - Biokhimiya 34:417-425, 1969. [Chl; in R, ab: E.]

1320 - **CHERNAVSKAYA, N.M., CHERNAVSKY, D.S.**: On oscillations in dark photosynthesis reactions. - Stud. biophys. 1 (Sonderh. 2): 83-90, 1966.

1321 - **CHERNAVSKAYA, N.M., NICHIPOROVICH, A.A.**: O garmonicheskom sochetanii usloviĭ osveshchennosti i azotnogo pitaniya. [On harmonic cooperation of conditions of illumination and nitrogen nutrition.] - In: Fotosinteziruyushchie Sistemy Vysokoĭ Produktivnosti. Pp. 169-177. Nauka, Moskva 1966. [Ps; in R.]

1322 - **CHERNAVSKIĬ, D.S., CHERNAVSKAYA, N.M.**: O kolebaniyakh v temnovykh reaktsiyakh fotosinteza. [Fluctuations in dark photosynthetic reactions.] - In: Kolebatel'nye Protsessy v Biologii i Khimicheskikh Sistemakh. Pp. 51-67. Nauka, Moskva 1967. [In R.]

1323 - **CHERNIĬ, N.E., SOLOV'EVA, Zh.V., FEDOROV, V.D., KONDRAT'EVA, E.N.**: Ul'trastruktura kletok dvukh vidov purpurnykh serobakteriĭ. [Cell ultrastructure of two species of purple sulphur bacteria.] - Mikrobiologiya 38: 479-484, 1969. [Chloroplast; in R.]

1324 - **CHERNOV, I.A.**: Lyuminestsentno-tsitokhimicheskie issledovaniya khloroplastov. [Luminescent and cytochemical studies of chloroplasts.] - In: Funktsional'nye Osobennosti Khloroplastov. Pp. 86-90. Kazan. Univ., Kazan 1969. [In R.]

1325 - **CHERNOV, I.A.**: Fotosintez izolirovannykh khloroplastov. [Photosynthesis of isolated chloroplasts.] - Uspekhi sovrem. Biol. 70: 227-238, 1970. [In R.]

1326 - **CHERNOV, I.A., SHAKIROVA, D.Z.**: Vliyanie povyshennoĭ temperatury na dinamiku intensivnosti fiksatsii $C^{14}O_2$ na svetu izolirovannymi khloroplastami gorokha. [Effect of increased temperature on the dynamics of the rate of $^{14}CO_2$ fixation by isolated pea chloroplasts in light.] - In: Trudy 3-eĭ Gorodskoĭ Nauchnoĭ Konferentsii Molodykh Uchenykh (Biol. Ser.). Pp. 75-78. Kazan 1967. [In R.]

1327 - **CHERNOV, I.A., SHAKIROVA, D.Z., GRISHINA, L.I.**: Fotosintez izolirovannykh khloroplastov posle deĭstviya submaksimal'nykh temperatur. [Photosynthesis of isolated chloroplasts after the action of sub-maximum temperatures.] - In: Funktsional'nye Osobennosti Khloroplastov. Pp. 69-74. Kazan. Univ., Kazan 1969. [In R.]

1328 - **CHERNOV, I.A., SIYANOVA, N.S., SHAKIROVA, D.Z.**: K metodike izucheniya fiksatsii CO_2 izolirovannymi khloroplastami i vliyaniya temperatury na ee intensivnost'. [On the method of CO_2 fixation by isolated chloroplasts and the effect of temperature on its rate.] - Fiziol. Rast. *14*: 172-175, 1967. [In R, ab: E.]

1329 - CHERNOV, I.A., TARCHEVSKIĬ, I.A.: Osobennosti vydeleniya khloroplastov pri izuchenii puti ugleroda v fotosinteze i vliyanie nekotorykh ekstremal'nykh faktorov
na fotosintez izolirovannykh khloroplastov. [On the peculiarities of isolation
of chloroplasts in the study of the pathway of carbon in photosynthesis and influence of extremal factors on photosynthesis of isolated chloroplasts.] - In:
KIRICHENKO, E.B. (ed.): Metody Vydeleniya Khloroplastov. Pp. 94-108. Pushchino-
na-Oke 1970. [In R, ab: E.]

1330 - CHERNYAD'EV, I.I., DOMAN, N.G.: K voprosu o putyakh ispol'zovaniya ugleroda v
fotosinteze *Rhodopseudomonas palustris*. [Carbon utilization in photosynthesis
of *Rhodopseudomonas palustris*.] - Biokhimiya *35*: 968-972,1970. [In R, ab: E.]

1331 - CHERNYAD'EV, I.I., KONDRAT'EVA, E.N.,DOMAN, N.G.: Assimilyatsiya uglekisloty *Rhodopseudomonas palustris*. [Carbon dioxide assimilation by *Rhodopseudomonas palustris*.] - Izv. Akad. Nauk SSSR, Ser. biol. *1969*: 670-675, 1969. [In R, ab: E.]

1332 - CHERNYAD'EV, I.I., KONDRAT'EVA, E.N., DOMAN, N.G.: Assimilyatsiya atsetata *Rhodopseudomonas palustris*. [Acetate assimilation in *Rhodopseudomonas palustris*.].
- Mikrobiologiya *39*: 24-29, 1970. [In R.]

1333 - CHERNYAD'EV, I.I., KONDRAT'EVA, E.N., DOMAN, N.G.: Ispol'zovanie formiata v
fotosinteze *Rhodopseudomonas palustris*. [Utilization of formiate in photosynthesis of *Rhodopseudomonas palustris*.] - Izv. Akad. Nauk SSSR, Ser. biol. *1970*:
895-898, 1970. [In R, ab: E.]

1334 - CHERNYAEVA, I.I., SAVEL'EVA, L.N.: Deĭstvie CO_2 i O_2 na temnovuyu reaktsiyu
prevrashcheniya ksantofillov. [Effect of CO_2 and O_2 on the dark reaction of
xanthophyll transformation.] - Sel'.-khoz. Biol. *1*: 583-587, 1966. [In R, ab:
E.]

1335 - CHERNYAEVA, I.I., ZINOV'EV, L.S.: Utochnenie velichiny okislitel'no-vosstanovitel'nogo potentsiala temnovoĭ reaktsii prevrashcheniya ksantofillov. [Verifi-
cation of value of the oxido-reduction potential of dark reaction of xanthophyll transformation.] - Byull. vsesoyuz. nauch.-issled. Inst. sel'.-khoz.
Mikrobiol. *1970 (14)*: 79-82, 1970. [In R.]

1336 - CHERRY, R.J.: Semiconduction and photoconduction of biological pigments. -
Quart. Rev. *22*: 160-178, 1968. [Chl, Car.]

1337 - CHERRY, R.J., CHAPMAN, D.: Photoconductivity of carotenoids. - Mol. Crys. *3*:
251-267, 1967.

1338 - CHESNOKOV, V.A., MIROSLAVOVA, S.A.: Obmen ugleroda v list'yakh *Bryophyllum daigremontianum* na svetu. [Carbon exchange in leaves of *Bryophyllum daigremontianum* in light.]. - Tr. petergof. biol. Inst. LGU *1970 (20)*: 225-248, 1970. [In
R.]

1339 - CHESNOKOV, V.A., SOLDATENKOV, S.V., STEPANOVA, A.M.: Zelenye rasteniya kak so-
birateli ugleroda. [Green plants as carbon accumulators.] - Vestn. leningrad.
Univ. *1968* (9): 73-78, 1968. [In R, ab: E.]

1340 - CHEVALIER, S., BACCOU, J.-C., SAUVAIRE, Y.: Influence d'un excès de calcium sur
la structure et le fonctionnement des chloroplastes du *Parietaria officinalis*
L. - Compt. rend. Acad. Sci. Paris, Sér. D *269*: 1653-1656, 1969.

1341 - CHIBA, Y., AIGA, I., IDEMORI, M., SATOH, Y., MATSUSHITA, K., SASA, T.: Studies
on chlorophyllase of *Chlorella protothecoides*. I. Enzymatic phytylation of
methyl chlorophyllide. - Plant Cell Physiol. *8*: 623-635, 1967.

1342 - CHIBA, Y., AIGA, I., TAKAGI, K., SASA, T.: Studies on chlorophyllase of *Chlorella protothecoides* in relation to the problem of chlorophyll formation. - In:
SHIBATA, K., TAKAMIYA, A., JAGENDORF, A.T., FULLER, R.C. (ed.): Comparative Biochemistry and Biophysics of Photosynthesis. Pp. 291-298. Univ. Tokyo Press,
Tokyo; Univ. Park Press, State College Pa. 1968.

1343 - CHIBISOV, A.K.: Izuchenie metodom impul'snogo fotoliza fotosensibilizirovannykh
pigmentami okislitel'no-vosstanovitel'nykh reaktsiĭ. [Impulse photolysis study
of oxido-reduction reactions photosensibilized by pigments.] - Biofizika *12*:
53-62, 1967. [Chl; in R.]

1344 - CHIBISOV, A.K.: A flash photolysis study of intermediates in photochemical reactions of chlorophyll. - Photochem. Photobiol. *10*: 331-347,1969.

1345 - CHIBISOV, A.K., BARASHKOV, B.I., KARYAKIN, A.V.: Spektrofotometricheskaya ustanovka s impul'snym fotovozbuzhdeniem dlya izucheniya pervichnykh protsessov fotosinteza. [Spectrophotometric device with impulse photoexcitation for studying primary processes of photosynthesis.] - Zh. prikl. Spektroskopii *12*: 243-247, 1970. [In R.]

1346 - CHIBISOV, A.K., KARYAKIN, A.V.: Issledovanie promezhutochnykh sostoyaniĭ v reaktsii fotovosstanovleniya khlorofilla. [Study of intermediary states in the reaction photosensibilized by chlorophyll.] - Biofizika *11*: 991-999, 1966. [In R.]

1347 - CHIBISOV, A.K., KARYAKIN, A.V.: Triplet-triplet energy transfer and chlorophyll photooxidation. - In: METZNER, H. (ed.): Progress in Photosynthesis Research. Vol. II. Pp. 757-762. Tübingen 1969.

1348 - CHIBISOV, A.K., KARYAKIN, A.V., DROZDOVA, N.N., KRASNOVSKII, A.A.: Izuchenie promezhutochnykh sostoyaniĭ v reaktsii fotookisleniya khlorofilla i ego analogov *p*-khinonom. [A study of intermediate conditions in the reaction of photooxidation of chlorophyll and its analogues by *p*-quinone.] - Dokl. Akad. Nauk SSSR *175*: 737-740, 1967. [In R.]

1349 - CHIBISOV, A.K., KARYAKIN, A.V., EVSTIGNEEV, V.B.: Pervichnye protsessy v reaktsii vzaimodeĭstviya khlorofilla s metilviologenom. [Primary processes in the interaction of chlorophyll and methyl viologen.] - Biofizika *11*: 983-990, 1966. [In R.]

1350 - CHIBISOV, A.K., KARYAKIN, A.V., ZUBRILINA, M.E.: Izuchenie promezhutochnykh sostoyaniĭ v fotosensibilizirovannoĭ khlorofillom reaktsii vosstanovleniya metilovogo krasnogo. [A study of intermediary conditions in a chlorophyll-photosensibilized reaction of methylene red reduction.] - Dokl. Akad. Nauk SSSR *170*: 198-201, 1966. [In R.]

1351 - CHIBISOV, A.K., KARYAKIN, A.V., ZUBRILINA, M.E.: Triplet-tripletnyĭ perenos energii s uchastiem khlorofilla. [Triplet-triplet energy transfer realized under participation of chlorophyll.] - Dokl. Akad. Nauk SSSR *177*: 468-470, 1967. [In R.]

1352 - CHIBISOV, A.K., KARYAKIN, A.V., ZUBRILINA, M.E.: Triplet-tripletnyĭ perenos energii i fotokislenie khlorofilla. [Triplet-triplet energy transfer and chlorophyll photooxidation.] - Biofizika *14*: 925-927, 1969. [In R., ab: E.]

1353 - CHICHESTER, C.O.: The biosynthesis of carotenoids. - Pure appl. Chem. *14*: 215-226, 1967. In: Carotenoids other than Vitamin *A*. Pp. 215-226. Butterworth, London 1967.

1354 - CHICHESTER, C.O., NAKAYAMA, T.O.M.: The biosynthesis of carotenoids and vitamin *A*. - In: BERNFELD, P.: Biogenesis of Natural Compounds. 2nd Ed. Pp. 641-678. Pergamon Press, Oxford-London-Edinburgh-New York-Toronto-Sydney-Paris-Braunschweig 1967.

1355 - CHIKOV, V.I.: Osobennosti izmereniya gazoobmena rasteniĭ s pomoshch'yu infrakrasnogo spektrometra IKS-21. [Features of determination of plant gas exchange by an infra-red spectrometer IKS-21.] - In: Funktsional'nye Osobennosti Khloroplastov. Pp. 109-112. Kazan. Univ. Kazan.,1969. [In R.]

1356 - CHINO, M., MITSUI, S.: [Effect of manganese toxicity on carbon assimilation of rice plants at the ripening stage.] - J. Sci. Soil Manure, Jap. [Nippon Dojo-Hiryogaku Zasshi] *39*: 320, 1968. [In Jap., ab: E.]

1357 - CHIRPUTKAR, M., JOSHI, G.V.: Studies in photosynthesis in viviparous seedlings of *Rhizophora mucronata*. - In: Proceedings of the 53rd Session of Indian Scientific Congress. Part III. Pp. 297. Chandigarh, India 1966.

1358 - CHITASHVILI, S.Sh.: Svetovye krivye fotosinteza osnovnykh lesoobrazuyushchikh drevesnykh porod gornykh lesov Gruzii. [Light curves of photosynthesis of main forest woody populations of mountain forests of Georgia.] - Bot. Zh. *51*: 720-723, 1966. [In R.]

1359 - CHITASHVILI, S.Sh.: Fotosintez sosnovykh i bukovykh drevostoev v svyazi s rubka-
mi ukhoda i razlichnoi intensivnosti. [Photosynthesis of pine and beech stands in
relation to intermediate cutting and cutting of different intensity.] - In:
SvetovoT Rezhim, Fotosintez i Produktivnost' Lesa. Pp. 167-179. Nauka, Moskva
1967. [In R.]

1360 - CHI-YING, H.: The role of photosynthesis and mineral nutrition in nitrate reductase in soybeans. - Taiwania 15: 189-197, 1970.

1361 - CHKANIKOV, D.I., KOSTINA, V.I.: Vliyanie galoidfenoksikislot na fotosinteticheskie protsessy. [Effect of haloid phenoxy acids on photosynthetic processes.] -
Khim. sel'.-Khoz. 6: 536-539, 1968. [In R.]

1362 - CHMORA, S.N.: Svetovye krivye fotosinteza v poseve kukuruzy. [Light curves of
photosynthesis in a maize canopy.] - Fotosinteziruyushchie Sistemy VysokoT Produktivnosti. Pp. 142-148. Nauka, Moskva 1966.

1363 - CHMORA, S.N., OYA, V.M.: Izmerenie svetovykh krivykh fotosinteza radiometricheskim metodom s primeneniem kamery-klina. [Measurement of light curves of photosynthesis by a radiometric method using a wedge-chamber.] - Fiziol. Rast. 14:
179-186, 1967. [In R, ab: E.]

1364 - CHMORA, S.N., OYA, V.M.: Izuchenie temperaturnoT zavisimosti fotosinteza lista.
[Study of the temperature dependence of leaf photosynthesis.] - Fiziol. Rast.
14: 603-611, 1967. [In R, ab: E.]

1365 - CHO, D.H.: Kinetics of the acid catalyzed pheophytinization of chlorophylls. -
Diss. Abstr. B 27: 716 B-717 B, 1966.

1366 - CHO, D.H., PARKS, L., ZWEIG, G.: Photoreduction of quinones by isolated spinach
chloroplasts. - Biochim. biophys. Acta 126: 200-206, 1966.

1367 - CHO, D.H., TOLLIN, G.: ESR studies of chlorophyll one-electron photochemistry:
kinetics and mechanism of reversible hydroquinone oxidation in solution. - Photochem. Photobiol. 8: 317-329, 1968.

1368 - CHO, F., GOVINDJEE: Fluorescence spectra of *Chlorella* in the 295-77°K range. -
Biochim. biophys. Acta 205: 371-378, 1970.

1369 - CHO, F., GOVINDJEE: Low-temperature (4-77°K) spectroscopy of *Chlorella*; temperature dependence of energy transfer efficiency. - Biochim. biophys. Acta 216:
139-150, 1970.

1370 - CHO, F., GOVINDJEE: Low-temperature (4-77°K) spectroscopy of *Anacystis*; temperature dependence of energy transfer efficiency. - Biochim. biophys. Acta 216:
151-161, 1970.

1371 - CHO, F., SPENCER, J., GOVINDJEE: Emission spectra of *Chlorella* at very low temperatures (-296° to -196°)- Biochim. biophys. Acta 126: 174-176, 1966.

1372 - CHOI, I.C., ARONOFF, S.: Photosynthetic transport using tritiated water. - Plant
Physiol. 41: 1119-1129, 1966.

1373 - CHOI, I.C., ARONOFF, S.: Apetiolar photosynthate translocation. - Plant Physiol.
41: 1130-1134, 1966.

1374 - CHOLLET, R., PAOLILLO, D.J. Jr.: Pigment concentration, plastid structure, &
photosynthetic capacity of a virescent corn mutant. - Plant Physiol. 46 (Suppl.):
22, 1970.

1375 - CHONAN, N.: [Studies on the photosynthetic tissues in the leaves of cereal crops.
II. Effect of shading on the mesophyll structure of the wheat leaves.] - Proc.
Crop. Sci. Soc. Jap. 35: 78-82, 1966. [In Jap., ab: E.]

1376 - CHONAN, N.: [Studies on the photosynthetic tissues in the leaves of cereal crops.
III. The mesophyll structure of rice leaves inserted at different levels of the
shoot.] - Proc. Crop Sci. Soc. Jap. 36: 291-296, 1967. [In Jap., ab: E.]

1377 - CHONAN, N.: [Studies on the photosynthetic tissues in the leaves of cereal crops.
IV. Effect of shading on the mesophyll structure of the rice leaves.] - Proc.
Crop. Sci. Soc. Jap. 36: 297-301, 1967. [In Jap., ab: E.]

1378 - CHONAN, N.: [Studies on the photosynthetic tissues in the leaves of cereal crops.

V. Comparison of the mesophyll structure among seedling leaves of cereal crops.]
- Proc. Crop Sci. Soc. Jap. *39*: 418-425, 1970. [In Jap., ab: E.]

1379 - CHONAN, N.: [Studies on the photosynthetic tissues in the leaves of cereal crops.
VI. Effect of nitrogen nutrition on the mesophyll structure of leaves in rice
varieties of different plant types.] - Proc. Crop Sci. Soc. Jap. *39*: 426-430,
1970. [In Jap., ab:E.]

1380 - CHOUDHURI, M.A., CHATTERJEE, S.K.: Seasonal changes in the levels of some cellular components in the abscission zone of *Coleus* leaves of different ages. -
Ann. Bot. *34*: 275-287, 1970. [Chl.]

1381 - CHOW, P.-C., HSEIN, K.-E., T'ANG, P.-S.: [Relationship between fluorescence decay and photophosphorylation in isolated wheat chloroplasts.] - Acta biochim.
biophys. sinica *6*: 225-230, 1966. [In Chin., ab: E.]

1382 - CHOW, P.C., YEH, Y.K., TANG, P.S.: [Reduction of 2,6-dichlorophenol-indophenol
(Hill reaction) by chloroplasts with different chlorophyll *a/b* ratios under monochromatic light.] - Acta bot. sinica *14*: 144-149, 1966. [In Chin., ab:E.]

1383 - CHRELASHVILI, M., GUGUSHVILI, N.: [Effect of anthocyanin on photosynthesis and
chlorophyll resistance.] - Vestn. gruz. bot. Obshch. Akad. Nauk Gruz. SSR *3*: 34-
37, 1966. [In Georg., ab: R.]

1384 - CHRELASHVILI, M.N.: [Effects of irradiance and temperature on photosynthesis in
some sempervirent plants.] - Vestn. gruz. bot. Obshch. Akad. Nauk Gruz. SSSR *3*:
73-78, 1966. [In Georg., ab: R.]

1385 - CHRELASHVILI, M.N.: Izmenenie khlorofillov *a* i *b* v ontogeneze lista. [Changes
of chlorophyll *a* and *b* during leaf ontogenesis.] - In: Merknian Mcenareta Fizio-
logia. [Fiziologiya Drevesnykh Rasteniĭ.] No. 2. Pp. 48-56. Mecniereba, Tbilisi
1966. [In R.]

1386 - CHRISTINE, L.: Étude de la vigueur hybride par l'emploi du carbone et du phosphore radioactifs. - In: Proceedings of the Symposium on Isotopes in Plant Nutrition and Physiology. Pp. 453-464. Int. at. Energy Agency, Vienna 1967. [Ps.]

1387 - CHU, C.C., KONG, L.: Photorespiration of sugar cane. - Taiwan Sugar exp. Sta.
annu. Rep. *1969*: 1-14, 1970.

1388 - CHU, T.C., CHU, E.J.-H.: Thin-layer chromatography of methyl esters of porphyrins, chlorins and related compounds with Eastman "Chromagram". - J. Chromatogr.
21: 46-51, 1966.

1389 - CHU, T.C., CHU, E.J.-H.: Rapid thin-layer chromatography of porphyrins and related compounds, and its application to the study of porphyrins. - J. chromatogr.
28: 475-478, 1967.

1390 - CHUA, N.-H., LEVINE, R.P.: The photosynthetic electron transport chain of *Chlamydomonas reinhardi*. VIII. The 520 nm light-induced absorbance change in the
wild-type and mutant strains. - Plant Physiol. *44*: 1-6, 1969.

1391 - CHUA, N.-H., LEVINE, R.P.: The 520 nm light-induced absorbance change in wild-
type and mutant strains of *Chlamydomonas reinhardi*. - In: METZNER, H. (ed.):
Progress in Photosynthesis Research. Vol. II. Pp. 978-990. Tübingen 1969.

1392 - CHUDNOVSKIĬ, A.F., MAKARYCHEV, I.N.: Fizicheskoe obosnovanie metodov kompleksnogo issledovaniya protsessov energo- i massoobmena v sisteme "rastenie-pochva-
vozdukh". [The physical background of methods for complex studying of energy
and mass exchange in the system "plant-soil-atmosphere".] - Vestn. sel'skokhoz.
Nauki (Moskva) *1969 (5)*: 118-124, 1969. [In R.]

1393 - CHULANOVSKAYA, M.V.: Fotosinteticheskiĭ koeffitsient u *Chlorella* v usloviyakh
razlichnoĭ temperatury. [Photosynthetic quotient in *Chlorella* under different
temperatures.] - Bot. Zh. *51*: 135-138, 1966. [In R.]

1394 - CHULANOVSKAYA, M.V., ZALENSKIĬ, O.V.: Svetozavisimoe pogloshchenie glyukozy kak
pokazatel' fotofosforilirovaniya in vivo. [Light-dependent absorption of glucose
as characteristics of photophosphorylation in vivo.] - In: KIRICHENKO, E.B.
(ed.): Metody Issledovaniya Fotofosforilirovaniya. Pp. 111-128. Pushchino-na-
Oke 1970. [In R.]

1395 - CHUMAKOV, V.M., YAGUZHINSKIĬ, L.S., KALMANSON, A.E.: O prirode temnovogo signala EPR v fotosinteziruyushchikh sistemakh. [Nature of the dark EPR signal in photosynthesizing systems.] - Biofizika *14*: 98-104, 1969. [In R, ab: E.]

1396 - CHUMAKOVSKIĬ, N.N., KRYSHNYAYA, S.V.: Vliyanie vodoobespechennosti na soderzhanie fitogormonov, khlorofillov i karotinoidov v vysokoroslom gorokhe. [Effect of water supply on the contents of phytohormones, chlorophylls and carotenoids in tall pea.] - In: Informatsionnyĭ Byulleten' 6. Sibir. Inst. Fiziol. i Biokhim. Rast. SO Akad. Nauk SSSR. Pp. 44-45. Irkutsk 1970. [In R.]

1397 - CÍCHA, V.: Klimatizované boxy v zemědělském výzkumu a experimentální botanice. [Air-conditioned cabinets in agricultural research and experimental botany.] - Zem. Technika (Praha) *14*: 49-66, 1968. [In Czech, ab: E.]

1398 - CLAES, H.: Maximal effectiveness of 670 mμ in the light dependent carotenoid synthesis in *Chlorella vulgaris*. - Photochem. Photobiol. *5*: 515-521, 1966.

1399 - CLAES, H.: Action spectrum of light-dependent carotenoid synthesis in *Chlorella vulgaris*. - In: GOODWIN, T.W. (ed.): Biochemistry of Chloroplasts. Vol. II. Pp. 441-444. Academic Press, London-New York 1967.

1400 - CLARK, J., BONGA, J.M.: Photosynthesis and respiration in black spruce *(Picea mariana)* parasitized by eastern dwarf mistletoe *(Arceuthobium pusillum)*. - Can. J. Bot. *48*: 2029-2031, 1970.

1401 - CLARKE, G.L.: Light measurements. - BioScience *18*: 965-966, 1968. [In sea.]

1402 - CLARKE, G.L., EWING, G.C., LORENZEN, C.J.: Spectra of backscattered light from the sea obtained from aircraft as a measure of chlorophyll concentration. - Science *167*: 1119-1121, 1970.

1403 - CLAUSS, H.: Effect of red and blue light on morphogenesis and metabolism of *Acetabularia mediterranea*. - In: BRACHET, J., BONOTTO, S. (ed.): Biology of *Acetabularia*. Pp. 177-191. Academic Press, New York-London 1970. [Ps.]

1404 - CLAUSS, H.: Der Einfluss von Rot- und Blaulicht auf die Hillaktivität von *Acetabularia*-Chloroplasten. - Planta *91*: 32-37, 1970.

1405 - CLAUSS, H., LÜTTKE, A., HELLMANN, F., REINERT, J.: Chloroplastenvermehrung in kernlosen Teilstücken von *Acetabularia mediterranea* und *Acetabularia cliftonii* und ihre Abhängigkeit von inneren Faktoren. - Protoplasma *69*: 313-329, 1970.

1406 - CLAYTON, R.K.: Physical processes involving chlorophylls in vivo. - In: VERNON, L.P., SEELY, G.R. (ed.): The Chlorophylls. Pp. 609-641. Academic Press, New York-London 1966.

1407 - CLAYTON, R.K.: Spectroscopic analysis of bacteriochlorophylls in vitro and in vivo. - Photochem. Photobiol. *5*: 669-677, 1966.

1408 - CLAYTON, R.K.: Fluorescence from major and minor bacteriochlorophyll components in vivo. - Photochem. Photobiol. *5*: 679-688, 1966.

1409 - CLAYTON, R.K.: Relations between photochemistry and fluorescence in cells and extracts of photosynthetic bacteria. - Photochem. Photobiol. *5*: 807-821, 1966.

1410 - CLAYTON, R.K.: Trapping of energy at photosynthetic reaction centers. - In: THOMAS, J.B., GOEDHEER, J.C. (ed.): Currents in Photosynthesis. Pp. 105-110. Donker, Rotterdam 1966.

1411 - CLAYTON, R.K.: The bacterial photosynthetic reaction center. - In: Energy Conversion by the Photosynthetic Apparatus. Brookhaven Symp. Biol. *19*: 62-70, 1967.

1412 - CLAYTON, R.K.: An analysis of the relations between fluorescence and photochemistry during photosynthesis. - J. theor. Biol. *14*: 173-186, 1967.

1413 - CLAYTON, R.K.: Some manifestations of energy storage in algae and chloroplasts. - Arch. Mikrobiol. *59*: 49-58, 1967.

1414 - CLAYTON, R.K.: Characteristics of prompt and delayed fluorescence from spinach chloroplasts. - Biophys. J. *9*: 60-76, 1969.

1415 - CLAYTON, R.K., SISTROM, W.R.: An absorption band near 800 mμ associated with P870 in photosynthetic bacteria. - Photochem. Photobiol. *5*: 661-668, 1966.

1416 - **CLAYTON, R.K., STRALEY, S.C.:** An optical absorption change that could be due to reduction of the primary photochemical electron acceptor in photosynthetic reaction centers. - Biochem. biophys. Res. Commun. *39*: 1114-1119, 1970.

1417 - **CLEMENT-METRAL, J.:** Analog computer simulation of the Blinks effect. - In: METZNER, H. (ed.): Progress in Photosynthesis Research. Vol. II. Pp. 1057-1063. Tübingen 1969.

1418 - **CLEMENT-METRAL, J., LAVOREL, J.:** Etude d'un modèle cinétique applicable aux transitoires de fluorescence de la chlorophylle et au "Jet" d'oxygène des chloroplastes isolés. - Photosynthetica *3*: 233-243, 1969.

1419 - **CLEWELL, A.F., SCHMID, G.H.:** Chlorophyll-deficient *Lespedeza procumbens*. - Planta *84*: 166-173, 1969.

1420 - **CLIJSTERS, H.:** On the photosynthetic activity of developing apple fruits. - Qual. Plant. Materiae veg. *19*: 129-140, 1969.

1421 - **CLIJSTERS, H.:** On the effect of light on carbon dioxide exchange in developing apple fruits. - In: METZNER, H. (ed.): Progress in Photosynthesis Research. Vol. I. Pp. 388-395. Tübingen 1969.

1422 - **CLOUD, P., GIBOR, A.:** The oxygen cycle. - Sci. Amer. *223*: 111-120, 1970.

1423 - **CLYDESDALE, F.M.:** Chlorophyllase activity in green vegetables with reference to pigment stability in thermal processing. - Diss. Abstr. B *27*: 1180-B, 1966.

1424 - **CLYDESDALE, F.M., FRANCIS, F.J.:** Chlorophyll changes in thermally processed spinach as influenced by enzyme conversion and pH adjustment. - Food Technol. *22* (6): 135-138, 1968.

1425 - **COBERN, D., HOBBS, J.S., LUCAS, R.A., MACKENZIE, D.J.:** Location of hydroperoxide groups in monohydroperoxides formed by chlorophyll-photosensitized oxidation of unsaturated esters. - J. chem. Soc. C *1966*: 1897-1902, 1966.

1426 - **COCKBURN, W.:** Pyrophosphate and isolated chloroplasts. - In: METZNER, H. (ed.): Progress in Photosynthesis Research. Vol. I. Pp. 267-271. Tübingen 1969.

1427 - **COCKBURN, W., BALDRY, C.W., WALKER, D.A.:** Oxygen evolution by isolated chloroplasts with carbon dioxide as the hydrogen acceptor. A requirement for orthophosphate or pyrophosphate. - Biochim. biophys. Acta *131*: 594-596, 1967.

1428 - **COCKBURN, W., BALDRY, C.W., WALKER, D.A.:** Photosynthetic induction phenomena in spinach chloroplasts in relation to the nature of the isolating medium. - Biochim. biophys. Acta *143*: 606-613, 1967.

1429 - **COCKBURN, W., BALDRY, C.W., WALKER, D.A.:** Some effects of inorganic phosphate on O_2 evolution by isolated chloroplasts. - Biochim. biophys. Acta *143*: 614-624, 1967.

1430 - **COCKBURN, W., WALKER, D.A., BALDRY, C.W.:** Photosynthesis by isolated chloroplasts. Reversal of orthophosphate inhibition by Calvin-cycle intermediates. - Biochem. J. *107*: 89-95, 1968.

1431 - **COCKBURN, W., WALKER, D.A., BALDRY, C.W.:** The isolation of spinach chloroplasts in pyrophosphate media. - Plant Physiol. *43*: 1415-1418, 1968.

1432 - **COCKSHULL, K.E.:** Effect of night-break treatment on leaf area and leaf dry weight in *Callistephus chinensis*. - Ann. Bot. *30*: 791-806, 1966. [Growth analysis.]

1433 - **COCKSHULL, K.E., van EMDEN, H.F.:** The effects of foliar applications of (2-chloroethyl)-trimethylammonium chloride on leaf area and dry matter production by the Brussels sprout plant. - J. exp. Bot. *20*: 648-657, 1969.

1434 - **COCUCCI, S., MARRE, E.:** Ricerche sull' invecchiamento in foglie recise. II. Sull' effetto di protezione degli zuccheri della luce e della cinetina sugli acidi ribonucleici e sulla capacitá fotosintetica. [Aging in cut leaves. II. Protective effect of sugar, light, and kinetin on RNA and photosynthetic capacity.] - Atti Accad. naz. Lincei, Rend., Cl. Sci. fis. mat. nat. *40*: 1095-1102, 1966. [In Ital.]

1435 - CODD, G.A., LORD, J.M., MERRETT, M.J.: The glycollate oxidising enzyme of algae.
       - FEBS Lett. *5*: 341-342, 1969.

1436 - COFFEY, M.D., MARSHALL, C., WHITBREAD, R.: The translocation of ^{14}C-labelled as-
       similates in tomato plants infected with *Alternaria solani* (ELL. and MART.)
       JONES and GROUT; - Ann. Bot. *34*: 605-615, 1970.

1437 - COFFMAN, R.E., STAVENS, B.W.: Solvent-induced EPR anisotropy change of the non-
       heme chromophore of spinach ferredoxin. - Biochim. biophys. Res. Commun. *41*:
       163-169, 1970.

1438 - COHEN, D.: The expected efficiency of water utilization in plants under differ-
       ent competition and selection regimes. - Israel J. Bot. *19*: 50-54, 1970. [Ps.]

1439 - COHEN, S.S.: Are/were mitochondria and chloroplasts microorganisms? - Amer. Sci.
       58: 281-289, 1970.

1440 - COHEN, W.S.: The effects of a naturally-occurring factor (from the leaves of
       Ricinus communis) on the light reactions of photosynthesis: model systems for
       the action of this factor. - Diss. Abstr. int. B *31*: 3177-B, 1970.

1441 - COHEN, W.S., NATHANSON, B., WHITE, J.E., BRODY, M.: Fatty acids as model systems
       for the action of *Ricinus* leaf extract on higher plant chloroplasts and algae.
       - Arch. Biochem. Biophys. *135*: 21-27, 1969.

1442 - COHEN, Y., ROTEM, J.: The relationship of sporulation to photosynthesis in some
       obligatory and facultative parasites. - Phytopathology *60*: 1600-1604, 1970.

1443 - COHEN-BAZIRE, G., LEFORT-TRAN, M.: Fixation of phycobiliproteins to photosynthe-
       tic membranes by glutaraldehyde. - Arch. Mikrobiol. *71*: 245-257, 1970.

1444 - COHEN-BAZIRE, G., SISTROM, W.R.: The procaryotic photosynthetic apparatus. - In:
       VERNON, L.P., SEELY, G.R. (ed.): The Chlorophylls. Pp. 313-341. Academic Press,
       New York-London 1966.

1445 - COINER, S., ORR, A.: Chlorophyll content in mutants of *Glycine Max* (L.) MERRIL
       during seedling development. - Proc. Iowa Acad. Sci. *76*: 90-93, 1970.

1446 - COLE, W.J., CHAPMAN, D.J., SIEGELMAN, H.W.: The structure of phycocyanobilin. -
       J. amer. chem. Soc. *89*: 3643-3645, 1967.

1447 - COLE, W.J., CHAPMAN, D.J., SIEGELMAN, H.W.: The structure and properties of
       phycocyanobilin and related bilatrienes. - Biochemistry *7*: 2929-2935, 1968.

1448 - COLE, W.J., Ô hEOCHA, C., MOSCOWITZ, A., KRUEGER, W.R.: The optical activity
       of urobilins derived from phycoerythrobilin. - Europe. J. Biochem. *3*: 202-207,
       1967.

1449 - COLLINS, B.G.: A modified Campbell Stokes sunshine recorder. - Meteor. Mag.
       97: 16-19, 1968.

1450 - COLLINS, B.G., PATTERSON, G.: A radiation recording system. - Aust. meteorol.
       Mag. *16 (4)*: 149-153, 1968.

1451 - COMBRES, J.C., BONHOMME, R., BALDY, C., GOILLOT, C., de PARCEVAUX, S.: Un pyra-
       nomètre linéaire destiné à la mesure du rayonnement sous couvert végétal. - In:
       Techniques d'Étude des Facteurs Physiques de la Biosphère. Pp. 59-70. INRA,
       Paris 1970.

1452 - CONNOR, D.J., CARTLEDGE, O.: Observed and calculated photosynthetic rates of
       Chloris gayana communities. - J. appl. Ecol. *7*: 353-362, 1970.

1453 - CONRAD, J.R.: The reversible photochemical oxidation of chlorophyll. - Diss.
       Abstr. B *28*: 4963-B, 1968.

1454 - CONSTANTOPOULOS, G.: Effect of manganese deficiency on the greening and the
       lipid composition of photoheterotrophic *Euglena gracilis* Z. - Plant Physiol. *43*
       (Suppl.): S 7, 1968.

1455 - CONSTANTOPOULOS, G.: Lipid metabolism of manganese-deficient algae. I. Effect
       of manganese deficiency on the greening and the lipid composition of *Euglena
       gracilis* Z. - Plant Physiol. *45*: 76-80, 1970.

1456 - CONSTANTOPOULOS, G., BLOCH, K.: Effect of light intensity on the lipid composition of *Euglena gracilis*. - J. biol. Chem. *242*: 3538-3542, 1967. [Chl, Hill reaction.]

1457 - CONSTANTOPOULOS, G., KENYON, C.N.: Release of free fatty acids and loss of Hill activity by aging spinach chloroplasts. - Plant Physiol. *43*: 531-536, 1968.

1458 - COOK, J.R.: Photosynthetic activity during the division cycle in synchronized *Euglena gracilis*. - Plant Physiol. *41*: 821-825, 1966.

1459 - COOK, J.R.: Quantitative measurement of paramylum in *Euglena gracilis*. - J. Protozool. *14*: 634-636, 1967.

1460 - COOKE, G.D.: The pattern of autotrophic succession in laboratory microcosms. - BioScience *17*: 717-721, 1967. [Ps, Chl.]

1461 - COOMBS, J., BALDRY, C.W.: CO_2 assimilation by chloroplasts illuminated on filter paper. - Nature *228*: 1349-1350, 1970.

1462 - COOMBS, J., LAURITIS, J.A., DARLEY, W.M., VOLCANI, B.E.: Studies on the biochemistry and fine structure of silica shell formation in diatoms. VI. Fine structure of colchicine-induced polyploids of *Navicula pelliculosa* (BREB.) HILSE; - Z. Pflanzenphysiol. *59*: 274-284, 1968. [Ps, Chl.]

1463 - COOMBS, J., SPANIS, C., VOLCANI, B.E.: Studies on the biochemistry and fine structure of silica shell formation in diatoms. Photosynthesis and respiration in silicon-starvation synchrony of *Navicula pelliculosa*. - Plant Physiol. *42*: 1607-1611, 1967.

1464 - COOMBS, J., WHITTINGHAM, C.P.: The effect of high partial pressures of oxygen on photosynthesis in *Chlorella* I. The effect on end products of photosynthesis. - Phytochemistry *5*: 643-651, 1966.

1465 - COOMBS, J., WHITTINGHAM, C.P.: The mechanism of inhibition of photosynthesis by high partial pressures of oxygen in *Chlorella*. - Proc. roy. Soc. (London) B *164*: 511-520, 1966.

1466 - COOPER, A.J.: Effects of shading and time of year on net assimilation rates of young glasshouse tomato plants. - Ann. appl. Biol. *59*: 85-90, 1967.

1467 - COOPER, C.S.: Relative growth of alfalfa and birdsfoot trefoil seedlings under low light intensity. - Crop Sci. *7*: 176-178, 1967. [Growth analysis.]

1468 - COOPER, C.S., MacDONALD, P.W.: Energetics of early seedling growth in corn *(Zea mays* L.). - Crop.Sci. *10*: 136-139, 1970.[Ps.]

1469 - COOPER, C.S., QUALLS, M.: Morphology and chlorophyll content of shade and sun leaves of two legumes. - Crop Sci. *7*: 672-673, 1967.

1470 - COOPER, R.B.: Photosynthesis and certain morphological characteristics of alfalfa as affected by potassium nutrition. - Diss. Abstr. B *27*: 2227-B, 1967.

1471 - COOPER, R.B., BLASER, R.E., BROWN, R.H.: Potassium nutrition effects on net photosynthesis and morphology of alfalfa. - Proc. Soil Sci. Soc. Amer. *31*: 231-235, 1967.

1472 - COOPER, R.L., BRUN, W.A.: Response of soybeans to a carbon dioxide-enriched atmosphere. - Crop Sci. *7*: 455-457, 1967. [Ps.]

1473 - COOPER, T.G., BENEDICT, C.R.: PEP carboxykinase exchange reaction in photosynthetic bacteria. - Plant Physiol. *43*: 788-792, 1968.

1474 - COOPER, T.G., FILMER, D., WISHNICK, M., LANE, M.D.: The active species of "CO_2" utilized by ribulose diphosphate carboxylase. - J. biol. Chem. *244*: 1081-1083, 1969.

1475 - COPE, B.T., SMITH, U., CRESPI, H.L., KATZ, J.J.: Studies on the identity of ordinary and deuterio-phycocyanins: end-groups, amino acid compositions, minimum molecular weights, peptide maps. - Biochim. biophys. Acta *133*: 446-453, 1967.

1476 - CORDUAN, G.: Autotrophe Gewebekulturen von *Ruta graveolens* und deren $^{14}CO_2$- Markierungsprodukte. - Planta *91*: 291-301, 1970.

1477 - CORKER, G.A.: The use of a nitroxide radical as a test reagent of the primary processes of photosynthesis. - Diss. Abstr. B *29*: 83-B-84-B, 1968.

1478 - CORKER, G.A., KLEIN, M.P., CALVIN, M.: Chemical trapping of a primary quantum conversion product in photosynthesis. - Proc. nat. Acad. Sci. U.S.A. *56*: 1365-1369, 1966.

1479 - CORKER, G.A., NICHOLSON, W.: Comparative kinetic behavior of the photoinduced EPR signal observed in whole cells and cell free preparations of *Rhodospirillum rubrum*. - Arch. Biochem. Biophys. *137*: 75-83, 1970.

1480 - CORNFORTH, I.S.: Chlorophyll compounds and nitrogen availability in West Indian soils. - Plant Soil *30*: 113-116, 1969.

1481 - CORNIC, G.: Evolution de l'activité photosynthétique de feuilles de rang différent chez le *Sinapis alba* en populations de différentes densités. - Compt. rend. Acad. Sci. Paris, Sér. D *268*: 1934-1937, 1969.

1482 - CORNIC, G.: Influence de la lumière sur la photorespiration. - Compt. rend. Acad. Sci. Paris, Sér. D *271*: 489-492, 1970.

1483 - CORNIC, G., MOUSSEAU, M.: Etude du dégagement de CO_2 à la lumière chez le Moutarde blanche(*Sinapis alba* L.). Influence de l'énergie lumineuse. - Compt. rend. Acad. Sci. Paris, Sér. D *269*: 1774-1776, 1969.

1484 - CORNIC, G., CHOPIN, G., MOUSSEAU, M.: Etude du dégagement de CO_2 à la lumière chez la Moutarde blanche (*Sinapis alba* L.). Influence du temps de photosynthèse. - Compt. rend. Acad. Sci. Paris, Sér. D *269*: 1194-1197, 1969.

1485 - CORNIC, G., MOUSSEAU, M., MONTENY, B.: Importance de la photorespiration dans le bilan photosynthétique au cours de la croissance foliaire. - Oecol. Plant. *5*: 355-363, 1970.

1486 - COST, H.B., GRAY, E.D.: Rapidly labeled RNA synthesis during morphogenesis. - Biochim. biophys. Acta *138*: 601-604, 1967. [Chl, car, Ps bacteria.]

1487 - COST, H.B., GRAY, E.D.: The effect of CO_2 on the biosynthesis of carotenoids in *Rhodopseudomonas spheroides*. - Biochim. biophys. Acta *177*: 118-123, 1969.

1488 - COST, K., BOLTON, J.R., FRENKEL, A.W.: The accelerating action of 5-methylphenazinium methyl sulfate on light induced absorbancy and ESR changes in *Rhodospirillum chromatophores*. - Photochem. Photobiol. *5*: 823-826, 1966.

1489 - COST, K., BOLTON, J.R., FRENKEL, A.W.: Comparative decay characteristics of the light generated free radical in chromatophores and chloroplasts. - Photochem. Photobiol. *10*: 251-258, 1969.

1490 - COST, K., FRENKEL, A.W.: Light-induced interactions of *Rhodospirillum rubrum* chromatophores with bromothymol blue. - Biochemistry *6*: 663-667, 1967.

1491 - COST, K., FRENKEL, A.W.: Comparative studies on phosphorylation and apparent proton changes in illuminated chromatophores. - In: SHIBATA, K. et al. (ed.): Comparative Biochemistry and Biophysics of Photosynthesis. Pp. 266-273. Univ. Tokyo Press, Tokyo; Univ. Park Press, State College, Pa. 1968.

1492 - COSTELLO, W.J., COBB, H.D.: The differential effect of low light intensities on photosynthesis and nitrogen uptake in *Chlorella pyrenoidosa*. - Plant Physiol. *46* (Suppl.): 40, 1970.

1493 - COSTES, C.: Biosynthèse du phytol des chlorophylles et du squalette tétratérpenique des caroténoïdes dans les feuilles vertes. - Phytochemistry *5*: 311-324, 1966.

1494 - COSTES, C.: Pigments caroténoïdes des pétales de la fleur de millepertuis, *Hypericum perforatum*. - Ann. Physiol. vég. *9*: 157-177, 1967.

1495 - COSTES, C.: Caroténoïdes et photosynthèse: variations induites de la teneur en pigments dans des folioles excisées de tomate. - Ann. Physiol. vég. *10*: 171-197, 1968.

1496 - COSTES, C.: Sur la présence d'anhydroeschscholtzxanthine dans les feuilles rouges de buis, *Buxus sempervirens*. - Bull. Soc. franç. Physiol. vég. *15*:55-70, 1969.

1497 - COSTES, C., DEROCHE, M.-E.: Hétérogénéité des liaisons des caroténoides dans les chloroplastes. - Bull. Soc. franç. Physiol. vég. *14*: 505, 1968.

1498 - COSTES, C., DEROCHE, M.-E., FERRON, F., CHARTIER, P.: L'utilisation du gaz carbonique et de la lumière dans les serres. - Bull. tech. Inform. *217*: 1-9, 1967.

1499 - COTTON, T.M., BALLSCHMITTER, K., KATZ, J.J.: Gas chromatographic determination of water in chlorophyll. - J. chromatogr. Sci. *8*: 546-549, 1970.

1500 - COUCH, R.W.: The effect of atrazine, bromacil, and diquat on carbon^{14} fixation in corn, cotton, and soybeans. - Diss. Abstr. B *27*: 1744-B, 1966.

1501 - COUCH, R.W., DAVIS, D.E.: Effect of atrazine, bromacil, and diquat on $^{14}CO_2$-fixation in corn, cotton and soybeans. - Weeds *14*: 251-255, 1966.

1502 - COUNIHAN, J.: An improved method of simulating an atmospheric boundary layer in a wind tunnel. - Atmos. Environm. *3*: 197-214, 1969.

1503 - COUPÉ, M., CHAMPIGNY, M.-L., MOYSE, A.: Sur la localisation intracellulaire de la nitrate réductase dans les feuilles et les racines d'Orge. - Physiol. vég. *5*: 271-291, 1967. [Chl.]

1504 - COWAN, I.R.: The interception and absorption of radiation in plant stands. - J. appl. Ecol. *5*: 367-379, 1968.

1505 - COWAN, I.R.: Mass, heat and momentum exchange between stands of plants and their atmospheric environment. - Quart. J. roy. meteorol. Soc. *94*: 523-544, 1968.

1506 - COWAN, I.R., MILTHORPE, F.L.: Physiological responses in relation to the environment within the plant cover. - In: ECKARDT, F.E. (ed.): Functioning of Terrestrial Ecosystems at the Primary Production Level. Pp. 107-130. UNESCO, Paris 1968. [Ps.]

1507 - COX, E.F.: Cyclic changes in transpiration of sunflower leaves in a steady environment. - J. exp. Bot. *19*: 167-175, 1968. [CO_2 in the leaf.]

1508 - COX, E.L., HUFFAKER, R.C.: Lower enzyme activities and pigment contents in chlorophyll mutants of barley. - Plant Physiol. *44* (Suppl.): 33, 1969.

1509 - COX, M.T., HOWARTH, T.T., JACKSON, A.H., KENNER, G.W.: Formation of the isocyclic ring in chlorophyll. - J. amer. chem. Soc. *91*: 1232-1233, 1969.

1510 - COX, R.M., FAY, P.: Special aspects of nitrogen fixation by blue-green algae. - Proc. roy. Soc. B *172*: 357-366, 1969. [Ps.]

1511 - CRAIG, I.W., CARR, N.G.: *C*-phycocyanin and allophycocyanin in two species of blue-green algae. - Biochem. J. *106*: 361-366, 1968.

1512 - CRAIG, I.W., GIBOR, A.: Biosynthesis of proteins involved with photosynthetic activity in enucleated *Acetabularia* sp. - Biochim. biophys. Acta *217*: 488-495, 1970.

1513 - CRAIG, I.W., GIBOR, A.: A direct light effect on maintaining photosynthetic activity of *Nitella* chloroplasts. - J. Cell Biol. *44*: 305-309, 1970.

1514 - CRAIGIE, J.S.: Some salinity-induced changes in growth, pigments, and cyclohexanetetrol content of *Monochrysis lutheri*. - J. Fish. Res. Board Can. *26*: 2959-2967, 1969.

1515 - CRAIGIE, J.S., McLACHLAN, J., ACKMAN, R.G., TOCHER, C.S.: Photosynthesis in algae. III. Distribution of soluble carbohydrates and dimethyl-β-propiothetin in marine unicellular *Chlorophyceae* and *Prasinophyceae*. - Can. J. Bot. *45*: 1327-1334, 1967.

1516 - CRAIGIE, J.S., McLACHLAN, J., MAJAK, W., ACKMAN, R.G., TOCHER, C.S.: Photosynthesis in algae. II. Green algae with special reference to *Dunaliella* spp. and *Tetraselmis* spp.- Can. J. Bot. *44*: 1247-1254, 1966.

1517 - CRAMER, W.A.: Low potential titration of the fluorescence yield changes in photosynthetic bacteria. - Biochim. biophys. Acta *189*: 54-59, 1969.

1518 - CRAMER, W.A., BUTLER, W.L.: Light-induced absorbance changes of two cytochrome *b* components in the electron transport system of spinach chloroplasts. - Biochim. biophys. Acta *143*: 332-339, 1967.

1519 - CRAMER, W.A., BUTLER, W.L.: Further resolution of chlorophyll pigments in photosystems 1 and 2 of spinach chloroplasts by low-temperature derivative spectroscopy. - Biochim. biophys. Acta *153*: 889-891, 1968.

1520 - CRAMER, W.A., BUTLER, W.L.: Potentiometric titration of the fluorescence yield of spinach chloroplasts. - Biochim. biophys. Acta *172*: 503-510, 1969.

1521 - CRANE, F.L., HENNINGER, M.D.: Function of quinones in photosynthesis. - Vitamines Hormones *24*: 489-517, 1966.

1522 - CRANE, F.L., HENNINGER, M.D., WOOD, P.M., BARR, R.: Quinones in chloroplasts. - In: GOODWIN, T.W. (ed.): Biochemistry of Chloroplasts. Vol. I. Pp. 133-151. Academic Press, London-New York 1966.

1523 - CREASY, L.L.: The role of low temperature in anthocyanin synthesis in "McIntosh" apples. - Proc. amer. Soc. hort. Sci. *93*: 716-724, 1968. [Chl.]

1524 - CRESPI, H.L., BOUCHER, L.J., NORMAN, G.D., KATZ, J.J., DOUGHERTY, R.C.: Structure of phycocyanobilin. - J. amer. chem. Soc. *89*: 3642-3643, 1967.

1525 - CRESPI, H.L., KATZ, J.J.: Exchangeable hydrogen in phycoerythrobilin. - Phytochemistry *8*: 759-761, 1969.

1526 - CRESPI, H.L., SMITH, U.H.: The chromophore-protein bonds in phycocyanin. - Phytochemistry *9*: 205-212, 1970.

1527 - CRESPI, H.L., SMITH, U., KATZ, J.J.: Phycocyanobilin. Structure and exchange studies by nuclear magnetic resonance and its mode of attachment in phycocyanin. A model for phytochrome. - Biochemistry *7*: 2232-2242, 1968.

1528 - CRIDDLE, R.S.: Protein and lipoprotein organization in the chloroplast. - In: GOODWIN, T.W. (ed.): Biochemistry of Chloroplasts. Vol. I. Pp. 203-231, Academic Press, London-New York 1966.

1529 - CRIDDLE, R.S.: Structural proteins of chloroplasts and mitochondria. - Annu. Rev. Plant Physiol. *20*: 239-252, 1969.

1530 - CRIDDLE, R.S., DAU, B., KLEINKOPF, G.E., HUFFAKER, R.C.: Differential synthesis of ribulosediphosphate carboxylase subunits. - Biochim. biophys. Res. Commun. *41*: 621-627, 1970.

1531 - CROFT, J.A., HOWDEN, M.E.H.: Chlorophyll *c* from *Sargassum flavicans* - Isolation and structure. - Phytochemistry *9*: 901-902, 1970.

1532 - CROFTS, A.R.: Uptake of ammonium ion by chloroplasts, and its relation to photophosphorylation. - Biochim. biophys. Res. Commun. *24*: 725-731, 1966.

1533 - CROFTS, A.R.: Amine uncoupling of energy transfer in chloroplasts. I. Relation to ammonium ion uptake. - J. biol. Chem. *242*: 3352-3359, 1967.

1534 - CROFTS, A.R.: Ammonium ion uptake by chloroplasts, and the light-energy state. - In: JÄRNEFELT, J. (ed.): Regulatory Functions of Biological Membranes. Pp. 247-263. Elsevier Publishing Co., Amsterdam-London-New York 1968.

1535 - CROFTS, A.R., DEAMER, D.W., PACKER, L.: Mechanisms of light-induced structural change in chloroplasts. II. The role of ion movements in volume changes. - Biochim. biophys. Acta *131*: 97-118, 1967.

1536 - CROOKSTON, R.K., MOSS, D.N.: The relation of carbon dioxide compensation and chlorenchymatous vascular bundle sheaths in leaves of dicots. - Plant Physiol. *46*: 564-567, 1970.

1537 - CRUDEN, D.L., COHEN-BAZIRE, G., STANIER, R.Y.: *Chlorobium* vesicles: the photosynthetic organelles of green bacteria. - Nature *228*: 1345-1347, 1970.

1538 - CRUDEN, D.L., STANIER, R.Y.: The characterization of *Chlorobium* vesicles and membranes isolated from green bacteria. - Arch. Mikrobiol. *72*: 115-134, 1970.

1539 - CSUPOR, L.: Das Phytol in vergilbten Blättern. - Planta med. *19*: 37-41, 1970.

1540 - CUKROVÁ, V., AVRATOVŠČUKOVÁ, N.: Photosynthetic activity, chlorophyll content and stomata characteristics in diploid and polyploid types of *Datura stramonium* L. - Photosynthetica *2*: 227-237, 1968.

1541 - CUMMINS, J.T., STRAND, J.A., VAUGHAN, B.E.: The movement of H^+ and other ions at the onset of photosynthesis in *Ulva*. - Biochim. biophys. Acta *173*: 198-205, 1969.

1542 - CUNNINGHAM, G.L., STRAIN, B.R.: An ecological significance of seasonal leaf variability in a desert shrub. - Ecology *50*: 400-408, 1969. [Ps.]

1543 - CUNNINGHAM, G.L., STRAIN, B.R.: Irradiance and productivity in a desert shrub. - Photosynthetica *3*: 69-71, 1969.

1544 - ČUPINA, T.: Primena radioaktivnich izotopa pri proučavanju problema fotosinteze. [Use of radioactive isotopes for studying photosynthesis.] - Poljoprivreda (Beograd) *15 (5)*: 16-23, 1967. [In Croatian.]

1545 - ČUPINA, T.: O metodima određivanja hlorofila *a* i *b* u biljnom materijalu. [On the methods of chlorophyll *a* and *b* determination in plant material.] - Savremena Poljoprivreda (Novi Sad) *17*: 1041-1049, 1969. [In Croatian.]

1546 - ČUPINA, T., ČURIĆ, R., GERIĆ, I., SARIĆ, M.: Proučavanje sadržaja hlorofila, šećera, slobodnih aminokiselina i NPK mineralnih elemenata kod biljaka kukuruza obolelih od "crvenila". [Studies of the content of chlorophyll, sugar, free amino acids and NPK by corn plants diseased with redness.] - Savremena Poljoprivreda (Novi Sad) *17*: 531-540, 1969. [In Croatian, ab: E.]

1547 - ČUPINA, T., JOCIĆ, B.: Uticaj količine i odnosa NPK mineralnih đubriva na dinamiku formiranja lisne površine i produktivnost fotosinteze pšenice. [Effect of amounts and ratios of NPK on the dynamics of formation of leaf area and productivity of wheat photosynthesis.] - Savremena Poljoprivreda (Novi Sad) *17* (11-12): 203-210, 1969. [In Croatian, ab: E.]

1548 - CURBY, W.A.: The importance of oxygen in the initiation of photosynthesis in *Phormidium persicinum*. - Bacteriol. Proc. *68*: 47, 1968.

1549 - CURBY, W.A., KING, H.D., MARIANI, H.A.: An effect of ionizing radiation on the initial photosynthetic rate of *Phormidium persicinum*. - Bacteriol. Proc. *66*: 96, 1966.

1550 - CURL, A.L.: The carotenoids of muskmelons. - J. Food Sci. *31*: 759-761, 1966.

1551 - CURTH, P., SUMPF, D.: Eine neue photoelektrische Messeinrichtung zur Bestimmung des Blattapparates von Rosettenpflanzen, insbesondere Zuckerrüben. - Biol. Zentralbl. *87*: 777-782, 1968.

1552 - CURTIS, K.K., SMITH, D.E.: Daily variation in chlorophyll content of corn seedlings. - Proc. Indiana Acad. Sci. *78*: 118-119, 1968.

1553 - CURTIS, P.E.: Varietal differences in photosynthetic reactions of the soybean (*Glycine max* (L.) MERRILL). - Diss. Abstr. int. B *30*: 27-B, 1969.

1554 - CURTIS, P.E., OGREN, W.L.: Varietal differences in soybean photosynthesis. - Plant Physiol. *43* (Suppl.): S 5, 1968.

1555 - CURTIS, P.E., OGREN, W.L., HAGEMAN, R.H.: Varietal effects in soybean photosynthesis and photorespiration. - Crop Sci. *9*: 323-327, 1969.

1556 - CUSANOVICH, M.A.: Light-induced electron transport in *Chromatium* st. D. - Diss. Abstr. B *28*: 470-B, 1967.

1557 - CUSANOVICH, M.A., BARTSCH, R.G.: A high potential cytochrome *c* from *Chromatium* chromatophores. - Biochim. biophys. Acta *189*: 245-255, 1969.

1558 - CUSANOVICH, M.A., BARTSCH, R.G., KAMEN, M.D.: Light-induced electron transport in *Chromatium* strain D. II. Light-induced absorbance changes in *Chromatium* chromatophores. - Biochim. biophys. Acta *153*: 397-417, 1968.

1559 - CUSANOVICH, M.A., BARTSCH, R.G., OLSON, J.M.: Light-induced reactions in *Chromatium* chromatophores under controlled redox conditions. - In: SHIBATA, K., TAKAMIYA, A., JAGENDORF, A.T., FULLER, R.C. (ed.): Comparative Biochemistry and Biophysics of Photosynthesis. Pp. 186-195. Univ. Tokyo Press, Tokyo; Univ. Park Press, State College, Pa. 1968.

84

1560 - CUSANOVICH, M.A., KAMEN, M.D.: Light-induced electron transport in *Chromatium* strain D. I. Isolation and characterization of *Chromatium* chromatophores. - Biochim. biophys. Acta *153*: 376-396, 1968.

1561 - CUSANOVICH, M.A., KAMEN, M.D.: Light-induced electron transfer in *Chromatium* strain D. III. Photophosphorylation by *Chromatium* chromatophores. - Biochim. biophys. Acta *153*: 418-426, 1968.

1562 - CUSHING, C.E.: Periphyton productivity and radionuclide accumulation in the Columbia River, Washington, U.S.A. - Hydrobiologia *29*: 125-139, 1967. [Chl.]

1563 - CUSHING, D.H., NICHOLSON, H.F.: Method of estimating algal production rates at sea. - Nature *212*: 310-311, 1966.

1564 - CZECZUGA, B.: Primary production of the green hydrosulphuric bacteria *Chlorobium limicola* NADS; *(Chlorobacteriaceae)*. - Photosynthetica *2*: 11-15, 1968.

1565 - CZECZUGA, B.: Primary production of the purple sulphuric bacteria, *Thiopedia rosea* WINOGR. *(Thiorhodaceae)*. - Photosynthetica *2*: 161-166, 1968.

1566 - CZECZUGA, B., CZERPAK, R.: Wpływ CCC i chlorowodorku dwuetyloaminy na rozwój niektórych gatunków *Chlorella*. [Effect of CCC and diethylamine hydrochloride on the development of some *Chlorella* species.] - Acta Soc. Bot. Pol. *36*: 321-335, 1967. [Chl; in Pol., ab: E.]

1567 - CZECZUGA, B., CZERPAK, R.: Studies on dyes found in *Chlorobium limicola* NADS. *(Chlorobacteriaceae)* from Wadolek lake. - Hydrobiologia *31*: 561-571, 1968. [Bacterioviridine.]

1568 - CZECZUGA, B., CZERPAK, R.: Investigations on vegetable pigments in post-glacial bed sediments of lakes. - Schweiz. Z. Hydrol. *30*: 217-231, 1968. [Chl, car.]

1569 - CZERPAK, R.: Porównanie działania chlorowodorku dwuetyloaminy i chlorku chlorocholiny na siewki *Pisum sativum* L. i *Hordeum sativum* L. [Comparison of the effect of diethylamine hydrochloride and CCC on seedlings of pea and barley.] - Rocz. Nauk rolnicz. A *94*: 617-628, 1968. [Chl; in Pol., ab: E, R.]

1570 - CZERPAK, R.: The effect of CCC and diethylamine hydrochloride on certain species of algae belonging to *Cyanophyceae, Chlorophyceae,* and *Diatomeae*. - Acta hydrobiol. *12*: 143-151, 1970. [Chl.]

1571 - CZOPEK, M.: Ecophysiological studies on photosynthesis and respiration of some plant species in meadow ecosystem. - Acta Soc. Bot. Pol. *36*: 73-86, 1967.

1572 - CZOPEK, M.: Photosynthesis and respiration of turions and vegetative fronds of *Spirodela polyrrhiza*. - Acta Soc. Bot. Pol. *36*: 87-96, 1967.

1573 - CZOPEK, M.: Metody pomiaru światła w ekofizjologii. [Methods of measuring light in ecophysiology.] - Wiadom. bot. *12*: 3-23, 1968. [In Pol.]

1574 - CZOPEK, M.: Photosynthetic production of some plant species in meadow ecosystem. - Acta Soc. Bot. Pol. *38*: 271-289, 1969.

1575 - CZOPEK, M., STARZECKI, W.: Methods for estimation of photosynthetic production of leaves of plants growing in natural ecosystems. - Bull. Acad. pol. Sci., Sér. Sci. Biol. *15*: 439-443, 1967.

1576 - CZOPEK, M., STARZECKI, W.: A field laboratory for photosynthesis measurement. - Bull. Acad. pol. Sci., Sér. Sci. biol. *18*: 657-662, 1970.

1577 - CZOPEK, M., STARZECKI, W., ŁAGISZ, J., MOTYKA, B.: An automatic registrator as a supplement of an electronic integrator of photosynthetically active radiation and their adaptation to ecophysiological studies. - Photosynthetica *1*: 65-68, 1967.

1578 - CZYGAN, F.-C.: Echinenon als Sekundär-Carotinoid einiger Grünalgen. - Z. Naturforsch. *21 b*: 197-198, 1966.

1579 - CZYGAN, F.-C.: Secondary carotenoids in green algae. - Tijdsskr. Kjemi Bergv. Metallurgi *26*: 125, 1966.

1580 - CZYGAN, F.-C.: Untersuchungen über die Wirkung des Hellgrünfaktors *h* auf die Zusammensetzung der Chloroplastenpigmente einiger Oenotheren. - Biol. Zentralbl. 85: 497-505, 1966.

1581 - CZYGAN, F.-C.: Biosynthese der Carotinoide. - Ber. deut. bot. Ges. 80: 627-644, 1967/8.

1582 - CZYGAN, F.-C.: Sekundär-Carotinoide in Grünalgen. I. Chemie, Vorkommen und Faktoren, welche die Bildung dieser Polyene beeinflussen. - Arch. Mikrobiol. 61: 81-102, 1968.

1583 - CZYGAN, F.-C.: Sekundär-Carotinoide in Grünalgen. II. Untersuchungen zur Biogenese. - Arch. Mikrobiol. 62: 209-236, 1968.

1584 - CZYGAN, F.-C.: Zum Vorkommen von Crustaxanthin (3,3',4,4'-Tetraoxi-β-carotin) und die Phoenicopteron (4-Oxo-α-carotin) in Aplanosporen von *Haematococcus pluvialis* FLOTOW em.WILLE. - Flora A 159: 339-345, 1968.

1585 - CZYGAN, F.-C.: Untersuchungen über den Stoffwechsel der Ketocarotinoide in *Adonis*-Arten. I. Pigment-Zusammensetzung nicht normal gefärbter Blütenblätter. - Planta 85: 35-41, 1969.

1586 - CZYGAN, F.-C.: Blutregen und Blutschnee: Stickstoffmangel-Zellen von *Haematococcus pluvialis* und *Chlamudomonas nivalis*. - Arch. Mikrobiol. 74: 69-76, 1970.

1587 - CZYGAN, F.-C.: Untersuchungen über die Bedeutung der Biosynthese von Sekundär-Carotinoiden als Artmerkmal bei Grünalgen. - Arch. Mikrobiol. 74: 77-81, 1970.

1588 - CZYGAN, F.-C., HEUMANN, W.: Die Zusammensetzung und Biogenese der Carotinoide in *Pseudomonas echinoides* und einigen Mutanten. - Arch. Mikrobiol. 57: 123-134, 1967.

1589 - CZYGAN, F.-C., KALB, K.: Untersuchungen zur Biogenese der Carotinoide in *Trentepohlia aurea*. - Z. Pflanzenphysiol. 55: 59-64, 1966.

1590 - CZYGAN, F.-C., KESSLER, E.: Nachweis von 3-Oxi-4.4'-dioxo-β-carotin in den Grünalgen *Chlorococcum wimmeri* und *Haematococcus* spec. - Z. Naturforsch. 22 b: 1085-1086, 1967.

1591 - CZYGAN, F.-C., WILLUHN, G.: Untersuchungen über Gehaltsänderungen der Lipochrome und Steroidalkaloide in rot- und gelbfrüchtigen Tomaten (*Lycopersicon esculentum* L.) verschiedener Reifegrade. - Planta med. 15: 404-415, 1967.

1592 - DADYKIN, V.P., POTAEVICH, E.V.: O propuskanii list'yami rastenii diffuznogo sveta. [Transmission of diffused radiation through leaves.] - Tr. Inst. Ekol. Rast. Zhivot. ural'. Fil. Akad. Nauk SSSR 1968 (62): 111-118, 1968. [Chl. in R.]

1593 - DAGET, N.: Variation de l'état d'oxydoréduction de la plastoquinone A dans une feuille aérienne éclairée en lumière monochromatique. - Compt. rend. Acad. Sci. Paris, Sér. D 368: 1928-1930, 1969.

1594 - DAIN, B.Ya., USACHEVA, M.N., ASHKINAZI, M.S., DOLIDZE, I.A.: O fotovosstanovlennykh formakh khlorofilla i feofitina. [Photoreduced forms of chlorophyll and pheophytin.] - Biofizika 14: 435-440, 1969. [In R, ab: E.]

1595 - DALAL, K.B., SALUNKHE, D.K., OLSON, L.E.: Physiological and biochemical changes in greenhouse-grown tomatoes (*Lycopersicon esculentum*). - J. Food Sci. 31: 461-467, 1966.[Chl, car.]

1596 - DALE, J.E.: Cell growth in expanding primary leaves of *Phaseolus*. - J. exp. Bot. 19: 322-332, 1968. [Chl.]

1597 - DALE, J.E., HEYES, J.K.: A virescent mutant of *Phaseolus vulgaris*; growth and pigment and plastid characters. - New Phytol. 69: 733-742, 1970.

1598 - DALE, J.E., MURRAY, D.: Photomorphogenesis, photosynthesis, and early growth of primary leaves of *Phaseolus vulgaris*. - Ann. Bot. 32: 767-780, 1968.

1599 - DALE, R.E., TEALE, F.W.J.: Spectroscopy of some phycobiliproteins. - In: THOMAS, J.B., GOEDHEER, J.C. (ed.): Currents in Photosynthesis. Pp. 169-175, Donker, Rotterdam 1966.

1600 - DALE, R.E., TEALE, F.W.J.: Number and distribution of chromophore types in native phycobiliproteins. - Photochem. Photobiol. *12*: 99-117, 1970.

1601 - DALEY, R.J., BROWN, S.R.: Growth kinetics, chlorophyll and photosynthetic capacity of a blue-green alga. - J. Phycol. *6* (Suppl.): 8, 1970.

1602 - DANCS, Zs., POZSÁR, B.I., FERENCZ, V.: Inhibition of CO_2 fixation and protein synthesis in apple leaves infected by *Venturia inaequalis*. - Acta agron. Acad. Sci. hung. *17*: 405-410, 1968.

1603 - DANIEL, P.: Chlorophyll und Carotin in Futterpflanzen. Problematik ihrer Bestimmung. - Wirtschaftseigene Futter *12*: 346-361, 1966.

1604 - DANIELL, J.W., CHAPPELL, W.E., COUCH, H.B.: Effect of sublethal and lethal temperatures on plant cells. - Plant Physiol. *44*: 1684-1689, 1969. [Chloroplasts.]

1605 - DARKANBAEV, T.B., SACHKOVA, O.P.: Vliyanie Mo i Mn na fiksatsiyu azota i fotosintez nekotorykh sinezelenykh vodoroslei. [Effect of Mo and Mn on nitrogen fixation and photosynthesis of some blue-green algae.] - Izv. Akad. Nauk Kaz. SSR, Ser. biol. *1970 (1)*: 26-29, 1970. [In R.]

1606 - DAS, B.C., LOUNASMAA, M., TENDILLE, C., LEDERER, E.: The structure of the plastoquinones *B* and *C*. - Biochem. biophys. Res. Commun. *26*: 211-215, 1967.

1607 - DAS, M., GOVINDJEE: A long-wave absorbing form of chlorophyll *a* responsible for the " red drop" in fluorescence at 298°K and the F_{723} band at 77°K. - Biochim. biophys. Acta *143*: 570-576, 1967.

1608 - DAS, M., RABINOWITCH, E., SZALAY, L.: Red drop in the quantum yield of fluorescence of sonicated algae. - Biophys. J. *8*: 1131-1137, 1968.

1609 - DAS, M., RABINOWITCH, E., SZALAY, L., PAPAGEORGIOU, G.: The "sieve effect" in *Chlorella* suspensions. - J. phys. Chem. *71*: 3543-3549, 1967.

1610 - DAS, O.A.K., GUPTA, A.B.: The effect of carbon dioxide concentration and light intensity on growth rate of *Scenedesmus obliquus*. - Labdev J. Sci. Technol. B *8*: 233-236, 1970. [Ps.]

1611 - DAS, V.S.R.: Photochemical events and energy conversion in photosynthesis. - Madras agr. J. *53*: 76-80, 1966.

1612 - DAS, V.S.R., REDDY, M.S.: Hill reaction activity of pepper fruit chloroplasts. - Curr. Sci. *36*: 272-273, 1967.

1613 - DASSIOU, C., AKOYUNOGLOU, G.: Effect of green and white flashes on the activity of the enzyme ribulose-diphosphate carboxylase in dark-grown barley plants. - Physiol. Plant. *22*: 570-574, 1969.

1614 - DASSIOU, C., AKOYUNOGLOU, G.: Effect of light on the RuDP-carboxylase activity in etiolated plants of different ages. - In: METZNER, H. (ed.): Progress in Photosynthesis Research. Vol. III. Pp. 1631-1635. Tübingen 1969.

1615 - DAUDA, K., HORVÁTH, M., VINCZE, H.: Pigment changes in bean and broad-bean. - Acta agron. Acad. Sci. hung. *17*: 445-448, 1968.

1616 - DAUKSHTA, A.S., VAINSHTEÏN, M.B.: Sravnenie fotosinteza fitoplanktona v razlichnykh ozerakh Latvii. [Comparison of phytoplankton photosynthesis in some lakes of Lithuania.] - Biol. vnutr. Vod, inf. Byull. *4*: 11-15, 1969. [In R.]

1617 - DAVENPORT, H.E., DODGE, A.D.: The effect of ultrasonic and heat treatment on some chloroplast reactions. - Phytochemistry *8*: 1849-1857, 1969.

1618 - DAVEY, E.W.: The photosynthetic and respiratory physiology of *Skeletonema costatum* (GREVILLE) CLEVE grown under simulated environmental conditions. - Diss. Abstr. int. B *3*: 4683-B, 1970.

1619 - DAVIDSON, J.L., MILTHORPE, F.L.: The effect of defoliation on the carbon balance in *Dactylis glomerata*. - Ann. Bot. *30*: 185-198, 1966. [Ps.]

1620 - DAVIES, A.G.: Iron, chelation and the growth of marine phytoplankton. I. Growth kinetics and chlorophyll production in cultures of the euryhaline flagellate *Dunaliella tertiolecta* under iron-limiting conditions. - J. mar. Biol. Ass. UK *50*: 65-86, 1970.

1621 - DAVIES, B.H.: The identification of carotenoids by their adsorption characteristics. - Biochem. J. *103*: 51 P-52 P, 1967.

1622 - DAVIES, B.H.: Alternative pathways of spirilloxanthin biosynthesis in *Rhodospirillum rubrum*. - Biochem. J. *116*: 101-110, 1970.

1623 - DAVIES, B.H., HOLMES, E.A.: The dehydrogenation sequence leading to coloured carotenoids in photosynthetic bacteria. - Biochem. J. *113*: 33 P-34 P, 1969.

1624 - DAVIES, B.H., HOLMES, E.A.: New pathways of methoxy carotenoid formation in *Rhodospirillum rubrum*. - Biochem. J. *113*: 34 P-35 P, 1969.

1625 - DAVIES, B.H., MATTHEWS, S., KIRK, J.T.O.: The nature and biosynthesis of the carotenoids of different colour varieties of *Capsicum annuum*. - Phytochemistry *9*: 797-805, 1970.

1626 - DAVIES, D.H.: A review of non-gravimetric methods of determining carbon dioxide and monoxide. - Talanta *16*: 1055-1065, 1969.

1627 - DAVIS, J.J., WILLARD,J.M., WOOD, H.G.: Phosphoenolpyruvate carboxytransphosphorylase. III. Comparison of the fixation of carbon dioxide and the conversion of phosphoenolpyruvate and phosphate into pyruvate and pyrophosphate. - Biochemistry *8*: 3127-3136, 1969.

1628 - DAVIS, R.G., ROBERSON, G.M., JOHNSON, W.C., WIESE, A.F.: A modified optical planimeter for measuring leaf area. - Agron. J. *58*: 106-107, 1966.

1629 - DAVIS, R.M. Jr., FILDES, R., BAKER, G.A., ZAHARA, M., MAY, D.M., TYLER, K.B.: A method of estimating the potential yield of a vegetable crop obtainable in a single harvest. - J. amer. Soc. hort. Sci. *95*: 475-480, 1970.

1630 - DAY, P.R., ZELITCH, I.: Can we make plants with more efficient photosynthesis? - Advanc. Frontiers Plant Sci. *20*: 4-5, 1968.

1631 - DAYNARD, T.B., HUNTER, R.B., TANNER, J.W., KANNENBERG, L.W.: An inexpensive space-integrating light meter for field crop research. - Can. J. Plant Sci. *49*: 231-234, 1969.

1632 - DEAMER, D.W., CROFTS, A.: Action of Triton X-100 on chloroplast membranes. Mechanisms of structural and functional disruption. - J. Cell Biol. *33*: 395-410, 1967.

1633 - DEAMER, D.W., CROFTS, A.R., PACKER, L.: Mechanisms of light-induced structural changes in chloroplasts. I. Light-scattering increments and ultrastructural changes mediated by proton transport. - Biochim. biophys. Acta *131*: 81-96, 1967.

1634 - DEAMER, D.W., PACKER, L.: Evidence for a H^+ ion mediated contractile system in chloroplasts. - Fed. Proc. *25*: 225, 1966.

1635 - DEAMER, D.W., PACKER, L.: Correlation of ultrastructure with light-induced ion transport in chloroplasts. - Arch. Biochem. Biophys. *119*: 83-97, 1967.

1636 - DEAMER, D.W., PACKER, L.: Light-dependent anion transport in isolated spinach chloroplasts. - Biochim. biophys. Acta *172*: 539-545, 1969.

1637 - DECHEV, G., TOMOVA, N., SECHENSKA, M.: On the possible enzymatic pathways of the electron to the molecular oxygen in electrontransport reactions of photosynthesis. - Dokl. bolg. Akad. Nauk *19*: 531-534, 1966.

1638 - DECKER, P.: Inverse assimilation and dissimilation, possible metabolic patterns in a methane and hydrogen containing atmosphere. - In: METZNER, H. (ed.): Progress in Photosynthesis Research. Vol. I. Pp. 458-463. Tübingen 1969.

1639 - DECKER, W.L.: The energy balance of a plant cover under high levels of productivity. - In: ECKARDT, F.E. (ed.): Functioning of Terrestrial Ecosystems at the Primary Production Level. Pp. 375-379. UNESCO, Paris 1968.

1640 - DECLEIRE, M.: Etude du reverdissement obtenu lors de la sterilisation de haricots prealablement conserves en saumure. - Rev. Ferment. Ind. aliment. *1966-67* *(21-22)*: 95-98, 1966-1967. [Chl derivatives.]

1641 - DECLEIRE, M.: Influence du traitement thermique sur le comportement des chloroplastes et de ses constituants. - Bull. INACOL *21*: 162-177, 1970.

1642 - DEGANI, H., SHAVIT, N.: Ion translocation and the contribution of a membrane potential to ATP formation in chloroplasts. - Israel J. Chem. *8*: 162 p, 1970.

1643 - DEHNER, T.R., CHAN, H.W.-S, CAPLE, M.B., CALVIN, M.: Tritium incorporation studies in photosynthetic bacteria. - Proc. nat. Acad. Sci. U.S.A. *58*: 1556-1559, 1967.

1644 - DEKOV, I., IGNATOV, G., DECHEV, G., STOYANOVA, E.: Sredy dlya podderzhivaniya maksimal'nogo fotofosforilirovaniya v khloroplastakh *Vicia faba*. [Media maintaining maximal photophosphorylation rate in chloroplasts of *Vicia faba* L.] - Fiziol. Rast. *17*: 871-876, 1970. [In R, ab: E.]

1645 - DELEENS, E., MOUSSEAU, M.: Relation entre le cycle journalier des échanges de CO_2 et la photorespiration chez le *Bryophyllum daigremontianum* BERGER (Crassulacée). - Compt. rend. Acad. Sci. Paris, Sér. D *271*: 1372-1375, 1970.

1646 - DELIEU, T., WALKER, D.A.: An illuminated constant temperature water bath for the study of photochemical reactions in biological systems. - Anal. Biochem. *16*: 160-166, 1966.

1647 - DELOSME, R.: Etude de l'induction de fluorescence des algues vertes et des chloroplastes au début d'une illumination intense. - Biochim. biophys. Acta *143*: 108-128, 1967.

1648 - DELRIEU, M.J.: Experimental and theoretical study of photosynthetic quantum yields in *Chlorella*. - In: METZNER, H. (ed.): Progress in Photosynthesis Research. Vol. II. Pp. 1110-1121. Tübingen 1969.

1649 - DELWICHE, C.C.: Energy relationships in soil biochemistry. - In: Soil Biochemistry. Pp. 173-193. M. Dekker, Inc., New York 1967. [Ps.]

1650 - DEMINA, N.S., IVANOV, I.D., MEDVEDEV, G.A.: O vzaimosvyazi protsessov azotfiksatsii, vydeleniya molekulyarnogo vodoroda i fosforilirovaniya u fotosinteziruyushchikh bakteriĭ. [Relation of processes of nitrogen fixation, molecular hydrogen production and phosphorylation in photosynthetic bacteria.] - Mikrobiologiya *38*: 428-431, 1969. [In R.]

1651 - DENCHEVA, A., VAKLINOVA, S.: The effect of the different nitrogen sources on photosynthesis products in *Scenedesmus obliquus*. - Dokl. bolg. Akad. Nauk *22*: 1079-1082, 1969.

1652 - DENCHEVA, A.V., RACHINSKIĬ, V.V.: Dinamika nakopleniya mechenogo ugleroda v sakharakh, aminokislotakh i organicheskikh kislotakh u rasteniĭ razlichnykh vidov v protsesse fotosinteza. [Dynamics of accumulation of labelled carbon in sugars, amino acids and organic acids in different plant species during photosynthesis.] - Izv. Timiryaz. sel'.-khoz. Akad. *1967* (2): 3-18, 1967. [In R, ab: E.]

1653 - DENISOVA, R.R., RYBAKOVA, M.I.: O izmenenii prochnosti svyazi pigmentov s belkom u ozimykh v protsesse perezimovki. [Changes in stability of pigment-protein complex in winter plants during wintering.] - Nauch. Tr. NII sel'.-khoz. Tsentr. Raĭonov nechernozemn. Zony *25* (2): 40-45, 1970. [In R.]

1654 - DENMEAD, O.T.: Comparative micrometeorology of a wheat field and a forest of *Pinus radiata*. - Agr. Meteorol. *6*: 357-371, 1969.

1655 - DENMEAD, O.T.: Transfer processes between vegetation and air: Measurement, interpretation and modelling. In: Prediction and Measurement of Photosynthetic Productivity. Pp. 149-164. PUDOC, Wageningen 1970.

1656 - DENNY, P., WEEKS, D.C.: Effects of light and bicarbonate on membrane potential in *Potamogeton schweinfurthii* (BENN.). - Ann. Bot. *34*: 483-496, 1970. [Ps.]

1657 - DENTICE DI ACCADIA, F., GRIBANOVSKI-SASSU, O., REYES LOZANO, N.: Blushing effect of some carbohydrate on the green alga *Dictyococcus cinnabarinus*. - Experientia *24*: 1177-1179, 1968.

1658 - DENTICE DI ACCADIA, F., GRIBANOVSKI-SASSU, O., ROMAGNOLI, A., TUTTOBELLO, L.: Isolation and identification of carotenoids produced by a green alga *(Dictyococcus cinnabarinus)* in submerged culture. - Biochem. J. *101*: 735-740, 1966.

1659 - DENTON, T.E.: The effect of a strontium replacement for calcium on carbohydrates and other products of photosynthesis in *Protosiphon botryoides*. - Diss. Abstr. B *27*: 590-B, 1966.

1660 - DERBY, S.B., RUBER, E.: Primary production: depression of oxygen evolution in algal cultures by organophosphorus insecticides. - Bull. Environm. Contam. Toxicol. *5*: 553-558, 1970.

1661 - DE REZENDE PINTO, M.C., CORTESAO, L.: On the structure of chloroplasts in *Phaeoceros laevis* (L.) PROSK. - Port. Acta biol., Ser. A *11*: 310-317, 1969.

1662 - DEROCHE, M.-E.: Extraction ménagée de chlorophylles à partir de chloroplastes lyophilisés. - Ann. Physiol. vég. *9*: 233-239, 1967.

1663 - DEROCHE, M.-E.: Sélection des molécules de chlorophylle *a* néoformées. - Bull. Soc. franç. Physiol. vég. *14*: 459-472, 1968.

1664 - DEROCHE, M.-E.: Hétérogénéité des pigments liposolubles des chloroplastes de Blé. - Physiol. vég. *7*: 335-389, 1969.

1665 - DEROCHE, M.-E.: Les protéines de structure des chloroplastes. - Ann. Physiol. vég. *11*: 441-461, 1969.

1666 - DEROCHE, M.-E., COSTES, C.: Hétérogénéité des chlorophylles. - Ann. Physiol. vég. *8*: 223-254, 1966.

1667 - DEROCHE, M.-E., COSTES, C.: Hétérogénéité du β-carotène dans les chloroplastes. - Ann. Physiol. vég. *8*: 259-269, 1966.

1668 - DEROCHE, M.-E., COSTES, C.: Heterogeneity of carotenoids in chloroplasts. - In: METZNER, H. (ed.): Progress in Photosynthesis Research. Vol. II. pp. 681-693, Tübingen 1969.

1669 - DEROCHE, M.-E., DUFOUR, P.: Etude de l'extinction de la scintillation due aux chlorophylles. - Bull. Soc. franç. Physiol. vég. *15*: 71-75, 1969.

1670 - DESCOMPS, S.: Effets de l'éclairement continue sur l'infrastructure des chloroplastes de *Vaucheria sessilis* (Xanthophycées). - Compt. rend. Acad. Sci. Paris, Sér. D *269*: 1955-1958, 1969.

1671 - DESIATI, L.: Il ruolo dell'idrogeno sul biochimismo della sintesi clorofilliana. [Role of hydrogen in biochemistry of chlorophyll synthesis.] - Corriere farm. *21*: 34-35, 1966. [In Ital.]

1672 - DESMORAS, J., GANTER, P., JAQUET, P., LAURENT, M.: Etudes sur le mode d'action du phénylcarbamoyloxy-2-N-éthylpropionamide, isomère D. Application au dosage des résidus dans les plantes. - Weed Res. *7*: 261-271, 1967. [Hill reaction.]

1673 - DESPAIN, D.G., BLISS, L.C., BOYER, J.S.: Carbon dioxide exchange in saguaro seedlings. - Ecology *51*: 912-914, 1970.

1674 - DESSUREAUX, L.: Four new genes responsible for chlorophyll deficiency in young alfalfa seedlings. - Can. J. Gen. Cytol. *9*: 408-411, 1967.

1675 - DETCHEV, G., SETCHENSKA, M., TOMOVA, N.: On the possible enzymatic pathways of the electron to the molecular oxygen in electron transport reactions of photosynthesis. - In: METZNER, H. (ed.): Progress in Photosynthesis Research. Vol. I. PP. 503-507. Tübingen 1969.

B1676 - Determination of Photosynthetic Pigments in Sea-Water. - Monographs on Oceanographic Methodology 1. UNESCO, Paris 1966.

1677 - DEUSER, W.G., DEGENS, E.T.: Carbon isotope fractionation in the system CO_2 (gas) - CO_2 (aqueous) - HCO_3^- (aqueous). - Nature *215*: 1033-1035, 1967.

1678 - DEVIDÉ, Z.: The effect of ionizing radiation and of ultraviolet light on developing plastids in primary leaves of etiolated bean seedlings. - In: Photochemistry and Photobiology in Plant Physiology. Europe. Photobiol. Symp. Hvar, Yugoslavia, 19th-22nd September 1967. Book of Abstracts. Pp. 9-12. Hvar 1967.

1679 - DEVIDÉ, Z.: Veränderungen im Chloroplasten-Feinbau nach Röntgenbestrahlung etiolierter Bohnenblätter. - Österreich. bot. Z. *116*: 444-453, 1969.

1680 - DEVIDÉ, Z.: Ultrastructural changes of plastids in ripe fruit of *Cucurbita pepo* cv. *ovifera*. - Acta bot. croat. *29*: 57-62, 1970.

1681 - DEVIDÉ, Z., WRISCHER, M.: Über den Einfluss von Atmungsstörungen auf die Diffe-
renzierung der Plastiden im Blattgewebe etiolierter Bohnenkeimlinge. - Z. Naturforsch. *22 b*: 447-450, 1967.

1682 - DEZHI, L., FARKASH, G.L.: Korrelatsiya mezhu aktivnost'yu oksidazy glikolevoĭ
kisloty i soderzhaniem khlorofilla v list'yakh yachmenya. [Correlation between
activity of glycolic acid oxidase and chlorophyll content in barley leaves.] -
Fiziol. Rast. *16*: 343-344, 1969. [In R.]

1683 - DIAMOND, J., SCHIFF, J.A.: Insensitivity of development of photosynthetic competence to doses of ultraviolet light which block green colony formation in *Euglena*. - Can. J. Bot. *48*: 1277-1283, 1970.

1684 - DIBROVA, V.S.: Deĭstvie tsinkovykh udobreniĭ na biokhimicheskie osobennosti kukuruzy pri razlichnom urovne obespechennosti tsinkom. [Effect of zinc fertilizers on biochemical properties of maize under various zinc supply.] - Fiziol.
Rast. *14*: 670-674, 1967. [In R, ab: E.]

1685 - DICKMANN, D.I.: Photosynthesis and respiration by developing leaves of *Populus
deltoides*. - Plant Physiol. *46* (Suppl.): 8, 1970.

1686 - DICKMANN, D.I., KOZLOWSKI, T.T.: Mobilization by *Pinus resinosa* cones and
shoots of C^{14}-photosynthate from needles of different ages. - Amer. J. Bot. *55*:
900-906, 1968.

1687 - DICKMANN, D.I., KOZLOWSKI, T.T.: Mobilization and incorporation of photoassimilated ^{14}C by growing vegetative and reproductive tissues of adult *Pinus resinosa* AIT. trees. - Plant Physiol. *45*: 284-288, 1970.

1688 - DICKMANN, D.I., KOZLOWSKI, T.T.: Photosynthesis by rapidly expanding green stro-
bili of *Pinus resinosa*. - Life Sci. II *9*: 549-552, 1970.

1689 - van DIE, J., LEEUWANGH, P., HOEKSTRA, S.M.R.: Translocation of assimilates in
Fritillaria imperialis L. 1. The secretion of ^{14}C-labelled sugars by the nectar-
ies in relation to phyllotaxis. - Acta bot. neerl. *19*: 16-23, 1970.

1690 - DIEHN, B.: Two perpendicularly oriented pigment systems involved in phototaxis
of *Euglena*. - Nature *221*: 366-367, 1969.

1691 - DIEHN, B., SEELY, G.R.: The oxidation of chlorophyll *a* in alcohols. - Biochim.
biophys. Acta *153*: 862-867, 1968.

1692 - DIEHN, B., TOLLIN, G.: Phototaxis in *Euglena*. IV. Effect of inhibitors of oxidative and photophosphorylation on the rate of phototaxis. - Arch. Biochem. Biophys. *121*: 169-177, 1967.

1693 - DIERS, L.: On the plastids, mitochondria, and other cell constituents during
oögenesis of a plant. - J. Cell Biol. *28*: 527-543, 1966.

1694 - DIERS, L.: Übertragung von Plastiden durch den Pollen bei *Antirrhinum majus*. -
Mol. gen. Genet. *100*: 56-62, 1967.

1695 - DIERS, L.: On the behaviour of the plastids during the development of the arche-
gonium and the egg cell in an archegoniate plant. - In: SIRONVAL, C. (ed.):
Le Chloroplaste, Croissance et Vieillissement. - Pp. 306-312. Masson, Paris 1967.

1696 - DIERS, L.: Origin of plastids: Cytological results and interpretations including
some genetical aspects. - In: Symposia of the Society for Experimental Biology.
XXIV: Control of Organelle Development. Pp. 129-145. Univ. Press, Cambridge
1970.

1697 - DIERS, L., SCHÖTZ, F.: Über die dreidimensionale Gestaltung des Thylakoidsystems
in den Chloroplasten. - Planta *70*: 322-343, 1966.

1698 - DIERS, L., SCHÖTZ, F.: Räumliche Beziehungen zwischen osmiophilen Granula und
Thylakoiden. - Z. Pflanzenphysiol. *58*: 252-265, 1968.

1699 - DIERS, L., SCHÖTZ, F.: Über ring- und schalenförmige Thylakoidbildungen in den
Plastiden. - Z. Pflanzenphysiol. *60*: 187-210, 1969.

1700 - DIERS, L., SCHÖTZ, F., BATHELT, H.: Zur Frage des Aufbaus von Thylakoiden aus

Vesikeln. - Planta *80*: 211-226, 1968.

1701 - DILCHER, D.L., PAVLICK, R.J., MITCHELL, J.: Chlorophyll derivatives in middle eocene sediments. - Science *168*: 1447-1449, 1970.

1702 - DILLEY, R.A.: Ion and water transport processes in spinach chloroplasts. - In: Energy Conversion by the Photosynthetic Apparatus. Brookhaven Symp. Biol. *19*: 258-280, 1967.

1703 - DILLEY, R.A.: Effect of poly-L-lysine on energy-linked chloroplast reactions. - Biochemistry *7*: 338-346, 1968.

1704 - DILLEY, R.A.: Observations on the structure of *Rhodopseudomonas viridis* as shown by freeze-etch. - In: METZNER, H. (ed.): Progress in Photosynthesis Research. Vol. I. Pp. 197-103. Tübingen 1969.

1705 - DILLEY, R.A.: Evidence for the requirement of H^+ ion transport in photophosphorylation by spinach chloroplast. - In: METZNER, H. (ed.): Progress in Photosynthesis Research. Vol. III. Pp. 1354-1360. Tübingen 1969.

1706 - DILLEY, R.A.: The effect of various energy-conversion states of chloroplasts on proton and electron transport. - Arch. Biochem. Biophys. *137*: 270-283, 1970.

1707 - DILLEY, R.A., PARK, R.B., BRANTON, D.: Ultrastructural studies of the light-induced chloroplast shrinkage. - Photochem. Photobiol. *6*: 407-412, 1967.

1708 - DILLEY, R.A., ROTHSTEIN, A.: Chloroplast membrane characteristics. - Biochim. Biophys. Acta *135*: 427-443, 1967.

1709 - DILLEY, R.A., SHAVIT, N.: On the relationship of H^+ transport to photophosphorylation in spinach chloroplasts. - Biochim. biophys. Acta *162*: 86-96, 1968.

1710 - DILLEY, R.A., VERNON, L.P.: Quantum requirement of the light-induced proton uptake by spinach chloroplasts. - Proc. nat. Acad. Sci. U.S.A. *57*: 395-400, 1967.

1711 - DILOVA, S., POPOV, K.: L'effet de l'obscurcissement préliminaire sur l'état des chlorophylles et l'activité photosynthétique des feuilles. - Stud. biophys. *5*: 97-104, 1967.

1712 - DILOVA, S., POPOV, K.: The influence of ascorbic acid on the dynamics of chlorophyll synthesis in barley plants in darkness. - Dokl. bolg. Akad. Nauk *23*: 735-738, 1970.

1713 - DILUNG, I., CHERNYUK, I.: Investigation of the reversible photoreactions of chlorophyll. - Abh. deut. Akad. Wiss. Berlin, Kl. Med. *1966*: 325-329, 1966.

1714 - DILUNG, I.I., IVNITSKAYA, I.N., KRASNOVA, V.A.: Obratimoe fotookislenie khlorofilla i ego rol' v protsessakh sensibilizatsii. [Reversible photooxidation of chlorophyll and its role in sensibilization processes.] - Fiziol. Biokhim. kul't. Rast. *1*: 197-201, 1969. [In R, ab: E.]

1715 - DILUNG, J.J.: The reversible photooxidation of chlorophyll and its role for the processes of sensibilization. - In: METZNER, H. (ed.): Progress in Photosynthesis Research. Vol. II. Pp. 746-749. Tübingen 1969.

1716 - DIMITROV, Kh.: V"rkhu dnevniya i sezonen khod na fotosintezata pri nyakoi gorsko-d"rvesni vidove. [Daily and seasonal course of photosynthesis in some mountain woody species.] - Gorkostop. Nauka (Sofia) *5 (3)*: 3-11, 1968. [In Bulg., ab: E, R.]

1717 - DION, M., NIGON, V.: Influence de la lumière sur la formation de chlorophylle et sur la multiplication cellulaire chez des populations étiolées d'*Euglena gracilis*. - Compt. rend. Acad. Sci. Paris, Sér. D *265*: 481-484, 1967.

1718 - DITMARS, W.E. Jr., VAN WINKLE, Q.: A comparative study of pheophytin a and pheophytin b monolayers. - J. phys. Chem. *72*: 39-45, 1968.

1719 - DJAVANCHIR, A.: Mise au point d'une chambre de transpiration pour mesurer la résistance stomatique. - Oecol. Plant. *5*: 301-318, 1970.

1720 - DJENDOV, C.: Încercări de mărire a rezistenței la săruri a plantelor de porumb și de sorg. [Increase of salt-resistance of corn and sorghum plants.] - Stud. Cercet. Biol., Ser. Bot. *18*: 471-475, 1966. [Ps, Chl; in Rum.]

1721 - van DOBBEN, W.H.: Physiology of growth in two *Senecio* species in relation to their ecological position. - Inst. Biol. Scheik. Onderz. Landbouwgewassen Wageningen Jaarb. *1967*: 75-83, 1967. [Ps.]

1722 - DOBBERSTEIN, R.H., STABA, E.J.: Chlorophyll production in Japanese mint suspension cultures. - Lloydia *29*: 50-57, 1966.

1723 - DOBBY, G.: Photosynthese-Bedingungen alles Lebens. - Scientia *103*: 26-35, 1968 [G.]; *103* (Suppl.): 17-24, 1968. [F.]

1724 - DÖBEL, P., SAGROMSKY, H.: Abhängigkeit der Chloroplastenstruktur und des Pigmentgehalts vom Nährstoffangebot bei der Tomatenmutante *venosa*. - Kulturpflanze *14*: 293-297, 1966.

1725 - DOBROSZ, M., GYESKŐ, A., H. NAGY, A., FALUDI-DÁNIEL, Á.: Hőmérsékleti viszonyok hatása mutáns kukorica csíranövények karotinoid-szintézisét kontrolláló gének expresszivitására. [The effect of temperature on the expressivity of genes controlling carotinoid-synthesis of mutant maize seedlings.] - Biol. Közlem. *15*: 3-9, 1967. [In Hung., ab: E, R.]

1726 - DOBRUNOV, L.G., CHUMINA, O.T., VERSHININA, K.T.: Izmeneniya fiziologo-biokhimicheskikh protsessov u kul'turnykh rasteniĭ pod vliyaniem razlichnogo rezhima mineral'nogo pitaniya. [Changes of physiological and biochemical processes of cultivated plants under varying mineral nutrition.] - In: Trudy I. Konferentsii Biokhimikov Respublik Sredneĭ Azii i Kazakhstana. Pp. 141-144. [In R.]

1727 - DODGE, A.D., WHITTINGHAM, C.P.: Photochemical activity of chloroplasts isolated from etiolated plants. - Ann. Bot. *30*: 711-719, 1966.

1728 - DODGE, J.D.: The fine structure of chloroplasts and pyrenoids in some marine dinoflagellates. - J. Cell Sci. *3*: 41-48, 1968.

1729 - DODGE, J.D.: Changes in chloroplast fine structure during the autumnal senescence of *Betula* leaves. - Ann. Bot. *34*: 817-824, 1970.

1730 - DOEMEL, W.N., BROCK, T.D.: The upper temperature limit of *Cyanidium caldarium*. - Arch. Mikrobiol. *72*: 326-332, 1970. [Ps.]

1731 - DÖHLER, G.: Transients in photosynthetic CO_2 uptake of *Chlorella* and *Anacystis*. - In: METZNER, H. (ed.): Progress in Photosynthesis Research. Vol. I. Pp. 531-535. Tübingen 1969.

1732 - DÖHLER, G., PRZYBYLLA, K.-R.: Untersuchung der Beziehung zwischen Pigmentzusammensetzung und CO_2-Gaswechsel bei der Blaualge *Anacystis nidulans*. - Planta *90*: 163-173, 1970.

1733 - DÖHLER, G., WEGMANN, K.: Untersuchungen zur Photosynthese-Induktion bei *Chlorella* mit Hilfe von radioaktivem CO_2. - Planta *89*: 266-274, 1969.

1734 - DOKE, N., HIRAI, T.: Effects of tobacco mosaic virus infection on photosynthetic CO_2 fixation and $^{14}CO_2$ incorporation into protein in tobacco leaves. - Virology *42*: 68-77, 1970.

1735 - DOLGOV, S.I., SHMIDT, G.N.: Analiz korrelyatsionnoĭ svyazi mezhdu urozhayami sel'skokhoyaĭstvennykh kul'tur i solnechnoĭ aktivnost'yu. [Analysis of correlation between crop yield and solar activity.] - Vestn. sel'.-khoz. Nauki *13* (*11*): 88-91, 1968. [In R, ab: E, F, G.]

1736 - DOLL, H.: Yield and variability of chlorophyll-mutant heterozygotes in barley. - Hereditas *56*: 255-276, 1966.

1737 - DOLPHIN, W.D.: Photoinduced carotenogenesis in chlorotic *Euglena gracilis*. - Plant Physiol. *46*: 685-691, 1970.

1738 - DOLZMANN, P.: Photosynthese-Reaktionen einiger Plastom-Mutanten von *Oenothera*. III. Strukturelle Aspekte. - Z. Pflanzenphysiol. *58*: 300-309, 1968.

1739 - DOLZMANN, P., ULLRICH, H.: Einige Beobachtungen über Beziehungen zwischen Chloroplasten und Mitochondrien im Palisadenparenchym von *Phaseolus vulgaris*. - Z. Pflanzenphysiol. *55*: 165-180, 1966.

1740 - DOMAN, N.G.: Biokhimicheskie reaktsii fotosinteza. [Biochemical reactions of photosynthesis.] - In: Fiziologiya Sel'skokhozyaĭstvennykh Rasteniĭ. Vol. 1. Pp. 207-266. Izdat. moskovskogo Univ., Moskva 1967. [In R.]

1741 - DOMAN, N.G.: Zavisimost'produktov fotosinteza ot usloviĭ osveshcheniya rasteniĭ. [Dependence of photosynthates on illuminance conditions of plants.] - In: Fiziologiya i Biokhimiya Zdorovogo i Bol'nogo Rasteniya. Pp. 293-300. Izdat. moskovskogo Univ., Moskva 1970. [In R.]

1742 - DOMAN, N.G., SHKOL'NIK, R.Ya., SPEKTOROV, K.S.: Produkty fotosinteticheskoĭ assimilyatsii $C^{14}O_2$ sinkhronnoĭ kul'turoĭ *Chlorella pyrenoidosa* na raznykh stadiyakh ee razvitiya. [Products of photosynthetic $^{14}CO_2$ assimilation in a synchronous culture of *Chlorella pyrenoidosa* in different phases of its development.] - Tr. mosk. Obshch. Ispyt. Prirody 24 (Biol. avtotrofnykh Mikroorganizmov): 68-80, 1966. [In R.]

1743 - DOMAN, N.G., SIMISKER, Ya.A.: Pathways of carbon photoassimilation by *Chloropseudomonas ethylicum*. - In: METZNER, H. (ed.): Progress in Photosynthesis Research. Vol. III. Pp. 1574-1578. Tübingen 1969.

1744 - DOMES, W., BERTSCH, A.: CO_2-Gaswechsel amphistomatischer Blätter. 2. Ein Vergleich von diffusivem CO_2 Austausch der beiden Blattepidermen von *Zea mays* mit dem im Porometer gemessenen viscosen Volumfluss. - Planta 86: 84-91, 1969.

1745 - DONNAT, P., BRIANTAIS, J.-M.: Variation du taux de manganèse induite par la lumière dans l'appareil photosynthétique. - Compt. rend. Acad. Sci. Paris, Sér. D 265: 21-24, 1967.

1746 - DONNAT, P., BRIANTAIS, J.-M.: Contenu et état du manganèse dans l'appareil photosynthétique. - Compt. rend. Acad. Sci. Paris, Sér. D 264: 2903-2906, 1967.

1747 - DONOHUE, H.V.: The interconversion of the oxygenated carotenoids. - Diss. Abstr. B 28: 470-B-471-B, 1967.

1748 - DONOHUE, H.V., NAKAYAMA, T.O.M., CHICHESTER, C.O.: Oxygen reactions of xanthophylls. - In: GOODWIN, T.W. (ed.): Biochemistry of Chloroplasts. Vol. II. Pp. 431-440, Academic Press, London-New York 1967.

1749 - DONZE, M., DUYSENS, L.N.M.: Forms of reaction center II as deduced from low temperature fluorescence changes. - In: METZNER, H. (ed.): Progress in Photosynthesis Research. Vol. II. Pp. 991-995. Tübingen 1969.

1750 - DÖRFLING, P., DUMMLER, W., MÜCKE, D.: Das Auftreten von Koproporphyrin in Kulturen von *Poteriochromonas stipitata* nach Inkubation mit 3-Amino-1,2,4-triazol (Amitrol). - Experientia 26: 728, 1970. [Chl.]

1751 - DÖRFLING, P., JONAS, L., DUMMLER, W., MÜCKE, D.: Veränderungen der Chloroplastenstruktur von *Poteriochromonas stipitata* nach Inkubation mit verschiedenen Amitrolkonzentrationen. - Z. allgem. Mikrobiol. 10: 363-365,1970.

1752 - DÖRING, G., BAILEY, J.L., KREUTZ, W., WEIKARD, J., WITT, H.T., VATER, J., RENGER, G., STIEHL, H.H., REINWALD, E., SIGGEL, U., RUMBERG, B., SEIFERT, K.: Some new results in photosynthesis. - Naturwissenschaften 55: 219-224, 1968.

1753 - DÖRING, G., BAILEY, J.L., WEIKARD, J., WITT, H.T.: The action of two chlorophyll-a_I-molecules in light reaction I of photosynthesis. - Naturwissenschaften 55: 219-220, 1968.

1754 - DÖRING, G., RENGER, G., VATER, J., WITT, H.T.: Properties of the photoactive chlorophyll-a_{II} in photosynthesis. - Z. Naturforsch. 24 B: 1139-1143, 1969.

1755 - DÖRING, G., STEIHL, H.H., WITT, H.T.: A second chlorophyll reaction in the electron chain of photosynthesis - Registration by the repetitive excitation technique. - Z. Naturforsch. 22 B: 639-644, 1967.

1756 - DORNHOFF, G.M., SHIBLES, R.M.: Varietal differences in net photosynthesis of soybean leaves. - Crop. Sci. 10: 42-45, 1970.

1757 - DOROKHOV, B.L.: Fotosintez. [Photosynthesis.] - In: Fiziologiya Sel'skokhozyaĭstvennykh Rasteniĭ. Vol. 10. Pp. 177-211. Izdat. moskov. Univ., Moskva 1968. [In R.]

1758 - DOROKHOV, B.L.: Izuchenie fotosinteza. [Studying photosynthesis.] - In: Issle-

dovaniya po Fiziologii i Biokhimii Rastenii v Moldavskoi SSR. Pp. 30-41. Akad.
Nauk Mold. SSR, Kishinev 1968. [In R.]

1759 - DOROKHOV, B.L.: Sortovaya spetsifika izmeneniya fotosinteticheskoi deyatel'nosti rastenii pri ikh razlichnom opylenii. [Cultivar specificity of photosynthetic activity of plants at various pollination.] - In: Fiziologiya i Biokhimiya
Sorta. Vol. I. Pp. 107-112. Irkutsk 1969. [In R.]

1760 - DOROKHOV, B.L., BARANINA, I.I.: Intensivnost' fotosinteza kolosa sovmestno s
verkhnei chast'yu steblya u pshenitsy. [Photosynthetic rate of ear together with
the upper part of stalk in wheat.] - In: Puti Povysheniya Intensivnosti i
Produktivnosti Fotosinteza. Vol. 3. Pp. 96-101. Naukova Dumka, Kiev 1969.
[In R.]

1761 - DOROKHOV, B.L., BARANINA, I.I.: Soderzhanie pigmentov v vegetativnykh i repro-
duktivnykh organakh ozimoi pshenitsy pri razlichnom mineral'nom pitanii. [Pig-
ment content in vegetative and reproductive organs of winter wheat under different mineral nutrition.] - In: Fotosinteticheskaya Deyatel'nost' Rastenii i
Vliyanie na nee Mineral'nogo Pitaniya. Pp. 3-38. Kishinev 1970. [In R.]

1762 - DOROKHOV, B.L., BARANINA, I.I.: Zimnii gazoobmen ozimoi pshenitsy pri razlichnom
mineral'nom pitanii. [Winter gas exchange of winter wheat under different mineral nutrition.] - In: Fotosintez i Pigmenty Osnovnykh Sel'sko-khozyaistvennykh
Rastenii Moldavii. Pp. 3-20. Kishinev 1970. [In R.]

1763 - DOROKHOV, B.L., BARANINA, I.I.: Vliyanie mineral'nogo pitaniya na soderzhanie
zheltykh pigmentov v list'yakh ozimoi pshenitsy. [Effect of mineral nutrition
on contents of yellow pigments in winter wheat leaves.] - In: Izuchenie Fotosinteza Odnoletnikh Rastenii. Pp. 4-23. Izdat. RIO Akad. Nauk Mold. SSR, Kishinev
1970. [In R.]

1764 - DOROKHOV, B.L., BARANINA, I.I., MAKHARINETS, S.N.: Fotokhimicheskaya aktivnost'
khloroplastov i soderzhanie khlorofillov u kukuruzy. [Photochemical activity of
chloroplasts and chlorophyll content in maize.] - Izv. Akad. Nauk Moldav. SSR
1966 (6): 59-61, 1966. [In R.]

1765 - DOROKHOV, B.L., BARANINA, I.I., MAKHARINETS, S.N.: Izmenenie intensivnosti fotosinteza u ozimoi pshenitsy pri razlichnom mineral'nom pitanii. [Changes in
photosynthetic rate of winter wheat under various mineral nutrition.] - Agrokhimiya *1966 (8)*: 109-116, 1966. [In R.]

1766 - DOROKHOV, B.L., BARANINA, I.I., MAKHARINETS, S.N.: O zimnem gazoobmene u ozimoi
pshenitsy. [Winter gas exchange in winter wheat.] - Fiziol. Rast. *13*: 162-164,
1966. [In R.]

1767 - DOROKHOV, B.L., BARANINA, I.I., MAKHARINETS, S.N.: Intensivnost' fotosinteza i
fotokhimicheskaya aktivnost' khloroplastov ozimoi pshenitsy pri razlichnom mineral'nom pitanii v osenne-zimnii period. [Photosynthetic rate and photochemical activity of winter wheat chloroplasts under different mineral nutrition
during the autumn-winter period.] - In: Mineral'nye Elementy i Mekhanizm Fotosinteza. Pp. 216-226. Kishinev 1969. [In R.]

1768 - DOROKHOV, B.L., GROZOV, D.N.: Vozdeistvie mineral'nogo pitaniya na fotosintez
vinogradnogo rasteniya. [Effect of mineral nutrition on photosynthesis of a vine
plant.] - Sadovodstvo, Vinogradarstvo Vinodelie Moldavii *1966 (11)*: 15-17,
1966. [In R.]

1769 - DOROKHOV, V.L., GROZOV, D.N.: Vliyanie mineral'nogo pitaniya na fotosintez vino-
gradnogo rasteniya. [Effect of mineral nutrition on vine photosynthesis.] - In:
Izuchenie Fotosinteza Vazhneishikh Sel'skokhozyaistvennykh Kul'tur Moldavii.
Pp. 31-47. Kartya Moldovenyaske, Kishinev 1968. [In R.]

1770 - DOROKHOV, B.L., IOVVA, E.P.: Assimilyatsionnaya poverkhnost'i intensivnost' fotosinteza u pervogo pokoleniya gibridov tomatov sorta No. 10. [Assimilatory
surface and photosynthetic rate in first generation hybrids of tomato cv. No.
10.] - Izv. Akad. Nauk SSSR, Ser. biol. *1966 (6)*: 62-66, 1966. [In R.]

1771 - DOROKHOV, B.L., IOVVA, E.P.: Izmeneniya ryada fiziologicheskikh protsessov u
pervogo pokoleniya gibridov tomatov. [Changes of a variety of physiological processes in first generation hybrids of tomato.] - In: Izuchenie Fotosinteza Vaz-

hneTshikh Sel'skokhozyaTstvennykh Kul'tur Moldavii. Pp. 107-128. Kartya Moldovenyaske, Kishinev 1968. [Ps, Chl; In R.]

1772 - DOROKHOV, B.L., LENNIK, Z.N.: Soderzhanie khlorofilla i fotokhimicheskaya aktivnost' izolirovannykh khloroplastov u kukuruzy pri perekrestnom opylenii.
[Chlorophyll content and photochemical activity of isolated chloroplasts of
cross-pollinated maize.] - In: Fotosinteticheskaya Deyatel'nost' RasteniT i
Vliyanie na nee Mineral'nogo Pitaniya. Pp. 114-125. Kishinev 1970. [In R.]

1773 - DOROKHOV, B.L., LENNIK, Z.N.: Soderzhanie pigmentov i fotokhimicheskaya aktivnost'khlorofilla u kukuruzy pri primenenii chuzherodnoT pyl'tsy. [Chlorophyll
content and photochemical activity of chlorophyll in foreign-pollinated maize.]
- In: Fotosintez i Pigmenty Osnovnykh Sel'sko-khozyaTstvennykh RasteniT Moldavii. Pp. 114-134. Kishinev 1970. [In R.]

1774 - DOROKHOV, B.L., MAKHARINETS, S.N.: Sostoyanie khlorofill-belkovo-lipoidnogo kompleksa v list'yakh ozimoT pshenitsy pri razlichnych usloviyakh mineral'nogo pitaniya. [State of chlorophyll-protein-lipid complex in leaves of winter wheat
under different mineral nutrition.] - Izv. Akad. Nauk Mold. SSR, Ser. biol.
khim. Nauk 1969 (6): 62-67, 1969. [In R.]

1775 - DOROKHOV, B.L., MAKHARINETS, S.N.: Vliyanie mineral'nogo pitaniya na fotokhimicheskuyu aktivnost' khloroplastov ozimoT pshenitsy. [Effect of mineral nutrition on photochemical activity of winter wheat chloroplasts.] - Izv. Akad.
Nauk Mold. SSR, Ser. biol. khim. Nauk 1970 (2): 47-51, 1970. [In R.]

1776 - DOROKHOV, B.L., MAKHARINETS, S.N.: Izuchenie fotokhimicheskoT aktivnosti khloroplastov u ozimoT pshenitsy. [Study of photochemical activity of winter wheat
chloroplasts.] - In: Fotosintez i Pigmenty Osnovnykh Sel'sko-khozyaTstvennykh
RasteniT Moldavii. Pp. 21-37. Kishinev 1970. [In R.]

1777 - DOROKHOV, B.L., NEVRYANSKAYA, A.D., LENNIK, Z.N.: Intensivnost' produktivnogo
fotosinteza, soderzhanie khlorofilla i uroven' okislitel'no-vosstanovitel'nykh
protsessov u kukuruzy. [Rate of productive photosynthesis, chlorophyll content
and level of oxido-reduction processes in maize.] - In: Izuchenie Fotosinteza
VazhneTshikh Sel'skokhozyaTstvennykh Kul'tur Moldavii. Pp. 83-106. Kartya Moldovenyaske, Kishinev 1968. [In R.]

1778 - DOROKHOV, B.L., SHISHKANU, G.V.: Izmenenie fiziologicheskikh i biologicheskikh
protsessov u tomatov v zavisimosti ot kachestva sveta i tipa opyleniya. [Changes
of physiological and biochemical processes in tomato related to light quality
and pollination type.] - In: Obmen Uglevodov Plodov i OvoshcheT v Ontogeneze.
Pp. 60-69. Kartya Moldovenyaske, Kishinev 1967. [In R.]

1779 - DOROKHOV, B.L., ZHAKOTE, A.G.: Fotokhimicheskaya aktivnost' khloroplastov u bo-
bovykh kul'tur pri nedostatochnosti mineral'nogo pitaniya. [Photochemical activity of chloroplasts of leguminous plants under insufficient mineral nutrition.]
- In: Mineral'nye Elementy i Mekhanizm Fotosinteza. Pp. 13-21. Kishinev 1969.
[In R.]

1780 - DOROKHOV, B.L., ZHAKOTE, A.G.: Pogloshchenie luchistoT energii list'yami fasoli
pri razlichnykh usloviyakh pitaniya. [Absorption of radiant energy by French
bean leaves under different nutrition.] - Fiziol. Biokhim. kul't. Rast. 2: 17-
20, 1970. [In R, ab: E.]

1781 - DOROKHOV, L.M., BARANINA, I.I., MAKHARINETS, S.N.: Vliyanie mineral'nogo pitaniya na fotosintez, nakoplenie suchogo veshchestva i urozhaT ozimoT pshenitsy i
yarovogo yachmenya.]Effect of mineral nutrition on photosynthesis, dry matter
accumulation and yield of winter wheat and spring barley.] - In: Izuchenie Fotosinteza VazhneTshikh Sel'skokhozyaTstvennykh Kul'tur Moldavii. Pp. 5-30. Kartya
Moldovenyaske, Kishinev 1968. [In R.]

1782 - DORONICHEV, N.I.: Opredelenie velichiny poverkhnosti khvoi adsorbtsionnym sposobom. [Determination of needle surface by an adsorption method.] - Lesovedenie 5:
93-95, 1969. [In R.]

1783 - DORONIN, A.N., KABANOVA, O.L., TIMOFEEV, S.A.: Tallievaya kolonka dlya opredeleniya rastvorennogo v vode kisloroda pri temperature do 80°. [Thallium
column for the determination of oxygen dissolved in water at $\leq$ 80°.] - Zh.
anal. Khim. 24: 274-276, 1969. [In R, ab: E.]

1784 - DOROSHENKO, V.F.: O svyazi formirovaniya kornevoĭ sistemy podsolnechnika s nali-
chiem pitatel'nykh veshchestv v pochve i fotosinteticheskoĭ deyatel'nost'yu.
[Relation between formation of root system of sunflower, presence of nutrients
in soil and photosynthetic activity.] - Fiziol. Rast. *16*: 250-254, 1969. [In R,
ab: E.]

1785 - DOROVSKAYA, I.F., BITAROV, M.I.: O nekotorykh osobennostyakh rosta i fotosinteza mazhlineĭnykh gibridov kukuruzy v usloviyakh orosheniya. [Some pecularities
of growth and photosynthesis of interline hybrids of irrigated maize.] - Zap.
tsent.-kavkaz.Otd. vsesoyuz.bot. Obshch. *2*: 77-80, 1967. [In R.]

1786 - DORRER, H.-D., FEDTKE, C., TREBST, A.: Intramolekulare Wasserstoffverschiebung
in der Hexosephosphatisomerase-Reaktion bei der photosynthetischen Stärkebildung
in *Chlorella*. - Z. Naturforsch. *21 b*: 557-562, 1966.

1787 - DORSCHEID, T.: Die Orientierung aktiver Moleküle des Chlorophyll *a* im Chloroplasten von *Mesotaenium caldariorum*. - Z. Pflanzenphysiol. *61*: 46-51, 1969.

1788 - DORSCHEID, T.: Die Temperaturabhängigkeit der positiven Chloroplastenphototaxis
bei *Mesotaenium caldariorum*. - Z. Pflanzenphysiol. *61*: 52-57, 1969.

1789 - DORSCHEID, T., WARTENBERG, A.: Chlorophyll als Photoreceptor bei der Schwachlichtbewegung des *Mesotaenium*-Chloroplasten. - Planta *70*: 187-192, 1966.

1790 - DOSTANOVA, R.Kh.: Vliyanie sernokislogo i khloristogo natriya na obmen plastidnykh pigmentov u rasteniĭ. [Effect of sodium sulphate and chloride on the metabolism of plastid pigments in plants.] - Fiziol. Rast. *13*: 614-622, 1966. [In R,
ab: E.]

1791 - DOSTANOVA, R.Kh.: Obmen pigmentov u rasteniĭ pri raznokachestvennom zasolenii
sredy. [Metabolism of pigments in plants under various salinization.] - In:
Trudy I. Konferentsii Biokhimikov Respublik Sredneĭ Azii i Kazakhstana. Pp. 157-
160. FAN, Tashkent 1967. [In R.]

1792 - DOUGHERTY, R.C., CRESPI, H.L., STRAIN, H.H., KATZ, J.J.: Nuclear magnetic resonance studies of plant biosynthesis. Bacteriochlorophyll. - J. amer. Chem. Soc.
88: 2854-2855, 1966.

1793 - DOUGHERTY, R.C., STRAIN, H.H., KATZ, J.J.: The use of fully deuterated pigments
to study their function. - In: GOODWIN, T.W. (ed.): Biochemistry of Chloroplasts.
Vol. II. Pp. 427-430. Academic Press, London-New York 1967.

1794 - DOUGHERTY, R.C., STRAIN, H.H., SVEC, W.A., UPHAUS, R.A., KATZ, J.J.: Structure
of chlorophyll *c*. - J. amer. chem. Soc. *88*: 5037-5038, 1966.

1795 - DOUGHERTY, R.C., STRAIN, H.H., SVEC, W.A., UPHAUS, R.A., KATZ, J.J.: The structure, properties and distribution of chlorophyll *c*. - J. amer. chem. Soc. *92*:
2826-2833, 1970.

1796 - DOVBYSH, K.P.: Pigmenty plastyd zernivok ozimogo yachmenyu. [Plastid pigments
of winter barley seeds.] - Ukr. bot. Zh. *26 (3)*: 73-76, 1969. [In Ukrain., ab:
E, R.]

1797 - DOVNAR, V.S.: Fotosintez i mineral'noe pitanie posevov kukuruzy. [Photosynthesis
and mineral nutrition of maize crops.] - Dokl. Akad. Nauk Belorus. SSR *10*: 492-
495, 1966. [In R.]

1798 - DOVNAR, V.S.: Nekotorye zakonomernosti izmeneniya produktivnosti fotosinteza i
optimal'noĭ ploshchadi list'ev u kukuruzy v Belorussii. [Some regularities of
changes in productivity of photosynthesis and optimum leaf area of maize in Belorussia.] - In: VazhneĭshIe Problemy Fotosinteza v Rastenievodstve. Pp. 298-
310. Kolos, Moskva 1970. [In R.]

1799 - DOWDELL, R.J., DODGE, A.D.: The photosynthetic capacity of pea leaves with a
controlled chlorophyll formation. - Planta *94*: 282-290, 1970.

1800 - DOWNES, R.W.: Effect of light intensity and leaf temperature on photosynthesis
and transpiration in wheat and sorghum. - Aust. J. biol. Sci. *23*: 775-782, 1970.

1801 - DOWNES, R.W.: Difference between tropical and temperate grasses in rates of
photosynthesis and transpiration. - In: Proceedings of the XI International
Grassland Congress. Pp. 527-530. Univ. Queensland Press, 1970.

1802 - **DOWNES, R.W., HESKETH, J.D.**: Enhanced photosynthesis at low oxygen concentrations: Differential response of temperate and tropical grasses. - Planta *78*: 79-84, 1968.

1803 - **DOWNEY, W.K., MURPHY, R.F., KEOGH, M.K.**: Separation of fatty acids, phospholipids and chloroplast pigments on Sephadex LH-20. - J. Chromatogr. *46*: 120-124, 1970.

1804 - **DOWNTON, J., BERRY, J., TREGUNNA, E.B.**: Photosynthesis: Temperate and tropical characteristics within a single grass genus. - Science *163*: 78-79, 1969.

1805 - **DOWNTON, W.J.S.**: Preferential C_4-dicarboxylic acid synthesis, the postillumination CO_2 burst, carboxyl transfer step, and grana configurations in plants with C_4- photosynthesis. - Can. J. Bot. *48*: 1795-1800, 1970.

1806 - **DOWNTON, W.J.S., BERRY, J.A., TREGUNNA, E.B.**: C_4-photosynthesis: Non-cyclic electron flow and grana development in bundle sheath chloroplasts. - Z. Pflanzenphysiol. *63*: 194-198, 1970.

1807 - **DOWNTON, W.J.S., BISALPUTRA, T., TREGUNNA, E.B.**: The distribution and ultrastructure of chloroplasts in leaves differing in photosynthetic carbon metabolism. II. *Atriplex rosea* and *Atriplex hastata (Chenopodiaceae)*. - Can. J. Bot. *47*: 915-919, 1969.

1808 - **DOWNTON, W.J.S., TREGUNNA, E.B.**: Carbon dioxide compensation - its relation to photosynthetic carboxylation reactions, systematics of the *Gramineae*, and leaf anatomy. - Can. J. Bot. *46*: 207-215, 1968.

1809 - **DOWNTON, W.J.S., TREGUNNA, E.B.**: Photorespiration and glycolate metabolism: a re-examination and correlation of some previous studies. - Plant Physiol. *43*: 923-929, 1968.

1810 - **DRABER, W., BÜCHEL, K.H., DICKORÉ, K., TREBST, A., PISTORIUS, E.**: Structure-activity correlation of 1,2,4-triazinones, a new group of photosynthesis inhibitors. - In: METZNER, H. (ed.): Progress in Photosynthesis Research. Vol. III. Pp. 1789-1795. Tübingen 1969.

1811 - **DRABER, W., DICKORÉ, K., BÜCHEL, K.H., TREBST, A., PISTORIUS, E.**: Struktur-Aktivität-Korrelation bei 1,2,4-Triazinonen, einer neuen Gruppe von Photosynthesehemmern. - Naturwissenschaften *55*: 446, 1968.

1812 - **DRABIKOWSKA, A.K.**: Biologiczne funkcje chinonów. [Biological functions of quinones.] - Postępy Biochem. *15*: 65-81, 1969. [Ps; in Pol., ab: E.]

1813 - **DRAGOVICH, A., KELLY, J.A. Jr., GOODELL, H.G.**: Hydrological and biological characteristics of Florida's west coast tributaries. - U.S. Fish. Wildlife Serv. Bull. *66*: 463-477, 1969. [Chl.]

1814 - **DRAPER, S.R.**: Lipid changes in senescing cucumber cotyledons. - Phytochemistry *8*: 1641-1647, 1969. [Chl.]

1815 - **DRAPER, S.R., SIMON, E.W.**: Lipid biosynthesis in senescing cotyledons of cucumber. - Phytochemistry *9*: 1997-2002, 1970. [Ps.]

1816 - **DRATZ, E.A.**: The geometry and electronic structure of biologically significant molecules as observed by natural and magnetic optical activity: I. Protein conformation and plasticity. II. Chlorophyll-chlorophyll interactions. III. Electronic structure of metal porphyrins. - Diss. Abstr. B *28*: 621-B-622-B, 1967.

1817 - **DRATZ, E.A., SCHULTZ, A.J., SAUER, K.**: Chlorophyll-chlorophyll interactions. - In: Energy Conversion by the Photosynthetic Apparatus. Brookhaven Symp. Biol. *19*: 303-318, 1967.

1818 - **DRAWERT, H.**: Der mikrospektralphotometrische Nachweis von Chlorophyll *b* in *Closterium* und einigen Mesotaeniaceen. - Ber. deut. bot. Ges. *79*: 403-406, 1966.

1819 - **DRAWERT, H.**: Untersuchung der Chlorophyllbleichung in lebenden Zellen von *Micrasterias rotata* durch kurzwelliges Licht mit dem Mikrospektralphotometer (UMSP I, C. Zeiss). - Mitt. Staatinst. allg. Bot. Hamburg *12*: 29-33, 1967.

1820 - **DRECHSLER, Z., NELSON, N., NEUMANN, J.**: Antimycin *A* as un uncoupler and electron transport inhibitor in photoreactions of chloroplasts. - Biochim. biophys.

Acta *189*: 65-73, 1969.

1821 - DRECHSLER, Z., NEUMANN, J., BEN-SHAUL, Y.: The effect of sonication on photoreactions in chloroplasts. - Israel J. Chem. *6*: 130 p, 1968.

1822 - DREGER, R.H., BRUN, W.A., COOPER, R.L.: Effect of genotype on the photosynthetic
rate of soybean (*Glycine max* (L.) MERR.). - Crop Sci. *9*: 429-431, 1969.

1823 - DREOSTI, I.E., de WIT, J.L., QUICKE, G.V.: Effect of saponification on the chromatographic separation of carotenoids from maize grain extracts. - S. Afr. J.
agr. Sci. *9*: 43-47, 1966.

1824 - DREW, E.A.: Photosynthesis and growth of attached marine algae down to 130 me-
ters in the Mediterranean. - In: MARGALEF, R. (ed.): Proceedings of the Sixth
International Seaweed Symposium. Pp. 151-159. Dirección gen. Pesca mar., Madrid
1969.

1825 - DREW, E.A., SMITH, D.C.: Studies in the physiology of lichens. VIII. Movement
of glucose from alga to fungus during photosynthesis in the thallus of *Peltige-
ra polydactyla*. - New Phytol. *66*: 389-400, 1967.

1826 - DREWS, G.: Zur Regulation der Tetrapyrrolsynthese bei *Rhodospirillum rubrum*. -
Z. Naturforsch. *21 b*: 1224-1229, 1966.

1827 - DREWS, G.: Nachweis der Zucker in den Thylakoiden von *Rhodospirillum rubrum* und
Rhodopseudomonas viridis. - Z. Naturforsch. *23 b*: 671-675, 1968.

1828 - DREWS, G., BIEDERMANN, M., OELZE, J.: Investigations of the thylakoid morphogenesis in *Rhodospirillum rubrum*. - In: METZNER, H. (ed.): Progress in Photosynthesis Research. Vol. I. Pp. 204-208. Tübingen 1969.

1829 - DREWS, G., BIEDERMANN, M., SCHÖN, G.: Zur Regulation der Synthese des Bakterien-
photosyntheseapparates. - Zentralbl. Bakteriol., Parasitenk., Infektionskr.,
Hyg. Abt. I. *205 (1-3)*: 38-41, 1967.

1830 - DREWS, G., LAMPE, H.-H., LADWIG, R.: Die Entwicklung des Photosyntheseapparates
in Dunkelkulturen von *Rhodopseudomonas capsulata*. - Arch. Mikrobiol. *65*: 12-28,
1969.

1831 - DRIESSCHE, R. VANDEN, WAREING, P.F.: Nutrient supply, dry-matter production and
nutrient uptake of forest tree seedlings. - Ann. Bot. *30*: 657-672, 1966.

1832 - DRIESSCHE, R. VANDEN, WAREING, P.F.: Dry-matter production and photosynthesis
in pine seedlings. - Ann. Bot. *30*: 673-682, 1966.

1833 - DRIESSCHE, T. VANDEN: Circadian rhythms in *Acetabularia*: photosynthetic capacity and chloroplast shape. - Exp. Cell Res. *42*: 18-30, 1966.

1834 - DRIESSCHE, T. VANDEN: The influence of constant light on the inulin content of
the chloroplasts in *Acetabularia mediterranea*. - In: METZNER, H. (ed.): Progress
in Photosynthesis Research. Vol. I. Pp. 450-457. Tübingen 1969.

1835 - DRIESSCHE, T. VANDEN: Circadian variation in ATP content in the chloroplasts
of *Acetabularia mediterranea*. - Biochim. biophys. Acta *205*: 526-528, 1970.

1836 - DROKOVA, I.G.: Suchasni uyavlennya pro biosyntez karotynoïdiv. [Present views
on the biosynthesis of carotenoids.] - Ukr. bot. Zh. *23* (2): 3-11, 1966. [In
Ukr., ab: E, R.]

1837 - DROKOVA, I.G.: Pigmenty dvokh shtamiv vodorosti *Dunaliella salina* TEOD. [Pig-
ments of two strains of the alga *Dunaliella salina* TEOD.] - Ukr. bot. Zh. *26 (6)*:
82-83, 1969. [In Ukr., ab: E.]

1838 - DROKOVA, I.G.: Deyaki karotynonosni shtamy vodorosti *Dunaliella salina* TEOD.
[Some strains of the alga *Dunaliella salina* TEOD. containing carotene.] - Ukr.
bot. Zh. *27*: 370-372, 1970. [In Ukr., ab: E, R.]

1839 - DROKOVA, I.G., DOVGORUKA, S.I.: Karotynoutvorennya u vodorosti *Dunaliella salina*
TEOD. pid vplyvom deyakikh dzherel vugletsyu. [Carotene synthesis in the alga
Dunaliella salina TEOD. under different carbon sources.] - Ukr. bot. Zh. *23* (1):
59-62, 1966. [In Ukr., ab: E, R.]

1840 - DROKOVA, I.G., KUZNETSOV, M.V.: Spoluky z C_5- izoprenoïdnoyu lankoyu na karotynoutvorennya u vodorosti *Dunaliella salina* TEOD. [Compounds with C$_5$-isprenoid cell and carotene formation in alga *Dunaliella salina* TEOD.] - Ukr. bot. Zh. *24* (6): 36-41, 1967. [In Ukrain., ab: E, R.]

1841 - DROKOVA, I.G., KUZNETSOV, M.V., POPOVA, R.Ts.: Vmist karotynu u vodorosti *Dunaliella salina* TEOD. pry vyroshchuvanni TT na ropi v laboratornych umovach. [Carotene content in the alga *Dunaliella salina* TEOD. grown on crude oil under laboratory conditions.] - Dopov. Akad. Nauk Ukr. RSR B *29*: 736-739, 1967. [In Ukr, ab: E, R.]

1842 - DROKOVA, I.G., POPOVA, R.Ts.: Vmist pigmentiv pry sumisnomu vyroshchuvanni dvokh shtamiv *Dunaliella salina* TEOD. [Pigment content in two strains of *Dunaliella salina* TEOD. cultivated together.] - Ukr. bot. Zh. *26* (2): 90-92, 1969. [In Ukr., ab: E, R.]

1843 - DROKOVA, I.G., POPOVA, R.Ts.: Vmist karotynu v vodorosti *Dunaliella salina* TEOD. v umovakh masovoÏ kul'tury. [Carotene content in the alga *Dunaliella salina* TEOD. in a mass culture.] - Ukr. bot. Zh. *26* (3): 17-20, 1969. [In Ukr., ab: E, R.]

1844 - DROZDOVA, N.N., KRASNOVSKIÏ, A.A.: Sravnitel'noe issledovanie tusheniya fluorestsentsii khlorofilla i ego analogov v rastvorakh i v khloroplastakh. [Comparative study of fluorescence quenching of chlorophyll and its analogues in solutions and in chloroplasts.] - Mol. Biol. (Moskva) *1*: 395-409, 1967. [In R, ab: E.]

1845 - DROZDOVA, N.N., KRASNOVSKIÏ, A.A.: Obratimoe fotookislenie agregirovannykh form bakteriokhlorofilla i khlorofilla khinonami. [Reversible photooxidation of aggregated forms of bacteriochlorophyll and chlorophyll by quinones.] - Dokl. Akad. Nauk SSSR *195*: 1222-1225, 1970. [In R.]

1846 - DROZDOVA, N.N., KRASNOVSKIÏ, A.A., OBOL'NIKOVA, E.A., SAMOKHVALOV, G.I.: Obratimoe fotookislenie bakteriokhlorofilla ubikhinonami. [Reversible photooxidation of bacteriochlorophyll by ubiquinones.] - Dokl. Akad. Nauk SSSR *183*: 221-224, 1968. [In R.]

1847 - DRUILHET, A.: Détermination de la diffusivité turbulente dans les premiers mètres au-dessus du sol à partir de la diffusion du thoron. - In: Techniques d'Etude des Facteurs Physiques de la Biosphère. Pp. 447-453. INRA, Paris 1970.

1848 - DRUMM, H.E., MARGULIES, M.M.: In vitro protein synthesis by plastids of *Phaseolus vulgaris*. IV. Amino acid incorporation by etioplasts and effect of illumination of leaves on incorporation by plastids. - Plant Physiol. *45*: 435-442, 1970. [Chl.]

1849 - DRURY, S., PARK, R.B.: The effect of leaf senescence on quantum efficiencies of photosynthetic light reactions. - Plant Physiol. *43* (Suppl.): S-29, 1968.

1850 - DRYAGINA, I.V.: Obluchenie plodovykh rasteniï. [Irradiation of fruit plants.] - Vest. mosk. Univ. *1966* (1): 46-60, 1966. [Chl; in R.]

1851 - DUBASH, P.J., REGE, D.V.: Chlorophyll formation in *Euglena gracilis* var. *bacillaris*: interference by analogues of purines, pyrimidines and amino acids. - J. gen. Microbiol. *48*: 283-292, 1967.

1852 - DUBASH, P.J., REGE, D.V.: Excretion of protoporphyrin IX by *Euglena*. - Biochim. biophys. Acta *141*: 209-211, 1967.

1853 - DUBASH, P.J., REGE, D.V.: Permanent bleaching of *Euglena* by Mg^{2+} starvation. - Biochim. biophys. Acta *136*: 185-187, 1967.

1854 - DUBASH, P.J., REGE, D.V.: Chlorophyll formation in *Euglena gracilis* var. *bacillaris*: interference by vitamin analogues. - J. gen. Microbiol. *51*: 127-135, 1968.

1855 - DUBASH, P.J., REGE, D.V.: Chlorophyll formation in *Euglena gracilis* var. *bacillaris:* Effect of enzyme inhibitors. - J. Protozool. *16*: 149-151, 1969.

1856 - DUBASH, P.J., REGE, D.V.: Chlorophyll formation in *Euglena gracilis* var. *bacillaris:* Effect of chelating agents. - J. Protozool. *17*: 349-351, 1970.

1857 - DUBOIS, J., DUJARDIN, D.: Croissance, respiration et thiols solubles de deux souches de tissus de carotte. - Compt. rend. Acad. Sci. Paris, Sér. D *267*: 1596-

1599, 1968. [Chl.]

1858 - DUBOVAYA, E.P.: Osobennosti postupleniya C^{14}-assimilyatov v etiolirovannuyu
       verkhushku rasteniĭ soi.[Some features of transfer of ^{14}C-assimilates into the
       etiolated apex of soybean plants.] - Dokl. Akad. Nauk SSSR 193: 951-954, 1970.
       [In R.]

1859 - DUBROV, A.P.: Vliyanie geomagnitnogo polya na fiziologicheskie protsessy u ras-
       teniĭ. [Effect of geomagnetic field on physiological processes in plants.] -
       Fiziol. Rast. 17: 836-842, 1970. [In R, ab: E.]

1860 - DUCET, G.: Diffusion de l'oxygène dans les tissues végétaux. - In: THOMAS, J.B.,
       GOEDHEER, J.C. (ed.): Currents in Photosynthesis. Pp. 449-453. Donker, Rotterdam
       1966.

1861 - DUCET, G.: Les cytochromes des végétaux supérieurs. - Bull. Soc. franc. Physiol.
       vég. 12: 69-79, 1966.

1862 - DUDKO, Z.G.: Izmenenie intensivnosti fotosinteza u kukuruzy v techenie vegetat-
       sionnogo perioda. [Changes in photosynthetic rate of maize during the vegetation
       period.]-Dokl. Akad. Nauk Belorus. SSR 10: 701-703, 1966. [In R.]

1863 - DUEKER, J., ARDITT, J.: Photosynthetic ^{14}CO$_2$ fixation by green *Cymbidium* (*Orchi-
       daceae)* flowers. - Plant Physiol. 43: 130-132, 1968.

1864 - DUGAN, R.E., RASSON, E., PORTER, J.W.: Separation of water-soluble steroid and
       carotenoid precursors by DEAE-cellulose column chromatography. - Anal. Biochem.
       22: 249-259, 1968.

1865 - DUGGER, W.M., Jr., TING, I.P.: Physiological and biochemical effect of air-pol-
       lution oxidants on plants. - In: STEELINK, C., RUNECKLES, V.C. (ed.): Recent Ad-
       vances in Phytochemistry. Vol. 3. Pp. 31-58. Appleton-Century Crofts, New York
       1970. [Ps.]

1866 - DUJARDIN, E., de KOUCHKOVSKY, Y., SIRONVAL, C.: Absence of photosystem II-pro-
       perties in leaves grown under a flash regime. - Photosynthetica 4: 223-227,
       1970.

1867 - DUJARDIN, E., SIRONVAL, C.: On the influence of daylength on spinach chloro-
       plasts. - Physiol. vég. 6: 311-335, 1968.

1868 - DUJARDIN, E., SIRONVAL, C.: The reduction of protochlorophyllide into chloro-
       phyllide. III. The phototransformability of the forms of the protochlorophylli-
       de-lipoprotein complex found in darkness. - Photosynthetica 4: 129-138, 1970.

1869 - DULIEU, H.: Étude de la stabilité d'une déficience chlorophyllienne induite chez
       le Tabac par traitement au méthane sulfonate d'éthyle. - Ann. Amél. Plant. 17:
       339-355, 1967.

1870 - DULIEU, H., BUGNON, F.: Chimères chlorophylliennes mériclines et ontogénie fo-
       liaire chez le Tabac (*Nicotiana Tabacum* L.). - Compt. rend. Acad. Sci. Paris,
       Sér. D 263: 1714-1717, 1966.

1871 - DUNAEVA, S.E.: Ul'trastruktura khloroplastov i sutochnoe respredelenie v nikh
       krakhmala u rasteniĭ, razlichayushchikhsya po stepeni razvitiya parenkhimnoĭ ob-
       kladki sosudistykh puchkov. [Chloroplast ultrastructure and diurnal distribution
       of starch in chloroplasts of plants differing in the development of bundle
       sheath parenchyma.] - Tsitologiya 12: 297-305, 1970. [In R, ab: E.]

1872 - DUNCAN, W.G.: A model for simulating photosynthesis and other radiation phenome-
       na in plant communities. - In: Proceedings of the 10th International Grassland
       Congress. Section 1, Paper No. 8. Pp. 120-125. Finland 1966.

1873 - DUNCAN, W.G.: Model building in photosynthesis. - In: SAN PIETRO, A., GREER,
       F.A., ARMY, T.J. (ed.): Harvesting the Sun. Pp. 309-314. Academic Press, New
       York-London 1967.

1874 - DUNCAN, W.G.: Cultural manipulation for higher yields. - In: EASTIN, J.D.,
       HASKINS, F.A., SULLIVAN, C.Y., van BAVEL, C.H.M. (ed.): Physiological Aspects
       of Crop Yield. Pp. 327-342. Amer. Soc. Agron. & Crop Sci. Soc. Amer., Madison,
       Wisc. 1969.

1875 - DUNCAN, W.G., BARFIELD, B.J.: Predicting effects of CO_2 enrichment with simulation models and a digital computer. - Trans. ASAE *13*: 246-248, 1970.

1876 - DUNCAN, W.G., HESKETH, J.D.: Net photosynthetic rates, relative leaf growth rates, and leaf numbers of 22 races of maize grown at eight temperatures. - Crop Sci. *8*: 670-674, 1968.

1877 - DUNCAN, W.G., LOOMIS, R.S., WILLIAMS, W.A., HANAU, R.: A model for simulating photosynthesis in plant communities. - Hilgardia *38*: 181-205, 1967.

1878 - DUNCAN, W.G., WILLIAMS, W.A., LOOMIS, R.S.: Tassels and the productivity of maize. - Crop. Sci. *7*: 37-39, 1967.

1879 - DUNCOMBE, W.G.: Effects of asymmetrical drying of paper chromatograms on the autoradiography of carbon-14. - J. Chromatogr. *36*: 557-559, 1968.

1880 - DUNCOMBE, W.G., RISING, T.J.: Scintillation counting of $^{14}CO_2$ from in vitro systems: A comparison of some trapping agents. - Anal. Biochem. *30*: 275-278, 1969.

1881 - DUNPHY, P.J., WHITTLE, K.J., PENNOCK, J.F.: Plastochromanol. - In: GOODWIN, T.W. (ed.): Biochemistry of Chloroplasts. Vol. I. Pp. 165-171. Academic Press, London-New York 1966.

1882 - DUNSTAN, W.M.: Photosynthesis in representative species of marine phytoplankton. - Diss. Abstr. Int. B *30*: 5581-B, 1970.

1883 - DUPUY, J., GERSTER, R.: Effet de la température sur l'évolution de la composition isotopique de l'oxygène photosynthétique dégagé par une suspension d'Euglènes enrichie en $H_2^{18}O$. - Compt. rend. Acad. Sci. Paris, Sér. D *271*: 1288-1291, 1970.

1884 - DURAND, R.: Utilisation des thermocouples. - In: Techniques d'Etude des Facteurs Physiques de la Biosphère. Pp. 131-141. INRA, Paris 1970.

1885 - DURANTON, J.: Biosynthèse de la chlorophylle *b* au cours du verdissement de plantules de *Zea Mays* étiolées. - Physiol. vég. *4*: 75-88,1966.

1886 - DURANTON, J.: Protéogenése et formation des chloroplastes au cours du verdissement de *Zea mays* L. - In: THOMAS, J.B., GOEDHEER, J.C. (ed.): Currents in Photosynthesis. Pp. 321-328. Donker, Rotterdam 1966.

1887 - DURANTON, J., BOQUET, M., GUIGNERY, G.: Comparative study of holochromic proteins in etiolated plastids and chloroplasts of *Zea mays* L. - In: METZNER, H. (ed.): Progress in Photosynthesis Research. Vol. I. Pp. 186-193. Tübingen 1969.

1888 - DURANTON, J., DUVAL, D.: Evolution du magnésium organique au cours du verdissement de feuilles etiolées de *Zea mays* L. - Bull. Soc. franç. Physiol. vég. *14*: 451 - 458, 1969.

1889 - DURANTON, J., LEFORT-TRAN,M.: Différentiation structurale et teneurs en protéines et en pigments chlorophylliens au cours du verdissement chez *Zea mays* L. - Bull. Soc. franç. Physiol. vég. *14*: 249-273, 1968.

1890 - DURQUETY, M.: Les anomalies morphologiques et pigmentaires chez *Vitis vinifera* L. (à suivre). - Progr. Agr. Vitic. *86*: 85-88, 114-115, 1969. [Chl.]

1891 - DURZAN, D.J.: Nitrogen metabolism of *Picea glauca*. III. Diurnal changes of amino acids, amides, protein, and chlorophyll in leaves of expanding buds. - Can. J. Bot. *46*: 929-937, 1968.

1892 - DUS, K., DE KLERK, H., BARTSCH, R.G., HORIO, T., KAMEN, M.D.: On the monotheme nature of cytochrome *c'(Rhodopseudomonas palustris)*. - Proc. nat. Acad. Sci. *57*: 367-370, 1967.

1893 - DUS, K., SLETTEN, K., KAMEN, M.D.: Cytochrome c_2 of *Rhodospirillum rubrum*. II. Complete amino acid sequence and phylogenetic relationships. - J. biol. Chem. *243*: 5507-5518, 1968.

1894 - DUST, J.V., SHINDALA, A.: Relationship of chlorophyll A to algal count and classification in oxidation ponds. - J. Water Pollut. Contr. Fed. *42*: 1362-1369, 1970.

1895 - DUTTON, L.P., KIHARA, T., CHANCE, B.: Early reactions in photosynthetic energy
       conservation: the photooxidation at liquid nitrogen temperatures of two cyto-
       chromes in chromatophores of *Rhodopseudomonas gelatinosa*. - Arch. Biochem. Bio-
       phys. *139*: 236-240, 1970.

1896 - DUVAL, D., DURANTON, J.: Effect of ethylenediaminetetraacetic acid on etiolated
       maize plastids. - Photochem. Photobiol. *12*: 423-428, 1970.

1897 - DÜVEL, D.: Elektrophoretische Untersuchungen des "Chlorophyll-Holochroms" von
       Nicotiana tabacum L. - Z. Naturforsch. *21 b*: 867-871, 1966.

1898 - DUYSEN, M.E.: The effect of two growth regulators on the initial photosynthetic
       reactions in isolated spinach chloroplasts. - Diss. Abstr. B *27*: 1745-B, 1966.

1899 - DUYSENS, L.N.M.: Kinetics of components of the photochemical reaction center in
       purple bacteria. - In: THOMAS, J.B., GOEDHEER, J.C. (ed.): Currents in Photosyn-
       thesis. Pp. 263-271. Donker, Rotterdam 1966.

1900 - DUYSENS, L.N.M.: Primary photosynthetic reactions in relation to transfer of
       excitation energy. - In: Energy Conversion by the Photosynthetic Apparatus.
       Brookhaven Symp. Biol. *19*: 71-80, 1967.

1901 - DUYSENS, L.N.M.: Photobiological principles and methods. - In: HALLDAL, P. (ed.):
       Photobiology of Microorganisms. Pp. 1-16. Wiley-Interscience, London-New York-
       Sydney-Toronto 1970.

1902 - DUYSENS, L.N.M., AMESZ, J.: Photosynthesis. - In: FLORKIN, M., STOTZ, E.H. (ed.):
       Comprehensive Biochemistry. Vol. 27. Pp. 237-266. Elsevier, Amsterdam-London-
       New York 1967.

1903 - DUYSENS, L.N.M., TALENS, A.: Reactivation of reaction center II by a product of
       photoreaction I. - In: METZNER, H. (ed.): Progress in Photosynthesis Research,
       Vol. II. Pp. 1073-1081. Tübingen 1969.

1904 - DVORAKOVSKIĬ, M.S., CHURKINA, V.P.: Rost i fotosintez podrosta buka (*Fagus silva-
       tica* L.) v razlichnykh usloviyakh osveshchennosti i soderzhaniya uglekisloty.
       [Growth and photosynthesis of beech-undergrowth under various illuminance and
       CO_2 content.] - Nauch. Dokl. vyssh. Shkoly, biol. Nauki *11* (10): 59-65, 1968.
       [In R.]

1905 - DVORNIC, V., MOLEA, I., GAINA, R., DUMITRESCU, A.: Influenţa microelementelor
       Mo, B si Cu asupra dinamicii unor procese fiziologice la mazarea F_{53-54}. [Dyna-
       mics of some physiological processes in F_{53-54} peas as affected by Mo, B and Cu.]
       - Lucr. Stiint. Inst. Agron. "N. Balescu" Bucurest, Ser. A *12*: 289-298, 1969.
       [Ps, Chl; in Rum., ab: E, F, R.]

1906 - DWIVEDI, R.S.: Evaluation of the methods for measuring productivity in wheat
       (*Triticum aestivum* L.) and marvel-grass (*Dichanthium annulatum*(FORSK.) STAPF).
       - Indian J. agr. Sci. *40*: 81-88, 1970.

1907 - DWORAKOWSKI, J., HEJNOWICZ, Z., KOLODZIEJ, H.B., WESEŁOWSKI, J.: Angular dis-
       tribution of annihiliation quanta from leaf tissue in darkness and in light. -
       Photosynthetica *4*: 302-308, 1970.

1908 - DWYER, M.R., SMILLIE, R.M.: The role of storage carbohydrates in chloroplast bio-
       synthesis in *Euglena*. - Proc. aust. biochem. Soc. *3*: 19, 1970.

1909 - DWYER, M.R., SMILLIE, R.M.: A light-induced β-1,3-glucan breakdown associated
       with the differentiation of chloroplasts in *Euglena gracilis*. - Biochim. bio-
       phys. Acta *216*: 392-401, 1970.

1910 - DYKSTRA, G.F., GATHERUM, G.E.: Physiological variation of Scotch pine seedlings
       in relation to prevenance and nitrogen. - Iowa State J. Sci. *41*: 487-502, 1967.
       [Ps.]

1911 - DYKYJOVÁ, D.: Kontaktdiagramme als Hilfsmethode für vergleichende Biometrie,
       Allometrie und Produktionsanalyse von *Phragmites*-Ökotypen. - Rev. roum. Biol.,
       Sér. Zool. *14*: 107-119, 1969.

1912 - DYKYJOVÁ, D.: Comparative biometry of *Phragmites communis* ecotypes and its
       significance to investigation of reed stands productivity. - In: Productivity

of Terrestrial Ecosystems. Production Processes. Czech. nat. Comm. IBP, PT-PP
Report No. 1. Pp. 105-109, Praha 1970.

1913 - DYKYJOVÁ, D., KVĚT, J.: Comparison of biomass production in reedswamp communities growing in south Bohemia and south Moravia. - In: Productivity of Terrestrial Ecosystems. Production Processes. Czech. nat. Comm. IBP, PT-PP Report No.
1. Pp. 71-79. Praha 1970.

1914 - DYKYJOVÁ, D., ONDOK, J.P., PŘIBÁŇ, K.: Seasonal changes in productivity and vertical structure of reed-stands (Phragmites communis TRIN.). - Photosynthetica 4:
280-287, 1970.

1915 - DYKYJOVÁ, D., VÉBER, K., PŘIBÁŇ, K.: Photosynthetic production and growth of
reedswamp macrophytes under controlled conditions of mineral nutrition. - Annu.
Rep. algol. Lab. Třeboň 1967: 153-159, 1968.

1916 - DYKYJOVÁ, D., VÉBER, K., PŘIBÁŇ, K.: Production and root/shoot ratio of dominant
reedswamp species growing in outdoor summer hydroponic cultures. - In: Productivity of Terrestrial Ecosystems. Production Processes. Czech. nat. Comm. IBP, PT-
PP Report No. 1. Pp. 101-104. Praha 1970.

1917 - DYSON, P.W., HUMPHRIES, E.C.: Modification of growth habit of Majestic potato by
growth regulators at different times. - Ann. appl. Biol. 58: 171-182, 1966.
[NAR.]

1918 - DZHANGALIEV, A.D., UVAROV, Yu.P.: Izmeneniya soderzhaniya i raspredeleniya azota i kaliya v yabloni v svyazi s usloviyami proizrastaniya i intensivnost'yu fotosinteza. [Changes in content and distribution of nitrogen and potassium in apple tree related to growing conditions and photosynthetic rate.] - Izv. Akad.
Nauk Kaz. SSR, Ser. biol. 1966 (4): 31-36, 1966. [In R.]

1919 - DZHANGALIEV, A.D., UVAROV, Yu.P.: Soderzhanie khlorofilla v list'yakh yablon' v
zavisimosti ot usloviĭ ikh proizrastaniya. [Chlorophyll content in apple leaves
in relation to their growing conditions.] - Izv. Akad. Nauk Kaz. SSR, Ser. biol.
1966 (5): 15-19, 1966. [In R.]

1920 - DZHASHI, V.S., DEMETRADZE, T.Ya.: Materialy ob anatomicheskikh, fiziologicheskikh i biokhimicheskikh izmeneniyakh, vyzvannykh v chaĭnykh list'yakh, povrezh-
dennykh guseĭnitsami mnogoyadnoĭ listovertki. [Anatomical, physiological and
biochemical changes in tea leaves damaged by polyphagous leaf roller caterpil-
lar.] - Subtrop. Kul't. 2: 34-40, 1967. [Ps; in R.]

1921 - DZHAVRSHYAN, D.: Bystroe razdelenie osnovnykh pigmentov lista, mechennykh C^14 s
pomoshch'yu bumazhnoĭ khromatografii. [Rapid separation of main leaf pigments,
labelled with ^{14}C, by paper chromatography.] - In: Nauchnaya Aspir. Konferent-
siya po Geol.-Mineral., Geogr. i Biol.-Pochv. Naukam. Pp. 127-128. Kazan. Univ.,
Kazan 1966. [In R.]

1922 - DZHAVRSHYAN, D.M.: K voprosu o deĭstvii temperatury na opticheskie svoĭstva fotosinteticheskogo apparata. [On the effect of temperature on optical properties
of the photosynthetic apparatus.] - In: Funktsional'nye Osobennosti Khloroplastov. Pp. 19-26. Kazan. Univ., Kazan 1969. [In R.]

1923 - DZHAVRSHYAN, D.M.: Opticheskie svoĭstva list'ev rasteniĭ, poluchennykh iz obluchennykh semyan. [Optical properties of plant leaves obtained from irradiated
seeds.] - Biol. Zh. Armenii 23: 80-86, 1970. [In R, ab: Armen.]

1924 - DZHEMUKHADZE, K.M.: Gazoobmen (fotosintez i dykhanie). [Gas exchange (photosynthesis and respiration).] - In: RUBIN, B.A. (ed.): Fiziologiya Sel'skokhozyaĭstvennykh Rasteniĭ. Vol. 9. Fiziologiya Vinograda i Chaya. Pp. 535-560. Izdat.
mosk. Univ., Moskva 1970. [Tea plants; in R.]

1925 - EAGLES, C.F.: Apparent photosynthesis and respiration in populations of Lolium
perenne from contrasting climatic regions. - Nature 215: 100-101, 1967.

1926 - EAGLES, C.F.: The effect of temperature on vegetative growth in climatic races
of Dactylis glomerata in controlled environments. - Ann. Bot. 31: 31-39, 1967.
[Growth analysis.]

1927 - EAGLES, C.F.: Time changes of relative growth-rate in two natural populations
of Dactylis glomerata L. - Ann. Bot. 33: 937-946, 1969.

1928 - EAGLES, C.F., TREHARNE, K.J.: Photosynthetic activity of *Dactylis glomerata* L. in different light regimes. - Photosynthetica *3*: 29-38, 1969.

1929 - EASTIN, J.A.: C-14 labeled photosynthate export from fully expanded corn (*Zea mays* L.) leaf blades. - Crop Sci. *10*: 415-418, 1970.

1930 - EASTIN, J.A., GRITTON, E.T.: Leaf area development, light interception, and the growth of canning peas (*Pisum sativum* L.) in relation to plant population and spacing. - Agron. J. *61*: 612-615, 1969.

1931 - EASTIN, J.D., SULLIVAN, C.Y.: Carbon dioxide exchange in compact and semi-open sorghum inflorescences. - Crop Sci. *9*: 165-166, 1969.

1932 - EAVES, C.A., FORSYTH, F.R.: The influence of light, modified atmospheres and benzimidazole on Brussels sprout (*Brassica oleracea* var. *gemmifera*). - J. hort. Sci. *43*: 317-322, 1968. [Chl.]

1933 - EBBON, J.G., TAIT, G.H.: Studies on S-adenosylmethionine-magnesium protophorphyrin methyltransferase in *Euglena gracilis* strain Z. - Biochem. J. *111*: 573-582, 1969.

1934 - EBREY, T.G., CLAYTON, R.K.: Polarization of fluorescence from bacteriochlorophyll in castor oil, in chromatophores and as P870 in photosynthetic reaction centers. - Photochem. Photobiol. *10*: 109-117, 1969.

1935 - EBRINGER, L.: The action of nalidixic acid on *Euglena* plastids. - J. gen. Microbiol. *61*: 141-144, 1970. [Chl.]

1936 - EBRINGER, L., JURÁŠEK, A., KADA, R.: Antibiotics and apochlorosis. III. Effect of 5-nitrofurans on the chloroplast system of *Euglena gracilis*. - Folia microbiol. *12*: 151-156, 1967. [Chl.]

1937 - EBRINGER, L., KRKOŠKA, R., MAČOR, M., JURÁŠEK, A., KADA, R.: Furan derivatives. Their common molecular denominator responsible for bleaching of *Euglena gracilis*. - Arch. Mikrobiol. *57*: 61-67, 1967. [Chl.]

1938 - EBRINGER, L., MEGO, J.L., JURÁŠEK, A., KADA, R.: The action of streptomycins on the chloroplast system of *Euglena gracilis*. - J. gen. Microbiol. *59*: 203-209, 1969.

1939 - EBRINGER, L., MEGO, J.L., PODOVA, G.: Reversal of streptomycin bleaching of *Euglena gracilis* by mutagenic concentrations of hydroxylamine. - Biochem. biophys. Res. Commun. *29*: 571-575, 1967.

1940 - EBRINGER, L., NEMEC, P., SANTOVÁ, H., FOLTÍNOVÁ, P.: Changes of the plastid system of *Euglena gracilis* induced with streptomycin and dihydrostreptomycin. - Arch. Mikrobiol. *73*: 268-280, 1970.

1941 - ECHLIN, P.: Origins of photosynthesis. - Sci. J. (London) 2 (4): 42-47, 1966.

1942 - ECHLIN, P.: The photosynthetic apparatus in *Prokaryotes* and *Eukaryotes*.- In: CHARLES, H.P., KNIGHT, B.C.J.G. (ed.): Organization and Control in Prokaryotic and Eucaryotic Cells. (Symposium of the Society for General Microbiology No. 20.) Pp. 221-248. Cambridge Univ. Press, London 1970.

1943 - ECK, R.V., DAYHOFF, M.O.: Evolution of the structure of ferredoxin based on living relics of primitive amino acid sequences. - Science *152*: 363-366, 1966.

1944 - ECKARDT, F.E.: Le principe de la soufflerie climatisée, appliqué à l'étude des échanges gazeux de la couverture végétale. - Oecol. Plant. *1*: 369-399, 1966.

1945 - ECKARDT, F.E.: Mécanisme de la production primaire des écosystèmes terrestres sous climat méditerranéen. Recherches entreprises à Montpellier dans le cadre du Programme Biologique International. - Oecol. Plant. *2*: 367-393, 1967.

1946 - ECKARDT, F.E.: Techniques de mesure de la photosynthèse sur le terrain basées sur l'emploi d'enceintes climatisées. - In: ECKARDT, F.E. (ed.): Functioning of Terrestrial Ecosystems at the Primary Production Level. Pp. 289-319. Unesco, Paris 1968.

1947 - ECKARDT, F.E., MÉTHY, M., SAUVEZON, R.: Un spectroradiomètre-fluxmètre de photons pour l'étude du climat radiatif au sein de l'écosystème. - Oecol. Plant. *4*: 267-294, 1969.

1948 - EDELMAN, J.: Some observations on sucrose formation in detached leaf tissues.
       - In: COOMBS, J. (ed.): Photosynthesis in Sugar Cane. Pp. 79-81. Tate and Lyle
       Ltd., London 1969. [Chl.]

1949 - EDELMAN, J., SCHOOLAR, A.I.: Light as a major factor in chlorophyll destruction
       in sugar cane leaf tissue. - Z. Pflanzenphysiol. 60: 470-471, 1969.

1950 - EDMONDSON, W.T.: Phosphorus, nitrogen, and algae in Lake Washington after diver-
       sion of sewage. - Science 169: 690-691, 1970. [Chl.]

1951 - EDWARDS, G.E., BOVELL, C.R.: Characteristics of a light-dependent proton trans-
       port in cells of Rhodospirillum rubrum. - Biochim. biophys. Acta 172: 126-133,
       1969.

1952 - EDWARDS, G.E., LEE, S.S., CHEN, T.M., BLACK, C.C.: Carboxylation reactions and
       photosynthesis of carbon compounds in isolated mesophyll and bundle sheath cells
       of Digitaria sanguinalis (L.) SCOP. - Biochem. biophys. Res. Commun. 39: 389-
       395, 1970.

1953 - EDWARDS, H.H.: Biphasic inhibition of photosynthesis in powdery mildewed barley.
       - Plant Physiol. 45: 594-597, 1970.

1954 - EDWARDS, H.H., ALLEN, P.J.: Distribution of the products of photosynthesis be-
       tween powdery mildew and barley. - Plant Physiol. 41: 683-688, 1966.

1955 - EDWARDS, R.A., REUTER, F.H.: Pigment changes during the maturation of tomato
       fruit. - Food Technol. 19: 352-355, 357, 1967.

1956 - EFIMOV, Yu.P.: O vliyanii temperaturnoĭ obrabotki zheludeĭ na soderzhanie khlo-
       rofilla i karotinoidov v list'yakh odnoletnikh seyantsev duba. [Effect of heat-
       ing of acorns on content of chlorophyll and carotenoids in leaves of one-year-
       old oak seeelings.] - Nauch. Dokl. vyssh. Shkoly, biol. Nauki 1966 (3): 170-173,
       1966. [In R.]

1957 - EFIMOV, N.A.: Fotosinteticheski aktivnaya radiyatsiya na territorii SSSR. [Pho-
       tosynthetically active radiation on the U.S.S.R. territory.] - In: Fotosintezi-
       ruyushchie Sistemy Vysokoĭ Produktivnosti. Pp. 70-77. Nauka, Moskva 1966. [In R.]

1958 - EGGER, K.: Die Plastochinon-Analogen, ein kritischer Vergleich. - Ber. deut.
       bot. Ges. 79: (123)-(126), 1966.

1959 - EGGER, K.: Verbreitung und Funktion pflanzlicher Karotinoide. - Biol. Rundschau
       5: 112-124, 1967.

1960 - EGGER, K.: Die Farbwachse roter Paprikaschoten. - Ber. deut. bot. Ges. 81: 153-
       156, 1968.

1961 - EGGER, K.: Die Xanthophyllester der Sonnenblume. - Z. Naturforsch. 23 b: 733-735,
       1968.

1962 - EGGER, K.: Zur Identität von Taraxanthin und Luteinepoxid. - Planta 80: 65-76,
       1968.

1963 - EGGER, K., KLEINIG, H.: Carotinoide in den Früchten von Brachychilus horsfieldii.
       G.O. PETERS. - Z. Pflanzenphysiol. 55: 224-228, 1966.

1964 - EGGER, K., KLEINIG, H.: Die Ketocarotinoide in Adonis annua L. - II. Zur Struk-
       tur der Ester. - Phytochemistry 6: 437-440, 1967.

1965 - EGGER, K., KLEINIG, H.: Die Ketocarotinoide in Adonis annua L. - III. Vergleich
       mit synthetischen Substanzen. - Phytochemistry 6: 903-905, 1967.

1966 - EGGER, K., KLEINIG, H.: Die Plastochinonanalogen - ein kritischer Vergleich. -
       Z. Pflanzenphysiol. 56: 113-121, 1967.

1967 - EGGER, K., KLEINIG-VOIGT, H.: Carotinoide der Commelinacee Palisota barteri -
       ein neues β-Citraurin-Vorkommen. - Z. Naturforsch. 23 b: 1105-1108, 1968.

1968 - EGGER, K., NITSCHE, H., KLEINIG, H.: Diatoxanthin und Diadinoxanthin-Bestand-
       teile des Xanthophyllgemisches von Vaucheria und Botrydium. - Phytochemistry 8:
       1583-1585, 1969.

1969 - EGGER, K., SCHWENKER, U.: Lutein-Fettsäureester im Herbstlaub. - Z. Pflanzen-
       physiol. 54: 407-416, 1966.

1970 - EGLE, K.: Die Photosynthese der grünen Pflanzen. - Umschau Wiss. Tech. *66*: 549-557, 1966.

1971 - EGLE, K.: Photosynthese: Die Umwandlung des Sonnenlichts in chemische Energie durch grüne Pflanze. - In: WIELAND, T., PFLEIDERER, G. (ed.): Molekularbiologie. Pp. 157-178. Umschau Verlag, Frankfurt/M. 1967.

1972 - EGLE, K., FOCK, H.: Light respiration-correlations between CO_2 fixation, O_2 pressure and glycollate concentration. - In: GOODWIN, T.W. (ed.): Biochemistry of Chloroplasts. Vol. II. Pp. 79-87. Academic Press, London-New York 1967.

1973 - EGLI, D.B.: Photosynthetic rate of three soybean (*Glycine max* (L.) MERRILL) communities as related to carbon dioxide levels and solar radiation. - Diss. Abstr. int. B *30*: 2989-B, 1970.

1974 - EGLI, D.B., PENDLETON, J.W., PETERS, D.B.: Photosynthetic rate of three soybean communities as related to carbon dioxide levels and solar radiation. - Agron. J. *62*: 411-414, 1970.

1975 - EGNÉUS, H.: Primary photosynthetic induction phenomena in wheat. - Physiol. Plant. *20*: 463-476, 1967.

1976 - EGNÉUS, H.: Action spectra for the transient and the normal photosynthetic oxygen evolution in wheat leaves. - Physiol. Plant. *21*: 602-614, 1968.

1977 - EGNÉUS, H., SUNDQVIST, C.: An action spectrum for the transformation of ALA-protochlorophyllide to ALA-chlorophyllide in the wavelength region 605-675 nm. - Photosynthetica *4*: 81-83, 1970.

1978 - EHEART, M.S.: Fertilizer effects in broccoli. Fertilization effects on the chlorophylls, carotene, pH, total acidity, and ascorbic acid in broccoli. - J. agr. Food Chem. *14*: 18-20, 1966.

1979 - EHEART, M.S.: Variety, fresh storage, blanching solution and packaging effects on ascorbic-acid, total acids, pH and chlorophylls in broccoli. - Food Technol. *23*: 104-106, 1969.

1980 - EHMANN, A., JYUNG, W.: Effect of zinc nutrition on photosynthetic CO_2 fixation. - Plant Physiol. *44* (Suppl.): 12, 1969.

1981 - EICHENBERGER, W.: Trennung biologischer Partikel durch Zentrifugation im Dichtegradienten. - Chimia *23*: 85-94, 1969. [Chl.]

1982 - EICHENBERGER, W., GROB, E.C.: Enzymatic formation of steryl glycosides by homogenates and chloroplast preparations from lettuce and spinach leaves. - In: METZNER, H. (ed.): Progress in Photosynthesis Research. Vol. I. Pp. 338-344. Tübingen 1969.

1983 - EICHHORN, M.: Zur Stoffproduktion kontinuierlicher Kulturen von *Scenedesmus obliquus* (TURP.) KÖTZING im Dauerlicht bei Phosphat- und Nitrat-Limitation. - Flora A *159*: 494-506, 1969. [Chl.]

1984 - EÏDEL'MAN, Z.M.: Osnovnye etapy razvitiya predstavleniĭ o fotosinteticheskom fosforilirovanii. [Main stages of the development of views on photosynthetic phosphorylation.] - In: KIRICHENKO, E.B. (ed.): Metody Issledovaniya Fotofosforilirovaniya. Pp. 10-36. Pushchino-na-Oke 1970. [In R, ab: E.]

1985 - EÏDEL'MAN, Z.M., POPOVA, O.F.: O razvitii fotokhimicheskoĭ aktivnosti plastidy v protsesse zeleneniya. [Development of photochemical activity of the chloroplast during greening.] - Bot. Zh. *51*: 590-599, 1966. [In R.]

1986 - EIK, K., HANWAY, J.J.: Leaf area in relation to yield of corn grain. - Agron. J. *58*: 16-18, 1966.

1987 - EIMHJELLEN, K.E.: Photosynthetic bacteria and carotenoids from a sea sponge *Halichondrium panicea*. - Acta chem. scand. *21*: 2280-2281, 1967.

1988 - EIMHJELLEN, K.E., STEENSLAND, H., TRAETTEBERG, J.: A *Thiococcus* sp. nov. gen., its pigments and internal membrane system. - Arch. Mikrobiol. *59*: 82-92, 1967. [Chl, car.]

1989 - EINHELLIG, F.A., RICE, E.L., RISSER, P.G., WENDER, S.H.: Effects of scopoletin on growth, CO_2 exchange rates, and concentration of scopoletin, scopolin, and chlorogenic acids in tobacco, sunflower and pigweed. - Bull. Torrey bot. Club *97*: 22-33, 1970.

1990 - EÏNOR, L.O.: Uchastie tsitokhromov v fotosinteze. [Participation of cytochromes in photosynthesis.] - In: Puti Povysheniya Intensivnosti i Produktivnosti Fotosinteza. Vol. *3*. Pp. 110-127, Naukova Dumka, Kiev 1969. [In R.]

1991 - EÏNOR, L.O., DZYUBAK, O.I.: Vplyv neorganichnykh soleĭ ta organichnykh rozchynnykiv na aktyvnistʹ reaktsiĭ Khilla z khloroplastamy gorokhu. [Effect of inorganic salts and organic solvents on the activity of Hill reaction with pea chloroplasts.] - Ukr. bot. Zh. *23* (4): 3-10, 1966. [In Ukr., ab: E, R.]

1992 - EÏNOR, L.O., LUTSYSHYNA, O.G.: Vplyv ekzogennogo tvarynnogo tsytokhromu *c* na fotofosforylyuvannya izolʹovanykh khloroplastiv. [Effect of exogenous animal cytochrome *c* on photophosphorylation of isolated chloroplasts.] - Dopov. Akad. Nauk Ukr. RSR B *29*: 941-944, 1967. [In Ukr., ab: E, R.]

1993 - EÏNOR, L.O., RESHETNIKOVA, T.P.: Izuchenie usloviĭ fotoreduktsii ekzogennogo tsitokhroma *c* preparatami khloroplastov iz gorokha. [Study of photoreduction of exogenous cytochrome *c* with pea chloroplast preparations.] - In: Puti Povysheniya Intensivnosti i Produktivnosti Fotosinteza. Pp. 77-84. Naukova Dumka, Kiev 1966. [In R.]

1994 - ELAGIN, I.N.: Vliyanie kobalʹta na soderzhanie khlorofilla, intensivnostʹ fotosinteza i urozhaĭ grechikhi. [Effect of cobalt on chlorophyll content, photosynthetic rate and yield of buckwheat.] - Dokl. VASKhNIL *1970* (7): 22-23, 1970. [In R.]

1995 - EL-BALDRY, A.M., BASSHAM, J.A.: Chloroplast inorganic pyrophosphatase. - Biochim. biophys. Acta *197*: 308-316, 1970.

B1996 - Elementarnye Fotoprotsessy v Molekulakh. [Elementary Photoprocesses in Molecules.] - Nauka, Moskva - Leningrad 1966. [In R.]

1997 - ELEY, J.H. Jr.: Enhancement of photosynthesis in *Chlorella* and a kinetic model for photosynthesis. - Diss. Abstr. B *27*: 4275-B, 1967.

1998 - ELEY, J.H. Jr., MYERS, J.: Enhancement of photosynthesis by alternated light beams and a kinetic model. - Plant Physiol. *42*: 598-607, 1967.

1999 - EL-FOULY, M.M., ASHOUR, N.I.: Interaction effect of chlorocholine chloride and gibberellic acid on photosynthetic pigments content in leaves of cotton seedlings. - Biochem. Physiol. Pflanzen *161*: 225-230, 1970.

2000 - ELIASSON, L.: Dependence of root growth on photosynthesis in *Populus tremula*. - Physiol. Plant. *21*: 806-810, 1968.

2001 - ELLIOTT, J.M.: Effect of rates of ammonium and nitrate nitrogen on Bright tobacco in Ontario. - Tobacco Sci. *14*: 131-137, 1970; Tobacco *171* (2): 19-25, 1970. [Chl.]

2002 - ELLIS, R.J.: Chloroplast ribosomes: Stereospecificity of inhibition by chloramphenicol. - Science *163*: 477-478, 1969.

2003 - ELLIS, R.J.: Effects of acetate on the growth and chlorophyll content of *Golenkinia*. - J. Phycol. *6*: 364-368, 1970.

2004 - ELLIS, R.J.: Further similarities between chloroplast and bacterial ribosomes. - Planta *91*: 329-335, 1970.

2005 - ELLSWORTH, R.K.: Chromatographic separation of milligram quantities of protoporphyrin IX monomethyl ester from protoporphyrin IX and its dimethyl ester. - Anal. Biochem. *32*: 377-380, 1969.

2006 - ELLSWORTH, R.K.: Gas chromatographic determination of some maleimides produced by the oxidation of heme and chlorophyll *a*. - J. Chromatogr. *50*: 131-134, 1970.

2007 - ELLSWORTH, R.K., ARONOFF, S.: Investigations of the biogenesis of chlorophyll *a*. I. Purification and mass spectra of maleimides from the oxidation of chlorophyll and related compounds. - Arch. Biochem. Biophys. *124*: 358-364, 1968.

2008 - ELLSWORTH, R.K., ARONOFF, S.: Investigations on the biogenesis of chlorophyll *a*. II. Chlorophyllide *a* accumulation by a *Chlorella* mutant. - Arch. Biochem. Biophys. *125*: 35-39, 1968.

2009 - ELLSWORTH, R.K., ARONOFF, S.: Investigations on the biogenesis of chlorophyll a. III. Biosynthesis of Mg-vinylpheoporphine a_5 methylester from Mg-protoporphine IX monomethylester as observed in *Chlorella* mutants. - Arch. Biochem. Biophys. *125*: 269-277, 1968.

2010 - ELLSWORTH, R.K., ARONOFF, S.: Investigations of the biogenesis of chlorophyll a. IV. Isolation and partial characterization of some biosynthetic intermediates between Mg-protoporphine IX monomethyl ester and Mg-vinylpheoporphine a_5 obtained from *Chlorella* mutants. - Arch. Biochem. Biophys. *130*: 374-383, 1969.

2011 - ELLSWORTH, R.K., PERKINS, H.J.: Gas chromatographic determination of phytol. - Anal. Biochem. *17*: 521-525, 1966.

2012 - ELLSWORTH, R.K., PERKINS, H.J., DETWILLER, J.P., LIU, K.: On the enzymatic conversion of ^{14}C-labeled chlorophyll a to ^{14}C-labeled chlorophyll b. - Biochim. biophys. Acta *223*: 275-280, 1970.

2013 - ELLYARD, P., SAN PIETRO, A.: Photosynthetic reactions of *Euglena* chloroplasts. - Plant Physiol. *43* (Suppl.): S 28, 1968.

2014 - ELLYARD, P.W.: The Warburg effect, investigations using isolated spinach chloroplasts. - Diss. Abstr. B *29*: 470-B-471-B, 1969.

2015 - ELLYARD, P.W., GIBBS, M.: The effect of oxygen on photosynthesis in chloroplasts. - Plant Physiol. *42* (Suppl.): S-33, 1967.

2016 - ELLYARD, P.W., GIBBS, M.: Inhibition of photosynthesis by oxygen in isolated spinach chloroplasts. - Plant Physiol. *44*: 1115-1121, 1969.

2017 - ELLYARD, P.W., SAN PIETRO, A.: The Warburg effect in a chloroplast-free preparation from *Euglena gracilis*. - Plant Physiol. *44*: 1679-1683, 1969.

2018 - ELMORE, C.D., HESKETH, J.D., MURAMOTO, H.: A survey of rates of leaf growth, leaf aging and leaf photosynthetic rates among and within species. - J. Ariz. Acad. Sci. *4*: 215-219, 1967.

2019 - EL-SAYED SAYED, Z.: On the productivity of the Southwest Atlantic Ocean and the waters of the Antarctic Peninsula. - In: Biology of the Antarctic Seas: III. American Geophysical Union, Washington, D.C. Antarctic Res. Ser. *11*: 15-47, 1967.

2020 - EL-SHARKAWY, M., HESKETH, J., ELMORE, D.: Photosynthesis and transpiration among species. - Adv. Frontiers Plant Sci. *18*: 33-37, 1967.

2021 - EL-SHARKAWY, M.A., LOOMIS, R.S., WILLIAMS, W.A.: Apparent reassimilation of respiratory carbon dioxide by different plant species. - Physiol. Plant. *20*: 171-186, 1967.

2022 - EL-SHARKAWY, M.A., LOOMIS, R.S., WILLIAMS, W.A.: Photosynthetic and respiratory exchanges of carbon dioxide by leaves of the grain amaranth. - J. appl. Ecol. *5*: 243-251, 1968.

2023 - ELSTNER, E.: Kofaktor- und Enzymabhängigkeit photosynthetischer Reaktionen in verschiedenen Typen fragmentierter Chloroplasten. - Ber. deut. bot. Ges. *81*: 325-331, 1968.

2024 - ELSTNER, E.: The role of plastocyanin and cytochrome f in photosynthetic electron transport. - In: METZNER, H. (ed.): Progress in Photosynthesis Research. Vol. II. Pp. 1035-1041. Tübingen 1969.

2025 - ELSTNER, E., PISTORIUS, E., BÖGER, P., TREBST, A.: Zur Rolle von Plastocyanin und Cytochrom f im photosynthetischen Elektronentransport. - Planta *79*: 146-161, 1968.

2026 - ELSTNER, E.F., HEUPEL, A., VAKLINOVA, S.: Über die Hemmung des photosynthetischen Elektronentransports in isolierten Chloroplasten durch Hydroxylamin. - Z. Pflanzenphysiol. *62*: 173-183, 1970.

2027 - ELSTNER, E.F., HEUPEL, A., VAKLINOVA, S.: Über die Oxidation von Hydroxylamin durch isolierte Chloroplasten und die mögliche Funktion einer Peroxidase aus Spinatblättern bei der Oxidation von Ascorbinsäure und Glykolsäure. - Z. Pflanzenphysiol. *62*: 184-200, 1970.

2028 - EL-TABBAKH, A.E.: The effect of leaf area development, leaf photosynthetic rates and temperature on growth and dry matter accumulation in certain species of forage crops. - Diss. Abstr. B *28*: 3560-B-3561-B, 1968.

2029 - EL TINAY, A.H.: Degradation of beta-carotene by molecular oxygen. - Diss. Abstr. int. B *31*: 2515-B, 1970.

2030 - van EMDEN, H.F., COCKSHULL, K.E.: The effects of soil applications of (2-chloro-ethyl)-trimethylammonium chloride on leaf area and dry matter production by the Brussels sprout plant. - J. exp. Bot. *18*: 707-715, 1967.

2031 - EMERSON, R.G., COCKBURN, W., GIBBS, M.: Sucrose as a product of photosynthesis in isolated spinach chloroplasts. - Plant Physiol. *42*: 840-844, 1967.

2032 - EMLEN, J.T.: A rapid method for measuring arboreal canopy cover. - Ecology *48*: 158-160, 1967.

2033 - EMMETT, J.M., WALKER, D.A.: Thermal uncoupling in chloroplasts. - Biochim. biophys. Acta *180*: 424-425, 1969.

2034 - EMRICH, H.M., JUNGE, W., WITT, H.T.: An artificial indicator for electric phenomena in biological membranes and interfaces. - Naturwissenschaften *56*: 514-515, 1969.

2035 - EMRICH, H.M., JUNGE, W., WITT, H.T.: Further evidence for an optical response of chloroplast bulk pigments to a light induced electrical field in photosynthesis. - Z. Naturforsch. *24 b*: 1144-1146, 1969.

B2036 - Energy Conversion by the Photosynthetic Apparatus. (Brookhaven Symposia in Biology No. 19). Brookhaven nat. Lab., Upton, N.Y., 1967.

2037 - ENGEL, K.H., MEINL, G.: Untersuchungen über die Ertragsbildung als Grundlage für pflanzenzüchterische und pflanzenbauliche Massnahmen. - Deut. Akad. Landw.-Wiss. Berlin, Tag.-Ber. *82* (2): 27-41, 1966.

2038 - ENGEL, K.H., RAEUBER, A., MEINL, G.: Phenometric studies on cultivated plants (proposal for PP/IBP). - Photosynthetica *2*: 298-302, 1968.

2039 - ENGELBRECHT, A.H.P., WEIER, T.E.: Chloroplast development in the germinating safflower *(Carthamus tinctorius)* cotyledon. - Amer. J. Bot. *54*: 844-856, 1967.

2040 - ENGELBRECHT, L., BRÄUNIGER, H., KOINE, A.: Kininwirkungen verschiedener Benzimidazolderivate. - Flora A *158*: 109-113, 1967. [Chl.]

2041 - ENGST, R., BLAZOVICH, M., KNOLL, R.: Über das Vorkommen von Lindan in Möhren und seinen Einfluss auf den Carotingehalt. - Nahrung *11*: 389-399, 1967.

2042 - ENOCH, H., EHRLICH-ROGOZINSKY, S., AVRON, M., PATCHORNIK, A.: A new portable CO_2 gas analyser and its use in field measurements. - Agr. Meteorol. *7*: 255-262, 1970.

2043 - ENOCH, H., RYLSKI, I., SAMISH, Y.: CO_2 enrichment to cucumber, lettuce and sweet pepper plants grown in low plastic tunnels in a subtropical climate. - Israel J. agr. Res. *20*: 63-69, 1970. [Ps, growth.]

2044 - ENTSCH, B., SMILLIE, R.M.: The relationship between phytoflavin, ferredoxin and photosynthesis. - Proc. aust. biochem. Soc. *3*: 87, 1970.

2045 - ENZELL, C.R., FRANCIS, G.W., LIAAEN-JENSEN, S.: Mass spectrometric studies of carotenoids. I. Occurrence and intensity ratios of M-92 and M-106 peaks. - Acta chem. scand. *22*: 1054-1055, 1968.

2046 - ENZELL, C.R., FRANCIS, G.W., LIAAEN-JENSEN, S.: Mass spectrometric studies of carotenoids. 2. A survey of fragmentation reactions. - Acta chem. scand. *23*: 727-750, 1969.

2047 - EPPLEY, R.W.: An incubation method for estimating the carbon content of phytoplankton in natural samples. - Limnol. Oceanogr. *13*: 574-582, 1968.

2048 - EPPLEY, R.W., HOLMES, R.W., PAASCHE, E.: Periodicity in cell division and physiological behavior of *Ditylum brightwellii*, a marine planktonic diatom, during growth in light-dark cycles. - Arch. Mikrobiol. *56*: 305-323, 1967. [Ps, Chl.]

2049 - EPPLEY, R.W., HOLMES, R.W., STRICKLAND, J.D.H.: Sinking rates of marine phytoplankton measured with a fluorometer. - J. exp. mar. Biol. Ecol. *1*: 191-208, 1967.

2050 - EPPLEY, R.W., SLOAN, P.R.: Growth rates of marine phytoplankton: correlation
       with light absorption by cell chlorophyll a. - Physiol. Plant. *19*: 47-59, 1966.

2051 - ERASMUS, G.M.M., MEYNHARDT, J.T., STRYDOM, D.K.: The effect of silver leaf dis-
       ease *(Stereum purpureum)* on the respiration rate and photosynthetic properties
       of peach leaves. - S. Afr. J. agr. Sci. *11*: 375-382, 1968.

2052 - ERBY, W.A., ERNER, W.E., WALDE, R.A.: AP-20, a new fast action postemergent her-
       bicide. - Proc. Sea Weed Conf. *20*: 111-115, 1967. [Chl.]

2053 - ERDELI, G.S., ZVYAGINTSEV, V.I., CHUGUNOVA, N.G.: Fiziologicheskie osobennosti
       vliyaniya regulyatorov rosta raznogo tipa na fosfornyĭ obmen v list'yakh pod-
       solnechnika. [Physiological features of the effect of various types of growth
       regulators on phosphorus metabolism in sunflower leaves.] - Uch. Zap. mosk.
       obl. pedagog. Inst. *169* (3): 148-158, 1967. [Ps; in R.]

2054 - ERGASHEV, A., ABDURAKHMANOVA, Z.N., BELAN, N.F.: Vliyanie vysokogornoĭ ul'trafio-
       letovoĭ radiatsii na fotosinteticheskuyu assimilyatsiyu CO_2 i kharakter rasprede-
       leniya ugleroda C^{14} sredi osnovnykh produktov fotosinteza chiny. [Effect of
       high-mountain ultra-violet radiation on photosynthetic CO_2 assimilation and the
       character of ^{14}C distribution among basic products in peavine.] - In: Tezisy
       Dokladov Molodykh Uchenykh na Vtoroĭ Respublikanskoĭ Nauchnoĭ Konferentsii,
       Posvyashchennoĭ 50-letiyu VLKSM. P. 164. Dushanbe 1968. [In R.]

2055 - ERGASHEV, A., ABDURAKHMANOVA, Z.N., BELAN, N.F.: Fotosinteticheskaya assimilyat-
       siya $C^{14}O_2$ u *Lathyrus* M. v svyazi s deĭstviem ul'trafioletovoĭ radiatsii. [Pho-
       tosynthetic assimilation of $^{14}CO_2$ in *Lathyrus* M. as related to ultra-violet ra-
       diation.] - In: Tezisy Dokladov Vtorogo Vsesoyuznogo Biokhimicheskogo S'ezda,
       Sektsiya " Problemy Fotosinteza". Pp. 126-127. FAN, Tashkent 1969. [In R.]

2056 - ERGASHEV, A., ABDURAKHMANOVA, Z.N., BELAN, N.F., NASYROV, Yu.S.: Vliyanie vyso-
       kogornoĭ radiatsii na fotosintez rasteniĭ. [Effect of ultra-violet radiation on
       plant photosynthesis.] - In: Fotosintez i Ispol'zovanie Solnechnoĭ Radiatsii.
       Pp. 74-75. Dushanbe 1967. [In R.]

2057 - ERISMANN, K.H., BRUNOLD, C.: Die Probeentnahme in kinetischen Stoffwechselunter-
       suchungen mit Wasserlinsen *Lemna minor* L. (Lemnaceen). - Experientia *23*: 235-
       236, 1967.

2058 - ERISMANN, K.H., KIRK, M.R.: The influence of nitrogen on metabolic intermediates
       in steady-state photosynthesis by *Lemna minor* L. - In: METZNER, H. (ed.): Pro-
       gress in Photosynthesis Research. Vol. III. Pp. 1538-1545. Tübingen 1969.

2059 - ERISMANN, K.H., WEGNER, F.: Der Einfluss einer wachstumshemmenden Kinetinkonzen-
       tration auf Chlorophyllgehalt, Photosyntheserate und Stärkeproduktion von *Lemna
       minor* L. - Flora A *158*: 433-442, 1967.

2060 - ERMOLAEVA, E. Ya., KOZLOVA, N.A.: O roli list'ev pri perekhode rasteniĭ k tsve-
       teniyu. [Role of leaves during transition to flowering.] - In: Ontogenez Vysshikh
       Rasteniĭ. Pp. 148-154. Izdat. Arm. SSR, Erevan 1970. [Ps; in R.]

2061 - ERNST, W.: ATP als Indikator für die Biomass mariner Sedimente. - Oecologia *5*:
       56-60, 1970.

2062 - EROKHIN, E., KRASNOVSKIĬ, A.A.: Uchastie raznykh form bakterioviridina i bakte-
       riokhlorofilla v fotosinteze bakteriĭ i migratsiya energii mezhdu formami etikh
       pigmentov. [Participation of different forms of bacterioviridin and bacterio-
       chlorophyll in the photosynthesis of bacteria and the migration of energy between
       the forms of these pigments.] - Tr. mosk. Obshch. Ispyt. Prirody *24* (Biol. av-
       totrof. Mikroorg.): 108-114, 1966. [In R.]

2063 - EROKHIN, Yu.E.: Sostoyanie pigmentov v kletkakh fotosinteziruyushchikh bakteriĭ.
       [State of pigments in cells of photosynthetic bacteria.] - Stud. biophys. *5*:
       171-174, 1967. [In R, ab: E.]

2064 - EROKHIN, Yu.E., SINEGUB, O.A.: O molekulyarnoĭ organizatsii pigmentnoĭ sistemy
       purpurnykh fotosinteziruyushchikh bakteriĭ. [Molecular organization of pigment
       system of purple photosynthetic bacteria.] - Mol. Biol. (Moskva) *4*: 401-410,
       1970. [In R, ab: E.]

2065 - EROKHIN, Yu.E., SINEGUB, O.A.: Izmeneniya v spektrakh pogloshcheniya khromatoforov *Chromatium* pri deĭstvii detergentov i organicheskikh rastvoriteleĭ. [Alterations in absorption spectra of *Chromatium* chromatophores under the action of detergents and organic solvents.] - Mol. Biol. (Moskva) *4*: 541-550, 1970. [In R, ab: E.]

2066 - EROKHINA, L.G., KRASNOVSKIĬ, A.A.: Vliyanie denaturiruyushchikh vozdeĭstviĭ na spektry pogloshcheniya i fluorestsentsii fikoeritrina iz *Callithamnion rubosum.* [Effect of denaturing agents on absorption and fluorescence spectra of phycoerythrin from *Callithamnion rubosum.*] - Mol. Biol. (Moskva) *2*: 550-561, 1968. [In R, ab: E.]

2067 - ERYGIN, P.S.: Fotosintez rasteniĭ risa. [Photosynthesis of rice plants.] - In: RUBIN, B.A. (ed.): Fiziologiya Sel'skokhozyaĭstvennykh Rasteniĭ. Tom 5, Fiziologiya Kukuruza i Risa. Pp. 381-397. Izdat. mosk. Univ., Moskva 1969. [In R.]

2068 - ESCHRICH, W.: Translokation ^{14}C-markierter Assimilate im Licht und im Dunkeln bei *Vicia faba.* - Planta *70*: 99-124, 1966.

2069 - ESHEL, Y.: Mode of action of 1-(3-chloro-4-methylphenyl)-3-methyl-2-pyrrolidinone. - Weeds *15*: 147-149, 1967. [Ps.]

2070 - ESHEL, Y.: Effect of pyrazon on photosynthesis of various plant species. - Weed Res. *9*: 167-172, 1969.

2071 - ESHEL, Y., SOMPOLINSKY, D.: Selectivity of pyrazon and benzthiazuron in sugar beet. - Weed Res. *10*: 196-203, 1970. [Ps.]

2072 - ESHEL, Y., WARREN, G.F.: Postemergence action of CIPC. - Weeds *15*: 237-241, 1967. [Chl.]

2073 - ETANA, P.*, SHILO, M.: Spread of viruses attacking blue green algae in freshwater ponds and their interaction with *Plectonema boryanum.* - Bamidgeh *20* (3): 77-87, 1968. [Ps.]

2074 - ETHERINGTON, J.R.: Measurement of photosynthesis and transpiration in controlled environments with particular reference to microclimate control in leaf cuvettes. - Ann. Bot. *31*: 653-660, 1967.

2075 - ETHERINGTON, J.R.: Soil water and the growth of grasses. II. Effects of soil water potential on growth and photosynthesis of *Alopecurus pratensis.* - J. Ecol. *55*: 373-380, 1967.

2076 - ETIENNE, A.L.: Étude de l'étape thermique de l'émission photosynthétique d'oxygène par une méthode d'écoulement. - Biochim. biophys. Acta *153*: 895-897, 1968.

2077 - ETTL, H.: Development and differentiation of the lobate chloroplast in the genus *Chlamydomonas.* - Annu. Rep. algol. Lab. Třeboň *1968*: 69-73, 1969.

2078 - EVANS, G.C.: Model and measurement in the study of woodland light climates. - In: BAINBRIDGE, R., EVANS, G.C., RACKHAM, O. (ed.): Light as an Ecological Factor. Pp. 53-76. Blackwell sci. Publ., Oxford 1966.

2079 - EVANS, G.C.: Apparatus for surveying light climate in woodlands, including area and intensity of sunflecks. - In: BAINBRIDGE, R., EVANS, G.C., RACKHAM, O. (ed.): Light as an Ecological Factor. Pp. 418-420. Blackwell sci. Publ., Oxford 1966.

2080 - EVANS, G.C.: The spectral composition of light in the field. I. Its measurement and ecological importance. - J. Ecol. *57*: 109-125, 1969.

2081 - EVANS, L.T., RAWSON, H.M.: Photosynthesis and respiration by the flag leaf and components of the ear during grain development in wheat. - Austr. J. biol. Sci. *23*: 245-254, 1970.

2082 - EVANS, L.V.: Chloroplast morphology and fine structure in British fucoids. - New Phytol. *67*: 173-178, 1968.

2083 - EVANS, M.C.W.: Ferredoxin NAD$^+$ reductase and the photoreduction of NAD$^+$ by *Chlorobium thiosulfatophilum.* - In: METZNER, H. (ed.): Progress in Photosynthesis Research. Vol. III. Pp. 1474-1475. Tübingen 1969.

* Erroneous sequence of names of the first author in the original paper; the correct author's name is PADAN Etana.

2084 - EVANS, M.C.W., BUCHANAN, B.B., ARNON, D.I.: A new ferredoxin-dependent carbon reduction cycle in a photosynthetic bacterium. - Federat. Proc. *25*: 226, 1966.

2085 - EVANS, M.C.W., BUCHANAN, B.B., ARNON, D.I.: A new ferredoxin-dependent carbon reduction cycle in a photosynthetic bacterium. - Proc. nat. Acad. Sci. U.S.A. *55*: 928-934, 1966.

2086 - EVANS, M.C.W., HALL, D.O., BOTHE, H., WHATLEY, F.R.: The stoicheiometry of electron transfer by bacterial and plant ferredoxins. - Biochem. J. *109*: 45 P, 1968.

2087 - EVANS, M.C.W., HALL, D.O., BOTHE, H., WHATLEY, F.R.: The stoicheiometry of electron transfer by bacterial and plant ferredoxins. - Biochem. J. *110*: 485-489, 1968.

2088 - EVANS, M.C.W., WHATLEY, F.R.: Photosynthetic mechanisms in *Prokaryotes* and *Eukaryotes*. - In: CHARLES, H.P., KNIGHT, B.C.J.G. (ed.): Symposium of the Society for General Microbiology. No. 20: Organization and Control in Prokaryotic and Eukaryotic Cells. Pp. 203-220. Cambridge Univ. Press, London 1970.

2089 - EVANS, W.R.: Photosynthesis in *Euglena*. - In: BUETOW, D.E. (ed.): The Biology of *Euglena*. Vol. 2. Pp. 73-84. Academic Press, New York-London 1968.

2090 - EVERSON, R.G.: Bicarbonate equilibria and the apparent $Km(HCO_3^-)$ of isolated chloroplasts. - Nature *222*: 876, 1969.

2091 - EVERSON, R.G.: Carbonic anhydrase and CO_2 fixation in isolated chloroplasts. - Phytochemistry *9*: 25-32, 1970.

2092 - EVERSON, R.G., COCKBURN, W., ELLYARD, P.W., GIBBS, M.: Desaspidin and CO_2 fixation by spinach chloroplasts. - Plant Physiol. *41*: 1240-1241, 1966.

2093 - EVERSON, R.G., GIBBS, M.: Photosynthetic assimilation of carbon dioxide and acetate by isolated chloroplasts. - Plant Physiol. *42*: 1153-1154, 1967.

2094 - EVERSON, R.G., SLACK, C.R.: Distribution of carbonic anhydrase in relation to the C_4 pathway of photosynthesis. - Phytochemistry *7*: 581-584, 1968.

2095 - EVERTON, M., LORDS, J.L.: Respiration and photosynthesis in the thermophilic blue-green algae. - Proc. Utah Acad. Sci. Arts Lett. *44*: 416, 1967.

2096 - EVSTIGNEEV, V.B.: Issledovanie fotosensibilizatsii okislitel'no-vosstanovitel'-nykh reaktsiĭ khlorofillom i ego analogami elektrometricheskimi metodami. [Study of photosensibilization of oxido-reduction reactions by chlorophyll and its analogues by means of electrometric methods.] - In: Elementarnye Fotoprotsessy v Molekulakh. Pp. 243-266. Nauka, Moskva-Leningrad 1966. [In R.]

2097 - EVSTIGNEEV, V.B.: O vzaimodeĭstvii khlorofilla s aktseptorami elektrona. [On the interrelationship between chlorophyll and electron acceptors.] - In: Mekhanizmy Dykhaniya, Fotosinteza i Fiksatsii Azota. Pp. 264-275. Nauka, Moskva 1967. [In R.]

2098 - EVSTIGNEEV, V.B.: Fotokhimicheskaya stadiya protsessa fotosinteza i zapasanie energii. [Photochemical phase of photosynthesis and energy storage.] - Sel'.-khoz. Biol. *2*: 515-521, 1967. [In R, ab: E.]

2099 - EVSTIGNEEV, V.B.: Mekhanizm fotosensibilizatsii khlorofillom. [Mechanism of photosensibilization by chlorophyll.] - In: Bioenergetika i Biologicheskaya Spektrofotometriya. Pp. 141-148. Nauka, Moskva 1967. [In R.]

2100 - EVSTIGNEEV, V.B.: On the mechanism of the photosensitizing action of chlorophyll. - In: Seventh International Congress of Biochemistry. Abstracts V. P. 904. Tokyo 1967.

2101 - EVSTIGNEEV, V.B.: Raboty po fotosintezu v laboratoriyakh Frantsii. [Photosynthetic research in French laboratories.] - Vestn. Akad. Nauk SSSR *1968* (4): 84-89, 1968. [In R.]

2102 - EVSTIGNEEV, V.B.: Sur l'oxydation photochimique réversible de la chlorophylle en rapport avec le mécanisme de son action photosensibilisatrice sur les réactions d'oxydoréduction. - J. Chim. phys. Phys.-chim. biol. *65*: 1447-1456, 1968.

2103 - EVSTIGNEEV, V.B.: Chlorophyll as photochemical electron donor. - In: METZNER, H. (ed.): Progress in Photosynthesis Research. Vol. II. Pp. 733-745. Tübingen 1969.

2104 - EVSTIGNEEV, V.B.: O roli kislotno-osnovnogo ravnovesiya v srede, kak regulyato-
ra fotokhimicheskikh reaktsiĭ khlorofilla *in vitro* i *in vivo*. [The role of redox
balances in the medium as a regulator of photochemical reactions of chlorophyll
in vitro and *in vivo*.] - In: Tezisy Sektsionnykh Soobshcheniĭ. Vtoroĭ Vsesoyuznyĭ Biokhimicheskiĭ S'ezd. 19. Sektsiya: Problemy Fotosinteza. Pp. 34-35. FAN,
Tashkent 1969. [In R.]

2105 - EVSTIGNEEV, V.B.: Fotosintez. [Photosynthesis.] - In: Budushchee Nauki *3*: 217-
225, 1970. [In R.]

2106 - EVSTIGNEEV, V.B.: Biologicheskoe ispol'zovanie solnechnoĭ energii. [Biological
utilization of solar radiation.] - Geliotekhnika (Tashkent) *1970*: 70-75, 1970.
[Ps; in R.]

2107 - EVSTIGNEEV, V.B.: O fotokhimicheskom vzaimodeĭstvii khlorofilla i ego analogov s
aktseptorami elektrona. [Photochemical interrelations of chlorophyll and its
analogues with electron acceptors.] - In: Molekulyarnaya Fotonika. Pp. 178-199.
Nauka, Leningrad 1970. [In R.]

2108 - EVSTIGNEEV, V.B.: O vozmozhnoĭ roli kislotno-osnovnogo ravnovesiya kak faktora
regulyatsii fotokhimicheskikh reaktsiĭ khlorofilla *in vitro* i *in vivo*. [On the
role of acid-base balance as a regulation factor of photochemical reactions of
chlorophyll *in vitro* and *in vivo*.] - Biofizika *15*: 239-253, 1970. [In R, ab: E.]

2109 - EVSTIGNEEV, V.B.: Kratkiĭ ocherk razvitiya issledovaniĭ na urovne izolirovannykh
khloroplastov. Vstupitel'noe slovo k seminaru. [A brief story of studies with
isolated chloroplasts: introduction to a seminar.] - In: KIRICHENKO, E.B. (ed.):
Metody Vydeleniya Khloroplastov. Pp. 3-6. Pushchino-na-Oke 1970. [In R.]

2110 - EVSTIGNEEV, V.B.: O fotokhimicheskoĭ stadii protsessa fotosinteza. [Photochemic-
al phase of photosynthesis.] - In: Vazhneĭshie Problemy Fotosinteza v Rastenievodstve. Pp. 52-67. Kolos, Moskva 1970. [In R.]

2111 - EVSTIGNEEV, V.B., BEKASOVA, O.D.: O fotokhimicheskikh svoĭstvakh fikoeritrobilina. [Photochemical properties of phycoerythrobilin.] - Biofizika *11*: 249-257,
1966. [In R.]

2112 - EVSTIGNEEV, V.B., BEKASOVA, O.D.: Fotokhimicheskie svoĭstva *C*-fikotsianina.
[Photochemical properties of *C*-phycocyanin.] - Mol. Biol. (Moskva) *2*: 380-388,
1968. [In R, ab: E.]

2113 - EVSTIGNEEV, V.B., BEKASOVA, O.D.: Fotokhimicheskie svoĭstva fikotsianobilina.
[Photochemical properties of phycocyanobilin.] - Mol. Biol. (Moskva) *3*: 32-40,
1969. [In R, ab: E.]

2114 - EVSTIGNEEV, V.B., BEKASOVA, O.D.: Fotoelektrokhimicheskiĭ effekt plenok i fikoeritrina fikoeritrobilina. [Photoelectrochemical effect of phycoerythrin and
phycoerythrobilin films.] - Biofizika *15*: 807-815, 1970. [In R, ab: E.]

2115 - EVSTIGNEEV, V.B., BEKASOVA, O.D.: O fotokhimicheskikh svoĭstvakh biliproteidov
vodorosleĭ. [Photochemical properties of algae biliproteins.] - In: ANDREENKO,
S.S. (ed.): Fiziologiya i Biokhimiya Zdorovogo i Bol'nogo Rasteniya. Pp. 170-
184. Izdat. mosk. Univ., Moskva 1970. [In R.]

2116 - EVSTIGNEEV, V.B., CHERKASHINA, N.A.: O vydelenii khlorofilla *d* iz vodorosli *Grateloupia dichotoma*. [Isolation of chlorophyll *d* from the alga *Grateloupia dichotoma*.] - Biokhimiya *35*: 48-52, 1970. [In R, ab: E.]

2117 - EVSTIGNEEV, V.B., GAVRILOVA, V.A.: Ob elektrodno-aktivnoĭ okislennoĭ forme khlorofilla. [On the oxidized form of chlorophyll active on the electrode.] - Biofizika *11*: 593-600, 1966. [In R.]

2118 - EVSTIGNEEV, V.B., GAVRILOVA, V.A.: O promezhutochnykh stadiyakh obratimogo fotookisleniya khlorofilla *b*. [Intermediary stages of reversible photooxidation of
chlorophyll *b*.] - Dokl. Akad. Nauk SSSR *174*: 476-479, 1967. [In R.]

2119 - EVSTIGNEEV, V.B., GAVRILOVA, V.A.: Izmenenie kislotno-osnovnogo ravnosiya v
srede pri fotovosstanovlenii i fotookislenii khlorofilla i ego analogov. [Chan-
ges of acid-base equilibrium in a medium during photoreduction and photooxidation of chlorophyll and its analogues.] - Mol. Biol. (Moskva) *2*: 869-877, 1968.
[In R, ab: E.]

2120 - EVSTIGNEEV, V.B., GAVRILOVA, V.A.: O fotosensibilizatsii khlorofillom i ego ana-
       logami okislitel'no-vosstanovitel'nykh reaktsiĭ pri nalichii dvukh aktseptorov
       elektrona. [On the photosensibilization by chlorophyll and its analogues of the
       redox reactions in the presence of two electron acceptors.] - Dokl. Akad. Nauk
       SSSR 188: 219-222, 1969. [In R.]

2121 - EVSTIGNEEV, V.B., GAVRILOVA, V.A.: Ob obratimom fotookislenii ftalotsianina mag-
       niya v svyazi s izucheniem fotokhimii khlorofilla. [Reversible photooxidation of
       magnesium phthalocyanin related to chlorophyll photochemistry.] - Biofizika 14:
       43-50, 1969. [In R, ab: E.]

2122 - EVSTIGNEEV, V.B., GAVRILOVA, V.A., BEKASOVA, O.D.: O vzaimodeĭstvii khlorofilla
       s aktseptorom elektrona, obladayushchim sil'no-otritsatel'nym okislitel'no-voss-
       tanovitel'nym potentsialom. [Interaction of chlorophyll with an electron accep-
       tor having a strong negative oxidation-reduction potential.] - Biofizika 11: 584-
       592, 1966. [In R.]

2123 - EVSTIGNEEV, V.B., GAVRILOVA, V.A., OLOVYANISHNIKOVA, G.D.: Ob obratimom fotookis-
       lenii bakterioviridina s obrazovaniem elektrodno-aktivnoĭ okislennoĭ formy. [Re-
       versible photooxidation of bacterioviridine with formation of an oxidized form
       active on the electrode.] - Mol. Biol. (Moskva) 1: 59-66, 1967. [In R, ab: E.]

2124 - EVSTIGNEEV, V.B., GAVRILOVA, V.A., SADOVNIKOVA, N.A.: O fotookislenii khlorofil-
       la a s obrazovaniem elektrodno-aktivnoĭ okislennoĭ formy pigmenta. [Photooxida-
       tion of chlorophyll a with formation of an electrode-active oxidized form of the
       pigment.] - Biokhimiya 31: 1229-1236, 1966. [In R, ab: E.]

2125 - EVSTIGNEEV, V.B., MUKHIN, E.N.: Fotosintez, kak protsess preobrazovaniya solnech-
       noĭ energii v khimicheskuyu. [Photosynthesis as a process of transformation of
       solar energy into chemical energy.] - Dokl. vsesoyuz. Konf. Ispol'z. solnech.
       Energii, Sekts. S-7: 11-20, 1969. [In R.]

2126 - EVSTIGNEEV, V.B., OLOVYANISHNIKOVA, G.D., POPOVA, N.B., SADOVNIKOVA, N.A.: O
       vozmozhnosti fotokhimicheskogo vzaimodeĭstviya fosfodoksina s khlorofillom. [Pos-
       sibility of photochemical interaction of phosphodoxine with chlorophyll.] - Bio-
       fizika 13: 616-621, 1968. [In R.]

2127 - EVSTIGNEEV, V.B., OLOVYANISHNIKOVA, G.D., SADOVNIKOVA, N.A.: Vzeimodeĭstvie
       ubikhinona 30 (koenzim Q_6) s vozbuzhdennymi svetom khlorofillom i bakterioviri-
       dinom. [Interaction of ubiquinone 30 (coenzyme Q_6) with light-excited chlorophyll
       and bacterioviridin.] - Mol. Biol. (Moskva) 3: 41-48, 1969. [In R, ab: E.]

2128 - EVSTIGNEEV, V.B., PROKHOROVA, L.I.: Ob opredelenii khlorofillov a i b v smesi
       bez razdeleniya komponentov. [Determination of chlorophylls a and b in mixture
       without separation of the components.] - Biokhimiya 33: 286-295, 1968. [In R,
       ab: E.]

2129 - EVSTIGNEEV, V.B., SADOVNIKOVA, N.A., OLOVYANISHNIKOVA, G.D.: O fotokhimicheskom
       vzaimodeĭstvii khlorofilla s plastokhinonom. [Photochemical interaction of chlo-
       rophyll with plastoquinone.] - Mol. Biol. (Moskva) 2: 21-28, 1968. [In R, ab: E.]

2130 - EVSTIGNEEV, V.B., SADOVNIKOVA, N.A., OLOVYANISHNIKOVA, G.D.: O svoĭstvakh labil'-
       nykh fotookislennykh form khlorofilla b. - [Properties of labile photooxidized
       forms of chlorophyll b.] - Dokl. Akad. Nauk SSSR 187: 1184-1187, 1969. [In R.]

2131 - EVTUSHENKO, G.A., KARAKEEVA, R.K.: Vliyanie ingibitorov rosta na sostav zeleno-
       plastidnykh pigmentov v list'yakh ozimykh pshenits. [Effect of growth inhibitors
       on pigment composition of green plastids in leaves of winter wheats.] - Izv.
       Akad. Nauk Kirg. SSR 6: 40-47, 1968. [In R.]

2132 - EYTAN, G., OHAD, I.: Biogenesis of chloroplast membranes. VI. Cooperation be-
       tween cytoplasmic and chloroplast ribosomes in the synthesis of photosynthetic
       lamellar proteins during the greening process in a mutant of Chlamydomonas rein-
       hardi y-1. - J. biol. Chem. 245: 4297-4307, 1970.

2133 - EYTAN, G., OHAD, I.: Synthesis and assembly of photosynthetic lamellar proteins
       during the greening process in Chlamydononas reinhardi y-1 mutant. - Israel J.
       Chem. 8: 128 p, 1970.

2134 - FABIAN, I.: Der Chlorophyllgehalt der Sonnenblumenblätter bei P- und K-Mangel.
       - Rev. roum. Biol., Sér. Bot. 15: 337-344, 1970.

2135 - **FABIAN, I., TIŢU, H.**: Influenţa fosforului asupra ultrastructurii cloroplastelor din mezofilul de floarea-soarelui. [Effect of phosphorus on chloroplast ultrastructure in the mesophyll of sunflower.] - Stud. Cercet. Biol., Ser. bot. *21*: 321-324, 1969. [In Rum.]

2136 - **FABIAN-GALAN, G.**: Despre transportul asimilatelor la mazăre şi ardei în ontogeneză. [Transport of assimilates in pea and pepper during ontogenesis.] - Stud. Cercet. Biol., Ser. bot. *18*: 271-280, 1966. [In Rum.]

2137 - **FABIAN-GALAN, G.**: Despre transportul asimilatelor în decursul dezvoltării fructelor. [Transport of assimilates during the development of fruits.] - Stud. Cercet. Biol., Ser. bot. *19*: 151-158, 1967. [In Rum.]

2138 - **FABIAN-GALAN, G.**: Fotosinteza şi transportul asimilatelor în decursul coacerii fructelor la *Fragaria* sp. [Photosynthesis and transport of assimilates during the ripening of *Fragaria* fruit.] - Stud. Cercet. Biol., Ser. bot. *20*: 423-428, 1968. [In Rum., ab: E.]

2139 - **FABIAN-GALAN, G.**: On the transport of assimilated substances in the course of fruit growth in *Fragaria* sp. - Rev. roum. Biol., Sér. Bot. *13*: 35-39, 1968.

2140 - **FABIAN-GALAN, G.**: Fotosinteza - aspecte actuale. [Photosynthesis - recent aspects.] - Natura, Ser. biol. *21* (2): 3-10, 1969. [In Rum.]

2141 - **FABIAN-GALAN, G.**: Photosynthesis and transport of assimilated substances in plants with different types of metabolism. - Rev. roum. Biol., Sér. Bot. *14*: 301-307, 1969.

2142 - **FABIAN-GALAN, G., ATANASIU, L., SĂLĂGEANU, N.**: Organic substances produced by photosynthesis in lichens. - In: METZNER, H. (ed.): Progress in Photosynthesis Research. Vol. III. Pp. 1553-1558. Tübingen 1969.

2143 - **FAHMY, R., MOSTAFA, D., FOUAD, S.**: Effect of indoleacetic acid on the nature and properties of organic compounds assimilated in *Linum usitatissimum* cv. Giza 4. - U.A.R. J. Bot. *13*: 1-8, 1970. [Chl, car.]

2144 - **FAÏNZIL'BER, A.M., KARAPETYAN, A.O.**: Analiticheskie metody resheniya i primenenie elektronno-vychislitel'nykh mashin dlya postroeniya matematicheskikh modeleĭ fotosinteza i rosta mikroorganizmov. [Analytical methods of solution and use of computers for construction of mathematical models of photosynthesis and growth of microorganisms.] - Dokl. TSKhA (Moskva) *154*: 207-212, 1969. [In R.]

2145 - **FAIRBAIRN, J.W., EL-MASRY, S.**: The alkaloids of *Papaver somniferum* L. - VI. "Bound" morphine and seed development. - Phytochemistry *7*: 181-187, 1968. [Chl.]

2146 - **FALK, H.**: Rough thylakoids: Polysomes attached to chloroplast membranes. - J. Cell Biol. *42*: 582-587, 1969.

2147 - **FALK, H., KLEINIG, H.**: Feinbau und Carotinoide von *Tribonema (Xanthophyceae)*. - Arch. Mikrobiol. *61*: 347-362, 1968.

2148 - **FALK, H., LÜTTGE, U., WEIGL, J.**: Untersuchungen zur Physiologie plasmolysierter Zellen. II. Ionenaufnahme, O_2-Wechsel, Transport. - Z. Pflanzenphysiol. *54*: 446-462, 1966. [Ps.]

2149 - **FALK, R.H., BOGORAD, L.**: Immunological distinction between fraction I protein and protochlorophyllide holochrome. - Plant Physiol. *44*: 1669-1671, 1969.

2150 - **FALUDI-DÁNIEL, Á., AMESZ, J., NAGY, A.H.**: P700 oxidation and energy transfer in normal maize and in carotenoid-deficient mutants. - Biochim. biophys. Acta *197*: 60-68, 1970.

2151 - **FALUDI-DÁNIEL, Á., DÉZSI, L., FARKAS, G.L., PACSÉRY, M.**: Glycolic acid oxidase activity in normal leaves and chloroplast mutants of increased photosensitivity. - Stud. biophys. *5*: 111-116, 1967.

2152 - **FALUDI-DÁNIEL, Á., FRIDVALSZKY, L., GYURJÁN, I.**: Pigment composition and plastid structure in leaves of carotenoid mutants of maize. - Planta *78*: 184-195, 1968.

2153 - **FALUDI-DÁNIEL, Á., KUNHALMI, M., HAFIEK, A., GYURJÁN, I.**: Free and protein-bound amino acid levels in normal and chloroplast mutant corn leaves. - Ann. Univ. Sci. budapest. R. Eötvös nomin., Sect. biol. *8*: 69-75, 1966. [Chl, car.]

2154 - FALUDI-DÁNIEL, Á., LÁNG, F., FRADKIN, L.I.: The state of chlorophyll a in leaves of carotenoid mutant maize. - In: GOODWIN, T.W. (ed.): Biochemistry of Chloroplasts. Vol. I. Pp. 269-274. Academic Press, London-New York 1966.

2155 - FALUDI-DÁNIEL, Á., LÁNG, F., NAGY, A., FALUDI, B.: The inheritance of carotenoid types in maize. - Acta agron. Acad. Sci. hung. 16: 1-6, 1967.

2156 - FALUDI-DÁNIEL, Á., NAGY, A.H., GYESKO, A.: Chloroplast development in maize leaves at different light intensities. - Ann. Univ. Sci. budapest. R. Eötvös nomin., Sect. biol. $9-10$: 143-149, 1968.

2157 - FALUDI-DÁNIEL, Á., NAGY, A.H., NAGY, Á.: The ratio of chlorophyll a to chlorophyll b in normal and mutant maize leaves. - Acta bot. Acad. Sci. hung. 14: 17-27, 1968.

2158 - FALUDI-DÁNIEL, Á., NAGY, A.H., NAGY, Á.: Chlorophyll synthesis in normal and photosensitive maize leaves. - In: METZNER, H. (ed.): Progress in Photosynthesis Research. Vol. II. Pp. 592-598. Tübingen 1969.

2159 - FALUDY-DÁNIEL, Á., DUBRAVITSKI, D.: PigmentnyΓ sostav, struktura i fotosinteticheskaya sposobnost' khloroplastov mutantnykh list'ev kukuruzy. [Pigment content, structure and photosynthetic capacity of chloroplasts in mutant maize leaves.] - Fiziol. Rast. 14: 232-236, 1967..[In R, ab: E.]

2160 - FAN, H.N., CRAMER, W.A.: The redox potential of cytochromes b-559 and b-563 in spinach chloroplasts. - Biochim. biophys. Acta 216: 200-207, 1970.

2161 - FANOUS, M.A.: Test for drought resistance in pearl millet *(Pennisetum typhoideum)*. - Agron. J. 59: 337-340, 1967. [Chl.]

2162 - FARAPONOVA, G.P.: Issledovanie radiatsionnykh termoelementov, prednaznachennykh dlya izmereniya radiatsionnykh potokov v atmosfere. [Study of radiation thermoelements determined for measuring radiant fluxes in the atmosphere.] - In: Aktinometriya i Optika Atmosfery. Pp. 202-212. Valgus, Tallin 1968. [In R, ab: E.]

2163 - FARAPONOVA, G.P., TIMANOVSKAYA, R.G.: Polevye ispytaniya radiatsionnykh termoelementov. [Field tests of radiation thermoelements.] - In: Aktinometriya i Optika Atmosfery. Pp. 212-218. Valgus, Tallin 1968. [In R.]

2164 - FARINEAU, J.: Étude de la genèse des composés phosphorylés au sein de feuilles vertes illuminées dans des atmosphères de compositions variées. - Compt. rend. Acad. Sci. (Paris), Sér. D 263: 36-39, 1966.

2165 - FARINEAU, J.: Métabolisme de quelques composés phosphorylés et photophosphorylation *"in vivo"* chez les feuilles de Maïs. - Planta 85: 135-156, 1969.

2166 - FARINEAU, J.: Pool sizes of phosphorylated compounds in maize leaves and their relation to the composition of gazeous medium and wavelengths of light. - In: METZNER, H. (ed.): Progress in Photosynthesis Research. Vol. III. Pp. 1141-1148. Tübingen 1969.

2167 - FARINEAU, N.: Action de lumières colorées sur l'ultrastructure des plastes et sur la teneur en pigments et en lipides polaires de feuilles de plantules étiolées de Maïs. - Bull. Soc. franç. Physiol. vég. 14: 275-305, 1968.

2168 - FARINEAU, N.: Étude comparée de l'action exercée par la lumière blanche et par diverses lumières monochromatiques sur l'évolution ultrastructurale et la biosynthèse pigmentaire des étioplastes de feuille de Maïs. - Bull. Soc. bot. France 117: 27-41, 1970.

2169 - FARINEAU, N.: Sur une méthode d'isolement d'étioplastes en bon état structural, à partir de feuilles de plantules étiolées de Maïs. - Compt. rend. Acad. Sci. (Paris), Sér. D 271: 664-666, 1970.

2170 - FARRON, F.: Isolation and properties of a chloroplast coupling factor and heat-activated adenosine triphosphatase. - Biochemistry 9: 3823-3828, 1970.

2171 - FARRON, F.: The conversion of coupling factor 1 from spinach chloroplasts to an active ATPase by heat. - Diss. Abstr. int. B 31: 1125-B, 1970.

2172 - FARRON, F., RACKER, E.: Studies on the mechanism of the conversion of coupling factor 1 from chloroplasts to an active adenosine triphosphatase. - Biochemistry 9: 3829-3836, 1970.

2173 - **FASULO, M.P., DALL'OLIO, G.:** Effetto del Cycocel e dell'AMO-1618 sulla crescita ed il metabolismo de *Euglena gracilis* KLEBS. [Effect of CCC and AMO-1618 on growth and metabolism of *Euglena gracilis* KLEBS.] - Ann. Univ. Ferrara, Ser. IV - Bot. *3* (13): 133-137, 1968. [Ps; in Ital., ab: E, F.]

2174 - **FATT, I.:** The oxygen electrode: some special applications. - Ann. N.Y. Acad. Sci. *148*: 81-92, 1968.

2175 - **FATTAH, Q.A., WORT, D.J.:** Effect of light and temperature on stimulation of vegetative and reproductive growth of bean plants by naphthenates. - Agron. J. *62*: 576-577, 1970. [Ps.]

2176 - **FATTAH, Q.A., WORT, D.J.:** Metabolic responses of bush bean plants to naphthenate application. - Can. J. Bot. *48*: 861-866, 1970. [Ps.]

2177 - **FAUST, H., ZAHN, H., DIETZE, H.-J.:** Anwendung stabiler Isotope bei der Untersuchung des Mineralstoffhaushaltes. 2. Untersuchung zum Magnesium-Stoffwechsel bei Avena sativa L. mit Hilfe von ^{25}Mg. - Tagungsber. DAW *85* (Mineralstoffversorgung Pflanze Tier): 173-177, 1966.[Chl.]

2178 - **FAVALI, M.A., CONTI, G.G.:** Ultrastructural observations on the chloroplasts of basil plants either infected with different viruses or treated with 3-amino-1,2,4-triazole. - Protoplasma *70*: 153-166, 1970.

2179 - **FAVREAU, G., RAYMOND, Y.:** Étude préliminaire des pigments des sommités fleuries d'*Anaphalis margaritacea* L. - Naturaliste can. *94*: 63-71, 1967. [Chl.]

2180 - **FAY, P.:** Cell differentiation and pigment composition in *Anabaena cylindrica*. - Arch. Mikrobiol. *67*: 62-70, 1969.

2181 - **FAY, P.:** Photostimulation of nitrogen fixation in *Anabaena cylindrica*. - Biochim. biophys. Acta *216*: 353-356, 1970. [Ps, Chl.]

2182 - **FEDARAŬ, N.I., RAPTUNOVICH, E.S.:** Zmyanenne fotasinteza sasny pad uplyvam karanёvaŭ gubki. [Changes of pine photosynthesis induced by a root fungus.] - Vesci Akad. Navuk Belarus. SSR, Ser. biyal. Navuk *1968* (4): 42-47, 105, 1968. [In Beloruss., ab: R.]

2183 - **FEDERER, C.A., TANNER, C.B.:** Sensors for measuring light available for photosynthesis. - Ecology *47*: 654-657, 1966.

2184 - **FEDOROV, V.D.:** Fiziologicheskie osobennosti bakterial'nogo fotosinteza. [Physiological properties of bacterial photosynthesis.] - Trudy mosk. Obshch. Ispyt. Prirody *24* (Biol. avtotrof. Mikroorg.): 124-130, 1966. [In R.]

2185 - **FEDOROV, V.D., MAKSIMOV, V.N.:** Izuchenie protsessov pervichnoĭ produktivnosti vodoemov metodom planiruemykh dobavok biogennykh elementov. [Study of primary productivity processes in waters by the method of planned addition of biogenic elements.] - Nauch. Dokl. vyssh. Shkoly, biol. Nauki *10* (4): 132-142, 1967. [In R.]

2186 - **FEDOROV, V.D., MAKSIMOV, V.N., KHROMOV, V.M.:** Vliyanie sveta i temperatury na pervichnuyu produktsiyu nekotorykh odnokletochnykh zelenykh i diatomovykh vodorosleĭ. [Effect of light and temperature on primary production of some unicellular green algae and diatoms.] - Fiziol. Rast. *15*: 640-651, 1968. [In R, ab: E.]

2187 - **FEDOSEEV, A.P.:** Radiatsionnyĭ rezhim sukhikh i svezhikh dubrav yuzhnoĭ lesostepi USSR. [Radiation regime of dry and fresh oak forests of southern forest-steppe of the Ukr.S.S.R.] - In: Svetovoĭ Rezhim, Fotosintez i Produktivnost' Lesa. Pp. 65-76. Nauka, Moskva 1967. [Photointegrator of PhAR; in R.]

2188 - **FEDOSEEVA, G.P.:** Adaptatsiya fotosinteza u ogurtsov k vysokoĭ temperature v usloviyakh zashchishchennogo grunta. [Adaptation of cucumber photosynthesis to high temperature on protected ground.] - Uchen. Zap. ural'. gos. Univ., Ser. biol. *58*: 67-73, 1967. [In R.]

2189 - **FEDOSEEVA, G.P.:** Izmenenie fotosinteticheskogo metabolizma ugleroda u ogurtsa v protsesse adaptatsii k vysokoĭ temperature. [Changes in photosynthetic carbon metabolism in cucumber during adaptation to high temperature.]- Uchen. Zap. ural'. gos. Univ., Ser. biol. *113*: 121-130, 1970. [In R.]

2190 - **FEDTKE, C.:** Intramolecular hydrogen transfer in isomerisation reactions of sugar phosphates in the Calvin cycle. - In: METZNER, H. (ed.): Progress in Photosynthesis Research. Vol. III. Pp. 1597-1603. Tübingen 1969.

2191 - FEDYUSHIN, A.A., BEDENKO, V.P.: Nekotorye rezul'taty izmereniya pryamoĭ solnechnoĭ fotosinteticheski aktivnoĭ radiatsii spektral'nym fitoaktinometrom. [Some results of measurements of direct photosynthetically active solar radiation with a spectral phytoactinometer.] - Fiziol. Rast. 16: 756-759, 1969. [In R, ab: E.]

2192 - FEICHTMAYR, F., HEILBRONNER, E., NÜRRENBACH, A., POMMER, H., SCHLAG, J.: The dispersive interaction of non-polar solutes with non-polar solvents: Solvent effects on electronic spectra of carotenoids. - Tetrahedron 25: 5383-5408, 1969.

2193 - FEIERABEND, J.: Änderungen im Enzymsystem bei der Bildung des Photosynthese-Apparates von Roggenkeimlingen. - Ber. deut. bot. Ges. 79: (68)-(69), 1966.

2194 - FEIERABEND, J.: Enzymbildung in Roggenkeimlingen während der Umstellung von heterotrophem auf autotrophes Wachstum. - Planta 71: 326-355, 1966. [Ps - enzymes.]

2195 - FEIERABEND, J.: Regulationsvorgänge bei der Bildung von Photosyntheseenzymen. - Umschau Wiss. Tech. 67: 494-495, 1967.

2196 - FEIERABEND, J.: Der Einfluss von Cytokininen auf die Bildung von Photosyntheseenzymen in Roggenkeimlingen. - Planta 84: 11-29, 1969.

2197 - FEIERABEND, J.: Formation of the photosynthetic apparatus during germination and its control. - In: METZNER, H. (ed.): Progress in Photosynthesis Research. Vol. I. Pp. 280-283. Tübingen 1969.

2198 - FEIERABEND, J.: Characterization of cytokinin action on enzyme formation during the development of the photosynthetic apparatus in rye seedlings. Enzymes of the reductive and oxidative pentose phosphate cycles. - Planta 94: 1-15, 1970.

2199 - FEIERABEND, J.: Proteinsynthese und Enzymbildung in Keimlingen bei niedrigen Wachstumstemperaturen und ihre Beziehungen zum Cytokininhaushalt. - Z. Pflanzenphysiol. 62: 70-82, 1970.

2200 - FEIERABEND, J., BERGER, C., MEYER, A.: Spezifische Störung von Entwicklung und Enzymbildung der Plastiden höherer Pflanzen durch hohe Wachstumstemperaturen. - Z. Naturforsch. 24 b: 1641-1647, 1969. [Chl.]

2201 - FEIERABEND, J., PIRSON, A.: Die Wirkung des Lichts auf die Bildung von Photosyntheseenzymen in Roggenkeimlingen. - Z. Pflanzenphysiol. 55: 235-245, 1966.

2202 - FEIGE, B.: Beiträge zur Physiologie einheimischer Algen. 1. ^{14}C-Markierungsprodukte von drei Süsswasserrotalgen. - Z. Pflanzenphysiol. 63: 288-291, 1970.

2203 - FEIGE. B., GIMMLER. H., JESCHKE, W.D., SIMONIS, W.: Eine Methode zur dünnschichtchromatographischen Auftrennung von ^{14}C- und ^{32}P-markierten Stoffwechselprodukten. - J. Chromatogr. 41: 80-90, 1969.

2204 - FEIGE, B., SIMONIS, W.: Untersuchungen zur Physiologie der Flechte Cladonia convoluta (LAM.) P. COUT. I. Allgemeines und Methodik der Untersuchungen. - Flora A 160: 552-560, 1969. [Chl.]

2205 - FELTON, R.H., DOLPHIN, D., BORG, D.C., FAJER, J.: Cations and cation radicals of porphyrins and ethyl chlorophyllide a. - J. amer. chem. Soc. 91: 196-198, 1969.

2206 - FERREE, M.E., BARDEN, J.A.: The influence of strains and root stocks on photosynthesis, respiration and morphology of delicious apple trees. - HortScience 5: 304, 1970.

2207 - FERRETTI, J.J., GRAY, E.D.: Control of enzyme synthesis during adaptation in synchronously dividing populations of Rhodopseudomonas spheroides. - Biochem. biophys. Res. Commun. 29: 501-507, 1967. [Chl.]

2208 - FETTER, F., ALTMANN, H., PFISTERER, E.E.: Über die Zusammenhänge zwischen Eisenaufnahme und Farbstoffsynthese in synchronen Chlorellazellen. - Bodenkultur 19: 255-258, 1968. [Chl.]

2209 - FIALA, K.: Rhizome biomass and its relation to shoot biomass and stand pattern in eight clones of Phragmites communis TRIN. - In: Productivity of Terrestrial Ecosystems. Production Processes. Czechosl. nat. Comm. IBP, PT-PP Report No. 1. Pp. 95-98. Praha 1970.

2210 - **FIALA, K.**: Seasonal changes in the growth of the underground organs in *Typha latifolia* L. - In: Productivity of Terrestrial Ecosystems. Production Processes. Czechosl. nat. Comm. IBP, PT-PP Report No. 1. Pp. 99-100. Praha 1970. [Growth analysis.]

2211 - **FIALA, K., DYKYJOVÁ, D., KVĚT, J., SVOBODA, J.**: Methods of assessing rhizome and root production in reed-bed stands. - In: Methods of Productivity Studies in Root Systems and Rhizosphere Organisms. Pp. 36-47. Nauka, Leningrad 1968.

2212 - **FIASSON, J.-L., ARPIN, N., LEBRETON, P.**: Sur l'analyse qualitative et quantitative des caroténoîdes naturels. - Chim. anal. *51*: 227-236, 1969.

2213 - **FICHERA, P., D'ARRIGO, C.M.**: La clorosi ferrica da calcare nei terreni agrumeti della Sicilia orientale. III. - Prove di "pieno campo" sull'influenza dell' azoto e dell'irigazione nella formazione della clorofilla. [Iron chlorosis caused by lime in the citrus fruit cultures of East Sicily. III. Field study of the effects of nitrogen and irrigation on the chlorophyll formation.] - Agrochimica *14*: 332-340, 1970. [In Ital., ab: E, F, G, Span.]

2214 - **FICHERA, P., D'ARRIGO, C.M.**: La clorosi ferrica da calcare nei terreni agrumeti della Sicilia orientale. IV. - Prove di "pieno campo" sull'influenza dell' azoto e dell'irrigazione nella nutrizione minerale. [Iron chlorosis caused by lime in the citrus fruit cultures of East Sicily. IV. Field study of the effects of nitrogen and irrigation on mineral nutrition.] - Agrochimica *14*: 426-433, 1970. [Chl; in Ital., ab: E, F, G, Span.]

2215 - **FILIMONOVA, V.D., CHERNAVSKAYA, M.M.**: Intensivnost' solnechnoî radiatsii v parkovykh nasazhdeniyakh goroda Erevana. [Density of solar radiation in parks of the town Erevan.] - Biol. Zh. Armenii *21* (10): 68-71, 1968. [In R.]

2216 - **FILIPPOVA, L.A.**: O metodakh vydeleniya khloroplastov v nevodnuyu sredu. [Methods for isolation of chloroplasts in a non-aqueous medium.] - Fiziol. Rast. *14*: 1107-1112, 1967. [In R, ab: E.]

2217 - **FILIPPOVA, L.A., LAVRENETSKAYA, T.E.**: Ispol'zovanie krakhmala, obrazovannogo pri fotosinteze, v dykhanii list'ev rasteniî. [Use of starch produced in photosynthesis for respiration of plant leaves.] - Bot. Zh. *52*: 995-998, 1967. [In R.]

2218 - **FILIPPOVA, L.A., ZALENSKIÏ, O.V.**: O vnutrikletochnoî lokalizatsii organicheskikh veshchestv, obrazovannykh v protsesse fotosinteza, i ikh uchastii v dykhanii. [Intercellular localization of organic substances formed in photosynthesis and their participation in respiration.] - Bot. Zh. *52*: 1158-1162, 1967. [In R.]

2219 - **FILIPPOVA, L.A., ZALENSKIÏ, O.V.**: Intracellular localization of assimilates and their utilization in the process of respiration. - Photosynthetica *3*: 104-111, 1969.

2220 - **FILIPPOVICH, I.I., TONGUR, A.M., ALINA, B.A., OPARIN, A.I.**: Strukturnaya organizatsiya beloksinteziruyushcheî sistemy khloroplastov. [Structural organization of the protein-synthesizing system of chloroplasts.] - Biokhimiya *35*: 247-256, 1970. [In R, ab: E.]

2221 - **FILIPPOVSKIÏ, Yu.N., SEMENENKO, V.E., NICHIPOROVICH, A.A.**: K voprosu o raspredelenii luchistoî energii v suspenzii khlorelly. [On the distribution of radiant energy in a *Chlorella* suspension.] - In: Fotosinteziruyushchie Sistemy Vysokoî Produktivnosti. Pp. 193-204. Nauka, Moskva 1966. [In R.]

2222 - **FILIPPOVSKIÏ, Yu.N., SEMENENKO, V.E., NICHIPOROVICH, A.A.**: Opticheskie svoîstva suspenzii khlorelly pri deîstvii slozhnykh spektrov. [Optical properties of *Chlorella* suspension under the action of combined spectra.] - In: Fotosinteziruyushchie Sistemy Vysokoî Produktivnosti. Pp. 204-212. Nauka, Moskva 1966. [In R.]

2223 - **FILIPPOVSKIÏ, Yu.N., SEMENENKO, V.E., NICHIPOROVICH, A.A., LEBEDEV, V.M., TSOGLIN, L.N.**: Raspredelenie luchistoî energii v populyatsiyakh vodorosleî v svyazi s ikh fotosinteticheskoî produktivnost'yu. [Distribution of radiant energy in algae populations in relation to their photosynthetic productivity.] - In: Bioenergetika i Biologicheskaya Spektrofotometriya. Pp. 231-241. Nauka, Moskva 1967. [In R.]

2224 - FILMER, D.L., COOPER, T.G.: Effect of varying temperature and pH upon the pre-
dicted rate of "CO_2" utilization by carboxylases. - J. theor. Biol. *29*: 131-145,
1970.

2225 - FILNER, B., KLEIN, A.O.: Changes in enzymatic activities in etiolated bean seedling leaves after a brief illumination. - Plant Physiol. *43*: 1587-1596, 1968.
[RuDP-carboxylase *etc.*]

2226 - FIRENZUOLI, A.M., RAMPONI, G., VANNI, P., ZANOBINI, A.: Ferredoxin-NADP reductase from *Pinus pinea*. - Life Sci. *7*: 905-915, 1968.

2227 - FIRENZUOLI, A.M., VANNI, P., ZANOBINI, A.: Sulla ferredoxina-NADP reduttase: Lo-
calizzazione in *Pinus pinea*. [Localization of ferredoxin NADP reductase in *Pinus
pinea*.] - Boll. Soc. ital. Biol. Sper. *44*: 1201-1204, 1968. [In Ital., ab: E.]

2228 - FIRKET, H.: A very simple trick to produce controlled CO_2 concentrations in the
gas phase overlying cell cultures. - Experientia *25*: 671, 1969.

2229 - FIRSANOVA, G.N., ZHIVUKHINA, G.M.: Fiziologicheskie i morfologo-anatomicheskie
izmeneniya u rasteniĭ v zavisimosti ot mesta naneseniya gibberellina. [Physiolo-
gical and morphologo-anatomical changes in plants in dependence on the site of
gibberellin application.] - Fiziol. Rast. *16*: 861-864, 1969. [Ps; in R, ab: E.]

2230 - FISCHER, E., WIESSNER, W.: Die gegenseitige Beeinflussung von Acetat- und CO_2-
Assimilation im Licht bei *Euglena gracilis*. - Ber. deut. bot. Ges. *81*: 347-348,
1968.

2231 - FISCHER, K., METZNER, H.: On chlorophyll and pigment P 750 of *Anacystis nidulans*.
- In: METZNER, H. (ed.): Progress in Photosynthesis Research. Vol. II. Pp. 547-
551. Tübingen 1969.

2232 - FISHER, D.A.: Correlation of pigment content with ultrastructural detail of *in
vivo* chloroplasts of barley seedlings. - Plant Physiol. *43* (Suppl.): S-6, 1968.

2233 - FISHER, D.A., WEIER, T.E.: Pigment content correlated with ultrastructural detail in chloroplasts of barley (*Hordeum vulgare* L.) seedlings treated with various herbicides. - Plant Physiol. *42* (Suppl.): S-26, 1967.

2234 - FISHER, D.B.: Kinetics of C-14 translocation in soybean. II. Kinetics in the
leaf. - Plant Physiol. *45*: 114-118, 1970.

2235 - FISHER, D.B.: Kinetics of C-14 translocation in soybean. III. Theoretical considerations. - Plant Physiol. *45*: 119-125, 1970.

2236 - FISHER, F.E., KROME, W.H., COLBURN, B.E., OBERBACHER, M.F.: Better colored limes
following spray treatments. - Proc. Fla. State hort. Soc. *80*: 391-395, 1967.
[Chl.]

2237 - FISHER, R.R., GUILLORY, R.J.: Inhibition of the energy conservation reactions
of *Rhodospirillum rubrum* by Dio-9. - Biochim. biophys. Acta *143*: 654-656, 1967.

2238 - FISHER, R.R., GUILLORY, R.J.: A soluble factor related to the energy-linked
transhydrogenase reaction of *Rhodospirillum rubrum* chromatophores. - J. biol.
Chem. *244*: 1078-1079, 1969.

2239 - FISHER, R.R., GUILLORY, R.J.: Partial resolution of energy-linked reactions in
Rhodospirillum rubrum chromatophores. - FEBS Lett. *3*: 27-30, 1969.

B2240 - Fitoaktinometricheskie Issledovaniya Rastitel'nogo Pokrova.[Phytoactinometric
Studies of a Plant Community.] - Valgus, Tallin 1967.

2241 - FLATMARK, T., DUS, K., de KLERK, H., KAMEN, M.D.: Comparative study of physico-
chemical properties of two *c*-type cytochromes of *Rhodospirillum molischianum*. -
Biochemistry *9*: 1991-1996, 1970.

2242 - FLAUMENHAFT, E., UPHAUS, R.A., KATZ, J.J.: Isotope biology of ^{13}C. Extensive incorporation of highly enriched ^{13}C in the alga *Chlorella vulgaris*. - Biochim.
biophys. Acta *215*: 421-429, 1970. [Apparatus for culture of algae in a $^{13}CO_2$ atmosphere.]

2243 - FLEISCHMAN, D.E.: Chemiluminescence in photosynthetic bacteria. - In: METZNER,
H. (ed.): Progress in Photosynthesis Research. Vol. II. Pp. 952-955. Tübingen
1969.

2244 - FLEISCHMAN, D.E., CLAYTON, R.K.: The effect of phosphorylation uncouplers and electron transport inhibitors upon spectral shifts and delayed light emission of photosynthetic bacteria. - Photochem. Photobiol. 8: 287-298, 1968.

2245 - FLEISCHMAN, D.L., CYLDESDALE, F.M., FRANCIS, F.J.: Effect of magnesium carbonate and sodium phosphate on the extraction of chlorophyll-like pigments after thermal processing of spinach puree. - J. Milk Food Technol. 33: 456-459, 1970.

2246 - FLEMING, I.: Absolute configuration and the structure of chlorophyll. - Nature 216: 151-152, 1967.

2247 - FLEMING, I.: The absolute configuration and the structure of chlorophyll and bacteriochlorophyll. - J. chem. Soc. (London) C 1968: 2765-2770, 1968.

2248 - FLEMION, F., DENGLER, R.E., DENGLER, N.G., STEWART, K.D.: Ultrastructure of the shoot apices and leaves of normal and physiologically dwarfed peach seedlings. I. Plastid development. - Contrib. Boyce Thompson Inst. 23: 331-344, 1967.

2249 - FLETCHER, R.A.: Retardation of leaf senescence by benzyladenine in intact bean plants. - Planta 89: 1-8, 1969. [Chl.]

2250 - FLETCHER, R.A., HOFSTRA, G., ADEPIPE, N.O.: Effects of benzyladenine on bean leaf senescence and the translocation of ^{14}C-assimilates. - Physiol. Plant. 23: 1144-1148, 1970.

2251 - FLETCHER, R.A., OSBORNE, D.J.: Gibberellin, as a regulator of protein and ribonucleic acid synthesis during senescence in leaf cells of Taraxacum officinale. - Can. J. Bot. 44: 739-745, 1966. [Chl.]

2252 - FLINN, A.M., PATE, J.S.: A quantitative study of carbon transfer from pod and subtending leaf to the ripening seeds of the field pea (Pisum arvense L.) - J. exp. Bot. 21: 71-82, 1970.

2253 - FLORENZANO, G., BALLONI, W., MATERASSI, R.: Un dispositivo in colonna per il rifornimento di CO_2 nel sistema di coltura massiva all'aperto di microalghe. [Column equipment for CO_2 saturation of algal mass culture.] - Ann. microbiol. enzimol. 16: 15-24, 1966. [In Ital., ab: E.]

2254 - FLORIAN, A.: Influența desimii de cultivare și a unor îngrășăminte minerale asupra recoltei plantelor de ardei. [The influence of density and certain mineral fertilizers on the yield of green pepper.] - An. Univ. Bucur., Biol. veg. 18: 257-264, 1969. [In Rum., ab: F, R.]

2255 - FLOROV, R.Ï.: Opyt opredeleniya skorosti vozniknoveniya entropii v listovoĭ sisteme lesnykh drevesnykh porod. [Determination of the rate of entropy production in a leaf system of forest woody plants.] - Fiziol. Rast. 13: 688-694, 1966. [In R, ab: E.]

2256 - FLOROV, R.Ï.: Opyt opredeleniya skorosti vozniknoveniya entropii v listovoĭ sisteme rasteniĭ na osnove zakona dissipatsii energii. [An attempt to determine the rate of entropy development in a leaf system of plants on the basis of the law of energy dissipation.] - Dokl. Akad. Nauk SSSR 178: 241-243, 1968. [In R.]

2257 - FLOROV, R.Ï.: Opyt opredeleniya potentsial'noĭ produktivnosti estestvennykh lesnykh mestoproizrastaniĭ na osnove zakona dissipatsii energii. [An attempt to determine the potential productivity of natural forest cultures on the basis of the energy dissipation law.] - Dokl. Akad. Nauk SSSR 179: 736-738, 1968. [In R.]

2258 - FLOROV, R.Ï., STOYANOV, Zh.: Entropiya dissipatsii listovoĭ sistemy kak vyraziteĺ zatrat assimilyatov na svetu. [Entropy of dissipation of the leaf system, serving to express the expenditure of assimilates under illumination.] - Dokl. Akad. Nauk SSSR 183: 1219-1220, 1968. [In R.]

2259 - FOALE, M.A.: The growth of the young coconut palm (Cocos nucifera L.). I. The role of the seed and of photosynthesis in seedling growth up to 17 months of age. - Aust. J. agr. Res. 19: 781-789, 1968.

2260 - FOCK, H.: Die Lichtatmung der grünen Pflanzen. Eine kritische Darstellung der bisher erarbeiteten Ergebnisse. - Biol. Zentralblatt 89: 545-572, 1970.

2261 - FOCK, H., BECKER, J.D., EGLE, K.: Use of labeled carbon dioxide for separation of CO_2 evolution from true CO_2 uptake by photosynthesizing Amaranthus and sunflower leaves. - Can. J. Bot. 48: 1185-1189, 1970.

2262 - FOCK, H., EGLE, K.: Über die "Lichtatmung" bei grünen Pflanzen. 1. Die Wirkung von Sauerstoff und Kohlendioxyd auf den CO_2-Gaswechsel während der Licht- und Dunkelphase. - Beitr. Biol. Pflanzen 42: 213-239, 1966.

2263 - FOCK, H., EGLE, K.: Über die Beziehungen zwischen dem Glykolsäure-Gehalt und dem Photosynthese-Gaswechsel von Bohnenblättern. - Z. Pflanzenphysiol. 57: 389-397, 1967.

2264 - FOCK, H., EGLE, K., SCHAUB, H., HILGENBERG, W.: Der Einfluss des Sauerstoff-Partialdrucks auf die Radioaktivität der nach $^{14}CO_2$-Zufütterung entstehenden Photosynthese-Produkte. - Z. Pflanzenphysiol. 61: 261-263, 1969.

2265 - FOCK, H., KROTKOV, G.: Relation between photorespiration and glycolate oxidase activity in sunflower and red kidney bean leaves. - Can. J. Bot. 47: 237-240, 1969.

2266 - FOCK, H., KROTKOV, G., CANVIN, D.T.: Photorespiration in liverworts and leaves. - In: METZNER, H. (ed.): Progress in Photosynthesis Research. Vol. I. Pp. 482-487. Tübingen 1969.

2267 - FOCK, H., SCHAUB, H., HILGENBERG, W.: Über den Sauerstoff- und Kohlendioxidgaswechsel von *Chlorella* und *Conocephalum* während der Lichtphase. - Z. Pflanzenphysiol. 60: 56-63, 1968.

2268 - FOCK, H., SCHAUB, H., HILGENBERG, W., EGLE, K.: Über den Einfluss niedriger und hoher O_2-Partialdrucke auf den Sauerstoff- und Kohlendioxidumsatz von *Amaranthus* und *Phaseolus* während der Lichtphase. - Planta 86: 77-83, 1969.

2269 - FOCKE, R.: Physiologisch-genetische Untersuchungen zur Trockenmassebildung in Maisblättern unter Berücksichtigung von Stofftransport und Ertrag. - Züchter 37: 371-376, 1967. [Dry matter increment determination.]

2270 - FODA, Y.H., EL-WARAKI, A., ZAID, M.A.: Effect of blanching and dehydration on the conversion of chlorophyll to pheophytin in green beans. - Food Technol. 22: 119-120, 1968.

2271 - FOGG, G.E.: Photosynthesis and nitrogen fixation in blue-green algae. - Ceylon Ass. Adv. Sci. Proc. annu. Sess. 25: 204-212, 1970; Proc. 25th annu. Sess. Ceylon Ass. Adv. Sci. Pp. 1-9. Print. Press, Vidyodaya Univ. Ceylon, Nugegoda 1970.

2272 - FOJTÍK, L.: The use of nuclear radiation for leaf area measurement. - Photosynthetica 3: 316-319, 1969.

2273 - FOLLETT, R.F., SCHMEHL, W.S., VIETS, F.G. Jr.: Seasonal leaf area, dry weight, and sucrose accumulation by sugarbeets. - J. amer. Soc. Sugar Beet Technol. 16: 235-252, 1970.

2274 - FOMENKO, A.A.: Opticheskie svoĭstva list'ev sakharnoĭ svekly pri razlichnykh usloviyakh mineral'nogo pitaniya rasteniĭ. [Optical properties of sugar beet leaves under various mineral nutrition of plants.] - In: Fotosintez i UrozhaĭnostĨ Sel'skokhozyaĭstvennykh RasteniĨ. Pp. 48-51. Min. sel'. Khoz. SSSR, Kiev 1970. [In R.]

2275 - FOOTE, C.S.: Mechanisms of photosensitized oxidation. - Science 162: 963-970, 1968. [Car.]

2276 - FOOTE, C.S., CHANG, Y.C., DENNY, R.W.: Chemistry of singlet oxygen. X. Carotenoid quenching parallels biological protection. - J. amer. chem. Soc. 92: 5216-5218, 1970.

2277 - FOOTE, C.S., DENNY, R.W., WEAVER, L., CHANG, Y.C., PETERS, J.: Quenching of singlet oxygen. - Ann. N.Y. Acad. Sci. 171: 139-148, 1970. [Car.]

2278 - FOOTT, J.H., HEINICKE, D.R.: Whitewash found harmless in applications of walnut leaves. - Calif. Agr. 21: 2-3, 1967. [Ps.]

2279 - FORBUSH, B., KOK, B.: Reaction between primary and secondary electron acceptors of photosystem II of photosynthesis. - Biochim. biophys. Acta 162: 243-253, 1968.

2280 - FORD, M.A., THORNE, G.N.: Effect of CO_2 concentration on growth of sugar-beet, barley, kale, and maize. - Ann. Bot. N.S. 31: 629-644, 1967.

2281 - FORK, D.C.: Evidence for the participation of carotenoids in the photosynthesis of algae and in a higher plant. - In: METZNER, H. (ed.): Progress in Photosynthesis Research. Vol. II. Pp. 800-810. Tübingen 1969.

2282 - FORK, D.C.: Light-induced reactions of carotenoids in the yellow-green alga *Botrydiopsis*. - Carnegie Inst. Year Book *67*: 496-503, 1969.

2283 - FORK, D.C., AMESZ, J.: Energy transfer between photosynthetic units of system 1 in algae. - Biochim. biophys. Acta *143*: 266-268, 1967.

2284 - FORK, D.C., AMESZ, J.: Light-induced shifts in the absorption spectrum of carotenoids in red and brown algae. - Photochem. Photobiol. *6*: 913-918, 1967.

2285 - FORK, D.C., AMESZ, J.: Transfer of energy between reaction centers of photosystem 1 in algae. - Carnegie Inst. Year Book *66*: 155-160, 1968.

2286 - FORK, D.C., AMESZ, J.: Light-induced shifts in the absorption spectrum of carotenoids in red, brown and yellow-green algae and in a barley mutant. - Carnegie Inst. Year Book *66*: 160-165, 1968.

2287 - FORK, D.C., AMESZ, J.: Action spectra and energy transfer in photosynthesis. - Annu. Rev. Plant Physiol. *20*: 305-328, 1969.

2288 - FORK, D.C., AMESZ, J.: Spectrophotometric studies of the mechanism of photosynthesis. - In: GIESE, A.C. (ed.): Photophysiology. Vol. 5. Pp. 97-126. Academic Press, New York-London 1970.

2289 - FORK, D.C., AMESZ, J., ANDERSON, J.M.: Light-induced reactions of chlorophyll *b*. - Carnegie Inst. Year Book *65*: 473-479, 1967.

2290 - FORK, D.C., AMESZ, J., ANDERSON, J.M.: Light-induced reactions of chlorophyll *b* and P 700 in intact plants and chloroplast fragments. - In: Energy Conversion by the Photosynthetic Apparatus. Brookhaven Symp. Biol. *19*: 81-94, 1967.

2291 - FORK, D.C., HEBER, U.W.: Studies on electron-transport reactions of photosynthesis in plastome mutants of *Oenothera*. - Plant Physiol. *43*: 606-612, 1968.

2292 - FORK, D.C., HEBER, U.W., MICHEL-WOLWERTZ, M.-R.: Studies on the photosynthesis of plastome mutants of *Oenothera*. - Carnegie Inst. Year Book *67*: 503-505, 1969.

2293 - FORK, D.C., de KOUCHKOVSKY, Y.: The 518-mμ absorbance change and its relation to the photosynthetic process. - Photochem. Photobiol. *5*: 609-619, 1966.

2294 - FORK, D.C., de KOUCHKOVSKY, Y.: Light-induced spectroscopic changes in the 600-mμ region in leaves of a higher plant and in *Chlorella*. - Biochim. biophys. Acta *153*: 891-894, 1968.

2295 - FORK, D.C., de KOUCHKOVSKY, Y.: A comparative study of the light-induced carotenoid change and fluorescence in the chlorophyll-*b*-less alga *Botrydiopsis alpina* (*Xanthophyceae*). - Carnegie Inst. Year Book *68*: 587-595, 1970.

2296 - FORK, D.C., MANTAI, K.E.: The effect of ultraviolet irradiation on the carotenoid change, electron transport, and photosynthesis of *Botrydiopsis alpina*. - Carnegie Inst. Year Book *68*: 595-598, 1970.

2297 - FORK, D.C., URBACH, W.: Studies made *in vivo* on the role of plastocyanin in photosynthesis. - In: THOMAS, J.B., GOEDHEER, J.C. (ed.): Currents in Photosynthesis. Pp. 293-303. Donker, Rotterdam 1966.

2298 - FORREST, H.S., van BAALEN, C.: Microbiology of unconjugated pteridines. - Annu. Rev. Microbiol. *24*: 91-108, 1970. [Role in Ps.]

2299 - FORRESTER, M.L., KROTKOV, G., NELSON, C.D.: Effect of oxygen on photosynthesis, photorespiration and respiration in detached leaves. I. Soybean. - Plant Physiol. *41*: 422-427, 1966.

2300 - FORRESTER, M.L., KROTKOV, G., NELSON, C.D.: Effect of oxygen on photosynthesis, photorespiration and respiration in detached leaves. II. Corn and other monocotyledons. - Plant Physiol. *41*: 428-431, 1966.

2301 - FORSYTH, F.R., HALL, I.V.: Rates of photosynthesis and respiration in leaves of the cranberry with emphasis on rates at low temperatures. - Can. J. Plant Sci. *47*: 19-23, 1967.

2302 - FORTI, G.: Studies on NADPH-cytochrome *f* reductase of chloroplasts. - In: Energy

Conversion by the Photosynthetic Apparatus. Brookhaven Symp. Biol. *19*: 195-201, 1967.

2303 - FORTI, G.: The stoichiometry of NADP dependent photosynthetic phosphorylation. - Biochem. biophys. Res. Commun. *32*: 1020-1024, 1968.

2304 - FORTI, G., MELANDRI, B.A., SAN PIETRO, A., KE, B.: Studies on the photoreduction of ferredoxin and the ferredoxin-NADP reductase flavoprotein by chloroplasts fragments: effect of pyrophosphate. - Arch. Biochem. Biophys. *140*: 107-112, 1970.

2305 - FORTI, G., MEYER, E.M.: Effect of pyrophosphate on photosynthetic electron transport reactions. - Plant Physiol. *44*: 1511-1514, 1969.

2306 - FORTI, G., STURANI, E.: On the structure and function of reduced nicotinamide adenine dinucleotide phosphate-cytochrome f reductase of spinach chloroplasts. - Europe. J. Biochem. *3*: 461-472, 1968.

2307 - FORTI, G., ZANETTI, G.: Photooxidation of cytochrome f and of pyridine nucleotides by digitonin extracts of chloroplasts. - In: THOMAS, J.B., GOEDHEER, J.C. (ed.): Currents in Photosynthesis. Pp. 197-205. Donker, Rotterdam 1966.

2308 - FORTI, G., ZANETTI, G.: Nicotinamide adenine dinucleotide phosphate-cytochrome f reductase of chloroplasts. - In: GOODWIN, T.W. (ed.): Biochemistry of Chloroplasts. Vol. II. Pp. 523-529. Academic Press, London-New York 1967.

2309 - FORTI, G., ZANETTI, G.: The electron pathway of cyclic photophosphorylation. - In: METZNER, H. (ed.): Progress in Photosynthesis Research. Vol. III. Pp. 1213-1216. Tübingen 1969.

B2310 - Fotosinteticheskaya Deyatel'nost' Rasteniĭ i Vliyanie na Nee Mineral'nogo Pitaniya. [Photosynthetic Activity of Plants as Affected by Mineral Nutrition.] - Red.-Izd. Otd. Akad. Nauk Mold. SSR, Kishinev 1970. [In R.]

B2311 - Fotosinteticheskaya Deyatel'nost' Yabloni i Slivy v Usloviyakh Moldavii. [Photosynthetic Activity of Apple and Plum Trees in Moldavian Conditions.] - Red.-Izd. Otd. Akad. Nauk Mold. SSR, Kishinev 1970. [In R.]

B2312 - Fotosinteticheskaya Produktivnost' Rastitel'nogo Pokrova. [Photosynthetic Productivity of a Plant Stand.] - Akad. Nauk Est. SSR, Inst. Fiz. Astron., Tartu 1969. [In R, ab: E.]

B2313 - Fotosintez i Produktivnost' Rastitel'nogo Pokrova. [Photosynthesis and Productivity of a Plant Community.] - Inst. Phys. and Astron. Acad. Sci. Estonian SSR, Tartu 1968. [In R, ab: E.]

B2314 - Fotosintez, Mineral'noe Pitanie, Svetovoĭ Rezhim. [Photosynthesis, Mineral Nutrition, Light Regime.] - Zinatne, Riga 1970. [In R.]

B2315 - Fotosintēzes Pētīšana Sējumos (Metodes un to Izmantošana). [Photosynthesis Study in Stands.] - Zinatne, Rīga 1970. [In Latvian, ab: R.]

B2316 - Fotosinteziruyushchie Sistemy Vysokoĭ Produktivnosti. [Photosynthesizing Systems of High Productivity.] - Nauka, Moskva 1966. [In R.]

2317 - FOUSOVÁ, S., AVRATOVŠČUKOVÁ, N.: Hybrid vigour and photosynthetic rate of leaf disks in *Zea mays* L. - Photosynthetica *1*: 3-12, 1967.

2318 - FOWLER, C.F., SYBESMA, C.: Light- and chemically-induced oxidation-reduction reactions in chromatophore fractions of *Rhodospirillum rubrum*. - Biochim. biophys. Acta *197*: 276-283, 1970.

2319 - FOWLER, C.W., RASMUSSON, D.C.: Leaf area relationships and inheritance in barley. - Crop Sci. *9*: 729-731, 1969.

2320 - FOX, C.H.: Studies of the cultural physiology of the lichen alga *Trebouxia*. - Physiol. Plant. *20*: 251-262, 1967. [Ps.]

2321 - FRĄCKOWIAK, D.: Fizyko-chemiczne modele procesu fotosyntezy. [Physico-chemical models of the photosynthesis process.] - Postępy Fiz. *17*: 383-402, 1966. [In Pol., ab: E.]

2322 - FRĄCKOWIAK, D.: Fizyczne badania procesu fotosyntezy. [The physical research in photosynthesis.] - Postępy Biochem. *13*: 335-357, 1967. [In Pol., ab: E.]

2323 - FRĄCKOWIAK, D., GRABOWSKI, J.: Excitation energy transfer between biliproteins and chlorophyllide *a*. - Photosynthetica *4*: 236-242, 1970.

2324 - FRĄCKOWIAK, D., GRABOWSKI, J., STACHOWIAK-HANS, E.: Energy transfer between bile pigments and chlorophyll *in vivo*. - Photosynthetica *3*: 39-44, 1969.

2325 - FRĄCKOWIAK, D., JANUSZCZYK, L., PASZKOWSKI, W.: The influence of sonication on spectra of chlorophyll *a* solutions. - Bull. Acad. pol. Sci., Sér. Sci. math. astron. phys. *17*: 781-787, 1969.

2326 - FRĄCKOWIAK, D., KOJRO, Z.: Polarization spectra of chlorophyllide *a*. - In: Proceedings of International Conference on Luminescence. Pp. 301-304. Budapest 1966.

2327 - FRĄCKOWIAK, D., KOJRO, Z., KOZŁOWSKA, H.: Polarization of fluorescence of chlorophyllide *a* and *b*. - Bull. Acad. pol. Sci., Sér. Sci. math. astron. phys. *15*: 881-884, 1967.

2328 - FRĄCKOWIAK, D., KOZŁOWSKI, S.: Chlorophyllide *a* in isotropic and anisotropic medium. - Bull. Acad. pol. Sci., Sér. Sci. math. astron. phys. *15*: 421-426, 1967.

2329 - FRĄCKOWIAK, D., MANIKOWSKI, H., SALAMON, Z.: The spectral properties of chlorophyll *b* in the two phases system. - Acta phys. pol. *34*: 669-674, 1968.

2330 - FRĄCKOWIAK, D., MIEDZIEJKO, E., SURMA, S.: Energy transfer between chloroplast dyes. - In: Photochemistry and Photobiology in Plant Physiology. European Photobiol. Symp. Hvar, Yugoslavia, 19th-22nd September 1967. Book of Abstracts. Pp. 17-19. Hvar 1967.

2331 - FRĄCKOWIAK, D., MIEDZIEJKO, E., SURMA, S.: Quenching of fluorescence of chlorophyll in different media. - Stud. biophys. *5*: 183-188, 1967.

2332 - FRACKOWIAK, D., MURTY, N.R.: Different forms of potassium chlorophyllides. - In: THOMAS, J.B., GOEDHEER, J.C. (ed.): Currents in Photosynthesis. Pp. 9-15. Donker, Rotterdam 1966.

2333 - FRACKOWIAK, D., SALAMON, Z.: The protective action of carotenoids on fluorescence of chlorophyll *b*. - Photochem. Photobiol., *11*: 559-563, 1970.

2334 - FRĄCKOWIAK, D., SURMA, S.: Spectral properties of chlorophyllide-bilirubin solutions in glycol. - Photosynthetica *2*: 75-84, 1968.

2335 - FRADKIN, L.I., FALUDY-DANIEL', A., SHLYK, A.A.: Fluorestsentsiya khlorofilla *b* v ζ-karotinovom mutante kukuruzy. [Chlorophyll *b* fluorescence in the ζ-carotenic mutant of maize.] - Dokl. Akad. Nauk SSSR *182*: 1420-1423, 1968. [In R.]

2336 - FRADKIN, L.I., KALININA, L.M., SHLYK, A.A.: Rannee obrazovanie dlinnovolnovykh form khlorofilla v postetiolirovannykh list'yakh. [Early formation of long-wave chlorophyll forms in postetiolated leaves.] - Dokl. Akad. Nauk SSSR *194*: 201-204, 1970. [In R.]

2337 - FRADKIN, L.I., SHLYK, A.A.: O perenose energii mezhdu karotinoidami i zelenymi pigmentami. [On energy transfer between carotenoids and green pigments.] - In: Bioenergetika i Biologicheskaya Spektrofotometriya. Pp. 135-140. Nauka, Moskva 1967. [In R.]

2338 - FRADKIN, L.I., SHLYK, A.A., KALININA, L.M., FALUDI-DÁNIEL, Å.: Fluorescence studies on the reaction centres of chlorophyll biosynthesis at the early stages of greening. - Photosynthetica *3*: 326-337, 1969.

2339 - FRADKIN, L.I., SHLYK, A.A., KOLYAGO, V.M.: Temnovoĭ biosintez khlorofilla *b* u kratkovremenno osveshchennykh etiolirovannykh prorostkov. [Dark biosynthesis of chlorophyll *b* in briefly illuminated etiolated seedlings.] - Dokl. Akad. Nauk SSSR *171*: 222-225, 1966. [In R.]

2340 - FRADKIN, L.I., ZEN'KO, A.A., KOLYAGO, V.M.: Vliyanie infil'tratsii rasteniĭ glutarovym al'degidom na sostoyanie i nekotorye metabolicheskie reaktsii pigmentov. [Effect of glutaraldehyde infiltration of plants on the state and some metabolic reactions of pigments.] - In: Metabolizm i Stroenie Fotosinteticheskogo Apparata. Pp. 53-68. Nauka i Tekhnika, Minsk 1970. [In R.]

2341 - FRANCIS, C.A.: Modifications of some ozalid paper technique for measuring integrated light transmission values in the field. - Crop Sci. *10*: 321-322, 1970.

2342 - FRANCIS, C.A., RUTGER, J.N., PALMER, A.F.E.: A rapid method for plant leaf area estimation in maize (*Zea mays* L.) - Crop Sci. *9*: 537-539, 1969.

2343 - FRANCIS, F.J.: Pigment content and color in fruits and vegetables. - Food Technol. *23*: 32-36, 1969. [Car.]

2344 - FRANCIS, G.W., HERTZBERG, S., ANDERSEN, K., LIAAEN-JENSEN, S.: New carotenoid glycosides from *Oscillatoria limosa*. - Phytochemistry *9*: 629-635, 1970.

2345 - FRANCIS, G.W., LIAAEN-JENSEN,S.: Bacterial carotenoids. XXXIII. Carotenoids of *Thiorhodaceae*. 9. The structures of the carotenoids of the rhodopinal series. - Acta chem. scand. *24*: 2705-2712, 1970.

2346 - FRANCK, U.F., HOFFMANN, N.: Luminescence of chlorophyll and its relation to photosynthesis: new results of the Kautsky effect. - In: METZNER, H. (ed.): Progress in Photosynthesis Research. Vol. II. Pp. 899-904. Tübingen 1969.

2347 - FRANCK, U.F., HOFFMANN, A., ARENZ, H., SCHREIBER, U.: Chlorophyllfluoreszenz als Indikator der photochemischen Primärprozesse der Photosynthese. - Ber. Bunsenges. phys. Chem. *73*: 871-879, 1969.

2348 - FRANDSEN, N.O.: Die Plastidenzahl als Merkmal bei der Kartoffel. - Theor. appl. Gen. *38*: 153-167, 1968.

2349 - FRANK, R., SWITZER, C.M.: Effects of pyrazon on growth, photosynthesis, and respiration. - Weed Sci. *17*: 344-348, 1969.

2350 - FRANKIVS'KYĬ, V.Ya.: Pro stijkist' khlorofilu v pryrodi. [On the stability of chlorophyll in nature.] - Ukr. bot. Zh. *24*: 104-106, 1967. [In Ukr.]

2351 - FRANZISKET, L.: The ratio of photosynthesis to respiration of reef building corals during a 24 hour period. - Forma Functio *1*: 153-158, 1969.

2352 - FRASER, A.I.: Recording some aspects of a forest environment. - In: WADSWORTH, R.M., CHAPAS, L.C., RUTTER, A.J., SOLOMON, M.E., WARREN WILSON, J. (ed.): The Measurement of Environmental Factors in Terrestrial Ecology. Pp. 235-242. Blackwell Sci. Publ., Oxford-Edinburgh 1968.

2353 - FREDERICK, S.E., NEWCOMB, E.H.: Microbody-like organelles in leaf cells. - Science *163*: 1353-1355, 1969. [Chloroplast.]

2354 - FREDERICQ, H., de GREEF, J.A.: Influence, réversible par la lumière rouge-clair, de courts éclairements rouge-foncé à la fin de la photopériode, sur la croissance et la teneur en chlorophylles de *Marchantia polymorpha* L. - Photochem. Photobiol. *5*: 431-440, 1966.

2355 - FREDERICQ, H., de GREEF, J.: Red (R), far-red (FR) photoreversible control of growth and chlorophyll content in light-grown thalli of *Marchantia polymorpha* L. - Naturwissenschaften *53*: 337, 1966.

2356 - FREDERICQ, H., de GREEF, J.: Red (R), far-red (FR) photoreversible control of growth and chlorophyll content in light-grown thalli of *Marchantia polymorpha*. - Plant Physiol. *41* (Suppl.): XV, 1966.

2357 - FREDERICQ, H., de GREEF, J.: Morphogenesis and chlorophyll metabolism of *Marchantia polymorpha* L. in response to short irradiations of various spectral composition. - In: Photochemistry and Photobiology in Plant Physiology. European Photobiol. Symp. Hvar, Yugoslavia, 19th-22nd September 1967. Book of Abstracts. Pp. 137-140. Hvar 1967.

2358 - FREDERICQ, H., de GREEF, J.: Photomorphogenic and chlorophyll studies in the bryophyte *Marchantia polymorpha*. I. Effect of red, far-red irradiations in short and long-term experiments. - Physiol. Plant. *21*: 346-359, 1968.

2359 - FREDON, J.-J., BONNEMAIN, J.-L.: Transport du ^{14}C assimilé chez le Radis à divers stades de son développement. - Compt. rend. Acad. Sci. (Paris), Sér. D *270*: 354-357, 1970.

2360 - FREDRICKS, W.W.: Regulation of electron transport in photosynthesis. - Biochem. biophys. Res. Commun. *31*: 582-587, 1968.

2361 - FREDRICKS, W.W., KOHLMANN, J.M.: Inhibitors of the transhydrogenase activity of spinach ferredoxin-nicotinamide adenine dinucleotide phosphate reductase. - J. biol. Chem. *244*: 522-528, 1969.

2362 - FREDRICKSON, A.G., TSUCHIYA, H.M.: Utilization of the effects of intermittent illumination on photosynthetic microorganisms. - In: Prediction and Measurement of Photosynthetic Productivity. Pp. 519-541. PUDOC, Wageningen 1970.

2363 - FREEMAN, J.A., RENNEY, A.J., DRIEDIGER, H.: Influence of atrazine and simazine on leaf chlorophylls and fruit yield of raspberries. - Can. J. Plant Sci. 46: 454-455, 1966.

2364 - FREEMAN, J.M., SEDDON, S.A.: Thin-layer chromatography. - School Sci. Rev. 52: 372-376, 1970. [Chl as school demonstration.]

2365 - FREŸ-VISSLING, A.: Ul'trastruktura khloroplastov, obnaruzhennaya metodom zamorazhivaniya s protravlivaniem. [Chloroplast ultrastructure observed by the freeze-etching method.] - In: Funktsional'naya Biokhimiya Kletochnykh Struktur. Pp. 104-109. Nauka, Moskva 1970. [In R.]

2366 - FRENCH, C.S.: Chloroplast pigments. - In: GOODWIN, T.W. (ed.): Biochemistry of Chloroplasts. Vol. I. Pp. 377-386. Academic Press, London-New York 1966.

2367 - FRENCH, C.S.: Die photochemische Nutzung der sichtbaren Strahlung durch zwei Lichtreaktionen der Photosynthese. - Nova Acta leopoldina N.F. 31: 169-187, 1966.

2368 - FRENCH, C.S.: Kinetics of oxygen evolution. - In: THOMAS, J.B., GOEDHEER, J.C. (ed.): Currents in Photosynthesis. Pp. 285-292. Donker, Rotterdam 1966.

2369 - FRENCH, C.S.: Changes with age in the absorption spectrum of chlorophyll a in a diatom. - Arch. Mikrobiol. 59: 93-103, 1967.

2370 - FRENCH, C.S.: Fluorescence spectra for several forms of chlorophyll in $vivo$. - In: Abstracts Volume - 7th International Congress of Biochemistry, H-70. Tokyo 1967.

2371 - FRENCH, C.S.: Absorption spectra of chlorophyll a in algae. - Carnegie Inst. Year Book 66: 177-186, 1968.

2372 - FRENCH, C.S.: Biophysics of plastid pigments. - In: METZNER, H. (ed.): Progress in Photosynthesis Research. Vol. II. Pp. 877-880. Tübingen 1969.

2373 - FRENCH, C.S.: The forms of chlorophyll a in fractions of chloroplasts from different sources. - Carnegie Inst. Year Book 68: 578-587, 1970.

2374 - FRENCH, C.S., BROWN, J.S., PRAGER, L., LAWRENCE, M.: Analysis of spectra of natural chlorophyll complexes. - Carnegie Inst. Year Book 67: 536-546, 1969.

2375 - FRENCH, C.S., KOERPER, M.A.: Fluorescence spectra of photosynthetic pigments. - Carnegie Inst. Year Book 65: 492-498, 1967.

2376 - FRENCH, C.S., LAWRENCE, M.: A spectrophotometer primarily for light-scattering samples at low temperature. - Carnegie Inst. Year Book 66: 175-177, 1968.

2377 - FRENCH, C.S., MICHEL-WOLWERTZ, M.R., MICHEL, J.M., BROWN, J.S., PRAGER, L.K.: Naturally occurring chlorophyll types and their function in photosynthesis. - In: GOODWIN, T.W. (ed.): Porphyrins and Related Compounds. Pp. 147-162. Academic Press, London-New York 1968.

2378 - FRENCH, C.S., PRAGER, L.: The forms of chlorophyll a in plants. - Plant Physiol. 43 (Suppl.): S 20, 1968.

2379 - FRENCH, C.S., PRAGER, L.: Absorption spectra for different forms of chlorophyll. - In: METZNER, H. (ed.): Progress in Photosynthesis Research. Vol. II. Pp. 555-564. Tübingen 1969.

2380 - FRENCH, S.A.W., HUMPHRIES, E.C.: Persistent effects of seedling treatment on growth of sugar beet in pots. - Ann. appl. Biol. 64: 161-175, 1969. [Growth analysis.]

2381 - FRENKEL, A.W.: Multiplicity of electron transport reactions in bacterial photosynthesis. - Biol. Rev. Cambridge phil. Soc. 45: 569-616, 1970.

2382 - FRENKEL, A.W., COST, K.: Photosynthetic phosphorylation. - In: FLORKIN, M., STOTZ, E.H. (ed.): Comprehensive Biochemistry. Vol. 14. Pp. 397-423. Elsevier Publ. Co., Amsterdam-London-New York 1966.

2383 - FRENKEL, C., KLEIN, I., DILLEY, D.R.: Methods for the study of ripening and protein synthesis in intact pome fruits. - Phytochemistry 8: 945-955, 1969. [Chl.]

2384 - FRENYŐ, C., THAI DUY NINH: Rézionok hatása a fotoszintetikus gáztermelésre. [Effect of copper ions on photosynthetic gas exchange.] - Bot. Közlem. 57: 107-112, 1970. [In Hung., ab: G.]

2385 - FREY-WYSSLING, A.: Ontogeny of chloroplasts. - In: SIRONVAL, C. (ed.): Le Chloroplaste, Croissance et Vieillissement. Pp. 17-20. Masson, Paris 1967.

2386 - FREY-WYSSLING, A.: Structure and chemistry of plastids. - In: METZNER, H. (ed.): Progress in Photosynthesis Research. Vol. I. Pp. 368-373. Tübingen 1969.

2387 - FRIČ, F., HASPELOVÁ-HORVATOVICOVÁ, A.: Ein Beitrag zur schnellen quantitativen Bestimmung von ATP, ADP und AMP in grünen Pflanzenorganen. - Phytochemistry 6: 633-640, 1967.

2388 - FRIČ, F., HASPELOVÁ-HORVATOVIČOVÁ, A., PAULECH, C.: Contribution to the study of respiration and photosynthesis of barley infected by powdery mildew (Erysiphe graminis f. spec. hordei MARCHAL). - Advanc. Frontiers Plant Sci. 16: 103-111, 1966.

2389 - FRICK, H., JONES, R.F.: Effects of nalidixic acid on Lemna. - Plant Physiol. 43 (Suppl.): S 40, 1968. [Chl.]

2390 - FRIDECKÝ, A., CIGL'AR, J., KOSTREJOVÁ, E.: The influence of mechanical interventions on growth-analytical characteristics of the photosynthetic productivity of winter wheat (Košútska variety). - Acta fytotech. (Nitra) 18: 17-26, 1968.

2391 - FRIEDLANDER, M., NEUMANN, J.: Stimulation of photoreactions of isolated chloroplasts by serum albumin. - Plant Physiol. 43: 1249-1254, 1968.

2392 - FRIEND, D.J.C.: Net assimilation rate of wheat as affected by light intensity and temperature. - Can. J. Bot. 47: 1781-1787, 1969.

2393 - FRIEND, J., ACTON, G.J.: Oxidation of carotenoids by isolated sugar-beet chloroplasts. - In: GOODWIN, T.W. (ed.): Biochemistry of Chloroplasts. Vol. I. Pp. 431-435. Academic Press, London-New York 1966.

2394 - FRIEND, J., ACTON, G.J.: The oxidation of unsaturated lipids by isolated sugar beet chloroplasts. - In: Energy Conversion by the Photosynthetic Apparatus. Brookhaven Symp. Biol. 19: 485-490, 1967.

2395 - FRIEND, J., HAWCROFT, D.M.: Carotenoids and fatty acids as uncouplers of photophosphorylation in isolated chloroplasts. - Biochem. J. 104: 60 P, 1967.

2396 - FRIEND, J., OLSSON, R.: Inhibition of photosynthetic electron transport by ioxynil. - Nature 214: 942-943, 1967.

2397 - FRIEND, J., OLSSON, R., REDFEARN, E.R.: Oxidation-reduction reactions of endogenous plastoquinone in chloroplasts and digitonin-fractionated chloroplasts. - In: GOODWIN, T.W. (ed.): Biochemistry of Chloroplasts. Vol. II. Pp. 537-543. Academic Press, London-New York 1967.

2398 - FRIIS-NIELSEN, B.: Active leaf area index, a meteorological-plant physiological parameter for photosynthetic production. 1. Under conditions of optimal water supplies. - Kong. vet. Landbohojskole Aarsskr. 1966: 49-60, 1966.

2399 - FRIIS-NIELSEN, B.: Active leaf area index, a meteorological-plant-physiological parameter for photosynthetic production. 2. Under conditions of varying water supplies. - Kong. vet. Landbohojskole Aarskr. 1970: 144-152, 1970.

2400 - FRÖHLICH, H.: Storage of light energy and photosynthesis. - Nature 219: 743-744, 1968.

2401 - FRÖHLICH, H.: Proposed model experiments on the storage of light energy in photosynthesis. - Nature 221: 976, 1969.

2402 - FRY, K.E.: Some factors affecting the Hill reaction activity in cotton chloroplasts. - Plant Physiol. 45: 465-469, 1970.

2403 - FRY, K.E., WALKER, R.B.: A pressure-infiltration method for estimating stomatal opening in conifers. - Ecology 48: 155-157, 1967.

2404 - FRYDRYCH, J.: Fotosyntetická aktivita a produkční schopnost di-, tri- a tetraploidního zelí. [Photosynthetic activity and yielding capacity of di-, tri- and tetraploid cabbage.] - Věd. Práce výzk. Úst. zel. (Olomouc) 1966: 69-75, 1966. [In Czech, ab: E, G, R.]

2405 - FRYDRYCH, J.: Photosynthetic characteristics of diploid and tetraploid forms of *Brassica oleracea* var. *gongylodes* grown under different irradiance. - Photosynthetica 4: 139-145, 1970.

2406 - FUCHS, M., KANEMASU, E.T., KERR, J.P., TANNER, C.B.: Effect of viewing angle on canopy temperature measurements with infrared thermometers. - Agron. J. 59: 494-496, 1967.

2407 - FUCIKOVSKY, L.A.: Changes in pigment, phosporus-32, and starch in unifoliate leaves of soybeans infected by *Cercospora sojina*. - Phytopathology 56: 987, 1966. [Chl, car.]

2408 - FUCIKOVSKY, L.A.: Changes of chlorophyll and carotenoid content in the leaves of *Chenopodium nuttalia* naturally infected by *Peronospora farinosa*. - Plant Dis. Reporter 54: 762-763, 1970.

2409 - FUESS, F.W., TESAR, M.B.: Photosynthetic efficiency, yields, and leaf loss in alfalfa. - Crop Sci. 8: 159-163, 1968.

2410 - FUHRHOP, J.-H., MAUZERALL, D.: The one-electron oxidation of metalloporphyrins. - J. amer. chem. Soc. 91: 4174-4181, 1969. [Bacteriochlorophyll.]

2411 - FUJIMORI, E.: Bacteriochlorophyll pheophytinization in chromatophores and sub-chromatophores from *Rhodospirillum rubrum*. - Biochim. biophys. Acta 180: 360-367, 1969.

2412 - FUJIMORI, E.: pH-induced reversible changes in the absorption spectrum and photoactivity of bacteriochlorophyll in photosynthetic bacteria chromatophores. - Biochim. biophys. Acta 223: 444-446, 1970.

2413 - FUJIMORI, E., PECCI, J.: Dissociation and association of phycocyanin. - Biochemistry 5: 3500-3508, 1966.

2414 - FUJIMORI, E., PECCI, J.: Distinct subunits of phycoerythrin from *Porphyridium cruentum* and their spectral characteristics. - Arch. Biochem. Biophys. 118: 448-455, 1967.

2415 - FUJIMORI, E., PECCI, J.: Circular dichroism of single- and double-peaked phycoerythrin: Mercurial-induced changes. - Biochim. biophys. Acta 221: 132-134, 1970.

2416 - FUJIMORI, E., TAVLA, M.: Light-induced electron transfer between chlorophyll and hydroquinone and the effect of oxygen and β-carotene. - Photochem. Photobiol. 5: 877-887, 1966.

2417 - FUJIMORI, E., TAVLA, M.: Chlorophyll-photosensitized oxidation of hydroquinone and p-phenylenediamine derivatives. - Photochem. Photobiol. 8: 31-45, 1968.

2418 - FUJITA, Y.: [Photosynthetic efficiency and chromatic adaptation in *Tolypothrix tenuis*.] - Chem. Control Plants 1: 173-175, 1966. [In Jap.]

2419 - FUJITA, Y., MURANO, F.: Photochemical activities of sonicated lamellae of spinach chloroplasts with special reference to the action of CRS (cytochrome c reducing substance). - Plant Cell Physiol. 8: 269-281, 1967.

2420 - FUJITA, Y., MURANO, F.: Occurrence of back flow of electrons against the action of photochemical system I in sonicated lamellar fragments. - In: SHIBATA, K., TAKAMIYA, A., JAGENDORF, A.T., FULLER, R.C. (ed.): Comparative Biochemistry and Biophysics of Photosynthesis. Pp. 161-169. Univ. Tokyo Press, Tokyo; Univ. Park Press, State College, Pa. 1968.

2421 - FUJITA, Y., MYERS, J.: Cytochrome c redox reactions induced by photochemical system 1 in sonicated preparations of *Anabaena cylindrica*. - Arch. Biochem. Biophys. 113: 730-737, 1966.

2422 - FUJITA, Y., MYERS, J.: Comparative studies of cytochrome c redox reactions by photochemical lamellar preparations obtained from blue-green, red, and green algae, and spinach chloroplasts. - Arch. Biochem. Biophys. 113: 738-741, 1966.

2423 - FUJITA, Y., MYERS, J.: Some properties of the cytochrome c reducing substance, a factor for light-induced redox reaction of cytochrome *c* in photosynthetic lamellae. - Plant Cell Physiol. *7*: 599-606, 1966.

2424 - FUJITA, Y., MYERS, J.: Kinetic analysis of light-induced cytochrome *c* redox reactions in *Anabaena* lamellar fragments. - Arch. Biochem. Biophys. *119*: 8-15, 1967.

2425 - FUJITA, Y., TSUJI, T.: Photochemically active chromoprotein isolated from the blue-green alga *Anabaena cylindrica*. - Nature *219*: 1270-1271, 1968.

2426 - FUKAMI, T., HILDEBRANDT, A.C.: Growth and chlorophyll formation in edible green plant callus tissues *in vitro* on media with limited sugar supplements. - Bot. Mag. (Tokyo) 80: 199-212, 1967.

2427 - FUKSHANSKIĬ, L.Ya.: Model' upravlyayushchego svetovogo vozdeĭstviya realizuemaya v zhivoĭ tkani. [Model of regulating light action in living tissue.] - Biofizika *11*: 374-377, 1966. [Ps; in R.]

2428 - FULLER, R.C.: The comparative biochemistry of carbon dioxide fixation in photosynthetic bacteria. - In: METZNER, H. (ed.): Progress in Photosynthesis Research. Vol. III. Pp. 1589-1592. Tübingen 1969.

2429 - FULLER, R.C., NUGENT, N.A.: Pteridines and the function of the photosynthetic reaction center. - Proc. nat. Acad. Sci. U.S.A. *63*: 1311-1318, 1969.

2430 - FURLONG, C.E., PREISS, J.: The regulation of the biosynthesis of α-1,4, glucans in photosynthetic systems. - In: METZNER, H. (ed.): Progress in Photosynthesis Research. Vol. III. Pp. 1604-1617. Tübingen 1969.

2431 - GAASTRA, P.: De betekenis van geconditioneerde groeiruimtes voor het landbouwkundig onderzoek. [Importance of air-conditioned growth chambers for the agricultural research.] - Verwarming Ventilatie *1966* (8): 1-7, 1966. [Radiation sources; in Dutch, ab: E.]

2432 - GAASTRA, P.: Radiation measurements for investigations of photosynthesis under natural conditions. - In: ECKARDT, F.E. (ed.): Functioning of Terrestrial Ecosystems at the Primary Production Level. Pp. 467-478. Unesco, Paris 1968.

2433 - GAASTRA, P.: Climate rooms as a tool for measuring physiological parameters for models of photosynthetic systems. - In: Prediction and Measurement of Photosynthetic Productivity. Pp. 387-398. PUDOC, Wageningen 1970. [Radiation sources.]

2434 - GABBOTT, P.A.: Inhibition of photoreactions in isolated chloroplasts by 2-azido-4-alkylamino-6-alkylamino-*s*-triazines. - In: METZNER, H. (ed.): Progress in Photosynthesis Research. Vol. III. Pp. 1712-1727. Tübingen 1969.

2435 - GACHKOVSKIĬ, V.F.: Obratimoe izmenenie spektrov pogloshcheniya i fluorestsentsii adsorbatov ftalotsianina magniya i khlorofilla. [Reversible variation of absorption and fluorescence spectra of the adsorbates of magnesium phthalocyanine and chlorophyll.] - Dokl. Akad. Nauk SSSR *186*: 471-473, 1969. [In R.]

2436 - GAILHOFER, M.: Der Einfluss von Huminsäure auf die Aktivität einiger Enzyme in Tomatenstecklingen. - Phyton (Austria) *14*: 69-78, 1970. [Chl.]

2437 - GÅL, I.E.: Über die chemische Zusammensetzung von Capsicidin. - Z. Lebensm.-Unters.-Forsch. *132* (2): 82-84, 1966.

2438 - GALE, J., KOHL, H.C., HAGAN, R.M.: Mesophyll and stomatal resistances affecting photosynthesis under varying conditions of soil, water and evaporation demand. - Israel J. Bot. *15*: 64-71, 1966.

2439 - GALE, J., KOHL, H.C., HAGAN, R.M.: Changes in the water balance and photosynthesis of onion, bean and cotton plants under saline conditions. - Physiol. Plant. *20*: 408-420, 1967.

2440 - GALE, J., MANES, A., POLJAKOFF-MAYBER, A.: A rapidly equilibrating thermocouple contact thermometer for measurement of leaf-surface temperatures. - Ecology *51*: 521-525, 1970.

2441 - GALE, J., POLJAKOFF-MAYBER, A.: Plastic films on plants as antitranspirants. - Science *156*: 650-652, 1967. [Ps.]

2442 - GALE, J., POLJAKOFF-MAYBER, A.: Resistance to gas flow through the leaf and its significance to measurements made with viscous flow and diffusion porometers. - Israel J. Bot. *16*: 205-211, 1967.

2443 - GALE, J., POLJAKOFF-MAYBER, A.: Resistances to the diffusion of gas and vapor in leaves. - Physiol. Plant. *21*: 1170-1176, 1968.

2444 - GALE, J., POLJAKOFF-MAYBER, A.: Interrelations between growth and photosynthesis of salt bush (*Atriplex halimus* L.) grown in saline media. - Aust. J. biol. Sci. *23*: 937-945, 1970.

2445 - GALE, J., POLJAKOFF-MAYBER, A., KAHANE, I.: The gas diffusion porometer technique and its application to the measurement of leaf mesophyll resistance. - Israel J. Bot. *16*: 187-204, 1967.

2446 - GALEEVA, S.G., KURMANOVA, S.A.: Ob "endogennom" fotosinteticheskom fosforiliro-vanii v izolirovannykh khloroplastakh rastenii. [" Endogenous" photosynthetic phosphorylation in isolated plant chloroplasts.] - In: Funktsional'nye Osobennos-ti Khloroplastov. Pp. 56-59. Kazan. Univ., Kazan 1969. [In R.]

2447 - GALKINA, T.B., KASHKOVSKIĬ, I.I., KURAPOVA, O.A., LEBEDEVA, E.K., MELESHKO, G.I., UL'YANIN, Yu.N.: Nekotorye kharakteristiki rosta i gazoobmena vodorosli *Anacystis nidulans* v intensivnoĭ kul'ture. [Some characteristics of growth and gas exchange in the alga *Anacystis nidulans* in intensive culture.] - In: Problemy Kosmicheskoĭ Biologii. Vol. 7. Pp. 480-486. Nauka, Moskva 1967. [In R.]

2448 - GALMES, J.M.: Effets de l'amputation totale ou partielle des limbes chez le blé. - Compt. rend. Acad. Agr. France *52*: 106-112, 1966. [Ps.]

2449 - GALMICHE, H.M.: Photosynthetic incorporation of carbon from CO_2 into malic acid and phosphoglyceric acid in *Zea mays*. - In: COOMBS, J. (ed.): Photosynthesis in Sugar Cane. Pp. 36-39. Tate and Lyle Ltd., London 1969.

2450 - GALMICHE, J.M.: Incorporation du carbone du gaz carbonique dans l'acide phospho-glycérique de feuilles de tomates soumises à des éclairements de lumière mono-chromatique. - Compt. rend. Acad. Sci. (Paris), Sér. D *264*: 2093-2096, 1967.

2451 - GALMICHE, J.-M., GIRAULT, G., TYSZKIEWICZ, E., FIAT, R.: Photophosphorylation et translocation de protons. - Compt. rend. Acad. Sci. (Paris), Sér. D *265*: 374-377, 1967.

2452 - GAMAYUNOVA, M.S., BERDEREVSKAYA, N.D., DUBROVSKAYA, A.A.: Svoĭstva ferredoksina khloroplastov gorokha pri razlichnom obespechenii zhelezom. [Properties of ferre-doxin from pea chloroplasts at various iron supply.] - In: Mineral'nye Elementy i Mekhanizm Fotosinteza. Pp. 153-159. Kishinev 1969. [In R.]

2453 - GANCHARYK, M.M., IVANCHANKA, V.M.: Fotasintez raslin. [Plant photosynthesis.] - Vestsi Akad. Navuk Belarus. SSR, Ser. biyal. Navuk *1968* (5): 31-35, 1968. [De-velopment in Belorussia; in Belorus.]

2454 - GANCHARYK, M.M., KRUCHYNINA, S.S.: Dynamika fotasintezu ŭ bul'by ŭ pasadkakh roz-naĭ gushchyni. [Dynamics of photosynthesis in potato sown at different density.] - Vestsi Akad. Navuk Belarus. SSR, Ser. biyal. Navuk *1966* (2): 20-25, 1966. [In Belorus., ab: R.]

2455 - GANCHARYK, M.M., LYAGENCHANKA, B.I., TALANAVA, K.S., DAROZHKINA, L.M.: Fotasintez i vodny rezhym bul'by ŭ suvyazi z lishkam iona Cl⁻ u glebe. [Photosynthesis and water relations in potato as related to the surplus Cl^- ion in the soil.] - Vestsi Akad. Navuk Belarus. SSR, Ser. biyal. Navuk *1968* (5): 82-87, 141, 1968. [In Belorus., ab: R.]

2456 - GANCHARYK, M.M., LAGUN, L.P., KANANUCHANKA, M.V., SIDENKA, V.T.: Fotasintetych-naya dzeĭnasts' raslin chetyrokh sartoŭ bul'by pry roznykh ploshchakh zhyŭlennya. [Photosynthetic activity of plants of four potato cultivars at various nutrition area.] - Vestsi Akad. Navuk Belarus. SSR, Ser. biyal. Navuk *1969* (1): 57-62, 139-140, 1969. [In Belorus., ab: R.]

2457 - GANCHARYK, M.M., SHARTSYANIKINA, A.V.: Ab strukturnykh zmenakh khlaraplastaŭ pry prymyanenni khlorzmyashchayuchykh kaliĭnykh ugnaennyaŭ. [Structure changes of chloroplasts induced by chlorine containing potassium fertilizers.] - Vestsi

Akad. Navuk Belarus. SSR, Ser. biyal. Navuk *1970* (5): 5-8, 133, 1970. [In Belorus., ab: R.]

2458 - GANCHARYK, M.M., SIDENKA, V.T.: Vykarystanne fotasintetychna aktyŭnaĭ radyyatsyi pasadkami bul'by. [Use of photosynthetically active radiation by potato crops.] - Vestsi Akad. Navuk Belarus. SSR, Ser. biyal. Navuk *1969* (2): 25-30, 137-138, 1969. [In Belorus., ab: R.]

2459 - GANCHARYK, M.M., TSYARENTSEVA, M.U., DAROZHKINA, L.M.: Uplyŭ mikraelementaŭ saligorskikh soleĭ na fotasintez i razvitstsë raslin bul'by. [Effect of microelements from saligorsk salts on photosynthesis and development of potato plants.] - Vestsi Akad. Navuk Belarus. SSR, Ser. biyal Navuk *1966* (4): 5-10, 1966. [In Belorus., ab: R.]

2460 - GANESAN, R.: Photosynthesis in the *Cladophoraceae*. - Indian sci. Congr. Ass. Proc. *54*: 365-366, 1967.

2461 - GANNUTZ, T.P.: Photosynthesis and respiration of antarctic lichens. - Amer. J. Bot. *54*: 644, 1967.

2462 - GANTT, E.: Properties and ultrastructure of phycoerythrin from *Porphyridium cruentum*. - Plant Physiol. *44*: 1629-1638, 1969.

2463 - GANTT, E., CONTI, S.F.: Granules associated with the chloroplast lamellae of *Porphyridium cruentum*. - J. Cell Biol. *29*: 423-434, 1966.

2464 - GANTT, E., CONTI, S.F.: Phycobiliprotein localization in algae. - In: Energy Conversion by the Photosynthetic Apparatus. Brookhaven Symp. Biol. *19*: 393-405, 1967.

2465 - GANTT, E., CONTI, S.F.: Ultrastructure of blue-green algae. - J. Bacteriol. *97*: 1486-1493, 1969. [Chl.]

2466 - GAPOCHKA, L.D.: Izuchenie geterotrofnogo fotosinteza sinezelenoĭ vodorosli *Anacystis nidulans*. [Heterotrophic photosynthesis of the blue-green alga *Anacystis nidulans*.] - Nauch. Dokl. vyssh. Shkoly, biol. Nauki *10* (1): 152, 1967. [In R.]

2467 - GAPONENKO, V.I.: Zavisimost' assimilyatsionnykh chisel ot soderzhaniya khlorofilla v list'yakh raznogo vozrasta. [Dependence of assimilation numbers on chlorophyll content in leaves of different ages.] - Stud. biophys. *5*: 197-202, 1967. [In R, ab: E.]

2468 - GAPONENKO, V.I., NIKOLAEVA, G.N., BALEVA, E.F., SHEVCHUK, S.N., LOSITSKAYA, T.V.: Vliyanie detergentov na metabolizm khlorofillov *a* i *b* v zatemnennykh rasteniyakh. [Effect of detergents on metabolism of chlorophylls *a* and *b* in darkened plants.] - In: Metabolizm i Stroenie Fotosinteticheskogo Apparata. Pp. 118-133. Nauka i Tekhnika, Minsk 1970. [In R.]

2469 - GAPONENKO, V.I., NIKOLAEVA, G.N., STANISHEVSKAYA, E.M., LOSITSKAYA, T.V., SHEVCHUK, S.N.: O labilizatsii khlorofilla v zelenom rastenii. [Labilization of chlorophyll in a green plant.] - In: Metabolizm i Stroenie Fotosinteticheskogo Apparata. Pp. 134-143. Nauka Tekhnika, Minsk 1970. [In R.]

2470 - GAPONENKO, V.I., STAZHETSKIĬ, V.: Izmenenie intensivnosti fotosinteza i soderzhaniya khlorofilla u ryaski v svyazi s vozrastom i usloviyami osveshcheniya. [Change in photosynthetic rate and chlorophyll content in duckweed *(Lemna)* as related to age and illumination conditions.] - Fiziol. Rast. *16*: 993-1001, 1969. [In R, ab: E.]

2471 - GAPONENKOV, T.K., SERDECHNAYA, K.I.: Fotoelektrokolorimetricheskiĭ metod opredeleniya karotina i pektinovykh veshchestv. [Photoelectrocolorimetric method for determining carotene and pectin substances.] - Sel'.-khoz. Biol. *1*: 788-790, 1966. [In R, ab: E.]

2472 - GARBETT, K., GILLARD, R.D., KNOWLES, P.F., STANGROOM, J.E.: Cotton effects in plant ferredoxin and xanthenine oxidase. - Nature *215*: 824-828, 1967.

2473 - GARCIA, A., VERNON, L.P., KE, B., MOLLENHAUER, H.: Some structural and photochemical properties of *Rhodopseudomonas palustris* subchromatophore particles obtained by treatment with Triton X-100. - Biochemistry *7*: 319-325, 1968.

2474 - GARCIA, A., VERNON, L.P., KE, B., MOLLENHAUER, H.: Some structural and photochemical properties of *Rhodopseudomonas* species NHTC 133 subchromatophore particles obtained by treatment with Triton X-100. - Biochemistry 7: 326-332, 1968.

2475 - GARCIA, A., VERNON, L.P., MOLLENHAUER, H.: Properties of *Chromatium* subchromatophore particles obtained by treatment with Triton X-100. - Biochemistry 5: 2399-2407, 1966.

2476 - GARCIA, A., VERNON, L.P., MOLLENHAUER, H.: Properties of *Rhodospirillum rubrum* subchromatophore particles obtained by treatment with Triton X-100. - Biochemistry 5: 2408-2416, 1966.

2477 - GARCIA, E.H., GUECO, C.: Carotene content of some vegetables, fruits and root crops: I. The effect of dehydration and refrigeration on the carotene of squash (*Cucurbita maxima* DUCHESNE). - Philippine J. Plant Ind. 33: 85-91, 1968.

2478 - GARCIA-MORIN, M., UPHAUS, R.A., NORRIS, J.R., KATZ, J.J.: Interpretation of chlorophyll electron spin resonance spectra. - J. phys. Chem. 73: 1066-1070, 1969.

2479 - GARGAS, E.: Measurements of primary production, dark fixation and vertical distribution of the microbenthic algae in the Øresund. - Ophelia 8: 231-253, 1970.

2480 - GARNIER, J.: Une méthode de détection, par photographie, de souches d'Algues vertes émettant *in vivo* une fluorescence anormale. - Compt. rend. Acad. Sci. (Paris), Sér. D 265: 874-877, 1967.

2481 - GARNIER, J.: Screening and selection of unicellular green algae mutants for the study of the photosynthetic mechanisms. - In: Prediction and Measurement of Photosynthetic Productivity. Pp. 543-549. PUDOC, Wageningen 1970.

2482 - GARNIER, J., de KOUCHKOVSKY, Y., LAVOREL, J.: Sélection de mutants d'Algues vertes unicellulaires pour l'étude de la photosynthèse. - Compt. rend. Séances Soc. Biol. 162: 365-368, 1968.

2483 - GARNIER, J., MAROC, J.: Recherche de plusieurs transporteurs d'électrons, notamment des cytochromes b-559 et c-553, chez trois mutants non photosynthétiques de *Chlamydomonas reinhardi*. - Biochim. biophys. Acta 205: 205-219, 1970.

2484 - GARSIDE, C., RILEY, J.P.: The absorptivity of fucoxanthin. - Deep-Sea Res. 15: 627, 1968.

2485 - GARSIDE, C., RILEY, J.P.: A thin-layer chromatographic method for the determination of plant pigments in sea water and cultures. - Anal. chim. Acta 46: 179-191, 1969.

2486 - GÄRTEL, W.: Untersuchungen über den Chlorophyllgehalt reifender Weinbeeren. - Weinberg Keller 17: 67-78, 1970.

2487 - GÄRTNER, R.: Die Bewegung des *Mesotaenium*-Chloroplasten im Starklichtbereich. I. Zeit- und temperaturabhängige Sekundärprozesse. - Z. Pflanzenphysiol. 63: 147-161, 1970.

2488 - GÄRTNER, R.: Die Bewegung des *Mesotaenium*-Chloroplasten im Starklichtbereich. II. Aktionsdichroismus und Wechselwirkungen des Photoreceptors mit Phytochrom. - Z. Pflanzenphysiol. 63: 428-443, 1970.

2489 - GASANOV, R.A.: Kinetika vydeleniya kisloroda posle kratkovremennogo osveshcheniya list'ev elodei monokhromaticheskim svetom. [Kinetics of oxygen efflux after a short-term irradiation of *Elodea* leaves with monochromatic radiation.] - Dokl. Akad. Nauk Azerb. SSR 24 (12): 34-36, 1968. [In R.]

2490 - GASANOV, R.A., ABUTALYBOV, M.G.: Spektr deĭstviya effekta Emersona i vzaimodeĭstvie fotokhimicheskikh pigmentnykh sistem fotosinteza vysshikh rasteniĭ v induktsionnoĭ faze. [Action spectrum of Emerson effect and interactions of photochemical pigment systems of photosynthesis in higher plants in induction phase.] - Dokl. Akad. Nauk SSR 192: 911-914, 1970. [In R.]

2491 - GASANOV, R.A., LITVIN, F.F.: Primenenie opticheskoĭ spektroskopii i polyarografii dlya odnovremennogo issledovaniya spektrov pogloshcheniya list'ev rasteniĭ i spektrov deĭstviya fotosinteza. [Use of optical spectroscopy and polarography for simultaneous study of absorption spectra of leaves and action spectra of

photosynthesis.] - In: Materialy I. Zakavkazskoĭ Konferentsii po Fiziologii Rasteniĭ. Pp. 53-56. Izdat. Akad. Nauk Azerb. SSR, Baku 1967. [In R.]

2492 - GASPAR, T., XHAUFFLAIRE, A.: Action comparée de la 6-furfurylaminopurine et de
la 6-(γ,γ-diméthylallylamino) purine sur la croissance, l'activité peroxydasique,
la teneur en chlorophylles et en caroténoïdes. - Physiol. Plant. 21: 792-799,
1968.

2493 - GASSER, H.: A growth analysis of *Phleum pratense* and of *Dactylis glomerata* grown
in pure and mixed stands at two densities. - Bot. Gaz. 129: 351-361, 1968.

2494 - GASSMAN, M., BOGORAD, L.: Control of chlorophyll production in rapidly greening
bean leaves. - Plant Physiol. 42: 774-780, 1967.

2495 - GASSMAN, M., BOGORAD, L.: Studies on the regeneration of protochlorophyllide after brief illumination of etiolated bean leaves. - Plant Physiol. 42: 781-784,
1967.

2496 - GASSMAN, M., GRANICK, S., MAUZERALL, D.: A rapid spectral change in etiolated
Red Kidney bean leaves following phototransformation of protochlorophyllide. -
Biochem. biophys. Res. Commun. 32: 295-300, 1968.

2497 - GASSMAN, M., PLUSCEC, J., BOGORAD, L.: δ-aminolevulinic acid transaminase in
Chlorella vulgaris. - Plant Physiol. 43: 1411-1414, 1968. [Chl.]

2498 - GATES, D.M.: Transpiration and energy exchange. - Quart. Rev. Biol. 41: 353-
364, 1966. [Ps, gas diffusion resistances.]

2499 - GATES, D.M.: Sensing biological environments with a portable radiation thermometer. - Appl. Optics 7: 1803-1809, 1968.

2500 - GATES, D.M.: Energy exchange and ecology. - BioScience 18: 90-95, 1968.

2501 - GATES, D.M.: Energy exchange in the biosphere. - In: ECKARDT, F.E. (ed.): Functioning of Terrestrial Ecosystems at the Primary Production Level. Pp. 33-43.
UNESCO, Paris 1968.

2502 - GATES, D.M.: The ecology of the elfin forest in Puerto Rico. 4. Transpiration
rates and temperatures of leaves in cool humid environment. - J. Arnold Arbore-
tum 50: 93-98, 1969. [Diffusion resistances.]

2503 - GATES, D.W., GUDAUSKAS, R.T.: Preliminary studies on the effect of maize dwarf
mosaic virus on photosynthesis and respiration in corn. - Phytopathology 57:
459, 1967.

2504 - GATES, D.W., GUDAUSKAS, R.T.: Photosynthesis, respiration, and evidence of a metabolic inhibitor in corn infected with maize dwarf mosaic virus. - Phytopathology 59: 575-580, 1969.

2505 - GATHERUM, G.E., GORDON, J.C., BROERMAN, B.F.S.: Effects of clone and light intensity on photosynthesis, respiration and growth of aspen-poplar hybrids. -
Silvae Genet. 16: 128-132, 1967.

2506 - GATHERUM, G.E., GORDON, J.C., BROERMAN, B.F.S.: Physiological variation in
Scotch pine seedlings in relation to light intensity and provenance. - Iowa State
J. Sci. 42: 19-26, 1967.[Ps.]

2507 - GATHERUM, G.E;, GORDON, J.C., BROERMAN, B.F.S.: Physiological variation in
European black pine seedlings in relation to light intensity and provenance. -
Iowa State J. Sci. 42: 27-35, 1967. [Ps.]

2508 - GAUDET, J.J., DAMM, G.: Low cost, portable laboratory radiometer. - BioScience
18: 513-514, 1968.

2509 - GAUDILLÈRE, J.P.: Etude thermogravimétrique de chloroplastes lyophilisés. - Ann.
Physiol. vég. 11: 113-122, 1969.

2510 - GAUHL, E.: Differential photosynthetic performance of *Solanum dulcamara* ecotypes
from shaded and exposed habitats. - Carnegie Inst. Year Book 67: 482-487, 1969.

2511 - GAUHL, E.: Leaf factors affecting the rate of light-saturated photosynthesis in
ecotypes of *Solanum dulcamara*. - Carnegie Inst. Year Book 68: 633-636, 1970.

2512 - GAUHL, E., BJÖRKMAN, O.: Simultaneous measurements of the effect of oxygen concentration on water vapor and carbon dioxide exchange in leaves. - Planta *88*: 187-191, 1969.

2513 - GAUNT, J.K., SROWE, B.B.: Analysis and distribution of tocopherols and quinones in the pea plant. - Plant Physiol. *42*: 851-858, 1967. [Chl, car.]

2514 - GAUSMAN, H.W., ALLEN, W.A., CARDENAS, R.: Reflectance of cotton leaves and their structure. - Remote Sens. Environm. *1*: 19-22, 1969.

2515 - GAUSMAN, H.W., CARDENAS, R.: Effect of leaf pubescence of *Gynura aurantiaca* on light reflectance. - Bot. Gaz. *130*: 158-162, 1969. [Chl.]

2516 - GAVOROVÁ, E.: Príspevok k štúdiu priebehu fotosyntézy ovocných stromov počas vegetačného obdobia. [Contribution to the study of the activity of photosynthesis in fruit trees during their growing season.] - Ved. Práce výsk. Ústavu rastl. Výr. (Piešťany) *7*: 237-245, 1969. [In Slovak, ab: E, R.]

2517 - GAVRILENKO, V.F.: O mekhanizme preobrazovaniya energii i put'yakh prevrashcheniya ugleroda pri fotosinteze. [On the energy transformation mechanism and carbon conversion pathways in photosynthesis.] - Nauch. Dokl. vyssh. Shkoly, biol. Nauki *11* (10): 119-136, 1968. [In R.]

2518 - GAVRILENKO, V.F.: Metabolizm zhelezo-porfirinov i energeticheskiǐ obmen. [Metabolism of Fe-porphyrins and of energy.] - In: Fiziologiya i Biokhimiya Zdorovogo i Bol'nogo Rasteniya. Pp. 208-222. Izdat. mosk. Univ., Moskva 1970. [In R.]

2519 - GAVRILENKO, V.F., RUBIN, B.A.: Zakonomernosti biosinteza porfirinov v kornevoǐ sisteme. [Regularities of biosynthesis of porphyrins in the root system.] - Sel'.-khoz. Biol. *3*: 587-606, 1968. [Chl; in R, ab: E.]

2520 - GEDZ', S.M., MAKHACHEK, V.N., KAPROVA, M.I.: Dinamika nakopleniya pigmentov v rasteniyakh kartofelya i vliyanie na nee nekotorykh mikroelementov. [Dynamics of pigment accumulation in potato as influenced by some microelements.] - Mikroelem. sel'.Khoz. Med. *4*: 44-52, 1968. [In R.]

2521 - GEE, A.R., TRUSCOTT, T.G.: Fluorescence spectra of chlorophyll excited by a continuous gas laser. - Chem. Commun. *1968*: 839-841, 1968.

2522 - GEE, R., KYLIN, A., SALTMAN, P.: Enhancement of photophosphorylation and photoreduction by a chloroplast factor from spinach leaves. - Biochem. biophys. Res. Commun. *40*: 642-648, 1970.

2523 - GEE, R., SALTMAN, P., WEAVER, E.: Studies on three photosynthetic mutants of *Scenedesmus*. - Biochim. biophys. Acta *189*: 106-115, 1969.

2524 - GEE, R.W.: Structure and biochemical function of the photosynthetic apparatus. - Diss. Abstr. B *28*: 1783-B-1784-B, 1967.

2525 - GEIGER, D.R.: Effect of sink region cooling on translocation of photosynthate. - Plant Physiol. *41*: 1667-1672, 1966.

2526 - GEIGER, D.R., CATALDO, D.A.: Leaf structure and translocation in sugar beet. - Plant Physiol. *44*: 45-54, 1969. [Photosynthates.]

2527 - GEIGER, D.R., SAUNDERS, M.A., CATALDO, D.A.: Translocation and accumulation of translocate in the sugar beet petiole. - Plant Physiol. *44*: 1657-1665, 1969.

2528 - GEISSLHOFER, M., BURIAN, K.: Biometrische Untersuchungen im geschlossenen Schilfbestand des Neusiedler Sees. - Oikos *21*: 248-254, 1970. [Ps.]

2529 - GEJ, B.: Zmiany w zawartości chlorofilu *a* i *b* w liściach róznego wieku niektórych roślin dwuliściennych. [Changes in chlorophyll *a* and *b* content in leaves of different ages in some dicotyledonous plants.] - Acta Soc. Bot. Pol. *35*: 209-224, 1966. [In Pol., ab: E.]

B2530 - GEJ, B.: Wzrost i Intensywność Fotosyntezy Liści Róznego Wieku Niektórych Roślin Dwuliściennych. [Growth and Photosynthetic Rate in Leaves of Different Ages in Some Dicotyledonous Plants.] - Dział Wyd. SGGW, Warszawa 1967. [In Polish]

2531 - GEJ, B.: Changes in $^{14}CO_2$ fixation rate in different-aged leaves after decapitation of certain annual plants. - Bull. Acad. pol. Sci., Sér. Sci. biol. *18*: 585-589, 1970.

2532 - GEJ, B.: Wpływ symazyny i diuronu na $^{14}CO_2$-fotosyntezę oraz na wzrost niektórych roślin dwuliściennych. [Influence of simazine and diuron on $^{14}CO_2$ uptake and on the growth of some dicotyledonous plants.] - Rocz. glebozn. *21*: 265-277, 1970. [In Pol.]

2533 - GELLER, D.M.: Modification of light-induced absorption changes in *Rhodospirillum rubrum* extracts. - Fed. Proc. *25*: 737, 1966.

2534 - GELLER, D.M.: Correlation of light-induced absorbance changes with photophosphorylation in *Rhodospirillum rubrum* extracts. - J. biol. Chem. *242*: 40-46, 1967.

2535 - GELLER, D.M.: The effects of phenazine dyes and N,N,N',N'-tetramethyl-*p*-phenylenediamine upon light-induced absorbance changes and photophosphorylation in *Rhodospirillum rubrum* extracts. - J. biol. Chem. *244*: 971-980, 1969.

2536 - van GEMERDEN, H.: Growth measurements of *Chromatium* cultures. - Arch. Mikrobiol. *64*: 103-110, 1968. [Chl.]

2537 - GENCHEV, S.: Prouchvane v"rkhu plastidnite pigmenti pri luka (*Allium cepa* L.). II. Dinamika na khlorofila i karotinoidite prez ontogenezisa na luka. [Study of plastid pigments in onion (*Allium cepa* L.). II. Dynamics of chlorophyll and carotenoids during onion ontogenesis.] - Fiziol. Rast. (Sofiya) *1*: 11-23, 1970. [In Bulg.]

2538 - GENCHEV, S.: Prouchvane v"rkhu plastidnite pigmenti pri luka (*Allium cepa* L.). III. V"rkhu vr"zkata na khlorofila s katalazata, peroksidazata i askorbinovata kiselina v listata na luka. [Study of plastid pigments in onion (*Allium cepa* L.). III. Relationship of chlorophyll content and catalase, peroxidase and ascorbic acid in onion leaves.] - Fiziol. Rast. (Sofiya) *1*: 25-31, 1970. [In Bulg.]

2539 - GENEROZOVA, I.P.: Dinamika krakhmalonosnykh plastid v ontogeneze rasteniĭ. [Dynamics of starch-containing plastids in plant ontogenesis.] - In: Khloroplasty i Mitokhondrii. Pp. 157-163. Nauka, Moskva 1969. [In R.]

2540 - GENEROZOVA, I.P., TAGEEVA, S.V.: Genezis khloroplastov i mitokhondriĭ v ontogeneze rasteniĭ. [Genesis of chloroplasts and mitochondria in plant ontogenesis.] - In: Khloroplasty i Mitokhondrii. Pp. 146-156. Nauka, Moskva 1969. [In R.]

2541 - GENEVÈS, L.: Distribution infrastructurale d'activités adénosine triphosphatasique dans les cellules en interphase et en division du tissu sporifère d'*Hypnum rusciforme*, vers la fin de la phase de prolifération. - Compt. rend. Acad. Sci. (Paris), Sér. D *271*: 392-395, 1970.

2542 - GENEVOIS, L., BARAUD, J., BENITEZ, F.: Les caroténoïdes du Maïs. - Qual. Plant. Mater. veg. *13*: 78-85, 1966.

2543 - GENGENBACH, B.G., GORZ, H.J., HASKINS, F.A.: Genetic studies of induced mutants in *Melilotus alba*. II. Inheritance and complementation of chlorophyll-deficient mutants. - Crop Sci. *10*: 154-156, 1970.

2544 - GENKEL', P.A., KURKOVA, E.B., PRONINA, N.D.: O vliyanii obezvozhivaniya na khod fotosinteza u gomeogidrovykh i poĭkilogidrovykh rasteniĭ. [Effect of drying on the course of photosynthesis in homeohydrous and poikilohydrous plants.] - Fiziol. Rast. *17*: 1140-1146, 1970. [In R, ab: E.]

2545 - GENSE, M.-T., GUÉRIN-DUMARTRAIT, E., LECLERC, J.-C., MIHARA, S.: Synchronisation de *Porphyridium:* Evolution des quantités de pigments et de la capacité photosynthétique au cours du cycle biologique. - Phycologia *8*: 135-141, 1969.

2546 - GENTER, C.F., JONES, G.D., CARTER, M.T.: Dry matter accumulation and depletion in leaves, stems, and ears of maturing maize. - Agron. J. *62*: 535-537, 1970.

2547 - GEORGESCU, M.: Unele procese fiziologice la soiuri de *Vitis vinifera*, cu diferite epoci de maturitate. [Some physiological processes of *Vitis vinifera* cultivars with various maturity period.] - Lucrări ştiinţ. Inst. agron. "N. Bălcescu", Ser. B *12*: 535-542, 1969. [Ps, Chl; in Rum., ab: E, F,R.]

2548 - GEORGIEV, D.: Asimilirane pridvizhvane i razpredelenie na radioaktivniya v'glerod C^{14} pri prisadeni rasteniya ot p'pesh v'rkhu tikva. [Assimilation, transport and distribution of radioactive carbon ^{14}C in melon plants grafted on gourd.] - Rasteniev. Nauki *5* (1): 9-18, 1968. [In Bulg., ab: E, R.]

2549 - GEREBTZOFF, D., RAMAUT, J.L.: Essai de localisation histochimique du zinc chez
         Hordeum gr. *vulgare* et toxicité. - Physiol. Plant. *23*: 574-582, 1970.[Chl.]

2550 - GERGIS, M.S.: A colorless *Chlorella* mutant containing thylakoids. - Arch. Mikro-
         biol. *68*: 187-190, 1969.

2551 - GERHARDT, B.: Manganeffekte in photosynthetischen Reaktionen von *Anacystis*. -
         Ber. deut. bot. Ges. *79*: (63)-(68), 1966.

2552 - GERHARDT, B., SANTO, R.: Photophosphorylierung in einem zellfreien System aus
         Anacystis. - Z. Naturforsch. *21 b*: 673-678, 1966.

2553 - GERHARDT, B., WIESSNER, W.: On the light-dependent reactivation of photosynthe-
         tic activity by manganese. - Biochem. biophys. Res. Commun. *28*: 958-964, 1967.

2554 - GERIĆ, I.: The photosynthetic activity of corn in combined crop growing. - J.
         sci. agr. Res. *19*: 105-134, 1966.

2555 - GERIĆ, I.: Transformacija svetlosne energije u procesu fotosinteze. [Transforma-
         tion of radiant energy in photosynthesis.] - Savremena Poljoprivreda *16*: 644-
         650, 1968. [In Croat.]

2556 - GERIĆ, I.: Uticaj vodnog režima zemljišta na fotosintetičke procese kod šećerne
         repe. [Effect of soil water relations on photosynthetic processes in sugar beet.]
         - Zbornik rad. *6*: 63-75, 1968. [In Croat., ab: E.]

2557 - GERIĆ, I.: Uticaj mineralne ishrane na fotosintetičke procese u kasnijim fazama
         razvoja pšenice. [Effect of mineral nutrition on photosynthetic processes in
         later phases of wheat development.] - Savremena Poljoprivreda *17* (11-12): 211-
         221, 1969. [In Croat., ab: E.]

2558 - GERONIMO, J., HERR, J.W.: Ultrastructural changes of tobacco chloroplasts in-
         duced by Pyriclor. - Weed Sci. *18*: 48-53, 1970.

2559 - GEROSA, V.: Chimica comparativa dei carotinoidi delle alghe. [Chemical compari-
         son of algae carotenoids.] - Stud. trentini Sci. nat. B *43*: 159-171, 1966. [In
         Ital.]

2560 - GEROSA, V.: La natura chimica delle sostanze che provocano l'arrossamento del
         Lago di Tovel. 2ª Nota - Cromatografia su strato sottile. [Chemical nature of
         substances which redden the lake of Tovel. 2nd note - Thin-layer chromatography.]
         - Stud. trentini Sci. nat. B *43*: 145-158, 1966. [Car; in Ital., ab: E, G, F.]

2561 - GERSTER, R., DUPUY, J.: Echanges isotopiques et photosynthèse étudiés au moyen
         d'oxygène 18 chez *Euglena gracilis*. - Compt. rend. Acad. Sci.(Paris), Sér.D *269*:
         1764-1766, 1969.

2562 - GERSTER, R., GUERIN DE MONTGAREUIL, P.: Isotopic composition of the oxygen evol-
         ved by a maize leaf with $C^{18}O_2$ as a tracer. - In: COOMBS, J. (ed.): Photosynthe-
         sis in Sugar Cane. P. 40. Tate and Lyle Ltd., London 1969.

2563 - GERSTER, R., MICHEL, J.-P.: Echange isotopique entre le gaz carbonique marqué à
         l'oxygène 18 et l'eau d'une feuille aérienne. - Compt. rend. Acad. Sci. (Paris),
         Sér. D *266*: 1024-1027, 1968.

2564 - GERTH, E.: Ein äquidensitometrisches Verfahren zur Registrierung des zeitlichen
         Verlaufes der Lichtintensität. - Feingerätetechnik *16*: 558-564, 1967.

2565 - GERVAIS, C., TENDILLE, C., GABORIT, T.: Présence de plastochromanol dans les
         tissus végétaux chlorophylliens. - Ann. Physiol. vég. (Paris) *10*: 209-217, 1968.

2566 - GES', D.K., REUTSKAYA, L.N.: Vliyanie virusnoĭ infektsii na soderzhanie pigmen-
         tov v list'yakh tomatov i ogurtsov. [Effect of virus infection on pigment con-
         tent in leaves of tomato and cucumber.] - Vestsi Akad. Navuk Belarus. SSR, Ser.
         biyal. Navuk *1966* (3): 129-130, 1966. [In R.]

2567 - GEST, H.: Comparative biochemistry of photosynthetic processes. - Nature *209*:
         879-882, 1966.

2568 - GETZ, L.L.: A method for measuring light intensity under dense vegetation. -
         Ecology *49*: 1168-1169, 1968.

2569 - GEVEL', L.M.: Zavisimost' intensivnosti fluorestsentsii v induktsionnom periode ot nekotorykh fizicheskikh i fiziologicheskikh faktorov. [Dependence of the rate of fluorescence in the induction period on some physical and physiological factors.] - In: Bioenergetika i Biologicheskaya Spektrofotometriya. Pp. 149-156. Nauka, Moskva 1967. [In R.]

2570 - GEVEL', L.M., BELYANIN, V.N., EROSHIN, N.S., SID'KO, F.Ya.: Zavisimost' kontsen-tratsii pigmentov i intensivnosti fluorestsentsii ot uslovii kul'tivirovaniya mikrovodorosleĭ. [Dependence of concentration of pigments and intensity of fluorescence on cultivation conditions of microalgae.] - In: Nepreryvnoe Uprav-lyaemoe Kul'tivirovanie Mikroorganizmov. Pp. 95-101. Nauka, Moskva 1967. [In R.]

2571 - GHAI, B.S., SRIVASTAVA, A.K., GILL, K.S.: Inheritence of amount of chlorophyll in wheat, *Triticum aestivum* L. - Euphytica *18*: 403-405, 1969.

2572 - GHOSH, A.K., BROKER, T.R., OLSON, J.M.: A kinetic study of bacteriochlorophyll pheophytinization in the protein complex from a green photosynthetic bacterium. - Biochim. biophys. Acta *162*: 402-413, 1968.

2573 - GHOSH, A.K., GOVINDJEE: Transfer of the excitation energy in *Anacystis nidulans* grown to obtain different pigment ratios. - Biophys. J. *6*: 611-619, 1966.

2574 - GHOSH, A.K., GOVINDJEE, CRESPI, H.L., KATZ, J.J.: Fluorescence studies on deute-rated *Chlorella vulgaris*. - Biochim. biophys. Acta *120*: 19-22, 1966. [Chl.]

2575 - GHOSH, A.K., OLSON, J.M.: Effects of denaturants on the absorption spectrum of the bacteriochlorophyll-protein from the photosynthetic bacterium *Chloropseudo-monas ethylicum* . - Biochim. biophys. Acta *162*: 135-148, 1968.

2576 - GIACOMELLI, M., BELLI DONINI, M.L., CERVIGNI, T.: Effects of kinetin on chloro-phyll breakdown and protein levels in irradiated barley leaves. - Radiat. Bot. *7*: 375-384, 1967.

2577 - GIBBON, D., HOLLIDAY, R., MATTEI, F., LUPPI, G.: Crop production potential and energy conversion efficiency in different environments. - Exp. Agr. *6*: 197-204, 1970.

2578 - GIBBS, M.: Photosynthesis. - Annu. Rev. Biochem. *36*: 757-784, 1967.

2579 - GIBBS, M.: Control of photosynthesis by oxygen. - In: DAY, P.R. (ed.): How Crops Grow: A Century Later. Pp. 63-79. Bull. 708, Conn. agr. exp. Sta., New Haven, Conn. 1969.

2580 - GIBBS, M.: Photorespiration, Warburg effect and glycolate. - Ann. N.Y. Acad. Sci. *168*: 356-368, 1969.

2581 - GIBBS, N.: The inhibition of photosynthesis by oxygen. - Amer. Sci. *58*: 634-640, 1970.

2582 - GIBBS, M., BAMBERGER, E.S., ELLYARD, P.W., EVERSON, R.G.: Assimilation of carbon dioxide by chloroplast preparations. - In: GOODWIN, T.W. (ed.): Biochemistry of Chloroplasts. Vol. II. Pp. 3-38. Academic Press, London-New York 1967.

2583 - GIBBS, M., ELLYARD, P.W., LATZKO, E.: Warburg effect: control of photosynthesis by oxygen. - In: SHIBATA, K., TAKAMIYA, A., JAGENDORF, A.T., FULLER, R.C. (ed.): Comparative Biochemistry and Biophysics of Photosynthesis. Pp. 387-399. Univ. Tokyo Press, Tokyo; Univ. Park Press, State College, Pa. 1968.

2584 - GIBBS, M., LATZKO, E., EVERSON, R.G., COCKBURN, W.: Carbon mobilization by the green plant. - In: SAN PIETRO, A., GREER, F.A., ARMY, T.J. (ed.): Harvesting the Sun. Pp. 111-130. Academic Press, New York-London 1967.

2585 - GIBBS, M., LATZKO, E., HARVEY, M.J., PLAUT, Z., SHAIN, Y.: Photosynthesis in the algae. - Ann. N.Y. Acad. Sci. *175*: 541-554, 1970.

2586 - GIBBS, M., LATZKO, E., O'NEAL, D., HEW, C.-S.: Photosynthetic carbon fixation by isolated maize chloroplasts. - Biochem. biophys. Res. Commun. *40*: 1356-1361, 1970.

2587 - GIBBS, M., PLAUT, Z., LATZKO, E., ELEY, J. Jr., SCHACTER, B.: Carbon fixation by isolated chloroplasts. - In: COOMBS, J. (ed.): Photosynthesis in Sugar Cane. Pp. 49-65. Tate and Lyle Ltd., London 1969.

2588 - GIBBS, S.P.: The comparative ultrastructure of the algal chloroplast. - Ann. N.Y. Acad. Sci. *175*: 454-473, 1970.

2589 - GIBOR, A.: Phenotypic variations among chloroplasts of a single cell. - Science 155: 327-329, 1967.

2590 - GIBOR, A., HERRON, H.A.: Chloroplast inheritance. - In: BUETOW, D.E. (ed.): The Biology of *Euglena*. Vol. 2. Pp. 335-349. Academic Press, New York-London 1968.

2591 - GIBSON, J.: Control of the photosynthetic apparatus in bacteria. - In: SHIBATA, K., TAKAMIYA, A., JAGENDORF, A.T., FULLER, R.C. (ed.): Comparative Biochemistry and Biophysics of Photosynthesis. Pp. 324-331. Univ. Tokyo Press, Tokyo; Univ. Park Press, State College, Pa. 1968.

2592 - GIBSON, J., HART, B.A.: Localization and characterization of Calvin cycle enzymes in *Chromatium* strain D. - Biochemistry 8: 2737-2741, 1969.

2593 - GIBSON, J., MORITA, S.: Changes in adenine nucleotides of intact *Chromatium* D. produced by illumination. - J.Bacteriol. 93: 1544-1550, 1967.

2594 - GIBSON, J.F., HALL, D.O., THORNLEY, J.H.M., WHATLEY, F.R.: The iron complex in spinach ferredoxin. - Proc. nat. Acad. Sci. U.S.A. 56: 987-990, 1966.

2595 - GIBSON, J.F., HALL, D.O., THORNLEY, J.H.M., WHATLEY, F.R.: The iron complex in spinach ferredoxin. - In: Magnetic Resonance in Biological Systems. Proc. 2nd Int. Conf. June 1966. Stockholm, Swed. Wenner-Gren Center Inst. Symp. Ser. 9. Pp. 181-182. Pergamon Press, London-New York-Sydney 1967.

B2596 - GIESE, A.C. (ed.): Photophysiology. Current Topics. Vol. III. - Academic Press. New York-London 1968.

2597 - GIESE, A.C., GRAINGER, R.M.: Studies on the red and blue forms of the pigment of *Blepharisma*. - Photochem. Photobiol. 12: 489-503, 1970.

2598 - GIFFORD, R.M., MUSGRAVE, R.B.: Diffusion and quasi-diffusion resistances in relation to the carboxylation kinetics of maize leaves. - Physiol. Plant. 23: 1048-1056, 1970.

2599 - GILFILLAN, I.M., JONES, W.W.: Effect of iron and manganese deficiency on the chlorophyll, amino acid and organic acid status of leaves of *Macadamia*. - Proc. amer. Soc. hort. Sci. 93: 210-214, 1968.

2600 - GILL, D., KILPONEN, R.G., RIMAI, L.: Resonance Raman scattering of laser radiation by vibrational modes of carotenoid pigment molecules in intact plant tissues. - Nature 227: 743-744, 1970.

2601 - GILLBRICHT, M.: Die Beziehungen zwischen verschiedenen Messungen im Meer. - Helgoländer wiss. Meeresunters. 13: 193-202, 1966. [Chl.]

2602 - GILLER, Yu.E.: Spektral'nye svoĭstva i sostoyanie fotosinteticheskikh pigmentov v iskusstvennom pigment-belkovo-lipoidnom komplekse. [Spectral properties and state of photosynthetic pigments in an artificial pigment-protein-lipoid complex.] - Biofizika 13: 1006-1017, 1968. [In R, ab: E.]

2603 - GILLER, Yu.E.: Modelirovanie nativnogo sostoyaniya i fotokhimicheskikh svoĭstv fotosinteticheskikh pigmentov s pomoshch'yu iskusstvennykh pigment-belkovolipoidnykh kompleksov. [Modelling of natural composition and photochemical properties of photosynthetic pigments by means of artificial pigment-protein-lipoid complexes.] - In: Tezisy Dokladov Vtorogo Vsesoyuznogo Biokhimicheskogo S"ezda, Sektsiya "Problemy Fotosinteza." Pp. 40-41. FAN, Tashkent 1969. [In R.]

2604 - GILLER, Yu.E., KAS'YANENKO, A.G., KOLOTOVA, L.R., TOLIBEKOV, D.T., YUSUPOVA, G.A.: Sostav, sostoyanie i fotokhimicheskaya aktivnost' pigmentov plastid zhiznesposobnykh khlorofil'nykh mutantov *Arabidopsis thaliana*. [Composition, state and photochemical activity of plastid pigments of vital chlorophyll mutants of *Arabidopsis thaliana*.] - In: Fotosintez i Ispol'zovanie Energii Solnechnoĭ Radiatsii. Pp. 45-46. Dushanbe 1967. [In R.]

2605 - GILLER, Yu.E., KHAITOVA, L.T.: Ob opticheskikh svoĭstvakh iskusstvennogo pigment-lipoproteidnogo kompleksa. [Optical properties of an artificial pigment-lipoproteid complex.] - Dokl. Akad. Nauk Tadzh. SSR 9 (12): 32-36, 1966. [In R.]

2606 - GILLER, Yu.E., KRASICHKOVA, G.V., SAPOZHNIKOV, D.I.: O roli vody i sostoyaniya nositelya v obrazovanii razlichnykh spektral'nykh form khlorofilla v iskusstvennom pigmentbelkovolipoidnom komplekse. [On the role of water and carrier con-

dition in the formation of different spectral forms of chlorophyll in an artificial pigment-protein-lipoid complex.] - Dokl. Akad. Nauk SSSR *182*: 1230-1233,
1968. [In R.]

2607 - GILLER, Yu.E., KRASICHKOVA, G.V., SAPOZHNIKOV, D.I.: Spektral'nye svoĭstva i sostoyanie fotosinteticheskikh pigmentov v iskusstvennom vodorastvorimom pigmentbelkovolipoidnom komplekse. [Spectral properties and state of photosynthetic
pigments in an artificial water-soluble pigment-protein-lipoid complex.] - Biofizika *15*: 38-46, 1970. [In R, ab: E.]

2608 - GILLER, Yu.E., MUZYKA, V.I.: K mekhanizmu izmeneniĭ opticheskoĭ sistemy list'ev
rasteniĭ pod deĭstviem dlinnovolnovoĭ UF radiatsii. [Mechanism of changes in
the optical system of plant leaves affected by long-wave UV radiation.] - In:
Issledovaniya po Fotosintezu. Pp. 37-43. Akad. Nauk Tadzh. SSR, Dushanbe 1967.
[In R, ab: Tadzh.]

2609 - GILLER, Yu.E., NASYROV, Yu.S.: O deĭstvii ul'trafioletovoĭ radiatsii na opticheskie svoĭstva i fluorestsentsiyu list'ev rasteniĭ. [Effect of UV radiation
on optical properties and fluorescence of leaves.] - Izv. Akad. Nauk Tadzh.
SSR, Otd. biol. Nauk *1966* (3): 32-45, 1966. [In R, ab: Tadzh.]

2610 - GILLER, Yu.E., NASYROV, Yu.S., SAPOZHNIKOV, D.I., KHAITOVA, L.T., YUSUPOVA, G.A.:
Energeticheskoe vzaimodeĭstvie khlorofillov *a* i *b* v razbavlennykh rastvorakh.
[Energetic interaction of chlorophyll *a* and *b* in diluted solutions.] - In: Bioenergetika i Biologicheskaya Spektrofotometriya. Pp. 163-168. Nauka, Moskva
1967. [In R.]

2611 - GILLER, Yu.E., YUKHANANOVA, L.N.: O zashchitnom deĭstvii karotina protiv foto-
destruktsii khlorofilla v model'nykh sistemakh. [Protective action of carotene
against photodestruction of chlorophyll in model systems.] - Izv. Akad. Nauk
Tadzh. SSR, Otd. biol. Nauk *1969* (2): 47-53, 1969. [In R.]

2612 - GILLER, Yu.E., YUKHANANOVA, L.N.: Izuchenie feofitinizatsii i fotofeofitinizat-
sii khlorofilla *a* v sostave iskusstvennogo pigment-belkovogo kompleksa. [Pheo-
phytinization and photopheophytinization of chlorophyll *a* in an artificial pigment-protein complex.] - Biokhimiya *35*: 873-878, 1970. [In R.]

2613 - GILLER, Yu.E., YUKHANANOVA, L.N., KHAITOVA, L.T.: K voprosu o roli komponentov
pigmento-belkovo-lipoidnogo kompleksa v povyshenii fotokhimicheskoĭ ustoĭchivosti khlorofilla. [Role of components of the pigment-protein-lipoid complex in
increasing the photochemical resistance of chlorophyll.] - In: Trudy I. Konferentsii Biokhimikov Respublik Sredneĭ Azii i Kazakhstana. Pp. 90-94. FAN,
Tashkent 1967. [In R.]

2614 - GILLER, Yu.E., YUSUPOVA, G.A.: O fotokhimicheskoĭ aktivnosti khlorofilla v sostave iskusstvennogo pigment-belkovolipoidnogo kompleksa. [Photochemical activity of chlorophyll in an artificial pigment-protein-lipoid complex.] - Dokl. Akad.
Nauk Tadzh. SSR *12* (4): 63-67, 1969. [In R.]

2615 - GILLER, Yu.E., YUSUPOVA, G.A.: Ob obratimykh fotoprevrashcheniyakh khlorofilla
v sostave iskusstvennogo pigment-belkovogo kompleksa. [On the reversible photo-
transformations of chlorophyll in the artificial pigment-protein complex.]
- Dokl. Akad. Nauk SSSR *190*: 1470-1473, 1970. [In R.]

2616 - GILLER, Yu.E., YUSUPOVA, G.A., YUKHANANOVA, L.N., ASOEVA, L.M.: O deĭstvii ul'-
trafioletovoĭ radiatsii na fluorestsentsiyu khlorofilla v model'nykh sistemakh
i nepovrezhdennykh list'yakh. [Effect of UV on fluorescence of chlorophyll in
model systems and in intact leaves.] - Izv. Akad. Nauk Tadzh. SSR, Otd. biol.
Nauk *1968* (2): 24-30, 1968. [In R.]

2617 - GILLINGHAM, J.T.: Determination of carotene in fresh forages and silages following freeze-drying and grinding. - J. Ass. offic. anal. Chem. *50*: 828-830, 1967.

2618 - GIMINGHAM, C.H., MILLER, G.R.: Part II. Measurement of the primary production
of dwarf shrub heaths. - In: MILNER, C., HUGHES, R.E. (ed.): Methods for the
Measurement of the Primary Production of Grassland. IBP Handbook No. 6. Pp. 43-
51. Blackwell Scientific Publications, Oxford-Edinburgh 1968.

2619 - GIMMLER, H., SIMONIS, W., URBACH, W.: Differential effect of Dio-9 on non-cyclic
photophosphorylation and ^{14}C-fixation *in vivo*. - Naturwissenschaften *56*: 371-
372, 1969.

2620 - GIMMLER, H., URBACH, W., JESCHKE, W.D., SIMONIS, W.: Die unterschiedliche Wirkung von Disalicylidenpropandiamin auf die cyclische und nichtcyclische Photophosphorylierung *in vivo* sowie auf die ^{14}C-Markierung einzelner Photosyntheseprodukte. - Z. Pflanzenphysiol. *58*: 353-364, 1968.

2621 - GINGRAS, G.: Etude comparative, chez quelques algues, de la photosynthèse et de la photoréduction réalisée en présence d'hydrogène. - Physiol. vég. *4*: 1-65, 1966.

2622 - GINGRAS, G.: Pigments sensibilisateurs de la photoréduction. - In: THOMAS, J.B., GOEDHEER, J.C. (ed.): Currents in Photosynthesis. Pp. 187-195. Donker, Rotterdam 1966.

2623 - GINGRAS, G., JOLCHINE, G.: Isolation of a P_{870}-enriched particle from *Rhodospirillum rubrum*. - In: METZNER, H. (ed.): Progress in Photosynthesis Research. Vol. I. Pp. 209-216. Tübingen 1969.

2624 - GINGRAS, G., SAMSON, J.P.: The Haxo and Blinks electrode. A mathematical model. - Biophys. J. *10*: 1189-1205, 1970.

2625 - GINZBURG, D., PADAN, E., SHILO, M.: Effect of cyanophage infection on CO_2 photoassimilation in *Plectonema boryanum*. - J. Virol. *2*: 695-701, 1968.

2626 - GIRAUD, G.: L'ultrastructure de l'appareil photosynthétique. - In: THOMAS, J.B., GOEDHEER, J.C. (ed.): Currents in Photosynthesis. Pp. 329-338. Donker, Rotterdam 1966.

2627 - GIRAUD, G., LICHTLE, C., THOMAS, J.C.: Some aspects of the localization of phycobiliproteins in the *Rhodophyceae* and the *Cyanophyceae*. - Biochem. J. *119*: 15P-16P, 1970.

2628 - GIRAULT, G., GALMICHE, J.-M., ROUX, E.: Rôle des quinones et des carotènes dans deux réactions photochimiques se produisant au sein d'une suspension de chloroplastes. - Compt. rend. Acad. Sci. Paris, Sér. D.*269*: 2265-2267, 1969.

2629 - GIRAULT, G., TYSZKIEWICZ, E., GALMICHE, J.: Photophosphorylation and light-induced increase of pH. - In: METZNER, H. (ed.): Progress in Photosynthesis Research. Vol. III. Pp. 1347-1353. Tübingen 1969.

2630 - GIRNIK, D.V.: Intensivnost' fotosinteza odnogo iz massovykh vidov fitoplanktona Chernogo morya v zavisimosti ot osveshcheniya i temperatury vody. 1. [Photosynthetic rate of one of the mass species of phytoplankton in Black Sea as affected by irradiance and water temperature. 1.] - In: Puti Povysheniya Intensivnosti i Produktivnosti Fotosinteza. Pp. 160-163. Naukova Dumka, Kiev 1966. [In R.]

2631 - GIRNIK, D.V.: Intensivnost' fotosinteza massovykh vidov fitoplanktona Chernogo morya v zavisimosti ot osveshcheniya i temperatury vody. [Photosynthetic rate of mass species of phytoplankton in Black Sea as affected by irradiance and water temperature.] - Puti Povysheniya Intensivnosti i Produktivnosti Fotosinteza. Vol. 3. Pp. 213-218. Naukova Dumka, Kiev 1969. [In R.]

2632 - GIRSHOVICH, Yu.E., KOBAK, K.I.: Issledovanie fotosinteticheskoĭ deyatel'nosti agrofitotsenoza. [Photosynthetic activity of agrophytocenosis.] - Tr. gl. geofiz. Observ. *229*: 48-62, 1968. [In R.]

2633 - GIVAN, A.L., LEVINE, R.P.: The photosynthetic electron transport chain of *Chlamydomonas reinhardi*. VII. Photosynthetic phosphorylation by a mutant strain of *Chlamydomonas reinhardi* deficient in active P700. - Plant Physiol. *42*: 1264-1268, 1967.

2634 - GIVAN, A.L., LEVINE, R.P.: The photosynthetic electron transport chain of a mutant strain of *Chlamydomonas reinhardi* lacking P_{700} activity. - Biochim. biophys. Acta *189*: 404-410, 1969.

2635 - GIVAN, C.V., GIVAN, A.L., LEECH, R.M.: Photoreduction of α-ketoglutarate to glutamate by *Vicia faba* chloroplasts. - Plant Physiol. *45*: 624-630, 1970.

2636 - GLABISZEWSKI, J., PLOSZYNSKI, M., SZUMILAK, G., ZURAWSKI, H.: Influence of herbicides from S triazine group on dynamics of some biochemical processes of oats in test experiments. II. Researches on action of propazine, prometrine and prometone. - Pamiet. Pulawski *1967* (28): 45-61, 1967.[Chl.]

2637 - GLADYSHEV, N.P.: K metodike opredeleniya ploshchadi list'ev yabloni. [On the methods of determination of the area of apple-tree leaves.] - Bot. Zh. *54*: 1571-1575. 1969. [In R.]

2638 - GLADYSHEVA, O.M., ESAULENKO, G.P.: Khlorofill-belkovo-lipoidnyĭ kompleks v list'-yakh pshenitsy v usloviyakh vysokikh temperatur. [Chlorophyll-protein-lipoid complex in wheat leaves at high temperatures.] - In: Trudy I. Konferentsii Biokhimi-kov Respublik Sredneĭ Azii i Kazakhstana. Pp. 152-154. FAN, Tashkent 1967. [In R.]

2639 - GLADYSHEVA, O.M., ESAULENKO, G.P.: Prochnost' svyazi khlorofilla s lipoproteida-mi u gibridnogo i roditel'skikh sortov yarovoĭ pshenitsy. [Stability of chloro-phyll-lipoproteid bound in hybrid and parental cultivars of spring wheat.] - Tr. Inst. Bot. Akad. Nauk Kaz. SSR *25*: 14-22, 1968. [In R.]

2640 - GLADYSHEVA, O.M., ESAULENKO, G.P.: Sostoyanie khlorofill-belkovo-lipoidnogo kom-pleksa v list'yakh yarovoĭ pshenitsy. [State of chlorophyll-protein-lipoid complex in leaves of spring wheat.] - Izv. Akad. Nauk Kaz. SSR, Ser. biol. *1968* (1): 25-32, 1968. [In R.]

2641 - GLAGOLEVA, T.A., MUUK, E.L., IVANOVA, N.A.: Vliyanie ingibitorov fotofosforili-rovaniya na metabolizm ugleroda C^{14} u khlorelly. [Effect of inhibitors of photo-phosphorylation on the ^{14}C metabolism in *Chlorella*.] - Bot. Zh. *54*: 1965-1973, 1969. [In R, ab: E.]

2642 - GLAGOLEVA, T.A., ZALENSKIĬ, O.V.: O bioenergetike assimiliruyushchikh kletok *Chlorella pyrenoidosa* CHICK. [Bioenergetics of photosynthesizing cells of *Chlo-rella pyrenoidosa* CHICK.] - Bot. Zh. *51*: 1683-1693, 1966. [In R, ab: E.]

2643 - GLAGOLEVA, T., ZALENSKIĬ, O.V.: Bioenergetics of assimilating cells of *Chlorel-la pyrenoidosa* CHICK. I. Relative rates of photophosphorylation and oxidative phosphorylation in intact cells. - Photosynthetica *4*: 15-20, 1970.

2644 - GLAGOLEVA, T.A., ZALENSKIĬ, O.V.: Metody kosvennoĭ otsenki fotofosforilirovaniya *in vivo*. [Methods of indirect estimation of photophosphorylation *in vivo*.] - In: KIRICHENKO, E.B. (ed.): Metody Issledovaniya Fotofosforilirovaniya. Pp. 89-110. Pushchino-na-Oke 1970. [In R, ab: E.]

2645 - GLIKMAN, T.S., ZABRODA, O.V.: O vozmozhnosti fotookisleniya vody marganetssoder-zhashchim proizvodnym khlorofilla. [Possibility of photooxidation of water by manganese-containing derivative of chlorophyll.] - Biokhimiya *34*: 302-307, 1969. [In R, ab: E.]

2646 - GLIKMAN, T.S., ZAVGORODNAYA, L.N.: O prirode vosstanovlennykh form marganets (III)-soderzhashchikh proizvodnykh khlorofilla i marganets (III)-sul'foftalot-sianina. [Nature of reduced forms of Mn^{3+} containing derivatives of chlorophyll and Mn^{3+}-sulphophthalocyanine.] - Biofizika *15*: 913-915, 1970. [In R, ab: E.]

2647 - GLINKA, Z., KATCHANSKY, M.Y.: The effect of water potential on the CO_2 compen-sation point of maize and sunflower leaf tissue. - Israel J. Bot. *19*: 533-541, 1970.

2648 - GLOGOV, L.: Opredelyane intenzivnostta na fotosintezata v posevite. [Determina-tion of photosynthetic rate in field crops.] - Rasteniev"d. Nauki (Sofia) *5* (5): 3-12, 1968. [In Bulg., ab: E, R.]

2649 - GLOOSCHENKO, W.A.: Diel periodicity of chlorophyll *a* in the Gulf of Mexico. - Quart. J. Fla. Acad. Sci. *33*: 187-192, 1970.

2650 - GLOSER, J.: Some problems of the determination of stomatal aperture by the microrelief method. - Biol. Plant. *9*: 28-33, 1967.

2651 - GLOSER, J.: The dependence of CO_2-exchange on density of irradiation, tempera-ture and water saturation deficit in *Stipa* and *Bromus*. - Photosynthetica *1*: 171-178, 1967.

2652 - GLOSER, J.: A set of leaf chambers suitable for gas exchange measurements in grasses. - Photosynthetica *4*: 312-313, 1970.

2653 - GLOSER, J.: Photosynthetic activity of some meadow grasses. - In: Productivity of Terrestrial Ecosystems. Production Processes. Czechosl. nat. Comm. IBP, PT-PP Report No. 1. Pp. 51-52. Praha 1970.

2654 - GNANAM, A., JAGENDORF, A.T., RANALLETTI, M.-L.: Chloroplasts and bacterial amino acid incorporation: a further comment. - Biochim. biophys. Acta *186*: 205-213, 1969.

2655 - GNANAM, A., KAHN, J.S.: Biochemical studies on the induction of chloroplast development in *Euglena gracilis*. I. Nucleic acid metabolism during induction. - Biochim. biophys. Acta *142*: 475-485, 1967.

2656 - GNANAM, A., KAHN, J.S.: Biochemical studies on the induction of chloroplast development in *Euglena gracilis*. II. Protein synthesis during induction. - Biochim. biophys. Acta *142*: 486-492, 1967.

2657 - GNANAM, A., KAHN, J.S.: Biochemical studies on the induction of chloroplast development in *Euglena gracilis*. III. Ribosome metabolism associated with chloroplast development. - Biochim. biophys. Acta *142*: 493-499, 1967.

2658 - GNANAM, A., KULANDAIVELU, G.: Photosynthetic studies with leaf cell suspensions from higher plants. - Plant Physiol. *44*: 1451-1456, 1969. [Mesophyll cell suspensions.]

2659 - GNUSKIN, Yu.A.: Elektrokhimicheskoe opredelenie kontsentratsii kisloroda v ekstensivnoĭ kul'ture vodorosleĭ. [Electrochemical determination of oxygen concentration in an extensive culture of algae.] - Nauch. Dokl. vyssh. Shkoly, biol. Nauki *1970* (1): 119-122, 1970. [In R.]

2660 - GÖBEL, F.: Measurements of the radiation balance in suspensions of photosynthetic bacteria. - In: METZNER, H. (ed.): Progress in Photosynthesis Research. Vol. II. Pp. 1122-1127. Tübingen 1969.

2661 - GOCHOLASHVILI, M.M., ADEĬSHVILI, N.I.: Vliyanie razlichnykh vneshnikh faktorov na intensivnost' fotosinteza v list'yakh chaĭnogo rasteniya. [Effect of different environmental factors on photosynthetic rate of tea leaves.] - Subtrop. Kul't. *1970* (6): 58-67, 1970. [In R.]

2662 - GODISH, T.J.: Effect of hydrogen chloride gas on photosynthesis, respiration, transpiration and photosynthetic pigments of tomato cv. Bonny Best. - Diss. Abstr. int. B *31*: 3104-3105,1970.

2663 - GODNEŬ, Ts.M.: Prablema fotasintezu ŭ rabotakh K.A. Tsimirazeva. [Problem of photosynthesis in papers of K.A. Timiryazev.] - Vestsi Akad. Navuk Belarus. SSR, Ser. biyal. Navuk *1968* (3): 5-8, 1968. [In Belorus.,ab:R.]

2664 - GODNEŬ, Ts.M., ABUTALYBAŬ, M.G., GUMATAŬ, M.R.: Uplyŭ kol'kastsi i suadnosin zhaleza i margantsu na ŭtvarenne pigmentaŭ u khlaraplastakh. [Effect of amount and ratio of iron and manganese on pigment formation in chloroplasts.] - Vestsi Akad. Navuk Belarus. SSR, Ser. biyal. Navuk *1969* (3): 48-51, 139, 1969. [In Belorus., ab: R.]

2665 - GODNEŬ, Ts.M., SHABEL'SKAYA, E.F.: Dasledavanne sumachnykh vagannyaŭ pigmentaŭ u nekatorykh svetalyubivykh i tsenevynoslivykh raslin. [Study of daily variations in pigment content in some sun and shade plants.] - Vestsi Akad. Navuk Belarus. SSR, Ser. biyal. Navuk *1966* (2): 5-9, 1966. [In Belorus.,ab: R.]

2666 - GODNEV, T.N., AKULOVICH, N.K., DOMASH, V.I.: O predele otritsatel'noĭ temperatury dlya reaktsii perekhoda protokhlorofillovogo pigmenta v khlorofillovyĭ v obezvozhennykh list'yakh. [On the threshold of negative temperature for the transition reaction of the protochlorophyllic pigment into the chlorophyllic one in dehydrated leaves.] - Dokl. Akad. Nauk SSSR *177*: 225-228, 1967. [In R.]

2667 - GODNEV, T.N., AKULOVICH, N.K., DOMASH, V.I.: O vliyanii temperatury na sostoyanie protokhlorofill-golokhroma v vysushennykh i nativnykh etiolirovannykh list'-yakh yachmenya. [Effect of temperature on the state of protochlorophyll-holochrome in dried and fresh etiolated barley leaves.] - In: Fiziologo-biokhimicheskie Issledovaniya Rasteniĭ. Pp. 8-14. Nauka i Tekhnika, Minsk 1967. [In R.]

2668 - GODNEV, T.N., AKULOVICH, N.K., DOMASH, V.I., ORLOVSKAYA, K.I.: K voprosu o posledeĭstvii moshchnykh svetovykh impul'sov na ustoĭchivost' fotosinteticheskogo apparata. [Aftereffect of intensive light impulses on the resistance of photosynthetic apparatus.] - In: Issledovaniya po Fiziologii i Biokhimii Rasteniĭ. Pp. 6-9. Nauka i Tekhnika, Minsk 1966. [In R.]

2669 - GODNEV, T.N., AKULOVICH, N.K., DOMASH, V.I., ORLOVSKAYA, K.I.: Vliyanie nagreva-

niya na sostoyanie protokhlorofill-golokhroma etiolirovannykh prorostkov yachmenya v protsesse ego perekhoda v khlorofill-golokhrom. [Effect of heating on
the state of protochlorophyll-holochrome in etiolated barley seedlings during
its transformation into chlorophyll-holochrome.] - In: Fiziologo-Biokhimicheskie
Issledovaniya Rastenii. Pp. 6-11. Nauka i Tekhnika, Minsk 1968. [In R.]

2670 - GODNEV, T.N., AKULOVICH, N.K., ORLOVSKAYA, K.I.: O protokhlorofille obolochek
semyan tykvy raznogo vozrasta i o bakterioprotokhlorofille. [On protochlorophyll
in seed coats in gourd of different age and on bacterioprotochlorophyll.] - In:
Fotosintez i Pitanie Rastenii. Pp. 9-12. Nauka i Tekhnika, Minsk 1969. [In R.]

2671 - GODNEV, T.N., AKULOVICH, N.K., ORLOVSKAYA, K.I., DOMASH, V.I.: O vliyanii fito-
khromnoi sistemy na formirovanie pigmentov v tkanevoi kul'ture morkovi. [Effect
of phytochrome system on the pigment formation in tissue culture of carrot.] -
Dokl. Akad. Nauk SSSR *169*: 692-694, 1966. [In R.]

2672 - GODNEV, T.N., AKULOVICH, N.K., ORLOVSKAYA, K.I., RASKIN, V.I.: O protokhlorofill-
golokhrome etiolirovannykh list'ev *Quercus robur* L. [Protochlorophyll-holochrome
in etiolated leaves of *Quercus robur* L.] - Dokl. Akad. Nauk SSSR *179*: 465-467,
1968. [In R.]

2673 - GODNEV, T.N., AKULOVICH, N.K., RASKIN, V.I., DOMASH, V.I.: O deistvii sveta i
temperatury na reaktsiyu prevrashcheniya protokhlorofillovogo pigmenta v khloro-
fillovyi. [Effect of light and temperature on transformation of protochlorophyll
into chlorophyll.] - Stud. biophys. *5*: 25-29, 1967. [In R, ab: E.]

2674 - GODNEV, T.N., AKULOVICH, N.K., RASKIN, V.I., DOMASH, V.N.: K teorii smeshcheniya
temperaturnykh predelov reaktsii protokhlorofill → khlorofill pri obezvozhivanii
list'ev. [Theory of shift of temperature limits of the reaction protochlorophyll
→ chlorophyll during dehydration of leaves.] - Dokl. Akad. Nauk SSSR *185*: 1366-
1367, 1969. [In R.]

2675 - GODNEV, T.N., ARNAUTOVA, A.I., KHODASEVICH, E.V.: Ob ustoichivosti pigmentnoi
sistemy ozimykh rastenii k vozdeistviyu kholoda v osenne-zimnii period. [Resist-
ance of the pigment system of winter plants to cold during the autumn-winter
period.] - Dokl. Akad. Nauk Belorus. SSR *10*: 897-900, 1966. [In R.]

2676 - GODNEV, T.N., ARNAUTOVA, A.I., KHODASEVICH, E.V.: O sostoyanii khloroplastov ozimoi rzhi v osenne-zimnii period. [Chloroplast state in winter rye during autumn-
winter period.] - In: Fiziologo-biokhimicheskie Issledovaniya Rastenii. Pp. 3-8.
Nauka i Tekhnika, Minsk 1967. [In R.]

2677 - GODNEV, T.N., DOMASH, V.I., AKULOVICH, N.K.: K voprosu o vliyanii mnogokratnykh
impul'sov krasnogo sveta na nakoplenie pigmentov rastenii. [Effect of repeated
impulses of red radiation on accumulation of plant pigments.] - In: Issledovaniya
po Fiziologii i Biokhimii Rastenii. Pp. 3-6. Nauka i Tekhnika, Minsk 1966. [In
R.]

2678 - GODNEV, T.N., GALAKTIONOV, S.G., RASKIN, V.I.: K voprosu o stericheskikh usloviyakh reaktsii gidrovaniya atomov C₇ i C₈ 4-go pirrol'nogo yadra protokhlorofil-
lovykh pigmentov. [Steric conditions of the hydration reaction of atoms C_7 and
C_8 of the fourth pyrrolic ring of protochlorophyll pigments.] - Dokl. Akad. Nauk
SSSR *181*: 237-240, 1968. [In R.]

2679 - GODNEV, T.N., GUMMATOV, M.R.: [Effect of iron and manganese on the amount of
pigments and their resistance to acids in maize leaves.] - Dokl. Akad. Nauk
Azerb. SSR *23* (6): 49-52, 1967. [In Azerb., ab: R.]

2680 - GODNEV, T.N., KAKHNOVICH, L.V.: O deistvii predposevnogo oblucheniya semyan ul'-
trafioletovymi luchami. [Effect of pre-sowing ultraviolet irradiation of seeds.]
- Dokl. Akad. Nauk Belorus. SSR *10*: 695-697, 1966. [Chl, car; in R.]

2681 - GODNEV, T.N., KAKHNOVICH, L.V., ANTIPOVA, A.I.: Vliyanie sootnosheniya krasnogo
i sinego sveta lyuminestsentnykh lamp na nakoplenie pigmentov v list'yakh nekotorykh rastenii. [Effect of ratio of the red and blue radiation of fluorescent
tubes on the accumulation of pigments in leaves of some plants.] - Fiziol. Rast.
13: 602-606, 1966. [In R, ab: E.]

2682 - GODNEV, T.N., KHODASEVICH, E.V., AKULOVICH, N.K., SOKOL, V.I.: O biosinteze pigmentov v usloviyakh predel'nykh znachenii temperatur'nogo i svetovogo faktorov.

[Pigment biosynthesis in limit ranges of temperature and light.] - In: Upravlyaemyi Biosintez. Pp. 299-303. Nauka, Moskva 1966. [In R.]

2683 - GODNEV, T.N., KHODASEVICH, E.V., ARNAUTOVA, A.I.: O biosinteze pigmentov pri otritsatel'noi temperature i lishainikov i zimuyushchikh rastenii. [Pigment biosynthesis at negative temperature in lichens and in wintering plants.] - Dokl. Akad. Nauk SSSR *167*: 451-453, 1966. [In R.]

2684 - GODNEV, T.N., KHODASEVICH, E.V., ARNAUTOVA, A.I.: O sposobnosti rastenii raznykh sistematicheskikh grupp k biosintezu pigmentov pri otritsatel'noi temperature. [Ability of plants of different systematic groups for biosynthesis of pigments at negative temperature.] - In: Fiziologo-biokhimicheskie Issledovanya Rastenii. Pp. 11-15. Nauka i Tekhnika, Minsk 1968. [In R.]

2685 - GODNEV, T.N., KHODASEVICH, E.V., ARNAUTOVA, A.I.: O kharaktere sezonnykh izmenenii v soderzhanii i sootnoshenii pigmentov u khvoinykh v estestvennykh usloviyakh v svyazi s temperaturoi vozdukha. [Character of seasonal variations in pigment content and ratio in coniferous plants in natural conditions in relation to air temperature.] - Fiziol. Rast. *16*: 102-105, 1969. [In R, ab: E.]

2686 - GODNEV, T.N., KHODASEVICH, E.V., LYAKHNOVICH, Ya.P., SHABEL'SKAYA, E.F.: O nekotorykh aspektakh deistviya temperaturnogo i svetovogo faktorov na biosintez i sostoyanie pigmentov v zelenykh rasteniyakh. [Some aspects of action of temperature and light on biosynthesis and state of pigments in green plants.] - Stud. biophys. *5*: 31-36, 1967. [In R, ab: E.]

2687 - GODNEV, T.N., KONDRAT'EVA, E.N., USPENSKAYA, V.E.: O vozmozhnykh putyakh biosinteza bakterioviridina (khlorobium-khlorofilla). [Possible pathways of biosynthesis of bacterioviridin (chlorobium-chlorophyll).] - Izv. Akad. Nauk SSSR, Ser. biol. *1966*: 525-531, 1966. [In R, ab: E.]

2688 - GODNEV, T.N., LIPSKAYA, G.A., FARTOTSKAYA, I.K.: K voprosu o biosinteze khlorofilla pod deistviem ekstrakta kletok propionovokislykh bakterii. [Biosynthesis of chlorophyll under influence of a cell extract of propionic bacteria.] - Dokl. Akad. Nauk SSSR,*186*: 228-230, 1969. [In R.]

2689 - GODNEV, T.N., LYAKHNOVICH, Ya.P.: K voprosu deistviya termoimpul'sa na izmenenie chislennosti kletok i formirovanie pigmentov u nekotorykh protokokkovykh vodoroslei. [Action of thermoimpulse on the change in cell number and pigment formation in some protococcal algae.] - In: Issledovaniya po Fiziologii i Biokhimii Rastenii. Pp. 16-21. Nauka i Tekhnika, Minsk 1966. [In R.]

2690 - GODNEV, T.N., LYAKHNOVICH, Ya.P.: K voprosu rosta i nakopleniya pigmentov khlorelloi na ryade pitatel'nykh sred s dobavleniem brosovogo kartofel'nogo soka. [Growth and pigment accumulation in *Chlorella* cultivated on several nutrient media with addition of an inferior potato sap.] - In: Issledovaniya po Fiziologii i Biokhimii Rastenii. Pp. 24-29. Nauka i Tekhnika, Minsk 1966. [In R.]

2691 - GODNEV, T.N., LYAKHNOVICH, Ya.P.: O posledeistvii termoimpul'sa na formirovanie pigmentov v suspenzii khlorelly. [Aftereffect of a thermoimpulse on the pigment formation in a *Chlorella* suspension.] - In: Upravlyaemyi Biosintez. Pp. 175-178. Nauka, Moskva 1966. [In R.]

2692 - GODNEV, T.N., LYAKHNOVICH, Ya.P.: O vliyanii vysokikh temperatur na sostoyanie khlorofilla i zhiznedeyatel'nost' kletok khlorelly. [Effect of high temperature on the state of chlorophyll and on the vitality of *Chlorella* cells.] - In: Fiziologo-biokhimicheskie Issledovaniya Rastenii. Pp. 15-23. Nauka i Tekhnika, Minsk 1968. [In R.]

2693 - GODNEV, T.N., LYAKHNOVICH, Ya.P.: Vliyanie kislorodnogo rezhima na sostoyanie pigmentov u nekotorykh vodoroslei. [Effect of oxygen regime on the state of pigments in some algae.] - In: Fotosintez i Pitanie Rastenii. Pp. 13-18. Nauka i Tekhnika, Minsk 1969. [In R.]

2694 - GODNEV, T.N., RASKIN, V.I.: K voprosy o dinamike prevrashcheniya protokhlorofillida v khlorofillid. [Dynamics of transformation of protochlorophyllide into chlorophyllide.] - In: Fiziologo-biokhimicheskie Issledovaniya Rastenii. Pp. 14-16. Nauka i Tekhnika, Minsk 1967. [In R.]

2695 - GODNEV, T.N., RASKIN, V.I.: Ob aktivirovanii monokhromaticheskim svetom formy,

protokhlorofill-golokhroma 634. [Activation of the protochlorophyll-holochrome
634 form by monochromatic light.] - Dokl. Akad. Nauk SSSR *180*: 235-236, 1968.
[In R.]

2696 - GODNEV, T.N., RASKIN, V.I., AKULOVICH, N.K., ORLOVSKAYA, K.I.: Prevrashchenie
protokhlorofill-golokhroma v monokhromaticheskom svete pri nachal'nom preoblada-
nii formy 634. [Transformation of protochlorophyll-holochrome in monochromatic
light at the initial predominance of the form 634.] - Dokl. Akad. Nauk SSSR
182: 709-711, 1968. [In R.]

2697 - GODNEV, T.N., RASKIN, V.I., KALER, V.L.: Zavisimost' skorosti prevrashcheniya
protokhlorofillida v khlorofillid ot intensivnosti sveta. [Dependence of the
rate of transformation of protochlorophyllide into chlorophyllide on irradiance.]
- Dokl. Akad. Nauk SSSR *174*: 225-226, 1967. [In R.]

2698 - GODNEV, T.N., ROTFARB, R.M., GVARDIYAN, V.N.: Ob uchastii protoporfirina i gema-
tina v biosinteze fikotsianinov. [Participation of protoporphyrin and haematin
in phycocyanin biosynthesis.] - Dokl. Akad. Nauk SSSR *169*: 1191-1194, 1966.
[In R.]

2699 - GODNEV, T.N., ROTFARB, R.M., GVARDIYAN, V.N.: K voprosu ob uchastii valina v
obrazovanii karotinoidov. [Participation of valine in formation of carotenoids.]
- In: Fiziologo-biokhimicheskie Issledovaniya Rastenii. Pp. 3-6. Nauka i Tekhnika, Minsk 1968. [In R.]

2700 - GODNEV, T., SELGA, M.: Vliyanie dlinovolnovoi i korotkovolnovoi dopolnitel'noi
ul'trafioletovoi radiatsii na rost i nakoplenie pigmentov v rasteniyakh ogurtsov
i tomatov v usloviyakh kul'tury zakrytogo grunta. [Effect of long-wave and
short-wave supplemental ultraviolet radiation on growth and pigment accumulation
in cucumber and tomato plants under conditions of glass-covered ground.] - Latvijas PSR Zinātnu Akad. Vēstis *1966* (3): 58-66, 1966. [In R, ab: E.]

2701 - GODNEV, T.N., SHABEL'SKAYA, E.F.: Sostoyanie khlorofilla i aktivnost' nekotorykh oksidaz u rastenii s razlichnoi ustoichivost'yu k dlitel'nomu polnomu zatemneniyu. [Chlorophyll state and activity of some oxidases in plants of various
resistance to prolonged full darkening.] - Dokl. Akad. Nauk Belorus. SSR *10*:
411-413, 1966. [In R.]

2702 - GODNEV, T.N., SHABEL'SKAYA, E.F.: K voprosu ob uchastii vitamina B_{12} v biosinteze khlorofilla. [Participation of vitamin B_{12} in chlorophyll biosynthesis.] -
Dokl. Akad. Nauk SSSR *171*: 1227-1229, 1966. [In R.]

2703 - GODNEV, T.N., SHABEL'SKAYA, E.F.: K voprosu o formirovanii plastidnogo apparata
v ontogeneze lista sakharnoi svekly v estestvennykh usloviyakh. [Formation of
the plastid apparatus during ontogenesis of sugar beet leaf in natural conditions.]
- Dokl. Akad. Nauk Belorus. SSR *10*: 987-990, 1966. [In R.]

2704 - GODNEV, T.N., SHABEL'SKAYA, E.F.: O vliyanii dlitel'nogo zatemneniya na pigmenty i plastidnyi apparat nekotorykh svetolyubivykh i tenevynoslivykh rastenii.
[Effect of prolonged darkening on pigments and plastid apparatus in some sun
and shade plants.] - Fiziol. Rast. *14*: 451-455, 1967. [In R, ab: E.]

2705 - GODNEV, T.N., SHABEL'SKAYA, E.F., GVARDIYAN, V.N.: Aktivnost' proteaz v svyazi
s ustoichivost'yu plastidnogo apparata rastenii k prodolzhitel'nomu zatemneniyu.
[Protease activity in relation to the resistance of the plastid apparatus to
prolonged darkening.] - Dokl. Akad. Nauk Belorus. SSR *12*: 827-829, 1968. [In R.]

2706 - GODNEV, T.N., SHABEL'SKAYA, E.F., GVARDIYAN, V.N.: Strukturnye i funktsional'nye
izmeneniya v plastidakh rastenii pri prodolzhitel'nom zatemnenii. [Structural
and functional changes in plant plastids under long-term darkness.] - In: Fotosintez i Pitanie Rastenii. Pp. 19-26. Nauka i Tekhnika, Minsk 1969. [In R.]

2707 - GODNEV, T.N., SHLYK, A.A.: Biosintez i metabolizm khlorofilla. [Biosynthesis
and metabolism of chlorophyll.] - Vestn. Akad. Nauk SSSR *1966* (10): 36-39,
1966. [In R.]

2708 - GODNEV, T.N., SMIRNOVA, L.F.: Sravnitel'naya kharakteristika nekotorykh gibridov
kukuruzy i ikh roditel'skikh form po nakopleniyu pigmentov. [Comparative characteristics of some maize hybrids and their parental forms according to the accumulation of pigments.] - Dokl. Akad. Nauk Belorus. SSR *11*: 460-463, 1967. [In R.]

2709 - GODNEV, T.N., VECHER, A.S., KHODASEVICH, E.V., CHAÏKA, M.T., KALER, V.L., FEDYUN'KIN, D.V.: O vozrastnoĭ aktivnosti khloroplastov klubneĭ kartofelya. [Ontogenetic activity of chloroplasts in potato tubers.] - Dokl. Akad. Nauk SSSR *173*: 1215-1217, 1967. [In R.]

2710 - GODNEV, T.N., VLASENKO, N.E.: Vliyanie razlichnykh form azota na nakoplenie karotina v list'yakh sakharnoĭ svekly i kormovoĭ kapusty. [Effect of various nitrogen forms on carotene accumulation in sugar beet and kale leaves.] - Dokl. Akad. Nauk Belorus. SSR *10*: 489-491, 1966. [In R.]

2711 - GODZIEMBA-CZYŻ, J.: Characteristics of vegetative and resting forms in *Wolffia arrhiza* (L.) WIMM. II. Anatomy, physical and physiological properties. - Acta Soc. Bot. Pol. *39*: 421-443, 1970. [Ps, Chl.]

2712 - GOEDHEER, J.C.: Fluorescence polarization and other fluorescence properties of chloroplasts and cells in relation to molecular structure. - In: GOODWIN, T.W. (ed.): Biochemistry of Chloroplasts. Vol. I. Pp. 75-82. Academic Press, London-New York 1966.

2713 - GOEDHEER, J.C.: Visible absorption and fluorescence of chlorophyll and its aggregates in solution. - In: VERNON, L.P., SEELY, G.R. (ed.): The Chlorophylls. Pp. 147-184. Academic Press, New York-London 1966.

2714 - GOEDHEER, J.C.: Chlorophyll-protein complexes. Part I. Complexes derived from green plants. - In: VERNON, L.P., SEELY, G.R. (ed.): The Chlorophylls. Pp. 399-411. Academic Press, New York-London 1966.

2715 - GOEDHEER, J.C.: On the function of accessory pigments. - In: THOMAS, J.B., GOEDHEER, J.C. (ed.): Currents in Photosynthesis. Pp. 177-186. Donker, Rotterdam 1966.

2716 - GOEDHEER, J.C.: Fotosynthese. - Chem. Courant *65*: 301-306, 1966.

2717 - GOEDHEER, J.C.: Les changements du spectre d'absorption et de fluorescence au cours du verdissement et du vieillissement des plastes. - In: SIRONVAL, C. (ed.): Le Chloroplaste, Croissance et Vieillissement. Pp. 77-85. Masson, Paris 1967.

2718 - GOEDHEER, J.C.: On the low-temperature fluorescence spectrum of blue-green and red algae. - Biochim. biophys. Acta *153*: 903-906, 1968.

2719 - GOEDHEER, J.C.: Carotenoids in blue-green and red algae. - In: METZNER, H. (ed.): Progress in Photosynthesis Research. Vol. II. Pp. 811-817. Tübingen 1969.

2720 - GOEDHEER, J.C.: Energy transfer from carotenoids to chlorophyll in blue-green, red and green algae and greening bean leaves. - Biochim. biophys. Acta *172*: 252-265, 1969.

2721 - GOEDHEER, J.C.: On the pigment system of brown algae. - Photosynthetica *4*: 97-106, 1970.

2722 - GOEDHEER, J.C., SIERO, J.P.J.: Investigation of magnesium tetrabenzporphyrin. - I. Absorption and fluorescence in organic solution and aqueous medium. - Photochem. Photobiol. *6*: 509-520, 1967.

2723 - GOEDHEER, J.C., van der TUIN, A.K.: Decline in bacteriochlorophyll fluorescence induced by carotenoid absorption. - Biochim. biophys. Acta *143*: 399-407, 1967.

2724 - GOEDHEER, J.C., VERHÜLSDONK, C.A.H.: Fluorescence and phototransformation of protochlorophyll with etiolated bean leaves from -196 to +20 °C. - Biochem. biophys. Res. Commun. *39*: 260-266, 1970.

2725 - GOGIYA, V.T., IVANOVA, L.N.: Biokhimicheskaya kharakteristika nekotorykh form gerani. Soobshchenie II. O soderzhanii pigmentov plastid v list'yakh gerani. [Biochemical characteristics of some forms of *Geranium*. Report II. Content of plastid pigments in *Geranium* leaves.] - Tr. sukhum. opyt. Sta. efirnomaslich. Kul'tur *1970* (9): 133-139, 1970. [In R.]

2726 - GOGOTOV, I.N.: Vydelenie vodoroda i assimilyatsiya ugleroda purpurnymi bakteriyami v zavisimosti ot intensivnosti sveta. [Hydrogen efflux and carbon assimilation in purple bacteria in dependence on irradiance.] - Dokl. Akad. Nauk SSSR *183*: 954-956, 1968. [In R.]

2727 - GOGOTOV, I.N., KONDRAT'EVA, E.N.: Obrazovanie molekulyarnogo vodoroda zelenymi
       fotosinteziruyushchimi bakteriyami. [Formation of molecular hydrogen by green
       photosynthesizing bacteria.] - Dokl. Akad. Nauk SSSR *175*: 714-717, 1967. [In R.]

2728 - GOGOTOV, I.N., KONDRAT'EVA, E.N.: Ob usloviyakh obrazovaniya vodoroda *Rhodo-
       pseudomonas* sp. [On conditions of formation of hydrogen by *Rhodopseudomonas* sp.]
       - Izv. Akad. Nauk SSSR, Ser. biol. *1969* (1): 161-165, 1969. [In R, ab: E.]

2729 - GOGOTOV, I.N., NOVIKOVA, N.A.: Vydelenie vodoroda rastushchimi kul'turami pur-
       purnykh serobakteriĭ. [Hydrogen evolution by growing cultures of purple sulphur
       bacteria.] - Mikrobiologiya *37*: 19-25, 1968. [In R, ab: E.]

2730 - GOGUADZE, V.P., VITUL'SKAYA, N.V.: Ochistka karotinoidov i piretrovykh soedine-
       niĭ ot khlorofilla i drugikh ballastnykh veshchestv. [Purification of carote-
       noids and pyrethrum compounds from chlorophyll and other ballast compounds.] -
       Soobshch. Akad. Nauk Gruz. SSR *42*: 609-612, 1966. [In R, ab: Georgian.]

2731 - GOL'D, V.M.: O vzaimovliyanii otdel'nykh uchastkov spektra v protsesse fotosin-
       teza pri svetovom nasyshchenii. [Interaction of individual spectrum ranges in
       light saturated photosynthesis.] - Nauch. Dokl. vyssh. Shkoly, biol. Nauki *9*
       (4): 149-153, 1966. [In R.]

2732 - GOL'D, V.M.: Deĭstvie vikasola na intensivnost' i napravlennost' gazoobmena u
       elodei. [Effect of vikasol on the rate and direction of gas exchange in *Elodea*.]
       - Nauch. Dokl. vyssh. Shkoly, biol. Nauki *11* (5): 114-119, 1968. [In R.]

2733 - GOL'D, V.M.: Nekotorye ekologo-fiziologicheskie osobennosti kharaktera spektra
       deĭstviya fotosinteza pri svetovom nasyshchenii. [Some eco-physiological feat-
       ures of the character of the action spectrum of light saturated photosynthesis.]
       - Fiziol. Rast. *16*: 594-602, 1969. [In R, ab: E.]

2734 - GOL'D, V.M.: Deĭstvie vikasola na nekotorye opticheskie svoĭstva elodei i ego
       veroyatnoe uchastie v reaktsiyakh okislitel'nogo fosforilirovaniya. [Effect of
       vikasol on some optical properties of *Elodea* and its probable participation in
       reactions of oxidative phosphorylation.] - Fiziol. Rast. *16*: 303-307, 1969.
       [In R, ab: E.]

2735 - GOL'D, V.M.: O vzaimovliyanii vidimykh uchastkov spektra na intensivnost' foto-
       sinteza. [Mutual effect of visible ranges of spectrum on photosynthetic rate.]
       - In: Voprosy Fotosinteza. Vol. 2. Pp. 193-202. Izdat.Tomskogo Univ., Tomsk 1970.
       [In R.]

2736 - GOL'D, V.M.: O vliyanii infrakrasnoĭ radiatsii na intensivnost' fotosinteza,
       vyzvannogo vidimymi uchastkami spektra. [Effekt of infra-red radiation on pho-
       tosynthetic rate induced by visible spectrum ranges.] - Tr. nauch. issled. Inst.
       Biol. Biofiz. tomsk. Univ. *1*: 170-176, 1970. [In R.]

2737 - GOL'D, V.M., KOL'TSOVA, V.G.: Vliyanie 2,4-dinitrofenola, uglekislogo gaza i
       kofaktorov tsiklicheskikh protsessov fotosinteza na biosintez zelenykh pigmen-
       tov. [Effect of 2,4-dinitrophenol, carbon dioxide and cofactors of cyclic pro-
       cesses of photosynthesis on the biosynthesis of green pigments.] - Nauch.
       Dokl. vyssh. Shkoly, biol. Nauki *13* (10): 64-68, 1970. [In R.]

2738 - GOLDBERG, I., OHAD, I.: Development of photosynthetic membranes in a mutant of
       Chlamydomonas reinhardi y⁻. - Israel J. Chem. *6*: 132-p, 1968.

2739 - GOLDBERG, I., OHAD, I.: Biogenesis of chloroplast membranes. IV. Lipid and pig-
       ment changes during synthesis of chloroplast membranes in a mutant of *Chlamy-
       domonas reinhardi* y-1. - J. Cell Biol. *44*: 563-571, 1970.

2740 - GOLDBERG, I., OHAD, I.: Biogenesis of chloroplast membranes. V. A radioautogra-
       phic study of membrane growth in a mutant of *Chlamydomonas reinhardi* y-1. - J.
       Cell Biol. *44*: 572-591, 1970.

2741 - GOLDMAN, C.R.: Photosynthetic efficiency and diversity of a natural phytoplank-
       ton population in Castle Lake, California. - In: Prediction and Measurement of
       Photosynthetic Productivity. Pp. 507-517. PUDOC, Wageningen 1970.

2742 - GOLDMAN, C.R., MASON, D.T., HOBBIE, J.E.: Variations in photosynthesis in two
       shallow Antarctic lakes. - Int. Ver. theor. angew. Limnol. Verhandl. *17*: 414-418,
       1969.

2743 - GOLDSTEIN, J.M.: Study of biological pigments by single specimen derivative spectrophotometry. - Biophys. J. *10*: 445-461, 1970.

2744 - GOLDSWORTHY, A.: A simple apparatus for generating an air stream containing a constant concentration of $^{14}CO_2$. - J. exp. Bot. *17*: 147-150, 1966.

2745 - GOLDSWORTHY, A.: Experiments on the origin of CO_2 released by tobacco leaf segments in the light. - Phytochemistry *5*: 1013-1019, 1966.

2746 - GOLDSWORTHY, A.: Comparison of the kinetics of photosynthetic carbon dioxide fixation in maize, sugar cane and tobacco, and its relation to photorespiration. - Nature *217*: 62, 1968.

2747 - GOLDSWORTHY, A.: Riddle of photorespiration. - Nature *224*: 501-502, 1969.

2748 - GOLDSWORTHY, A.: Photorespiration. - Bot. Rev. *36*: 321-340, 1970.

2749 - GOLDSWORTHY, A., DAY, P.R.: A simple technique for the rapid determination of plant CO_2 compensation points. - Plant Physiol. *46*: 850-851, 1970.

2750 - GOLDSWORTHY, A., DAY, P.R.: Further evidence for reduced role of photorespiration in low compensation point species. - Nature *228*: 687-688, 1970.

2751 - GOLDTHWAITE, J.J., LAETSCH, W.M.: Regulation of senescence in bean leaf discs by light and chemical growth regulators. - Plant Physiol. *42*: 1757-1762, 1967. [Chl.]

2752 - GOLDTHWAITE, J.J., LAETSCH, W.M.: Control of senescence in *Rumex* leaf discs by gibberellic acid. - Plant Physiol. *43*: 1855-1858, 1968. [Chl.]

2753 - GOLINKA, P.I.: Vliyanie obrezki vinogradnykh kustov na razvitie fotosinteticheskogo apparata list'ev. [Effect of cutting of vine shrubs on the development of photosynthetic apparatus of leaves.] - Fiziol. Rast. *13*: 607-613, 1966. [In R, ab: E.]

2754 - GOLOD, M.G.: Vzaemozv'yazok mizh vmistom pigmentiv i aktyvnistyu khlorofilazy u vnutrishnikh obolonkakh nasinnya garbuza v protsesi Togo rozvytku. [Correlation between pigment content and chlorophyllase activity in inner coats of pumpkin seed during its development.] - Ukr. bot. Zh. *23* (6): 26-31, 1966. [In Ukr., ab: E, R.]

2755 - GOLOD, M.G.: Aktyvnist' khlorofilazy shchodo protokhlorofilu. [Chlorophyllase activity for protochlorophyll.] - Ukr. bot. Zh. *24* (3): 3-7, 1967. [In Ukr., ab: E, R.]

2756 - GOLOD, M.G.: Osobennosti sostoyaniya, biosinteza i fotokhimicheskoT aktivnosti protokhlorofilla. [Peculiarities of the state, biosynthesis, and photochemical activity of protochlorophyll.] - In: Tezisy Dokladov II Vsesoyuznogo Biokhimicheskogo S"ezda. 19. Sektsiya: Problemy Fotosinteza. Pp. 61-62. FAN, Tashkent 1969. [In R.]

2757 - GOLOD, M.G., SUD'ĬNA, O.G.: Spektral'ni ta fotosensybilizatsiĬni vlastyvosti protokhlorofilu v model'nykh systemakh. [Spectral and photosensibilizing properties of protochlorophyll in model systems.] - Ukr. bot. Zh. *25* (3): 15-21, 1968. [In Ukr.]

2758 - GOLOVIN, V.V., MIGUNOV, V.S.: Opredelenie ploshchadi lista soi po parametram. [Leaf area determination in soybean by means of parameters.] - Vestn. sel'.-khoz. Nauki *13* (12): 90-91, 1968. [In R.]

2759 - GOLUBEV, V.N., MAKHAEVA, L.V.: Kalorymetrychne vyvchennya produktyvnosti trav'yano-napivchagarnychkovykh roslyn ta ugrupovan' Kryms'koĬ yaĬly. [Calorimetric determination of productivity of herbaceous-semishrub plants and cenoses in Crimea.] - Ukr. bot. Zh. *26* (3): 99-105, 1969. [In Ukr., ab: E, R.]

2760 - GOLUBKOVA, B.M., KISLYAKOVA, T.E., BOGACHEVA, I.I., KUZNETSOVA, L.I., KUDRYAVTSEVA, L.F.: Struktura i funktsiya fotosinteticheskogo apparata u rasteniĬ razlichnykh sistematicheskikh grupp. [Structure and function of the photosynthetic apparatus in plants of various systematic groups.] - In: Khloroplasty i Mitokhondrii. Pp. 74-88. Nauka, Moskva 1969. [In R.]

2761 - GONCHARIK, M.N.: Osobennosti ottoka assimilyatov u rasteniya kartofelya. [Features of photosynthates efflux in a potato plant.] - Uch. Zap. Tartu.gos. Univ. *185*: 394-401, 1966. [In R.]

2762 - GONCHARIK, M.N., IVANCHENKO, V.I.: Vliyanie formy udobreniT na intensivnost' fotosinteza i urozhaT sel'skokhozyaTstvennykh kul'tur. [Effect of fertilizer form on photosynthetic rate and yield of agricultural crops.] - In: VazhneTshie Problemy Fotosinteza v Rastenievodstve. Pp. 171-183. Kolos, Moskva 1970. [In R.]

2763 - GONCHARIK, M.N., IVANCHENKO, V.M.: Fotokhimicheskaya aktivnost' izolirovannykh khloroplastov khlorotravlennykh i intaktnykh rasteniT kartofelya. [Photochemical activity of isolated chloroplasts in chlorine-poisoned and intact potato plants.] - Fiziol. Rast. *13*: 429-432, 1966. [In R, ab: E.]

2764 - GONCHARIK, M.N., IVANCHENKO, V.M.: O khlornom khloroze. [Chlorine chlorosis.] - In: Fiziologo-biokhimicheskie Issledovaniya RasteniT. Pp. 38-45. Nauka i Tekhnika, Minsk 1967. [In R.]

2765 - GONCHARIK, M.N., KOZLOVA, A.P.: Vliyanie ionov khlora na soderzhanie khlorofilla v list'yakh grechikhi. [Effect of chlorine ions on chlorophyll content in buckwheat leaves.] - Vestsi Akad. Navuk Belarus. SSR, Ser. biyal. Navuk *1970* (5): 118-120, 1970. [In R.]

2766 - GONCHARIK, M.N., KRUCHININA, S.S.: O soderzhanii pigmentov u kartofelya v posadkakh raznoT gustoty. [Pigment content in potato in stands of various density.] - In: Issledovaniya po Fiziologii i Biokhimii RasteniT. Pp. 45-52. Nauka i Tekhnika, Minsk 1966. [In R.]

2767 - GONCHARIK, M.N., LEGENCHENKO, B.I., IVANCHENKO, V.M.: O prichinakh depressii fotosinteza u rasteniT kartofelya v usloviyakh izbytka khloridov v pochve. [Causes of photosynthesis depression in potato plants under surplus of chlorides in soil.] - In: Fiziologo-biokhimicheskie Issledovaniya RasteniT. Pp. 23-29. Nauka i Tekhnika, Minsk 1970. [In R.]

2768 - GONCHARIK, M.N., MIKUL'SKAYA, S.A.: DeTstvie khlora na fotosinteticheskiT apparat sakharnoT svekly. [Effect of chlorine on the photosynthetic apparatus of sugar beet.] - Agrokhimiya *1966* (3): 103-110, 1966. [In R.]

2769 - GONCHARIK, M.N., MIKUL'SKAYA, S.A.: Vliyanie khloridov kaliTnykh soleT na fotosinteticheskuyu funktsiyu i khozyaTstvennuyu produktivnost' sakharnoT svekly. [Effect of chlorides of potassium salts on photosynthetic function and agricultural productivity of sugar beet.] - In: Issledovaniya po Fiziologii i Biokhimii RasteniT. Pp. 52-62. Nauka i Tekhnika, Minsk 1966. [In R.]

2770 - GONCHARIK, M.N., URBANOVICH, T.A.: O deTstvii Cl⁻ na fotosinteticheskuyu aktivnost' khloroplastov. [Effect of Cl⁻ on photochemical activity of chloroplasts.] - Dokl. Akad. Nauk Belorus. SSR *14*: 761-763, 1970. [In R.]

2771 - GONCHAROVA, N.V., EVSTIGNEEV, V.B.: Metodicheskie aspekty issledovaniya fotofosforilirovaniya na beskletochnykh preparatakh fotosinteziruyushchikh bakteriT. [Methodical aspects of studying photophosphorylation on non-cellular preparations of photosynthetic bacteria.] - In: KIRICHENKO, E.B. (ed.): Metody Issledovaniya Fotofosforilirovaniya. Pp. 49-68. Pushchino-na-Oke 1970. [In R, ab: E.]

2772 - GOOD, N., IZAWA, S., HIND, G.: Uncoupling and energy transfer inhibition in photophosphorylation. - Current Topics Bioenerg. *1*: 75-112, 1966.

2773 - GOOD, N.E., WINGET, G.D., WINTER, W., CONNOLLY, T.N., IZAWA, S., SINGH, R.M.M.: Hydrogen ion buffers for biological research. - Biochemistry *5*: 467-477, 1966. [Ps.]

2774 - GOODCHILD, D.J., HIGHKIN, H.R., BOARDMAN, N.K.: The fine structure of chloroplasts in a barley mutant lacking chlorophyll *b*. - Exp. Cell Res. *43*: 684-688, 1966.

2775 - GOODENOUGH, U.W., ARMSTRONG, J.J., LEVINE, R.P.: Photosynthetic properties of ac-31, a mutant strain of *Chlamydomonas reinhardi* devoid of chloroplast membrane stacking. - Plant Physiol. *44*: 1001-1012, 1969.

2776 - GOODENOUGH, U.W., LEVINE, R.P.: Chloroplast ultrastructure in mutant strains of *Chlamydomonas reinhardi* lacking components of the photosynthetic apparatus. - Plant Physiol. *44*: 990-1000, 1969.

2777 - GOODENOUGH, U.W., LEVINE, R.P.: Chloroplast structure and function in *ac-20*, a mutant strain of *Chlamydomonas reinhardi*. III. Chloroplast ribosomes and membrane organization. - J. Cell Biol. *44*: 547-562, 1970.

2778 - GOODENOUGH, U.W., LEVINE, R.P.: The genetic activity of mitochondria and chloroplasts. - Sci. Amer. *223* (5): 22-29, 1970.

2779 - GOODMAN, P.J.: Effect of varying plant populations on growth and yield of sugar beet. - Agr. Progr. *41*: 89-107, 1966. [Growth analysis.]

2780 - GOODMAN, P.J.: Physiological analysis of the effects of different soils on sugar beet crops in different years. - J. appl. Ecol. *5*: 339-357, 1968. [Growth analysis.]

2781 - GOODWIN, T.W.: The carotenoids. - In: SWAIN, T. (ed.): Comparative Phytochemistry. Pp. 121-137. Academic Press, London 1966.

B2782 - GOODWIN, T.W. (ed.): Biochemistry of Chloroplasts. Volume I. - Academic Press, London-New York 1966. Volume II. - Academic Press, London-New York 1967.

2783 - GOODWIN, T.W.: Terpenoids and chloroplast development. - In: GOODWIN, T.W. (ed.): Biochemistry of Chloroplasts. Vol. II. Pp. 721-733. Academic Press, London-New York 1967.

2784 - GOODWIN, T.W.: The biological significance of terpenoids in plants. - In: PRIDHAM, J.B. (ed.): Terpenoids in Plants. (Proc. Phytochemical Group on Terpenoids in Plants. Aberystwyth 1966.) Pp. 1-23. Academic Press, London-New York 1967. [Chl, Car.]

2785 - GOODWIN, T.W.: The mechanism and regulation of carotenoid biosynthesis in chloroplasts. - Stud. biophys. *5*: 1-6, 1967.

B2786 - GOODWIN, T.W. (ed.): Porphyrins and Related Compounds. - Academic Press, London-New York 1968.

2787 - GOODWIN, T.W.: Recent developments in the biosynthesis of carotenoids. - J. sci. ind. Res. *27*: 103-105, 1968.

2788 - GOODWIN, T.W.: Carotenoid biosynthesis in chloroplasts. - In: METZNER, H. (ed.): Progress in Photosynthesis Research. Vol. II. Pp. 669-674. Tübingen 1969.

2789 - GOPAL, N.H., RAO, I.M.: Effect of boron toxicity on some leaf constituents in groundnut (*Arachis hypogaea* L.) plants. - Andhra agr. J. (India) *15*: 21-24, 1968. [Chl.]

2790 - GORCHAKOV, V.V., SINYUKHIN, A.M.: Avtomaticheskaya registratsiya gazovogo obmena i elektrofiziologicheskikh kharakteristik tkaneĭ lista vysshikh rasteniĭ. [Automatic recording of gas exchange and electrophysiological characteristics of leaf tissues in plants.] - In: Izmeritel'naya Tekhnika v Sel'skom Khozyaĭstve. Pp. 176-182. Moskva 1967. [In R.]

2791 - GORCHEIN, A.: The relation between the pigment content of isolated chromatophores and that of the whole cell in *Rhodopseudomonas spheroides*. - Proc. roy. Soc. (London) Ser. B, biol. Sci. *170*: 247-254, 1968.

2792 - GORCHEIN, A.: The nature of the internal fine structure of *Rhodopseudomonas spheroides* as determined by the study of cell fragments. - Proc. roy. Soc. (London), Ser. B, biol. Sci. *170*: 255-263, 1968.

2793 - GORCHEIN, A., NEUBERGER, A., TAIT, G.H.: Adaptation of *Rhodopseudomonas spheroides* from aerobic to semianaerobic conditions. - In: GOODWIN, T.W. (ed.): Biochemistry of Chloroplasts. Vol. II. Pp. 411-420. Academic Press, London-New York 1967. [Bacteriochlorophyll.]

2794 - GORCHEIN, A., NEUBERGER, A., TAIT, G.H.: The isolation and characterization of subcellular fractions from pigmented and unpigmented cells of *Rhodopseudomonas spheroides*. - Proc. Roy. Soc. (London), Ser. B - biol. Sci. *170*: 229-246, 1968.

2795 - GORDETSKIĬ, A.V., IL'ENKO, E.V.: Ispol'zovanie zheleza seyantsami i sazhentsami yabloni pri raznykh usloviyakh pitaniya azotom. [Utilization of iron by apple seedlings and nursery-treated plants under different nitrogen supply.] - Sel'skokhoz. Biol. *5*: 31-36, 1970. [In R, ab: E.]

2796 - GORDON, J.C.: Photosynthesis, respiration and growth of Scotch pine seedlings. - Diss. Abstr. B *27*: 1011-B, 1966.

2797 - GORDON, J.C.: Effect of shade on photosynthesis and dry weight distribution in yellow birch (*Betula alleghaniensis* BRITTON) seedlings. - Ecology *50*: 924-927, 1969.

2798 - GORDON, J.C., GATHERUM, G.E.: Photosynthesis and growth of selected Scotch pine populations. - Silva fenn. *2*: 183-194, 1968.

2799 - GORDON, J.C., GATHERUM, G.E.: Photosynthesis and growth of selected Scotch pine seed sources. - U.S. Forest Serv. Res. Pap. NC *23*: 20-23, 1968.

2800 - GORDON, J.C., GATHERUM, G.E.: Effect of environmental factors and seed source on CO_2 exchange of Scotch-pine seedlings. - Bot. Gaz. *130*: 5-9, 1969.

2801 - GORDON, J.C., LARSON, P.R.: Seasonal course of photosynthesis, respiration and distribution of ^{14}C in young *Pinus resinosa* trees as related to wood formation. - Plant Physiol. *43*: 1617-1624, 1968.

2802 - GORELOVA, Z.P., NIKOLAEV, B.A.: Vliyanie povyshennykh temperatur i obezvozhivaniya na ATF-aznuyu aktivnost' khloroplastov. [Effect of increased temperature and dehydration on ATP-ase activity of chloroplasts.] - In: Funktsional'nye Osobennosti Khloroplastov. Pp. 65-68. Kazan. Univ., Kazan 1969. [In R.]

2803 - GOREN, R.: The effect of fluometuron on the behaviour of citrus leaves. - Weed Res. *9*: 121-135, 1969. [Ps, Chl.]

2804 - GOREN, R., MONSELISE, S.P.: Some physiological effects of triazines on citrus trees. - Weeds *14*: 141-144, 1966. [Chl.]

2805 - GORHAM, E., SANGER, J.: Caloric values of organic matter in woodland, swamp, and lake soils. - Ecology *48*: 492-494, 1967. [Also plants, algae.]

2806 - GORHAM, E., SANGER, J.: Plant pigments in woodland soils. - Ecology *48*: 306-308, 1967.

2807 - GORID'KO, I.V.: Dinamika intensivnosti fotosinteza list'ev kartofelya pod vliyaniem kobal'ta. [Dynamics of photosynthetic rate in potato leaves as affected by cobalt.] - Fiziol. Rast. *16*: 405-407, 1969. [In R.]

2808 - GÖRING, H., HOFFMANN, P.: Chlorophyllgehalt und Photosyntheseintensität bei Heterosishybriden und ihren Elternformen. - Biol. Zentralbl. *85*: 289-303, 1966.

2809 - GORLENKO, V.M.: Fotosinteziruyuyushchie serobakterii vodoemov yuzhnoĭ chasti Krymskogo poluostrova. [Photosynthetic sulphur bacteria in water reservoirs of the southern part of the Crimea peninsula.] - Mikrobiologiya *37*: 745-748, 1968. [In R, ab: E.]

2810 - GORLENKO, V.M., ZHILINA, T.N.: Izuchenie tonkoĭ struktury zelenykh serobakteriĭ shtamma SK-413. [Fine structure of green sulphur bacteria of the strain SK-413.] Mikrobiologiya *37*: 1052-1056, 1968. [In R, ab: E.]

2811 - GORMAN, D.S., LEVINE, R.P.: Photosynthetic electron transport chain of *Chlamydomonas reinhardi*. IV. Purification and properties of plastocyanin. - Plant Physiol. *41*: 1637-1642, 1966.

2812 - GORMAN, D.S., LEVINE, R.P.: Photosynthetic electron transport chain of *Chlamydomonas reinhardi*. V. Purification and properties of cytochrome 553 and ferredoxine. - Plant Physiol. *41*: 1643-1647, 1966.

2813 - GORMAN, D.S., LEVINE, R.P.: Photosynthetic electron transport chain of *Chlamydomonas reinhardi*. VI. Electron transport in mutant strains lacking either cytochrome 553 or plastocyanin. - Plant Physiol. *41*: 1648-1656, 1966.

2814 - GORSHKOV, V.K.: Issledovanie migratsii energii v tverdykh rastvorakh khlorofilla i nekotorykh drugikh veshchestv metodom polyarizovannoĭ lyuminestsentsii pri 290 i 60 °K. [Migration of energy in solid solutions of chlorophyll and some other substances studied by means of polarized luminescence at 290 and 60 °K.] - Biofizika *14*: 28-33, 1969. [In R, ab: E.]

2815 - GORSHKOVA, L.M., LEBEDEV, S.I.: O fotosinteticheskoĭ produktivnosti rasteniĭ konopli raznogo pola. [Photosynthetic productivity of hemp plants of different sex.] - In: Fotosintez i UrozhaĭInost' Sel'skokhozyaĭstvennykh RasteniĭI. Pp. 58-64. Min. sel'. Khoz. SSSR, Kiev 1970. [In R.]

2816 - GORYA, V.S., RAZMERITSA, D.M.: Izmenenie soderzhaniya pigmentov v list'yakh ku-kuruzy v zavisimosti of rezhima osveshcheniya. [Change in pigment content in maize leaves in dependence on light regime.] - In: Materialy IV Konferentsii Mo-lodykh Uchenykh Moldavii, 1964. Sektsiya Fiziologii, Biokhimii i Genetiki Raste-niĭI. Pp. 41-43, Kishinev 1966. [In R.]

2817 - GORYSHINA, T.K.: Rannevesenniĭ fotosintez perezimovavshikh list'ev dubravnykh travyanistykh rasteniĭI. [Early-spring photosynthesis of over-wintered leaves of herbaceous plants from an oak forest.] - Bot. Zh. 54: 919-923, 1969. [In R, ab: E.]

2818 - GORYSHINA, T.K., MITINA, M.B.: O nekotorykh osobennostyakh fotosinteza i dykha-niya rannevesennikh efemeroidov dubovogo lesa. [Some features of photosynthesis and respiration of early-spring ephemeroids from an oak forest.] - In: Svetovoĭ Rezhim, Fotosintez i Produktivnost' Lesa. Pp. 270-273. Nauka, Moskva 1967. [In R.]

2819 - GOSTIMSKIĬ, S.A.: Fotosinteticheskiĭ mutant *Pisum sativum*. [Photosynthetic mu-tant of *Pisum sativum*.] - Genetika *1966* (4): 80-85, 1966. [In R.]

2820 - GOTO, K., HIGUCHI, M., SAKAI, H., KIKUCHI, G.: Differential inhibition of in-duced syntheses of δ-amino levulinate synthetase and bacteriochlorophyll in dark aerobically grown *Rhodopseudomonas spheroides*. - J. Biochem. (Tokyo) *61*: 186-192, 1967.

2821 - GOTTWALD, S.: Experimentelle Apparaturen zur Herstellung von Sauerstoffgradien-ten. - Arch. Hydrobiol. *68*: 143-150, 1970.

2822 - GOUD, J.V., MURALEEDHARAN NAYAR, K.: Effects of irradiation on seedlings of methi *(Trigonella foenum graecum)*. - Mysore J. agr. Sci. *2*: 53-56, 1968. [Chl.]

2823 - GOULDER, R.: Day-time variations in the rates of production by two natural com-munities of submerged freshwater macrophytes. - J. Ecol. *58*: 521-528, 1970.

2824 - GOVINDJEE: Fluorescence studies on algae, chloroplasts and chloroplast frag-ments. - In: THOMAS, J.B., GOEDHEER, J.C. (ed.): Currents in Photosynthesis. Pp. 93-103. Donker, Rotterdam 1966.

2825 - GOVINDJEE: Transformation of light energy into chemical energy: Photochemical aspects of photosynthesis. - Crop Sci. *7*: 551-560, 1967.

2826 - GOVINDJEE, BAZZAZ, M.: On the Emerson enhancement effect in the ferricyanide Hill reaction in chloroplast fragments. - Photochem. Photobiol. *6*: 885-894, 1967.

2827 - GOVINDJEE, DÖRING, G., GOVINDJEE, R.: The active chlorophyll a_{II} in suspensions of lyophilized and Tris-washed chloroplasts. - Biochim. biophys. Acta *205*: 303-306, 1970.

2828 - GOVINDJEE, MUNDAY, J.C. Jr., PAPAGEORGIOU, G.: Fluorescence studies with algae: changes with time and preillumination. - In: Energy Conversion by the Photosyn-thetic Apparatus. Brookhaven Symp. Biol. *19*: 434-445, 1967. [Chl.]

2829 - GOVINDJEE, PAPAGEORGIOU, G., RABINOWITCH, E.: Chlorophyll fluorescence and pho-tosynthesis. - In: GUILBAULT, G.G. (ed.): Fluorescence Theory, Instrumentation and Practice. Pp. 511-564. M. Dekker, Inc., New York 1967.

2830 - GOVINDJEE, YANG, L.: Structure of the red fluorescence band in chloroplasts. - J. gen. Physiol. *49*: 763-780, 1966.

2831 - GOVINDJEE, R., GOVINDJEE, LAVOREL, J., BRIANTAIS, J.M.: Fluorescence character-istics of lyophilized maize chloroplasts suspended in buffer. - Biochim. bio-phys. Acta *205*: 361-370, 1970.

2832 - GOVINDJEE, R., RABINOWITCH, E., GOVINDJEE: Maximum quantum yield and action spectra of photosynthesis and fluorescence in *Chlorella*. - Biochim. biophys. Acta *162*: 539-544, 1968.

2833 - GOVINDJEE, R., SYBESMA, C.: Light-induced reduction of pyridine nucleotide and its relation to light-induced electron transport in whole cells of *Rhodospiril-lum rubrum*. - Biochim. biophys. Acta *223*: 251-260, 1970.

2834 - GRACE, J., WOOLHOUSE, H.W.: A physiological and mathematical study of the growth and productivity of a *Calluna-Sphagnum* community. I. Net photosynthesis of *Calluna vulgaris* L. HULL. - J. appl. Ecol. *7*: 363-381, 1970.

2835 - GRADYUSHKO, A.T., SEVCHENKO, A.N., SOLOV'EV, K.N., TSVIRKO, M.P.: Molekulyarnaya energetika khlorofilla i porfirinov. [Molecular energetics of chlorophyll and porphyrins.] - Izv. Akad. Nauk SSSR, Ser. fiz. *34*: 636-640, 1970. [In R.]

2836 - GRADYUSHKO, A.T., SEVCHENKO, A.N., SOLOVYOV, K.N., TSVIRKO, M.P.: Energetics of photophysical processes in chlorophyll-like molecules. - Photochem. Photobiol. *11*: 387-400, 1970.

2837 - GRAHAM, D., GRIEVE, A.M., SMILLIE, R.M.: Phytochrome as the primary photoregulator of the synthesis of Calvin cycle enzymes in etiolated pea seedlings. - Nature *218*: 89-90, 1968.

2838 - GRAHAM, D., HATCH, M.D., SLACK, C.R., SMILLIE, R.M.: Light-induced formation of enzymes of the C_4-dicarboxylic acid pathway of photosynthesis in detached leaves. - Phytochemistry *9*: 521-532, 1970.

2839 - GRAHAM, D., WHITTINGHAM, C.P.: The path of carbon during photosynthesis in *Chlorella pyrenoidosa* at high and low carbon dioxide concentrations. - Z. Pflanzenphysiol. *58*: 418-427, 1968.

2840 - GRAINGER, J.: An approach by computer to the prediction of crop yields. - Pest Articles News Summ. B. Plant Dis. *14*: 347-352, 1968.

2841 - GRANICK, S.: The heme and chlorophyll biosynthetic chain. - In: GOODWIN, T.W. (ed.): Biochemistry of Chloroplasts. Vol. II. Pp. 373-410. Academic Press, London-New York 1967.

2842 - GRANICK, S.: Differentiation of the erythrocyte and chloroplast: Examples of a phasing principle in developmental biology. - Fed. Proc. *29*: 729, 1970.

2843 - GRANICK, S., GASSMAN, M.: Rapid regeneration of protochlorophyllide$_{650}$. - Plant Physiol. *45*: 201-205, 1970.

2844 - GRANICK, S., GIBOR, A.: The DNA of chloroplasts, mitochondria, and centrioles. - Prog. nucleic Acid Res. mol. Biol. *6*: 143-186, 1967.

2845 - GRANT, B.R.: The effect of carbon dioxide concentration and buffer system on nitrate and nitrite assimilation by *Dunaliella tertiolecta*. - J. gen. Microbiol. *54*: 327-336, 1968. [Ps.]

2846 - GRANT, B.R., ATKINS, C.A., CANVIN, D.T.: Intracellular location of nitrate reductase and nitrite reductase in spinach and sunflower leaves. - Planta *94*: 60-72, 1970. [Ps.]

2847 - GRANT, B.R., CANVIN, D.T.: The effect of nitrate and nitrite on oxygen evolution and carbon-dioxide assimilation and the reduction of nitrate and nitrite by intact chloroplasts. - Planta *95*: 227-246, 1970.

2848 - GRANT, B.R., WHATLEY, F.R.: Some factors affecting the onset of cyclic photophosphorylation. - In: GOODWIN, T.W. (ed.): Biochemistry of Chloroplasts. Vol. II. Pp. 505-521. Academic Press, London-New York 1967.

2849 - GRAY, E.D.: Studies on the adaptive formation of photosynthetic structures in *Rhodopseudomonas spheroides*. I. Synthesis of macromolecules. - Biochim. biophys. Acta *138*: 550-563, 1967.

2850 - GRAY, I.K., RUMSBY, M.G., HAWKE, J.C.: The variations in linolenic acid and galactolipid levels in *Graminae* species with age of tissue and light environment. - Phytochemistry *6*: 107-113, 1967. [Ps, Chl.]

2851 - GREBINSKIĬ, S.O., PALANITSA, R.P.: Vliyanie gibberellina na soderzhanie khlorofilla v list'yakh i v khloroplastakh. [Effect of gibberellin on chlorophyll content in leaves and in chloroplasts.] - Fiziol. Rast. *17*: 175-176, 1970. [In R.]

2852 - GRECHUKHINA, O.A., BEZSHKURAYA, Yu.G., VALIKHANOVA, G.Zh.: Vliyanie vnekornevoĭ i kornevoĭ podkormki azotom na soderzhanie pigmentov v list'yakh rasteniĭ. [Effect of extra-root and root nitrogen supply on pigment content in plant leaves.] - Uch. Zap. Tartu. gos. Univ. *185*: 387-394, 1966. [In R.]

2853 - De GREEF, J., FREDERICQ, H.: Photomorphogenic and chlorophyll studies in the bryophyte *Marchantia polymorpha*. II. Photobiological responses to terminal irradiations with different red/far-red ratios. - Physiol. Plant. *22*: 462-468, 1969.

2854 - De GREEF, J.A., CAUBERGS, R.: Chlorophyll *c* in *Vaucheria*. - Naturwissenschaften *57*: 673-674, 1970.

2855 - GREEN, W.G.E., ISRAELSTAM, G.F.: Kinetics of nicotinamide adenine dinucleotides during dark-light transients in *Chlorella*. - Physiol. Plant. *23*: 217-231, 1970.

2856 - GREENE, G.L., PROAÑO, V.A.: Gibberellin-induced paling in leaves of a dwarf bean mutant. - Turrialba *18*: 70-72, 1968. [Chl.]

2857 - GREENE, R.W.: Symbiosis in sacoglossan opisthobranchs: Functional capacity of symbiotic chloroplasts. - Mar. Biol. *7*: 138-142, 1970. [Ps, Chl.]

2858 - GREENE, R.W.: Symbiosis in sacoglossan opisthobranchs: symbiosis with algal chloroplasts. - Malacologia *10*: 357-368, 1970. [Chl, Car.]

2859 - GREENE, R.W.: Symbiosis in sacoglossan opisthobranchs: translocation of photosynthetic products from chloroplast to host tissue. - Malacologia *10*: 369-380, 1970.

2860 - GREGORY, R.P.F.: Inhibitory effects of uncoupling agents on systems I and II of photosynthesis electron transport in chloroplast. - Biochem. J. *112*: 10 P, 1969.

2861 - GREIG, J.K., MOTES, J.E., AL-TIKRITI, A.S.: Effect of nitrogen levels and micronutrients on yield, chlorophyll and mineral content of spinach. - Proc. amer. Soc. hort. Sci. *92*: 508-515, 1968.

2862 - GREVTSOVA, A.T.: Intensyvnist' deyakikh fiziologichnykh protsesiv u odnorichnykh siyantsiv duba zvyachaĩnogo.[Rate of several physiological processes in one-year oak seedlings.] - Visn. sil's'kogospod. Nauk *6*: 84-85, 1968. [Ps; in Ukr.]

2863 - GRIFFIN, D.M., NAIR, N.G., BAXTER, R.I., SMILES, D.E.: Control of gaseous environment of organisms using a diffusion column technique. - J. exp. Bot. *18*: 518-525, 1967.

2864 - GRIFFITHS, D.A., GRIFFITHS, D.J.: The fine structure of autotrophic and heterotrophic cells of *Chlorella vulgaris* (Emerson strain). - Plant Cell Physiol. *10*: 11-19, 1969. [Chl.]

2865 - GRIFFITHS, D.J.: The pyrenoid. - Bot. Rev. *36*: 29-58, 1970.

2866 - GRIFFITHS, M., PERROTT, P.S., EDMONDSON, W.T.: Oscillaxanthin in the sediment of Lake Washington. - Limnol. Oceanogr. *14*: 317-326, 1969.

2867 - GRIFFITHS, W.T.: 'Plastoquinone *B*' - some structural studies. - Biochem. biophys. Res. Commun. *25*: 596-602, 1966.

2868 - GRIFFITHS, W.T., THRELFALL, D.R., GOODWIN, T.W.: Observations on the nature and biosynthesis of terpenoid quinones and related compounds in tobacco shoots. - Europe. J. Biochem. *5*: 124-132, 1968.

2869 - GRIFFITHS, W.T., WALLWORK, J.C., PENNOCK, J.F.: Presence of a series of plastoquinones in plants. - Nature *211*: 1037-1039, 1966.

2870 - GRIGOROV, L.N., KONONENKO, A.A., RUBIN, A.B.: Ob obratimom fotoindutsirovannom okislenii vnutrikletochnykh tsitokhromov fotosinteziruyushchikh purpurnykh bakteriĩ *Rhodopseudomonas* species pri temperature zhidkogo azota. [Reversible photoinduced oxidation of intracellular cytochromes in photosynthetic purple bacteria *Rhodopseudomonas* sp. at the temperature of liquid nitrogen.] - Izv. Akad. Nauk SSSR, Ser. Biol. *1969*: 448-451, 1969. [In R, ab: E.]

2871 - GRIGOROV, L.N., KONONENKO, A.A., RUBIN, A.B.: Issledovanie nizkotemperaturnykh fotoindutsirovannykh okislitel'no-vosstanovitel'nykh reaktsiĩ tsitokhromov v kletkakh serobakterii *Rhodopseudomonas* sp. [Studies of the low temperature light-induced oxidation-reduction reactions of cytochromes in sulphur bacteria *Rhodopseudomonas* sp.] - Mol. Biol. (Moskva) *4*: 483-490, 1970. [In R, ab: E.]

2872 - GRIME, J.P.: Shade avoidance and shade tolerance in flowering plants. - In: BAINBRIDGE, R., EVANS, G.V., RACKHAM, O. (ed.): Light as an Ecological Factor. Pp. 187-207. Blackwell sci. Publ., Oxford 1966. [Ps.]

2873 - GRIMES, D.W., CARTER, L.M.: A linear rule for direct nondestructive leaf area
       measurements. - Agron. J. *61*: 477-479, 1969.

2874 - GRIMME, L.H.: Photosynthese und Photoreduktion von *Chlorella fusca* unter Chlo-
       rid-Mangel. - Ber. deut. bot. Ges. *83*: 481-483, 1970.

2875 - GRIMME, L.H., KESSLER, E.: Chloride effect on photosynthesis and photoreduction
       in *Chlorella*. - Naturwissenschaften *57*: 133-134, 1970.

2876 - GRINENKO, V.V.: O znachenii vodnogo rezhima i prisposobitel'nom metabolizme,
       ustoĭchivosti i produktivnosti rasteniĭ. [Significance of water relations and
       accommodation metabolism, resistance, and productivity of plants.] - Uch. Zap.
       Tartu.gos. Univ. *185*: 155-164, 1966. [In R.]

2877 - GRINENKO, V.V.: O vozmozhnostyakh povysheniya fotosinteticheskogo potentsiala
       plodovykh nasazhdeniĭ. [Possibilities of increasing photosynthetic potential in
       stands of fruit trees.] - In: Vazhneĭshie Problemy Fotosinteza v Rastenievodstve.
       Pp. 263-272. Kolos, Moskva 1970. [In R.]

2878 - GRISHINA, G.S., BELL, L.N., BUKINA, G.S.: Primenenie amperometricheskogo meto-
       da dlya issledovaniya obmena kisloroda na svetu. [Application of the amperomet-
       ric technique for investigation of oxygen metabolism in light.] - Fiziol. Rast.
       13: 737-744, 1966. [In R, ab: E.]

2879 - GRISHINA, G.S., VOSKRESENSKAYA, N.P.: Sravnitel'naya spektral'naya effektivnost'
       nasyshchayushchikh intensivnosteĭ sveta v reaktsii Melera. [Comparative spectral
       effectivity of saturating irradiances in the Mehler reaction.] - Fiziol. Rast.
       13: 942-948, 1966. [In R, ab: E.]

2880 - GRISHINA, G.S., VOSKRESENSKAYA, N.P.: Effect of oxygen and ascorbate on the
       electron transport chain of photosynthesis. - In: METZNER, H. (ed.): Progress
       in Photosynthesis Research. Vol. III. Pp. 1262-1267. Tübingen 1969.

2881 - GROB, E.C., RUFENER, J.: Influence of sugar containing nutrients on ultrastruc-
       ture and photosynthesis activity of *Spirodela oligorrhiza* chloroplasts. - In:
       METZNER, E. (ed.): Progress in Photosynthesis Research. Vol. I. Pp. 55-62.
       Tübingen 1969.

2882 - GROB, E.C., SEILER, J.: Beitrag zur Charakterisierung der Chlorophyllase. -
       Chimia *21*: 466-468, 1967.

2883 - GRODZINSKIĬ, D.M., LUTSISHINA, E.G.: Primenenie teorii misheni dlya izucheniya
       fotosinteticheski aktivnykh edinits khloroplastov. [Application of the target
       theory to the study of photosynthetically active units of chloroplasts.] - Bio-
       fizika *14*: 276-279, 1969. [In R, ab: E.]

2884 - GRODZINSKIĬ, D.M., LUTSISHINA, E.G.: Primenenie teorii misheni pri izuchenii
       fotosinteticheski aktivnykh edinits khloroplastov. I. Reaktsiya Khilla s ferrit-
       sianidom. [Utilization of target theory in the study of photosynthetically active
       units of chloroplasts. I. Hill reaction with ferricyanide.] - Tsitol. Genet. *3*:
       440-443, 1969. [In R.]

2885 - GRODZYNS'KYĬ, D.M., GULYAEV, B.I., MAKARENKO, K.I., MANUĬL'SKYĬ, V.D.: Osoblyvos-
       ti gazoobminu lyestya tsukrovykh buryakiv pry rizkomu poslablenni intensyvnosti
       svitla. [Features of gas exchange of a sugarbeet leaf at a rapid decrease in ir-
       radiance.] - Dopovidi Akad. Nauk URSR *1969* B: 560-563, 573, 1969. [In Ukr., ab:
       R.]

2886 - GROESCHEL, E.C., NELSON, A.I., STEINBERG, M.P.: Changes in color and other cha-
       racteristics of green beans stored in controlled refrigerated atmospheres. - J.
       Food Sci. *31*: 488-496, 1966.

2887 - GROGAN, C.O., BLESSIN, C.W.: Characterization of major carotenoids in yellow
       maize lines of differing pigment concentration. - Crop Sci. *8*: 730-732, 1968.

2888 - GROMET-ELHANAN, Z.: The relation of cyclic and noncyclic electron flow patterns
       with reduced indophenols to photophosphorylation. - Israel J. Chem. *4* (1a): 79,
       1966.

2889 - GROMET-ELHANAN, Z.: Inhibitors of photophosphorylation by chromatophores of
       Rhodospirillum rubrum. - Israel J. Chem. *5* (4A): 97, 1967.

2890 - **GROMET-ELHANAN, Z.**: The relationship of cyclic and non-cyclic electron flow patterns with reduced indophenols to photophosphorylation. - Biochim. biophys. Acta *131*: 526-537, 1967.

2891 - **GROMET-ELHANAN, Z.**: Energy-transfer inhibitors and electron transport inhibitors in chloroplasts. - Arch. Biochem. Biophys. *123*: 447-456, 1968.

2892 - **GROMET-ELHANAN, Z.**: The inhibition of photoreactions of chloroplasts by ioxynil. - Biochem. biophys. Res. Commun. *23*: 28-31, 1968.

2893 - **GROMET-ELHANAN, Z.**: Inhibitors of photophosphorylation and photoreduction by chromatophores from *Rhodospirillum rubrum*. - Arch. Biochem. Biophys. *131*: 299-305, 1969.

2894 - **GROMET-ELHANAN, Z.**: Two types of cyclic electron transport in isolated chloroplasts. - In: METZNER, H. (ed.): Progress in Photosynthesis Research. Vol. III. Pp. 1197-1202. Tübingen 1969.

2895 - **GROMET-ELHANAN, Z.**: Differences in sensitivity to valinomycin and nonactin of various phosphorylating and photoreducing systems of *Rhodospirillum rubrum* chromatophores. - Biochim. biophys. Acta *223*: 174-182, 1970.

2896 - **GROMET-ELHANAN, Z., AVRON, M.**: Desaspidin: a nonspecific uncoupler of photophosphorylation. - Plant Physiol. *41*: 1231-1236, 1966.

2897 - **GROMET-ELHANAN, Z., REDLICH, N.**: Diaminodurene-induced plastocyanin dependent oxygen uptake and its relation to photophosphorylation in isolated lettuce chloroplasts. A comparison of the systems using either water or ascorbate as the electron donors. - Europe. J. Biochem. *17*: 523- 528, 1970.

2898 - **GROMOVA, T.P., OSADCHIÏ, A.**: Vliyanie torfo-mineral'nykh shchelochnykh kompostov na dinamiku nakopleniya khlorofilla v list'yakh kartofelya. [Effect of peat-mineral alkaline composts on the dynamics of accumulation of chlorophyll in potato leaves.] - Tr. vologodsk. molochn. Inst. *56*: 105-111, 1968. [In R.]

2899 - **GROOT, J.**: The use of silicone rubber plastic for replicating leaf surfaces. - Acta bot. neerl. *18*: 703-708, 1969.

2900 - **GROSS, E., DILLEY, R.A., SAN PIETRO, A.**: Control of electron flow in chloroplasts by cations. - Arch. Biochem. Biophys. *134*: 450-462, 1969.

2901 - **GROSS, E., SAN PIETRO, A.**: Interaction of uncouplers and energy transfer inhibitors with high-energy states of chloroplasts. - Arch. Biochem. Biophys. *131*: 49-56, 1969.

2902 - **GROSS, E., SHAVIT, N., SAN PIETRO, A.**: Synthalin: an inhibitor of energy transfer in chloroplasts. - Arch. Biochem. Biophys. *127*: 224-228, 1968.

2903 - **GROSS, E.L., PACKER, L.**: Ion transport and conformational changes in spinach chloroplast grana. I. Osmotic properties and divalent cation-induced volume changes. - Arch. Biochem. Biophys. *121*: 779-789, 1967.

2904 - **GROSS, E.L., PACKER, L.**: Ion transport and conformational changes in spinach chloroplast grana. II. Light-induced changes. - Arch. Biochem. Biophys. *122*: 237-245, 1967.

2905 - **GROSS, J.A., SHEFNER, A.M., BECKER, M.J.**: Distribution of chlorophylls in chloroplast fragments. - Nature *209*: 615-616, 1966.

2906 - **GROSS, J.A., STROZ, R.**: Photostimulation of carotenoid biosynthesis in a non-photosynthetic *Euglena* mutant. - Plant Physiol. *44* (Suppl.): 41, 1969.

2907 - **GROSS, J.A., WHITFIELD, M.D.**: Wavelength dependence of electron flow and oxygen evolution in isolated chloroplasts: a possible role for carotenoids. - Biochem. biophys. Res. Commun. *40*: 1216-1223, 1970.

2908 - **GROSS, R.E., DUGGER, W.M. Jr.**: Responses of *Chlamydomonas reinhardtii* to peroxyacetyl nitrate. - Environm. Res. *2*: 256-266, 1969. [Ps, Chl, Car.]

2909 - **GROSS, R.E., PUGNO, P., DUGGER, W.M.**: Observations on the mechanism of copper damage in *Chlorella*. - Plant Physiol. *46*: 183-185, 1970. [Chl, car.]

2910 - **GROSSWEINER, L.I.**: The study of labile states of biological molecules with flash photolysis. - Advances Radiat. Biol. *2*: 83-133, 1966. [Chl.]

2911 - GROZDINSKIĬ, D.M., KHODOS, V.N.: O stepeni geterogennosti fondov fosfornykh metabolitov v list'yakh gorokha. [Degree of heterogeneity of pools of phosphorus metabolites in pea leaves.] - In: Puti Povysheniya Intensivnosti i Produktivnosti Fotosinteza. Pp. 51-55. Naukova Dumka, Kiev 1967. [Chi; in R.]

2912 - GROZOV, D.N.: Dinamika nakopleniya pigmentov v list'yakh yabloni v zavisimosti ot sortovykh osobennosteĭ i formirovaniya krony. [Dynamics of pigment accumulation in apple leaves in dependence on cultivar features and forming of the crown.] - In: Fotosinteticheskaya Deyatel'nost' Yabloni i Slivy v Usloviyakh Moldavii. Pp. 39-61. Kishinev 1970. [In R.]

2913 - GROZOV, D.N.: Soderzhanie zheltykh pigmentov v list'yakh vinograda v zavisimosti ot usloviĭ mineral'nogo pitaniya. [Content of yellow pigments in vine leaves in dependence on mineral nutrients supply.] - In: Fotosinteticheskaya Deyatel'nost' Rasteniĭ i Vliyanie na nee Mineral'nogo Pitaniya. Pp. 55-68. Kishinev 1970. [In R.]

2914 - GROZOV, D.N., DOROKHOV, B.L.: Prirost list'ev i odnoletnikh pobegov vinograda pri razlichnom mineral'nom pitanii. [Growth of leaves and one-year shoots of vine under different mineral nutrition.] - Sadovodstvo, Vinogradarstvo Vinodelie Moldavii 1968 (12): 14-16, 1968. [In R.]

2915 - GRÜNHAGEN, H.H., WITT, H.T.: Primary ionic events in the functional membrane of photosynthesis. Umbelliferone as indicator for pH changes in one turn-over. - Z. Naturforsch. 25 b: 373-386, 1970.

2916 - GRUNWALD, C.: Sterol distribution in intracellular organelles isolated from tobacco leaves. - Plant Physiol. 45: 663-666, 1970. [Chi.]

2917 - GUBAR', G.D., KREĬTSBERG, O.E., KRISTKALNE, S.Kh.: Vliyanie urovnya mineral'nogo pitaniya na fotosinteticheskuyu deyatel'nost' rasteniĭ v protsesse formirovaniya urozhaya. [Effect of mineral nutrition on photosynthetic activity of plants during yield formation.] - Latvijas PSR Zinãtnu Akad. Vēstis 1966.(7): 53-64, 1966. [In R, ab: E, Latv.]

2918 - GUBAR', G.D., VOĬTSEKHOVICH, Z.V., GROSA, V.F.: Svetovye krivye fotosinteza v zavisimosti ot urovnya mineral'nogo pitaniya i svetovogo rezhima vyrashchivaniya rasteniĭ+ [Light curves of photosynthesis as related to mineral nutrition and light regime of plant cultivation.] - In: Fotosintez, Mineral'noe Pitanie, Svetovoĭ Rezhim. Pp. 31-50. Zinatne, Riga 1970. [In R.]

2919 - GUBARE, G.: Augu fotosintētiskā darbība un tās nozīme ražas veidošanās likumsakarību novērtēšanā. [Plant photosynthetic activity and its use for assessing yield formation.] - In: Fotosintēzes Pētīšana Sējumos. Pp. 7-20. Zinatne, Rīgā 1970. [In Latvian, ab: E.]

2920 - GUÉRIN de MONTGAREUIL, P., ANDRÉ, M., SEIMANDI, N.: Capacités en oxygène photosynthétique dans le cas d'une feuille aérienne. - Compt. rend. Acad. Sci. (Paris), Sér. D 265: 485-488, 1967.

2921 - GUÉRIN de MONTGAREUIL, P., GERSTER, R.: Plastoquinones et oxygène-18. - In: THOMAS, J.B., GOEDHEER, J.C. (ed.): Currents in Photosynthesis. Pp. 401-408. Donker, Rotterdam 1966.

2922 - GUÉRIN-DUMARTRAIT, E.: Quelques effets du 3-amino-1,2,4-triazol sur Chlorella pyrenoidosa CHICK. Recherches sur l'action de l'aminotriazol sur la synthèse des acides nucléiques chez des Chlorelles cultivées en cultures synchrones. - Physiol. vég. 4: 135-193, 1966. [Chi.]

2923 - GUÉRIN-DUMARTRAIT, E.: Étude, en cryodécapage, de la morphologie des surfaces lamellaires chloroplastiques de Chlorella pyrenoidosa, en cultures synchrones. - Planta 80: 96-109, 1968.

2924 - GUÉRIN-DUMARTRAIT, E., MIHARA, S., MOYSE, A.: Composition de Chlorella pyrenoidosa, structure des cellules et de leurs lamelles chloroplastiques, en fonction de la carence en azote et de la levée de carence. - Can. J. Bot. 48: 1147-1154, 1970.

2925 - GUÉRIN-DUMARTRAIT, E., MOYSE, A.: Structure des lamelles chloroplastiques de Chlorella pyrenoidosa, soit au cours d'une culture synchrone, soit en fonction de la carence en azote puis de la levée de carence. - Soc. Phycol. Fr. Bull. 1970 (15): 74-79, 1970.

2926 - GUÉRIN-DUMARTRAIT, E., SARDA, C., LACOURLY, A.: Sur la structure fine du chloroplaste de *Porphyridium* sp. (LEWIN). - Compt. rend. Acad. Sci. (Paris), Sér. D
270: 1977-1979, 1970.

2927 - GUILLOT-SOLOMON, T., DOUCE, R., SIGNOL, M.: Relations entre les modifications
de l'ultrastructure plastidiale, la teneur en pigments et la composition en lipides polaires de feuilles de maïs traitées par l'aminotriazole. - Bull. Soc.
fr. Physiol. vég. *13*: 63-79, 1967.

2928 - GULANYAN, S.A., ANDRIANOV, V.K., KURELLA, G.A., LITVIN, F.F.: Elektrometriches-
kiĭ metod nepreryvnoĭ registratsii obmena uglekislogo gaza pri fotosinteze.
[Electrometric method of continuous recording of CO_2 exchange in photosynthesis.]
- Nauch. Dokl. vyssh. Shkoly, biol. Nauki *1970* (9): 106-111, 1970. [In R.]

2929 - GULLVÅG, B.M.: Fine structure of the plastids and possible ways of distribution
of the chloroplast products in some spores of *Archegoniatae*. - Phytomorphology
18: 520-535, 1969.

2930 - GULYAEV, B.A.: Struktura spektrov pogloshcheniya nativnykh form fotosinteticheskikh pigmentov i modelirovanie ikh spektral'nykh svoĭstv. [Structure of absorption spectra of native forms of photosynthetic pigments and modelling of their
spectral properties.] - Nauch. Dokl. vyssh. Shkoly, biol. Nauki *10* (1): 146,
1967. [In R.]

2931 - GULYAEV, B.A., LITVIN, F.F.: Pervaya i vtoraya proizvodnye spektra pogloshcheniya
khlorofilla i soprovozhdayushchikh pigmentov v kletkakh vysshikh rasteniĭ vodorosleĭ pri 20°. [First and second derivatives of absorption spectrum of chlorophyll
and accompanying pigments in cells of higher plants and in algae at 20 °C.] -
Biofizika *15*: 670-680, 1970. [In R, ab: E.]

2932 - GULYAEV, B.I.: Nekotorye voprosy izucheniya radiatsionnogo rezhima posevov v
svyazi s fotosintezom. [Some problems of the study of radiation regime of stands
in relation to photosynthesis.] - In: Puti Povysheniya Intensivnosti i Produktivnosti Fotosinteza. Pp. 177-185. Naukova Dumka, Kiev 1966. [In R.]

2933 - GULYAEV, B.I.: Dannye o raspredelenii FAR v razlichnykh posevakh i otsenka vliyaniya na produktivnost' fotosinteza. [Data on PhAR distribution in different
stands and evaluation of its effect on the photosynthetic productivity.] - In:
Aktinometriya i Optika Atmosfery. Pp. 282-283. Valgus, Tallin 1968. [In R.]

2934 - GULYAEV, B.I., LARIN, A.P.: Chasovyĭ khid fotosyntezu roslyn pry zmini umov ko-
renevogo zhyvlennya. [Time course of photosynthesis after a change in conditions
of root nutrition.] - Ukr. biokhim. Zh. *25*: 98-101, 1968. [In Ukr.]

2935 - GULYAEV, B.I., LAVRENTOVICH, D.I., MANUIL'SKIĬ, V.D., OKANENKO, A.S.: Radiatsionnyĭ rezhim i fotosintez posevov. [Radiation regime and photosynthesis in
stands.] - In: Puti Povysheniya Intensivnosti i Produktivnosti Fotosinteza. Pp.
82-96. Naukova Dumka, Kiev 1967. [In R.]

2936 - GULYAEV, B.I., LAVRENTOVICH, D.I., MITROFANOV, B.A., MANUIL'SKIĬ, V.D.: Spektral'nye kharakteristiki list'ev rasteniĭ i posevov kukuruzy. [Spectral characteristics of plant leaves and of maize stands.] - In: Puti Povysheniya Intensivnosti i Produktivnosti Fotosinteza. Pp. 97-108. Naukova Dumka, Kiev 1967.
[In R.]

2937 - GULYAEV, B.I., LITVIN, F.F.: K voprosu o edinoĭ sisteme agregirovannykh (polimernykh) form fotosinteticheskikh pigmentov v kletkakh vysshikh rasteniĭ, vodorosleĭ i bakteriĭ. [United system of aggregated (polymeric) forms of photosynthetic pigments in cells of higher plants, algae and bacteria.] - Biofizika *12*:
845-854, 1967. [In R.]

2938 - GULYAEV, B.I., MANUÏL'SKIĬ, V.D.: Vplyv rukhu prodykhiv na fotosyntez ta trans-
piratsiyu. [Effect of stomata movement on photosynthesis and transpiration.] -
Dopov. Akad. Nauk Ukr. SSR *32*: 364-367, 384, 1970. [In Ukr., ab: E, R.]

2939 - GULYAEV, B.I., MANUIL'SKIĬ, V.D., OKANENKO, A.S.: Otsenka pogreshnosteĭ izmereniya intensivnosti fotosinteza gazometricheskim metodom. [Estimation of error
of measuring photosynthetic rate by the gasometric method.] - Fiziol. Biokhim.
kul't. Rast. *2*: 34-40, 1970. [In R, ab: E.]

2940 - GULYAEV, B.I., MITROFANOV, B.A., MANUIL'SKIĬ, V.D.: Osobennosti raspredeleniya fotosinteticheski aktivnoĭ radiatsii v poseve ozimoĭ pshenitsy. [Features of PhAR distribution in a winter wheat stand.] - In: Puti Povysheniya Intensivnosti i Produktivnosti Fotosinteza. Vol. 3. Pp. 87-95. Naukova Dumka, Kiev 1969. [In R.]

2941 - GULYAEV, V.I., MANUIL'SKIĬ, V.D.: Zavisimost' intensivnosti fotosinteza ot intensivnosti fotosinteticheski aktivnoĭ radiatsii (FAR) pri razlichnoĭ vlazhnosti pochvy. [Dependence of photosynthetic rate on irradiance with PhAR at different soil moisture.] - In: Puti Povysheniya Intensivnosti i Produktivnosti Fotosinteza. Vol. 3. Pp. 165-170, Naukova Dumka, Kiev 1969. [In R.]

2942 - GUMINETSKIĬ, S.G., KIREEVA, V.V.: Zavisimost' opticheskoĭ aktivnosti list'ev rasteniĭ ot ugla padeniya napravlennogo obIucheniya. [Dependence of optical activity of plant leaves on angle of incidence of directed radiation.] - Nauch. Dokl. vyssh. Shkoly, biol. Nauki *1966* (2): 93-98, 1966. [In R.]

2943 - GUMINETSKIĬ, S.G., RVACHEV, V.P.: Opredelenie spektrov pogloshcheniya otdel'nykh kletok list'ev rasteniĭ. [Determination of absorption spectra of individual cells of leaves.] - Zh. prikl. Spektroskop. *5*: 674-680, 1966. [In R, ab: E.]

2944 - GUMINETSKIĬ, S.G., RVACHEV, V.P.: Ob istinnom spektre pogloshcheniya smesi pigmentov v zhivykh kletkakh list'ev rasteniĭ. [Actual absorption spectrum of a mixture of pigments in living leaf cells.] - Zh. prikl. Spektroskop. *5*: 73-80, 1966. [In R, ab: E.]

2945 - GUNAR, I.I.: Fotosintez, dykhanie i mineral'noe pitanie rasteniĭ. [Photosynthesis, respiration, and mineral nutrition in plants.] - Zemledelie (Moskva) *1968* (6): 2-7, 1968. [In R.]

2946 - GUNNING, B.E.S., JAGOE, M.P.: The prolamellar body. - In: GOODWIN, T.W. (ed.): Biochemistry of Chloroplasts. Vol. II. Pp. 655-676. Academic Press, London-New York 1967.

2947 - GUPTA, S.B.: Chlorophyll variegation caused by somatic elimination of an alien chromosomal fragment in *Nicotiana tabacum*. - Can. J. Genet. Cytol. *10*: 106-111, 1968.

2948 - GUPTA, S.B.: Chlorophyll variegation caused by unstable behaviour of an alien chromosome in hybrid derivatives of *Nicotiana* species. - Genetica *39*: 193-208, 1968.

2949 - GURINOVICH, G.P.: Kvantovaya effektivnost' elementarnykh stadiĭ fotokhimicheskikh reaktsiĭ khlorofilla. [Quantum efficiency of elementary stages of chlorophyll photochemical reactions.] - In: Molekulyarnaya Fotonika. Pp. 221-231. Nauka, Leningrad 1970. [In R.]

2950 - GURINOVICH, G.P., BYTEVA, I.M.: O mekhanizme reaktsii fotovosstanovleniya khlorofilla i ego analogov. [Mechanism of reaction of photoreduction of chlorophyll and its analogues.] - Biofizika *15*: 602-607, 1970. [In R, ab: E.]

B2951 - GURINOVICH, G.P., SEVCHENKO, A.N., SOLOV'EV, K.N.: Spektroskopiya Khlorofilla i Rodstvennykh Soedineniĭ. [Spectroscopy of Chlorophyll and Related Compounds.] - Nauka i Tekhnika, Minsk 1968. [In R.]

2952 - GURINOVICH, G.P., STRELKOVA, T.I.: O mekhanizme assotsiatsii khlorofilla i ego analogov. [Mechanism of association of chlorophyll and its analogues.] - Biofizika *13*: 782-792, 1968. [In R, ab: E.]

2953 - GUSEĬNOV, S.G.: Vliyanie predposevnogo oblucheniya semyan khlopchatnika na fotosintez i dykhanie. [Effect of presowing irradiation of cotton seeds on photosynthesis and respiration.] - Dokl. Akad. Nauk AzerbaĬdzh. SSR *1966* (3): 71-74, 1966. [In R, ab: AzerbaĬdzh.]

2954 - GUSEĬNOV, S.G.: Vliyanie ioniziruyushchikh izlucheniĭ na soderzhanie khlorofilla, askorbinovoĭ kisloty i sukhogo veshchestva v list'yakh khlopchatnika. [Effect of ionizing radiation on the content of chlorophyll, ascorbic acid, and dry matter in cotton leaves.] - Dokl. Akad. Nauk AzerbaĬdzh. SSR *23* (6): 45-48, 1967. [In R, ab: AzerbaĬdzh.]

2955 - GUSEĬNOVA, G.I.: Ispol'zovanie solnechnoĭ energii rasteniyami v usloviyakh oranzherei. [Solar energy utilization by greenhouse plants.] - In: Fotosintez i Ispol'zovanie Energii Solnechnoĭ Radiatsii. Pp. 11-12. Dushanbe 1967. [In R.]

2956 - GUSEÏNOVA, G.I.: Ispol'zovanie solnechnoĭ radiatsii rasteniyami v usloviyakh oranzherei v osenne-zimniĭ period. [Solar energy utilization by greenhouse plants in autumn-winter period.] - Izv. Akad. Nauk Tadzh. SSR, Otd. biol. Nauk *1969* (2): 60-65, 1969. [In R.]

2957 - GUSEV, M.V.: Pigmenty sine-zelenykh vodorosleĭ. [Pigments of blue-green algae.] - In: Biologiya Sine-zelenykh Vodorosleĭ. Vol. 2. Pp. 88-109. Moskov. Univ., Moskva 1969. [In R.]

2958 - GUSEV, M.V., NIKITINA, K.A., KORZHENEVSKAYA, T.G.: Metabolicheski aktivnye sferoplasty sinezelenykh vodorosleĭ. [Metabolically active spheroplasts of blue-green algae.] - Mikrobiologiya *39*: 862-868, 1970. [Ps, Chl; in R.]

2959 - GUSEV, M.V., SHENDEROVA, L.V., KONDRAT'EVA, E.N.: Vliyanie kontsentratsii kisloroda na rost i vyzhivaemost' fotosinteziruyushchikh bakteriĭ. [Effect of oxygen concentration on growth and survival in photosynthetic bacteria.] - Mikrobiologiya *39*: 562-566, 1970. [In R, ab: E.]

2960 - GUSEV, M.V., ZHEVNER, V.D., SHESTAKOV, S.V.: Khromaticheskie izmeneniya u sinezelenykh vodorosleĭ. [Chromatographic changes in blue-green algae.] - Nauch. Dokl. vyssh. Shkoly, biol. Nauki *10* (1): 154-155, 1967. [Chl, car; in R.]

2961 - GUSEVA, A.I.: Ovodnennost' tkaneĭ i soderzhanie khlorofilla v list'yakh gaploidov kukuruzy. [Water and chlorophyll content in the leaves of maize haploids.] - In: Apomiksis i Selektsiya. Pp. 258-261. Nauka, Moskva 1970. [In R.]

2962 - GUYOMARC'H, C.: Influence des pH alcalins et des ions bicarbonates sur la photosynthèse de *Chlorella vulgaris* BEIJ. - Bull. Soc. sci. Bretagne, Sci. math. phys. nat. *45*: 113-126, 1970.

2963 - GUYOMARC'H, C., VILLERET, S.: Influence comparée du gas carbonique dissous et des solutions bicarbonatées sur la croissance de l'algue *Chlorella vulgaris* BEIJ. - Bull. Soc. sci. Bretagne, Sci. math. phys. nat. *40*: 193-206, 1965 (1966).

2964 - GUZ, A.F., VOLYNETS, A.P.: Formirovanie i rabota fotosinteticheskogo apparata u sortov l'na-dolguntsa pri khimicheskoĭ propolke. [Formation and activity of photosynthetic apparatus in *Linum* cultivars under chemical weeding.] - In: Fiziologo-biokhimicheskie Issledovaniya Rasteniĭ. Pp. 61-69. Nauka i Tekhnika, Minsk 1967. [In R.]

2965 - GVARDIYAN, V.N.: Sostoyanie plastidnogo apparata kartofelya v usloviyakh prodolžitel' nogo zatemneniya i pri vosstanovlenii normal'nogo svetovogo rezhima. [State of plastid apparatus of the potato plant under prolonged darkening and restoration of the normal light regime.] - In: Tezisy IV Nauchnoĭ Konferentsii Molodykh Uchenykh po Sovremennym Problemam Biologii. P. 1. Minsk 1970. [In R.]

2966 - GVOZDYKIVS'KA, A.T., BERSHTEĬN, B.I.: Vyvchennya roli fosforylyuvannya v fotosyntezi u vyshchykh roslyn. [Role of phosphorylation in photosynthesis of higher plants.] - Ukr. bot. Zh. *25* (6): 8-14, 1968. [In Ukr., ab: E, R.]

2967 - GYLDENHOLM, A.O.: Macromolecular physiology of plastids. V. On the nucleic acid metabolism during chloroplast development. - Hereditas *59*: 142-168, 1968.

2968 - GYLDENHOLM, A.O., WHATLEY, F.R.: The onset of photophosphorylation in chloroplasts isolated from developing bean leaves. - New Phytol. *67*: 461-468, 1968.

2969 - GYURJÁN, I., KEVE, T.: $^{14}CO_2$ incorporation by normal and mutant maize leaves at different light intensity. - Stud. biophys. *5*: 117-122, 1967.

2970 - GYURJÁN, I., KEVE, T., FALUDI-DÁNIEL, Á., ANDA, S.: $^{14}CO_2$ assimilation of normal and chloroplast mutant leaves at different light intensities. - Acta biol. Acad. Sci. hung. *20*: 325-334, 1969.

2971 - GYURJÁN, I., LÁNG, F., PACSÉRY, M.: Normális és mutáns kukoricalevelek $^{14}CO_2$ asszimilációja különböző megvilágításo viszonyok között. [$^{14}CO_2$ assimilation in leaves of normal and mutant maize plants under different irradiance.] - Biol. Közl. *14*: 97-106, 1966. [In Hung., ab: E, R.]

2972 - GYURJÁN, I., RAKOVÁN, J.N., FALUDI-DÁNIEL, Á.: Chlorophyll differentiation and $^{14}CO_2$ fixation in normal and mutant maize leaves. - In: METZNER, H. (ed.): Progress in Photosynthesis Research. Vol. I. Pp. 63-72. Tübingen 1969.

2973 - HAAPALA, H.: Accumulation and disintegration of starch in the chloroplasts of *Stellaria media* grown under constant illumination. - Physiol. Plant. *21*: 866-871, 1968.

2974 - HAAPALA, H.: Starch metabolism of chloroplasts as influenced by light. - Aquilo, Ser. bot. *8*: 42-65, 1969.

2975 - HAAPALA, H.: Photosynthesis and starch metabolism of chloroplasts during prolonged illumination. - Planta *86*: 259-266, 1969.

2976 - HABER, A.H., THOMPSON, P.J., WALNE, P.L., TRIPLETT, L.L.: Nonphotosynthetic retardation of chloroplast senescence by light. - Plant Physiol. *44*: 1619-1628, 1969.

2977 - HABER, A.H., WALNE, P.L.: Actions of acute gamma radiation on excised wheat leaf tissue. I. Radioresistance of senescence and establishing criteria of death from studies of chlorophyll loss. - Radiat. Bot. *8*: 389-397, 1968.

2978 - HABERDITZL, W., PREIDEL, J., STÖSSER, R.: Elektronen-Spin-Resonanz des Ferredoxins aus *Chlorella*. - Z. Chem. *9*: 344 A, 1969.

2979 - HABERMANN, H.M.: Light-inhibited leaf development in a white mutant: resemblance to effects of 2-thiouracil in normally pigmented *Helianthus annuus*. - Physiol. Plant. *19*: 122-127, 1966.

2980 - HABERMANN, H.M.: Reversal of copper inhibition in chloroplast reactions by manganese. - Plant Physiol. *44*: 331-336, 1969.

2981 - HABERMANN, H.M., HANDEL, M.A., McKELLAR, P.: Kinetics of chloroplast-mediated photooxidation of diketogulonate. - Photochem. Photobiol. *7*: 211-224, 1968.

2982 - HABESHAW, D.: The effect of light on the translocation from sugar-beet leaves. - J. exp. Bot. *20*: 64-71, 1969.

2983 - HABIG, W., RACUSEN, D.: The effect of light on ribulose diphosphate carboxylase in corn leaves. - Can. J. Bot. *47*: 1051-1054, 1969.

2984 - HADFIELD, W.: Leaf temperature, leaf pose and productivity of the tea bush. - Nature *219*: 282-284, 1968.

2985 - HADLEY, E.B.: Physiological ecology of *Pinus ponderosa* in southwestern North Dakota. - Amer. Midland Naturalist *81*: 289-314, 1969. [Ps.]

2986 - HADLEY, E.B., LEVIN, D.A.: Physiological evidence of hybridization and reticulate evolution in *Phlox maculata*. - Amer. J. Bot. *56*: 561-570, 1969. [Ps.]

2987 - HAFEZ, M.G.A., YOUNIS, M.E.: Studies in stomatal behaviour: The effects of moving, normal and carbon dioxide - free air on the stomata of some plants. - Physiol. Plant. *22*: 332-337, 1969.

2988 - HAFEZ, M.G.A., YOUNIS, M.E.: Studies in stomatal behaviour. Effects of moving carbon dioxide free and still air on the stomata of some plants. - Beitr. Biol. Pflanzen *47*: 1-9, 1970. [Porometer.]

2989 - HAGER, A.: Die Zusammenhänge zwischen lichtinduzierten Xanthophyll-Umwandlungen und Hill-Reaktion. - Ber. deut. bot. Ges. *79*: (94)-(107), 1966.

2990 - HAGER, A.: Untersuchungen über die lichtinduzierten reversiblen Xanthophyllumwandlungen an *Chlorella* und *Spinacia*. - Planta *74*: 148-172, 1967.

2991 - HAGER, A.: Untersuchungen über die Rückreaktionen im Xanthophyll-Cyclus bei *Chlorella*, *Spinacia* und *Taxus*. - Planta *76*: 138-148, 1967.

2992 - HAGER, A.: Lichtbedingte pH-Erniedrigung in einem Chloroplasten-Kompartiment als Ursache der enzymatischen Violaxanthin- → Zeaxanthin-Umwandlung; Beziehungen zur Photophosphorylierung. - Planta *89*: 224-243, 1969.

2993 - HAGER, A.: Ausbildung von Maxima im Absorptionsspektrum von Carotinoidem im Bereich um 370 nm; Folgen für die Interpretation bestimmter Wirkungsspektren. - Planta *91*: 38-52, 1970.

2994 - HAGER, A., MEYER-BERTENRATH, T.: Die Isolierung und quantitative Bestimmung der Carotinoide und Chlorophylle von Blättern, Algen und isolierten Chloroplasten mit Hilfe dünnschichtchromatographischer Methoden. - Planta *89*: 198-217, 1966.

2995 - HAGER, A., MEYER-BERTENRATH, T.: Beziehungen zwischen Absorptionsspektrum und Konstitution bei Carotinoiden von Algen und höheren Pflanzen. - Ber.deut. bot. Ges. *80*: 426-436, 1967.

2996 - HAGER, A., MEYER-BERTENRATH, T.: Die Identifizierung der an Dünnschichten getrennten Carotinoide grüner Blätter und Algen. - Planta *76*: 149-168, 1967.

2997 - HAGER, A., PERZ, H.: Veränderung der Lichtabsorption eines Carotinoids im Enzym (De-epoxidase)-Substrat (Violaxanthin)-Komplex. - Planta *93*: 314-322, 1970.

2998 - HAGER, A., STRANSKY, H.: Das Carotinoidmuster und die Verbreitung des lichtinduzierten Xanthophyllcyclus in verschiedenen Algenklassen. I. Methoden zur Identifizierung der Pigmente. - Arch. Mikrobiol. *71*: 132-163, 1970.

2999 - HAGER, A., STRANSKY, H.: Das Carotinoidmuster und die Verbreitung des lichtinduzierten Xanthophyllcyclus in verschiedenen Algenklassen. III. Grünalgen. - Arch. Mikrobiol. *72*: 68-83, 1970.

3000 - HAGER, A., STRANSKY, H.: Das Carotinoidmuster und die Verbreitung des lichtinduzierten Xanthophyllcyclus in verschiedenen Algenklassen. V. Einzelne Vertreter der *Cryptophyceae, Euglenophyceae, Bacillariophyceae, Chrysophyceae* und *Phaeophyceae*. - Arch. Mikrobiol. *73*: 77-89, 1970.

3001 - HAGIN, R.D., LINSCOTT, D.L., ROBERTS, R.N., DAWSON, J.E.: Carotenoid pigments in plants. Major interfering substances in determining 2,4-D, a metabolite of 2,4-DB. - J. agr. Food Chem. *14*: 630-632, 1966.

3002 - HAHN, L.W., MILLER, J.H.: Light dependence of chloroplast replication and starch metabolism in the moss *Polytrichum commune*. - Physiol. Plant. *19*: 134-141, 1966.

3003 - HAIDAK, D.J., MATHEWS, C.K., SWEENEY, B.M.: Pigment protein complex from *Gonyaulax*. - Science *152*: 212-213, 1966.

3004 - HAIGHT, T.H., KUEHNERT, C.C.: The effect of colored light on pigment synthesis in cultured fern leaf primordia. - Physiol. Plant. *23*: 704-714, 1970.

3005 - HÁJKOVÁ, A., KVĚT, J.: Analysis of primary productivity in two types of inundated meadows. - In: Productivity of Terrestrial Ecosystems. Production Processes. Czechosl. nat. Comm. IBP, PT-PP Report No. 1. Pp. 47-50. Praha 1970.

3006 - HALFACRE, R.G., BARDEN, J.A., ROLLINS, H.A. Jr.: Effects of Alar on morphology, chlorophyll content and net CO_2 assimilation rate of young apple trees. - Proc. amer. Soc. hort. Sci. *93*: 40-52, 1968.

3007 - HALL, D.O., EVANS, M.C.W., GIBSON, J.F., JOHNSON, C.E., WHATLEY, F.R.: Some properties of ferredoxins. - In: METZNER, H. (ed.): Progress in Photosynthesis Research. Vol. III. Pp. 1433-1443. Tübingen 1969.

3008 - HALL, D.O., GIBSON, J.F., WHATLEY, F.R.: Electron spin resonance spectra of spinach ferredoxin. - Biochem. biophys. Res. Commun. *23*: 81-84, 1966.

3009 - HALL, D.O., TELFER, A.: The effect of sulphate and sulphite on photophosphorylation by spinach chloroplasts. - In: METZNER, H. (ed.): Progress in Photosynthesis Research. Vol. III. Pp. 1281-1287. Tübingen 1969.

3010 - HALL, D.O., WHATLEY, F.R.: The chloroplast. - In: ROODYN, D.B. (ed.): Enzyme Cytology. Pp. 181-237. Academic Press, London-New York 1967.

3011 - HALL, I.V., FORSYTH, F.R., LOCKHART, C.L., AALDERS, L.E.: Effect of time of day, a parasitic fungus and a genetic mutation on rate of photosynthesis in the lowbush blueberry. - Can. J. Bot. *44*: 529-533, 1966.

3012 - HALL, T.C.: Protein, amino acid and chlorophyll metabolism during the ontogeny of snap beans. - Proc. amer. Soc. hort. Sci. *93*: 379-387, 1968.

3013 - HALLDAL, P.: Induction phenomena and action spectra analyses of photosynthesis in ultraviolet and visible light studied in green and blue green algae, and in isolated chloroplast fragments. - Z. Pflanzenphysiol. *54*: 28-44, 1966.

3014 - HALLDAL, P.: Ultraviolet action spectra in algology. A review. - Photochem. Photobiol. *6*: 445-460, 1967. [Also Ps, Chl.]

3015 - HALLDAL, P.: Photosynthetic capacities and photosynthetic action spectra of endozoic algae of the massive coral Favia. - Biol. Bull. *134*: 411-424, 1968.

3016 - HALLDAL, P.: Automatic recording of action spectra of photobiological processes, spectrophotometric analyses, fluorescence measurements and recording of the first derivative of the absorption curve in one simple unit. - Photochem. Photobiol. *10*: 23-34, 1969.

3017 - HALLDAL, P.: The photosynthetic apparatus of microalgae and its adaptation to environmental factors. - In: HALLDAL, P. (ed.): Photobiology of Microorganisms. Pp. 17-55. Wiley-Interscience, London-New York-Sydney-Toronto 1970.´

3018 - HALLIER, U.W.: Untersuchungen des Photosyntheseapparates bei Plastommutanten von *Oenothera*. - Ber. deut. bot. Ges. *79*: (69)-(71), 1966.

3019 - HALLIER, U.W.: On the use of ^{14}C and ^{32}P in locating genetically caused defects in photosynthesis of some plastid mutants. - In: Proceedings of the Symposium on Isotopes in Plant Nutrition and Physiology. 5-9 Sept. 1966, Vienna, Austria. Pp. 235-245. Int. at. Energy Agency, Vienna 1967.

3020 - HALLIER, U.W.: Photosynthese-Reaktionen einiger Plastom-Mutanten von *Oenothera*. II. Die Bildung von ATP und NADPH. - Z. Pflanzenphysiol. *58*: 289-299, 1968.

3021 - HALLIER, U.W., HEBER, U., STUBBE, W.: Photosynthese-Reaktionen einiger Plastom-Mutanten von *Oenothera*. I. Der reduktive Pentosephosphatzyklus. - Z. Pflanzenphysiol. *58*: 222-239, 1968.

3022 - HALLIER, U.W., PARK, R.B.: Photosystems I and II in chemically fixed *Anacystis nidulans* and *Chlorella pyrenoidosa*. - Plant Physiol. *43* (Suppl.): S 20, 1968.

3023 - HALLIER, U.W., PARK. R.B.: Photosynthetic light reactions in chemically fixed *Anacystis nidulans, Chlorella pyrenoidosa,* and *Porphyridium cruentum*. - Plant Physiol. *44*: 535-539, 1969.

3024 - HALLIER, U.W., PARK, R.B.: Photosynthetic light reactions in chemically fixed spinach thylakoids. - Plant Physiol. *44*: 544-546, 1969.

3025 - HALLOIN, J.M., WALKER, J.C., de ZOETEN, G.A., GAARD, G.: Effects of tentoxin on plastids of cucumber and cabbage. - Phytopathology *59*: 1028-1029, 1969. [Chl.]

3026 - HALLOIN, J.M., de ZOETEN, G.A., GAARD, G., WALKER, J.C.: The effects of tentoxin on chlorophyll synthesis and plastid structure in cucumber and cabbage. - Plant Physiol. *45*: 310-314, 1970.

3027 - HAMED, M.G.E.: Zur Gefriertrocknung von grünen Bohnen. - Z. Lebensm.-Unters. Forsch. *131*: 144-149, 1966/67. [Chl.]

3028 - HAMMER, L.: Salzgehalt und Photosynthese bei marinen Pflanzen. - Mar. Biol. *1*: 185-190, 1968.

3029 - HAMMER, L.: "Free space-photosynthesis" in the algae *Fucus virsoides* and *Laminaria saccharina*. - Mar. Biol. *4*: 136-138, 1969.

3030 - HAND, D.W., BOWMAN, G.E.: Carbon dioxide assimilation measurement in a controlled environment glasshouse. - J. agr. eng. Res. *14*: 92-99, 1969.

3031 - HANNAH, L.C., TOMES, M.L.: Effects of tomato flesh pigment mutant genes on leaf chlorophylls and carotenoids. - J. amer. Soc. hort. Sci. *95*: 503-507,1970.

3032 - HANNAN, H.H., DORRIS, T.C.: Succession of a macrophyte community in a constant temperature river. - Limnol. Oceanogr. *15*: 442-453, 1970. [Ps.]

3033 - HANSCH, C.: Theoretical considerations of the structure-activity relationship in photosynthesis inhibitors. - In: METZNER, H. (ed.): Progress in Photosynthesis Research, Vol. III. Pp. 1685-1692. Tübingen 1969.

3034 - HANSCH, C., DEUTSCH, E.W.: The structure-activity relationship in amides inhibiting photosynthesis. - Biochim. biophys. Acta *112*: 381-391, 1966.

3035 - HANSEN, H., LINDHARD, J.: Bidrag til tørstofbestemmelsens metodik. [Contribution to the method of dry matter determination.] - T. Planteavl. *73*: 94-98, 1969. [In Danish, ab: E.]

3036 - HANSEN, P.: ^{14}C-studies on apple trees. I. The effect of the fruit on the translocation and distribution of photosynthates. - Physiol. Plant. *20*: 382-391, 1957.

3037 - HANSEN, P.: ^{14}C-studies on apple trees. II. Distribution of photosynthates from top and base leaves from extension shoots. - Physiol. Plant. *20*: 720-725, 1967.

3038 - HANSEN, P.: ^{14}C-studies on apple trees. III. The influence of season on storage and mobilization of labelled compounds. - Physiol. Plant. *20*: 1103-1111, 1967.

3039 - HANSEN, P.: ^{14}C-studies on apple trees. IV. Photosynthate consumption in fruits in relation to the leaf-fruit ratio and to the leaf-fruit position. - Physiol. Plant. *22*: 186-198, 1969.

3040 - HANSEN, P.: ^{14}C-studies on apple trees. V. Translocation of labelled compounds from leaves to fruit and their conversion within the fruit. - Physiol. Plant. *23*: 564-573, 1970.

3041 - HANSEN, P.: ^{14}C-studies on apple trees. VI. The influence of the fruit on the photosynthesis of the leaves, and the relative photosynthetic yields of fruits and leaves. - Physiol. Plant. *23*: 805-810, 1970.

3042 - HARADA, H.: Retardation of the senescence of *Rumex obtusifolius* L. leaves by growth retardants. - Plant Cell Physiol. *7*: 701-703, 1966. [Chl.]

3043 - HARAGUCHI, N., SHIMIZU, S.: Photosynthetic activities in tobacco plants. II. Relationship between photosynthetic activity and chlorophyll content. - Bot. Mag. (Tokyo) *83*: 411-418, 1970.

3044 - HARASHIMA, K.: [Biological oxidation of carotenes.] - Yukagaku [J. Jap. Oil Chemists' Soc.] *16*: 491-498, 1967. [In Jap.]

3045 - HARDING, H., WILLIAMS, P.H., McNABOLA, S.S.: Chlorophyll changes, photosynthesis, and ultrastructure of chloroplasts in *Albugo candida* induced "Green Islands" on detached *Brassica juncea* cotyledons. - Can. J. Bot. *46*: 1229-1234, 1968.

3046 - HARDON, J.J., WILLIAMS, C.N., WATSON, I.: Leaf area and yield in the oil palm in Malaya. - Exp. Agr. *5*: 25-32, 1969.

3047 - HARDWICK, K., LUMB, H., WOOLHOUSE, H.W.: A chamber suitable for measurement of gas exchange by leaves under controlled conditions. - New Phytol. *65*: 526-531, 1966.

3048 - HARDWICK, K., WOOD, M., WOOLHOUSE, H.W.: Photosynthesis and respiration in relation to leaf age in *Perilla frutescens* (L.) BRITT. - New Phytol. *67*: 79-86, 1968.

3049 - HARDWICK, K., WOOLHOUSE, H.W.: Foliar senescence in *Perilla frutescens* (L.) BRITT. - New Phytol. *66*: 545-552, 1967. [Chl.]

3050 - HARDY, P.J.: Selective diffusion of basic and acidic products of CO_2 fixation into the transpiration stream in grapevine. - J. exp. Bot. *20*: 856-862, 1969.

3051 - HARDY, S.I., CASTELFRANCO, P.A., REBEIZ, C.A.: Effect of the hypocotyl hook on greening in etiolated cucumber cotyledons. - Plant Physiol. *46*: 705-707, 1970.

3052 - HARLEY, J.L.: The importance of micro-organisms to colonising plants. - Bot. Soc. Edinburgh Trans. *41*: 65-70, 1970. [Ps.]

3053 - HARNISCHFEGER, G.: Changes in appearance, volume and activity during the early stages of disintegration in isolated chloroplasts. - Planta *92*: 164-177, 1970.

3054 - HARNISCHFEGER, G., GAFFRON, H.: Transient color sensitivity of the Hill reaction during the disintegration of chloroplasts. - Planta *89*: 385-388, 1969.

3055 - HARNISCHFEGER, G., GAFFRON, H.: Transient light effects in the Hill reaction of disintegrating chloroplasts *in vitro*. - Planta *93*: 89-105, 1970.

3056 - HARRIS, G.P., SCOTT, M.A.: Studies on the glasshouse carnation: Effects of light and temperature on the growth and development of the flower. - Ann. Bot. *33*: 143-152, 1969. [Ps.]

3057 - HARRIS, J.B., NAYLOR, A.W.: Changes in the etiolated tobacco leaf during greening: I. Structure of the chloroplast. - Tobacco Sci. *11*: 168-172, 1967; Tobacco *165* (23): 26-30, 1967.

3058 - HARRIS, J.B., NAYLOR, A.W.: Changes in the etiolated tobacco leaf during greening. II. Composition of subcellular fractions. - Tobacco Sci. *12*: 25-33, 1968.

3059 - HARRIS, J.B., NAYLOR, A.W.: Changes in the etiolated tobacco leaf during greening. III. The effect of organic metabolites and light intensity on chlorophyll and carotenoid production. - Tobacco Sci. *12*: 170-176, 1968; Tobacco *167* (11): 24-29, 1968.

3060 - HARRIS, J.B., NAYLOR, A.W.: Changes in the etiolated tobacco leaf during greening. IV. Carbon dioxide fixation. - Tobacco Sci. *14*: 69-72, 1970; Tobacco *170*
(20): 23-26, 1970.

3061 - HARRIS, R.C., KIRK, J.T.O.: Control of chloroplast formation in *Euglena graci-*
lis: antagonism between carbon and nitrogen sources. - Biochem. J. *106*: 34 P,
1968.

3062 - HARRIS, R.C., KIRK, J.T.O.: Control of chloroplast formation in *Euglena graci-*
lis. Antagonism between carbon and nitrogen sources. - Biochem. J. *113*: 195-205,
1969.

3063 - HARRIS, W.M.: Chromoplasts of tomato fruits. III. The high-delta tomato. - Bot.
Gaz. *131*: 163-166, 1970.

3064 - HARRIS, W.M., SPURR, A.R.: Chromoplasts of tomato fruits. I. Ultrastructure of
low-pigment and high-beta mutants. Carotene analyses. - Amer. J. Bot. *56*: 369-
379, 1969.

3065 - HARRIS, W.M., SPURR, A.R.: Chromoplasts of tomato fruits. II. The red tomato.
- Amer. J. Bot. *56*: 380-389, 1969.

3066 - HARRISS, R.C., WHITE, D.B., MACFARLANE, R.B.: Mercury compounds reduce photosynthesis by plankton. - Science *170*: 736-737, 1970.

3067 - HARTLEY, W.R., WEISS, C.M.: Light intensity and the vertical distribution of algae in tertiary oxidation ponds. - Water Res. *4*: 751-763, 1970. [Chl, Car.]

3068 - HARTMAN, R.T.: Photosynthetic carbon fixation in aquatic vascular plants as
measured by radiocarbon tracer techniques. - In: Proceedings of the Symposium
on Isotopes in Plant Nutrition and Physiology. 5-9 Sept. 1966, Vienna, Austria.
Pp. 111-125. Int. at. Energy Agency, Vienna 1967.

3069 - HARTMAN, R.T., BROWN, D.L.: Changes in internal atmosphere of submersed vascular
hydrophytes in relation to photosynthesis. - Ecology *48*: 252-258, 1967.

3070 - HARTMANN, C., DURAND, B.: Sur les activités respiratoires et photosynthétiques
de *Mercurialis annua* L. (2 n = 16), espèce dioïque.- Compt. rend. Acad. Sci.
(Paris), Sér. D *269*: 1762-1763, 1969.

3071 - HARTT, C.E.: Translocation in colored light. - Plant Physiol. *41*: 369-372, 1966.
[Ps.]

3072 - HARTT, C.E.: Effect of moisture supply upon translocation and storage of ^{14}C in
sugarcane. - Plant Physiol. *42*: 338-346, 1967.

3073 - HARTT, C.E.: Effect of potassium deficiency upon translocation of ^{14}C in attached blades and entire plants of sugarcane. - Plant Physiol. *44*: 1461-1469, 1969.

3074 - HARTT, C.E.: Effect of potassium deficiency upon translocation of ^{14}C in detached blades of sugarcane. - Plant Physiol. *45*: 183-187, 1970.

3075 - HARTT, C.E.: Effect of nitrogen deficiency upon translocation of ^{14}C in sugarcane. - Plant Physiol. *46*: 419-422, 1970.

3076 - HARTT, C.E., BURR, G.O.: Factors affecting photosynthesis in sugar cane. - In:
Proceedings of the 12th ISSCT Congress, Puerto Rico 1965. Pp. 590-608. Elsevier
Publ. Co., Amsterdam 1967.

3077 - HARTT, C.E., KORTSCHAK, H.P.: Radioactive isotopes in sugarcane physiology. -
In: Proceedings of the 12th ISSCT Congress, Puerto Rico 1965. Pp. 647-662. Elsevier Publ. Co., Amsterdam 1967. [Ps.]

3078 - HARTT, C.E., KORTSCHAK, H.P.: Translocation of ^{14}C in the sugarcane plant during
the day and night. - Plant Physiol. *42*: 89-94, 1967.

3079 - HARVEY, M.J., BROWN, A.P.: Nicotinamide cofactors of intact chloroplasts isolated on a sucrose density gradient. - Biochim. biophys. Acta *172*: 116-125, 1969.

3080 - HARVEY, M.J., BROWN, A.P.: Uptake and light activated esterification of ^{32}P by
isolated chloroplasts. - Biochim. biophys. Acta *180*: 520-528, 1969.

3081 - HARVEY, M.J., GIBBS, M.: Distribution of photosynthetic products between chloroplasts and incubation medium. - Plant Physiol. *46* (Suppl.): 6, 1970.

3082 - HASEGAWA, K., MACMILLAN, J.D., MAXWELL, W.A., CHICHESTER, C.O.: Photosensitized bleaching of β-carotene with light at 632.8 nm from a continuous-wave gas laser. - Photochem. Photobiol. *9*: 165-169, 1969.

3083 - HASPEL-HORVATOVIČ, E.: Further proof of direct oxygen transfer by carotenoids in respiration and photosynthesis. - Nature *209*: 1135, 1966.

3084 - HASPEL-HORVATOVIČ, E.: Pathophysiologischer Beitrag zur Interaktion oxydierter und reduzierter Xanthophyllformen. - Stud. biophys. *5*: 239-246, 1967.

3085 - HASPEL-HORVATOVIČ, E., PAULECH, C.: Changes of oxidized and reduced carotenoids, of the respiration and the photosynthesis of barley (*Hordeum sativum* L.) in connection with its resistance against powdery mildew. - In: METZNER, H. (ed.): Progress in Photosynthesis Research. Vol. I. Pp. 396-401. Tübingen 1969.

3086 - HASPELOVÁ-HORVATOVIČOVÁ, A.: Genetika chlorofylového aparátu a dedičnost' pigmentových abnormít. [Genetics of chlorophyll apparatus and heredity of pigment abnormities.] - Acta Fac. Rerum natur. Univ. Comeniae *12* (Genet. 1): 99, 1966. [In Slovak.]

3087 - HASPELOVÁ-HORVATOVIČOVÁ, A.: Dünnschichtchromatographie zur Isolierung des Luteinisomers Zeaxanthin. - Biológia (Bratislava) *25*: 439-444, 1970.

3088 - HASPELOVÁ-HORVATOVIČOVÁ, A., POSTULKOVÁ, M.: Quantitative Berechnung von Assimilationspigmenten mit Hilfe des Elektronenrechners GIER in der Sprache ALGOL. - Biológia (Bratislava) *24*: 783-793, 1969.

3089 - HASPELOVÁ-HORVATOVIČOVÁ, A., POSTULKOVÁ, M.: Auswertung biologischer Versuchsergebnisse durch Students *t*-Test mit Hilfe des Elektronenrechners GIER in der Sprache ALGOL. - Biológia (Bratislava) *25*: 251-254, 1970.

3090 - HATCH, M.D.: Chemical energy costs for CO_2 fixation by plants with differing photosynthetic pathways. - In: Prediction and Measurement of Photosynthetic Productivity. Pp. 215-220. PUDOC, Wageningen 1970.

3091 - HATCH, M.D., SLACK, C.R.: Photosynthesis by sugar-cane leaves. A new carboxylation reaction and the pathway of sugar formation. - Biochem. J. *101*: 103-111, 1966.

3092 - HATCH, M.D., SLACK, C.R.: The participation of phosphoenolypyruvate synthetase in photosynthetic CO_2 fixation of tropical grasses. - Arch. Biochem. Biophys. *120*: 224-225, 1967.

3093 - HATCH, M.D., SLACK, C.R.: New enzyme for the interconversion of pyruvate and phosphopyruvate and its role in the dicarboxylic acid pathway of photosynthesis. - Biochem. J. *106*: 141-146, 1968.

3094 - HATCH, M.D., SLACK, C.R.: NADP-specific malate dehydrogenase and glycerate kinase in leaves and evidence for their location in chloroplasts. - Biochem. biophys. Res. Commun. *34*: 589-593, 1969.

3095 - HATCH, M.D., SLACK, C.R.: Studies on the mechanism of activation and inactivation of pyruvate, phosphate dikinase. A possible regulatory role for the enzyme in the C_4 dicarboxylic acid pathway of photosynthesis. - Biochem. J. *112*: 549-558, 1969.

3096 - HATCH, M.D., SLACK, C.R.: Photosynthetic CO_2-fixation pathways. - Annu. Rev. Plant Physiol. *21*: 141-162, 1970.

3097 - HATCH, M.D., SLACK, C.R.: The C_4-dicarboxylic acid pathway of photosynthesis. - In: REINHOLD, L., LIWSCHITZ, Y. (ed.): Progress in Phytochemistry. Vol. 2. Pp. 35-106. Intersci. Publ., London-New York-Sydney-Toronto 1970.

3098 - HATCH, M.D., SLACK, C.R., BULL, T.A.: Light-induced changes in the content of some enzymes of the C_4-dicarboxylic acid pathway of photosynthesis and its effect on other characteristics of photosynthesis. - Phytochemistry *8*: 697-706, 1969.

3099 - HATCH, M.D., SLACK, C.R., JOHNSON, H.S.: Further studies on a new pathway of photosynthetic carbon dioxide fixation in sugar-cane and its occurrence in other plant species. - Biochem. J. *102*: 417-422, 1967.

3100 - HATTORI, A., UESUGI, I.: Ferredoxin-dependent photoreduction of nitrate and nitrite by subcellular preparations of *Anabaena cylindrica*. - In: SHIBATA, K., TAKAMIYA, A., JAGENDORF, A.T., FULLER, R.C. (ed.): Comparative Biochemistry and Biophysics of Photosynthesis. Pp. 201-205. Univ. Tokyo Press, Tokyo; Univ. Park Press, State College, Pa. 1968.

3101 - HAUPT, W.: Die Inversion der Schwachlichtbewegung des *Mougeotia*-Chloroplasten: Versuche zur Kinetik der Phytochrom-Umwandlung. - Z. Pflanzenphysiol. *54*: 151-160, 1966.

3102 - HAUPT, W.: Phototaxis in plants. - Int. Rev. Cytol. *19*: 267-299, 1966. [Chloroplasts.]

3103 - HAUPT, W.: VIII. Bewegungen. - Fortschr. Bot. *29*: 160-169, 1967. [Also stomata.]

3104 - HAUPT, W.: Die Orientierungsbewegungen der Chloroplasten. - Biol. Rundschau *6*: 121-136, 1968.

3105 - HAUPT, W.: VIII. Bewegungen. - Fortschr. Bot. *31*: 164-171, 1969. [Also chloroplasts.]

3106 - HAUPT, W.: Chloroplastenbewegung bei *Mougeotia*: Vergleich der Induktionswirkung von Blaulicht und Rotlicht. - Ber. deut. bot. Ges. *83*: 201, 1970.

3107 - HAUPT, W., GÄRTNER, R.: Die Chloroplasten-Orientierung von *Mesotaenium* in starkem Licht. - Naturwissenschaften *53*: 411, 1966.

3108 - HAUPT, W., HEYMANN, N.: Versuche zur Mechanik der Chloroplastenbewegung von *Mougeotia*. - Z. Pflanzenphysiol. *57*: 68-71, 1967.

3109 - HAUPT, W., KRÖGER, B., LAUX, A.: Aktionsdichroismus der Chloroplastenbewegung von *Mougeotia* im Blaulicht. - Naturwissenschaften *56*: 642, 1969.

3110 - HAUPT, W., SCHOLZ, A.: Nachweis des Linseneffektes bei der Chloroplastenorientierung von *Hormidium flaccidum*. - Naturwissenschaften *53*: 388, 1966.

3111 - HAUPT, W., SCHÖNBOHM, E.: Light-oriented chloroplast movements. - In: HALLDAL, P. (ed.): Photobiology of Microorganisms. Pp. 283-307. Wiley-Interscience, London-New York-Sydney-Toronto 1970.

3112 - HAUPT, W., SEITZ, K.: Fördernde und hemmende Dunkelrot-Wirkungen bei der phytochrominduzierten Chloroplastenbewegung von *Mougeotia*. - Z. Pflanzenphysiol. *56*: 102-103, 1967.

3113 - HAUPT, W., WEISENSEEL, M.: Die Temperaturabhängigkeit der Chloroplastenbewegungen bei *Lemna trisulca*. - Naturwissenschaften *53*: 411-412, 1966.

3114 - HAUPT, W., WEISENSEEL, M.: Chloroplastenbewegung bei *Lemna trisulca* in polarisiertem Licht. - Naturwissenschaften *54*: 48-49, 1967.

3115 - HAUPT, W., WIRTH, H.: Nachweis einer Schraubenstruktur in der *Mougeotia*-Zelle. - Plant Cell Physiol. *8*: 541-543, 1967. [Chloroplast.]

3116 - HAUSKA, G.A., McCARTY, R.E., OLSON, J.S.: The relation of the light-induced increase in absorbance at 518 nm to photophosphorylation in digitonin subchloroplast particles. - FEBS Letters *7*: 151-156, 1970.

3117 - HAUSKA, G.A., McCARTY, R.E., RACKER, E.: The site of phosphorylation associated with photosystem I. - Biochim. biophys. Acta *197*: 206-218, 1970.

3118 - HAVELKA, J., ROZTOČIL, J.: Integrátor fotosynteticky aktivního záření. [Integrator of photosynthetically active radiation.] - Jemná Mech. Opt. *13*: 341-344, 1968. [In Czech, ab: E, G, R.]

3119 - HAYASHI, K.: Efficiencies of solar energy conversion in rice varieties. - In: Photosynthesis and Utilization of Solar Energy. Level III Experiments. Pp. 39-42. Jap. nat. Subcomm. for PP (JPP), Tokyo 1968.

3120 - HAYASHI, K.-I.: Response of net assimilation rate to different intensity of sunlight of rice varieties. - Proc. Crop Soc. Jap. *37*: 528-533, 1968.

3121 - HAYES, D.J.: The sensitivity of the thermistor katharometer. - J. sci. Instrum. (J. Phys. E.) Ser. 2, *1*: 761-764, 1968.

3122 - HAYYIM, G.B., AVRON, M.: Cytochrome *b* of isolated chloroplasts. - Europe. J. Biochem. *14*: 205-213, 1970.

3123 - HAZAMA, K., MURAKAMI, T.: [Carbon dioxide content of air in a sericultural experiment station.] - Acta Sericologia [Sanshi Kenkyu] *64*: 18-25, 1967. [Ps; in Jap.]

3124 - HEALEY, F.P.: The carotenoids of four blue-green algae. - J. Phycol. *4*: 126-129, 1968.

3125 - HEALEY, F.P.: The mechanism of hydrogen evolution by *Chlamydomonas moewusii*. - Plant Physiol. *45*: 153-159, 1970. [Ps.]

3126 - HEALEY, F.P., COOMBS, J., VOLCANI, B.E.: Changes in pigment content of the diatom *Navicula pelliculosa* (BRÉB.) HILSE in silicon-starvation synchrony. - Arch. Mikrobiol. *59*: 131-142, 1967.

3127 - HEARN, A.B.: The growth and performance of cotton in a desert environment: II. Dry matter production. - J. agr. Sci. *73*: 75-86, 1969.

3128 - HEATH, O.V.S.: Light measurements in plant growth investigations. - Nature *210*: 752-753, 1966.

B3129 - HEATH, O.V.S.: The Physiological Aspects of Photosynthesis. - Stanford Univ. Press, Stanford, Calif. 1969.

3130 - HEATH, O.V.S., MANSFIELD, T.A.: The movements of stomata. - In: WILKINS, M.B. (ed.): The Physiology of Plant Growth and Development. Pp. 301-332. McGraw-Hill, London 1969. [Ps.]

3131 - HEATH, O.V.S., McCREE, K.J.: Light measurements in plant growth investigations. - Nature *210*: 752-753, 1966.

3132 - HEATH, O.V.S., MEIDNER, H.: Compensation points and carbon dioxide enrichment for lettuce grown under glass in winter. - J. exp. Bot. *18*: 746-751, 1967.

3133 - HEATH, O.V.S., ORCHARD, B.: Carbon assimilation at low carbon dioxide levels. - II. The processes of apparent assimilation. - J. exp. Bot. *19*: 176-192, 1968.

3134 - HEATH, R.L.: Kinetics studies on the fluorescence quencher in isolated chloroplasts. - Biophys. J. *10*: 1173-1188, 1970.

3135 - HEATH, R.L., HIND, G.: The role of Cl^- in photosynthesis. II. The effect of Cl^- upon fluorescence. - Biochim. biophys. Acta *172*: 290-299, 1969.

3136 - HEATH, R.L., HIND, G.: The role of chloride ion in photosynthesis. IV. Studies on the low temperature fluorescence emission spectrum. - Biochim. biophys. Acta *180*: 414-416, 1969.

3137 - HEATH, R.L., HIND, G.: On the functional site of manganese in photosynthetic oxygen evolution. - Biochim. biophys. Acta *189*: 222-233, 1969.

3138 - HEATH, R.L., PACKER, L.: Photoperoxidation in isolated chloroplasts. I. Kinetics and stoichiometry of fatty acid peroxidation. - Arch. Biochem. Biophys. *125*: 189-198, 1968.

3139 - HEATH, R.L., PACKER, L.: Photoperoxidation in isolated chloroplasts. II. Role of electron transfer. - Arch. Biochem. Biophys. *125*: 850-857, 1968.

3140 - HEATH, R.L., PACKER, L.: Steady-state fluorescence of spinach chloroplasts and electron flow. - Arch. Biochem. Biophys. *125*: 1019-1022, 1968.

3141 - HEBER, U.: Freezing injury and uncoupling of phosphorylation from electron transport in chloroplasts. - Plant Physiol. *42*: 1343-1350, 1967.

3142 - HEBER, U.: Freezing injury in relation to loss of enzymes activities and protection against freezing. - Cryobiology *5*: 188-201, 1968. [Ps.]

3143 - HEBER, U.: Conformational changes of chloroplasts induced by illumination of leaves *in vivo*. - Biochim. biophys. Acta *180*: 302-319, 1969.

3144 - HEBER, U.: Control of photosystem I-mediated cyclic electron transfer by photosystem II and electron acceptors. - In: METZNER, H. (ed.): Progress in Photosynthesis Research. Vol. II. Pp. 1082-1090. Tübingen 1969.

3145 - HEBER, U.: Inhibition of cyclic electron transfer *in vivo* by red light and by oxygen. - Carnegie Inst. Year Book *67*: 525-528, 1969.

3146 - HEBER, U.: Adenylattransport zwischen Chloroplasten und Cytoplasma. - Ber. deut. bot. Ges. *83*: 447-450, 1970.

3147 - HEBER, U.: Flow of metabolites and compartmentation phenomena in chloroplasts. - In: Int. Symp. Stofftransport und Stoffverteilung in Zellen höherer Pflanzen. Band b. Pp. 151-184. Abhandl. deut. Akad. Wiss., Berlin 1970. [Also Ps.]

3148 - HEBER, U.: Protein capable of protecting chloroplast membranes against freezing. - In: The Frozen Cell. Pp. 175-188. Ciba, London 1970.

3149 - HEBER, U., FRENCH, C.S.: Effects of oxygen on the electron transport chain of photosynthesis. - Planta 79: 99-112, 1968.

3150 - HEBER, U., HALLIER, U.W., HUDSON, M.A.: Untersuchungen zur intrazellulären Verteilung von Enzymen und Substraten in der Blattzelle. II. Lokalisation von Enzymen des reduktiven und des oxydativen Pentosephosphat-Zyklus in den Chloroplasten und Permeabilität der Chloroplasten-Membran gegenüber Metaboliten. - Z. Naturforsch. 22 b: 1200-1215, 1967.

3151 - HEBER, U., KEMPFLE, M.: Proteine als Schutzstoffe gegenüber dem Gefriertod der Zelle. - Z. Naturforsch. 25 b: 834-842, 1970. [Ps.]

3152 - HEBER, U., SANTARIUS, K.A.: Pyridinnucleotide in Chloroplasten und Zytoplasma von Blattzellen im Licht und im Dunkeln. - In: THOMAS, J.B., GOEDHEER, J.C. (ed.): Currents in Photosynthesis. Pp. 393-400. Donker, Rotterdam 1966.

3153 - HEBER, U., SANTARIUS, K.A.: Direct and indirect transfer of ATP and ADP across the chloroplast envelope. - Z. Naturforsch. 25 b: 718-728, 1970.

3154 - HEBER, U., SANTARIUS, K.A., HUDSON, M.A., HALLIER, U.W.: Untersuchungen zur intrazellulären Verteilung von Enzymen und Substraten in der Blattzelle. I. Intrazellulärer Transport von Zwischenprodukten der Photosynthese im Photosynthese-Gleichgewicht und im Dunkel-Licht-Dunkel-Wechsel. - Z. Naturforsch. 22 b: 1189-1199, 1967.

3155 - HEBER, U.W.: Transport metabolites in photosynthesis. - In: GOODWIN, T.W. (ed.): Biochemistry of Chloroplasts. Vol. II. Pp. 71-78. Academic Press, London and New York 1967.

3156 - HECHT-BUCHHOLZ, C.: Über die Dunkelfärbung des Blattgrüns bei Phosphormangel. - Z. Pflanzenernähr. Bodenkunde 118: 12-22, 1967. [Chl.]

3157 - HEFTEL, W.R.: Characterization of the primary photochemical apparatus of green plants. - Diss. Abstr. int. B 30: 4530-B-4531-B, 1970.

3158 - HEICHEL, G.H., MUSGRAVE, R.B.: Relation of CO_2 compensation concentration to apparent photosynthesis in maize. - Plant Physiol. 44: 1724-1728, 1969.

3159 - HEICHEL, G.H., MUSGRAVE, R.B.: Varietal differences in net photosynthesis of Zea mays L. - Crop Sci. 9: 483-486, 1969.

3160 - HEICHEL, G.H., MUSGRAVE, R.B.: Photosynthetic response to drought in maize. - Philippine Agr. 54: 102-114, 1970.

3161 - HEIDT, L.J.: The path of oxygen from water to molecular oxygen. - J. chem. Educ. 43: 623-636, 1966. [Ps.]

3162 - HEILMAN, M.D., GONZALEZ, C.L., SWANSON, W.A., RIPPERT, W.J.: Adaptation of a linear transducer for measuring leaf thickness. - Agron. J. 60: 578-579, 1968.

3163 - HEINICKE, D.R.: The effect of natural shade on photosynthesis and light intensity in Red Delicious apple trees. - Proc. amer. Soc. hort. Sci. 88: 1-8, 1966.

3164 - HEINICKE, D.R., FOOT, J.W.: The effect of several phosphate insecticides on photosynthesis of Red Delicious apple leaves. - Can. J. Plant Sci. 46: 589-591, 1966.

3165 - HELDT, H.W.: Adenine nucleotide translocation in spinach chloroplasts. - FEBS Letters 5: 11-14, 1969.

3166 - HELDT, H.W., RAPLEY, L.: Unspecific permeation and specific uptake of substances in spinach chloroplasts. - FEBS Letters 7: 139-142, 1970.

3167 - HELDT, H.W., RAPLEY, L.: Specific transport of inorganic phosphate, 3-phosphoglycerate and dihydroxyacetonephosphate, and of dicarboxylates across the inner membrane of spinach chloroplasts. - FEBS Letters 10: 143-148, 1970.

3168 - HELLEBUST, J.A.: Light plants. - In: KINNE, O. (ed.): Marine Ecology. Vol. I. Pp. 125-158. Wiley-Interscience, New York 1970. [Ps.]

3169 - HELLBUST, J.A., TERBORGH, J.: Effects of environmental conditions on the rate of photosynthesis and some photosynthetic enzymes in *Dunaliella tertiolecta* BUTCHER. - Limnol. Oceanogr. *12*: 559-567, 1967.

3170 - HELLEBUST, J.A., TERBORGH, J., McLEOD, G.C.: The photosynthetic rhythm of *Acetabularia crenulata*. II. Measurements of photoassimilation of carbon dioxide and the activities of enzymes of the reductive pentose cycle. - Biol. Bull. *133*: 670-678, 1967.

3171 - HELLMUTH, E.O.: A method of determining true values for photosynthesis and respiration under field conditions. - Flora B *157*: 265-286, 1967.

3172 - HELMS, J.A.: Summer net photosynthesis of ponderosa pine in its natural environment. - Photosynthetica *4*: 243-253, 1970.

3173 - HELMY, F.M., HACK, M.H., YAEGER, R.G.: Comparative lipid biochemistry. - VI. Lipids of green and etiolated *Euglena gracilis* and of *Blastocrithidia culicis*. - Comp. Biochem. Physiol. *23*: 565-567, 1967. [Chl, car.]

3174 - HELTNE, J., BONNETT, H.T.: Chloroplast development in isolated roots of *Convolvulus arvensis* (L.). - Planta *92*: 1-12, 1970.

3175 - van HEMERT, P., KILBURN, D.G., RIGHELATO, R.C., van WEZEL, A.L.: A steam-sterilizable electrode of the galvanic type for the measurement of dissolved oxygen. - Biotechnol. Bioeng. *11*: 549-560, 1969.

3176 - HENDRICH, W.: Struktura chloroplastów a biochemia fotosyntezy. [Structure of chloroplasts and biochemistry of photosynthesis.] - Postępy Biochem. *13*: 311-333, 1967. [In Pol., ab: E.]

3177 - HENDRICH, W.: Widma IR chlorofilu a w roztworach zasad organicznych. [Infrared spectra of chlorophyll a in solutions of organic hydroxides.] - Roczn.Chem. *41*: 743-751, 1967. [In Pol., ab: E, R.]

3178 - HENDRICH, W.: Nowy produkt fotoredukcji chlorofilu a. [New product of chlorophyll a photoreduction.] - Roczn. Chem. *42*: 1715-1723, 1968. [In Pol., ab: E,R.]

3179 - HENDRICH, W.: Reversible photoreduction of chlorophyll a to a compound, Ch_aH_2610, with an absorption maximum at 610 nm. - Biochim. biophys. Acta *162*: 265-270, 1968.

3180 - HENDRICH, W.: Photoreduction of chlorophyll b in the presence of benzylamide. - Acta biochem. pol. *16*: 111-118, 1969.

3181 - HENDRICH, W.: Photoreduction of chlorophyll a in the presence of morpholine. - Photosynthetica *4*: 228-235, 1970.

3182 - HENDRICKS, S.B.: Light in plant life. - In: SAN PIETRO, A., GREER, F.A., ARMY, T.J. (ed.): Harvesting the Sun. Pp. 1-14. Academic Press, New York-London 1967. [Ps.]

3183 - HENDRICKS, S.B.: How light interacts with living matter. - Sci. Amer. *219*: 174-184, 186, 1968. [Also Ps, Chl.]

3184 - HENDRIX, J.E.: Labeling pattern of translocated stachyose in squash. - Plant Physiol. *43*: 1631-1636, 1968.[Ps.]

3185 - HENNINGER, M.D., BARR, R., CRANE, F.L.: Plastoquinone B. - Plant Physiol. *41*: 696-700, 1966.

3186 - HENNINGER, M.D., CRANE, F.L.: Electron transport in chloroplasts. I. A combined requirement for plastoquinones A and C for photoreduction of 2,6-dichloroindophenol. - J. biol. Chem. *241*: 5190-5196, 1966.

3187 - HENNINGER, M.D., CRANE, F.L.: Electron transport in chloroplasts: a new redox protein, rubimedin. - Biochem. biophys. Res. Commun. *24*: 386-390, 1966.

3188 - HENNINGER, M.D., CRANE, F.L.: Electron transport in chloroplasts. III. The role of plastoquinone C. - J. biol. Chem. *242*: 1155-1159, 1967.

3189 - HENNINGER, M.D., CRANE, F.L.: Plastoquinone C involved in electron transport in chloroplasts. - Nature *214*: 921-922, 1967.

3190 - HENNINGER, M.D., GELARDI, C., CRANE, F.L.: Rubimedin particles on chloroplast
        lamellae. - Exp. Cell Res.*44*: 655-658, 1966.

3191 - HENNINGER, M.D., MAGREE, L., CRANE, F.L.: Distribution of plastoquinones in
        fractionated chloroplasts. - Biochim. biophys. Acta *131*: 119-126, 1967.

3192 - HENNINGSEN, K.W.: An action spectrum for vesicle dispersal in bean plastids. -
        In: THOMAS, J.B., GOEDHEER, J.C. (ed.): Currents in Photosynthesis. Pp. 441-447.
        Donker, Rotterdam 1966.

3193 - HENNINGSEN, K.W.: An action spectrum for vesicle dispersal in bean plastids. -
        In: GOODWIN, T.W. (ed.): Biochemistry of Chloroplasts. Vol. II. Pp. 453-457.
        Academic Press, London-New York 1967.

3194 - HENNINGSEN, K.W.: Spectral shifts of chlorophyllous pigments associated with re-
        arrangements of plastid membranes. - Ann. Acad. Sci. fenn. Ser. A IV Biol. *128*:
        39-40, 1968.

3195 - HENNINGSEN, K.W.: Macromolecular physiology of plastids. VI. Changes in membrane
        structure associated with shifts in the absorption maxima of the chlorophyllous
        pigments. - J. Cell Sci. *7*: 587-621, 1970.

3196 - HENNINGSEN, K.W., BOYNTON, J.E.: The physiology and ultrastructure of barley
        mutants at loci controlling the development of the lamellar systems in chloro-
        plasts. - Stud. biophys. *5*: 89-90, 1967.

3197 - HENNINGSEN, K.W., BOYNTON, J.E.: Macromolecular physiology of plastids. VII. The
        effect of a brief illumination on plastids of dark-grown barley leaves. - J.
        Cell Sci. *5*: 757-793, 1969.

3198 - HENNINGSEN, K.W., BOYNTON, J.E.: The physiology and chloroplast structure of
        mutants at loci controlling chlorophyll synthesis in barley. - In: METZNER, H.
        (ed.): Progress in Photosynthesis Research. Vol. I. P. 73. Tübingen 1969.

3199 - HENNINGSEN, K.W., BOYNTON, J.E.: Macromolecular physiology of plastids. VIII.
        Pigment and membrane formation in plastids of barley greening under low light
        intensity. - J. Cell Biol. *44*: 290-304, 1970.

3200 - HENRY-HISS, Y.: Contribution à l'étude de la nutrition et du métabolisme à la
        lumière et à l'obscurité chez *Gonium octonarium* POCOCK. - Compt. rend. Acad.
        Sci.(Paris),Sér. D *269*: 1517-1520, 1969. [Ps.]

3201 - HERRMANN, R.G.: Die Plastidenpigmente einiger Desmidiaceen. - Protoplasma *66*:
        357-368, 1968.

3202 - HERRMANN, R.G.: Carotin-cis-trans-Isomere unter den Plastidenpigmenten des Le-
        bermooses *Sphaerocarpos*. - Z. Naturforsch. *23 b*: 1496-1499, 1968.

3203 - HERRMANN, R.G.: Are chloroplasts polyploid? - Exp. Cell Res. *55*: 414-416, 1969.

3204 - HERRMANN, R.G.: Multiple amounts of DNA related to the size of chloroplasts.
        I. An autoradiographic study. - Planta *90*: 80-96, 1970.

3205 - HERRMANN, R.G., KOWALLIK, K.V.: Multiple amounts related to the size of chloro-
        plasts. II. Comparison of electron-microscopic and autoradiographic data. -
        Protoplasma *69*: 365-372, 1970.

3206 - HERRON, H., MAUZERALL, D.: The functional development of chloroplasts in a mu-
        tant of *Chlorella* using the oxygen luminometer. - Fed. Proc. *27*: 827, 1968.

3207 - HERRON, H.A., MAUZERALL, D.: The light saturation curve of photosynthesis. -
        Biochim. biophys. Acta *205*: 312-314, 1970.

3208 - HERSCOVICI, A., SHAVIT, N.: The effect of ATP on coupled electron flow and
        phosphorylation in chloroplasts. - Israel J. Chem. *8*: 164 p, 1970.

3209 - HERTZBERG, S., LIAAEN JENSEN, S.: The carotenoids of blue-green algae. I. The
        carotenoids of *Oscillatoria rubescens* and an *Athrospira* sp. - Phytochemistry
        5: 557-563, 1966.

3210 - HERTZBERG, S., LIAAEN JENSEN, S.: The carotenoids of blue-green algae. II.
        The carotenoids of *Aphanizomenon flos-aquae*. - Phytochemistry *5*: 565-570, 1966.

3211 - HERTZBERG, S., LIAAEN JENSEN, S.: The carotenoids of blue-green algae. III.

A comparative study of mutochrome and flavacin. - Phytochemistry *6*: 1119-1126, 1967.

3212 - HERTZBERG, S., LIAAEN JENSEN, S.: The structure of myxoxanthophyll. - Phytochemistry *8*: 1259-1280, 1969.

3213 - HERTZBERG, S., LIAAEN JENSEN, S.: The structure of oscillaxanthin. - Phytochemistry *8*: 1281-1292, 1969.

3214 - HESKETH, J.: Enhancement of photosynthetic CO_2 assimilation in the absence of oxygen, as dependent upon species and temperature. - Planta *76*:371-374, 1967.

3215 - HESKETH, J., BAKER, D.: Light and carbon assimilation by plant communities. - Crop. Sci. *7*: 285-293, 1967.

3216 - HESKETH, J.D.: Effects of light and temperature during plant growth on subsequent leaf CO_2 assimilation rates under standard conditions. - Aust. J. biol. Sci. *21*: 235-241, 1968.

3217 - HESKETH, J.D., BAKER, D.N.: Relative rates of leaf expansion in seedlings of species with differing photosynthetic rates. - J. Arizona Acad. Sci. *5*: 216-221, 1969.

3218 - HESKETH, J.D., BAKER, D.N.: The relationship between leaf anatomy and photosynthetic CO_2 assimilation among and within species. - In: Prediction and Measurement of Photosynthetic Productivity. Pp. 317-322. PUDOC, Wageningen 1970.

3219 - HESLOP-HARRISON, J.: Structural features of the chloroplast. - Sci. Progr. *54*: 519-541, 1966.

3220 - HESS, J.L., TOLBERT, N.E.: Glycolate, glycine, serine, and glycerate formation during photosynthesis by tobacco leaves. - J. biol. Chem. *241*: 5705-5711, 1966.

3221 - HESS, J.L., TOLBERT, N.E.: Rate of formation of phosphoglycerate and glycolate during photosynthesis. - Fed. Proc. *25*: 226, 1966.

3222 - HESS, J.L., TOLBERT, N.E.: Glycolate pathway in algae. - Plant Physiol. *42*: 371-379, 1967.

3223 - HESS, J.L., TOLBERT, N.E.: Changes in chlorophyll *a/b* ratio and products of $^{14}CO_2$ fixation by algae grown in blue or red light. - Plant Physiol. *42*: 1123-1130, 1967.

3224 - HESS, J.L., TOLBERT, N.E., PIKE, L.M.: Glycolate biosynthesis by *Scenedesmus* and *Chlorella* in the presence or absence of $NaHCO_3$. - Planta *74*: 278-285, 1967. [Ps.]

3225 - HEVESI, J., SINGHAL, G.S.: Temperature dependence of the relationship between the absorption and the emission spectra of chlorophyll *a* and its derivatives. - Acta biochim. biophys. Acad. Sci. hung. *3*: 454, 1968.

3226 - HEVESI, J., SINGHAL, G.S.: Relation between the absorption and the emission spectra of chlorophyll *a* and its derivatives at room and low temperature. - Spectrochim. Acta A *25*: 1751-1758, 1969.

3227 - HEW, C.-S., GIBBS, M.: A study of chloroplasts of corn, sorghum and sugar cane. - Plant Physiol. *44* (Suppl., Abstr.): 10, 1969.

3228 - HEW, C.-S., GIBBS, M.: Light-induced O_2 evolution, triphosphopyridine nucleotide reduction, and phosphorylation by chloroplasts of maize, sugarcane, and sorghum. - Can. J. Bot. *48*: 1265-1269, 1970.

3229 - HEW, C.S., KROTKOV, G.: Effect of temperature on apparent photosynthesis, CO_2 evolution in light and in darkness by attached leaves of sunflower, soybean and egg plants. - Plant Physiol. *42* (Suppl.): S 47, 1967.

3230 - HEW, C.S., KROTKOV, G.: Effect of oxygen on the rates of CO_2 evolution in light and in darkness by photosynthesizing and non-photosynthesizing leaves. - Plant Physiol. *43*: 464-466, 1968.

3231 - HEW, C.-S., KROTKOV, G., CANVIN, D.T.: Determination of the rate of CO_2 evolution by green leaves in light. - Plant Physiol. *44*: 662-670, 1969.

3232 - HEW, C.-S., KROTKOV, G., CANVIN, D.T.: Effects of temperature on photosynthesis and CO_2 evolution in light and darkness by green leaves. - Plant Physiol. *44*: 671-677, 1969.

3233 - HEW, C.S., NELSON, C.D., KROTKOV, G.: Hormonal control of translocation of photosynthetically assimilated ^{14}C in young soybean plants. - Amer. J. Bot. *54*: 252-256, 1967.

3234 - HEYTLER, P.G.: Polarographic measurement of respiration and photosynthesis. - Fed. Proc. *28*: 533, 1969.

3235 - HICKS, B.B.: Measurement of atmospheric fluxes by eddy correlation. - Aust. meteorol. Mag. *17*: 173-175, 1969.

3236 - HICKS, B.B.: A simple instrument for the measurement of Reynolds stress by eddy correlation. - J. appl. Meteorol. *8*: 825-827, 1969.

3237 - HICUGHI, M., KIKUCHI, G.: Induced formation of ribulose 1,5-diphosphate carboxylase in *Rhodopseudomonas spheroides* with particular concern to its relation with chromatophore formation. - Plant Cell Physiol. *10*: 149-160, 1969.

3238 - HIESEY, W.M., BJÖRKMAN, O., NOBS, M.A.: Light-saturated rates of photosynthesis in *Mimulus cardinalis*. - Carnegie Inst. Year Book *65*: 461-464, 1967.

3239 - HIESEY, W.M., NOBS, M.A., BJÖRKMAN, O.: Photosynthetic rates of *M. lewisii* and *M. cardinalis* in comparison with their F_1 hybrid. - Carnegie Inst. Year Book *65*: 464-468, 1967.

3240 - HIESEY, W.M., NOBS, M.A., BJÖRKMAN, O.: Photosynthetic rates of *Mimulus* races and hybrid derivatives. - Carnegie Inst. Year Book *66*: 214-216, 1968.

3241 - HIESEY, W.M., NOBS, M.A., BJÖRKMAN, O.: Photosynthesis of an amphiploid *Mimulus* in comparison with its progenitors. - Carnegie Inst. Year Book *67*: 489-491, 1969.

3242 - HIGHKIN, H.R., BOARDMAN, N.K., GOODCHILD, D.J.: Photosynthetic studies on a pea mutant deficient in chlorophyll *b*. - Plant Physiol. *42*: S 35, 1967.

3243 - HIGHKIN, H.R., BOARDMAN, N.K., GOODCHILD, D.J.: Photosynthetic studies on a pea-mutant deficient in chlorophyll. - Plant Physiol. *44*: 1310-1320, 1969.

3244 - HIGUCHI, M., OHBA, T., SAKAI, H., KURASHIMA, Y., KIKUCHI, G.: Change of metabolic stability of δ-aminolevulinate synthetase of *Rhodopseudomonas spheroides* cells probably related to changes of intracellular oxidation-reduction state. - J. Biochem. (Tokyo) *64*: 795-805, 1968. [Chl.]

3245 - HILDRETH, W.W.: Laser-activated electron transport in a *Chlamydomonas* mutant. - Plant Physiol. *43*: 303-312, 1968.

3246 - HILDRETH, W.W.: Laser-induced kinetics of cytochrome oxidation and the 518 mµ absorption change in spinach leaves and chloroplasts. - Biochim. biophys. Acta *153*: 197-202, 1968.

3247 - HILDRETH, W.W.: The 520-nm absorption change in barley and a chlorophyll *b*-deficient mutant. - Arch. Biochem. Biophys. *139*: 1-8, 1970.

3248 - HILDRETH, W.W., AVRON, M., CHANCE, B.: Laser activation of rapid changes in spinach chloroplasts and *Chlorella*. - Plant Physiol. *41*: 983-991, 1966.

3249 - HILL, A.C.: Air quality standards for fluoride vegetation effects. - APCA J. *19*: 331-336, 1969. [Ps.]

3250 - HILL, A.C., BENNETT, J.H.: Inhibition of apparent photosynthesis by nitrogen oxides. - Atmos. Environm. *4*: 341-348, 1970.

3251 - HILL, A.C., LITTLEFIELD, N.: Ozone. Effect on apparent photosynthesis, rate of transpiration, and stomatal closure in plants. - Environm. Sci. Technol. *3*: 52-56, 1969.

3252 - HILL, E.R., PUTALA, E.C., VENGRIS, J.: Atrazine-induced ultrastructural changes of barnyardgrass chloroplasts. - Weed Sci. *16*: 377-380, 1968.

3253 - HILL, H.A.O., MACFARLANE, A.J., MANN, B.E., WILLIAMS, R.J.P.: Molecular complexes of vitamin K_3 with a porphyrin and chlorophyll. - Chem. Commun. *1968* (3): 123-124, 1968.

3254 - HILL, H.M., ROGERS, L.J.: Conversion of lycopene into β-carotene by chloroplasts of higher plants. - Biochem. J. *113*: 31 P-32 P, 1969.

3255 - HILL, H.Z., EPSTEIN, H.T., SCHIFF, J.A.: Studies of chloroplast development in *Euglena*. XIV. Sequential interactions of ultraviolet light and photoreactivating light in green colony formation. - Biophys. J. *6*:135-144, 1966.

3256 - HILL, H.Z., SCHIFF, J.A., EPSTEIN, H.T.: Studies of chloroplast development in *Euglena*. XIII. Variation of ultraviolet sensitivity with extent of chloroplast development. - Biophys. J. *6*: 125-133, 1966.

3257 - HILL, H.Z., SCHIFF, J.A., EPSTEIN, H.T.: Studies of chloroplast development in *Euglena*. XV. Factors influencing the decay of photoreactivability of green colony formation. - Biophys. J. *6*: 373-383, 1966.

3258 - HILL, R.: The growth of our knowledge of photosynthesis. - In: NEEDHAM, J. (ed.): The Chemistry of Life. Pp. 1-14. Cambridge Univ. Press, New York 1970.

3259 - HILL, R., BENDALL, D.S.: Oxidation-reduction potentials in relation to components of the chloroplast. - In: GOODWIN, T.W. (ed.): Biochemistry of Chloroplasts. Vol. II. Pp. 559-564. Academic Press, London-New York 1967.

3260 - HILLER, R.G.: Light-driven scattering changes and increased 515 nm absorbance changes associated with fatty acid inhibition of photosynthesis in *Chlorella*. - Biochim. biophys. Acta *172*: 546-552, 1969.

3261 - HILLER, R.G.: Transients in the photosynthetic carbon reducation cycle produced by iodoacetic acid and ammonium chloride. - J. exp. Bot. *21*: 628-638, 1970.

3262 - HILLER, R.G., GREENWAY, H.: Effects of low water potentials on some aspects of carbohydrate metabolism in *Chlorella pyrenoidosa*. - Planta *78*: 49-59, 1968.

3263 - HILLER, R.G., WHATLEY, F.R.: Photosynthesis- the present state of our knowledge. - Advance Sci. *23*: 643-651, 1967.

3264 - HILLIARD, J.H., WEST, S.H.: Starch accumulation associated with growth reduction at low temperatures in a tropical plant. - Science *168*: 494-496, 1970.

3265 - HILTON, J.L., SCHAREN, A.L., JOHN, S.B.St., MORELAND, D.E., NORRIS, K.H.: Modes of action of pyridazinone herbicides. - Weed Sci. *17*: 541-547, 1969. [Ps, Chl.]

3266 - HIND, G.: Effect of desaspidin on photosynthetic phosphorylation and related processes. - Nature *210*: 703-708, 1966.

3267 - HIND, G.: Photo-oxidation of desaspidin sensitized by chlorophyll. - Plant Physiol. *41*: 1237-1239, 1966.

3268 - HIND, G.: Light-induced changes in cytochrome *b*-559 in spinach chloroplasts. - Photochem. Photobiol. *7*: 369-375, 1968.

3269 - HIND, G.: The site of action of plastocyanin in chloroplasts treated with detergent. - Biochim. biophys. Acta *153*: 235-240, 1968.

3270 - HIND, G., HEATH, R.L., IZAWA, S.: The role of chloride ion in photosynthesis. - In: METZNER, H. (ed.): Progress in Photosynthesis Research. Vol. II. Pp. 1022-1026. Tübingen 1969.

3271 - HIND, G., NAKATANI, H.Y.: Determination of the concentration and the redox potential of chloroplast cytochrome 559. - Biochim. biophys. Acta *216*: 223-225, 1970.

3272 - HIND, G., NAKATANI, H.Y., IZAWA, S.: Electron transfer in chloroplasts treated with detergent. - Fed. Proc. *26*: 731, 1967.

3273 - HIND, G., NAKATANI, H.Y., IZAWA, S.: The role of Cl^- in photosynthesis. I. The Cl^- requirement of electron transport. - Biochim. biophys. Acta *172*: 277-289, 1969.

3274 - HIND, G., OLSON, J.M.: Light-induced changes in cytochrome b_6 in spinach chloroplasts. - In: Energy Conversion by the Photosynthetic Apparatus. Brookhaven Symp. Biol. *19*: 188-194, 1967.

3275 - HIND, G., OLSON, J.M.: Electron-transport pathways in photosynthesis. - Annu. Rev. Plant Physiol. *19*: 249-282, 1968.

3276 - HINES, G.D., ELLSWORTH, R.K.: Methyl chlorophyllide *a* as a probable intermediate in the chlorophyll *a* pathway. - Plant Physiol. *44*: 1742-1744, 1969.

3277 - HIPP, B.W., COWLEY, W.R., GERARD, C.J., SMITH, B.A.: Influence of solar radiation and date of planting on yield of sweet *Sorghum*. - Crop Sci. *10*: 91-92, 1970. [Ps.]

3278 - HIRASAWA, R., MUKAIBO, T., HASEGAWA, H., ODAN, N., MARUYAMA, T.: Study on the kinetics of fast electrode processes with pulse technique and electron spin resonance methods. - J. phys. Chem. *72*: 2541-2547, 1968.

3279 - HIRAYAMA, K., HIRANO, R.: Influence of high temperature and residual chlorine on marine phytoplankton. - Mar. Biol. *7*: 205-213, 1970. [Ps.]

3280 - HIRAYAMA, O.: Lipids and lipoprotein complex in photosynthetic tissues. II. Pigments and lipids in blue-green algae, *Anacystis nidulans*. - J. Biochem. (Tokyo) *61*: 179-185, 1967.

3281 - HIRAYAMA, O.: Lipids and lipoprotein complex in photosynthetic tissues. III. Characterization of lamellar fractions isolated from a blue-green alga, *Anacystis nidulans*. - Agr. biol. Chem. *31*: 947-952, 1967.

3282 - HIRAYAMA, O.: Lipids and lipoprotein complex in photosynthetic tissues. Part IV. Lipids and pigments of photosynthetic bacteria. - Agr. biol. Chem. *32*: 34-41, 1968.

3283 - HIRAYAMA, O., OIDO, H.: [Changes of lipid and pigment compositions in spinach leaves during their storage.] - J. agr. chem. Soc. Jap. *43*: 423-428, 1969. [In Jap., ab: E.]

3284 - HIRAYAMA, O., SUZUKI, T.: Lipids and lipoprotein complex in photosynthetic tissues. Part V. A comparison of lipid and pigment patterns in light-grown and dark-grown plant leaves. - Agr. biol. Chem. *32*: 549-554, 1968.

3285 - HIROI, T., KOYAMA, H.: Photosynthetic characteristics of various greenhouse plants. - In: Photosynthesis and Utilization of Solar Energy. Level III Experiments. Pp. 91-92. Jap. nat. Subcomm. for PP (JPP), Tokyo 1968.

3286 - HIROI, T., MONSI, M.: Dry-matter economy of *Helianthus annuus* communities grown at varying densities and light intensities. - J. Fac. Sci., Univ. Tokyo, Sec. III *9*: 241-285, 1966.

3287 - HIROMI, K., ONO, S., ITOH, S., NAGAMURA, T.: A versatile stopped-flow apparatus for the measurement of changes in absorption, fluorescence and optical rotation. - J. Biochem. (Tokyo) *64*: 897-900, 1968.

3288 - HIROSE, H., KUMANO, S.: Spectroscopic studies on the phycoerythrins from Rhodophycean algae with special reference to their phylogenetical relations. - Bot. Mag. (Tokyo) *79*: 105-113, 1966.

3289 - HIROSE, H., KUMANO, S., MADONO, K.: Spectroscopic studies on phycoerythrins from Cyanophycean and Rhodophycean algae with special reference to their phylogenetical relations. - Bot. Mag. (Tokyo) *82*: 197-203, 1969.

3290 - HIRSCHAUER, M., PRIOUL, J.-L.: Essai de restitution du relief sur des clichés de répliques obtenues par cryodécapage. - Compt. rend. Acad. Sci. Paris, Sér. D *268*: 3064-3067, 1969. [Chloroplast.]

3291 - HISATAKE, M., SATAKE, H., SHITOMI, H.: [Chlorophyll changes in brined cucumber.] - J. Food Sci. Technol. (Tokyo) [Nippon Shokuhin Kogyo Gakkai-Shi] *17*: 187-192, 1970. [In Jap., ab: E.]

3292 - HIYAMA, T., NISHIMURA, M., CHANCE, B.: Interaction of photosynthetic and respiratory electron transfer chains in *Chlamydomonas*. - Plant Physiol. *43*: S 28, 1968.

3293 - HIYAMA, T., NISHIMURA, M., CHANCE, B.: Determination of carotenes by thin-layer chromatography. - Anal. Biochem. *29*: 339-342, 1969.

3294 - HIYAMA, T., NISHIMURA, M., CHANCE, B.: Energy and electron transfer system of *Chlamydomonas reinhardi*. I. Photosynthetic and respiratory cytochrome systems of the pale green mutant. - Plant Physiol. *44*: 527-534, 1969.

3295 - HIYAMA, T., NISHIMURA, M., CHANCE, B.: Energy and electron transfer systems of
Chlamydomonas reinhardi. II. Two cyclic pathways of photosynthetic electron transfer in the pale green mutant. - Plant Physiol. *46*: 163-168, 1970.

3296 - HO, L.C., PEEL, A.J.: Transport of ^{14}C-labelled assimilates and ^{32}P-labelled
phosphate in *Salix viminalis* in relation to phyllotaxis and leaf age. - Ann. Bot.
33: 743-751, 1969. [Ps.]

3297 - HOARAU, J.: La photoréduction du flavine mononucléotide et l'activité de la nitrate réductase foliaire. - Photosynthetica *2*: 135-148, 1968.

3298 - HOARE, D.S., HOARE, S.L., SMITH, A.J.: Assimilation of organic compounds by
blue-green algae and photosynthetic bacteria. - In: METZNER, H. (ed.): Progress
in Photosynthesis Research. Vol. III. Pp. 1570-1573. Tübingen 1969.

3299 - HOBBS, A.P.: Gas analysis. - Anal. Chem. *34*: 91R-98R, 1966. [Also O_2, CO_2.]

3300 - HOCH, G., KNOX, R.S.: Primary processes in photosynthesis. - In: GIESE, A.C.
(ed.): Photophysiology. Vol. III. Pp. 225-251. Academic Press, New York-London
1968.

3301 - HOCHAPFEL, A., BERCHET, D., VIOVY, R.: Modifications de la chlorophylle *a* en
présence de polypeptides et d'amines. - Biochim. biophys. Acta *222*: 180-190,
1970.

3302 - HOCHAPFEL, A., JOURNEAUX, R., VIOVY, R.: Inclusions de pigments dans des cris-
taux: modifications du spectre d'absorption. Influence de la température. -
Compt. rend. Acad. Sci. (Paris), Sér. C *264*: 1792-1795, 1967. [Chl.]

3303 - HOCHAPFEL, A., JOURNEAUX, R., VIOVY, R.: Existence et nature des formes agrégées
de la chlorophylle *a* dans les solvants binaires. I. Etude des spectres d'absorption. - J. Chim. phys. Phys.-chim. biol. *66*: 1467-1473, 1969.

3304 - HODÁŇOVÁ, D.: Photosynthetic rate and chlorophyll content as related to leaf age
and ontogenesis of sugar-beet plants. - In: Productivity of Terrestrial Ecosystems. Production Processes. Czechosl. nat. Comm. IBP, PT-PP Report No. 1. Pp.
175-177. Praha 1970.

3305 - HODÁŇOVÁ, D.: Canopy structure and photosynthetic rate as related to different
leaf age. - In: Productivity of Terrestrial Ecosystems. Production Processes.
Czechoslov. nat. Comm. IBP, PT-PP Report No. 1. Pp. 179-180. Praha 1970.

3306 - HODGES, J.D.: Patterns of photosynthesis under natural environmental conditions.
- Ecology *48*: 234-242, 1967.

3307 - HODGES, J.D., SCOTT, D.R.M.: Photosynthesis in seedlings of six conifer species
under natural environmental conditions. - Ecology *49*: 973-981, 1968.

3308 - HODGKINSON, K.C., VEALE, J.A.: The distribution of photosynthate within lucerne
as influenced by illumination. - Aust. J. biol. Sci. *19*: 15-21, 1966.

3309 - HODGSON, G.L.: Physiological and ecological studies in the analysis of plant environment. XIII. A comparison of the effects of seasonal variations in light
energy and temperature on the growth of *Helianthus annuus* and *Vicia faba* in the
vegetative phase. - Ann. Bot. *31*: 291-308, 1967.

3310 - HOFFMAN, F.M., MILLER, J.H.: An endogenous rhythm in the Hill-reaction activity
of tomato chloroplasts. - Amer. J. Bot. *53*:543-548, 1966.

3311 - HOFFMAN, G.J., PHENE, C.J., RAWLINS, S.L.: Microchamber for studying plant response to environmental factors. - Trans. A.S.A.E. *12*: 598-601, 1969. [Ps.]

3312 - HOFFMANN, P.: Der Chloroplast - eine biochemische und funktionelle Einheit. -
Stud. biophys. *5*: 149-156, 1967.

3313 - HOFFMANN, P.: Zur Physiologie der Gibberellinsäurewirkung bei Weizenkeimpflanzen.
- Wiss. Z. Univ. Rostock, math-nat. Reihe *16*: 609-610, 1967. [Chl.]

3314 - HOFFMANN, P.: Zur Physiologie der Photosynthese bei höheren Pflanzen. - Bot.
Stud. (Jena) *18*: 1-151, 1968.

3315 - HOFFMANN, P.: Gegenwartsprobleme und Aufgaben der Photosyntheseforschung. -
Biol. Schule *11*: 456-463, 1968.

3316 - HOFFMANN, P.: Pigmentgehalt und Gaswechsel von *Myrothamnus*-Blättern nach Austrocknung und Wiederaufsättigung. - Photosynthetica *2*: 245-252, 1968.

3317 - HOFFMANN, P.: Kinetinbedingte Verzögerung des lichtinduzierten Chlorophyllabbaus bei der Grünalge *Monostroma balticum* (ARESCH.). - Biol. Rundschau *7*: 72-73, 1969.

3318 - HOFFMANN, P.: Pigmentphysiologische Untersuchungen an isolierten Kinetin-behandelten Weizenprimärblättern. - Wiss. Z. Ernst-Moritz-Univ. Greifswald, math.-nat. Reihe *19*: 55-68, 1970.

3319 - HOFFMANN, P.: Versuche zur künstlichen Photosynthese. - Wiss. Fortschr. *20*: 41, 1970.

3320 - HOFFMANN, P., ALBRECHT, E.: Vergleichende Untersuchungen über Ascorbinsäuregehalt, Chlorophyllkonzentration und Gaswechselintensität bei höheren Pflanzen. - Z. Pflanzenphysiol. *55*: 292-295, 1966.

3321 - HOFFMANN, P., IRMLER, R.: Vitamin K_1-Gehalt und Photosynthese bei höheren Pflanzen. - Ber. deut. bot. Ges. *79*: 71-86, 1966.

3322 - HOFFMANN, P., KARSTEN, U.: Zur Frage der Chlorophyllbildung in pflanzlichen Tumoren.- Planta *69*: 96-100, 1966.

3323 - HOFFMANN, P., RATHSACK, R.: Zur Cytokinin-Wirkung bei Blaualgen. - Biochem. Physiol. Pflanzen *161*: 95-96, 1970. [Chl.]

3324 - HOFFMANN, P., SACHERT, H.: Der Einfluss von Harnstoff auf die Entwicklung von *Salicornia brachystachya* G.F.W. MEYER - ein Beitrag zum Halophytenproblem. - Ber. deut. bot. Ges. *80*: 437-446, 1967. [Chl.]

3325 - HOFFMANN, P., TICHÁ, I.: Der Gaswechsel von *Phaseolus vulgaris*- und *Pisum sativum*-Keimpflanzen nach der Entfernung des Wurzelsystems. - Photosynthetica *3*: 73-78, 1969.

3326 - HOFFMANN, P., WALTER, G.: Der Einfluss von Chloramphenicol und Streptomycin auf die Entwicklung des photosynthetischen Apparates bei Weizenkeimpflanzen. - Biol. Zentralbl. *89*: 163-200, 1970.

3327 - HOFFMANN, P., WERNER, D.: Zur Spektrophotometrischen Chlorophyllbestimmung unter besonderer Berücksichtigung verschiedener Gerätetypen. - Jenaer Rundschau *11*: 300-303, 1966 [G.], Jena Rev. *11*: 300-303, 1966 [E.], lenskoe Obozrenie *11*: 260-263, 1966 [R.]

3328 - HOFSTRA, G.: Effects of extending the day length with low-intensity light on the growth of wheat and cocksfoot. - Aust. J. biol. Sci. *22*: 333-341, 1969. [Ps.]

3329 - HOFSTRA, G., HESKETH, J.D.: Effects of temperature on the gas exchange of leaves in the light and dark. - Planta *85*: 228-237, 1969.

3330 - HOFSTRA, G., HESKETH, J.D.: The effect of temperature on stomatal aperture in different species. - Can. J. Bot. *47*: 1307-1310, 1969.

3331 - HOFSTRA, G., NELSON, C.D.: A comparative study of translocation of assimilated ^{14}C from leaves of different species. - Planta *88*: 103-112, 1969.

3332 - HOFSTRA, G., NELSON, C.D.: The translocation of photosynthetically assimilated ^{14}C in corn. - Can. J. Bot. *47*: 1435-1442, 1969.

3333 - HOGETSU, D., MIYACHI, S.: Effect of oxygen on the light-enhanced dark carbon dioxide fixation in *Chlorella* cells. - Plant Physiol. *45*: 178-182, 1970.

3334 - HOLDEN, M.: Chlorophyll bleaching systems in leaves. - Phytochemistry *9*: 1771-1777, 1970.

3335 - HOLLIES, M.A.: Effect of shade on the structure and chlorophyll content of Arabica coffee leaves. - Exp. Agr. *3*: 183-190, 1967.

3336 - HOLM, O., BOOTH, C.R.: The measurement of adenosine triphosphate in the ocean and its ecological significance. - Limnol. Oceanogr. *11*: 510-519, 1966. [ATP as measure of biomass.]

3337 - HOLMES, R.W.: Description and evaluation of methods for determining incident solar radiation, submarine daylight, chlorophyll *a*, and primary production. Used by Scripps Tuna oceanography research program in the Eastern Tropical Pacific. -

U.S. Fish Wildl. Serv. Bur. Comm. Fish. Spec. Sci. Rep. Fish. *564*: 1-31, 1968.

3338 - HOLMES, R.W., WILLIAMS, P.M., EPPLEY, R.W.: Red water in La Jolla Bay, 1964-1966. - Limnol. Oceanogr. *12*: 503-512, 1967. [Ps, Chl.]

3339 - HOLMGREN, P.: A device to procure air mixtures with accurate carbon dioxide concentrations. - Lantbrukshögsk. Ann. *34*: 219-224, 1968. [For Ps measurement.]

3340 - HOLMGREN, P.: Comparative studies on photosynthesis in plants native to habitats differing in level of irradiance. - Acta Univ. Upsaliensis, Abstr. Uppsala Diss. Sci. *117*: 1-13, 1968.

3341 - HOLMGREN, P.: Leaf factors affecting light-saturated photosynthesis in ecotypes of *Solidago virgaurea* from exposed and shaded habitats. - Physiol. Plant. *21*: 676-698, 1968.

3342 - HOLMGREN, P., JARVIS, P.G.: Carbon dioxide efflux from leaves in light and darkness. - Physiol. Plant. *20*: 1045-1051, 1967.

3343 - HOLM-HANSEN, P.: Ecology, physiology, and biochemistry of blue-green algae. - Annu. Rev. Microbiol. *22*: 47-70, 1968. [Ps.]

3344 - HOLM-HANSEN, O.: Determination of microbial biomass in ocean profiles. - Limnol. Oceanogr. *14*: 740-747, 1969. [Chl.]

3345 - HOLOWINSKY, A.W., SCHIFF, J.A.: Events surrounding the early development of *Euglena* chloroplasts. I. Induction of preillumination. - Plant Physiol. *45*: 339-347, 1970.

3346 - HOLT, A.S.: Recently characterized chlorophylls. - In: VERNON, L.P., SEELY, G.R. (ed.): The Chlorophylls. Pp. 111-118. Academic Press, New York-London 1966.

3347 - HOLT, A.S., PURDIE, J.W., WASLEY, J.W.F.: Structures of *Chlorobium* chlorophylls (660). - Can. J. Chem. *44*: 88-93, 1966.

3348 - von HOLT, C., von HOLT, M.: Transfer of photosynthetic products from *Zooxanthellae* to coelenterate hosts. - Compar. Biochem. Physiol. *24*: 73-81, 1968.

3349 - HOLT, S.C., CONTI, S.F., FULLER, R.C.: Effect of light intensity on the formation of the photochemical apparatus in the green bacterium *Chloropseudomonas ethylicum*. - J. Bacteriol. *91*: 349-355, 1966.

3350 - HOLT, S.C., CONTI, S.F., FULLER, R.C.: Photosynthetic apparatus in the green bacterium *Chloropseudomonas ethylicum*. - J. Bacteriol. *91*: 311-323, 1966.

3351 - HOLT, S.C., STERN, A.I.: The effect of 3-(3,4-dichlorophenyl)-1,1-dimethylurea on chloroplast development and maintenance of *Euglena gracilis*. I. Ultrastructural characterization of light-grown cells by the techniques of thin sectioning of freeze-etching. - Plant Physiol. *45*: 475-483, 1970.

3352 - HOLT, S.C., TRÜPER, H.G., TAKÁCS, B.J.: Fine structure of *Ectothiorhodospira mobilis* strain 8113 thylakoids: Chemical fixation and freeze-etching studies. - Arch. Mikrobiol. *62*: 111-128, 1968.

3353 - HOLTON, R.W., MYERS, J.: Water-soluble cytochromes from a blue-green alga. I. Extraction, purification, and spectral properties of cytochromes c (549, 552, and 554, *Anacystis nidulans*). - Biochim. biophys. Acta *131*: 362-374, 1967.

3354 - HOLTON, R.W., MYERS, J.: Water-soluble cytochromes from a blue-green alga. II. Physicochemical properties and quantitative relationships of cytochromes c (549, 552, and 554 *Anacystis nidulans*). - Biochim. biophys. Acta *131*: 375-384, 1967.

3355 - HOLUBOWICZ, T., BOE, A.A.: Development of cold hardiness in apple seedlings treated with gibberellic acid and abscisic acid. - J. amer. Soc. hort. Sci. *94*: 661-664, 1969. [Ps.]

3356 - HOMANN, P.H.: Studies on the manganese of the chloroplast. - Plant Physiol. *42*: 997-1007, 1967.

3357 - HOMANN, P.H.: The activity of chloroplasts in relation to their structure and manganese content. - Plant Physiol. *42*: S 35, 1967.

3358 - HOMANN, P.H.: Effects of manganese on the fluorescence of chloroplasts. - Biochem. biophys. Res. Commun. *33*: 229-234, 1968.

3359 - HOMANN, P.H.: Effect of manganese and vanadium deficiency on the photosynthetic system of green plants. - Plant Physiol. *43*: S 13, 1968.

3360 - HOMANN, P.H.: Fluorescence properties of chloroplasts from manganese deficient and mutant tobacco. - Biochim. biophys. Acta *162*: 545-554, 1968.

3361 - HOMANN, P.H.: Cation effects on the fluorescence of isolated chloroplasts. - Plant Physiol. *44*: 932-936, 1969.

3362 - HOMANN, P.H.: Fluorescence studies on tobacco chloroplasts deficient in photosystem II. - In: METZNER, H. (ed.): Progress in Photosynthesis Research. Vol. II. Pp. 932-937. Tübingen 1969.

3363 - HOMANN, P.H., SCHMID, G.H.: Photosynthetic reactions of chloroplasts with unusual structures. - Plant Physiol. *42*: 1619-1632, 1967.

3364 - HOMANN, P.H., SCHMID, G.H., GAFFRON, H.: Structure and photochemistry in tobacco chloroplasts. - In: SHIBATA, K., TAKAMIYA, A., JAGENDORF, A.T., FULLER, R.C. (ed.): Comparative Biochemistry and Biophysics of Photosynthesis. Pp. 50-56. Univ. Tokyo Press, Tokyo; Univ. Park Press, State College, Pa. 1968.

3365 - HOMMA, S.: [Phytotoxicity of 3-amino-1,2,4-triazole for mulberry plants.] - J. Sericult. Sci. Jap. *38*: 123-130, 1969. [Ps, Chl; in Jap., ab: E.]

3366 - HONDA, S.I., HONGLADAROM, T., LATIES, G.G.: A new isolation medium for plant organelles. - J. exp. Bot. *17*: 460-472, 1966. [Chloroplasts.]

3367 - HONEYCUTT, R.C., KROGMANN, D.W.: A light-dependent oxygen-reducing system from *Anabaena variabilis*. - Biochim. biophys. Acta *197*: 267-275, 1970.

3368 - HONGLADAROM, T., HONDA, S.I.: Reversible swelling and contraction of isolated spinach chloroplasts. - Plant Physiol. *41*: 1686-1694, 1966.

3369 - HONGLADAROM, T., HONDA, S.I., WILDMAN, S.G.: Swelling of spinach chloroplasts induced by HCHO and $KMnO_4$, an effect partially hidden by the wide size range of chloroplasts. - Plant Cell Physiol. *9*: 159-168, 1968.

3370 - HONGLADAROM-HONDA, T., HONDA, S.I.: Size distributions of chloroplasts in living mesophyll cells. - In: XIth International Bot. Congress, Abstracts of the Papers. P. 94. Washington 1969.

3371 - HONJO, T., HANAOKA, T.: [Diurnal fluctuations of photosynthetic rate and pigment contents in marine phytoplankton.] - J. oceanogr. Soc. Jap. *25*: 182-190, 1969. [In Jap., ab: E.]

3372 - HOOBER, J.K., SIEKEVITZ, P., PALADE, G.E.: Formation of chloroplast membranes in *Chlamydomonas reinhardi* y-1. Effects of inhibitors of protein synthesis. - J. biol. Chem. *244*: 2621-2631, 1969.

3373 - HOOD, W., CARR, N.G.: Association of NAD and NADP linked glyceraldehyde-3-phosphate dehydrogenase in the blue-green alga, *Anabaena variabilis*. - Planta *86*: 250-258, 1969.

3374 - HOOL, G.: Wirkung von Antibiotika auf Wachstum und Ionenaufnahme bei *Zea mays* L. - Ber. schweiz. bot. Ges. *77*: 210-256, 1967. [Chl.]

3375 - HORANIC, G.E., GARDNER, F.E.: An improved method of making epidermal imprints. - Bot. Gaz. *128*: 144-150, 1967. [Also stomata.]

3376 - HORIO, T., BARTSCH, R.G., KAKUNO, T., KAMEN, M.D.: Two reduced nicotinamide adenine dinucletide dehydrogenases from the photosynthetic bacterium, *Rhodospirillum rubrum*. - J. biol. Chem. *244*: 5899-5909, 1969.

3377 - HORIO, T., KAMEN, M.D.: Bacterial cytochromes: II. Functional aspects. - Annu. Rev. Microbiol. *24*: 399-428, 1970.

3378 - HORIO, T., NISHIKAWA, K., HORIUTI, Y., KAKUNO, T.: Mode of coupling of the phosphorylation system to the electron transfer system in *Rhodospirillum rubrum* chromatophores. - In: SHIBATA, K., TAKAMIYA, A., JAGENDORF, A.T., FULLER, R.C. (ed.): Comparative Biochemistry and Biophysics of Photosynthesis. Pp. 408-424. Univ. Tokyo Press, Tokyo; Univ. Park Press, State College, Pa. 1968.

3379 - HORIO, T., NISHIKAWA, K., OKAYAMA, S., HORIUTI, Y., YAMAMOTO, N., KAKUTANI, Y.: The requirement of ubiquinone-10 for an ATP-forming system and an ATPase system of chromatophores from *Rhodospirillum rubrum*. - Biochim. biophys. Acta *153*: 913-916, 1968.

3380 - HORIO, T., NISHIKAWA, K., YAMASHITA, J.: Synthesis and possible character of a high-energy intermediate in bacterial photophosphorylation. - Biochem. J. *98*: 321-329, 1966.

3381 - HORIO, T., von STEDINGK, L.-V., BALTSCHEFFSKY, H.: Photophosphorylation in presence and absence of added adenosine diphosphate in chromatophores from *Rhodospirillum rubrum*. - Acta chem. scand. *20*: 1-10, 1966.

3382 - HORIUTI, Y., NISHIKAWA, K., HORIO, T.: Oxidation-reduction potential-dependent adenosine triphosphatase activity of chromatophores from *Rhodospirillum rubrum*. - J. Biochem. (Tokyo) *64*: 577-587, 1968.

3383 - HORTON, A.A., HALL, D.O.: 2-amino-1,1,3-tricyanopropene: a new inhibitor of oxygen evolution in photosynthesis. - Biochim. biophys. Acta *131*: 201-203, 1967.

3384 - HORTON, A.A., HALL, D.O.: Determining the stoichiometry of photosynthetic phosphorylation. - Nature *218*: 386-388, 1968.

3385 - HORTON, A.A., PACKER, L.: Effect of tetraphenylboron on light-induced uptake of monovalent cations by chloroplasts. - Arch. Biochem. Biophys. *128*: 820-823, 1968.

3386 - HORVÁTH, I.: The effect of the spectral composition of light on the quantity of the photosynthetic pigments and on the proportion of components. - Acta biol. (Szeged) *12*: 25-34, 1966.

3387 - HORVÁTH, I.: The effect of the spectral composition of light on the light absorption of the photosynthetic pigment complex, on the quantity of pigments and on the proportion of components. - Stud. biophys. *5*: 67-70, 1967.

3388 - HORVÁTH, I., SZÁSZ, K.: Effect of spectral composition of light on the accumulation of carbohydrates and nitrogen compounds. A probable interpretation. - In: METZNER, H. (ed.): Progress in Photosynthesis Research. Vol. III. Pp. 1675-1677. Tübingen 1969.

3389 - HORVÁTH, I., SZÁSZ, K., GARAY, A.: A fotoszintézis és a növényi szervesanyagprodukció. [Photosynthesis and organic matter production.] - Bot. Közlem. *56*: 71-75, 1969. [In Hung., ab: E.]

3390 - HORVÁTH, L., KISS, B., POZSÁR, B.: Ammóniumnitrát- és karbamid-műtrágyázás hatása a Bezosztája 1. búza-csíranövények fotoszintetikus széndioxid-fixálásának serkentésére. [Stimulating effect of ammonium nitrate and urea fertilization on the photosynthetic carbon dioxide fixation by the wheat variety Bezostaya 1.] - Bot. Közlem. *56*: 237-239, 1969. [Ps, Chl; in Hung., ab: E.]

3391 - HORVÁTH, L., POZSÁR, B.I.: Cation-dependent effect of chloride on the photosynthetic carbon dioxide fixation by bean leaves. - Acta agron. Acad. Sci. hung. *19*: 331-332, 1970.

3392 - HORVÁTH, L., POZSÁR, B.: Az Agronit hatása a fotoszintetikus széndioxid-fixálás intenzitásának serkentésére, bablevelekben. [Stimulating effect of Agronit on the rate of carbon dioxide fixation in bean leaves.] - Agrobotanika *12*: 127-129, 1970 (1971). [In Hung., ab: E.]

3393 - HORVÁTH, M., LÁSZTITY, D.: Pigment changes in etiolated barley leaves. - Acta agron. Acad. Sci. hung. *15*: 119-125, 1966.

3394 - HORVÁTH, M., LÁSZTITY, D.: Effect of kinetin on the pigment content of barley leaves. - Acta agron. Acad. Sci. hung. *16*: 393-396, 1967.

3395 - HORVÁTH, M., LONTAI, I.: Investigation on physiological changes in roots and shoots as a result of a herbicide treatment. (II). Development of the pigment content of shoots. - Acta biol. (Szeged) *16*: 95-98, 1970.

3396 - HORVÁTHné MÉSZÁROS, M., D.-POSCH, M.: A hónapos retek (*Raphanus sativus* L. convar. *radicula* (PERS.) DC.) pigmenttartalmának változása. [Changes in pigment content of radish (*Raphanus sativus* L. convar. *radicula* (PERS.) DC.)]- Bot. Közlem. *56*: 117-120, 1969. [In Hung., ab: G.]

3397 - HOSEMANN, R., KREUTZ, W.: On the tertiary structure of the protein layers of chloroplasts. - Naturwissenschaften *53*: 298-304, 1966.

3398 - HOSHIAI, T.: [Ecological observations of the colored layer of the sea ice at Syowa station.] - Antarctic Res. *34*: 60-72, 1969. [Chl; in Jap.]

3399 - HOSHINA, S., NISHIDA, K.: Electron microscopic observations of the balloon-formation of isolated spinach chloroplasts. - Experientia *26*: 1275-1276, 1970.

3400 - HOSHINO, M., NISHIMURA, S., OKUBO, T.: Studies on the assimilation and translocation of $^{14}CO_2$ in Ladino clover. III. Uptake and distribution of ^{14}C by the plants in the different stages of regrowth. - Proc. Crop Sci. Soc. Jap. *36*: 269-274, 1967.

3401 - HOSHINO, T., MATSUSHIMA, S., TOMITA, T., KIKUCHI, T.: [Analysis of yield-determining process and its application to yield-prediction and culture improvement of lowland rice. LXXXVIII. Combined effects of air-temperature and water-temperature in seedling periods on the characteristics of seedlings of rice plants (In case of seedlings treated for different number of days during which the seedlings attained an identical leaf-age).] - Proc. Crop Sci. Soc. Jap. *38*: 273-278, 1969. [Net assimilation rate; in Jap., ab: E.]

3402 - HOTTA, R., HARAGUCHI, N., SHIMIZU, S.: Metabolism of chlorophyll in higher plants. II. Action of proteolytic enzymes on a chlorophyll-protein complex. - Bot. Mag. (Tokyo) *81*: 347-355, 1968.

3403 - HOTTA, R., SHIMIZU, S., TAMAKI, E.: Photosynthetic activities in tobacco plants. I. Cytochrome f from tobacco leaves. - Bot. Mag. (Tokyo) *80*: 23-26, 1967.

3404 - HOUGH, L., STACEY, B.E.: Variation in the allitol content of *Itea* plants during photosynthesis. - Phytochemistry *5*: 171-175, 1966.

3405 - HOUSSIER, C., SAUER, K.: Optical properties of the protochlorophyll pigments. I. Isolation, characterization, and infrared spectra. - Biochim. biophys. Acta *172*: 476-491, 1969.

3406 - HOUSSIER, C., SAUER, K.: Optical properties of the protochlorophyll pigments. II. Electronic absorption, fluorescence, and circular dichroism spectra. - Biochim. biophys. Acta *172*: 492-502, 1969.

3407 - HOUSSIER, C., SAUER, K.: Circular dichroism and magnetic circular dichroism of the chlorophyll and protochlorophyll pigments. - J. amer. chem. Soc. *92*: 779-791, 1970.

3408 - HOUTMAN, T.J.: Repeat measurements of temperature, salinity, and ^{14}C depletion at an ocean station. - New Zealand J. Sci. *9*: 457-471, 1966.

3409 - HOWARD, A., HAIGH, M.V.: Chloroplast aberrations in irradiated fern spores. - Mutat. Res. *6*: 263-280, 1968.

3410 - HOWARD, J.A.: Spectral energy relations of isobilateral leaves. - Aust. J. biol. Sci. *19*: 757-766, 1966. [Chl.]

3411 - HOWARD, P.J.A.: A method for the estimation of carbon dioxide evolved from the surface of soil in the field. - Oikos *17*: 267-271, 1966.

3412 - HOWELL, J.A., FREDRICKSON, A.G., TSUCHIYA, H.M.: Optimal and dynamic characteristics of a continuous photosynthetic algal gas exchanger. - Chem. Eng. Progr. Symp. Ser. *62*: 56-68, 1966.

3413 - HOWELL, J.A., TSUCHIYA, H.M., FREDRICKSON, A.G.: Continuous synchronous culture of photosynthetic microorganisms. - Nature *214*: 582-584, 1967.

3414 - HOWELL, S.H., MOUDRIANAKIS, E.N.: Function of the "quantasome" in the photosynthesis: structure and properties of membrane-bound particle active in the dark reactions of photophosphorylation. - Proc. nat. Acad. Sci. U.S.A. *58*: 1261-1268, 1967.

3415 - HOWELL, S.H., MOUDRIANAKIS, E.N.: Hill reaction site in chloroplast membranes: Non-participation of the quantasome particle in photoreduction. - J. mol. Biol. *27*: 323-333, 1967.

3416 - HOWES, C.D., BATRA, P.P.: Mechanism of photoinduced carotenoid synthesis. Further studies on the action spectrum and other aspects of carotenogenesis. - Arch. Biochem. Biophys. *137*: 175-180, 1970.

3417 - HOWES, C.D., STERN, A.I.: Photophosphorylation during chloroplast development in red kidney bean. I. Characterization of the mature system and the effect of BSA and sulfhydryl reagents. - Plant Physiol. 44: 1515-1522, 1969.

3418 - HOWMILLER, R., WEINER, A.: A limnological study of a mangrove lagoon in the Galapagos. - Ecology 49: 1184-1186, 1968. [Chl.]

3419 - HOY, H.: Chlorophyll in the sea off South Africa. - S. Afr. Div. Sea Fish. Fish. Bull. 6: 1-9, 1970.

3420 - HOYT, P.B.: Chlorophyll-type compounds in soil. I. Their origin. - Plant Soil 25: 167-180, 1966.

3421 - HOYT, P.B.: Chlorophyll-type compounds in soil. II. Their decomposition. - Plant Soil 25: 313-328, 1966.

3422 - HOYT, P.B.: Chlorophyll-type compounds in soil. III. Their significance in arable soils. - Plant Soil 26: 5-13, 1967.

3423 - HOZUMI, K., KIRITA, H.: Estimation of the rate of total photosynthesis in forest canopies. - Bot. Mag. (Tokyo) 83: 144-151, 1970.

3424 - HOZUMI, K., YODA, K., KIRA, T.: Production ecology of tropical rain forests in southwestern Cambodia. II. Photosynthetic production in an evergreen seasonal forest. - In: KIRA, T., IWATA, K. (ed.): Nature and Life in Southeastern Asia. Vol. VI. Pp. 57-81. Jap. Soc. Promotion Sci., Tokyo 1969.

3425 - HOZUMI, K., YODA, K., KOKAWA, S., KIRA, T.: Production ecology of tropical rain forests in southwestern Cambodia. I. Plant biomass. - In: KIRA, T., IWATA, K. (ed.): Nature and life in Southeastern Asia. Vol. VI. Pp. 1-51. Jap. Soc. Promotion Sci., Tokyo 1969.

3426 - HOZYO, Y., KOBAYASHI, H.: Tracer studies on the behaviour of photosynthetic products during the grain ripening stage of six-rowed barley plant (Hordeum sativum, JESSEN). - Nogyo Gijutsu Kenkyusho Hokoku [Bull. nat. Inst. agr. Sci. (Jap.)] Ser. D 20: 35-77, 1969.

3427 - HUANG, J.-C.: Effects of toxic organics on photosynthetic reoxygenation. - Diss. Abstr. 28: 4156 B, 1968.

3428 - HUANG, J.-C., GLOYNA, E.F.: Effect of organic compounds on photosynthetic oxygenation. I. Chlorophyll destruction and suppression of photosynthetic oxygen production. - Water Res. 2: 347-366, 1968.

3429 - HUANG, J.-C., GLOYNA, E.F.: Effect of organic compounds on photosynthetic oxygenation. II. Design modification for waste stabilization ponds. - Water Res. 2: 459-469, 1968.

3430 - HUBER, W., de FEKETE, M.A.R., ZIEGLER, H.: Enzyme des Stärkeumsatzes in Bündelscheiden- und Palisadenchloroplasten von Zea mays L. - Planta 87: 360-364, 1969.

3431 - HUBÍK, E.: The influence of water and soda-water sprinkling and increased CO_2 concentration in the canopy on photosynthetic production of wheat varieties. - In: Productivity of Terrestrial Ecosystems. Production Processes. Czechosl. nat. Comm. IBP, PT-PP Report No. 1. Pp. 215-216. Praha 1970.

3432 - HUDÁK, J.: Der Einfluss von Bor auf die Ultrastruktur der Chloroplasten. - Naturwissenschaften 57: 458, 1970.

3433 - HUDOCK, G.A., BART, C.: Responses of a mutant strain of Chlamydomonas reinhardi to prolonged organotrophic growth. - Plant Physiol. 42: 186-190, 1967. [Chl.]

3434 - HUDOCK, G.A., BART, C.: Isolation of a strain of Chlamydomonas reinhardi with altered control of glyceraldehyde 3-phosphate dehydrogenase. - J. Protozool. 16: 597-598, 1969.

3435 - HUDOCK, G.A., KIVIC, P.A., BART, C.: Synthetic requirement of chloroplast development in Chlamydomonas reinhardi. - J. Protozool. 15: 678-679, 1968.

3436 - HUFFAKER, R.C., COX, E.L., KLEINKOPF, G.E., STANFORD, E.H.: Regulation of synthesis of chlorophyll, carotene, ribulose-1,5-diP carboxylase and phosphoribulokinase in a temperature-sensitive chlorophyll mutant of Medicago sativa. - Physiol. Plant. 23: 404-411, 1970.

3437 - HUFFAKER, R.C., OBENDORF, R.L., KELLER, C.J., KLEINKOPF, G.E.: Effect of light
intensity on photosynthetic carboxylative phase enzymes and chlorophyll synthesis in greening leaves of *Hordeum vulgare* L. - Plant Physiol. *41*: 913-918, 1966.

3438 - HUFFAKER, R.C., RADIN, T., KLEINKOPF, G.E., COX, E.L.: Effect of mild water
stress on enzymes of nitrate assimilation and of the carboxylative phase of
photosynthesis in barley. - Crop Sci. *10*: 471-474, 1970.

3439 - HUGHES, A.P.: The importance of light compared with other factors affecting plant
growth. - In: BAINBRIDGE, R., EVANS, G.C., RACKHAM, O. (ed.): Light as an Ecological Factor. Pp. 121-146. Blackwell sci. Publ., Oxford 1966. [Ps.]

3440 - HUGHES, E.E., DORKO, W.D.: Direct mass spectrometric determination of atmospheric carbon dioxide. - Anal. Chem. *40*: 866-869, 1968.

3441 - HUGHES, M.K., LINCOLN, E.: A simple integrator for use with solarimeters. -
Oikos *20*: 161-165, 1969.

3442 - HÜLSEN, W.: Eine vollautomatische Probenverbrennungsanlage für die Messtechnik
von H-3 und C-14 im Flüssigszintillations-Spektrometer. - Experientia *26*: 1406-
1407, 1970.

3443 - HUMPHREY, G.F.: The concentration of chlorophylls *a* and *c* in the South-West Pacific Ocean. - Aust. J. mar. Freshwater Res. *21*: 1-10, 1970.

3444 - HUMPHREY, G.F., JITTS, H.R.: The measurement of photosynthesis in discrete spectral bands. - In: Proceedings of Conference on Instrumentation for Plant Environment Measurements, Aspendale 1966. Pp. 14-15. Soc. Instrum. Technol., Mel-
bourne 1966.

3445 - HUMPHREY, G.F., KERR, J.D.: Seasonal variations in the Indian ocean along 110° E.
III. Chlorophylls *a* and *c*. - Aust. J. mar. Freshwater Res. *20*: 55-64, 1969.

3446 - HUMPHREY, G.F., SUBBA RAO, D.V.: Photosynthetic rate of the marine diatom *Cylin-
drotheca closterium*. - Aust. J. mar. Freshwater Res. *18*: 123-127, 1967.

3447 - HUMPHREY, G.F., WOOTTON, M.: Comparison of the techniques used in the determination of phytoplankton pigments. - In: Determination of Photosynthetic Pigments
in Sea-Water. Pp. 37-63. UNESCO, Paris 1966.

3448 - HUMPHREYS, L.R., ROBINSON, A.R.: Subtropical grass growth. I. Relationship between carbohydrate accumulation and leaf area in growth. - Queensland J. agr.
anim. Sci. *23*: 211-259, 1966. [Ps.]

3449 - HUMPHRIES, E.C.: Internal control of rate of leaf production in sugar beet. -
Physiol. Plant. *19*: 827-829, 1966.

3450 - HUMPHRIES, E.C.: The dependence of photosynthesis on carbohydrate sinks: current
concepts. - Proc. 1st int. Symp. trop. Root Crops. Univ. West Indies, II: 34-45.
St. Augustin, Trinidad 1967.

3451 - HUMPHRIES, E.C.: The effect of growth regulators, CCC and B9, on protein and
total nitrogen of bean leaves *(Phaseolus vulgaris)* during development. - Ann.
Bot. *32*: 497-507, 1968. [Ps, Chl.]

3452 - HUMPHRIES, E.C., BOND, W.: Experiments with CCC on wheat: effects of spacing,
nitrogen and irrigation. - Ann. appl. Biol. *64*: 375-384, 1969. [Ps, Chl.]

3453 - HUMPHRIES, E.C., DYSON, P.W.: Effect of growth inhibitor, N-dimethylaminosucci-
namic acid (B9), on potato plants in the field. - Europe. Potato J. *10*: 116-126,
1967. [Growth analysis.]

3454 - HUMPHRIES, E.C., DYSON, P.W.: Effects of growth regulators, CCC and B9, on some
potato varieties. - Ann. appl. Biol. *60*: 333-341, 1967. [Ps, Chl.]

3455 - HUMPHRIES, E.C., FRENCH, S.A.W.: Photosynthesis in sugar beet depends on root
growth. - Planta *88*: 87-90, 1969.

3456 - HUMPHRIES, E.C., FRENCH, S.A.W.: Effect of seedling treatment on growth and
yield of sugar beet in the field. - Ann. appl. Biol. *64*: 385-393, 1969. [Ps.]

3457 - HUNT, L.A.: Use of the Cionco model to obtain further information on the nature
of leaf boundary layers. - Can. J. Bot. *46*: 177-178, 1968. [Ps.]

3458 - HUNT, L.A., CHRISTIE, B.R.: Determination of stomatal numbers, stomatal lengths, and dry weight increments of detached bromegrass leaves. - Can. J. Plant Sci. *49*: 597-602, 1969.

3459 - HUNT, L.A., COOPER, J.P.: Productivity and canopy structure in seven temperate forage grasses. - J. appl. Ecol. *4*: 437-458, 1967.

3460 - HUNT, L.A., IMPENS, I.I., LEMON, E.R.: Preliminary wind tunnel studies of the photosynthesis and evapotranspiration of forage stands. - Crop Sci. *7*: 575-578, 1967.

3461 - HUNT, L.A., IMPENS, I.I., LEMON, E.R.: Estimates of the diffusion resistance of some large sunflower leaves in the field. - Plant Physiol. *43*: 522-526, 1968.

3462 - HUNT, L.A., MOORE, C.E., WINCH, J.E.: Light attenuation coefficient and productivity in "Vernal" alfalfa. - Can. J. Plant Sci. *50*: 469-474, 1970.

3463 - HURD, R.G.: Effect of CO_2-enrichment on the growth of young tomato plants in low light. - Ann. Bot. *32*: 531-542, 1968. [Ps.]

3464 - HURD, R.G., REES, A.R.: Transmission error in the photometric estimation of leaf area. - Plant Physiol. *41*: 905-906, 1966.

3465 - HURDUC, N., NASTASIA, I.: La dynamique de la concentration des pigments foliaires sur la verticale chez la mais. - Physiol. Plant Rom. (Bucureşti) *1970*: 17-21, 1970.

3466 - HURDUC, N., ŞTEFAN, V.: Influenţa îngrăşămintelor asupra activităţii fotosintetice a plantelor de porumb. [Effect of fertilizers on the photosynthetic activity of maize plants.] - Probl. agr. (Bucureşti) *1966* (9): 27-36, 1966. [In Rum., ab: E, F, R.]

3467 - HURTER, J., BERÜTER, J., BOSSHARDT, H.P.: Zur Resistenz der Graminee *Imperata cylindrica* L. gegenüber dem herbiziden Wirkstoff Simazin. - Experientia *24*: 217, 1968. [Ps.]

3468 - HUSÁK, Š., KVĚT, K.: Productive structure of *Phragmites communis* and *Typha angustifolia* stands after cutting at two different levels. - In: Productivity of Terrestrial Ecosystems. Production Processes. Czechosl. nat. Comm. IBP, PT-PP Report No. 1. Pp. 117-119. Praha 1970.

3469 - HÜSEMANN, W.: Der Einfluss verschiedener Lichtqualitäten auf Chlorophyllgehalt und Wachstum von Gewebekulturen aus *Crepis capillaris* (L.) WALLR. - Plant Cell Physiol. *11*: 315-322, 1970.

3470 - HUSHMAN, L.J.: A common regression for estimating dry matter content of carbohydrate-containing commodities from specific gravity. - Amer. Potato J. *46*: 234-238, 1969.

3471 - HUSSAINY, S.U.: Studies on the limnology and primary production of a tropical lake. - Hydrobiologia *30*: 335-352, 1967. [Ps, Chl.]

3472 - HUTCHINSON, T.C.: A physiological study of *Teucrium scorodonia* ecotypes which differ in their susceptibility to lime-induced chlorosis and iron-deficiency chlorosis. - Plant Soil *18*: 81-105, 1968. [Chl.]

3473 - HUTCHINSON, T.C.: Lime chlorosis as a factor in seedling establishment on calcareous soils. II. The development of leaf water deficit in plants showing lime-chlorosis. - New Phytol. *69*: 143-157, 1970. [Chl.]

3474 - HUTCHINSON, T.C.: Lime chlorosis in seedling establishment on calcareous soils. III. The ability of green and chlorotic plants fully to reverse large leaf water-deficits. - New Phytol. *69*: 261-268, 1970. [Chl.]

3475 - HUTH, W.: Enzyme in grünen Einzellern in Abhängigkeit von der Kohlenstoffversorgung. - Flora A *158*: 58-87, 1967. [Chl.]

3476 - HUTNER, S.H., ZAHALSKY, A.C., AARONSON, S., SMILLIE, R.M.: Resemblances between chloroplasts and mitochondria inferred from flagellates inhibited with the carcinogens 4-nitroquinoline N-oxide and ethionine. - In: GOODWIN, T.W. (ed.): Biochemistry of Chloroplasts. Vol. II. Pp. 703-720. Academic Press, London-New York 1967.

3477 - HUXLEY, A.F.: A theoretical treatment of the reflexion of light by multilayer structures. - J. exp. Biol. *48*: 227-246, 1968.

3478 - HUZISIGE, H., ISIMOTO, M., INOUE, H.: A new factor required for oxygen evolution. - In: SHIBATA, K., TAKAMIYA, A., JAGENDORF, A.T., FULLER, R.C. (ed.): Comparative Biochemistry and Biophysics of Photosynthesis. Pp. 170-178. Univ. Tokyo Press, Tokyo; Univ. Park Press, State College, Pa. 1968.

3479 - HUZISIGE, H., USIYAMA, H., KIKUTI, T., AZI, T.: Purification and properties of the photoactive particle corresponding to photosystem II. - Plant Cell Physiol. *10*: 441-455, 1969.

3480 - HUZISIGE, H., WADA, Y., SUNAGUTI, H., OHMORI, M.: Physiological studies on tobacco plants. I. Relationship between photosynthetic activity and growth stage, with special reference to the effect of topping-treatment. - Bot. Mag. (Tokyo) *79*: 722-732, 1966.

3481 - HYSON, P.: The tungsten wire temperature sensor. - J. appl. Meteorol. *7*: 684-690, 1968.

3482 - IBRAGIMOV, A.P., MARFINA, K.G.: Vliyanie razlichnykh doz gamma-luchěi Co^{60} na metabolizm ugleroda C^{14}, pogloshchennogo v protsesse fotosinteza. [Effect of different doses of gamma radiation from ^{60}Co on the metabolism of ^{14}C absorbed in photosynthesis.] - Radiobiologiya *7*: 922-925, 1967. [In R.]

3483 - ICHIKI, K., YOKOI, H., YUMURA, Y.: [Effect of wilting on nutrients and water absorption by the rice plant: I. Water absorption in the process of recovery from wilting.] - Bull. Tokai-Kinki nat. agr. exp. Sta. *18*: 407-418, 1969. [Ps; in Jap., ab: E.]

3484 - ICHIMURA, S., NAGASAWA, S., TANAKA, T.: On the oxygen and chlorophyll maxima found in the metalimnion of a mesotrophic lake. - Bot. Mag. (Tokyo) *81*: 1-10, 1968.

3485 - ICHIMURA, S.E.: Phytoplankton photosynthesis. - In: JACKSON, D.F. (ed.): Algae, Man, and the Environment. Pp. 103-120. Syracuse Univ. Press, Syracuse, N.Y. 1968.

3486 - IDLE, D.B.: The measurement of apparent surface temperature. - In: WADSWORTH, R.M., CHAPAS, L.C., RUTTER, A.J., SOLOMON, M.E., WARREN WILSON, J. (ed.): The Measurement of Environmental Factors in Terrestrial Ecology. Pp. 47-57. Blackwell sci. Publ., Oxford-Edinburgh 1968.

3487 - IDLE, D.B.: Scanning electron microscopy of leaf surface replicas and the measurement of stomatal aperture. - Ann. Bot. *33*: 75-76, 1969.

3488 - IDLE, D.B.: The calculation of transpiration rate and diffusion resistance of a single leaf from micrometeorological information subject to errors of measurement. - Ann. Bot. *34*: 159-176, 1970.

3489 - IDSO, S.B.: A holocoenotic analysis of environment-plant relationships. With special emphasis being given to the calculation of net photosynthesis, transpiration, and sensible heat exchange. - Tech. Bull. agr. exp. Sta., Univ. Minnesota (St. Paul) *264*: 1-147, 1968.

3490 - IDSO, S.B.: An analysis of the heating coefficient concept. - J. appl. Meteorol. *7*: 716-717, 1968.

3491 - IDSO, S.B.: Atmospheric- and soil-induced water stresses in plants and their effects on transpiration and photosynthesis. - J. theor. Biol. *21*: 1-12, 1968.

3492 - IDSO, S.B.: The photosynthetic response of plants to their environment: A holocoenotic method of analysis. Volume 1: Theory. Volume 2: Application. Volume 3: Data and computer output. - Diss. Abstr. *28*: 3557-B, 1968.

3493 - IDSO, S.B.: A theoretical framework for the photosynthetic modeling of plant communities. - Advanc. Frontiers Plant Sci. *23*: 91-118, 1969.

3494 - IDSO, S.B.: The relative sensitivities of polyethylene shielded net radiometers for short and long wave radiation. - Rev. sci. Instrum. *41*: 939-943, 1970.

3495 - IDSO, S.B., BAKER, D.G.: Method for calculating the photosynthetic response of a crop to light intensity and leaf temperature by an energy flow analysis of the meteorological parameters. - Agron. J. *59*: 13-21, 1967.

3496 - IDSO, S.B., BAKER, D.G.: Relative importance of reradiation, convection, and transpiration in heat transfer from plants. - Plant Physiol. *42*: 631-640, 1967.

3497 - IDSO, S.B., BAKER, D.G.: The naturally varying energy environment and its effects upon net photosynthesis. - Ecology *49*: 311-316, 1968.

3498 - IDSO, S.B., BAKER, D.G., BLAD, B.L.: Relations of radiation fluxes over natural surfaces. - Quart. J. roy. meteorol. Soc. *95*: 244-257, 1969.

3499 - IDSO, S.B., BAKER, D.G., BLAD, B.L.: Reply to discussion of "Relations of radiation fluxes over natural surfaces". - Quart. J. roy. meteorol. Soc. *96*: 765-767, 1970.

3500 - IDSO, S.B., BAKER, D.G., GATES, D.M.: The energy environment of plants. - Advance. Agron. *18*: 171-218, 1966.

3501 - IDSO, S.B., de WIT, C.T.: Light relations in plant canopies. - Appl. Optics *9*: 177-184, 1970.

3502 - IGNAT'EVSKAYA, M.A., RAÏKOV, N.I.: Vliyanie kontsentratsii sredy na fiziologiyu *Anacystis nidulans* v usloviyakh intensivnoī kul'tury. [Effect of the concentration of nutrient medium salts on the physiology of *Anacystis nidulans* under conditions of intense culture.] - Fiziol. Rast. *14*: 634-643, 1967. [Ps; in R, ab: E.]

3503 - IGNATOV, G., DETCHEV, G.: Comparing the temperature dependences of the cyclic photophosphorylation and the light-induced uptake of oxygen by isolated chloroplasts from *Vicia faba* L. - Dokl. bolg. Akad. Nauk *23*: 109-112, 1970.

3504 - IGNJATOVIC, L.R.: Effect of photosynthesis on oxygen saturation. - J. Water Pollut. Contr. Fed. *40*: R 151-R 161, 1968.

3505 - IKEDA, T.: Analytical studies on the structure of prolamellar body. - Bot. Mag. (Tokyo) *81*: 517-527, 1968.

3506 - IKEDA, T.: Changes in fine structure of prolamellar body in relation to the formation of the chloroplast. - Bot. Mag. (Tokyo) *83*: 1-9, 1970.

3507 - IKEGAMI, I., KATOH, S., TAKAMIYA, A.: Nature of heme moiety and oxidation-reduction potential of cytochrome 558 in *Euglena* chloroplasts. - Biochim. biophys. Acta *162*: 604-605, 1968.

3508 - IKEGAMI, I., KATOH, S., TAKAMIYA, A.: Light-induced changes of b-type cytochromes in the electron transport chain of *Euglena* chloroplasts. - Plant Cell Physiol. *11*: 777-791, 1970.

3509 - IKEHARA, N., SUGAHARA, K.: Inactivation of the Hill reaction in spinach chloroplasts by pre-treatment of Tris buffer in the light. - Bot. Mag. (Tokyo) *82*: 271-277, 1969.

3510 - IKEHARA, N., URIBE, E.G.: A pH dependent alteration of photosystem II activity in tris washed chloroplasts. - FEBS Letters *9*: 321-323, 1970.

3511 - IKEMORI, M.: The relation of leaf age to the translocation of ^{14}C and ^{32}P in *Halophila ovalis*. - Rec. oceanogr. Works Jap. *10*: 157-171, 1970.

3512 - IKEMORI, M.: [Relation of calcium uptake to photosynthetic activity as a factor controlling calcification in marine algae.] - Bot. Mag. (Tokyo) *83*: 152-162, 1970. [In Jap., ab: E.]

3513 - IKEMORI, M., NAKANO, A., NISHIDA, K.: Changes in photosynthetic and respiratory activity during the spore formation of *Padina arborescens*. - Annu. Rep. Noto mar. Lab. *8*: 1-13, 1968.

3514 - IKEMORI, M., NISHIDA, K.: [Effects of cations on photosynthesis of brown and green algae.] - Annu. Rep. Noto mar. Lab. *6*: 1-8, 1966. [In Jap.]

3515 - IKEMORI, M., NISHIDA, K.: Inorganic carbon source and the inhibitory effect of diamox on the photosynthesis of marine algae, *Ulva pertusa*. - Annu. Rep. Noto mar. Lab. *7*: 1-5, 1967.

3516 - IKEMORI, M., NISHIDA, K.: [Translocation of assimilates in *Sargassum*, and photo-
       synthetic and respiratory activity in various parts of the fronds.] - Annu. Rep.
       Noto mar. Lab. *7*: 7-13, 1967. [In Jap.]

3517 - IKONOMOVA, E.: Vliyanie molibdena v"rkhu usvoyavaneto i izpolzuvaneto na azota
       ot rasteniyata. [The effect of molybdenum on nitrogen uptake and utilization by
       plants.] - Pochvozn. Agrokhim. (Sofiya) *4* (6): 29-35, 1969. [Chl; in Bulg., ab:
       E, R.]

3518 - IKUSIMA, I.: Ecological studies on the productivity of aquatic plant communities.
       II. Seasonal changes in standing crop and productivity of a natural submerged
       community of *Vallisneria denseserrulata*. - Bot. Mag. (Tokyo) *79*: 7-19, 1966.

3519 - IKUSIMA, I.: Ecological studies on the productivity of aquatic plant communities.
       III. Effect of depth on daily photosynthesis in submerged macrophytes. - Bot.
       Mag. (Tokyo) *80*: 57-67, 1967.

3520 - IKUSIMA, I.: Ecological studies on the productivity of aquatic plant communities.
       IV. Light condition and community photosynthesic production. - Bot. Mag. (Tokyo)
       83: 330-341, 1970.

3521 - IL'INYKH, Z.G.: Reaktsiya fotosinteza na uroven' azotnogo pitaniya. [Response
       of photosynthesis on nitrogen supply.] - Uch. Zap. ural'.gos. Univ. Ser. biol.
       58: 50-57, 1967. [In R.]

3522 - IL'INYKH, Z.G., KONDRAT'EVA, A.A.: Svyaz' produktivnosti rasteniĭ s velichinoĭ
       intensivnosti fotosinteza i kharakterom raspredeleniya assimilyatov. [Relation
       of plant productivity with photosynthetic rate and character of photosynthates
       distribution.] - Uch. Zap. ural'. Univ. *113*: 163-173, 1970. [In R.]

3523 - IL'KUN, G.M., PANKRAT'EV, V.V., TARASENKO, S.A., MIRONOVA, A.S., MIKHAÏLENKO,
       L.A.: Puti povysheniya gazostoĭkosti rasteniĭ. [Ways of increasing gas resist-
       ance of plants.] - In: Puti Povysheniya Intensivnosti i Produktivnosti Fotosin-
       teza. Pp. 124-133. Naukova Dumka, Kiev 1967. [In R.]

3524 - IL'YASHUK, E.M., OKANENKO, A.S.: Intensivnost' ottoka C^{14}-assimilyatov iz
       list'ev sakharnoĭ svekly. [Rate of efflux of ^{14}C assimilates from sugar-beet
       leaves.] - Fiziol. Biokhim. kul't. Rast. *2*: 176-180, 1970. [In R, ab: E.]

3525 - IL'YASHUK, E.M., OKANENKO, A.S.: Vliyanie kaliya na peredvizhenie fotosinte-
       ticheski assimilirovannoĭ C^{14}O$_2$ u sakharnoĭ svekly. [Effect of potassium on the
       translocation of photosynthetically assimilated ^{14}CO$_2$ in sugar beet.] - Fiziol.
       Rast. *17*: 445-451, 1970. [In R, ab: E.]

3526 - IMAMALIEV, A., AKBAROV, K.: Posledeĭstvie defoliantov na nekotorye fiziologi-
       cheskie protsessy u khlopchatnika. [After-effect of defoliants on some physiol-
       ogical processes in cotton.] - Agrokhimiya *1967* (9): 69-73, 1967. [Ps, chl; in
       R.]

3527 - IMHOFF, V.: Identification d'inositol et de galactinol dans les chloroplastes
       de *Lolium italicum*. - Compt. rend. Acad. Sci.(Paris), Sér. D *270*: 2441-2443, 1970.

3528 - IMHOFF, V., BOURDU, R.: Distribution de glucides solubles non phosphorylés et
       de l'inositol dans les chloroplastes et les cellules foliaires de *Lolium itali-
       cum*. - Physiol. vég. *8*: 649-659, 1970. [Chl.]

3529 - IMHOFF, V., ROCHER, J.-P.: Une méthode d'estimation rapide d'osides cétosiques
       dans les chloroplastes. - Compt. rend. Acad. Sci.(Paris), Sér. D *268*: 3071-3073,
       1969.

3530 - IMPENS, I., LEMEUR, R., MOERMANS, R.: Spatial and temporal variation of net ra-
       diation in crop canopies. - Agr. Meteorol. *7*: 335-337, 1970.

3531 - INAMDAR, J.A., PATEL, R.C.: A new technique for making plant epidermal imprints
       using various domestic adhesives. - J. Microscop. *90*: 269-272, 1969.

3532 - INAMDAR, J.A., PATEL, R.C., BHATT, D.C.: A new and easy method of making im-
       prints of plant tissues: sectional and surface. - Z. wiss. Mikrosk. mikrosk.
       Tech. *70*: 140-145, 1970.

3533 - INCOLL, L.D.: A convenient assay for 14-carbon in leaf tissue. - Plant Physiol.
       44 (Suppl.): 42, 1969.

3534 - INCOLL, N.C., WRIGHT, W.H.: A field technique for measuring photosynthesis using 14-carbon dioxide. - Conn. agr. exp. Sta., spec. Bull. Soils *30/100*: 1-10, 1969.

3535 - INHOFFEN, H.H.: Zur Chemie des Chlorophylls und des Hämins. - Naturwissenschaften *55*: 457-462, 1968.

3536 - INHOFFEN, H.H., BIERE, H.: Zur weiteren Kenntnis des Chlorophylls und des Hämins XI. Ein Weg zum Protochlorophyll. - Tetrahedron Lett. *1966* (42): 5145-5149, 1966.

3537 - INHOFFEN, H.H., BLIESENER, K., BROCKMANN, H. Jr.: Zur weiteren Kenntnis des Chlorophylls und des Hämins, XIV (1) Substituierte Deuteroporphyrine. - Tetrahedon Lett. *1967* (8): 727-730, 1967.

3538 - INHOFFEN, H.H., BROCKMANN, H. Jr., BLIESENER, K.-M.: Zur weiteren Kenntnis des Chlorophylls und des Hämins, XXX. Photoprotoporphyrine und ihre Umwandlung in Spirographis- sowie Isospirographis-porphyrin. - J. Liebigs Ann. Chem. *730*: 173-185, 1969.

3539 - INHOFFEN, H.H., BUCHLER, J.W.: Zur weiteren Kenntnis des Chlorophylls und des Hämins, XVIII (1). Octaäthylporphinato-aluminium-hydroxid. - Tetrahedon Lett. *1968* (17): 2057-2059, 1968.

3540 - INHOFFEN, H.H., BUCHLER, J.W., THOMAS, R.: Zur weiteren Kenntnis des Chlorophylls und des Hämins, XXV (1). 3,4,7,8-Tetrahydro-octaäthylporphin ("Bacterio-octaäthylchlorin"). - Tetrahedron Lett. *1969* (14): 1141-1144, 1969.

3541 - INHOFFEN, H.H., BUCHLER, J.W., THOMAS, R.: Zur weiteren Kenntnis des Chlorophylls und des Hämins, XXVI (1). *cis*- und *trans*-7,8-Dihydro-octaäthylporphin (epimere Octaäthylchlorine). - Tetrahedron Lett. *1969* (14): 1145-1148, 1969.

3542 - INHOFFEN, H.H., FUHRHOP, J.-H., VOIGT, H., BROCKMANN, H. Jr.: Zur weiteren Kenntnis des Chlorophylls und des Hämins, VI. Formylierung der *meso*-Kohlenstoffatome von Alkylsubstituierten Porphyrinen. - J. Liebigs Ann. Chem. *695*: 133-143, 1966.

3543 - INHOFFEN, H.H., GOSSAUER, A.: Zur weiteren Kenntnis des Chlorophylls und des Hämins, XXIII. *meso*-Acetoxylierung des Gemini-monoketons vom Octaäthyl-porphin. - J. Liebigs Ann. Chem. *723*: 135-148, 1969.

3544 - INHOFFEN, H.H., JÄGER, P., MÄHLHOP, R., MENGLER, C.-D.: Zur weiteren Kenntnis des Chlorophylls und des Hämins, XII. Elektrochemische Reduktionen an Porphyrinen und Chlorinen, IV. - J. Liebigs Ann. Chem. *704*: 188-207, 1967.

3545 - INHOFFEN, H.H., NOLTE, W.: Zur weiteren Kenntnis des Chlorophylls und des Hämins, XXIV. Oxidative Umlagerungen am Octaäthylporphin zu Geminiporphin-polyketonen. - J. Liebigs Ann. Chem. *725*: 167-176, 1969.

3546 - INKINA, A.G.: Vliyanie vnutrisutochnogo sootnosheniya sveta i temnoty na kharakter razvitiya, soderzhanie pigmentov i rastvorimykh uglevodov u zlakovykh rasteniĭ. [Effect of diurnal light-darkness ratio on the characteristics of development, and content of pigments and soluble hydrocarbons in cereals.] - Sb. Rab. mol. Uch. vsesoyuz. sel'.-genet. Inst. *1969*: 123-125, 1969. [In R.]

3547 - INOUE, E.: The CO_2-concentration profile within crop canopies and its significance for the productivity of plant communities. - In: ECKARDT, F.E. (ed.): Functioning of Terrestrial Ecosystems at the Primary Production Level. Pp. 359-366. UNESCO, Paris 1968.

3548 - INOUE, E., UCHIJIMA, Z., SAITO, T., ISOBE, S., UEMURA, K.: The "assimitron", a newly devised instrument for measuring CO_2 flux in the surface air layer. - J. agr. Meteorol. *25*: 165-172, 1969.

3549 - INOUE, E., UCHIJIMA, Z., UDAGAWA, T., HORIE, T., KOBAYASHI, K.: [Studies of energy and gas exchange within crop canopies (2) - CO_2 flux within and above a corn plant canopy.] - J. agr. Meteorol. *23*: 165-176, 1968. [In Jap., ab: E.]

3550 - INOUE, E., UCHIJIMA, Z., UDAGAWA, T., HORIE, T., KOBAYASHI, K.: CO_2 environment and CO_2-exchange within a corn canopy. - In: Photosynthesis and Utilization of Solar Energy. Level III Experiments. Pp. 1-8. Jap. nat. Subcomm. for PP (JPP), Tokyo 1968.

3551 - IONESCU, P.: Contribuţii la studiul unor aspecte privind intensitatea fotosintezei la viţa de vie. [Study of photosynthetic rate in vine.] - Stud. Cercet. Biol., Ser. bot. *22*: 147-152, 1970. [In Rum., ab: E.]

3552 - IONESOVA, A.S.: O zelenom pigmente v semenakh pustynnykh rasteniĭ. [Green pigment in seeds of desert plants.] - Uzb. biol. Zh. 12 (2): 24-27, 1968. [In R,
ab: E, Uzb.]

3553 - ĬORDANOV, I.: Vliyanie kinetina na intensivnost' i produkty fotosinteza list'ev
makhorki, obrabotannykh povyshennymi temperaturami. [The effect of kinetin on
the rate and products of photosynthesis in leaves of Nicotiana rustica subjected
to high temperature.] - Fiziol. Rast. 16: 1008-1013, 1969. [In R, ab: E.]

3554 - ĬORDANOV, I., POPOV, K.: Vliyanie fiziologicheskogo sostoyaniya rasteniĭ na intensivnost' fotosinteza list'ev fasoli i na fotokhimicheskuyu aktivnost' izolirovannykh khloroplastov. [Effect of physiological state of plants on photosynthetic rate in Phaseolus leaves and on photochemical activity of isolated chloroplasts.] - Stud. biophys. 5: 123-130, 1967. [In R, ab: E.]

3555 - ĬORDANOV, I., POPOV, K., CHICHEV, P.: Izpolzuvane na ^{14}C pri izuchavane intenzivnostta i produktite na fotosintezata. [Use of ^{14}C in studies of photosynthetic
rate and products.] - In: P"rva Nats. Konf. Ispolz. Ionizir. L"cheniya i Izotop.
Biol. i Selsk. Stop., 1966. Pp. 59-72. Sofiya 1969. [In Bulg., ab: E, R.]

3556 - ĬORDANOV, I., POPOV, K., CHICHEV, P.: Svetlinni krivi na fotosintezata i razpredelenie na ^{14}C v lista s razlichno fiziologichno s"stoyanie. [Light curves of
photosynthesis and distribution of ^{14}C in leaves of a different physiological
state.] - Izv. Inst. Fiziol. Rast. "Metodiĭ Popov" b"lg. Akad. Nauk 16: 109-124,
1970. [In Bulg., ab: R, E.]

3557 - ĬORDANOV, I., SHOPOVA, K.: Fotosintetichna deĭnost na tsarevitsata v zavisimost
ot usloviyata na otglezhdane. II. Izmenenie na s"d"rzhanieto na khlorofil i ka-
rotinoidi v listata na tsarevitsata pod vliyanie na g"stotata na poseva, vodos-
nabdyavaneto i toreneto. [Photosynthetic activity of maize in dependence on cultivation conditions. II. Changes in the content of chlorophyll and carotenoids
in maize leaves as affected by stand density, water supply, and fertilization.]
- Rast. Nauki (Sofiya) 7: 19-27, 1970. [In Bulg.]

3558 - ĬORDANOV, I.T.: Preobrazuvane na energiyata v"v fotosintichniya protses. [Energy
transformation in photosynthesis.] - Biol. Khim. (Sofiya) 13 (4): 1-6, 1970. [In
Bulg.]

3559 - IOVVA, E.P.: Izmenenie soderzhaniya zelenykh pigmentov u roditel'skikh i gibrid-
nykh rasteniĭ tomatov. [Change of content of green pigments in parental and hybrid tomato plants.] - In: Fotosintez i Pigmenty Osnovnykh Sel'skokhozyaĭstvennykh Rasteniĭ Moldavii. Pp. 81-91. Kishinev 1970. [In R.]

3560 - IOVVA, E.P.: Proyavlenie geterozisa u tomatov v zavisimosti ot tipa opyleniya i
usloviĭ mineral'nogo pitaniya roditeleĭ. [Heterosis realization in tomatoes in
dependence on pollination type and mineral nutrition of parents.] - In: Fotosinteticheskaya Deyatel'nost' Rasteniĭ i Vliyanie na nee Mineral'nogo Pitaniya. Pp.
81-101. Kishinev 1970. [Ps; in R.]

3561 - IRBE, K.I., ROMANOVSKAYA, O.I.: Vliyanie fenazona na fotosinteticheskie protsessy v khloroplastakh rasteniĭ. [Effect of fenazon on photosynthetic processes in
plant chloroplasts.] - Izv. Akad. Nauk Latv. SSR 1970 (9): 20-27, 1970. [In R.]

3562 - IRIZARRY, H., ELLISON, J.H., ORTON, P.: Inheritance of persistent-green color
in Asparagus officinalis L. - Proc. amer. Soc. hort. Sci. 87: 274-278, 1968.
[Chl, car.]

3563 - IRVINE, J.E.: Photosynthesis in sugarcane varieties under field conditions. -
Crop Sci. 7: 297-300, 1967.

3564 - IRVINE, J.E.: Evidence for photorespiration in tropical grasses. - Physiol.
Plant. 23: 607-612, 1970.

3565 - ISAEV, P.I., LIBERMAN, E.A., SAMUILOV, V.D., SKULACHEV, V.P., TSOFINA, L.M.:
Conversion of biomembrane-produced energy into electric form. III. Chromatophores of Rhodospirillum rubrum. - Biochim. biophys. Acta 216: 22-29, 1970.

3566 - ISHINGALIEVA, M.K., KOTSUR, N.V.: Vliyanie ponizhennoĭ temperatury v zone korneĭ na formirovanie fotosinteticheskogo apparata prorostkov pshenitsy. [Effect
of low temperature in root zone on the formation of photosynthetic apparatus in
wheat seedlings.] - Rost Ustoĭchivost' Rast. 4: 248-252, 1968. [In R.]

3567 - ISHIZUKA, Y.: Engineering for higher yields. - In: EASTIN, J.D., HASKINS, F.A., SULLIVAN, C.Y., van BAVEL, C.H.M. (ed.): Physiological Aspects of Crop Yield. Pp. 15-25. Amer. Soc. Agron. & Crop Sci. Soc. Amer., Madison, Wisc. 1969. [Growth analysis.]

3568 - ISLER, O., RÜEGG, R., SCHWIETER, U.: Carotenoids as food colourants. - Pure appl. Chem. *14*: 245-263, 1967.

3569 - ISOBE, S.: Theory of the light distribution and photosynthesis in canopies of randomly dispersed foliage area. - Bull. nat. Inst. agr. Sci. Ser. A *16*: 1-25, 1969.

3570 - ISOE, S., HYEON, S.B., SAKAN, T.: Photo-oxygenation of carotenoids. I. The formation of dihydroactinidiolide and β-ionone from β-carotene. - Tetrahedon Lett. *1969* (4): 279-281, 1969.

3571 - ISRAEL, H.W., MAPES, M.O., STEWARD, F.C.: Pigments and plastids in cultures of totipotent carrot cells. - Amer. J. Bot. *56*: 910-917, 1969.

3572 - ISRAEL, H.W., STEWARD, F.C.: The fine structure and development of plastids in cultured cells of *Daucus carota*. - Ann. Bot. *31*: 1-18, 1967.

B3573 - Issledovaniya po Fotosintezu. [Photosynthesis Studies.] - Akad. Nauk Tadzh. SSR, Inst. Fiziol. Biofiz. Rast., Dushanbe 1967. [In R, ab: Tadzh.]

3574 - ISTATKOV, S.: Nyakoi osobennosti pri usvoyavaneto na CO_2 i obmyanata na v"glerodnite s"edineniya v rasteniya ot samooprashenii linii i kheterozisni kombinatsii tsarevitsa. [Some features of CO_2 uptake and transformation into carbon compounds in maize plants of self-pollinated lines and of heterosis combinations.] - In: P"rva Nats. Konf. Ispolz. Ionizir. L"cheniya i Izotop. Biol. i Selsk. Stop., 1966. Pp. 47-58. Sofiya 1969. [In Bulg.]

3575 - ISTRATI, L.N., SHISHKANU, G.V., POPUSHOÏ, I.S.: Fotosinteticheskaya deyatel'nost' list'ev slivy pri razlichnoĭ stepeni porazheniya vertitsillezom. [Photosynthetic activity of plum leaves infected to various degrees by *Verticillium*.] - In: Fotosinteticheskaya Deyatel'nost' Yabloni i Slivy v Usloviyakh Moldavii. Pp. 114-150. Kishinev 1970. [In R.]

3576 - ITÔ, A.: [Geometrical structure of rice canopy and penetration of direct solar radiation.] - Proc. Crop Sci. Soc. Jap. *38*: 355-363, 1969. [In Jap., ab: E.]

3577 - ITOH, M., YAMASHITA, K., NISHI, T., KONISHI, K., SHIBATA, K.: The site of manganese function in photosynthetic electron transport system. - Biochim. biophys. Acta *180*: 509-519, 1969.

3578 - IVANCHANKA, V.M., LYAGENCHANKA, B.I., GANCHARYK, N.M.: Ustanoŭka dlya vyvuchennya fotasintezu ŭ strumeni mechanaĭ vuglekistlaty. [Apparatus for measuring photosynthesis in a stream of labeled CO_2.] - Vestsi Akad. Navuk Belarus. SSR, Ser. biyal. Navuk *1970* (3): 28-31, 134, 1970. [In Belorus., ab: E.]

3579 - IVANCHENKO, V.M.: Popytka modelirovaniya vzaimosvyazi struktury i funktsiĭ khloroplastov. [Trial to simulate interrelation between chloroplast structure and functions.] - In: Tezist IV Nauchnoĭ Konferentsii Molodykh Uchennykh po Sovremennym Problemam Biologii. Pp. 70-73. Minsk 1970. [In R.]

3580 - IVANCHENKO, V.M., GONCHARIK, M.N.: Voprosy kinetiki fotosinteza v svyazi s vodnym rezhimom assimilyatsionnoĭ tkani. [Kinetics of photosynthesis in relation to water relations of the assimilatory tissue.] - In: Tezisy Dokladov. Biokhimiya Rastenii i Mikroorganizmov. Vol. 1. P. 4ĭ. Akad. Nauk Beloruss. SSR, Minsk 1968.

3581 - IVANCHENKO, V.M., GONCHARIK, M.N., KRUCHININA, S.S., MARSHAKOVA, M.I., URBANOVICH, T.A., DOROZHKINA, L.N., TALANOVA, K.S.: O korrelyatsii fotosinteticheskikh protsessov s ob"emnymi izmeneniyami khloroplastov in vivo. [Correlation of photosynthetic processes with volume changes of chloroplasts in vivo.] - In: II Vsesoyuznyĭ Biokhimicheskiĭ S"ezd. Sektsiya 19. Problemy Fotosinteza. Pp. 17-18. FAN, Tashkent 1969.

3582 - IVANCHENKO, V.M., KRUCHININA, S.S.: Nabukhanie belkov kak prichina ob"emnykh izmeneniĭ khloroplastov. [Protein swelling as the cause of volume changes in chloroplasts.] - Fiziol. Rast. *16*: 780-785, 1969. [In R, ab: E.]

3583 - IVANCHENKO, V.M., KRUCHININA, S.S., GONCHARIK, M.N.: O sokratitel'noǐ sposobnosti glitserinovykh modeleǐ khloroplastov. [Shrinkability of glycerol models of
chloroplasts.] - In: Fotosintez i Pitanie Rasteniǐ. Pp. 43-47. Nauka i Tekhnika,
Minsk 1968.

3584 - IVANCHENKO, V.M., KRUCHININA, S.S., URBANOVICH, T.A., MARSHAKOVA, M.I.,
DOROZHKINA, L.N., TALANOVA, K.S.: O korrelatsii mezhdu ob"emom khloroplastov,
reaktsieǐ Khilla, fotosinteticheskim fosforilirovaniem i assimilyatsieǐ CO_2 v
svyazi s vodnym defitsitom lista. [Correlation between chloroplast volume, Hill
reaction, photophosphorylation and CO_2 assimilation in relation to leaf water
deficit.] - In: Fiziologo-biokhimicheskie Issledovaniya Rasteniǐ. Pp. 12-17.
Nauka i Tekhnika, Minsk 1970. [In R.]

3585 - IVANCHENKO, V.M., KRUCHININA, S.S., URBANOVICH, T.A., MARSHAKOVA, M.I.,
LEGENCHENKO, B.I., DOROZHKINA, L.N.: Khloroplasty kak dinamicheskaya biokhimicheskaya sistema. [Chloroplasts as a dynamic biochemical system.] - In: Materialy IV Biokhimicheskoǐ Konferentsii Pribaltiǐskikh Respublik i Belorusskoǐ SSR.
Pp. 343. Akad. Nauk Litov. SSR, Vil'nyus 1970.

3586 - IVANCHENKO, V.M., LEGENCHENKO, B.I., GONCHARIK, M.N.: O "fenomene Brilliant".
[On the "Brilliant phenomenon".] - Fiziol. Rast. 15: 1070-1073, 1968. [In R.]

3587 - IVANCHENKO, V.M., URBANOVICH, T.A., MARSHAKOVA, M.I., GONCHARIK, M.N.: Sezonnaya dinamika reaktsii Khilla i fotofosforilirovaniya v svyazi s razlichnymi usloviyami vodoobespechennosti rasteniǐ ovsa. [Seasonal dynamics of Hill reaction
and photophosphorylation in oat plants in relation to various conditions of
water supply.] - Dokl. Akad. Nauk Belorus. SSR 13: 936-938, 1969. [In R.]

3588 - IVANOǓ, A.F., RAKHTSEENKA, L.I., MAǏSEENKA, A.I.: Intensiǔnasts' fotosintezu ǔ
seyantsaǔ drevavykh raslin pry roznaǐ kol'kastsi ǔ glebe alyuminiyu i margantsu. [Photosynthetic rate in seedlings of woody plants at different aluminium
and magnesium content in soil.] - Vestsi Akad. Navuk Belarus. SSR, Ser. biyal.
Navuk 1969 (5): 11-15, 129, 1969. [In Belorus., ab: E.]

3589 - IVANOV, A.G.: O vozmozhnosti peredvizheniya khlorofilla v rastenii. [Possibility
of chlorophyll translocation in plants.] - Fiziol. Rast. 16: 139-140, 1969. [In
R.]

3590 - IVANOV, I.D., DEMINA, N.S.: Fiksatsiya molekulyarnogo azota v svyazi s elektron-
donornoǐ sistemoǐ dykhaniya i fotosintezom. [Fixation of molecular nitrogen as
related to electron-donor system of respiration and photosynthesis.] - Izv.
Akad. Nauk SSSR, Ser. biol. 31: 115-120, 1966. [In R, ab: E.]

3591 - IVANOVA, E.V.: Vliyanie raznykh sootnosheniǐ mezhdu fosforom i zhelezom v pitatel'noǐ srede na pogloshchenie i raspredelenie etikh elementov v fasoli. [Effect
of different P/Fe ratios in nutrient medium on their absorption and distribution
in French bean.] - Vestn. mosk. Univ., Ser. 6 - Biol. Pochvoved. 25 (6): 87-91,
1970. [Chl; in R.]

3592 - IVANOVA, M.G., CHERYATNIKOVA, T.A.: Chistaya produktivnost' fotosinteza kukuruzy gibrida Bukovinskiǐ 3 v raznykh usloviyakh vyrashchivaniya. [Net photosynthetic productivity of the maize hybrid Bukovskiǐ 3 in different growing conditions.] - Izv. Timiryaz. sel'.-khoz. Akad. 1968 (3): 30-35, 1968. [In R.]

3593 - IVANOVA, N.A.: Kinetika fotosinteticheskogo metabolizma ugleroda u khlorelly pri
raznykh temperaturakh. [Kinetics of photosynthetic carbon metabolism in *Chlorella* at varying temperature.] - Uch. Zap. ural'. gos. Univ., Ser. biol. 58: 73-76,
1967. [In R.]

3594 - IVANOVA, N.A.: Fotosinteticheskiǐ metabolizm diskov sireni pri dlitel'nom khra-
nenii. [Photosynthetic metabolism of lilac discs during prolonged storage.] -
Uch. Zap. ural'.Univ. 113: 107-113, 1970. [In R.]

3595 - IVANOVA, N.A.: Vliyanie ukoreneniya na fotosinteticheskiǐ metabolizm izolirovannykh list'ev tomatov. [Effect of rooting on photosynthetic metabolism of isolated tomato leaves.] - Uch. Zap. ural'. Univ. 113: 114-120, 1970. [In R.]

3596 - IVNITSKAYA, I.N., MISHCHENKO, V.A., DILUNG, I.I.: Ob uchastii produkta obratimogo fotookisleniya khlorofilla v protsessakh sensibilizatsii. [Participation of
a product of reversible chlorophyll photooxidation in sensibilization processes.]
- Biofizika 13: 329-331, 1968. [In R.]

3597 - IWAKI, H.: Net production and photosynthesis of a cultivated turf of *Zoysia matrella*. - In: Photosynthesis and Utilization of Solar Energy, Level III Experiments. Pp. 60-64. Jap. nat. Subcomm. for PP (JPP), Tokyo 1968.

3598 - IWAKI, H., TAKADA, K., MONSI, M.: Studies on the dry matter production of *Solidago altissima* community I. The plant biomass and annual net production. - Bot. Mag. (Tokyo) *82*: 215-225, 1969.

3599 - IWAMURA, T., KUWASHIMA, S.: DNA species in chloroplasts of *Chlorella*. - In: SHIBATA, K., TAKAMIYA, A., JAGENDORF, A.T., FULLER, R.C. (ed.): Comparative Biochemistry and Biophysics of Photosynthesis. Pp. 354-359. Univ. Tokyo Press, Tokyo; Univ. Park Press, State College, Pa. 1968. [Chl.]

3600 - IZAWA, S.: Effect of Hill reaction inhibitors on photosystem I. - In: SHIBATA, K., TAKAMIYA, A., JAGENDORF, A.T., FULLER, R.C. (ed.): Comparative Biochemistry and Biophysics of Photosynthesis. Pp. 140-147. Univ. Tokyo Press, Tokyo; Univ. Park Press, State College, Pa. 1968.

3601 - IZAWA, S.: High concentration effect of Hill reaction inhibitors on isolated chloroplasts. The effect on the absorption spectra of chloroplast fragments and artificial chlorophyll *a* colloid. - In: METZNER, H. (ed.): Progress in Photosynthesis Research. Vol. III. Pp. 1742-1751. Tübingen 1969.

3602 - IZAWA, S.: Photoreduction of 2,6-dichlorophenolindophenol by chloroplasts with exogenous Mn^{2+} as electron donor. - Biochim. biophys. Acta *197*: 328-331, 1970.

3603 - IZAWA, S.: The relation of post-illumination ATP formation capacity (X_E) to H^+ accumulation in chloroplasts. - Biochim. biophys. Acta *223*: 165-173, 1970.

3604 - IZAWA, S., CONNOLLY, T.N., WINGET, G.D., GOOD, N.E.: Inhibition and uncoupling of photophosphorylation in chloroplasts. - In: Energy Conversion by the Photosynthetic Apparatus. Brookhaven Symp. Biol. *19*: 169-187, 1967.

3605 - IZAWA, S., GOOD, N.E.: Effect of salts and electron transport on the conformation of isolated chloroplasts. I. Light-scattering and volume changes. - Plant Physiol. *41*: 533-543, 1966.

3606 - IZAWA, S., GOOD, N.E.: Effect of salts and electron transport on the conformation of isolated chloroplasts. II. Electron microscopy. - Plant Physiol. *41*: 544-552, 1966.

3607 - IZAWA, S., GOOD, N.E.: The stoichiometric relation of phosphorylation to electron transport in isolated chloroplasts. - Biochim. biophys. Acta *162*: 380-391, 1968.

3608 - IZAWA, S., GOOD, N.E.: Effect of *p*-chloromercuribenzoate (PCMB) and mercuric ion on chloroplast photophosphorylation. - In: METZNER, H. (ed.): Progress in Photosynthesis Research. Vol. III. Pp. 1288-1298. Tübingen 1969.

3609 - IZAWA, S., HEATH, R.L., HIND, G.: The role of chloride ion in photosynthesis. III. The effect of artificial electron donors upon electron transport. - Biochim. biophys. Acta *180*: 388-398, 1969.

3610 - IZAWA, S., HIND, G.: The kinetics of the pH rise in illuminated chloroplast suspensions. - Biochim. biophys. Acta *143*: 377-390, 1967.

3611 - IZAWA, S., WINGET, G.D., GOOD, N.E.: Phlorizin, a specific inhibitor of photophosphorylation and phosphorylation-coupled electron transport in chloroplasts. Biochem. biophys. Res. Commun. *22*: 223-226, 1966.

3612 - IZHAR, S.: Physiological and genetical studies of the net carbon dioxide exchange by individual intact leaves of several dry bean varieties (*Phaseolus vulgaris* L.). - Diss. Abstr. B *27*: 4246-B-4247-B, 1967.

3613 - IZHAR, S., WALLACE, D.H.: Studies of the physiological basis for yield differences. III. Genetic variation in photosynthetic efficiency of *Phaseolus vulgaris* L. - Crop Sci. *7*: 457-460, 1967.

3614 - IZHAR, S., WALLACE, D.H.: Effect of night temperature on photosynthesis of *Phaseolus vulgaris* L. - Crop Sci. *7*: 546-547, 1967.

B3615 - Izuchenie Fotosinteza Vazhneīshikh sel'skokhozyaīstvennykh Kul'tur Moldavii. [Studying Photosynthesis in the Most Important Moldavian Crops.] - Kartya Moldovenyaske, Kishinev 1968. [In R.]

3616 - JACKSON, A.H., KENNER, G.W.: Recent developments in porphyrin chemistry. -
In: GOODWIN, T.W. (ed.): Porphyrins and Related Compounds. Pp. 3-18. Academic
Press, London-New York 1968.

3617 - JACKSON, J.B., CROFTS, A.R.: Energy-linked reduction of nicotinamide adenine
dinucleotides in cells of *Rhodospirillum rubrum*. - Biochem. biophys. Res.
Commun. *32*: 908-915, 1968.

3618 - JACKSON, J.B., CROFTS, A.R.: The high energy state in chromatophores from
Rhodopseudomonas spheroides. - FEBS Letters *4*: 185-189, 1969.

3619 - JACKSON, W.A.: Comments on water and CO_2 transport in the photosynthetic process. - In: SAN PIETRO, A., GREER, F.A., ARMY, T.J. (ed.): Harvesting the Sun.
Pp. 249-254. Academic Press, New York-London 1967.

3620 - JACKSON, W.A., VOLK, R.J.: Role of potassium in photosynthesis and respiration. - In: Role of Potassium in Agriculture. Proc. Symp. Muscle Shoals. Pp.
109-146. Alabama 1968.

3621 - JACKSON, W.A., VOLK, R.J.: Oxygen uptake by illuminated maize leaves. - Nature
222: 269-271, 1969.

3622 - JACKSON, W.A., VOLK, R.J.: Photorespiration. - Annu. Rev. Plant Physiol. *21*:
385-432, 1970.

3623 - JACOBI, G.: Die Desintegration von isolierten Chloroplasten und die Beurteilung
von Chloroplastenfragmenten. - Ber. deut. bot. Ges. *79*: (72)-(81), 1966.

3624 - JACOBI, G.: The functional and structural state of chloroplast fragments. - In:
THOMAS, J.B., GOEDHEER, J.C. (ed.): Currents in Photosynthesis. Pp. 339-347.
Donker, Rotterdam 1966.

3625 - JACOBI, G.: Die photochemische Aktivität von Ultraschall-desintegrierten isolierten Chloroplasten. - Z.Pflanzenphysiol. *57*: 255-268, 1967.

3626 - JACOBI, G.: Die Fragmentation isolierter Chloroplasten. II. Isolation und Charakterisierung von zwei unterschiedlichen Photosystem-I-Partikeln. - Z. Pflanzenphysiol. *61*: 203-217, 1969.

3627 - JACOBI, G.: Die funktionelle Charakterisierung aktiver Substrukturen aus voriso-
lierten Granastapeln. - Ber. deut. bot. Ges. *83*: 451-463, 1970.

3628 - JACOBI, G., LEHMANN, H.: Die Fragmentation isolierter Chloroplasten. I. Die
funktionelle und strukturelle Beurteilung von Fragmenten Ultraschall-behandel-
ter Chloroplasten. - Z. Pflanzenphysiol. *59*: 457-476, 1968.

3629 - JACOBI, G., LEHMANN, H.: Photochemical activities of chloroplast fragments. -
In: METZNER, H. (ed.): Progress in Photosynthesis Research. Vol. I. Pp. 159-173.
Tübingen 1969.

3630 - JACOBSON, A.B.: A procedure for isolation of proplastids from etiolated maize
leaves. - J. Cell Biol. *38*: 238-244, 1968. [Chl.]

3631 - JACOBY, B., DAGAN, J.: Effects of [6]N-benzyladenine on primary leaves of intact
bean plants and on their sodium absorption capacity. - Physiol. Plant. *23*: 397-
403, 1970. [Chl.]

3632 - JACQUARD, P.: Comparaison du rythme saisonnier de croissance de deux grami-
nées pérennes: *Festuca arundinacea* SCHREB. (cv. S. 170) et *Phleum pratense* L.
(cv. Mélusine). - Ann. Amelior. Plant. (Paris) *20*: 45-77, 1970. [Ps.]

3633 - JACQUINOT, L.: La nutrition carbonée du mil (*Pennisetum typhoides* STAPF et HUBB.).
I. - Migrations des assimilats carbonés durant la formation des grains. - Agron.
trop. *25*: 1088-1095, 1970.

3634 - JAGELS, R.: Photosynthetic apparatus in *Selaginella*. I. Morphology and photosynthesis under different light and temperature regimes. - Can. J. Bot. *48*:
1843-1852, 1970.

3635 - JAGELS, R.: Photosynthetic apparatus in *Selaginella*. II. Changes in plastid ultrastructure and pigment content under different light and temperature regimes.
- Can. J. Bot. *48*: 1853-1860, 1970.

3636 - JAGENDORF, A.T.: Acid-base transitions and phosphorylation by chloroplasts. -
Fed. Proc. *26*: 1361-1369, 1967.

3637 - JAGENDORF, A.T.: Photosynthesis - light and energy conversion. - In: Proceedings of the XVII International Horticultural Congress (1966). Vol. 3. P. 393. College Park, Md. 1967.

3638 - JAGENDORF, A.T.: The chemiosmotic hypothesis of photophosphorylation. - In: SAN PIETRO, A., GREER, F.A., ARMY, T.J. (ed.): Harvesting the Sun. Pp. 69-78. Academic Press, New York-London 1967.

3639 - JAGENDORF, A.T., URIBE, E.: ATP formation caused by acid-base transition of spinach chloroplasts. - Proc. nat. Acad. Sci. U.S.A. 55:: 170-177, 1966.

3640 - JAGENDORF, A.T., URIBE, E.: Photophosphorylation and the chemi-osmotic hypothesis. - In: Energy Conversion by the Photosynthetic Apparatus. Brookhaven Symp. Biol. 19: 215-245, 1967.

3641 - JAIN, M.L.: Biochemical definition of yellow-virescent and light-green suppressor mutations in barley. - Genetics 54: 813-818, 1966. [Chl.]

3642 - JAIN, T.C., MISRA, D.K.: Leaf area estimation by linear measurements in *Ricinus communis*. - Nature 212: 741-742, 1966.

3643 - JAKRLOVÁ, J.: Plant production, chlorophyll content and its vertical distribution in inundated meadows. - Photosynthetica 1: 199-205, 1967.

3644 - JAMES, A.T., NICHOLS, B.W.: Lipids of photosynthetic systems. - Nature 210: 372-375, 1966.

3645 - JANÁČ, J.: The accuracy of the differential measurement of small CO_2 concentration differences with the infrared gas analyzer. - Photosynthetica 4: 302-308, 1970.

3646 - JANES, B.E.: Effect of carbon dioxide, osmotic potential of nutrient solution, and light intensity on transpiration and resistance to flow of water in pepper plants. - Plant Physiol. 45: 95-103, 1970.

3647 - JANKIEWICZ, L.S., ANTOSZEWSKI, R., KLIMOWICZ, E.: Distribution of labelled assimilates within a young apple tree after supplying $^{14}CO_2$ to a leaf or shoot. - Biol. Plant. 9: 116-121, 1967. [Ps.]

3648 - JANKIEWICZ, L.S., PLICH, H., ANTOSZEWSKI, R.: Preliminary studies on the translocation of ^{14}C-labelled assimilates and $^{32}PO^{3-}$ towards the gall evoked by *Cynips (Diplolepis) Quercus-folii* L. on oak leaves. - Marcellia (Strasburg) 36: 163-172, 1970.

3649 - JANUSZCZYK, L., PASZKOWSKI, W., FRĄCKOWIAK, D.: The influence of sonication on spectra of chlorophyll a solutions. - Bull. Acad. pol. Sci., Sér. Sci. math. astron. phys. 17: 781-787, 1969.

3650 - JARVIS, P.G.: Characteristics of the photosynthetic apparatus derived from its response to natural complexes of environmental factors. - In: Prediction and Measurement of Photosynthetic Productivity. Pp. 353-367. PUDOC, Wageningen 1970.

3651 - JARVIS, P.G., SLATYER, R.O.: A controlled-environment chamber for studies of gas exchange by each surface of a leaf. - CSIRO Aust. Div. Land Res. tech. Pap. 29: 2-16, 1966.

3652 - JARVIS, P.G., SLATYER, R.O.: The role of the mesophyll cell wall in leaf transpiration. - Planta 90: 303-322, 1970. [N_2O porometer.]

3653 - JAVORNICKÝ, P.: On the utilization of light by fresh-water phytoplankton. - Arch. Hydrobiol. 39 (Suppl. 1/2): 68-85, 1970. [Ps.]

3654 - JAYANGOUDAR, I.S.: Bacterial photosynthesis in the oxidation ponds of Ahmedabad, India. - Water Sewage Works 115: 380-383, 1968.

3655 - JEFFERS, D.L. Jr.: Field studies on photosynthesis in soybeans. - Diss. Abstr. B 29: 19-B, 1968.

3656 - JEFFERS, D.L., SHIBLES, R.M.: Some effects of leaf area, solar radiation, air temperature, and variety on net photosynthesis in field-grown soybeans. - Crop Sci. 9: 762-764, 1969.

3657 - JEFFREY, S.W.: Two spectrally distinct components in preparations of chlorophyll c. - Nature 220: 1032-1033, 1968.

3658 - JEFFREY, S.W.: Photosynthetic pigments of the phytoplankton of some coral reef
       waters. - Limnol. Oceanogr. *13*: 350-355, 1968. [PC, TLC.]

3659 - JEFFREY, S.W.: Pigment composition of *Siphonales* algae in the brain coral *Favia*.
       - Biol. Bull. *135*: 141-148, 1968.

3660 - JEFFREY, S.W.: Quantitative thin-layer chromatography of chlorophylls and caro-
       tenoids from marine algae. - Biochim. biophys. Acta *162*: 271-285, 1968.

3661 - JEFFREY, S.W.: Properties of two spectrally different components in chlorophyll
       c preparations. - Biochim. biophys. Acta *177*: 456-467, 1969.

3662 - JEFFREY, S.W., ALLEN, M.B.: A paper chromatographic method for the separation
       of phytoplankton pigments at sea. - Limnol. Oceanogr. *12*: 533-537, 1967.

3663 - JEFFREY, S.W., HAXO, F.T.: Photosynthetic pigments of symbiotic dinoflagellates
       (zooxanthellae) from corals and clams. - Biol. Bull. *135*: 149-165, 1968.

3664 - JEFFREY, S.W., SHIBATA, K.: Some spectral characteristics of chlorophyll *c* from
       Tridacna crocea zooxanthellae. - Biol. Bull. *136*: 54-62, 1969.

3665 - JEFFREY, S.W., ULRICH, J., ALLEN, M.B.: Some photochemical properties of chloro-
       plast preparations from the chrysomonad *Hymenomonas* sp. - Biochim. biophys.
       Acta *112*: 35-44, 1966.

3666 - JEN, J.J., MACKINNEY, G.: On the photodecomposition of chlorophyll *in vitro* -
       I. Reaction rates. - Photochem. Photobiol. *11*: 297-302, 1970.

3667 - JEN, J.J., MACKINNEY, G.: On the photodecomposition of chlorophyll *in vitro* -
       II. Intermediates and breakdown products. - Photochem. Photobiol. *11*: 303-308,
       1970.

3668 - JEN, J.J.-S.: Photodecomposition of chlorophyll. - Diss. Abstr. int. B *30*:
       4490-B, 1970.

3669 - JENNINGS, V.-M., SHIBLES, R.M.: Genotypic differences in photosynthetic contri-
       butions of plant parts to grain yield in oats. - Crop Sci. *8*: 173-175, 1968.

3670 - JENSEN, A.: Algal carotenoids. V. Isofucoxanthin - a rearrangement product of
       fucoxanthin. - Acta chem. scand. *20*: 1728-1730, 1966.

3671 - JENSEN, A.: Carotenoids of Norwegian brown seaweeds and of seaweed meals. -
       Norsk Inst. Tang- Tareforsk. Rep. *31*: 1-138, 1966.

3672 - JENSEN, R.G., BASSHAM, J.A.: Conditions for obtaining photosynthetic carbon com-
       pound photosynthesis with isolated chloroplasts comparable to *in vivo* in rates
       and products. - Plant Physiol. *41* (Suppl.): LVII-LVIII, 1966.

3673 - JENSEN, R.G., BASSHAM, J.A.: Photosynthesis by isolated chloroplasts. - Proc.
       nat. Acad. Sci. U.S.A. *56*: 1095-1101, 1966.

3674 - JENSEN, R.G., BASSHAM, J.A.: Photosynthesis by isolated chloroplasts. II. Ef-
       fects of addition of cofactors and intermediate compounds. - Biochim. biophys.
       Acta *153*: 219-226, 1968.

3675 - JENSEN, R.G., BASSHAM, J.A.: Photosynthesis by isolated chloroplasts. III.
       Light activation of the carboxylation reaction. - Biochim. biophys. Acta *153*:
       227-234, 1968.

3676 - JENSEN, R.G., SEELY, G.R., VERNON, L.P.: Photochemistry of a water-soluble poly-
       meric derivative of chlorophyll. - J. phys. Chem. *70*: 3307-3314, 1966.

3677 - JENSEN, S.G.: Metabolism in barley yellow dwarf virus-infected barley. - Phyto-
       pathology *56*: 883, 1966. [Chl.]

3678 - JENSEN, S.G.: Photosynthesis, respiration and other physiological relationships
       in barley infected with barley yellow dwarf virus. - Phytopathology *59*: 204-208,
       1968.

3679 - JERLOV, N.G.: Aspects of light measurement in the sea. - In: BAINBRIDGE, R.,
       EVANS, G.C., RACKHAM, O. (ed.): Light as an Ecological Factor. Pp. 91-98. Black-
       well sci. Publ., Oxford 1966.

3680 - JESCHKE, W.D.: Über einige Zusammenhänge zwischen Photosynthese und Anionenauf-
       nahme bei *Elodea*. - Ber. deut. bot. Ges. *79*: (121)-(123), 1966.

3681 - JESCHKE, W.D.: Die cyclische und die nichtcyclische Photophosphorylierung als Energiequellen der lichtabhängigen Chloridionenaufnahme bei *Elodea*. - Planta *73*: 161-174, 1967.

3682 - JESCHKE, W.D.: On the connexion between electron transport and ion transport. - In: MOTHES, K., MÜLLER, E., NELLES, A., NEUMANN, D. (ed.): Transport and Distribution of Matter in Cells and Higher Plants. Abhandl. deut. Akad. Wiss. Berlin *4 a*: 127-143, 1968. [Ps.]

3683 - JESCHKE, W.D.: Der Influx von Kaliumionen bei Blättern von *Elodea densa*, Abhängigkeit vom Licht, von der Kaliumkonzentration und von der Temperatur. - Planta *91*: 111-128, 1970. [Ps.]

3684 - JESCHKE, W.D.: Lichtabhängige Veränderungen des Membranpotentials bei Blattzellen von *Elodea densa*. - Z. Pflanzenphysiol. *62*: 158-172, 1970. [Chloroplast.]

3685 - JESCHKE, W.D., GIMMLER, H., SIMONIS, W.: Incorporation of ^{32}P and ^{14}C into photosynthetic products of *Ankistrodesmus braunii* as affected by X-rays. - Plant Physiol. *42*: 380-386, 1967.

3686 - JESCHKE, W.D., SIMONIS, W.: Effects of carbon dioxide and light quality upon the light-dependent Cl^- uptake in *Elodea densa*. - In: Photochemistry and Photobiology in Plant Physiology. European Photobiol. Symp. Hvar, Yugoslavia, 19th-22nd September 1967. Book of Abstracts. Pp. 101-104. Hvar 1967. [Ps.]

3687 - JESCHKE, W.D., SIMONIS, W.: Effect of CO_2 on photophosphorylation *in vivo* as revealed by the light-dependent Cl^- uptake in *Elodea densa*. - Z. Naturforsch. *22 b*: 873-876, 1967.

3688 - JESCHKE, W.D., SIMONIS, W.: Über die Wirkung von CO_2 auf die lichtabhängige Cl^--Aufnahme bei *Elodea densa*: Regulation zwischen nichtcyclischer und cyclischer Photophosphorylierung. - Planta *88*: 157-171, 1969.

3689 - JEŠKO, T.: Príspevok k meraniu listovej plochy polárnym planimetrom. [Contribution to measuring of leaf area by means of polar planimeter.] - Biológia (Bratislava) *21*: 904-908, 1966. [In Slovak, ab: G, R.]

3690 - JEŠKO, T., LUKAČOVIČ, A., HEINRICHOVÁ, K.: Einfluss einer Mischung der Mikroelemente auf einige Prozesse der Zuckerrübe im Verlauf der Vegetation. I. Änderungen in der Intensität der Photosynthese und der Wasserhaushalts in der Pflanze. - Biológia (Bratislava) *21*: 405-420, 1966.

3691 - JEWISS, O.R., WOLEDGE, J.: The effect of age on the rate of apparent photosynthesis in leaves of tall fescue (*Festuca arundinacea* SCHREB.). - Ann. Bot. *31*: 661-671, 1967.

3692 - JEYARAJAN, R., BAKHSHI, J.C., NAURIYAL, J.P.: Biochemical differences in healthy and declining sweet-orange trees. 1. Relation of tree conditions to pigment content, elemental composition, and enzyme activity. - J. Res., Punjab agr. Univ. *7*: 183-187, 1970.

3693 - JI, T.H., BENSON, A.A.: Association of lipids and proteins in chloroplast lamellar membrane. - Biochim. biophys. Acta *150*: 686-693, 1968.

3694 - JI, T.H., HESS, J.L., BENSON, A.A.: Studies on chloroplast membrane structure. I. Association of pigments with chloroplast lamellar protein. - Biochim. biophys. Acta *150*: 676-685, 1968. [Preparation of β-carotenoprotein complex.]

3695 - JI, T.H., HESS, J.L., BENSON,A.A.:The nature of β-carotene association in chloroplast lamellae. - In: SHIBATA, K., TAKAMIYA, A., JAGENDORF, A.T., FULLER, R.C. (ed.): Comparative Biochemistry and Biophysics of Photosynthesis. Pp. 36-49. Univ. Tokyo Press, Tokyo; Univ. Park Press, State College, Pa. 1968.

3696 - JIMENEZ-SAENZ, E.: Importancia de la viá del glicolato en la fysiologiá de la plantas. [Importance of glycolate pathway in plant physiology.] - Turrialba *17*: 35-39, 1967. [In Span.]

3697 - JITTS, H.R.: Seasonal variations in the Indian Ocean along 110° E. IV. Primary production. - Aust. J. mar. Freshwater Res. *20*: 65-75, 1969.

3698 - JODO, S.: [Stomatal movement and water relations in crops. I. Performance test on a newly improved recording porometer.] - Proc. Crop Sci. Soc. Jap. *39*: 431-439, 1970. [In Jap., ab: E.]

3699 - JOHN, J., SUBBARAYAN, C., CAMA, H.R.: Carotenoids in 3 stages of ripening of mango. - J. Food Sci. *35*: 262-265, 1970.

3700 - JOHN, J.B.St.: Determination of ATP in *Chlorella* with the luceferin-luciferase enzyme system. - Anal. Biochem. *37*: 409-416, 1970.

3701 - JOHNSON, B.E., BRUN, W.E.: The effect of simulated catterpillar damage on CO_2 fixation and translocation in banana leaves. - Physiol. Plant. *19*: 417-421, 1966.

3702 - JOHNSON, C.E., ELSTNER, E., GIBSON, J.F., BENFIELD, G., EVANS, M.C.W., HALL, D.O.: Mössbauer effect in the ferredoxin of *Euglena*. - Nature *220*: 1291-1293, 1968.

3703 - JOHNSON, C.E., HALL, D.O.: Mössbauer effect study of the state of iron in spinach ferredoxin. -Nature *217*: 446-448, 1968.

3704 - JOHNSON, E.J.: Occurrence of the adenosine monophosphate inhibition of carbon dioxide fixation in photosynthetic and chemosynthetic autotrophs. - Arch. Biochem. Biophys. *114*: 178-183, 1966.

3705 - JOHNSON, E.J., BRUFF, B.S.: The effect of metal ions and nucleoside phosphates on ATP-dependent CO_2 fixation in fractions of *Spinacea oleracea*. - Fed. Proc. 25: 738, 1966.

3706 - JOHNSON, E.J., BRUFF, B.S.: Chloroplast integrity and ATP-dependent CO_2 fixation in *Spinacea oleracea*. - Plant Physiol. *42*: 1321-1328, 1967.

3707 - JOHNSON, H.S., HATCH, M.D.: Distribution of the C_4-dicarboxylic pathway of photosynthesis and its occurrence in dicotyledonous plants. - Phytochemistry *7*: 375-380, 1968.

3708 - JOHNSON, H.S., HATCH, M.D.: The C_4-dicarboxylic acid pathway of photosynthesis. Identification of intermediates and products and quantitative evidence for the route of carbon flow. - Biochem. J. *114*: 127-134, 1969.

3709 - JOHNSON, H.S., HATCH, M.D.: Properties and regulation of leaf nicotinamide-adenine dinucleotide phosphate-malate dehydrogenase and 'malic' enzyme in plants with the C_4-dicarboxylic acid pathway of photosynthesis. - Biochem. J. *119*: 273-280, 1970.

3710 - JOHNSON, L.B., SCHAFER, J.F., LEOPOLD, A.C.: Nutrient mobilization in leaves by *Puccinia recondita*. - Phytopathology *56*: 799-803, 1966. [Chl.]

3711 - JOHNSON, L.D.: Oscillator strengths in magnesium porphin. - Diss. Abstr. B *28*: 4241-B, 1968.

3712 - JOHNSON, P.L., KELLEY, J.J. Jr.: Dynamics of carbon dioxide and productivity in an Arctic biosphere. - Ecology 51: 73-80, 1970. [Ps.]

3713 - JOHNSON, R.E.: Comparison of methods for estimating cotton leaf area. - Agron. J. *59*: 493-494, 1967.

3714 - JOHNSON, U.G., TOGASAKI, R.K., LEVINE, R.P.: A mutant strain of *Chlamydomonas* defective in chloroplast ribosome production. - Plant Physiol. *43*: S 6, 1968. [Ps.]

3715 - JOHNSTON, C.S., COOK, J.P.: A preliminary assessment of the techniques for measuring primary production in macrophytic marine algae. - Experientia *24*: 1176-1177, 1968.

3716 - JOHNSTON, T.J.: Field studies of light relationships as affecting seed yields and photosynthesis of individual leaves within canopies of soybeans (*Glycine max* (L.) MERRILL). - Diss. Abstr. B *29*: 448-B-449-B, 1969.

3717 - JOHNSTON, T.J., PENDLETON, J.W., PETERS, D.B., HICKS, D.R.: Influence of supplemental light on apparent photosynthesis, yield, and yield components of soybeans (*Glycine max* L.). - Crop Sci. *9*: 577-581, 1969.

3718 - JOHRI, M.M.: Electron transport and photophosphorylation in chloroplasts. - Botanica (India) *17* (2): 23-28, 1966.

3719 - JOLCHINE, G., MOYSE, A.: Action conjuguée de D_2O et du Triton X-100 sur les complexes pigmentaires de *Rhodospirillum rubrum*. - Physiol. vég. *4*: 195-220, 1966.

3720 - JOLCHINE, G., REISS-HUSSON, F., KAMEN, M.D.: Active center fractions from *Rhodopseudomonas spheroides* strain Y. (DE KLERK). - Proc. nat. Acad. Sci. U.S.A. *64*: 650-653, 1969.

3721 - JOLIOT, A.: Actions du chlorométhylurée et de l'hydroxylamine sur la réaction photochimique d'émission d'oxygène (système II). - Biochim. biophys. Acta *126*: 587-590, 1966.

3722 - JOLIOT, A.: Cinétique des deux étapes photochimiques de la photosynthèse. - Physiol. vég. *6*: 235-254, 1968.

3723 - JOLIOT, P.: Oxygen evolution of algae illuminated by modulated light. - In: Energy Conversion by the Photosynthetic Apparatus. Brookhaven Symp. Biol. *19*: 418-433, 1967.

3724 - JOLIOT, P.: Diskussionsbemerkung. - In: 19. Colloquium der Gesellschaft für Biologische Chemie vom 24.-27. April 1968 in Mosbach/Baden. Pp. 307-312. Springer-Verlag, Berlin-Heidelberg-New York 1968. [Ps.]

3725 - JOLIOT, P.: Kinetic studies of photosystem II in photosynthesis. - Photochem. Photobiol. *8*: 451-463, 1968.

3726 - JOLIOT, P., BARBIERI, G., CHABAUD, R.: Un nouveau modèle des centres photochimiques du Système II. - Photochem. Photobiol. *10*: 309-329, 1969.

3727 - JOLIOT, P., DELOSME, R., JOLIOT, A.: Étude de la reaction photochimique liée à l'émission d'oxygène (système II) chez *Chlorella pyrenoidosa*. - In: THOMAS, J.B., GOEDHEER, J.C. (ed.): Currents in Photosynthesis. Pp. 359-366. Donker, Rotterdam 1966.

3728 - JOLIOT, P., HOFNUNG, M., CHABAUD, R.: Étude de l'émission d'oxygène par des algues soumises a un éclairement modulé sinusoidalement. - J. Chim. phys. *63*: 1423-1441, 1966.

3729 - JOLIOT, P., JOLIOT, A.: A polarographic method for detection of oxygen production and reduction of Hill reagent by isolated chloroplasts. - Biochim. biophys. Acta *153*: 625-634, 1968.

3730 - JOLIOT, P., JOLIOT, A., KOK, B.: Analysis of the interactions between the two photosystems in isolated chloroplasts. - Biochim. biophys. Acta *153*: 635-652, 1968.

3731 - JOLLIFFE, P.A., TREGUNNA, E.B.: Effect of temperature, CO_2 concentration, and light intensity on oxygen inhibition of photosynthesis in wheat leaves. - Plant Physiol. *43*: 902-906, 1968.

3732 - JOLLIFFE, P.A., TREGUNNA, E.B.: Estimation of the CO_2 free exchange pool size in wheat and corn leaves. - Can. J. Bot. *47*: 1506-1508, 1969.

3733 - JONES, C.B., ELLSWORTH, R.K.: Evidence for the presence of protochlorophyllase in etiolated wheat seedlings. - Plant Physiol. 44: 1478-1480, 1969.

3734 - JONES, C.W., VERNON, L.P.: Nicotinamide-adenine dinucleotide photoreduction in *Rhodospirillum rubrum* chromatophores. - Biochim. biophys. Acta *180*: 149-164, 1969.

3735 - JONES, I.D., BENNETT, L.S., WHITE, R.C.: Thin layer chromatography for identification of chlorophylls and derivatives. - J. Elisha Mitchell sci. Soc. *82*: 90, 1966.

3736 - JONES, I.D., BENNETT, L.S., WHITE, R.C.: Recording of thin layer chromatograms on Polacolor film under ultraviolet. - J. Chromatogr. *30*: 622-625, 1967. [Chl, pheophytins, pheophorbides.]

3737 - JONES, I.D., WHITE, R.C., GIBBS, E., DENARD, C.D.: Absorption spectra of copper and zinc complexes of pheophytins and pheophorbides. - Agr. Food Chem. *16*: 80-83, 1968.

3738 - JONES, L.J.: The physiology of production in *Phaseolus*. - Eucarpia Congr. Ass. Europe. Amelior. Plant. *5*: 465-484, 1969. [Ps.]

3739 - JONES, L.W.: Two quantum-hit requirement for delayed light emission from photosynthetic green algae. - Proc. nat. Acad. Sci. U.S.A. *58*: 75-80, 1967.

3740 - JONES, L.W., KOK, B.: Photoinhibition of chloroplast reactions. I. Kinetics and action spectra. - Plant Physiol. *41*: 1037-1043, 1966.

3741 - JONES, L.W., KOK, B.: Photoinhibition of chloroplast reactions. II. Multiple effects. - Plant Physiol. *41*: 1044-1049, 1966.

3742 - JONES, M.B., MANSFIELD, T.A.: A circadian rhythm in the level of the carbon dioxide compensation point in *Bryophyllum* and *Coffea*.- J. exp. Bot. *21*: 159-163, 1970.

3743 - JONES, O.T.G.: A protein-protochlorophyll complex obtained from inner seed coats of *Cucurbita pepo*. The resolution of its two pigment groups into true protochlorophyll and a pigment related to bacterial protochlorophyll. - Biochem. J. *101*: 153-160, 1966.

3744 - JONES, O.T.G.: Haem synthesis by isolated chloroplasts. - Biochem. biophys. Res. Commun. *28*: 671-674, 1967.

3745 - JONES, O.T.G.: Intermediates in chlorophyll biosynthesis in *Rhodopseudomonas spheroides*: effects of substrates and inhibitors. - Phytochemistry *6*: 1355-1362, 1967.

3746 - JONES, O.T.G.: Biosynthesis of chlorophylls. - In: GOODWIN, T.W. (ed.): Porphyrins and Related Compounds. Pp. 131-145. Academic Press, London-New York 1968.

3747 - JONES, O.T.G., WHALE, F.R.: The oxidation and reduction of pyridine nucleotides by *Rhodopseudomonas spheroides* and *Chlorobium thiosulfatophilum*. - Arch. Mikrobiol. *72*: 48-59, 1970.

3748 - JONES, P.C.T.: The effect of light, temperature, and anaesthetics on ATP levels in the leaves of *Chenopodium rubrum* and *Phaseolus vulgaris*. - J. exp. Bot. *21*: 58-63, 1970.

3749 - JONES, R.: Estimating productivity and apparent photosynthesis from differences in consecutive measurements of total living plant parts of an Australian heathland. - Aust. J. Bot. *16*: 589-602, 1968.

3750 - JONES, R., HODGKINSON, K.C., RIXON, A.J.: Growth and productivity in rangeland species of *Atriplex*. - In: JONES, R. (ed.): The Biology of *Atriplex*. Pp. 31-42. CSIRO, Canberra 1970.

3751 - JONES, R.F., CHEN, J.H.: Amino acid stimulation of carbon dioxide fixation in cell-free extracts of *Chlamydomonas reinhardtii*. - Plant Physiol. *46*: 761-762, 1970.

3752 - JONES, R.J., HAYDOCK, K.P.: Yield estimation of tropical and temperature pasture species using an electronic capacitance meter. - J. agr. Sci. *75*: 27-36, 1970.

3753 - JONES, R.J., MANSFIELD, T.A.: Increases in the diffusion resistances of leaves in a carbon dioxide-enriched atmosphere. - J. exp. Bot. *21*: 951-958, 1970.

3754 - JONSSON, M., PETTERSSON, E.: Isoelectric focusing of plant pigments. - Sci. Tools *15* (1): 2-6, 1968.

3755 - JORDAN, C.F.: Derivation of leaf-area index from quality of light on the forest floor. - Ecology *50*: 663-666, 1969.

3756 - JORDAN, J.M.: A photographic study of stomatal response of ten higher plants to seven concentrations of carbon dioxide at three temperatures. - Diss. Abstr. int., Ser. B *30*: 2491-B-2492-B, 1970.

3757 - JORDAN, R.B., ODELL, A.L.: Purification of carbon dioxide for mass spectral analysis by gas-liquid chromatography. - Anal. Chem. *39*: 681-682, 1967. [For ^{18}O analysis.]

3758 - JØRGENSEN, E.G.: Photosynthetic activity during the life cycle of synchronous *Skeletonema* cells. - Physiol. Plant. *19*: 789-799, 1966.

3759 - JØRGENSEN, E.G.: The adaptation of plankton algae. II. Aspects of the temperature adaptation of *Skeletonema costatum*. - Physiol. Plant. *21*: 423-427, 1968. [Ps.]

3760 - JØRGENSEN, E.G.: The adaptation of plankton algae. IV. Light adaptation in different algal species. - Physiol. Plant. *22*: 1307-1315, 1969. [Ps, Chl.]

3761 - JØRGENSEN, E.G.: The adaptation of plankton algae. V. Variation in the photosynthetic characteristics of *Skeletonema costatum* cells grown at low light intensity. - Physiol. Plant. *23*: 11-17, 1970.

3762 - JOSHI, G.V.: Photosynthesis in marine plants of Bombay. - In: KRISHNAMURTHY, V.

(ed.): Proceedings of the 1st Seminar Sea, Salt, Plants. Pp. 256-267. Bhavnagar, India 1967.

3763 - JOSHI, G.V., MISHRA, S.D.: Photosynthesis and mineral metabolism in senescent leaves of *Clerodendron inerme* GAERTN. - Indian. J. exp. Biol. *8*: 41-43. 1970.

3764 - JOURNEAUX, R., HOCHAPFEL, A.,VIOVY, R.: Existence et nature des formes agrégées de chlorophylle *a* dans les solvants binaires. II. Étude des spectres de fluorescence. - J. Chim. phys. Phys.-chim. biol. *66*: 1474-1478, 1969.

3765 - JOUSSAUME, M., BOURDU, R.: Conversion of carbon-14-labelled orotic acid into pyrimidine nucleotides by chloroplasts. - Nature *210*: 1363-1364, 1966.

3766 - JOY, K.W.: Carbon and nitrogen sources for protein synthesis and growth of sugar-beet leaves. - J. exp. Bot. *18*: 140-150, 1967. [Ps.]

3767 - JOZWIAK, Z., LEYKO, W.: Ratio of adenine nucleotides to other nucleotides of spinach. - In: METZNER, H. (ed.): Progress in Photosynthesis Research. Vol. I. Pp. 345-351. Tübingen 1969.

3768 - JUNG, J., DRESSEL, J., RIEHLE, G.: Der Einfluss von N,N-Dimethyl-(2-Chloräthyl)-hydrazoniumchlorid (CHM) auf den Gehalt an N, P, K, Mg und Blattpigmenten von Weizen. - Z. Pflanzenphysiol. *62*: 343-351, 1970.

3769 - JUNGE, W.: The critical electric potential difference for photophosphorylation. Its relation to the chemiosmotic hypothesis and to the triggering requirements of the ATP-ase system. - Europe. J. Biochem. *14*: 582-592, 1970.

3770 - JUNGE, W., EMRICH, H.M., WITT, H.T.: The indication of a light induced electrical field by pigments incorporated in chloroplast membranes. - In: SNELL, F., WOLKEN, J., IVERSON, G., LAM, J.(ed.): Physical Principles of Biological Membranes. Pp. 383-396. Gordon and Breach sci. Publ., New York-London-Paris 1970.

3771 - JUNGE, W., REINWALD, E., RUMBERG, B., SIGGEL, U., WITT, H.T.: Further evidence for a new function unit of photosynthesis. - Naturwissenschaften *55*: 36-37, 1968.

3772 - JUNGE, W., RUMBERG, B., SCHRÖDER, H.: The necessity of an electric potential difference and its use for photophosphorylation in short flash groups. - Europe. J. Biochem. *14*: 575-581, 1970.

3773 - JUNGE, W., SCHLIEPHAKE, W.D., WITT, H.T.: Experimental evidence for the chemiosmotic hypothesis. - In: METZNER, H. (ed.): Progress in Photosynthesis Research. Vol. III. Pp. 1383-1391. Tübingen 1969.

3774 - JUNGE, W., WITT, H.T.: On the ion transport system of photosynthesis. - Investigations on a molecular level. - Z. Naturforsch. *23*: 244-254, 1968.

3775 - KABANOVA, Yu.G.: Izuchenie mineral'nogo pitaniya morskogo fitoplanktona. [Study of mineral nutrition of marine phytoplankton.] - Okeanologiya *7*: 495-504, 1967. [Chl; in R.]

3776 - KÁBELOVÁ, J., MINÁŘ, J., TICHÝ, V.: Vliv některých humusových frakcí na obsah lipofilních barviv v řase *Scenedesmus quadricauda* (TURP.) BRÉB.[Effect of some humus fractions on the content of lipophilic pigments in the alga *Scenedesmus quadricauda* (TURP.) BRÉB.] - Spisy přírodověd. Fak. Univ. Brno *1969* (4): 139-148, 1969. [In Czech.]

3777 - KACHARAVA, N.F.: Vliyanie ul'trafioletovoĭ radiatsii na fotosintez svetolyubivykh rasteniĭ. [Effect of ultraviolet radiation on photosynthesis of sun and shade plants.] -Soobshch. Akad. Nauk Gruz. SSR *43*: 723-727, 1966. [In R.]

3778 - KACPERSKA-PALACZ, A.E.: Chlorophyll content in tomato leaves as a measure of amitrol uptake by different root zones. - Photosynthetica *1*: 193-198, 1967.

3779 - KAFALIEVA, D.N.: Electron-spin resonance studies of photo-induced free radicals in native chloroplasts. - Dokl. bolg. Akad. Nauk *20*: 141-144, 1967.

3780 - KAFALIEVA, D.N., BLYUMENFEL'D, L.A., SOLOV'EV, I.S., LIVSHITS, V.A., DARMANYAN, A.P.: Relaksatsionnye kharakteristiki fotoindutsirovannykh paramagnitnykh tsentrov v khloroplastakh *Vicia faba*. [Relaxation characteristics of photoinduced para-

magnetic centres in chloroplasts of *Vicia faba*.] - Biofizika *14*: 1117-1119,
1969. [In R, ab: E.]

3781 - KAHN, A.: Developmental physiology of bean leaf plastids. II. Negative contrast
electron microscopy of tubular membranes in prolamellar bodies. - Plant Physiol.
43: 1769-1780, 1968.

3782 - KAHN, A.: Developmental physiology of bean leaf plastids. III. Tube transformation and protochlorophyll(ide) photoconversion by a flash irradiation. - Plant
Physiol. *43*: 1781-1785, 1968.

3783 - KAHN, A., BOARDMAN, N.K., THORNE, S.W.: Energy transfer between protochlorophyllide molecules: Evidence for multiple chromophores in the photoactive protochlo-
rophyllide-protein complex *in vivo* and *in vitro*. - J. mol. Biol. *48*: 85-101,
1970.

3784 - KAHN, J.S.: Photophosphorylation by isolated chloroplasts of *Euglena gracilis*.
- Biochem. biophys. Res. Commun. *24*: 329-33, 1966.

3785 - KAHN, J.S.: Chlorotri-*n*-butyltin. An inhibitor of photophosphorylation in isolated chloroplasts. - Biochim. biophys. Acta *153*: 203-210, 1968.

3786 - KAHN, J.S.: Absence of a common intermediate pool among individual enzyme chains
of the energy-conservation pathway in chloroplasts of *Euglena gracilis*. - Biochem. J. *116*: 55-60, 1970.

3787 - KAÏBIYAÏNEN, T.M., PETROV, V.E.: O mikrokalorimetricheskom izmerenii energii,
zapasaemoĭ fotosinteziruyushcheĭ vodorosl'yu *Chlorella* r. 84. [Microcalorimetric
determination of energy accumulated by *Chlorella* strain 84.] - In: Funktsional'-
nye Osobennosti Khloroplastov. Pp. 9-13. Kazan. Univ., Kazan 1969. [In R.]

3788 - KAÏBIYAÏNEN, T.M., PETROV, V.E., NIKOLAEV, V.: Vliyanie povyshennykh temperatur
na velichinu zapasaemoĭ energii i gazoobmen *Chlorella* r. 84 T. [Effect of increased temperature on the amount of stored energy and gas exchange of *Chlorella*
strain 84 T.] - In: Funktsional'nye Osobennosti Khloroplastov. Pp. 14-18. Kazan.
Univ., Kazan 1969. [In R.]

3789 - KAÏRYUKSHTIS, L.A.: Ratsional'noe ispol'zovanie solnechnoĭ energii kak faktor
povysheniya produktivnosti listvenno-elovykh nasazhdeniĭ. [Effective utilization
of solar energy as factor of increasing the productivity of mixed deciduous
forest with fir.] - In: Svetovoĭ Rezhim, Fotosintez i Produktivnost' Lesa. Pp.
151-166. Nauka, Moskva 1967. [In R.]

3790 - KAJA, H.: Elektronenmikroskopische Untersuchungen zum Feinbau der Chromatophoren
von *Mougeotia* spec. - Z. Naturforsch. *21 b*: 379-384, 1966.

3791 - KAKHNOVICH, L.V.: O nakoplenii pigmentov i izmenenii plastidnogo apparata v
list'yakh redisa i salata v zavisimosti ot spektral'nogo sostava sveta. [Pig-
ment accumulation and changes of the pigment apparatus in leaves of radish and
lettuce in relation to spectral composition of light.] - In: Botanika Issledovaniya. Vol. 9. Pp. 14-19. Nauka i Tekhnika, Minsk 1967. [In R.]

3792 - KAKHNOVICH, L.V., KLIMOVICH, A.S.: Fotokhimicheskaya aktivnost' khloroplastov
pri vozdeĭstvii sinego i krasnogo sveta. [Photochemical activity of chloroplasts
under action of blue and red light.] - Dokl. Akad. Nauk Beloruss. SSR *14*: 664-
667, 1970. [In R.]

3793 - KAKIE, T.: Effect of phosphorus deficiency on the photosynthetic carbon dioxide
fixation-products in tobacco plants. - Soil Sci. Plant Nutrition *15*: 245-251,
1969.

3794 - KAKIE, T.: [Phosphorus metabolism in tobacco plants. Relationship between ethanol soluble phosphorus compounds concerning photosynthesis and maturity of tobacco leaves.] - Bull. Hatano Tobacco exp. Sta. *63*: 43-50, 1969. [In Jap., ab:
E.]

3795 - KALACHEVA, V. Ya., KIRSHTEÏNE, B.E., BIRYUZOVA, V.I., KOSTYCHEVA, L.I., SISAKYAN,
N.M.: Ochistka mitokhondriĭ iz prorostkov gorokha ot fragmentov khloroplastov
tsentrifugirovaniem v gradiente plotnosti. - [Purification of pea seedlings mitochondria from chloroplast fragments by density gradient centrifugation.] -
Biokhimiya *32*: 520-526, 1967. [In R, ab: E.]

3796 - KALBE, L., SCHULZE, H.-A.: Zur Primärproduktion des Kummerower Sees. - Wiss.
       Z. Univ. Rostock, math.-naturwiss. Reihe *18*: 763-767, 1969.

3797 - KALBERER, P.P., BUCHANAN, B.B., ARNON, D.I.: Rates of photosynthesis by isolated
       chloroplasts. - Proc. nat. Acad. Sci. U.S.A. *57*: 1542-1549, 1967.

3798 - KALER, V.L., FEDYUN'KIN, D.V.: Vliyanie dlitel'nogo zatemneniya na razmery khlo-
       roplastov i lyuminestsentnye kharakteristiki zelenogo lista. [Effect of prolon-
       ged darkening on chloroplast dimensions and luminescence characteristics of the
       green leaf.] - In: Fiziologo-biokhimicheskie Issledovaniya Rasteniĭ. Pp. 34-38.
       Nauka i Tekhnika, Minsk 1967. [In R.]

3799 - KALER, V.L., FEDYUN'KIN, D.V., GODNEV, T.N.: Obrazovanie khlorofilla v list'yakh
       Tradescantia guianensis v temnote. [Chlorophyll formation in leaves of *Trades-
       cantia guianensis* in darkness.] - Dokl. Akad. Nauk SSSR *170*: 469-471, 1966. [In
       R.]

3800 - KALER, V.L., PODCHUFAROVA, G.M.: Vliyanie sveta na skorost' resinteza proto-
       khlorofillida v etiolirovannykh prorostkakh yachmenya. [Effect of light on rate
       of protochlorophyllide resynthesis in etiolated barley seedlings.] - In: Issle-
       dovaniya po Fiziologii i Biokhimii Rasteniĭ. Pp. 62-68. Nauka i Tekhnika, Minsk
       1966. [In R.]

3801 - KALER, V.L., PODCHUFAROVA, G.M.: Ustanovka dlya zhidkost'-zhidkostnoĭ mikroek-
       straktsii pigmentov pri raspredelenii "kislykh" i fitolizirovannykh form. [Ap-
       paratus for liquid-liquid microextraction of pigments in separation of "acid"
       and phytolized forms.] - In: Issledovaniya po Fiziologii i Biokhimii Rasteniĭ.
       Pp. 68-71. Nauka i Tekhnika, Minsk 1966. [In R.]

3802 - KALER, V.L., PODCHUFAROVA, G.M.: Svetovoe upravlenie skorost'yu resinteza pro-
       tokhlorofillida v etiolirovannykh prorostkakh yachmenya. [Rate of light driven
       protochlorophyllide resynthesis in etiolated barley seedlings.] - In: Fiziologo-
       biokhimicheskie Issledovaniya Rasteniĭ. Pp. 27-34. Nauka i Tekhnika, Minsk 1967.
       [In R.]

3803 - KALER, V.L., PODCHUFAROVA, G.M.: Lokalizatsiya upravlyayushchego zvena v siste-
       me biosinteza khlorofilla. [Localization of regulatory link in the system of
       chlorophyll biosynthesis.] - Tr. mosk. Obshchestv. Ispyt. Prirody, Otd. biol.
       28: 185-189, 1968. [In R.]

3804 - KALER, V.L., PODCHUFAROVA, G.M.: Sootnoshenie mezhdu kolichestvom prevrativshe-
       gosya protokhlorofillida i skorost'yu ego resinteza v etiolirovannykh prorost-
       kakh yachmenya. [Relation of the amount of transformed protochlorophyllide and
       rate of its resynthesis in etiolated barley seedlings.] - In: Fotosintez i Pita-
       nie Rasteniĭ. Pp. 27-31. Nauka i Tekhnika, Minsk 1969. [In R.]

3805 - KALER, V.L., PODCHUFAROVA, G.M.: Aktivnost' khlorofillazy v etiolirovannykh ras-
       teniyakh. [Chlorophyllase activity in etiolated plants.] - In: Fotosintez i Pi-
       tanie Rasteniĭ. Pp. 32-35. Nauka i Tekhnika, Minsk 1969. [In R.]

3806 - KALER, V.L., PODCHUFAROVA, G.M., SERGEEV, A.A.: Kolebaniya kontsentratsii proto-
       khlorofillida posle zatemneniya zelenykh prorostkov yachmenya. [Fluctuations in
       protochlorophyllide concentration after darkening of green barley seedlings.] -
       In: FRANK, G.M. (ed.): Kolebatel'nye Protsessy v Biologicheskikh i Khimicheskikh
       Sistemakh. Pp. 128-134. Nauka, Moskva 1967. [In R.]

3807 - KALER, V.L., PODCHUFAROVA, G.M., SERGEYEV, A.A.: Some results of the study of
       chlorophyll biosynthesis control mechanisms, using an analog computer. - In:
       METZNER, H. (ed.): Progress in Photosynthesis Research. Vol. II. Pp. 599-605.
       Tübingen 1969.

3808 - KALER, V.L., SERGEEV, A.A., SKACHKOV, N.M.: Proizvodnaya spektrofotometriya bio-
       logicheskikh ob'ektov na spektrofotometre SF-10. [Derivative spectrophotometry
       of biological objects on the spectrophotometer SF-10.] - In: Bioenergetika i
       Biologicheskaya Spektrofotometriya. Pp. 244-248. Nauka, Moskva 1967. [In R.]

3809 - KALER, V.L., SIGALOV, G.G.: Uproshchennaya model' obrazovaniya khlorofilla pri ego
       biosinteze v etiolirovannykh prorostkakh. [Simplified model of chlorophyll form-
       ation during its biosynthesis in etiolated seedlings.] - In: Kibernetika v Ras-
       tenievodstve. Vyp. 95. Pp. 230-240. Moskva 1967. [In R.]

3810 - KALFF, J.: Primary production rates and the effect of some environmental factors
       on algal photosynthesis in small arctic tundra ponds. - Diss. Abstr. 26: 6331,
       1966.

3811 - KALININA, L.M.: K voprosu o prirode polosy 470 nm v spektrakh vozbuzhdeniya
       fluorestsentsii khlorofilla v rasteniyakh. [Nature of the 470 nm band in fluor-
       escence excitation spectra of chlorophyll in leaves.] - In: Metabolizm i Stroe-
       nie Fotosinteticheskogo Apparata. Pp. 82-88. Nauka i Tekhnika, Minsk 1970. [In
       R.]

3812 - KALISHEVICH, S.V., LIPSKAYA, G.A., ANTIPOVA, A.I.: Vliyanie razlichnogo kachest-
       va aeratsii pitatel'nogo rastvora na izmenenie fotosinteziruyushchego apparata
       list'ev sakharnoi svekly. [Effect of different aeration of nutrient solution on
       changes of photosynthesizing apparatus of sugar beet leaves.] - Dokl. Akad. Nauk
       Beloruss. SSR 12: 155-158, 1968. [In R.]

3813 - KALLIS, A.: Koeffitsienty pogloshcheniya FAR rastitel'nym pokrovom na raznykh
       shirotakh. [Coefficients of PhAR absorption by plant communities at different
       latitudes.] - In: Voprosy Effektivnosti Fotosinteza. Pp. 44-63. Tartu 1969. [In
       R, ab: E.]

3814 - KALTOFEN, H.: Relationales Modell des vegetativen Pflanzenwachstums in Abhängig-
       keit von Stickstoffernährung und Photosynthese. - Biol. Rundschau 8: 391-401,
       1970.

3815 - KALTWASSER, H., STUART, T.S., GAFFRON, H.: Light-dependent hydrogen evolution by
       Scenedesmus. - Planta 89: 309-322, 1969.

3816 - KALUDIN, K., KALUDIN, I.:Prouchvaniya v"rkhu karotinnoto s"d"rzhanie na listata
       na nyakoi iglolistni d"rvesni vidove. [Carotene content in leaves of certain co-
       niferous tree species.] - Gorskostop. Nauka 4 (3): 81-88, 1967. [In Bulg., ab:
       E, R.]

3817 - KAMEN, M.: The future of photosynthesis. - In: SAN PIETRO, A., GREER, F.A.,
       ARMY, T.J. (ed.): Harvesting the Sun. Pp. 333-341. Academic Press, New York-
       London 1967.

3818 - KAMEN, M.D., HORIO, T.: Bacterial cytochromes: I. Structural aspects. - Annu.
       Rev. Biochem. 39: 673-700, 1970.[Ps.]

3819 - KAMENTSEVA, I.E.: TeploustoTchivost' fotosinteza i dykhaniya vesennikh i let-
       nikh list'ev *Pulmonaria obscura* DUMORT. [Thermostability of photosynthesis and
       respiration in vernal and estival leaves of *Pulmonaria obscura* DUMORT.] - Dokl.
       Akad. Nauk SSSR 186: 968-970, 1969. [In R.]

3820 - KAMP NIELSEN, L.: The influence of copper on the photosynthesis and growth of
       Chlorella pyrenoidosa. - Dan. Tidsskr. Farm. 43: 249-254, 1969.

3821 - KAMRIN, M.: Changes in the capacity of algae to fluoresce during steady-state
       photosynthesis. - Biochim. biophys. Acta 126: 262-268, 1966.

3822 - KANAI, R., SIMONIS, W.: Einbau von ^{32}P in verschiedene Phosphatfraktionen, be-
       sonders Polyphosphate, bei einzelligen Grünalgen *(Ankistrodesmus braunii)* im
       Licht und im Dunkeln. - Arch. Mikrobiol. 62: 56-71, 1968.

3823 - KANAI, R., SIMONIS, W.: Über den Einfluss der Vorbelichtung auf die anschliess-
       ende Phosphorylierung im Dunkeln bei einzelligen Grünalgen *(Ankistrodesmus brau-
       nii)*. - Arch. Mikrobiol. 63: 29-40, 1968.

3824 - KANAZAWA, T., KANAZAWA, K., KIRK, M.R., BASSHAM, J.A.: Regulation of photosyn-
       thesis carbon metabolism in synchronously growing *Chlorella pyrenoidosa*. - Plant
       Cell Physiol. 11: 149-160, 1970.

3825 - KANAZAWA, T., KANAZAWA, K., KIRK, M.R., BASSHAM, J.A.: Difference in nitrate re-
       duction in "light" and "dark" stages of synchronously grown *Chlorella pyrenoido-
       sa* and resultant metabolic changes. - Plant Cell Physiol. 11: 445-452, 1970.
       [Ps.]

3826 - KANAZAWA, T., KIRK, M.R., BASSHAM, J.A.: Regulatory effects of ammonia on car-
       bon metabolism in photosynthesizing *Chlorella pyrenoidosa*. - Biochim. biophys.
       Acta 205:401-408, 1970.

3827 - **KANDELER, R.:** The role of photophosphorylation in flower initiation of the long-day plant *Lemna gibba*. - In: Photochemistry and Photobiology in Plant Physiology. Europe. Photobiol. Symp. Hvar, Yugoslavia, 19th-22nd September 1967. Book of Abstracts. P. 45. Hvar 1967.

3828 - **KANDLER, O.:** Biosynthesis of poly- and oligosaccharides during photosynthesis in green plants. - In: SAN PIETRO, A., GREER, F.A., ARMY, T.J. (ed.): Harvesting the Sun. Pp. 131-152. Academic Press, New York-London 1967.

3829 - **KANDLER, O., ELBERTZHAGEN, H., HABERER-LIESENKÖTTER, I.:** Rate-limiting factors in the photosynthesis of isolated chloroplasts. - In: GOODWIN, T.W. (ed.): Biochemistry of Chloroplasts. Vol. II. Pp. 39-51. Academic Press, London-New York 1967.

3830 - **KANDLER, O., SENSER, M.:** Differences in the pattern of products of photosynthesis after short term photosynthesis, in $^{14}CO_2$, of *Oryza* and *Sorghum*. - In: COOMBS, J. (ed.): Photosynthesis in Sugar Cane. Pp. 30-32. Tate and Lyle Ltd., London 1969.

3831 - **KANDLER, O., TANNER, W.:** Die Photoassimilation von Glucose als Indikator für die Lichtphosphorylierung *in vivo*. - Ber. deut. bot. Ges. *79*: (48)-(58), 1966.

3832 - **KANEMASU, E.T., TANNER, C.B.:** Stomatal diffusion resistance of snap beans. I. Influence of leaf-water potential. - Plant Physiol. *44*: 1547-1552, 1969.

3833 - **KANEMASU, E.T., TANNER, C.B.:** Stomatal diffusion resistance of snap beans. II. Effect of light. - Plant Physiol. *44*: 1542-1546, 1969.

3834 - **KANEMASU, E.T., THURTELL, G.W., TANNER, C.B.:** Design, calibration and field use of a stomatal diffusion porometer. - Plant Physiol. *44*: 881-885, 1969.

3835 - **KANEMATSU,H., NIIYA, I., IMAMURA, M., KAWAKITA, H.:** [Studies on the quantitative determination of β-carotene and vitamin A in margarine. I. Separation and determination of β-carotene.] - Vitamins (Kyoto) [Vitamin] *33*: 52-56, 1966. [In Jap., ab: E.]

3836 - **KANESHIRO, T., ZWEIG, G.:** Effect of diquat (1,1'-ethylene-2,2'-dipyridylium dibromide) on the formation and photoreactions of chromatophores from *Rhodospirillum rubrum*. - Biochim. biophys. Acta *126*: 225-233, 1966.

3837 - **KANIVETS, V.I., OKANENKO, A.S.:** Vliyanie kal'tsiya na intensivnost' fotosinteza i elementy obmena veshchestv u sakharnoĭ svekly. [Effect of calcium on photosynthetic rate and elements of metabolism in sugar beet.] - Fiziol. Biokhim. kul't. Rast. *2*: 499-504, 1970. [In R, ab: E.]

3838 - **KANNANGARA, C.G.:** The formation of ribulose diphosphate carboxylase protein during chloroplast development in barley. - Plant Physiol. *44*: 1533-1537, 1969.

3839 - **KANNANGARA, C.G., van WYK, D., MENKE, W.:** Immunological evidence for the presence of latent Ca^{2+} dependent ATP-ase and carboxydismutase on the thylakoid surface. - Z. Naturforsch. *25 b*: 613-618, 1970.

3840 - **KANWAR, J.S., BHAMBOTA, J.R.:** Effect of different water tables and salinity levels on the chlorophyll content and chemical composition of leaves of sweet orange *(Citrus sinensis)*. - Indian J. agr. Sci. *38*: 238-243, 1968.

3841 - **KANWISHER, J.W.:** Photosynthesis and respiration in some seaweeds. - In: Some Contemporary Studies in Marine Science. Pp. 407-420. Allen and Unwin, London 1966; Hafner Publ. Co., New York 1966.

3842 - **KANWISHER, J.W., WAINWRIGHT, S.A.:** Oxygen balance in some reef corals. - Biol. Bull. *133*: 318-390, 1967.

3843 - **KAO, O., BERNS, D.S.:** The monomer molecular weight of *C*-phycocyanin. - Biochem. biophys. Res. Commun. *33*: 457-462, 1968.

3844 - **KAPITANCHUK, V.A., NATSIK, V.I., SEMENOV, A.G.:** Sostoyanie i perspektivy razvitiya issledovaniĭ po probleme "Fotosintez" na Ukraine. [State and perspectives of development on photosynthesis studies in the Ukraine.] - Fiziol. Rast. *14*: 562-564, 1967. [In R.]

3845 - **KAPLAN, J.H., JAGENDORF, A.T.:** Further studies on chloroplast adenosine triphosphatase activation by acid-base transition. - J. biol. Chem. *243*: 972-979, 1968.

3846 - KAPLAN, J.H., URIBE, E., JAGENDORF, A.T.: ATP hydrolysis caused by acid-base transition of spinach chloroplasts. - Arch. Biochem. Biophys. *120*: 365-370, 1967.

3847 - KAPLER, R., NEKRASOV, L.I.: Sensibilizatsiya reaktsii vosstanovleniya metilovogo krasnogo adsorbirovannym khlorofillom a i b. [Sensibilization of reduction of methyl-red by adsorbed chlorophyll a and b.] - Biofizika *11*: 420-426, 1966. [In R.]

3848 - KAPLER, R., NEKRASOV, L.I.: Spektral'nye i paramagnitnye svoĭstva adsorbtsionnykh sloev khlorofillov a i b i sostoyanie pigmentov v etikh sloyakh. [Spectral and paramagnetic properties of adsorption layers of chlorophyll a and b and state of pigments in these layers.] - Nauch. Dokl. vyssh. Shkoly, biol. Nauki *13* (3): 133, 1970. [In R.]

3849 - KARABANOV, I.A.: K voprosu o deĭstvii gibberellina na soderzhanie khlorofilla v rasteniyakh. [Action of gibberellin on chlorophyll content in plants.] - Fiziol. Rast. *15*: 1068-1070, 1968. [In R.]

3850 - KARABANOV, I.A.: Vliyanie gibberellina na vodnyĭ rezhim i assimilyatsionnuyu deyatel'nost' list'ev chernoĭ smorodiny v svyazi s mineral'nym pitaniem. [Effect of gibberellin on water relations and assimilatory activity of black currant leaves in relation with mineral nutrition.]- Nauch. Dokl. vyssh. Shkoly, biol. Nauki *12* (1): 74-80, 1969. [In R.]

3851 - KARABASHEV, G.S.: Fotometr dlya issledovaniya spektral'nykh funktsiĭ oslableniya obluchennosti v more. [Photometer for the measurement of irradiance spectral attenuation functions in the sea.] - Okeanologiya *6*: 886-891, 1966. [In R, ab: E.]

3852 - KARANOV, E., IVANOVA, Ĭ., VASILEV, G.: Tsitokininovaya aktivnost' i khimicheskoe stroenie nekotorykh proizvodnykh N-allil-N'-feniltiomocheviny. [Cytokinin activity and chemical structure of some derivatives of N-allyl-N'-phenylthiourea.] - Fiziol. Rast. *15*: 430-435, 1968. [Chl; in R, ab: E.]

3853 - KARANOV, E., NIKOLOV, K., POPOV, K.Y.: Influence of some growth regulators on the aging and state of plastide pigments in leaves of different age. - Dokl. bolg. Akad. Nauk *23*: 1155-1158, 1970.

3854 - KARANOV, E., VASSILEV, G.: Cytokinin activity and chemical structure of some derivatives of N-normal-butyl-N'-phenylthiourea. Retardation of the chlorophyll breakdown. - Dokl. Akad. sel'skokhoz. Nauk Bolg. *3*: 343-348, 1970.

3855 - KARANOV, E.N., KIMENOV, G.P.: Comparative testing of certain growth regulators for retarding the destruction of the chlorophyll and of the proteins and the state of the water in detached leaves of *Phaseolus vulgaris*. - Dokl. bolg. Akad. Nauk *23*: 1425-1428, 1970.

3856 - KARANOV, E.N., NEICHEVA, M.: The influence of CCC, Alar and Phosfone-D and their interaction with kinetin for delaying the destruction of chlorophyll and carotenoids in discs of leaves of *Raphanus sativa* of different age. - Dokl. bolg. Akad. Nauk *23*: 1541-1544, 1970.

3857 - KARAPETYAN, N.V.: Light-induced transformations of bacteriochlorophyll *in vivo*. - In: METZNER, H. (ed.): Progress in Photosynthesis Research. Vol. II. Pp. 778-785. Tübingen 1969.

3858 - KARAPETYAN, N.V., KRAKHMALEVA, I.N., KRASNOVSKIĬ, A.A.: Deĭstvie temperaturnoĭ inaktivatsii na differential'nye spektry pogloshcheniya purpurnykh fotosinteziruyushchikh bakteriĭ. [Effect of thermal inactivation on differential absorption spectra of purple photosynthetic bacteria.] - Dokl. Akad. Nauk SSSR *171*: 1201-1204, 1966. [In R.]

3859 - KARAPETYAN, N.V., KRASNOVSKIĬ, A.A.: Issledovanie svetovykh prevrashcheniĭ pigmentov v kletkakh fotosinteziruyushchikh bateriĭ metodom differentsial'noĭ spektrofotometrii. [Light transformations of pigments in cells of photosynthetic bacteria studied by differential spectrophotometry.] - Tr. mosk. Obshchestv. Ispyt. Prirody *24* (Biol. Avtotrof. Mikroorganizmov): 94-107, 1966. [In R.]

3860 - KARAPETYAN, N.V., KRASNOVSKIĬ, A.A.: Issledovanie spektral'nykh svoĭstv i svetovykh prevrashcheniĭ bakteriokhlorofilla v purpurnykh bateriyakh *Rhodothece* sp. [Spectral properties and light transformations of bacteriochlorophyll in purple bacteria *Rhodothece* sp.] - Dokl. Akad. Nauk SSSR *180*: 989-992, 1968. [In R.]

3861 - **KARDO-SYSOEVA, E.K., GILLER, Yu.E.:** Vliyanie ul'trafioletovogo oblucheniya na tkani lista v usloviyakh pamirskogo vysokogor'ya. [Effect of ultraviolet radiation on leaf tissues in Pamir.] - In: Problemy Botaniki 9 (Rastitel'nyT Mir VysokogoriT SSSR i Voprosy ego Ispol'zovaniya): 363-369, 1967. [Chl; in R.]

3862 - **KARIMOV, Kh.Kh., CHERNER, R.I.:** O biologicheskoT produktivnosti rastitel'nykh soobshchestv Yuzhnogo Tadzhikistana v svyazi s zimneT vegetatsieT i letnim pokoem. [Biological productivity of plant communities of South Tadzhikistan in relation to winter vegetation and summer rest.] - In: Obshchie Teoreticheskie Problemy BiologicheskoT Produktivnosti. Pp. 110-115. Nauka, Leningrad 1969. [In R.]

3863 - **KARIMOV, Kh.Kh., CHERNER, R.I., RAKHMONOV, A.:** Ispol'zovanie energii solnechnoT radiatsii zimnevegetiruyushchimi kul'turami v usloviyakh GissarskoT doliny Tadzhikistana. [Use of solar radiant energy by crops growing in winter in the Gissar valley of Tadzhikistan.] - In: Fotosintez i Ispol'zovanie Energii SolnechnoT Radiatsii. Pp. 14-15. Dushanbe 1967. [In R.]

3864 - **KARIMOV, Kh.Kh., CHERNER, R.I., RAKHMONOV, A.:** Ispol'zovanie energii solnechnoT radiatsii zimnevegetiruyushchimi kul'turami v usloviyakh GissarskoT doliny Tadzhikistana. [Use of solar radiant energy by crops growing in winter in the Gissar valley of Tadzhikistan.] - Izv. Akad. Nauk Tadzh. SSR, Otd. biol. Nauk 1969 (2): 32-37, 1969. [In R.]

3865 - **KARIMOV, Kh.Kh., CHERNER, R.I., RAKHMONOV, A.:** Ispol'zovanie solnechnoT radiatsii zimnevegetiruyushchimi bobovo-zlakovymi kul'turami v sredneT Azii. [Use of solar energy by leguminous crops growing in winter in Central Asia.] - Sel'-skokhoz. Biol. 5: 767-770, 1970. [In R.]

3866 - **KARIMOV, Yu.Yu., USMANBEKOV, N.Ya.:** Vliyanie izmeneniya vodnogo rezhima i pitaniya na fiziologicheskie protsessy v rasteniyakh kartofelya. [Effect of changes in water regime and nutrition on physiological processes in potato plants.] - Uzbeksk. biol. Zh. 14 (5): 80-81, 1970. [Ps; in R.]

3867 - **KARIMOVA, M.A., ROMANOVA, A.K., DOMAN, N.G.:** Karboksilaza ribuleza-1,5-difosfata,ee svoTstva i rol' v fotosinteze. [Ribulose 1,5-diphosphate carboxylase, its characteristics and role in photosynthesis.] - In: Fotosintez i Ispol'zovanie Energii SolnechnoT Radiatsii. P. 54. Dushanbe 1967. [In R.]

3868 - **KARIMOVA, M.A., SHKOL'NIK, R.Ya., DOMAN, N.G.:** Karboksidismutaza i soputstvuyushchie eT fermenty iz list'ev gorokha, lyutserny i rezushki talya (Arabidopsis thaliana). [Carboxydismutase and accompanying enzymes from leaves of pea, alfalfa and Arabidopsis thaliana.] - In: Issledovaniya po Fotosintezu. Pp. 66-77. Akad. Nauk Tadzh. SSR, Dushanbe 1967. [In R, ab: Tadzh.]

3869 - **KARLANDER, E.P., KRAUSS,R.W.:** Responses of heterotrophic cultures of Chlorella vulgaris BEYERINCK to darkness and light. I. Pigment and pH changes. - Plant Physiol. 41: 1-6, 1966.

3870 - **KARLANDER, E.P., KRAUSS,R.W.:** The laser as a light source for the photosynthesis and growth of Chlorella vannielii. - Biochim. biophys. Acta 153: 312-314, 1968.

3871 - **KARLANDER, E.P., PATTERSON, G.W.:** Chlorophyll bleaching, fluorescence and absorption shifts in Chlorella sorokiniana promoted by NaCl. - Plant Physiol. 43: S-21, 1968.

3872 - **KARLISH, S., AVRON, M.:** Light-induced proton uptake in chloroplasts - relevance to theories of energy coupling in photophosphorylation. - Israel J. Chem. 5: 98 p, 1967.

3873 - **KARLISH, S., AVRON, M.:** The effect of ion permeability-inducing agents on ion transport in chloroplasts. - In: TAGER, J.M., PAPA, S., QUAGLIARIELLO, E., SLATER, E.C. (ed.): Electron Transport and Energy Conservation. Pp. 431-448. Adriatica Ed., Bari 1970.

3874 - **KARLISH, S.J.D., AVRON, M.:** Relevance of proton uptake induced by light to the mechanism of energy coupling in photophosphorylation. - Nature 216: 1107-1109, 1967.

3875 - **KARLISH, S.J.D., AVRON, M.:** Analysis of light-induced proton uptake in isolated chloroplasts. - Biochim. biophys. Acta 153: 878-888, 1968.

3876 - KARLISH, S.J.D., AVRON, M.: Dinitrophenol and valinomycin as uncouplers in iso-
       lated chloroplasts. - FEBS Letters 1: 21-24, 1968.

3877 - KARLISH, S.J.D., AVRON, M.: The relevance of light-induced proton uptake to the
       mechanism of energy coupling in photophosphorylation. - In: SHIBATA, K., TAKA-
       MIYA, A., JAGENDORF, A.T., FULLER, R.C. (ed.): Comparative Biochemistry and Bio-
       physics of Photosynthesis. Pp. 214-221. Univ. Tokyo Press, Tokyo; Univ. Park
       Press, State College, Pa. 1968.

3878 - KARLISH, S.J.D., AVRON, M.: On the effect of phosphorylating conditions on pho-
       toinduced changes in light scattering. - Photosynthetica 3: 79-82, 1969.

3879 - KARLISH, S.J.D., SHAVIT, N., AVRON, M.: On the mechanism of uncoupling in chlo-
       roplasts by ion-permeability inducing agents. - Europe. J. Biochem. 9: 291-298,
       1969.

3880 - KARLSTAM, B., ALBERTSSON, P.-Å.: Demonstration of three classes of spinach chlo-
       roplasts by counter-current distribution. - FEBS Letters 5: 360-363, 1969.

3881 - KARLSTAM, B., ALBERTSSON, P.-Å.: A spectrophotometric method for the quantita-
       tive estimation of intact (Class I) chloroplasts. - Biochim. biophys. Acta 216:
       220-222, 1970.

3882 - KARMANOV, V.G., RYABOVA, E.P.: Pribor dlya izmereniya temperatury rasteniĭ.
       [Device for measuring plant temperature.] - Byul. nauch.-tech. Inform. agron.
       Fiz. (Leningrad) 12: 24-27, 1968. [In R.]

3883 - KARMANOV, V.G., SOLOV'EV, E.V., MUKHIN, V.P.: Pribor dlya izmereniya kontsentrat-
       sii dvuokisi ugleroda po teploprovodnosti s poluprovodnikovymi termosoprotivle-
       niyami. [Apparatus for measuring CO_2 concentration by determination of heat con-
       ductance with semiconductive thermistors.] - Sb. Tr. agron. Fiz. 24 (Voprosy eksp.
       Biofiz. Kibernetiki Rast.): 67-73, 1969. [In R.]

3884 - KARPILOV, Yu.S.: Obrazovanie i metabolizm aminokislot pri fotosinteze. [Formation
       and metabolism of amino acids during photosynthesis.] - In: Puti Povysheniya In-
       tensivnosti i Produktivnosti Fotosinteza. Pp. 58-76. Naukova Dumka, Kiev 1966.
       [In R.]

3885 - KARPILOV, Yu.S.: Uchastie azota i fosfora v reaktsiyakh fotosinteza. [Participa-
       tion of nitrogen and phosphorus in photosynthetic reactions.] - In: Mineral'nye
       Elementy i Mekhanizm Fotosinteza. Pp. 3-12. Kishinev 1969. [In R.]

3886 - KARPILOV, Yu.S.: Osobennosti funktsiĭ i struktury fotosinteticheskogo apparata
       nekotorykh vidov rasteniĭ tropicheskogo proiskhozhdeniya. [Peculiarities of the
       functions and structure of the photosynthetic apparatus in some species of
       plants of tropical origin.] - Tr. mold. nauch.-issled. Inst. orosh. Zemled.
       Ovoshchevod. 11: 3-34, 1969. [In R, ab: E.]

3887 - KARPILOV, Yu.S.: Fotodykhanie list'ev kukuruzy. [Photorespiration of maize
       leaves.] - In: Kooperativnyĭ Fotosintez Kserofitov. Tr. mold. nauch.-issled.
       Inst. orosh. Zemled. Ovoshchevod. 11 (3): 46-64, 1970. [In R, ab: E.]

B3888 - KARPILOV, Yu.S.: Fotosintez Kserofitov (Evolyutsionnye Aspekty). [Photosynthesis
       of xerophytes (Evolutionary Aspects).] - Akad. Nauk Mold. SSSR, Kishinev 1970.
       [In R.]

3889 - KARPILOV, Yu.S.: Ob odnoĭ gruppe kserorezistentnykh rasteniĭ, skhodnykh po ana-
       tomii lista, fotosintezu i vodnomu rezhimu. [On one group of xeroresistent
       plants similar in the leaf anatomy, photosynthesis and water relations.] - In:
       Kooperativnyĭ Fotosintez Kserofitov. Tr. mold. nauch.-issled. Inst. orosh. Zem-
       led.Ovoshchevod. 11 (3): 3-17, 1970. [In R, ab: E.]

3890 - KARPILOV, Yu.S., BIL', K.Ya.: Sootnoshenie dvukh tipov khloroplastov v list'-
       yakh kukuruzy pri razlichnom azotno-fosfornom pitanii rasteniĭ. [Relationship
       of two types of chloroplasts in corn leaves in the case of a different nitrogen-
       phosphorus nutrition of plants.] - Dokl. Akad. Nauk SSSR 193: 1198-1200, 1970.
       [In R.]

3891 - KARPILOV, Yu.S., BIL', K.Ya., MALYSHEV, O.G., KARNAUKHOV, V.N.: Osobennosti fo-
       tosistem khloroplastov v mezofille i parenkhimnykh obkladkakh provodyashchikh
       puchkov list'ev u odnodol'nykh i dvudol'nykh rasteniĭ. [Properties of photosys-

tems of chloroplasts in mesophyll and bundle sheath cells in leaves of monocotyledonous and dicotyledonous plants.] - In: Kooperativnyĭ Fotosintez Kserofitov. Tr. mold. nauch.-issled. Inst. orosh. Zemled. Ovoshchevod. *11* (3): 25-32, 1970. [In R, ab: E.]

3892 - KARPILOV, Yu.S., BRIK, P.L.: Osobennosti stroeniya i funktsional'noĭ deyatel'-nosti fotosinteticheskogo apparata kukuruzy pri razlichnoĭ stepeni azotnogo i fosfornogo golodaniya. [Peculiarities of structure and function of the photosynthetic apparatus of maize under various levels of nitrogen and phosphorus starvation.] - In: Mineral'nye Elementy i Mekhanizm Fotosinteza. Pp. 160-168. Kishinev 1969. [In R.]

3893 - KARPILOV, Yu.S., KOTOVA, N.F.: Vliyanie usloviĭ azotno-fosfornogo pitaniya na obrazovanie produktov fotosinteza u gorokha i fasoli pri razlichnoĭ intensivnosti sveta. [Effect of nitrogen and phosphorus nutrition on formation of photosynthates in pea and bean at various irradiances.] - In: Puti Povysheniya Intensivnosti i Produktivnosti Fotosinteza. Vol. 3. Pp. 39-48. Naukova Dumka, Kiev 1969. [In R.]

3894 - KARPILOV, Yu.S., MALYSHEV, O.G.: Uglerodnyĭ metabolizm fotosinteza v list'yakh kukuruzy. [Carbon metabolism of photosynthesis in maize leaves.] - In: Kooperativnyĭ Fotosintez Kserofitov. Tr. mold. nauch.-issled. Inst. orosh. Zemled. Ovoshchevod. *11* (3): 33-45, 1970. [In R, ab: E.]

3895 - KARPILOV, Yu.S., MASLOVA, N.F., RUSSU, L.: Osobennosti pigmentnogo sostava list'-ev i fotosinteza v zavisimosti ot proiskhozhdeniya i skorospelosti sortov. [Peculiarities of pigment composition in leaves and photosynthesis in relation to origin and earliness of cultivars.] - In: Fiziologiya i Biokhimiya Sorta. No. 1. Pp. 113-118. Irkutsk 1969. [In R.]

3896 - KARPILOV, Yu.S., NEDOPEKINA, I.F., KOTOVA, N.F.: O nekotorykh osobennostyakh puti ugleroda v fotosinteze u razlichnykh vidov rasteniĭ. [Some peculiarities of carbon pathways in photosynthesis of various plant species.] - Tr. mold. nauch.-issled. Inst. orosh. Zemled. Ovoshchevod. *8* (1): 19-34, 1968. [In R.]

3897 - KARPILOV, Yu.S., NEDOPEKINA, I.F., KOTOVA, N.F., ZARVANSKAYA, E.I.: Uchastie azota i fosfora v biokhimicheskikh reaktsiyakh fotosinteza v svyazi s ikh osobennostyami u razlichnykh vidov ovoshchnykh rasteniĭ. [Participation of nitrogen and phosphorus in biochemical reactions of photosynthesis in relation to their peculiarities in various species of fruit trees.] - Tr. mold. nauch.-issled. Inst. orosh. Zemled. Ovoshchevod. *9*: 181-187, 1968. [In R, ab: E.]

3898 - KARPILOV, Yu.S., PRISTUPA, N.A., BIL', K.Ya., BRIK, P.L.: Stroenie fotosinteticheskogo apparata list'ev u rasteniĭ s kooperativnym fotosintezom. [Formation of photosynthetic apparatus in leaves of plants with co-operation photosynthesis.] - In: Kooperativnyĭ Fotosintez Kserofitov. Tr. mold. nauch.-issled. Ins. orosh. Zemled. Ovoshchevod. *11* (3): 18-24, 1970. [In R, ab: E.]

3899 - KARPOV, V.L., BYKOV, O.D., SELEZNEV, Yu.M.: Tekhnika vyrashchivaniya khlorelly v atmosfere $C^{14}O_2$ vysokoĭ udel'noĭ radioaktivnosti. [Technique for *Chlorella* cultivation in an atmosphere of $^{14}CO_2$ of high specific activity.] - Fiziol. Rast. *16*: 169-174, 1969. [In R, ab: E.]

3900 - KARPUSHKIN, L.T.: Opredelenie vertikal'nogo uglekislotnogo profilya v posevakh. [Determination of vertical profile of CO_2 in canopies.] - In: Fotosinteziruyushchie Sistemy Vysokoĭ Produktivnosti. Pp. 149-156. Nauka, Moskva 1966. [In R.]

3901 - KARTASHOVA, E.R.: Dykhatel'nye sistemy list'ev, razlichayushchikhsya po soderzhaniyu khlorofilla. [Respiratory systems of leaves differing in chlorophyll content. - Nauch. Dokl. vyssh. Shkoly, biol. Nauki *12* (6): 77-83, 1969. [In R.]

3902 - KARU, A.E., MOUDRIANAKIS, E.N.: Fractionation and comparative studies of enzymes in aqueous extracts of spinach chloroplasts. - Arch. Biochem. Biophys. *129*: 655-671, 1969.

3903 - KARVÉ, A.D., MISHAL, B.D.: The phenomenon of overshoot in the case of stomatal opening. - Naturwissenschaften *53*: 280, 1966.

3904 - KARYAGIN, Yu.G., TOLSTENKO, L.A.: Vliyanie udobreniĭ na urozhaĭ i nekotorye fiziologo-biokhimicheskie protsessy v rasteniyakh soi v svyazi s problemoĭ uchas-

tiya kluben'kovykh bakteriĭ v azotofiksatsii. [Effect of fertilizers on yield
and some physiological and biochemical processes in soybean as related to the
nitrogen-fixing activity of nodule bacteria.] - Tr. kaz. nauch.-issled. Inst.
Zemled. *1970* (9/10): 252-258, 1970. [Chi; in R.]

3905 - KARYAKIN, A.V., CHIBISOV, A.K.: Tripletnye sostoyaniya pigmentov i ikh rol' v
fotoreaktsiyakh. [Triplet state of pigments and their role in photoreactions.]
- In: Elementarnye Fotoprotsessy v Molekulakh. Pp. 296-313. Nauka, Moskva -
Leningrad 1966. [In R.]

3906 - KARYAKIN, A.V., CHMUTINA, L.A.: Issledovanie vzaimodeĭstviya khlorofillina (*a +
b*) s glitsininom pri razlichnykh znacheniyakh pH. [Interaction between chlorophyllin (*a + b*) and glycinine at various pH.] - Biokhimiya *31*: 3-7, 1966. [In
R, ab: E.]

3907 - KARYAKIN, A.V., MURADOVA, G.A., SAENKO, G.N.: O sostoyanii vody v veshchestvakh,
prinimayushchikh uchastie v protsesse fotosinteza. [Water state in substances
participating in photosynthesis.] - Biofizika *14*: 240-244, 1969. [In R, ab: E.]

3908 - KARYAKIN, A.V., NIKITINA, A.A., CHMUTINA, L.A.: Issledovanie vzaimodeĭstviya
belka s krasitelyami i khlorofillom po infrakrasnym spektram pogloshcheniya.
[Interaction of protein with dyes and chlorophyll studied by infra-red absorption spectra.] - Biofizika *12*: 344-345, 1967. [In R.]

3909 - KASEMIR, H., MOHR, H.: Die Wirkung von Phytochrom und Actinomycin *D* auf die Chlorophyll A-Synthese von Senfkeinlingen (*Sinapis alba* L.). - Planta *73*: 187-197,
1967.

3910 - KASIM, M.: Beitrag zur qualitativen Analyse der Carotinoide von Kartoffeln. -
Nahrung *11*: 405-409, 1967.

3911 - KASIM, M.: Untersuchungen über die Zusammensetzung der Kartoffel-Carotinoide. -
Nahrung *11*: 411-415, 1967.

3912 - KASK, K.: O deĭstvii alkiliruyushchikh soedineniĭ u chereshni v M_1. [Action of
alkylating substances in M_1 of cherry.] - Izv. Akad. Nauk Est. SSR, Ser. biol.
17: 181-186, 1968. [Chi; in R.]

3913 - KASPERBAUER, M.J., HIATT, A.J.:Photoreversible control of leaf shape and chlorophyll content in *Nicotiana tabacum* L. - Tobacco *162* (8): 30-33, 1966; Tobacco
Sci. *10*: 29-32, 1966.

3914 - KASSNER, R.J., KAMEN, M.D.: The photoreduction of spinach ferredoxin in the presence of porphyrin and an electron donor. - Proc. nat. Acad. Sci. U.S.A. *58*:
2445-2450, 1967.

3915 - KASSNER, R.J., KAMEN, M.D.: Trace metal composition of photosynthetic bacteria.
- Biochim. biophys. Acta *153*: 270-278, 1968.

3916 - KAS'YANENKO, A.G.: Deĭstvie geneticheskikh faktorov na fotosinteticheskiĭ apparat. [Effect of genetic factors on the photosynthetic apparatus.] - In: Issledovaniya po Fotosintezu. Pp. 77-96. Akad. Nauk Tadzh. SSR, Dushanbe 1967. [In
R, ab: Tadzh.]

3917 - KAS'YANENKO, A.G.: Indutsirovanie oblucheniem khlorofil'nykh mutantov u *Arabidopsis thaliana* i indeks fotosinteza u nekotorykh iz nikh. [Chlorophyll mutants
of *Arabidopsis thaliana* and chlorophyll index in some mutants.] - In: Experimental'nyĭ Mutagenez u Sel'skokhozyaĭstvennykh Rasteniĭ i ego Ispol'zovanie v Selektsii. Tr. mosk. Obshch. Ispyt. Prirody *23*: 284-288, 1967. [In R, an: E.]

3918 - KAS'YANENKO, A.G.: Mutatsionnye sistemy fotosinteticheskogo apparata. [Mutation
systems of the photosynthetic apparatus.] - In: Fotosintez i Ispol'zovanie Energii Solnechnoĭ Radiatsii. Pp. 55-56. Dushanbe 1967. [In R.]

3919 - KAS'YANENKO, A.G.: Vliyanie geneticheskikh faktorov na soderzhanie fotosinteticheskikh pigmentov *Arabidopsis thaliana*. [Effect of genetic factors on the content of photosynthetic pigments in *Arabidopsis thaliana*.] - In: Trudy I. Konferentsii Biokhimikov Respublik Sredneĭ Azii i Kazakhstana. Pp. 101-104. FAN,
Tashkent 1967. [In R.]

3920 - KAS'YANENKO, A.G., GILLER, Yu.E.: K kharakteristike fotosinteticheskogo apparata khlorofil'nykh mutantov *Arabidopsis thaliana* (L.). [Characteristics of pho-

tosynthetic apparatus of chlorophyll mutants of *Arabidopsis thaliana* (L.).] - Genetika *1967* (6): 42-49, 1967. [In R.]

3921 - KAS'YANENKO, A.G., NASYROV, Yu.S.: O detstvii geneticheskikh faktorov na foto-sinteticheskiĭ apparat *Arabidopsis thaliana* (L.) HEYNH. [Action of genetic factors on photosynthetic apparatus of *Arabidopsis thaliana* (L.) HEYNH.] - Fiziol. Rast. *15*: 422-429, 1968. [In R, ab: E.]

3922 - KAS'YANENKO, A.G., SMOLINA, Z.A.: Morfologicheskaya kharakteristika khloroplas-tov u khlorofil'nykh mutantov. [Morphological characteristics of chloroplasts in chlorophyll mutants.] - In: Tezisy Dokladov Konferentsii Molodykh Uchenykh. Pp. 46-47. Dushanbe 1966. [In R.]

3923 - KAS'YANENKO, A.G., TIMOFEEV-RESOVSKIĬ, N.V.: O neskol'kikh interesnykh "khloro-fil'nykh" mutatsiyakh u *Arabidopsis thaliana* (L.) HEYNH. [Some interesting "chlo-rophyll" mutations in *Arabidopsis thaliana* (L.) HEYNH.] - Byul. mosk. Obshch. Ispyt. Prir. Otd. biol. *72*: 100-105, 1967. [In R.]

3924 - KASZTORI, R., TYUPINA, T.: A Cu, B, Mn, és Zh katása a kloroplaszt festékanya-gainak tartalmára a búza fejlődésínek egyes fázisaiban. [Coloring matter con-tent of chloroplasts as affected by Cu, B, Mn, and Zn in the individual phases of the development of wheat.] - Agrokém. Talajtan *16*: 161-168, 1967. [In Hung.]

3925 - KATANA, Kh.: Soderzhanie khlorofilla v list'yakh yabloni v zavisimosti ot voz-rasta plodovykh organov. [Chlorophyll content in apple leaves in relation to age of fruit organs.] - Dokl. mosk. sel'.-khoz. Akad. K.A. Timiryazeva *125*: 33-37, 1966. [In R.]

3926 - KATAYAMA, M., BENSON, A.A.: α-linolenate and photosynthetic activity in *Chlorel-la prototheoides*. - Plant Physiol. *42*: 308-313, 1967.

3927 - KATAYAMA, T.: [Comparative biochemistry of carotenoids in algae. - II. On caro-tenoids in *Codium intricatum* and their biosynthesis.] - Bull. Jap. Soc. Sci. Fish. *32*: 610-620, 1966. [In Jap., ab: E.]

3928 - KATAYAMA, Y., SHIDA, S.: [Studies on the contamination and the variation of crop plants caused by radioactivity: Leaf pigment variants found in γ-rayed rice and barley progenies.] - Bull. Fac. Agr. Univ. Miyazaki *14*: 369-376, 1967. [In Jap.]

3929 - KATAYAMA, Y., SHIDA, S.: Studies on the change of chlorophyll *a* and *b* contents due to projected materials and some environmental conditions. - Cytologia *35*: 171-180, 1970.

3930 - KATES, J.R., JONES, R.F.: Pattern of CO_2 fixation during vegetative development and gametic differentiation in *Chlamydomonas reinhardtii*. - J. cell. Physiol. *67*: 101-105, 1966.

3931 - KATO, T., YASUOKA, K.-I.: [Fundamental studies on the apparent photosynthesis for cucumber plants grown under vinyl films.] - Res. Rep. Kochi Univ. agr. Sci. *19*: 1-13, 1970. [In Jap.]

3932 - KATOH, S., IKEGAMI, I., TAKAMIYA, A.: Effects of hydroxylamine on electron-transport system in chloroplasts. - Arch. Biochem. Biophys. *141*: 207-218, 1970.

3933 - KATOH, S., SAN PIETRO, A.: The role of plastocyanin in NADP photoreduction by chloroplasts. - In: PEISACH, J., AISEN, P., BLUMBERG, W.E. (ed.): Biochemistry of Copper. Pp. 407-422. Academic Press, New York-London 1966.

3934 - KATOH, S., SAN PIETRO, A.: Inhibitory effect of salicylaldoxime on chloroplast photooxidation-reduction reactions. - Biochem. biophys. Res. Commun. *24*: 903-908, 1966.

3935 - KATOH, S., SAN PIETRO, A.: Activities of chloroplast fragments. I. Hill reac-tion and ascorbate-indophenol photoreductions. - J. biol. Chem. *241*: 3575-3581, 1966.

3936 - KATOH, S., SAN PIETRO, A.: The role of *C*-type cytochrome in the Hill reaction with *Euglena* chloroplasts. - Arch. Biochem. Biophys. *118*: 488-496, 1967.

3937 - KATOH, S., SAN PIETRO, A.: Photooxidation and reduction of cytochrome-552 and NADP photoreduction by *Euglena* chloroplasts. - Arch. Biochem. Biophys. *121*: 211-219, 1967.

3938 - KATOH, S., SAN PIETRO, A.: Ascorbate-supported NADP photoreduction by heated
       Euglena chloroplasts. - Arch. Biochem. Biophys. *122*: 144-152, 1967.

3939 - KATOH, S., SAN PIETRO, A.: A comparative study of the inhibitory action on the
       oxygen-evolution system of various chemical and physical treatments of *Euglena*
       chloroplasts. - Arch. Biochem. Biophys. *128*: 378-386, 1968.

3940 - KATOH, S., SAN PIETRO, A.: Photoreaction of chloroplasts: NADP photoreduction
       by *Euglena* chloroplasts. - In: SHIBATA, K., TAKAMIYA, A., JAGENDORF, A.T.,
       FULLER, R.C. (ed.): Comparative Biochemistry and Biophysics of Photosynthesis.
       Pp. 148-160. Univ. Tokyo Press, Tokyo; Univ. Park Press, State College, Pa.
       1968.

3941 - KATRUSHENKO, I.V.: Fotosintez podrosta eli vo vtorichnykh soobshchestvakh yuzh-
       noĭ taĭgi. [Photosynthesis of fir undergrowth in secondary communities of south-
       ern taiga.] - In: Svetovoĭ Rezhim, Fotosintez i Produktivnost' Lesa. Pp. 237-
       242. Nauka, Moskva 1967. [In R.]

3942 - KATZ, J.J.: Coordination properties of magnesium in chlorophyll from IR and NMR
       spectra. - Developm. appl. Spectroscopy *6*: 201-218, 1968.

3943 - KATZ, J.J., BALLSCHMITER, K.: Wechselwirkungen zwischen Chlorophyll und Wasser.
       - Angew. Chem. *80*: 283-284, 1968.

3944 - KATZ, J.J., BALLSCHMITER, K.: Chlorophyll-water interaction. - Angew. Chem.
       (Int. Ed.) *7*: 286-287, 1968.

3945 - KATZ, J.J., BALLSCHMITER, K., GARCIA-MORIN, M., STRAIN, H.H., UPHAUS, R.A.:
       Electron paramagnetic resonance of chlorophyll-water aggregates. - Proc. nat.
       Acad. Sci. U.S.A. *60*: 100-107, 1968.

3946 - KATZ, J.J., CRESPI, H.L.: Deuterated organisms: cultivation and uses. - Science
       151: 1187-1194, 1966. [Application to biosynthesis of pigments and photosynthe-
       sis.]

3947 - KATZ, J.J., DOUGHERTY, R.C., BOUCHER, L.J.: Infrared and nuclear magnetic reso-
       nance spectroscopy of chlorophyll. - In: VERNON, L.P., SEELEY, G.R. (ed.): The
       Chlorophylls. Pp. 185-251. Academic Press, New York-London 1966.

3948 - KATZ, J.J., DOUGHERTY, R.C., CRESPI, H.L., STRAIN, H.H.: Nuclear magnetic reso-
       nance studies of plant biosynthesis. A bacteriochlorophyll isotope mirror expe-
       riment. - J. amer. chem. Soc. *88*: 2856-2857, 1966.

3949 - KATZ, J.J., NORMAN, G.D., SVEC, W.A., STRAIN, H.H.: Chlorophyll diastereoisomers.
       The nature of chlorophylls a' and b' and evidence for bacteriochlorophyll epi-
       mers from proton magnetic resonance studies. - J. amer. chem. Soc. *90*: 6841-6845,
       1968.

3950 - KATZ, J.J., STRAIN, H.H., LEUSSING, D.L., DOUGHERTY, R.C.: Chlorophyll-ligand
       interactions from nuclear magnetic resonance studies. - J. amer. chem. Soc. *90*:
       784-791, 1968.

3951 - KAUL, R.: Relative growth rates of spring wheat, oats, and barley under polyethy-
       lene glycol-induced water stress. - Can. J. Plant Sci. *46*: 611-617, 1966.

3952 - KAUL, R.N., ROY, R.D.: Chlorophyll stability index, a suitable criterion for
       rapid screening of tree provenance in arid zones. - Experientia *23*: 37-38, 1967.

3953 - KAUSCH, W.: Neuere Untersuchungen zur Frage: Wie unterscheiden sich Sonnen- und
       Schattenblätter der Blutbuche? - Umschau Wiss. Tech. *1968*: 373-374, 1968. [Ps.]

3954 - KAWABE, K., NAEMURA, M., MATSUKAWA, Y.: Emission spectrum of chlorophyll-a solu-
       tion excited by argon ion laser. - Technol. Rep. Osaka Univ. *20*: 665-672, 1970.

3955 - KAWABE, K., YOSHINO, K., INUISHI, Y.: Q-switching of ruby laser by natural chlo-
       rophyll and the influence of solvents. - J. phys. Soc. Jap. *24*: 966, 1968.

3956 - KAWAMURA, T., TAGUCHI, S.: [Amount of pigments and daily production of phyto-
       plankton occurring in acidified lake Toya.] - Bull. Fac. Fish. Hokkaido Univ. *21*:
       201-209, 1970. [In Jap., ab: E.]

3957 - KAWANABE, S., SAKAI, H., FUZIWARA, K.: [Dry matter production in ladino clover
       sward.] - Proc. Crop Sci. Soc. Jap. *38*: 327-332, 1969. [Growth analysis; in
       Jap., ab: E.]

3958 - KAWANABE, S., SASAKI, H.: Dry matter production in ladino clover sward. - In: Photosynthesis and Utilization of Solar Energy. Level III Experiments. Pp. 55-59. Jap. nat. Subcomm. for PP (JPP), Tokyo 1968.

3959 - KAWANABE, S., USHIYAMA, M.: [Comparison of dry matter production in forage grass species.] - Proc. Crop Sci. Soc. Jap. *39*: 84-89, 1970. [In Jap., ab: E.]

3960 - KAWASHIMA, N.: Comparative studies on fraction 1 protein from spinach and tobacco leaves. - Plant Cell Physiol. *10*: 31-40, 1969.

3961 - KAWASHIMA, N., MITAKE, T.: Studies on protein metabolism in higher plants. Part VI. Changes in ribulose diphosphate carboxylase activity and fraction 1 protein content in tobacco leaves with age. - Agr. biol. Chem. *33*: 539-543, 1969.

3962 - KAWASHIMA, R.: [Studies on the leaf orientation-adjusting movement in soybean plants. I. The leaf orientation-adjusting movement and light intensity on leaf surface.] - Proc. Crop Sci. Soc. Jap. *38*: 718-729, 1969. [In Jap., ab: E.]

3963 - KAWASHIMA, R.: [Studies on the leaf orientation-adjusting movement in soybean plants. II. Fundamental pattern of the leaf orientation-adjusting movement and its significance for the dry matter production.] - Proc. Crop Sci. Soc. Jap. *38*: 730-742, 1969. [In Jap., ab: E.]

3964 - KAYE, S., KOENCY, J.E.: Spectrophotometric method for determining oxygen in gases. - Anal. Chem. *41*: 1491-1493, 1969.

3965 - KAZARYAN, V.O.: Rol' kornevoĭ sistemy v intensifikatsii fotosinteza. [Role of root system in intensification of photosynthesis.] - In: Vazhneĭshie Problemy Fotosinteza v Rastenivodstve. Pp. 153-160. Kolos, Moskva 1970. [In R.]

3966 - KAZARYAN, V.O., DAVTYAN, V.A.: O zavisimosti aktivnosti fotosinteza rasteniĭ ot moshchnosti i metabolicheskoĭ deyatel'nosti korneĭ. [Dependence of plant photosynthetic activity on the extent and metabolism of roots.] - Biol. Zh. Armenii *20* (11): 49-58, 1967. [In R.]

3967 - KAZARYAN, V.O., DAVTYAN, V.A.: Ob izmenenii aktivnosti fotosinteza list'ev pod deĭstviem faktorov, vliyayushchikh na kornevuyu sistemu. [Variation of photosynthetic rate induced by factors affecting root system.] - Fiziol. Rast. *14*: 860-865, 1967. [In R, ab: E.]

3968 - KAZARYAN, V.O., KHURSHUDYAN, P.A., KARAPETYAN, K.A.: O vnutrennikh faktorakh ontogeneticheskogo zatukhaniya rosta vysshikh rasteniĭ. [Internal factors of ontogenetic growth inhibition in higher plants.] - Biol. Zh. Armenii *21* (11): 30-38, 1968. [Ps; in R.]

3969 - KAZARYAN, V.V.: O vliyanii moshchnosti kornevoĭ sistemy na fotosinteticheskuyu aktivnost' list'ev i nakoplenie khlorofilla. [Effect of extent of root system on photosynthetic activity of leaves and accumulation of chlorophyll.] - Dokl. Akad. Nauk Arm. SSR *42*: 304-308, 1966. [In R, ab: Armen.]

3970 - KE, B.: Optical rotatory dispersion of pigment-containing particles from green plants and photosynthetic bacteria. - In: THOMAS, J.B., GOEDHEER, J.C. (ed.): Currents in Photosynthesis. Pp. 149-156. Donker, Rotterdam 1966.

3971 - KE, B.: Some properties of chlorophyll monolayers and crystalline chlorophyll. - In: VERNON, L.P., SEELY, G.R. (ed.): The Chlorophylls. Pp. 253-279. Academic Press, New York-London 1966.

3972 - KE, B.: Chlorophyll-protein complexes. Part III. Optical rotatory dispersion of chlorophyll-containing particles from green plants and photosynthetic bacteria. - In: VERNON, L.P., SEELEY, G.R. (ed.): The Chlorophylls. Pp. 427-436. Academic Press, New York-London 1966.

3973 - KE, B.: Photoreduction sites for 2,6-dichlorophenolindophenol in chloroplasts. - Plant Physiol. *42*: 1310-1312, 1967.

3974 - KE, B.: Nature of the primary electron acceptor in bacterial photosynthesis. - Biochim. biophys. Acta *172*: 583-585, 1969.

3975 - KE, B., BREEZE, R.H., GREEN, M.: Adaptation of the Cary recording spectrophotometer to circular dichroism measurements. - Anal. Biochem. *25*: 181-191, 1968. [Chromatophores.]

3976 - KE, B., CHANEY, T.H., REED, D.W.: The electrostatic interaction between the reac-
       tion-center bacteriochlorophyll derived from *Rhodopseudomonas spheroides* and mam-
       malian cytochrome *c* and its effect on light-activated electron transport. - Bio-
       chim. biophys. Acta *216*: 373-383, 1970.

3977 - KE, B., GREEN, M., VERNON, L.P., GARCIA, A.F.: Some optical properties of a
       "carotenoid complex" derived from *Rhodospirillum rubrum*. - Biochim. biophys. Ac-
       ta *162*: 467-469, 1968.

3978 - KE, B., IMSGARD, F., KJØSEN, H., LIAAEN-JENSEN, S.: Electronic spectra of caro-
       tenoids at 77 °K. - Biochim. biophys. Acta *210*: 139-152, 1970.

3979 - KE, B., KATOH, S., SAN PIETRO, A.: Light-induced rapid absorption changes during
       photosynthesis. VII. Some reactions in sonicated chloroplasts. - Biochim. bio-
       phys. Acta *131*: 538-547, 1967.

3980 - KE, B., NGO, E.: Light-induced rapid absorption changes during photosynthesis.
       VIII. Cytochrome and bacteriochlorophyll reactions in *Rhodospirillum rubrum*
       cells. - Biochim. biophys. Acta *143*: 319-331, 1967.

3981 - KE, B., SELISKAR, C., BREEZE, R.: Pigment composition and optical rotary disper-
       sion of chloroplast fractions obtained by detergent action. - Plant Physiol. *41*:
       1081-1082, 1966.

3982 - KE, B., SPERLING, W.: Evidence for the presence of ordered aggregates in chloro-
       phyll *a* monolayers. - In: Energy Conversion by the Photosynthetic Apparatus.
       Brookhaven Symp. Biol. *19*: 319-327, 1967.

3983 - KE, B., VERNON, L.P.: Fluorescence of the subchloroplast particles obtained by
       the action of Triton X-100. - Biochemistry *6*: 2221-2226, 1967.

3984 - KE, B., VERNON, L.P., GARCIA, A., NGO, E.: Coupled photooxidation of bacterio-
       chlorophyll P890 and photoreduction of ubiquinone in a photochemically active
       subchromatophore particle derived from *Chromatium*. - Biochemistry *7*: 311-318,
       1968.

3985 - KEAY, J., TURTON, A.G., CAMPBELL, N.A.: Some effects of nitrogen and phosphorus
       fertilization of *Pinus pinaster* in Western Australia. - Forest Sci. *14*: 408-417,
       1968. [Ps, Chl.]

3986 - KECK, R.W.: Compositional and functional observations in a soybean chloroplast
       mutant. - Plant Physiol. *44* (Suppl.): 11, 1969.

3987 - KECK, R.W.: The role of plastoquinone in a soybean mutant exhibiting high photo-
       synthetic capacity. - Plant Physiol. *46* (Suppl.): 41, 1970.

3988 - KECK, R.W., DILLEY, R.A., ALLEN, C.F., BIGGS, S.: Chloroplast composition and
       structure differences in soybean mutant. - Plant Physiol. *46*: 692-698, 1970.

3989 - KECK, R.W., DILLEY, R.A., KE, B.: Photochemical characteristics in soybean mu-
       tant. - Plant Physiol. *46*: 699-704, 1970.

3990 - KEERBERG, O., VYARK, E., KEERBERG, Kh., PYARNIK, T.: O regulyatsii fotosinteti-
       cheskoǐ assimilyatsii CO$_2$ usloviyami osveshcheniya. [Control of photosynthetic
       CO$_2$ assimilation by irradiance.] - In: Na Puti k Obnovleniyu Zemli. Pp. 280-
       296. Valgus, Tallin 1968. [In R.]

3991 - KEERBERG, O.F., VYARK, E.Ya., KEERBERG, K.I., PYARNIK, T.R.: Issledovanie kine-
       tiki vklyucheniya C^{14} v produkty fotosinteza u list'ev fasoli. [Kinetics of ^{14}C
       incorporation into photosynthates in *Phaseolus* leaves.] - Dokl. Akad. Nauk SSSR
       195: 238-241, 1970. [In R.]

3992 - KEISTER, D.L., MINTON, N.J.: Energy-linked reactions in photosynthetic bacteria.
       III. Further studies on energy-linked nicotinamide-adenine dinucleotide reduct-
       ion by *Rhodospirillum rubrum* chromatophores. - Biochemistry *8*: 167-173, 1969.

3993 - KEISTER, D.L., MINTON, N.J.: Energy-linked reactions in photosynthetic bacteria.
       IV. Interaction of the photochemical and respiratory systems of *Rhodospirillum
       rubrum*. - In: METZNER, H. (ed.): Progress in Photosynthesis Research. Vol. III.
       Pp. 1299-1305. Tübingen 1969.

3994 - KEISTER, D.L., MINTON, N.J.: Energy-linked reactions in photosynthetic bacteria.
       V. Relation of the light-induced proton uptake to photophosphorylation in *R.
       rubrum* chromatophores. - Proc. nat. Acad. Sci. U.S.A. *63*: 489-495, 1969.

3995 - **KEISTER, D.L., MINTON, N.J.**: K^+-independent effects of valinomycin in photosynthetic systems. - J. Bioenerg. *1*: 367-377, 1970.

3996 - **KEISTER, D.L., YIKE, N.J.**: Studies on an energy-linked pyridine nucleotide transhydrogenase in photosynthetic bacteria. I. Demonstration of the reaction in *Rhodospirillum rubrum*. - Biochem. biophys. Res. Commun. *24*: 519-525, 1966.

3997 - **KEISTER, D.L., YIKE, N.J.**: Energy-linked reactions in photosynthetic bacteria. I. Succinate-linked ATP-driven NAD^+ reduction by *Rhodospirillum rubrum* chromatophores. - Arch. Biochem. Biophys. *121*: 415-422, 1967.

3998 - **KEISTER, D.L., YIKE, N.J.**: Energy-linked reactions in photosynthetic bacteria. II. The energy-dependent reduction of oxidized nicotinamide-adenine dinucleotide phosphate by reduced nicotinamide-adenine dinucleotide in chromatophores of *Rhodospirillum rubrum*. - Biochemistry *6*: 3847-3857, 1967.

3999 - **KELLER, C.J., HUFFAKER, R.C.**: Evidence for *in vivo* light-induced synthesis of ribulose-1,5-diP carboxylase and phosphoribulokinase in greening barley leaves. - Plant Physiol. *42*: 1277-1283, 1967.

4000 - **KELLER, E.C., Jr., MATTONI, R.H.T.**: Preliminary studies of the dynamics of algal populations in high rate photosynthetic reactors. - Advanc. Frontiers Plant Sci. *16*: 137-160, 1966.

4001 - **KELLER, J., BACHOFEN, R.**: Reactions of spinach chloroplasts with added Mn^{2+}. - In: METZNER, H. (ed.): Progress in Photosynthesis Research. Vol. II. Pp. 1013-1021. Tübingen 1969.

4002 - **KELLER, T.**: Über den Einfluss von transpirationshemmenden Chemikalien (Antitranspirantien) auf Transpiration, CO_2-Aufnahme und Wurzelwachstum von Jungfichten. - Forstwiss. Centralbl. *85*: 65-79, 1966.

4003 - **KELLER, T.**: The influence of fertilization on gaseous exchange of forest tree species. - In: HOLOPAINEN, V. (ed.): Colloquium on Forest Fertilization. Pp. 65-79. Int. Potash Inst., Berne 1967.

4004 - **KELLER, T.**: Die Wirkung einer Bodenabdeckung (Mulchung) im Forstpflanzgarten auf den Gaswechsel junger Fichten. - Forstwiss. Centralbl. *87*: 1-8, 1968.

4005 - **KELLER, T.**: Influence de la nutrition minérale sur les échanges gazeux des arbres de fôret. - Phosphore Agr. 22 (50): 25-37, 1968.

4006 - **KELLER, T.**: Nettoassimilation, Spross- und Wurzelatmung junger Pappeln bei unterschiedlicher Ernährung. - In: Klimaresistenz, Photosynthese und Stoffproduktion. Tagungsber. deut. Akad. Landwirtschaftswiss. Berlin *100*: 233-253, 1968.

4007 - **KELLER, T.**: Laborversuche über den Einfluss von Antitranspirantien auf den Gaswechsel junger Coniferen. - Schweiz. Z. Forstwesen *120*: 32-43, 1969. [Ps.]

4008 - **KELLER, T.**: Über die Assimilation einer junger Arve im Winterhalbjahr. - Bündnerwald *23*: 49-54, 1970.

4009 - **KELLER, T.**: Wuchsleistung, Gaswechsel, Überlebensprozente und Schneeschimmelpilzbefall gedüngter Ballenpflanzen an der oberen Waldgrenze. - Mitt. schweiz. Anst. forst. Versuchswes. *46*: 1-32, 1970.

4010 - **KELLER, T.**: Über den Einfluss organischer Bodenabdeckungen im Forstpflanzgarten auf die Wuchsleistung von Verschulfichten sowie auf Bodentemperatur und Bodenfeuchtigkeit. - Mitt. schweiz. Anst. forst. Versuchswes. *46*: 33-65, 1970. [Productivity.]

4011 - **KELLER, T., PREIS, H.**: Der Bleigehalt von Fichtennadeln als Indikator einer verkehrsbedingten Luftverunreinigung. - Schweiz. Z. Forstwes. *118*: 143-162, 1967. [Ps.]

4012 - **KELLER, T., ZUBER, R.**: Über die Bleiaufnahme und die Bleiverteilung in jungen Fichten. - Forstwiss. Centralbl. *89*: 20-26, 1970. [Ps.]

4013 - **KELLY, A.R., PORTER, G.**: Model systems for photosynthesis. I. Energy transfer and light harvesting mechanisms. - Proc. roy. Soc. London *A* 315: 149-161, 1970.

4014 - **KELLY, D.P.**: Problems of the autotrophic micr-organisms. - Sci. Progr. (Oxford) *55* (217): 35-51, 1967. [Ps and chemoautotrophy of bacteria and algae.]

4015 - KELLY, J., SAUER, K.: Functional photosynthetic unit sizes for each of the two
light reactions in spinach chloroplasts. - Biochemistry 7: 882-890, 1968.

4016 - KELLY, J.J.: The role of the pigment array in the photochemistry and kinetics
of photosynthesis. - Diss. Abstr. B 29: 3240-B-3241-B, 1969.

4017 - KELLY, J.M., PORTER, G.: The interaction of photo-excited chlorophyll a with
duroquinone, α-tocopherylquinone and vitamin K_1. - Proc. roy. Soc. London A 319:
319-329, 1970.

4018 - KEMMERER, A.J.: A method to determine fertilization requirements of a small
sport fishing lake. - Trans. amer. Fish.Soc. 97: 425-428, 1968. [Ps.]

4019 - KEMMERER, A.J., GLUCKSMAN, J., STEWART, P.A., McCONNELL, W.J.: Some productivity relations in seven fishing impoundments in eastern Arizona. - J. Arizona
Acad. Sci. 5 (2): 80-85, 1968. [Ps.]

4020 - KEREFOVA, M.L., EL'MESOV, A.M.: Nekotorye fiziologicheskie osobennosti rosta i
razvitiya otdel'nykh podvidov kukuruzy na rannich etapakh ontogeneza v svyazi
so srokami seva. [Some physiological peculiarities of growth and development of
individual maize subspecies in early ontogenetic phases in relation to sowing
time.] - Uch. Zap. Kabardino-Balkarsk. Univ. 35: 56-64, 1967. [Chl; in R.]

4021 - KERESZTES, Å., FALUDI-DÁNIEL, Å.: Changes in the structure, pigment content and
photosynthetic activity of normal and mutant chloroplasts of developing Trades-
cantia-leaves. - In: Septième Congrès International de Microscopie Électronique.
Pp. 181-182. Grenoble 1970.

4022 - KERESZTES-NAGY, S., MARGOLIASH, E.: Preparation and characterization of alfalfa
ferredoxin. - J. biol. Chem. 241: 5955-5966, 1966.

4023 - KERESZTES-NAGY, S., PERINI, F., MARGOLIASH, E.: Primary structure of alfalfa
ferredoxin. - J. biol. Chem. 244: 981-995, 1969.

4024 - KERR, J.D., SUBBA RAO, D.V.: Extraction of chlorophyll a from Nitzschia closterium by grinding. - In: Determination of Photosynthetic Pigments in Sea-water.
Pp. 65-69. UNESCO, Paris 1966.

4025 - KESHEVA, A., SLONOV, L.: Vliyanie vnekornevoĭ podkormki mikroelementami na
urozhaĭ i kachestvo vinograda Kabardino-Balkarii. [Effect of foliar nutrition
by microelements on yield and quality of vine in Kabardino-Balkaria.] - Vestn.
mosk. Univ. Ser. 6 - Biol., Pochvoved. 25 (6): 92-95, 1970. [Ps, chl; in R.]

4026 - KESSLER, B., SPIEGEL, S., ZOLOTOV, Z.: Control of leaf senescence by growth retardants. - Nature 213: 311-312, 1967. [Chl.]

4027 - KESSLER, E.: Physiologische und biochemische Beiträge zur Taxonomie der Gattung
Chlorella. III. Merkmale von 8 autotrophen Arten. - Arch. Mikrobiol. 55: 346-357,
1967. [Car.]

4028 - KESSLER, E.: Effect of hydrogen adaptation on fluorescence in normal and manganese-deficient algae. - Planta 81: 264-273, 1968.

4029 - KESSLER, E.: Effect of manganese deficiency on growth and chlorophyll content
of algae with and without hydrogenase. - Arch. Mikrobiol. 63: 7-10, 1968.

4030 - KESSLER, E.: Effect of manganese deficiency on fluorescence in algae adapted to
hydrogen. - In: METZNER, H. (ed.): Progress in Photosynthesis Research. Vol. II.
Pp. 938-942. Tübingen 1969.

4031 - KESSLER, E.: Photosynthesis, photooxidation of chlorophyll and fluorescence of
normal and manganese-deficient Chlorella with and without hydrogenase. - Planta
92: 222-234, 1970.

4032 - KESSLER, E., CZYGAN, F.-C.: Physiologische und biochemische Beiträge zur Taxonomie der Gattung Chlorella. II. Untersuchungen an Mutanten. - Arch. Mikrobiol.
54: 37-45, 1966. [Car.]

4033 - KESSLER, E., CZYGAN, F.-C.: Physiologische und biochemische Beiträge zur Taxonomie der Gattungen Ankistrodesmus und Scenedesmus. I. Hydrogenase, Sekundär-Carotinoide und Gelatine-Verflüssung. - Arch. Mikrobiol. 55: 320-326, 1967.

4034 - KESSLER, E., CZYGAN, F.-C.: The effect of iron supply on the activity of nitrate
and nitrite reduction in green algae. - Arch. Mikrobiol. 60: 282-284, 1968. [Chl.]

4035 - KESSLER, E., CZYGAN, F.-C., FOTT, B., NOVÁKOVÁ, M.: Über *Halochlorella rubescens* DANGEARD. - Arch. Protistenk. *110*: 462-468, 1968. [Car.]

4036 - KETCHUM, P.A., HOLT, S.C.: Isolation and characterization of the membranes from *Rhodospirillum rubrum*. - Biochim. biophys. Acta *196*: 141-161, 1970. [Chl.]

4037 - KETSKHOVELI, E.N.: O fotosinteticheskoĭ deyatel'nosti zelenykh plastid pobegov. [Photosynthetic activity of green plastids of shoots.] - Vestn. gruz. bot. Obshch. Akad. Nauk Gruz. SSR *2*: 40-49, 1967.

4038 - KEY, A., PARKER, D., DAVIES, R.: Use of epoxy resin in oxygen electrodes. - Phys. Med. Biol. *15*: 569-572, 1970.

4039 - KEYS, A.J.: The intracellular distribution of free nucleotides in the tobacco leaf. Formation of adenosine 5'-phosphate from adenosine 5'-triphosphate in the chloroplasts. - Biochem. J. *108*: 1-8, 1968.

4040 - KEYS, A.J.: WHITTINGHAM, C.P.: Nucleotide metabolism in chloroplast and non-chloroplast components of tobacco leaves. - In: METZNER, H. (ed.): Progress in Photosynthesis Research. Vol. I. Pp. 352-358. Tübingen 1969. [Chl.]

4041 - KHACHIDZE, O.T., TKHELIDZE, P.A., MATIKASHVILI, I.A.: Obrazovanie aminokislot v list'yakh vinogradnoĭ lozy pri fotosinteze. [Formation of amino acids in vine leaves during photosynthesis.] - Soobshch. Akad. Nauk Gruz. SSR *48*: 85-90, 1967. [In R.]

4042 - KHACHIDZE, O.T., TKHELIDZE, P.A., PKHAKADZE, N.V.: Raspredelenie pogloshchenno-go pri fotosinteze C^{14} vo fraktsiyakh uglevodov, organicheskikh kislot i amino-kislot v list'yakh vinogradnoĭ lozy. [Distribution of ^{14}C absorbed during photo-synthesis between fractions of saccharides, organic acids and amino acids in vine leaves.] - Soobshch. Akad. Nauk Gruz. SSR *44*: 317-324, 1966. [In R.]

4043 - KHADASEVICH, E.V., MEL'NIKAVA, L.M., GODNEŬ, Ts.M.: Ab stane khlaraplastaŭ u sasny zvychaĭnaĭ z zhaŭtseyuchaĭ u asenne-zimovy peryyad iglitsaĭ. [State of chloroplasts in pine with needles yellowing in autumn and winter.] - Vestsi Akad. Navuk BSSR, Ser. biyal. Navuk *1970* (3): 50-53, 136, 1970. [In Belorus., ab: R.]

4044 - KHAITOVA, L.T., LOGINOVA, N.P., GILLER, Yu.E.: Opticheskie svoĭstva i fotokhimi-cheskaya ustoĭchivost' fotosinteticheskikh pigmentov v iskusstvennom fitokhrom-proteĭdnom kompletse i komplekse Lyubimenko. [Optical characteristics and photo-chemical stability of photosynthetic pigments in an artificial phytochrome-pro-teid complex and Lubimenko complex.] - In: Tezisy Dokladov Konferentsii Molodykh Uchenykh. P. 50. Dushanbe 1966. [In R.]

4045 - KHAN, A.A., SAGAR, G.R.: Distribution of ^{14}C-labelled products of photosynthesis during the commercial life of the tomato crop. - Ann. Bot. *30*: 727-743, 1966.

4046 - KHAN, A.A., SAGAR, G.R.: Translocation in tomato: the distribution of the pro-ducts of photosynthesis of the leaves of a tomato plant during the phase of fruit production. - Hort. Res. *7*: 61-69, 1967.

4047 - KHAN, A.A., SAGAR, G.R.: Alteration of the pattern of distribution of photosyn-thetic products in the tomato by manipulation of the plant. - Ann. Bot. *33*: 753-762, 1969.

4048 - KHAN, A.A., SAGAR, G.R.: Changing patterns of distribution of the products of photosynthesis in the tomato plant with respect to time and to the age of a leaf. - Ann. Bot. *33*: 763-779, 1969.

4049 - KHAN, M.A., TSUNODA, S.: Evolutionary trends in leaf photosynthesis and related leaf characters among cultivated wheat species and its wild relatives. - Jap. J. Breed. *20*: 133-140, 1970.

4050 - KHAN, M.A., TSUNODA, S.: Leaf photosynthesis and transpiration under different levels of air flow rate and light intensity in cultivated wheat species and its wild relatives. - Jap. J. Breed. *20*: 305-314, 1970.

4051 - KHAN, M.A., TSUNODA, S.: Differences in leaf photosynthesis and leaf transpira-tion rates among six commercial wheat varieties of West Pakistan. - Jap. J. Breed. *20*: 344-350, 1970.

4052 - KHAU van KIEN, L.: Étude histochimique nouvelle au niveau des chloroplastes cellulaires. Essais de caractérisation et de localisation des plastoquinones. -
Ann. histochim. *11*: 51-61, 1966.

4053 - KHAU van KIEN, L.: New histochemical study at cellular chloroplast level: char-
acterization and localization of plastoquinones. - In: GOODWIN, T.W. (ed.):
Biochemistry of Chloroplasts. Vol. II. Pp. 545-550. Academic Press, London-New
York 1967.

4054 - KHAZANOV, V.S., SHISHOV, D.M., TSEL'NIKER, Yu.L.: Izmerenie radiatsii pod pologom lesa. [Measurement of radiation under the forest canopy.] - In: Svetovoĭ
Rezhim, Fotosintez i Produktivnost' Lesa. Pp. 36-47. Nauka, Moskva 1967. [In R.]

4055 - KHAZANOV, V.S., TSEL'NIKER, Yu.L.: Izmerenie fotosinteticheski aktivnoĭ radiatsii v lesu s pomoshch'yu lyuksmetra. [Measurement of photosynthetically active
radiation in forest using a luxmeter.] - In: Aktinometriya i Optika Atmosfery.
Pp. 382-385. Valgus, Tallin 1968. [In R.]

4056 - KHEĬN, Kh.Ya., NICHIPOROVICH, A.A.: Izmerenie svetovykh krivykh fotosinteza intaktnykh list'ev radiometricheskim metodom. [Light curves of photosynthesis in
intact leaves measured by radiometric technique.] - Fiziol. Rast. *17*: 1284-1290,
1970. [In R, ab: E.]

4057 - KHITRYUK, L.A.: Rol' list'ev razlichnykh yarusov i vozrastov v formirovanii
urozhaya morkovi. [The effect of different leaves and their age on yield of
carrots.] - Sb. nauch. Tr. beloruss. sel'.-khoz. Akad. *64*: 17-23, 1970. [In R.]

B4058 - Khloroplasty i Mitokhondrii. [Chloroplasts and Mitochondria.] - Nauka, Moskva
1969. [In R.]

4059 - KHLYUSTOVA, T.M.: Produkty fotosinteza list'ev shpinata i izvlechennykh iz nikh
khloroplastov. [Photosynthates of spinach leaves and isolated chloroplasts.] -
In: Funktsional'nye Osobennosti Khloroplastov. Pp. 75-77. Kazan. Univ., Kazan
1969. [In R.]

4060 - KHLYUSTOVA, T.M.: Deĭstvie povyshennoĭ temperatury na fotosinteticheskiĭ metabolizm ugleroda v izolirovannykh khloroplastakh. [Effect of increased temperature on photosynthetic carbon metabolism in isolated chloroplasts.] - In: Funktsional'nye Osobennosti Khloroplastov. Pp. 78-80. Kazan. Univ., Kazan 1969. [In R.]

4061 - KHOBOT'EV, V.G., KAPKOV, V.I.: Vliyanie polimetallicheskikh rud na vydelenie i
pogloshchenie kisloroda v protsesse fotosinteza i dykhaniya protokokkovykh vodorosleĭ. [Effect of polymetallic ores on oxygen efflux and absorption during photosynthesis and respiration of protococcous algae.] - Nauch. Dokl. vyssh. Shkoly,
biol. Nauki *11* (4): 82-85, 1968. [In R.]

4062 - KHODASEVICH, E.V., ARNAUTOVA, A.I., GODNEV, T.N.: Formirovanie i sostoyanie fonda pigmentov i plastid v ontogeneze lista khvoĭnykh. [Formation and state of
pigments and plastids during ontogenesis of needles.] - In: Metabolizm i Stroenie Fotosinteticheskogo Apparata. Pp. 152-163. Nauka i Tekhnika, Minsk 1970.
[In R.]

4063 - KHODASEVICH, E.V., GODNEV, T.N., SIDOROVA, T.V.: K metodike razdeleniya i radio-
khimicheskoĭ ochistki al'fa- i beto-karotinov. [Method of separation and radio-
chemical purification of α- and β-carotenes.] - In: Issledovaniya po Fiziologii
i Biokhimii Rasteniĭ. Pp. 13-16. Nauka i Tekhnika, Minsk 1966. [In R.]

4064 - KHODOS, V.M., KHOMLYAK, M.M.: Vydilennya khloroplastiv z lystkiv tsukrovogo bu-
ryaka. [Isolation of chloroplasts from sugar beet leaves.] - Ukr. biokhim. Zh.
39: 444-448, 1967. [In Ukr., ab: R, E.]

4065 - KHODOS, V.N.: Razdelenie fosforilirovannykh soedineniĭ iz list'ev gorokha i
sakharnoĭ svekly na kolonkakh ionoobmennykh smol. [Separation of phosphorylated
compounds from pea and sugar beet leaves on ionex columns.] - In: Puti Povysheniya Intensivnosti i Produktivnosti Fotosinteza. Vol. 2. Pp. 191-197. Naukova
Dumka, Kiev 1967. [In R.]

4066 - KHODZHAEV, A.S., SABIROV, B.: O pigmentnom sostave semyan saksaula. [Pigment
content of saxaul seeds.] - Dokl. Akad. Nauk Uzb. SSR *1968* (10): 49-50, 1968.
[In R.]

4067 - **KHODZHAEV,D.Kh.:** Vliyanie mikroelementov na soderzhanie khlorofilla i prochnost'
khlorofill-belkovo-lipoidnogo kompleksa khlopchatnika na rannikh fazakh razvitiya. [Effect of microelements on chlorophyll content and stability of the chlorophyll-protein-lipoid complex of cotton at early phases of development.] -
Dokl. Akad. Nauk Uzb. SSR *1966* (8): 61-63, 1966. [In R.]

4068 - **KHODZHAEV,D.Kh.:** Vliyanie nekotorykh mikroelementov na nakoplenie khlorofilla v
rasteniyakh khlopchatnika. [Effect of some microelements on chlorophyll accumulation in cotton plants.] - Nauch. Dokl. vyssh. Shkoly, biol. Nauki *9:* 161-
163, 1966. [In R.]

4069 - **KHOLMOGOROV, V.E.:** Issledovanie metodom EPR temnovykh i fotokhimicheskikh reaktsiĭ tetrapirrol'nykh pigmentov s otritsatel'nymi molekulyarnymi ionami v rastvore. [EPR study of dark and photochemical reactions of tetrapyrrol pigments with
negative molecular ions in solution.] - Biofizika *15:* 983-992, 1970. [In R; ab: E.]

4070 - **KHOLMOGOROV, V.E., SIDOROV, A.N.:** Spektral'noe issledovanie vzaimodeĭstviya khlorofilla s otritsatel'nymi ionami antratsena i piridina. [Spectral study of chlorophyll interaction with negative anthracene and pyridine ions.] - Dokl. Akad.
Nauk SSSR *178:* 897-900, 1968. [In R.]

4071 - **KHOSSEÏN, M.M.:** Vliyanie intensivnosti, kachestva sveta i fotoperiodizma na fotosinteticheskuyu produktivnost' rasteniĭ fasoli. [Effect of intensity and quality of light and photoperiodism on photosynthetic productivity of bean plants.]
- In: Fotosintez i Urozhaĭnost' Sel'skokhozyaĭstvennykh Rasteniĭ. Pp. 96-102.
Min. sel'. Khoz. SSSR, Kiev 1970. [In R.]

4072 - **KHRYANIN, V.N.:** Sravnitel'noe izuchenie deĭstviya tekhnicheskogo i kristalliches-
kogo gibberellina na konoplyu. [Comparison of effect of technical and crystalline
gibberelin on hemp plants.] - Nauch. Dokl. vyssh. Shkoly, biol. Nauki *11* (9):
71-74, 1968. [Ps, Chl; In R.]

4073 - **KHUDAIRI, A.K.:** Chlorophyll degradation by light in leaf discs in the presence
of sugar. - Physiol. Plant. *23:* 613-622, 1970.

4074 - **KHUDYAK, M.I., POPOVA, A.F., GOLOD, M.G., ONANKO, L.K., SHCHAPOVA, I.S.:** Pro ko-
relyatyvnyĭ zv'yazok mizh embriogenezom i nayavnistyu zelenykh pigmentiv u nasin-
ni garbuziv. [Correlation between embryogenesis and appearance of green pigments
in melon seeds.] - Ukr. bot. Zh. *27:* 116-119, 1970. [In Ukr.]

4075 - **KHVEDELIDZE, M.A., LOMSADZE, M.Sh., SHARASHIDZE, N.B., CHRELASHVILI, M.N.:** Mag-
nitnyĭ effekt pri fotosinteze. [Magnesium effect in photosynthesis.] - Soobshch.
Akad. Nauk Gruz. SSR *51:* 693-696, 1968. [In R, ab: Georg.]

4076 - **KIBALENKO, A.P.:** Vliyanie bora na strukturu i funktsiyu khloroplastov sakharnoĭ
svekly. [Boron effect on structure and function of sugar beet chloroplasts.] -
In: Khloroplasty i Mitokhondrii. Pp. 183-189. Nauka, Moskva 1969. [In R.]

4077 - **KIBALENKO, A.P., SILAEVA, A.M.:** Vliyanie bora na fotosinteticheskiĭ apparat
sakharnoĭ svekly. [Boron effect on the photosynthetic apparatus of sugar beet.]
- In: Materialy ko II Vsesoyuznomu Simpoziumu po Primeneniyu Elektronnoĭ Mikros-
kopii v Botanicheskikh Issledovaniyakh. Pp. 73-75. Naukova Dumka, Kiev 1967.
[In R.]

4078 - **KICHIGIN, A.A.:** Ob izemeniyakh soderzhaniya karotina v pochkakh i list'yakh drevesnykh i kustarnikovykh rasteniĭ v khode ontogeneza. [Changes in carotene content in buds and leaves of trees and bushes during their ontogenesis.] - Uch.
Zap. leningrad. gos. ped. Inst. A.I. Gertsena *421:* 13-31, 1970. [In R.]

4079 - **KICHIGIN, A.A.:** Vliyanie usloviĭ aeratsii i temperatury na protsessy prorasta-
niya i nakopleniya karotina v rasteniyakh. [Effect of aeration and temperature
on germination and carotene accumulation in plants.] - Uch. Zap. leningrad.
gos. ped. Inst. A.I. Gertsena *421:* 95-104, 1970. [In R.]

4080 - **KIEFER, D., STRICKLAND, J.D.H.:** A comparative study of photosynthesis in sea-
water samples incubated under two types of light attenuator. - Limnol. Oceanogr. *15:* 408-412, 1970.

4081 - **KIERAS, F.J., HASELKORN, R.:** Properties of ribulose-1,5-diphosphate carboxylase
(carboxydismutase) from Chinese Cabbage and photosynthetic microorganisms. -
Plant Physiol. *43:* 1264-1270, 1968.

4082 - KIERMAYER, O.: Hemmung der Kern- und Chloroplastenmigration von *Micrasterias* durch Colchizin. - Naturwissenschaften *55*: 299-300, 1968.

4083 - KIHARA, T., CHANCE, B.: Cytochrome photooxidation at liquid nitrogen temperature in photosynthetic bacteria. - Biochim. biophys. Acta *189*: 116-124, 1969.

4084 - KIHARA, T., DUTTON, P.L.: Light-induced reactions of photosynthetic bacteria. I. Reactions in whole cells and in cell-free extracts at liquid nitrogen temperatures. - Biochim. biophys. Acta *205*: 196-204, 1970.

B4085 - KIKNADZE, G.S.: Vliyanie Rezhima Osveshcheniya na Kinetiku Fluorestsentsii Khlorofilla v List'yakh Nekotorykh Vysshikh Rastenii. [Effect of Radiation Regime on Kinetics of Chlorophyll Fluorescence in Leaves of Some Higher Plants.] - Inst. Biofiz. Akad. Nauk SSSR, Pushchino-na-Oke 1970. [In R.]

4086 - KIKUCHI, G.: Studies on the induced syntheses of bacteriochlorophyll and δ-aminolevulinate synthetase in *Rhodopseudomonas spheroides* with special reference to chromatophore formation. - In: SHIBATA, K., TAKAMIYA, A., JAGENDORF, A.T., FULLER, R.C. (ed.): Comparative Biochemistry and Biophysics of Photosynthesis. Pp. 313-323. Univ. Tokyo Press, Tokyo; Univ. Park Press, State College, Pa. 1968.

4087 - KIM, W.S.: Complete fractionation of bacteriochlorophyll and its degradation products. - Biochim. biophys. Acta *112*: 392-402, 1966.

4088 - KIM, W.S.: Copper replacement of magnesium in the chlorophylls and bacteriochlorophyll. - Z. Naturforsch. *22 b*: 1054-1061, 1967.

4089 - KIM, Y.D.: The conformational stability of bacteriochlorophyll-protein complex isolated from a green photosynthetic bacterium. II. The action of detergents on the bacteriochlorophyll-protein complex. - Arch. Biochem. Biophys. *140*: 354-361, 1970.

4090 - KIM, Y.D., KE, B.: Conformational stability of the bacteriochlorophyll-protein complex isolated from the green photosynthetic bacterium *Chloropseudomonas ethylicum*. I. The conformational states at acid, neutral, and alkaline pH values. - Arch. Biochem. Biophys. *140*: 341-353, 1970.

4091 - KIMURA, M.: Ecological and physiological studies on the vegetation of Mt. Shimagare: VII. Analysis of production processes of young *Abies* stand based on the carbohydrate economy. - Bot. Mag. (Tokyo) *82*: 6-19, 1969.

4092 - KIMURA, M.: Analysis of production processes of an undergrowth of subalpine *Abies* forest, *Pteridophyllum racemosum* population 1. Growth, carbohydrate economy and net production. - Bot. Mag. (Tokyo) *83*: 99-108, 1970.

4093 - KIMURA, M.: Analysis of production processes of an undergrowth of subalpine *Abies* forest, *Pteridophyllum racemosum* population 2. Respiration, gross production and economy of dry matter. - Bot. Mag. (Tokyo) *83*: 304-311, 1970.

4094 - KING, D.L.: The role of carbon in eutrophication. - J. Water Pollut. Contr. Fed. *42*: 2035-2051, 1970.

4095 - KING, R.W., EVANS, L.T.: Photosynthesis in artificial communities of wheat, lucerne, and subterranean clover plants. - Aust. J. biol. Sci. *20*: 623-635, 1967.

4096 - KING, R.W., WARDLAW, I.F., EVANS, L.T.: Effect of assimilate utilization on photosynthetic rate in wheat. - Planta *77*: 261-276, 1967.

4097 - KINRAIDE, W.T.B., AHMADJIAN, V.: The effects of usnic acid on the physiology of two cultured species of the lichen alga *Trebouxia* PUYM. - Lichenologist (Oxford) *4*: 234-247, 1970. [Ps, Chl.]

4098 - KIRA, T.: A rational method for estimating total respiration of trees and forest stands. - In: ECKARDT, F.E. (ed.): Functioning of Terrestrial Ecosystems at the Primary Production Level. Pp. 399-407. UNESCO, Paris 1968.

4099 - KIRA, T., SHINOZAKI, K., HOZUMI, K.: Structure of forest canopies as related to their primary productivity. - Plant Cell Physiol. *10*: 129-142, 1969. [Ps, growth analysis.]

4100 - KIRENSKIĬ, L.V., TERSKOV, I.A., GITEL'ZON, I.I., LISOVSKIĬ, G.N., KOVROV, B.G., OKLADNIKOV, Yu.N.: Nepreryvnaya kul'tura mikrovodorosleĭ v kachestve zvena zamknutoĭ ekologicheskoĭ sistemy. [Continuous culture of microalgae as component of the closed ecological system.] - Kosm. Biol. Med. *1* (4): 19-22, 1967. [In R.]

4101 - KIRENSKIĬ, L.V., TERSKOV, I.A., GITEL'ZON, I.I., LISOVSKIĬ, G.M., KOVROV, B.G., SID'KO, F.Ya., OKLADNIKOV, Yu.N., ANTONYUK, M.P., BELYANIN, V.N., RERBERG, M.S.: Gazoobmen mezhdu chelovekom i kul'turoĭ mikrovodorosleĭ v 30-sutochnom eksperimente. [Gas exchange between man and microalgae culture in a 30-day experiment.] - Kosm. Biol. Med. *1* (4): 23-28, 1967. [Ps; in R.]

4102 - KIRICHENKO, E.B.: Vklyuchenie S^{35} v razlichnye belki khloroplastov. [^{35}S incorporation into various chloroplast proteins.] - Fiziol. Rast. *14*: 15-20, 1967. [Chl; in R, ab: E.]

4103 - KIRICHENKO, E.B.: Vydelenie plastid v organicheskikh sredakh i issledovanie ikh funktsional'noĭ deyatel'nosti. [Chloroplast isolation in organic media and the study of their functional activity.] - In: KIRICHENKO, E.B. (ed.): Metody Vydeleniya Khloroplastov. Pp. 18-32. Pushchino-na-Oke 1970. [In R, ab: E.]

4104 - KIRICHENKO, E.B.: Vydelenie plastid tsentrifugirovaniem v gradiente plotnosti organicheskoĭ sredy. [Chloroplast isolation by centrifugation in the density gradient of organic media.] - In: KIRICHENKO, E.B. (ed.): Metody Vydeleniya Khloroplastov. Pp. 109-115. Pushchino-na-Oke 1970. [In R, ab: E.]

4105 - KIRIENKO, I.M., LEBEDEV, S.I.: Deĭstvie vnekornevoĭ podkormki mineral'nymi elementami na fotosinteticheskuyu produktivnost' rasteniĭ sakharnoĭ svekly. [Effect of extra-root mineral nutrition on photosynthetic productivity of sugar beet plants.] - In: Mineral'nye Elementy i Mekhanizm Fotosinteza. Pp. 227-234. Kishinev 1969. [In R.]

4106 - KIRITA, M.: [The effect of germanium dioxide, a diatom-eliminating chemical, on the growth of free-living conchocelis of *Porphyra yezoensis*.] - Bull. jap. Soc. phycol. *18*: 167-170, 1970. [Ps; in Jap.]

4107 - KIRK, J.T.O.: Nature and function of chloroplast DNA. - In: GOODWIN, T.W. (ed.): Biochemistry of Chloroplasts. Vol. I. Pp. 319-320. Academic Press, London-New York 1966. [Car.]

4108 - KIRK, J.T.O.: Studies on the dependence of chlorophyll synthesis on protein synthesis in *Euglena gracilis*, together with a nomogram for determination of chlorophyll concentration. - Planta *78*: 200-207, 1968.

4109 - KIRK, J.T.O.: Biochemical aspects of chloroplast development. - Annu. Rev. Plant Physiol. *21*: 11-42, 1970.

4110 - KIRK, J.T.O.: Failure to detect effects of cycloheximide on energy metabolism in *Euglena gracilis*. - Nature *226*: 182, 1970. [Chl.]

4111 - KIRK, J.T.O., JUNIPER, B.E.: The ultrastructure of the chromoplasts of different colour varieties of *Capsicum*. - In: GOODWIN, T.W. (ed.): Biochemistry of Chloroplasts. Vol. II. Pp. 691-701. Academic Press, London-New York 1967.

4112 - KIRK, J.T.O., KEYLOCK, M.J.: Control of chloroplast formation in *Euglena gracilis:* Dependence of rate of chlorophyll synthesis on previous nutritional history of cells. - Biochem. biophys. Res. Commun. *28*: 927-931, 1967.

4113 - KIRK, J.T.O., READE, J.A.: The action spectrum of photosynthesis in *Euglena gracilis* at different stages of chloroplast development. - Aust. J. biol. Sci. *23*: 33-41, 1970.

B4114 - KIRK, J.T.O., TILNEY-BASSETT, R.A.E.: The Plastids. Their Chemistry, Structure, Growth and Inheritance. - W.H. Freeman and Co., London-San Francisco 1967.

4115 - KIRSHIN, I.K.: O fotosinteticheskoĭ deyatel'nosti posevov mnogoletnikh trav. [Photosynthetic activity of communities of perennial grasses.] - Uch. Zap. ural'. gos. Univ., Ser. biol. *58*: 105-113, 1967. [In R.]

4116 - KIRYATSEVA, O.F., LEBEDEV, S.T.: Pro rol' pigmentov plastyd u protsesi rostu roslyn. [Role of plastid pigments in plant growth process.] - In: Pidvyshchennya Vrozhaĭnosti Sil's'kogospod. Kul'tur. Pp. 196-203. Urozhaĭ, Kiev 1968. [In Ukr.]

4117 - KISAKI, T., TOLBERT, N.E.: Glycolate and glyoxylate metabolism by isolated peroxisomes or chloroplasts. - Plant Physiol. *44*: 242-250, 1969.

4118 - KISAKI, T., TOLBERT, N.E.: Glycine as a substrate for photorespiration. - Plant Cell Physiol. *11*: 247-258, 1970.

4119 - KISELEV, B.A., KOZLOV, Yu.N., EVSTIGNEEV, V.B.: K voprosu o polyarografii khlo-
         rofilla. [On chlorophyll polarography.] - Biofizika 15: 594-601, 1970. [In R,
         ab: E.]

4120 - KISELEV, V.E.: Kharakter deĭstviya 2,4-D i smeseĭ s azotnokislym ammoniem na so-
         derzhanie belka i khlorofilla v rasteniyakh. [Character of action of 2,4-D and
         its mixtures with ammonium nitrate on the amounts of protein and chlorophyll in
         plants.] - In: Fiziologicheskie Mekhanizmy Regulyatsii Prisposoblenii i Ustoĭ-
         chivosti u Rasteniĭ. Pp. 105-110. Nauka, Novosibirsk 1966. [In R.]

4121 - KISLYAKOVA, T.E., GOLUBKOVA, B.M., BOGACHEVA, I.I.: Vzaimosvyaz' struktūry i
         funktsii fotosinteticheskogo apparata v ontogeneze kartofelya. [Relation of
         structure and function of the photosynthetic apparatus in potato ontogenesis.]
         - Fiziol. Rast. 14: 5-14, 1967. [In R, ab: E.]

4122 - KISLYAKOVA, T.E., GOLUBKOVA, B.M., KUZNETSOVA, L.I.: Vliyanie khloramfenikola
         na strukturu i funktsiyu fotosinteticheskogo apparata kartofelya. [Effect of
         chloramphenicol on structure and function of the photosynthetic apparatus in
         potato.] - In: Khloroplasty i Mitokhondrii. Pp. 173-182. Nauka, Moskva 1969.
         [In R.]

4123 - KISS, A.S., HORVÁTH, L., POZSÁR, B.: Az Agronit serkentő hatása a Bezostaja I
         őszi búza levelének magnéziumtartalmára és fotoszintetikus szén-dioxid-fixálá-
         sára. [Stimulating effect of Agronit on magnesium content and photosynthetic
         carbon dioxide fixation in leaves of the winter wheat cv. Bezostaya 1.] - Búza-
         termesztési Kísérletek 1960-1970: 105-110, 1970. [In Hung, ab: R, E.]

4124 - KISS, B., KISS, A.S., POZSÁR, B.: A technikai ammóniumklorid hatása a fotoszin-
         tetikus széndioxidfixálásra és a fehérjeszintézisre, szudánifű-levelekben. [Ef-
         fect of technical ammonium chloride on photosynthetic carbon dioxide fixation
         and protein synthesis in leaves of sudangrass.] - Takarmánybázis 10: 19-26,
         1970. [In Hung, ab: E, F, G, R.]

4125 - KISSER, J., NĘTRUP-AUST, H.: Eine neue empfindliche Methode zur indirekten Fest-
         stellung des Öffnungszustandes der Spaltöffnungen. - Ber. deut. bot. Ges. 80:
         157-166, 1967.

4126 - KITLAEV, B.N., TARUSOV, B.N.: Nizkotemperaturnye vspyshki fotosinteticheskoĭ
         lyuminestsentsii rasteniĭ. [Low-temperature flashes of photosynthetic lumines-
         cence in plants.] - Dokl. Akad. Nauk SSSR 195: 725-727, 1970. [In R.]

4127 - KLEE, R., STEUBING, L.: Studien über das Interzellularvolumen von Laubblättern.
         - Ber. deut. bot. Ges. 80: 416-425, 1967.

4128 - KLEESE, R.A.: Photophosphorylation in barley. - Crop Sci. 6: 524-527, 1966.

4129 - KLEIN, R.M.: Packaged sunshine. - Garden J. (New York) 16: 122-127, 173-178,
         1966. [Also Ps.]

4130 - KLEIN, R.M., CRONQUIST, A.: A consideration of the evolutionary and taxonomic
         significance of some biochemical, micromorphological, and physiological charac-
         ters in the Thallophytes. - Quart. Rev. Biol. 42: 105-296, 1967. [Chl, Car, Ps.]

4131 - KLEIN, S., NEUMAN, J.: The greening of etiolated bean leaves and the development
         of chloroplast fine structure in absence of photosynthesis. - Plant Cell Physiol.
         7: 115-123, 1966.

4132 - KLEIN, W., FRENZ, W., STEINBERG, B.: Photometrische Blattflächenmessungen. - An-
         gew. Bot. 46: 285-292, 1970.

4133 - KLEINHOFS, A.: Ribulose-diphosphate carboxylase activity in induced barley mu-
         tants. - In: DOYLE, E. (ed.): Induced Mutations in Plants. Pp. 101-108. Int. at.
         Energy Agency, Vienna 1969.

4134 - KLEINHOFS, A., SHUMWAY, L.K.: Correlation of ribulose 1,5-diphosphate carboxyla-
         se activity with chlorophyll content and ultrastructure in induced mutants of
         Hordeum vulgare. - Biochem. Genet. 3: 485-492, 1969.

4135 - KLEINIG, H.: Die Bildung von Sekundärcarotinoiden in Acetabularia mediterranea. -
         Ber. deut. bot. Ges. 79:(126)-(130), 1966.

4136 - KLEINIG, H.: Anteraxanthin-Zeaxanthin-Umwandlung in Vaucheria sessilis (Xantho-
         phyceae). - Planta 75: 73-76, 1967.

4137 - KLEINIG, H.: Sekundärcarotinoide in der Grünalge *Sphaeroplea*. - Z. Naturforsch. *22 b*: 977-979, 1967.

4138 - KLEINIG, H.: Carotenoids of siphonous green algae: A chemo-taxonomical study. - J. Phycol. *5*: 281-284, 1969.

4139 - KLEINIG, H.: The structure of siphonaxanthin. - Tetrahedron Letters *1969*: 5139-5142, 1969.

4140 - KLEINIG, H., CZYGAN, F.-C.: Lipids of *Protosiphon (Chlorophyta)*. I. Carotenoids and carotenoid esters of five strains of *Protosiphon botryoides* (KÜTZ.) KLEBS. - Z. Naturforsch. *24 b*: 927-930, 1969.

4141 - KLEINIG, H., EGGER, K.: Carotinoide der *Vaucheriales Vaucheria* und *Botrydium (Xanthophyceae)*. - Z. Naturforsch. *22 b*: 868-872, 1967.

4142 - KLEINIG, H., EGGER, K.: Ketocarotinoidester in *Acetabularia mediterranea* LAM. - Phytochemistry *6*: 611-619, 1967.

4143 - KLEINIG, H., EGGER, K.: Zur Struktur von Siphonaxanthin und Siphonein, den Hauptcarotinoiden siphonaler Grünalgen. - Phytochemistry *6*: 1681-1686, 1967.

4144 - KLEINIG, H., NIETSCHE, H.: Carotinoidester-Muster in gelben Blütenblättern. - Phytochemistry *7*: 1171-1175, 1968.

4145 - KLEINIG, H., WRISCHER, M.: Die Feinstruktur von *Acetabularia*-Chloroplasten bei Sekündärcarotinoid-Bildung. - Z. Pflanzenphysiol. *58*: 248-251, 1968.

4146 - KLEINKOPF, G.E., HUFFAKER, R.C., MATHESON, A.: A simplified purification and some properties of ribulose 1,5-diphosphate carboxylase from barley. - Plant Physiol. *46*: 204-207, 1970.

4147 - KLEINKOPF, G.E., HUFFAKER, R.C., MATHESON, A.: Light-induced *de novo* synthesis of ribulose 1,5-diphosphate carboxylase in greening leaves of barley. - Plant Physiol. *46*: 416-418, 1970.

4148 - KLEMME, J.-H.: Hydrogenase and photosynthetic electron transport in chromatophores from the facultative phototroph, *Rhodopseudomonas capsulata*. - In: METZNER, H. (ed.): Progress in Photosynthesis Research. Vol. III. Pp. 1492-1503. Tübingen 1969.

4149 - KLEMME, J.-H.: Studies on the mechanism of NAD-photoreduction by chromatophores of the facultative phototroph, *Rhodopseudomonas capsulata*. - Z. Naturforsch. *24 b*: 67-76, 1969.

4150 - KLEMME, J.-H., SCHLEGEL, H.G.: Lichtabhängige Pyridinnucleotid-Reduktion mit molekularem Wasserstoff durch subzelluläre Photopigment-Partikel aus *Rhodopseudomonas capsulata*. - Z. Naturforsch. *22 b*: 899-900, 1967.

4151 - KLEMME, J.-H., SCHLEGEL, H.G.: Cyclic photophosphorylation by chromatophores of the facultative phototroph, *Rhodopseudomonas capsulata*. - Arch. Mikrobiol. *63*: 154-169, 1968.

4152 - de KLERK, H., GOVINDJEE, KAMEN, M.D., LAVOREL, J.: Age and fluorescence characteristics in some species of *Athiorhodaceae*. - Proc. nat. Acad. Sci. U.S.A. *62*: 972-978, 1969.

4153 - KLEUSER, D., BÜCHER, H.: Elektrochromie von Chlorophyll-*a* und Chlorophyll-*b* in monomolekularen Filmen. - Z. Naturforsch. *24 b*: 1371-1374, 1969.

4154 - KLEVANTSOVA, V.A., POKROVSKAYA, I.A.: O vliyanii okolosolnechnoĭ radiatsii na pokazaniya pirgeliometrov s raznymi dlinami trubok. [Influence of sun's oreol radiation on readings of pyrheliometers with various tube lengths.] - In: Aktinometriya i Optika Atmosfery. Pp. 226-229. Valgus, Tallin 1968. [In R, ab: E.]

4155 - KLIMAKHIN, G.I., FIRSOV, I.P.: Kharakteristika ust'ichnogo apparata di-, tri-, tetraploidnykh form svekly. [Characteristic of stomatal apparatus of di-, tri-, and tetraploid forms of sugar beet.] - Dokl. mosk. sel'skokhoz. Akad. K.A. Timiryazeva *136*: 55-59, 1968. [In R.]

4156 - KLIMASHEVSKIĬ, E.L.: O neodinakovoĭ ustoĭchivosti raznykh sortov kukuruzy k kislotnosti pitatel'nogo rastvora. [Different resistance of various maize cultivars to acidity of nutrient medium.] - Agrokhimiya *1966* (4): 98-106, 1966. [Ps, Chl; in R.]

4157 - KLIMASHEVSKIĬ, E.L.: O fiziologicheskikh osobennostyakh assimilyatsionnogo appa-
       rata i kornevykh sistem raznykh sortov kukuruzy v svyazi s neodinakovoĭ ustoĭchi-
       vost'yu ikh k kislomu pH sredy. [Physiological specificity of the assimilating
       apparatus and root systems of different maize varieties in relation to their
       reaction to the acidic pH of the media.] - Dokl. VASKHNIL 1966 (11): 15-17, 1966.
       [Ps; in R.]

4158 - KLIMASHEVSKIĬ, E.L.: Spetsifika fotosinteticheskoĭ deyatel'nosti genotipov kul'-
       turnykh rasteniĭ v svyazi s urovnem kornevogo pitaniya. [Features of photosyn-
       thetic activity of crop plant genotypes in relation to root nutrition.] - In:
       Mineral'nye Elementy i Mekhanizm Fotosinteza. Pp. 187-200. Kishinev 1969. [In R.]

4159 - KLIMASHEVSKIĬ, E.L., BAGAUTDINOVA, R.I.: Fotosintez rasteniĭ pri razlichnom kor-
       nevom pitanii. [Plant photosynthesis under various root nutrition.] - Sel'.-khoz.
       Biol. 3: 218-226, 1968. [In R, ab: E.]

4160 - KLIMASHEVSKIĬ, E.L., ZHURAVLEV, Yu.N., POPOVA, Z.S.: K fiziologicheskomu analizu
       geneticheski raznokachestvennykh sortov Zea mays L., neodinakovo ustoĭchivykh k
       povyshennoĭ kislotnosti sredy. [Physiological analysis of genetically heteroge-
       neous cultivars of maize unequally resistant to increased acidity of medium.] -
       Fiziol. Rast. 15: 343-351, 1968. [Ps; in R, ab: E.]

4161 - KLIMOVITSKAYA, Z.M., BESKINSKAYA, E.P.: Izuchenie reaktsii Khilla v svyazi s oso-
       bennostyami pitaniya rasteniĭ makro- i mikroelementami. [Study of Hill reaction
       in relation to peculiarities of plant nutrition with macro- and microelements.]
       - In: Mikroelementy v Sel'skom Khozyaĭstve i Meditsine, Vol. 3. Pp. 19-26. Nau-
       kova Dumka, Kiev. 1967. [In R.]

4162 - von KLITZING, L., SCHWEIGER, H.G.: A method for recording the circadian rhythm
       of the oxygen balance in a single cell of Acetabularia mediterranea. - Protoplas-
       ma 67: 327-332, 1969.

4163 - KLOCHKOVA, M.P.: Spektral'nye svoĭstva list'ev rasteniĭ, vyrashchennykh na is-
       kusstvennom osveshchenii. [Spectral properties of leaves of plants grown under
       artificial irradiation.] - Sb. Tr. agr. Fiz. (Leningrad) 15: 55-61, 1968. [In R.]

4164 - KLOCHKOVA, M.P.: O fluorestsentsii list'ev vysshikh rasteniĭ v zelenoĭ chasti
       spektra. [Fluorescence of leaves of higher plants in the green part of the spec-
       trum.] - Sb. Tr. agron. Fiz. (Leningrad) 15: 62-68, 1968. [In R.]

4165 - KLOCHKOVA, M.P.: Ustanovka dlya izmereniya spektrov fluorestsentsii list'ev ras-
       teniĭ. [Apparatus for measuring fluorescence spectra of plant leaves.] - Sb. Tr.
       agron. Fiz. (Leningrad) 15: 168-171, 1968. [In R.]

4166 - KLOCHKOVA, M.P.: Issledovanie pigmentov list'ev perilly v zavisimosti ot inten-
       sivnosti i spektral'nogo sostava sveta. [Study of pigments of Perilla leaves in
       relation to irradiance and spectral composition of radiation.] - Sb. Tr. agron.
       Fiz. (Leningrad) 21: 85-90, 1970. [In R.]

4167 - KLOFAT, W., HANNIG, K.: Elektrophoretische Isolierung von Chloroplasten. - Hoppe-
       Seyler's Z. physiol. Chem. 348: 739-741, 1967.

4168 - KLOFAT, W., HANNIG, K.: Elektrophoretische Trennung von Chloroplastenfragmenten
       mit unterschiedlichem Verhältnis von Chlorophyll a : Chlorophyll b. - Hoppe-
       Seyler's Z. physiol. Chem. 348: 1332-1334, 1967.

4169 - KLUGE, M.: Untersuchungen über den Gaswechsel von Bryophyllum während der Licht-
       periode. I. Zum Problem der CO_2-Abgabe. - Planta 80: 255-263, 1968.

4170 - KLUGE, M.: Untersuchungen über den Gaswechsel von Bryophyllum während der Licht-
       periode. II. Beziehungen zwischen dem Malatgehalt des Blattgewebes und der CO_2-
       Aufnahme. - Planta 80: 359-377, 1968.

4171 - KLUGE, M.: Über den CO_2-Gaswechsel von Bryophyllum. - Ber. deut. bot. Ges. 82:
       25-28, 1969.

4172 - KLUGE, M.: Zur Analyse des CO_2-Austausches von Bryophyllum. I. Messung der Än-
       derung des Mengenverhältnisses einiger Phosphatverbindungen im Blattgewebe während
       bestimmter Phasen der Licht-Dunkel-Periode. - Planta 85: 160-170, 1969.

4173 - KLUGE, M.: Veränderliche Markierungsmuster bei $^{14}CO_2$-Fütterung von Bryophyllum
       tubiflorum zu verschiedenen Zeitpunkten der Hell/Dunkelperiode. I. Die $^{14}CO_2$-Fi-
       xierung unter Belichtung. - Planta 88: 113-129, 1969.

4174 - KLUGE, M., FISCHER, K.: Über Zusammenhänge zwischen dem CO_2-Austausch und der Abgabe von Wasserdampf durch *Bryophyllum daigremontianum* BERG. - Planta *77*: 212-223, 1967.

4175 - KLYACHENKO, V.I.: Vliyanie kal'tsiya i magniya na strukturu organoidov rastitel'-noi kletki. [Effect of calcium and magnesium on organoid structure of plant cell.] - In: Fotosintez i UrozhaTnost' Sel'skokhozyaTstvennykh RasteniT. Pp. 72-77. Min. sel'. Khoz. SSSR, Kiev 1970. [Chl; in R.]

4176 - KLYUCHNIKOV, L.Yu., BAGAEVA, M.V.: Fiziologicheskie reaktsii khvoTnykh dreves-nykh porod na obrabotku 2,4-D v smeshannykh molodnyakakh. [Physiological response of woody needle plants to 2,4-D treatment in mixed young communities.] - Fiziol. Rast. *16*: 443-446, 1966. [Ps, Chl; in R, ab: E.]

4177 - KNAFF, D.B., ARNON, D.I.: Light-induced oxidation of a chloroplast *b*-type cyto-chrome at -189°C. - Proc. nat. Acad. Sci. U.S.A. *63*: 956-962, 1969.

4178 - KNAFF, D.B., ARNON, D.I.: Spectral evidence for a new photoreactive component of the oxygen-evolving system in photosynthesis. - Proc. nat. Acad. Sci. U.S.A. *63*: 963-969, 1969.

4179 - KNAFF, D.B., ARNON, D.I.: A concept of three light reactions in photosynthesis by green plants. - Proc. nat. Acad. Sci. U.S.A. *64*: 715-722, 1969.

4180 - KNAFF, D.B., ARNON, D.I.: Contrasting requirement for plastocyanin in the photo-oxidation of chloroplast cytochromes. - Biochim. biophys. Acta *223*: 201-204, 1970.

4181 - KNAVEL, D.E.: Influence of growth retardants on growth, nutrient content and yield of tomato plants grown at various fertility levels. - J. amer. Soc. hort. Sci. *94*: 32-35, 1969. [Chl.]

4182 - KNIGA, M.I., KNIGA, N.M., NASONOVA, M.G., SHEVCHENKO, I.M.: Vliyanie fotosinte-za na produktivnost' kukuruzy v zavisimosti ot udobreniya pochvy. [Influence of photosynthesis on maize productivity in dependence on soil fertilization.] - Dokl. VASKHNIL *1970* (2): 11-13, 1970. [In R.]

4183 - KNIGHT, G.J., PRICE, C.A.: Measurements of s-ρ coordinates during the development of *Euglena* chloroplasts. - Biochim. biophys. Acta *158*: 283-285, 1968.

4184 - KNIPL, Ya.S., KULAEVA, O.N.: DeTstvie kumarina i sinteticheskikh retardantov rosta na soderzhanie khlorofilla i belka v otrezkakh list'ev yachmenya v temnote i na svetu. [Effect of coumarin and synthetic growth retardants on chlorophyll and protein content in barley leaf segments in light and darkness.] - Fiziol. Rast. *17*: 14-22, 1970. [In R, ab: E.]

4185 - KNIPL, Ya.S., KULAEVA, O.N.: Vliyanie sinteticheskikh retardantov rosta i kuma-rina na sintez i raspad RNK i belka v srezannykh list'yakh v protsesse ikh sta-reniya. [Effect of synthetic growth retardants and coumarin on synthesis and decomposition of RNA and proteins during ageing of detached leaves.] - Biokhimiya *35*: 1219-1229, 1970. [Chl; in R, ab: E.]

4186 - KNOBLOCH, K.: Photosynthetische Sulfid-Oxydation grüner Pflanzen. I. - Planta *70*: 73-86, 1966.

4187 - KNOBLOCH, K.: Photosynthetische Sulfid-Oxydation grüner Pflanzen. II. Wirkung von Stoffwechselinhibitoren. - Planta *70*: 172-186, 1966.

4188 - KNOBLOCH, K.: Über die Rolle von Schwefelwasserstoff in der Photosynthese ein-zelliger Grünalgen. - Ber. deut. bot. Ges. *79*: (115)-(118), 1966.

4189 - KNOBLOCH, K.: Sulphide oxidation via photosynthesis in green algae. - In: METZ-NER, H. (ed.): Progress in Photosynthesis Research. Vol. II. Pp. 1032-1034. Tübingen 1969.

4190 - KNOP, J.V., FUHRHOP, J.-H.: Die Reaktivitäten des Porphins, Chlorins, Bacterio-chlorins und Phlorins. Ladungsdichten, freie Valenzen und Aussenelektronendich-ten. - Z. Naturforsch. *25 b*: 729-734, 1970.

4191 - KNOTH, R., HAGEMANN, R.: Phänotypische Normalisierung der Mutante *albina* von *Lycopersicon pimpinellifolium* (JUSL.) MILL. - Biochem. Physiol. Pflanzen *161*: 106-132, 1970. [Chl.]

4192 - KNOWLES, F.C., PON, N.G.: On the structure of ribulose 5-phosphate as an intermediate of the photosynthetic pentose phosphate cycle. - J. amer. chem. Soc.
90: 6536-6537, 1968.

4193 - KNOWLES, R.E., LIVINGSTON, A.L., NELSON, J.W., KOHLER, G.O.: Xanthophyll and
carotene storage stability in commercially dehydrated and freeze-dried alfalfa.
- J. agr. Food Chem. *16*: 654-658, 1968.

4194 - KNOX, R.S.: On the theory of trapping of excitation in the photosynthetic unit.
- J. theor. Biol. *21*: 244-259, 1968.

4195 - KNOX, R.S.: Storage of light energy and photosynthesis. - Nature *221*: 263-264,
1969.

4196 - KNOX, R.S.: Thermodynamics and the primary processes of photosynthesis. - Biophys. J. *9*: 1351-1362, 1969.

4197 - KNUTSEN, G.: Effects of phenylethyl alcohol on *Chlorella pyrenoidosa*. - Physiol.
Plant. *19*: 142-151, 1966. [Ps, Chl.]

4198 - KNYPL, J.S.: Coumarin, phosfon D and CCC - the inhibitors of chlorophyll and
protein degradation in senescing leaf tissue of kale. - Flora A *158*: 230-240,
1967.

4199 - KNYPL, J.S.: Retardation of chlorophyll degradation in *Zea mays* by coumarin,
Phosfon D, and CCC. - Naturwissenschaften *54*: 146, 1967.

4200 - KNYPL, J.S.: Inhibition of chlorophyll synthesis by growth retardants and coumarin, and its reversal by potassium. - Nature *224*: 1025-1026, 1969.

4201 - KNYPL, J.S.: The control of RNA, protein and chlorophyll synthesis in senescing
leaf tissue of kale by coumarin and growth retardants. - Flora A *160*: 217-233,
1969.

4202 - KNYPL, J.S.: Complementary action of potassium and benzylaminopurine on growth,
chlorophyll, protein and RNA synthesis in cucumber cotyledons. - Curr. Sci.
39: 534-535, 1970.

4203 - KNYPL, J.S.: Control of chlorophyll synthesis by coumarin and plant growth retarding chemicals. - Acta Soc. Bot. Pol. *39*: 321-332, 1970.

4204 - KNYPL, J.S.: Inhibition of chlorophyll synthesis by growth retarding chemicals
and coumarin in detached cotyledons of pumpkin. - Biochem. Physiol. Pflanzen
161: 1-13, 1970.

4205 - KNYPL, J.S., RENNERT, A.: Działanie inhibitorów syntezy kwasów nukleinowych i
białek na wzrost i syntezę chlorofilu w izolowanych liścieniach ogórka. [Effect
of inhibitors of nucleic acid and protein synthesis on growth and chlorophyll
synthesis in isolated cucumber leaves.] - Zesz. nauk. Univ. Łódzk. Ser. 2, *1970*
(37): 77-96, 1970. [In Pol.]

4206 - KNYPL, J.S., RENNERT, A.: Stimulation of growth and chlorophyll synthesis in detached cotyledons of cucumber by potassium. Z. Pflanzenphysiol. *62*: 97-107, 1970.

4207 - KOBAK, K.I.: Izuchenie proizvoditel'nosti fitotsenozov aerodinamicheskim metodom.
[Aerodynamic method of studying productivity of phytocenoses.] - In: Soobshcheniya
po Anatomii i Fiziologii Drevesnykh Rasteniĭ. Pp. 67-71. Leningrad. lesotekh.
Akad. Kirova, Leningrad 1967. [In R, ab: E.]

4208 - KOBAK, K.I.: Uglekislota vozdukha kak kharakteristika atmosfery lesnogo biogeot-
senoza. [Air carbon dioxide as characteristics of atmosphere of forest biogeno-
sis.] - In: Svetovoĭ Rezhim, Fotosintez i Produktivnost' Lesa. Pp. 180-199. Nauka,
Moskva 1967. [In R.]

4209 - KOBLENTS-MISHKE, O.I., OCHAKOVSKIĬ, Yu.E.: Ob izmereniyakh sveta pri izuchenii
pervichnoĭ produktsii v more. [Light measurements in primary production studies
in the sea.] - Okeanologiya *1966*: 535-542, 1966. [In R, ab: E.]

4210 - KOBLET, W.: Wanderung von Assimilaten in Rebtrieben. - Schweiz. Z. Obst- Weinbau
105: 501-508, 1969. [Ps.]

4211 - KOCH, W.: Die Temperaturabhängigkeit der Photosynthese. - Mitt. Staatsforstver-
waltung Bayerns *1967* (36): 1-7, 1967.

4212 - KOCH, W.: Neue Bemühungen um eine standortgerechte Erfassung des Gaswechsels unserer Waldbäume. - Deut. Akad. Landwirtschaftswissensch. Berlin, Tagungsber. *100*: 171-176, 1968.

4213 - KOCH, W.: Untersuchungen über die Wirkung von CO_2 auf die Photosynthese einiger Holzgewächse unter Laboratoriumsbedingungen. - Flora B *158*: 402-428, 1969.

4214 - KOCH, W., KLEIN, E., WALZ, H.: Neuartige Gaswechsel-Messanlage für Pflanzen in Laboratorium und Freiland. - Siemens-Z. *42*: 392-404, 1968.

4215 - KOCH, W., WALZ, H.: Neuer Wasserdampfabscheider mit Peltierkühlung als vielseitiges Zusatzgerät bei Gaswechselmessungen. - Naturwiss. Rundschau *19*: 163, 1966.

4216 - KOCH, W., WALZ, H.: Kleinklimaanlage zur Messung des pflanzlichen Gaswechsels. Ein neuartiges Verfahren der Feuchteregelung und Transpirationsmessung. - Naturwissenschaften *54*: 321-322, 1967.

4217 - KOCHERZHENKO, I.E., MAÏKO, T.K.: Fotoperiodizm i osobennosti anatomicheskoĭ struktury seyantsev drevesnykh rasteniĭ. [Photoperiodism and some features of anatomical structure of seedlings of woody plants.] - Fiziol. Rast. *15*: 63-73, 1968. [Chl; in R, ab: E.]

4218 - KOCHUBEÏ, S.M.: Ob interpretatsii ul'trafioletovogo spektra β-karotina. [Interpretation of ultra-violet spectrum of β-carotene.] - Zh. prikl. Spektroskop. *8*: 166-168, 1968. [In R.]

4219 - KOCHUBEÏ, S.M.: Spektr fluorestsentsii khlorofilla *a* v khloroplastakh pri 4°K. [Fluorescence spectrum of chlorophyll *a* in chloroplasts at 4 K]. - Dokl. Akad. Nauk SSSR *192*: 446-448, 1970. [In R.]

4220 - KOCHUBEÏ, S.M., KUCHERENKO, V.P.: Spektral'nye svoĭstva supernatantov, poluchennykh pri tsentrifugirovanii khloroplastov, obrabotannykh poverkhnostno-aktivnymi veshchestvami. [Spectral properties of supernatants in centrifugates of chloroplasts treated with surface-active substances.] - Biofizika *14*: 628-633, 1969. [In R, ab: E.]

4221 - KOCHUBEÏ, S.M., MANUIL'SKAYA, S.V., OSTROVSKAYA, L.K.: O vozmozhnykh putyakh issledovaniya tsentrov I fotokhimicheskoĭ sistemy. [Possible ways of studying centres of photosystem I.] - In: Problemy Biofotokhimii. Pp. 33-34. Izdat. moskov. gos. Univ., Moskva 1970. [In R.]

4222 - KOCHUBEÏ, S.M., REÏNGARD, T.A.: Vliyanie sostava smesi dlya vydeleniya fragmentov khloroplastov na ikh spektral'nye i fotokhimicheskie svoĭstva. [Effect of different composition of the buffer mixture used for extracting fragments of the chloroplasts on their spectral and photochemical characteristics.] - Biokhimiya *35*: 868-872, 1970. [In R, ab: E.]

4223 - KOF, E.M.: Deĭstvie metabolicheskikh ingibitorov na protsessy obrazovaniya khlorofilla i flavonoidnykh ingibitorov rosta pri zelenenii rasteniĭ. [Effect of metabolic inhibitors on processes of chlorophyll formation and flavonoid growth inhibitors during the greening of plants.] - Dokl. Akad. Nauk SSSR *192*: 676-679, 1970. [In R.]

4224 - KOHL, D.H., WOOD, P.M.: On the molecular identity of ESR signal II observed in photosynthetic systems: The effect of heptane extraction and reconstitution with plastoquinone and deuterated plastoquinone. - Plant Physiol. *44*: 1439-1445, 1969.

4225 - KOHL, D.H., WRIGHT, J.R., WEISSMAN, M.: Electron spin resonance studies of free radicals derived from plastoquinone, α- and γ-tocopherol and their relation to free radicals observed in photosynthetic materials. - Biochim. biophys. Acta *180*: 536-544, 1969.

4226 - KOHLER, G.O., KNOWLES, R.E., LIVINGSTON, A.L.: An improved analytical procedure for the determination of xanthophyll. - J. Ass. offic. anal. Chem. *50*: 707-711, 1967.

4227 - KOHN, H.I., McLEOD, G.C., WRIGHT, K.A.: Inhibition of chlorophyll *a* and *b* synthesis in *Chlamydomonas reinhardi* mutant strain y-2 by ionizing radiation. - Rad. Bot. *7*: 123-128, 1967.

4228 - KOJIĆ, M.: Über die Assimilationsleistung bei Roggen und Gerste unter feldmässigen Bedingungen. - Ber. deut. bot. Ges. *81*: 437-441, 1968.

4229 - KOJIMA, C., KITADA, K.: [A simple integrating light meter and its application
to the continuous measurement of insolation (intensity of illumination) in several stands.] - J. Jap. Forest. Soc. *49*: 69-72, 1967. [In Jap.]

4230 - KOK, B.: Concentration and normal potential of primary photooxidants and reductants in photosynthesis. - In: THOMAS, J.B., GOEDHEER, J.C. (ed.): Currents in
Photosynthesis. Pp. 383-392. Donker, Rotterdam 1966.

4231 - KOK, B.: Photosynthesis - physical aspects. - In: SAN PIETRO, A., GREER, F.A.,
ARMY, T.J. (ed.): Harvesting the Sun. Pp. 29-48. Academic Press, New York-
London 1967.

4232 - KOK, B.: Photosynthesis. - In: WILKINS, M.B. (ed.): The Physiology of Plant
Growth and Development. Pp. 333-379. McGraw-Hill, London 1969.

4233 - KOK, B., CHENIAE, G.M.: Kinetics and intermediates of the oxygen evolution step
in photosynthesis. - Curr. Topics Bioenerg. *1*: 1-47, 1966.

4234 - KOK, B., FORBUSH, B., McGLOIN, M.: Cooperation of charges in photosynthetic O_2
evolution. I. A linear four step mechanism. - Photochem. Photobiol. *11*: 457-475,
1970.

4235 - KOK, B., JOLIOT, P., GLOIN, M.P.: Electron transfer between the photoacts. -
In: METZNER, H. (ed.): Progress in Photosynthesis Research. Vol. II. Pp. 1042-
1056. Tübingen 1969.

4236 - KOK, B., MALKIN, S., OWENS, O., FORBUSH, B.: Observations on the reducing side
of the O_2-evolving photoact. - In: Energy Conversion by the Photosynthetic Apparatus. Brookhaven Symp. Biol. *19*: 446-459, 1967.

4237 - KOK, B., RURAINSKI, H.J.: Long-wave absorption and emission bands in chloroplast
fragments. - Biochim. biophys. Acta *126*: 584-587, 1966.

4238 - KOLESNIKOV, P.A.: K voprosu ob uchastii substrat-O_2-oksidoreduktaz v razlozhenii
vody v protsesse fotosinteza. [Participation of substrate-O_2-oxidoreductases in
photolysis of water.] - In: Mekhanizmy Dykhaniya, Fotosinteza i Fiksatsii Azota.
Pp. 275-280. Nauka, Moskva 1967. [In R.]

4239 - KOLESNIKOV, P.A.: Ferredoksin i plastotsianin v tsepi fotosinteticheskogo perenosa elektronov. [Ferredoxin and plastocyanin in the photosynthetic electron transfer chain.] - In: Mekhanizmy Dykhaniya, Fotosinteza i Fiksatsii Azota. Pp. 289-
291. Nauka, Moskva 1967. [In R.]

4240 - KOLESNIKOV, P.A., PETRACHENKO, E.I., ZORE, S.V., MUTUSKIN, A.A., PSHENOVA, K.V.:
Glikolatno-glioksalatnyi tsikl i kislorodnyi obmen khloroplastov. [Glycolate-
glyoxylate pathway and oxygen exchange of chloroplasts.] - Fiziol. Rast. *17*:
496-501, 1970. [In R, ab: E.]

4241 - KOLESNIKOV, V.A., AGAFONOV, N.V., KHRYPOVA, N.Kh.: Razmeshchenie assimilyatsionnogo apparata v krone yabloni. [Distribution of the assimilation apparatus in
crown of apple trees.] - Izv. timiryaz. sel'skokhoz. Akad. *1970* (3): 153-158,
1970. [Leaf area; in R, ab: E.]

4242 - KOLESNIKOV, V.A., PIL'SHCHIKOV, F.N.: Vliyanie plantazhnoi vspashki na rost
kornei i nakoplenie khlorofilla v list'yakh yabloni.[Effect of deep ploughing
on root growth and chlorophyll accumulation in apple leaves.] - Izv. timiryaz.
sel'skokhoz. Akad. (Moskva) *1968* (6): 152-158, 1968. [In R.]

4243 - KOLLER, D.: Characteristics of the photosynthetic apparatus derived from its
response to natural complexes of environmental factors. - In: Prediction and
Measurement of Photosynthetic Productivity. Pp. 283-294. PUDOC, Wageningen 1970.

4244 - KOLLER, D.: The partitioning of resistances to photosynthetic CO_2 uptake in the
leaf. - New Phytol. *69*: 971-981, 1970.

4245 - KOLLER, H.R., NYQUIST, W.E., CHORUSH, I.S.: Growth analysis of the soybean community. - Crop Sci. *10*: 407-412, 1970.

4246 - KOLLMANN, R.: Autoradiographischer Nachweis der Assimilat-Transportbahn im se-
kundären Phloem von *Metasequoia glyptostroboides*. - Z. Pflanzenphysiol. *56*: 401-
409, 1967.

4247 - KOLOMIETS, N.G.: Dnevnaya i sezonnaya dinamika soderzhaniya khlorofilla i karotina v list'yakh gorokha. [Diurnal and seasonal dynamics of chlorophyll and carotene content in pea leaves.] - Tr. kharkovsk. sel'.-khoz. Inst. *57* (94): 77-80, 1966. [In R.]

4248 - **KOLOTOVA, L.R.**: Lyutein-epoksid kak promezhutochnoe soedinenie pri prevrashche-niyakh ksantofillov. [Lutein-epoxide as intermediary compound in xanthophyll transformations.] - In: Tezisy Dokladov Konferentsii Molodykh Uchenykh. P. 45. Dushanbe 1966. [In R.]

4249 - **KOLOTOVA, L.R.**: K voprosu ob usloviyakh obrazovaniya lyutein-epoksida. [Conditions of formation of lutein-epoxide.] - Dokl. Akad. Nauk Tadzh. SSR *10* (7): 53-55, 1967. [In R.]

4250 - **KOLOTOVA, L.R.**: Deĭstvie temperatury na soderzhanie lyutein-epoksida. [Effect of temperature on the content of lutein-epoxide.] - Izv. Akad. Nauk Tadzh. SSR, Otd. biol. Nauk *1967* (3): 87-88, 1967. [In R.]

4251 - **KOLOTOVA, L.R., SAPOZHNIKOV, D.I.**: Deĭstvie serovodoroda na izmenenie soderzhaniya lyutein-epoksida. [Effect of hydrogen sulfide on changes of lutein-epoxide content.] - Dokl. Akad. Nauk Tadzh. SSR *10* (8): 57-59, 1967. [In R, ab: Tadzh.]

4252 - **KOLOTOVA, L.R., TOLIBEKOV, D.**: Izmeneniya soderzhaniya lyutein-epoksida v go-mogenate. [Changes in lutein epoxide content in homogenate.] - Dokl. Akad. Nauk Tadzh. SSR *9* (5) : 31-33, 1966. [In R.]

4253 - **KOLOVU, M.**: Izmeneniya na produktivnostta na fotosintezata v khoda na individual-noto razvitie na rasteniya ot pshenitsa sort Yubileĭna III. [Changes in the productivity of photosynthesis during ontogenesis of wheat plants cv. Yubileĭna III.] - Rasteniev"d. Nauki (Sofia) *5* (2): 3-12, 1968. [In Bulg., ab: E, R.]

4254 - **KOL'TSOVA, T.I.**: Opredelenie ob"ema i poverkhnosti kletok fitoplanktona. [Determination of volume and surface of phytoplankton cells.] - Nauch. Dokl. vyssh. Shkoly, biol. Nauki *13* (6): 114-119, 1970. [In R.]

4255 - **KOMÁREK, J., PŘIBYL, S.**: Heat of combustion in the biomass of the alga *Scene-desmus quadricauda* during its ontogenetic cycle. - Nature *219*: 635-636, 1968.

4256 - **KOMARETSKAYA, E.N.**: Ob izemenenii soderzhaniya fotosinteticheskikh pigmentov list'ev v techenie sutok. [Diurnal changes in the level of photosynthetic pigments in leaves.] - In: Fotosintez i Urozhaĭnost' Sel'skokhozyaĭstvennykh Rasteniĭ. Pp. 127-132. Min. sel'. Khoz. SSSR, Kiev 1970. [In R.]

4257 - **KOMARNITSKIĬ, P.A., LEBEDEV, S.I.**: Fiziologo-biokhimicheskie izmeneniya v poch-kakh chereshni v period pokoya. [Physiological and biochemical changes in dorm-ant cherry buds.] - In: Fotosintez i Urozhaĭnost' Sel'skokhozyaĭstvennykh Rasteniĭ. Pp. 118-126. Min. sel'.Khoz.SSR, Kiev 1970. [Chl, Car; in R.]

4258 - **KOMISSAROV, D.A., SHTEĬNVOL'F, L.P.**: Intensivnost' fotosinteza podrosta eli v raznykh ekologicheskikh usloviyakh. [Photosynthetic rate of spruce undergrowth in various ecological conditions.] - In: Svetovoĭ Rezhim, Fotosintez i Produk-tivnost' Lesa. Pp. 243-254. Nauka, Moskva 1967. [In R.]

4259 - **KOMISSAROV, G.G.**: Ob odnoĭ osobennosti stroeniya khloroplasta, obespechivayush-cheĭ vysokuyu effektivnost' pogloshcheniya svetovoĭ energii. [Peculiarity of chloroplast structure ensuring high efficiency of radiant energy absorption.] - In: Bioenergetika i Biologicheskaya Spektrofotometriya. Pp. 241-243. Nauka, Moskva 1967. [In R.]

4260 - **KOMISSAROV, G.G.**: O vozmozhnosti fotoelektricheskogo mekhanizma razlozheniya vody pri fotosinteze. [Possibility of photoelectric mechanism of water split-ting in photosynthesis.] - In: Mekhanizmy Dykhaniya, Fotosinteza i Fiksatsii Azota. Pp. 286-289. Nauka, Moskva 1967. [In R.]

4261 - **KOMISSAROV, G.G.**:O vozmozhnosti fotoelektricheskogo mekhanizma razlozheniya vo-dy pri fotosinteze. [Possibility of photoelectric mechanism of water splitting in photosynthesis.] - Biofizika *12*: 558-560, 1967. [In R.]

4262 - **KOMISSAROV, G.G., SHUMOV, Yu.S.**: Izuchenie fotovol'taicheskogo effekta v plen-kakh β-karotina. [Photovoltaic effect in β-carotene films.] - Biofizika *13*: 421-427, 1968. [In R.]

4263 - KOMISSAROV, G.G., SHUMOV, Yu.S.: O vozmozhnosti fotoelektroliza vody v modelyakh
         soderzhashchikh fotosinteticheskie pigmenty, i v zelenom liste. [Possibility of
         water photoelectrolysis in models with photosynthetic pigments, and in a green
         leaf.] - Dokl. Akad. Nauk SSSR *182*: 1226-1229, 1968. [In R.]

4264 - KOMISSAROV, G.G., SHUMOV, Yu.S., ATAMANCHUK, L.M.: Izmenenie velichiny i kineti-
         ki fotopotentsiala fotosinteticheskikh pigmentov s pomoshch'yu razlichnykh doba-
         vok. [Change in the value and kinetics of photopotential of photosynthetic pig-
         ments by means of different additions.] - Biofizika *13*: 324-325, 1968. [In R.]

4265 - KOMISSAROV, G.G., SHUMOV, Yu.S., BORISEVICH, Yu.E.: Fotovol'taicheskaya batareya
         - funktsional'naya model' khloroplasta. [Photovoltaic battery as a function model
         of the chloroplast.] - Dokl. Akad. Nauk SSSR *187*: 670-673, 1969. [In R.]

4266 - KOMOV, S.V.: Tipy otvetnykh reaktsiĭ (po skorosti fiksatsii $C^{14}O_2$) u rasteniĭ
         ovsa, vyrashchennogo pri razlichnykh usloviyakh kornevogo pitaniya. [Types of
         response reactions (according to the rate of $^{14}CO_2$ fixation) in oat plants grown
         at various root nutrition.] - Uch. Zap. ural'.gos. Univ.,Ser. biol. *58*: 35-39,
         1967. [In R.]

4267 - KOMOV, S.V.: Otsenka dinamicheskoĭ organizatsii fotosinteza po usvoeniyu CO_2 pri
         stupenchatom, impul'snom i chastotnom rezhimakh osveshcheniya. [Evaluation of
         dynamic organization of photosynthesis from CO_2 assimilation at step-wise, im-
         pulse and frequency irradiation.] - Uch. Zap. ural'.gos. Univ. *113*: 20-44, 1970.
         [In R.]

4268 - KOMOV, S.V., NOSOVA, I.P.: Issledovanie zavisimosti induktsionnykh yavleniĭ (po
         pogloshcheniyu $C^{14}O_2$) ot fiziologicheskogo sostoyaniya rasteniĭ. [Relation of
         induction phenomena (in $^{14}CO_2$ absorption) to the physiological state of plants.]
         - Uch. Zap. ural'. gos. Univ.,Ser. biol. *58*: 27-34, 1967. [In R.]

4269 - KONDRAT'EVA, E.N.: Fotosinteziruyushchie organizmy v svyazi s evolyutsieĭ foto-
         sinteza. [Photosynthetic organisms in relation to evolution of photosynthesis.]
         - Tr. mosk. Obshch. Ispyt. Prirody, Otd. biol. *24*: 26-37, 1966. [In R.]

4270 - KONDRAT'EVA, E.N.: Metabolizm ugleroda u zelenykh fotosinteziruyushchikh bakte-
         riĭ. [Carbon metabolism of green photosynthetic bacteria.] - Nauch. Dokl. vyssh.
         Shkoly, biol. Nauki *10* (10): 77-94, 1967. [In R.]

4271 - KONDRATEVA, E.N., GOGOTOV, I.N.: Production of hydrogen by green photosynthetic
         bacteria *(Chloropseudomonas)*. - Nature *221*: 83-84, 1969.

4272 - KONDRAT'EVA, E.N., GOGOTOV, I.N.: Vydelenie i potreblenie molekulyarnogo vodo-
         roda *Chloropseudomonas*. [Efflux and absorption of molecular hydrogen in *Chloro-
         pseudomonas*.] - Mikrobiologiya *38*: 938-944, 1969. [In R.]

4273 - KONDRAT'EVA, E.N., KRASIL'NIKOVA, E.N., TROTSENKO, Yu.A., GOGOTOV, I.N.: O put-
         yakh ispol'zovaniya nekotorykh soedineniĭ ugleroda zelenymi bakteriyami. [Path-
         ways of utilization of some carbon compounds by green bacteria.] - Nauch. Dokl.
         vyssh. Shkoly, biol. Nauki *13* (3): 141, 1970. [In R.]

4274 - KONDRAT'EVA, E.N., NESTEROV, A.I., GOGOTOV, I.N.: Fotosintez u purpurnykh i zele-
         nykh bakteriĭ pri deĭstvii monokhromaticheskogo sveta raznykh dlin voln. [Photo-
         synthesis in purple and green bacteria under monochromatic irradiation with va-
         rious wavelengths.] - Nauch. Dokl. vyssh. Shkoly, biol. Nauki *13* (12): 69-77,
         1970. [In R.]

4275 - KONDRAT'EVA, E.N., RUBIN, L.B.: O fiziologo-biokhimicheskikh svoĭstvakh fotosin-
         teziruyushchikh bakteriĭ *Ectothiorhodospira shaposhnikovii*. [Physiological and
         biochemical properties of photosynthetic bacterium *Ectothiorhodospira shaposhni-
         kovii*.] - In: Fiziologiya i Biokhimiya Zdorovogo i Bol'nogo Rasteniya. Pp. 273-
         292. Izd. mosk. Univ., Moskva 1970. [In R.]

4276 - KONDRAT'EVA, E.N., TROTSENKO, Yu.A., VORONINA, O.I.: Fiziologicheskie osobennos-
         ti razlichnykh shtammov *Chloropseudomonas*. [Physiological peculiarities of va-
         rious strains of *Chloropseudomonas*.] - Izv. Akad. Nauk SSSR, Ser. biol. *1968*:
         218-226, 1968. [Chi; in R, ab: E.]

4277 - KONDRAT'EVA, E.N., USPENSKAYA, V.E.: Biosintez zelenymi bakteriyami porfirinov
         i bakterioviridina 660 (bakteriokhlorofilla *C*). [Biosynthesis of porphyrins
         and bacterioviridin 660 (bacteriochlorophyll *C*) in green bacteria.] - Nauch.
         Dokl. vyssh. Shkoly, biol. Nauki *11* (4): 97-110, 1968. [In R.]

4278 - de KONING, H.W., AGHION, J.: Ètude des complexes de pigments chlorophylliens
       d'*Euglena gracilis* au cours de la croissance et à l'état adulte. - Plant Cell
       Physiol. *8*: 129-139, 1967.

4279 - de KONING, H.W., JEGIER, Z.: A study of the effects of ozone and sulfur dioxide
       on the photosynthesis and respiration of *Euglena gracilis*. - Atmos. Environm.
       2: 321-326, 1968.

4280 - de KONING, H.W., JEGIER, Z.: Quantitative relation between ozone concentration
       and reduction of photosynthesis of *Euglena gracilis*. - Atmos. Environm. *2*: 615-
       616, 1968.

4281 - de KONING, H., JEGIER, Z.: Effect of ozone on pyridine nucleotide reduction
       and phosphorylation of *Euglena gracilis*. - Arch. environm. Health *18*: 913-916,
       1969.

4282 - de KONING, H., JEGIER, Z.: Effect of aldehydes on photosynthesis and respiration
       of *Euglena gracilis*. - Arch. environm. Health *20*: 720-722, 1970.

4283 - de KONING, H.W., JEGIER, Z.: Effects of sulfur dioxide and ozone on *Euglena
       gracilis*. - Atmos. Environm. *4*: 357-361, 1970.

4284 - KONISHI, K., OGAWA, T., ITOH, M., SHIBATA, K.: Minor carotenoid components in
       the chloroplasts of higher plants. - Plant Cell Physiol. *9*: 519-527, 1968.

4285 - KONONENKO, A.A.: O prirode nekotorykh fotoindutsirovannykh reaktsiĭ perenosa
       elektronov v kletkakh purpurnykh bakteriĭ *Rhodopseudomonas (Thiorhodaceae)*. [Nature
       of some photoinduced reactions of electron transfer in cells of purple bacteria
       Rhodopseudomonas (Thiorhodaceae).] - Nauch. Dokl. vyssh. Shkoly, biol. Nauki
       13 (3): 139, 1970. [In R.]

4286 - KONONENKO, A.A., ANDREĬTSEV, A.P., RUBIN, A.B.: Vysokochuvstvitel'nyĭ different-
       sial'nyĭ spektrofotometr dlya issledovaniya nekotorykh pervichnykh stadiĭ foto-
       sinteza. [Differential spectrophotometer of high sensibility for studying some
       primary phases of photosynthesis.] - Nauch. Dokl. vyssh. Shkoly, biol. Nauki
       10 (8): 138-144, 1967. [In R.]

4287 - KONOVALOV, I.N., SAAKOV, V.S.: Posledeĭstvie ponizhennykh temperatur na biosin-
       tez i metabolizm karotinoidov. [After-effect of low temperatures on biosynthesis
       and metabolism of carotenoids.] - Tr. Inst. Ekol. Rast. Zhivot. *62*: 94-103,
       1968. [In R.]

4288 - KONYAEV, N.F.: Matematicheskiĭ metod opredeleniya ploshchadi list'ev rasteniĭ.
       [Mathematical method of determination of plant leaf area.] - Dokl. VASKHNIL
       1970 (9): 5-6, 1970. [In R.]

4289 - KORDUNYANU, N.V.: Fotosintez list'ev vinograda v zavisimosti ot sistemy vedeniya
       kustov. [Photosynthesis of vine leaves as affected by shrub formation.] - Sado-
       vod. Vinograd. Vinodel. Mold. *25* (7): 57-60, 1970. [In R.]

4290 - KOREN, L.E., HUTNER, S.H.: High yield media for photosynthesizing *Euglena graci-
       lis*. - J. Protozool. *14* (Suppl.): 17, 1967.

4291 - KORNEEVA, G.A., SEREBROVSKAYA, K.B.: Fotoaktivnost' khlorofilla v vodnykh sis-
       temakh amfoternogo PAV-letsitina. [Photoactivity of chlorophyll in aqueous sys-
       tems of amphoterous PAV-lecithin.] - Dokl. Akad. Nauk SSSR *187*: 1188-1190, 1969.
       [In R.]

4292 - KORNHER, A.: Über den Einfluss einer Deckfrucht auf das Wachstum von Wiesenschwin-
       gel (*Festuca pratensis* HUDS.) und Rotklee (*Trifolium pratense* L.) mit besonderer
       Berücksichtigung von Beschattung und Stickstoffdüngung. - Lantbrukshögskolans
       Ann. *36*: 273-322, 1970.

4293 - KORNHER, A., RODSKJER, N.: Über die Bestimmung der Globalstrahlung in Pflanzen-
       beständen. - Flora B *157*: 149-164, 1967.

4294 - KORNHER, A., RODSKJER, N.: Ein photoelektrisches Planimeter zur Bestimmung von
       Blattflächen nach einem Kompensationsverfahren. - Angew. Bot. *42*: 263-269, 1969.

4295 - KORNHER, A., RODSKJER, N.: Über die Globalstrahlungsverhältnisse in der Marginal-
       zone verschiedener Getreidebestände. - Lantbrukshögskolans Ann. *36*: 337-350,
       1970.

4296 - KORNILOV, A.A.: Osobennosti fotosinteza zernobobovykh kul'tur. [Peculiarities of photosynthesis of pulse crops.] - In: Vazhnefshie Problemy Fotosinteza v Rastenievodstve. Pp. 221-234. Kolos, Moskva 1970. [In R.]

4297 - KORNYUSHENKO, G.A., POPOVA, I.A.: Sravnitel'naya kharakteristika metodov bumazhnof i tenkoslofnof khromatografii karotinoidov zelenogo lista. [Comparison of methods of paper and thin-layer chromatography of carotenoids from green leaves.] - Fiziol. Rast. 17: 1277-1283, 1970. [In R, ab: E.]

4298 - KORNYUSHENKO, G.A., SAPOZHNIKOV, D.I.: Metodika opredeleniya karotinoidov zelenogo lista s pomoshch'yu tonkoslofnof khromatografii. [Method of determining carotenoids of green leaf using thin-layer chromatography.] - In: Metody Kompleksnogo Izucheniya Fotosinteza. Tr. VNII Rastenievod. (Leningrad) 40 (Suppl.): 181-192, 1969. [In R.]

4299 - KOROTAEV, M.M., KUSTOV, V.V., MELESHKO, G.I., MIKHAÏLOV, V.I., SHEPELEV, E.Ya.: O vliyanii nekotorykh gazoobraznykh primesef obitaemof atmosfery na fotosinteticheskuyu deyatel'nost' khlorelly. [Effect of some gas admixtures of atmosphere on photosynthetic activity of *Chlorella*.] - In: Problemy Kosmicheskof Biologii. Vol. 7. Pp. 475-480. Nauka, Moskva 1967. [In R.]

4300 - KORSHUNOVA, V.S., KRENDELEVA, T.E., RUBIN, A.B.: Ob otnositel'nof effektivnosti dvukh pigmentnykh sistem khloroplastov v protsessakh fotosinteticheskogo fosforilirovaniya s nekotorymi kofaktorami. [Relative efficiency of two pigment systems of chloroplasts in photosynthetic phosphorylation with some cofactors.] - Biokhimiya 32: 980-987, 1967. [In R, ab: E.]

4301 - KORSHUNOVA, V.S., KRENDELEVA, T.E., RUBIN, A.B.: O svyazi transporta elektronov s protsessom fotofosforilirovaniya v izolirovannykh khloroplastakh. [Coupling of electron transport with photophosphorylation in isolated chloroplasts.] - Biokhimiya 34: 359-366, 1969. [In R, ab: E.]

4302 - KORTSCHAK, H.P.: Photosynthesis in sugar cane and related species. - In: COOMBS, J. (ed.): Photosynthesis in Sugar Cane. Pp. 18-29. Tate and Lyle Ltd., London 1969.

4303 - KORTSCHAK, H.P., FORBES, A.: The effects of shade and age on the photosynthesis rate of sugarcane. - In: METZNER, H. (ed.): Progress in Photosynthesis Research. Vol. I. Pp. 383-387. Tübingen 1969.

4304 - KORTSCHAK, H.P., HARTT, C.E.: The effects of varied conditions on carbon dioxide fixation in sugarcane leaves. - Naturwissenschaften 53: 253, 1966.

4305 - KORTSCHAK, H.P., NICKELL, L.G.: Calvin-type carbon dioxide fixation in sugarcane stalk parenchyma tissue. - Plant Physiol. 45: 515-516, 1970.

4306 - KOSHKIN, V.A.: Potentsial'naya intensivnost' fotosinteza sortov yarovof pshenitsy razlichnykh rafonov vozdelyvaniya. [Potential photosynthetic rates of cultivars of spring wheat in different cultivation areas.] - Sb. Tr. Aspirantov molodykh nauch. Sotrudn. vsesoyuz. nauch.-issled. Inst. Rastenivod. 10 (14): 61-70, 1969. [In R.]

4307 - KOSHKIN, V.A., BYKOV, O.D.: Temperaturnye krivye potentsial'nof intensivnosti fotosinteza yarovof pshenitsy razlichnogo proiskhozhdeniya. [Temperature curves of potential photosynthetic rate of spring wheat of different provenience.] - Dokl. vsesoyuz. Akad. sel'.-khoz. Nauk 1970 (12): 10-11, 1970. [In R.]

4308 - KOSOBOKOV, G.I., CHERNETSOVA, E.A., VYATLEVA, T.I., STANKO, S.A.: Regulirovanie svetovykh uslovif pri pomoshchi polimernykh plenok s zadannymi opticheskimi svofstvami. [Control of light conditions by means of polymer films of certain optical properties.] - Vestn. sel'.-khoz. Nauki 13 (7): 11-15, 1968. [In R, ab: E, G, F.]

4309 - KOSOBOKOV, G.I., STANKO, S.A., VYATLEVA, T.I., SUBBOTIN, A.A.: Posledefstvie svetoimpul'snof obrabotski semyan na razvitie i produktivnost' ovoshchnykh rastenif. [After-effect of light impulse seed treatment on the development and productivity of vegetables.] - In: Svetoimpul'snoe Obluchenie Rastenif (Tr. Lab. evolyuts. i ekol. Fiziol. B.A. Kellera 6). Pp. 141-155. Nauka, Moskva 1967. [In R.]

4310 - KOSSOVICH, N.L.: Fotosintez i produktivnost' 45-letnikh elef v elovo-listvennom

drevostoe v rezul'tate rubok ukhoda 6-letneĭ davnosti. [Photosynthesis and productivity of 45 year old spruces in mixed spruce-deciduous forests as a result
of intermediate cutting 6 years ago.] - In: SvetovoĭRezhim, Fotosintez i Produktivnost' Lesa. Pp. 129-150. Nauka, Moskva 1967. [In R.]

4311 - KOSTKOVÄ, H.: Remarks on the use of maize in comparative growth analytical investigations of photosynthetic production. - Photosynthetica 2: 212-214, 1968.

4312 - KOSTYUK, M.D.: Pro vbyrannya i rozpodil vugletsyu -14 pri riznykh umovakh kore-
nevogo zhyvlennya kukurudzy. [Absorption and distribution of ^{14}C at various root
nutrition of maize.] - Dopovidi Akad. Nauk URSR B 1967: 743-746, 1967. [In Ukr.,
ab: E, R.]

4313 - KOTAKA, S., KRUEGER, A.P.: Air ion effects on EDTA-induced bleaching in green
barley leaves. - Int. J. Biometeorol. 12: 331-342, 1968. [Chl.]

4314 - KOTAKA, S., KRUEGER, A.P.: Some observations on the bleaching effect of ethylenediaminetetraacetic acid on green barley leaves. - Plant Physiol. 44: 809-815,
1969. [Chl.]

4315 - KOTAKA, S., KRUEGER, A.P., ANDRIESE, P.C.: Bleaching effect of EDTA on green
barley leaves. - Plant Physiol. 43: S 4, 1968. [Chl.]

4316 - KOTAKA, S., KRUEGER, A.P., ANDRIESE, P.C.: The effect of air ions on light-induced swelling and dark-induced shrinking of isolated chloroplasts. - Int. J.
Biometeorol. 12: 85-92, 1968.

4317 - KOTELEVETS, O.S.: Ob osobennostyakh prisposobleniya rasteniĭ k usloviyam buko-
viny. [Peculiarities of plant adaptation to a beech community.] - Tr. bot.
Inst. V.L. Komarova Akad. Nauk SSSR, Ser. 4 - eksp. Bot. 19: 23-38, 1967. [Chl;
in R.]

4318 - van KOTEN-HERTOGS, M., WESSELS, J.S.C.: Ferredoxin-stimulated photoreduction of
2,4-dinitrophenol with solubilized chlorophyll a. - In: THOMAS, J.B., GOEDHEER,
J.C. (ed.): Currents in Photosynthesis. Pp. 207-216. Donker, Rotterdam 1966.

4319 - KOTLYAR, V.Z., GULYAEV, B.I., LYUBIN'SKIĬ, M.A.: Do pytannya pro porivnyal'ne
vyvchennya dennoĭ dynamiki sukhoĭ rechovyny, intensyvnosti fotosyntezu ta vmistu vody v nevidokremlenykh lystkakh u zv'yazku z riznoyu zabezpechenistyu roslyn vodoyu ta mineral'nym zhyvlennyam. [Comparison of daily dynamics of dry
matter, photosynthetic rate and water content in leaves in relation to different water supply and mineral nutrition of plants.] - Ukr. bot. Zh. 25 (4): 73-
78, 1968. [In Ukr., ab: E, R.]

4320 - KOTOVA, K.A.: Sravnitel'naya kharakteristika iskhodnykh form dikikh vidov kartofelya i ikh poliploidov. [Comparison of initial forms of wild potato species
and their polyploids.] - Zap. leningrad. sel'skokhoz. Inst. 117 (3): 46-50,
1968. [Ps; in R.]

4321 - KOTOVA, N.F., KARPILOV, Yu.S.: Produkty fotosinteza gorokha i vliyanie na ikh
obrazovanie usloviĭ azotno-fosfornogo pitaniya. [Products of pea photosynthesis
and effects of nutrition with nitrogen and phosphorus on their formation.] -
Tr. mold. nauch.-issled. Inst. oroshaem. Zemled. Ovoshchevod. 8: 35-39, 1968.
[In R.]

4322 - KOTOVA, N.F., KARPILOV, Yu.S.: Vliyanie azotnoĭ i fosfornoĭ nedostatochnosti na
produkty fotosinteza fasoli v zavisimosti ot intensivnosti sveta. [Effect of nitrogen and phosphorus deficiency on photosynthates of French bean in relation to
irradiance.] - Tr. mold. nauch.-issled. Inst. oroshaem. Zemled. Ovoshchevod. 10:
23-28, 1969. [In R.]

4323 - KOTSUR, N.V.: Termoreguliruemaya germeticheskaya kamera dlya izucheniya temperaturnoĭ zavisimosti fotosinteza rasteniĭ s primeneniem mechenoĭ uglekisloty.
[Thermo-regulated hermetic chamber for studying temperature relations of plant
photosynthesis using labeled carbon dioxide.] - In: Puti Povysheniya Intensivnosti i Produktivnosti Fotosinteza. Vol. 2. Pp. 185-190. Naukova Dumka, Kiev
1967. [In R.]

4324 - KOTSUR, N.V.: Metodika opredeleniya parametrov IKSS pri konstruirovanii appara-
tury i postanovka model'nykh opytov. [Method of determining parameters of IKSS
for apparatus construction and model experiments.] - In: Svetoimpul'snoe Oblu-

chenie RasteniĬ. (Tr. evolyuts. ekol. Fiziol. B.A. Kellera 6). Pp. 178-193. Nauka, Moskva 1967. [In R.]

4325 - KOTSUR, N.V., LYUBINSKIĬ, N.A.: VozdeĬstvie na semena intensivnym svetom. [Effect of intense radiation on seeds.] - In: Puti Povysheniya Intensivnosti i Produktivnosti Fotosinteza. Pp. 129-135. Naukova Dumka, Kiev 1966. [Ps; In R.]

4326 - KOTSUR, N.V., OKANENKO, A.S.: DeĬstvie intensivnogo osveshcheniya semyan na rost i fotosintez rasteniĬ. [Effect of intense irradiation of seeds on growth and photosynthesis of plants.] - In: Svetoimpul'snoe Obluchenie RasteniĬ (Tr. Lab. evolyuts. ekol. Fiziol. B.A. Kellera 6). Pp. 99-108. Nauka, Moskva 1967. [In R.]

4327 - de KOUCHKOVSKY, Y.: Étude, au spectrographe de masse, des échanges rapides d'oxygène effectués à la lumière par des chloroplastes isolés. - Compt. rend. Acad. Sci. Paris, Sér. D 262: 919-922, 1966.

4328 - de KOUCHKOVSKY, Y.:Étude de l'interaction des deux systèmes photochimiques de la photosynthèse. - In: THOMAS, J.B., GOEDHEER, J.C. (ed.): Currents in Photosynthesis. Pp. 367-374. Donker, Rotterdam 1966.

4329 - de KOUCHKOVSKY, Y.: Nature et role des pigments intervenant dans le transfert d'énergie et le transport d'électrons en photosynthèse. - Bull. Soc. franç. Physiol. vég. 14: 409-450, 1968.

4330 - de KOUCHKOVSKY, Y.: Relationships between light-induced absorption changes, fluorescence and oxygen evolution in photosynthesis. - In: METZNER, H. (ed.): Progress in Photosynthesis Research. Vol. II. Pp. 959-970. Tübingen 1969.

4331 - de KOUCHKOVSKY, Y.: Ionic environment and fluorescence changes in photosynthesizing cells. - In: GREGORY, J.G. (ed.): 8th International Congress of Biochemistry. Switzerland 1970. Abstracts. Symp. No. 4. Biol. Oxidation and Bioenergetics. 1. Photosynthesis. Pp. 139-141. Staples Print. Ltd., Rochester, Kent 1970.

4332 - de KOUCHKOVSKY, Y., JOLIOT, P.: Cinétique des échanges d'oxygène et de la fluorescence des chloroplastes isolés. - Photochem. Photobiol. 6: 567-587, 1967.

4333 - de KOUCHKOVSKY, Y., PROVENDIER-GIOT, F.: Effects of photoreactions I and II on the electron carrier revealed by the photosynthetic oxygen burst. - In: Abstracts Volume - Seventh International Congress of Biochemistry. H-86. Tokyo 1967.

4334 - KOUKOL, J., DUGGER, W.M. Jr., PALMER, R.L.: Inhibitory effect of peroxyacetyl nitrate on cyclic photophosphorylation by chloroplasts from Black Valentine bean leaves. - Plant Physiol. 42: 1419-1422, 1967.

4335 - KOUSALOVÁ, I.: Studies of internal factors affecting the yield of winter wheat. - In: Productivity of Terrestrial Ecosystems. Production Processes. Czechosl. nat. Comm. IBP, PT-PP Report No. 1. Pp. 205-207. Praha 1970.

4336 - KOVALEVSKAYA, R.Z., OSTAPENYA, A.P.: Nablyudeniya za soderzhaniem khlorofilla a v sestone poverkhnostnogo sloya morskikh vod. [Chlorophyll a content in the seston of surface water layers of sea.] - Okaneologiya 1966: 849-852, 1966. [In R.]

4337 - KOVROV, B.G., BELYANIN, V.N.: Zavisimost' rosta mikrovodorosleĬ ot kontsentratsii uglekisloty. [Dependence of growth of microalgae on CO_2 concentration.] - Dokl. Akad. Nauk SSSR 182: 705-708, 1968. [In R.]

4338 - KOVTUN, I.I.: Dinamika pigmentov v list'yakh ozimoĬ pshenitsy v ontogeneze. [Pigment dynamics in leaves of winter wheat during ontogenesis.] - Nauch.-tekh. Byul. mironovsk. nauch.-issled. Inst. Selektsii Semenovod. Pshenitsy 1970: 27-30, 1970. [In R.]

4339 - KOWALLIK, U., KOWALLIK, W.: Eine wellenlängeabhängige Atmungssteigerung während der Photosynthese von Chlorella. - Planta 84: 141-157, 1969.

4340 - KOWALLIK, W.: Einfluss verschiedener Lichtwellenlängen auf die Zusammensetzung von Chlorella in Glucosekultur bei gehemmter Photosynthese. - Planta 69: 292-295, 1966.

4341 - KOWALLIK, W.: Chlorophyll-independent photochemistry in algae. - In: Energy Conversion by the Photosynthetic Apparatus. Brookhaven Symp. Biol. 19: 467-477, 1967.

4342 - **KOWALLIK, W.**: Der Einfluss von Licht auf die Atmung von *Chlorella* bei gehemmter Photosynthese. - Planta *86*: 50-62, 1969.

4343 - **KOZHEVNIKOVA, N.F., STANKO, S.A.**: Vliyanie predposevnoī obrabotki semyan kukuruzy v peremennom elektricheskom pole na nekotorye fiziologicheskie protsessy rasteniī. [Effect of pre-sowing treatment of maize seeds in intermittent electric field on some physiological processes of plants.] - Elektron. Obrabotka Mater. *2*: 70-76, 1966. [Ps, Chl; in R.]

4344 - **KOZHOVA, O.M.**: O valovoī i chistoī produktsiī perifitonnykh i planktonnykh vodorosleī. [Gross and net production of periphytonic and planktonic algae.] - Dokl. Akad. Nauk SSSR *195*: 965-968, 1970. [In R.]

4345 - **KOZLOWSKI, T.T., KELLER, T.**: Food relations of woody plants. - Bot. Rev. *32*: 293-382, 1966. [Ps.]

4346 - **KOZLOWSKI, T.T., SASAKI, S.**: Importance of cotyledons to early development of *Pinus resinosa* seedlings. - Amer. J. Bot. *55*: 730, 1968. [Ps.]

4347 - **KOZLOVSKY, D.G;**: A critical evaluation of the trophic level concept. I. Ecological efficiencies. - Ecology *49*: 48-60, 1968. [Ps.]

4348 - **KOZYREV, B.P.**: Vysokochuvstvitel'nyī neselektivnyī piranometr s chernoī priemnoī poverkhnost'yu i s ksenonovym napolneniem. [Highly sensitive non-selective xenon-filled pyranometer with black receiving surface.] - In: Aktinometriya i Optika Atmosfery. Pp. 170-177. Valgus, Tallin 1968. [In R, ab: E.]

4349 - **KOZYREV, B.P.**: Kompensirovannyī termoelektricheskiī balansomer s beloī i blestyashcheī priemnymi poverkhnost'yami, zashchishchennymi ot vozdushnykh potokov polusferami iz KRS-5. [Compensated thermoelectrical net radiometer with white and polished receiving surfaces protected with KRS-5 hemispheres.] - In: Aktinometriya i Optika Atmosfery. Pp. 178-185. Valgus, Tallin 1968. [In R, ab: E.]

4350 - **KOZYREV, B.P.**: Termoelektricheskiī fitopiranometr s polusfericheskimi fil'trami iz stekol BS-8 i KS-19. [Thermoelectrical phytopyranometer with hemispherical filters BS-8 and KS-19.] - In: Aktinometriya i Optika Atmosfery. Pp. 185-192. Valgus, Tallin 1968. [In R, ab: E.]

4351 - **KOZYREV, B.P.**: Termoelektricheskie fitopiranometry dlya absolyutnykh izmereniī fotosinteticheski aktivnoī radiatsii (FAR). [Thermoelectric phytopyranometers for absolute measurements of photosynthetically active radiation (PhAR).] - Fiziol. Rast. *17*: 861-870, 1970. [In R, ab: E.]

4352 - **KOZYREV, B.P., BUCHENKOV, V.A., VASILEVSKAYA, L.M., PARAMONOV, A.I.**: Metody i rezul'taty issledovaniī parametrov novykh aktinometricheskikh priborov sistemy LETI i pokrytiī poverkhnosteī. [Methods and results of investigation of parameters of new actinometric instruments and receiving surfaces produced in the Leningrad Electrotechnical Institute.] - In: Aktinometriya i Optika Atmosfery. Pp. 193-202. Valgus, Tallin 1968. [In R, ab: E.]

4353 - **KRAAN, G.P.B., AMESZ, J., VELTHUYS, B.R., STEEMERS, R.G.**: Studies on the mechanisms of delayed and stimulated fluorescence of chloroplasts. - Biochim. biophys. Acta *223*: 129-145, 1970.

4354 - **KRAAYENHOF, R.**: "State 3-State 4 transition" and phosphate potential in "Class I" spinach chloroplasts. - Biochim. biophys. Acta *180*: 213-215, 1969.

4355 - **KRAAYENHOF, R.**: Quenching of uncoupler fluorescence in relation to the "energized state" in chloroplasts. - FEBS Letters *6*: 161-165, 1970.

4356 - **KRAAYENHOF, R., GROOT, G.S.P., van DAM, K.**: The reversibility of photophosphorylation in different classes of spinach chloroplasts. - FEBS Letters *4*: 125-128, 1969.

4357 - **KRAFT, V.A., DOMAN, N.G., VASILEVA, Z.A.**: Vliyanie defoliantov na nekotorye produkty fotosinteticheskoī assimilyatsii uglekisloty. [Effect of defoliants on some products of photosynthetic assimilation of carbon dioxide.] - Fiziol. Rast. *13*: 595-601, 1966. [In R, ab: E.]

4358 - **KRANZ, A.R.**: Genome mutation and assimilation efficiency. - In: LANDA, Z. (ed.): Mechanism of Mutation and Inducing Factors. Pp. 419-421. Academia, Praha 1966.

4359 - KRANZ, A.R.: Stoffproduktion und Assimilationsleistung in der Evolution der
       Kulturpflanzen. - Biol. Zentralbl. *85*: 597-626, 1966.

4360 - KRANZ, A.R.: Stoffproduktion und Assimilationsleistung in der Evolution der
       Kulturpflanzen. II. Versuchsergebnisse und zusammenfassende Diskussion. - Biol.
       Zentralbl. *85*: 681-734, 1966.

4361 - KRANZ, A.R.: Assimilationsleistung und Evolution des Weizens. - Ber. deut. bot.
       Ges. *80*: 119-123, 1967.

4362 - KRANZ, A.R.: Genetic and ontogenetic correlations between leaf pigment and
       assimilation efficiency in mutants of *Arabidopsis thaliana*. - Arabidopsis Inf.
       Serv. (Göttingen) *4*: 20-21, 1967.

4363 - KRANZ, A.R.: Veränderungen der Assimilationsleistung in der Kulturpflanzenevo-
       lution. - Umschau Wissensch. Tech. *1967* (10): 326, 1967.

4364 - KRANZ, A.R.: Endogene und exogene Beeinflussung der apparenten Strahlungsenergie-
       nutzung annueller Pflanzen. - Angew. Bot. *41*: 271-278, 1968. [Chl.]

4365 - KRASICHKOVA, G.V., SAPOZHNIKOV, D.I.: Osushchestvlenie reaktsii dezepoksidatsii
       violaksantina v iskusstvennom vodnorastvorimom komplekse. [De-epoxidation of
       violaxanthin in an artificial water-soluble complex.] - Dokl. Akad. Nauk Tadzh.
       SSR *11* (7): 59-60, 1968. [In R.]

4366 - KRASIL'NIKOVA, E.N.: Izotsitratliaznaya aktivnost' fotosinteziruyushchikh bak-
       terii v zavisimosti ot prisutstviya uglekisloty. [Isocitrate lyase activity of
       photosynthesizing bacteria as a factor of the presence of carbon dioxide.] -
       Nauch. Dokl. vyssh. Shkoly, biol. Nauki *13* (12): 78-81, 1970. [In R.]

4367 - KRASNOVSKII, A.A.: Fotokhimiya khlorofilla i ego analogov. [Photochemistry of
       chlorophyll and its analogues.] - In: Elementarnye Fotoprotsessy v Molekulakh.
       Pp. 213-242. Nauka, Moskva-Leningrad 1966. [In R.]

4368 - KRASNOVSKII, A.A.: Pervichnye protsessy fotosinteza rastenii. [Primary processes
       of plant photosynthesis.] - In: OPARIN, A.I. (ed.): Fiziologiya Sel'skokhozyaist-
       vennykh Rastenii. Vol. 1. Pp. 149-206. Izdat. mosk. Univ., Moskva 1967. [In R.]

4369 - KRASNOVSKII, A.A.: Fotokhimiya khlorofilla i molekulyarnaya organizatsiya pig-
       mentnoi sistemy organizmov. [Chlorophyll photochemistry and molecular organiza-
       tion of pigment system of organisms.] - In: Funktsional'naya Biokhimiya Kletoch-
       nykh Struktur. Pp. 15-38. Nauka, Moskva 1970. [In R.]

4370 - KRASNOVSKII, A.A., BRIN, G.P.: Narushenie reaktsii Khilla deistviem nagrevaniya,
       rastvoritelei i detergentov; usloviya reaktivatsii. [Disturbance of Hill reac-
       tion by heating, solvents and detergents; reactivation conditions.] - Dokl.
       Akad. Nauk SSSR *179*: 726-729, 1968. [In R.]

4371 - KRASNOVSKII, A.A., BYSTROVA, M.I.: Perestroika agregirovannykh form khlorofilla
       i bakteriokhlorofilla. [Reconstruction of aggregated forms of chlorophyll and
       bacteriochlorophyll.] - Dokl. Akad. Nauk SSSR *174*: 480-483, 1967. [In R.]

4372 - KRASNOVSKII, A.A., BYSTROVA, M.I.: Fotokhimicheskie svoistva agregirovannykh
       form khlorofilla i ego analogov. [Photochemical properties of aggregated forms
       of chlorophyll and its analogues.] - Dokl. Akad. Nauk SSSR *182*: 211-213, 1968.
       [In R.]

4373 - KRASNOVSKII, A.A., BYSTROVA, M.I., LANG, F.: Issledovanie fotovosstanovleniya
       protokhlorofilla do khlorofilla v rastvore. [Photoreduction of protochlorophyll
       to chlorophyll in solution.] - Dokl. Akad. Nauk SSSR *194*: 1441-1444, 1970. [In R.]

4374 - KRASNOVSKII, A.A., BYSTROVA, M.I., MAL'GOSHEVA, I.N.: Issledovanie agregatsii
       bakterioviridina po spektram pogloshcheniya v infrakrasnoi i vidimoi oblastyakh.
       [Bacterioviridin aggregation studied by absorption spectra in infra-red and
       visible regions.] - Dokl. Akad. Nauk SSSR *189*: 885-888, 1969. [In R.]

4375 - KRASNOVSKII, A.A., BYSTROVA, M.I., PAKSHINA, E.V.: Vliyanie atoma magniya v mole-
       kule pigmenta na spektral'nye svoistva agregirovannykh form analogov khlorofilla.
       [Effect of magnesium atom in the pigment molecule on spectral properties of ag-
       gregated forms of chlorophyll analogues.] - Dokl. Akad. Nauk SSSR *167*: 691-694,
       1966. [In R.]

4376 - **KRASNOVSKIĬ, A.A., DROZDOVA, N.N.:** Sravnitel'noe issledovanie tusheniya fluorestsentsii khlorofilla i ego analogov; deĬstvie karotina na effekt tusheniya. [Com-
parative study of fluorescence quenching of chlorophyll and its analogues; the
influence of carotene on the quenching effect.] - Dokl. Akad. Nauk SSSR *166*:
223-226, 1966. [In R.]

4377 - **KRASNOVSKIĬ, A.A., DROZDOVA, N.N.:** Tushenie fluorestsentsii khlorofilla i ego
analogov metilviologenom. [Fluorescence quenching of chlorophyll and its analogues
by methyl viologen.] - Dokl. Akad. Nauk SSSR *167*: 928-930, 1966. [In R.]

4378 - **KRASNOVSKIĬ, A.A., DROZDOVA, N.N.:** DeĬstvie sinego i krasnogo sveta na reaktsiyu
obratimogo okisleniya bakteriokhlorofilla i khlorofilla khinonami; fotoaktivat-
siya okislennykh form pigmentov. [Effect of blue and red light on the reaction
of reversible oxidation of bacteriochlorophyll and chlorophyll by quinones;
photoactivation of oxidized forms of pigments.] - Dokl. Akad. Nauk SSSR *188*:
1384-1386, 1969. [In R.]

4379 - **KRASNOVSKIĬ, A.A., DROZDOVA, N.N., BOKUCHAVA, E.M.:** Stupenchatoe fotookislenie
bakteriokhlorofilla. Spektry fluorestsentsii i pogloshcheniya promezhutochnykh
form. [Stepwise photooxidation of bacteriochlorophyll. Fluorescence and absorption spectra of intermediary forms.] - Dokl. Akad. Nauk SSSR *190*: 464-467, 1970.
[In R.]

4380 - **KRASNOVSKIĬ, A.A., DROZDOVA, N.N., SAPOZHNIKOVA, I.M.:** Usloviya obratimogo i
neobratimogo fotookisleniya bakteriokhlorofilla. [Conditions for reversible and
irreversible photooxidation of bacteriochlorophyll.] - Dokl. Akad. Nauk SSSR
177: 1225-1228, 1967. [In R.]

4381 - **KRASNOVSKIĬ, A.A., EROKHINA, L.G.:** Usloviya tusheniya i vozgoraniya fluorestsentsii fikoeritrina. [Conditions for quenching and induction of phycoerythrin fluorescence.] - Dokl. Akad. Nauk SSSR *183*: 470-473, 1968. [In R.]

4382 - **KRASNOVSKIĬ, A.A., EROKHINA, L.G.:** Issledovanie vzaimodeĬstviya khlorofilla s
fikoeritrinom i fikotsianinom. [Interaction between chlorophyll, phycoerythrin
and phycocyanin.] - Dokl. Akad. Nauk SSSR *188*: 957-960, 1969. [In R.]

4383 - **KRASNOVSKIĬ, A.A., EROKHINA, L.G.:** Vliyanie denaturiruyushchikh vozdeĬstviĬ na
fotokhimicheskie svoĬstva fikoeritrina i fikotsianina. [Effect of denaturating
treatments on the photochemical properties of phycoerythrin and phycocyanin.] -
Dokl. Akad. Nauk SSSR *193*: 1415-1418, 1970. [In R.]

4384 - **KRASNOVSKIĬ, A.A., FEDENKO, E.P., LANG, F., KONDRAT'EVA, E.N.:** Spektrofluorome-
triya pigmentov iskhodnogo shtamma i protokhlorofil'nykh mutantov *Rhodopseudomonas palustris*. [Spectrofluorometry of pigments of the initial strain and of
protochlorophyll mutants of *Rhodopseudomonas palustris*.] - Dokl. Akad. Nauk
SSSR *190*: 218-221, 1970. [In R.]

4385 - **KRASNOVSKIĬ, A.A., LUGANSKAYA, A.N.:** Rol' kisloroda pri fotosensibilizirovannom
khlorofillom vosstanovlenii metilviologena i drugikh krasiteleĬ. [Role of oxygen
in the reduction of methylviologene and other dyes, photosensibilized by chlorophyll.] - Dokl. Akad. Nauk SSSR *183*: 1441-1444, 1968. [In R.]

4386 - **KRASNOVSKIĬ, A.A., MIKHAĬLOVA, E.S.:** Okislitel'no-vosstanovitel'nye prevrashcheniya tsitokhroma *c*, fotosensibilizirovannye khlorofillom v vodnykh rastvorakh
detergentov. [Redox transformations of cytochrome *c* photosensibilized by chlorophyll in water solutions of detergents.] - Dokl. Akad. Nauk SSSR *185*: 938-941,
1969. [In R.]

4387 - **KRASNOVSKIĬ, A.A., MIKHAĬLOVA, E.S.:** Aktiviruyushchee deĬstvie flavinovykh ko-
fermentov na fotosensibilizirovannye khlorofillom prevrashcheniya tsitokhroma
c. [Activating effect of flavin co-enzymes on cytochrome *c* transformations photosensibilized by chlorophyll.] - Dokl. Akad. Nauk SSSR *194*: 953-956, 1970.
[In R.]

4388 - **KRASNOVSKIĬ, A.A., PAKSHINA, E.V., SAPOZHNIKOVA, I.M.:** Sravnenie fotookisleniya
bakterioviridina i bakteriokhlorofilla. Reaktsii fotoproduktov s vosstanovitel-
yami. [Comparison of photooxidation of bacterioviridin and bacteriochlorophyll.
Reactions of photoproducts with reducing agents.] - Dokl. Akad. Nauk SSSR *172*:
727-730, 1967. [In R.]

4389 - KRASNOVSKIĬ, A.A., SAPOZHNIKOVA, I.M.: Fotokhimicheskaya reaktsiya khlorofillov s tiomochevinoĭ i kislorodom. [Photochemical response of chlorophyll to thiourea and oxygen.] - Dokl. Akad. Nauk SSSR *169*: 695-698, 1966. [In R.]

4390 - KRASNOVSKIĬ, A.A., SHAPOSHNIKOVA, M.G.: Fluorometricheskiĭ metod opredeleniya feofitina v list'yakh rasteniĭ. [Fluorometric method of pheophytin determination in plant leaves.] - Fiziol. Rast. *17*: 436-439, 1970. [In R, ab: E.]

4391 - KRASNOVSKIĬ, A.A. Jr., LITVIN, F.F.: Khemilyuminestsentsiya khlorofilla i drugikh pigmentov pri fotookislenii. [Chemiluminescence of chlorophyll and other pigments in the course of photooxidation.] - Mol. Biol. (Moskva) *1*: 699-712, 1967. [In R, ab: E.]

4392 - KRASNOVSKIĬ, A.A. Jr., LITVIN, F.F.: Termokhemilyuminestsentsiya rastvorov khlorofilla i ego analogov posle osveshcheniya pri nizkoĭ temperature. [Thermochemiluminescence of solutions of chlorophyll and its analogues after a low-temperature illumination.] - Mol. Biol. (Moskva) *3*: 282-293, 1969. [In R, ab: E.]

4393 - KRASNOVSKIĬ, A.A. Jr., LITVIN, F.F.: Fotoindutsirovannaya termokhemilyuminestsentsiya analogov khlorofilla i krasiteleĭ. [Photoinduced thermochemiluminescence of analogues of chlorophyll and dyes.] - Dokl. Akad. Nauk SSSR *194*: 197-200, 1970. [In R.]

4394 - KRASNOVSKIĬ, A.A. Jr., LITVIN, F.F.: Khemilyuminestsentsiya pri fotookislenii pigmentov fotosinteziruyushchikh bakteriĭ. [Chemiluminescence during photooxidation of pigments of photosynthetic bacteria.] - Nauch. Dokl. vyssh. Shkoly, biol. Nauki *13* (3): 136-137, 1970. [In R.]

4395 - KRASNOVSKY, A.A.: The participation of chlorophyll and bacteriochlorophyll in photosynthetic hydrogen transfer. - Stud. biophys. *5*: 165-170, 1967.

4396 - KRASNOVSKY, A.A.: The principles of light energy conversion in photosynthesis: photochemistry of chlorophyll and the state of pigments in organisms. - In: METZNER, H. (ed.): Progress in Photosynthesis Research. Vol. II. Pp. 709-727. Tübingen 1969.

4397 - KRATKY, B.A., WARREN, G.F.: Reversal of the action of a photosynthetic inhibitor by activated carbon. - Plant Physiol. *44* (Suppl.): 12, 1969.

4398 - KRAŬCHANKA, L.U.: Da pytannya ab uplyve naftavaga rostavaga rechyva na intensiŭnasts' fotasintezu i rost bakavykh parastkaŭ u nekatorykh drevavykh raslin. [Effect of rock-oil growth substance on photosynthetic rate and growth of lateral branches of some woody plants.] - Vestsi Akad. Navuk Belarus. SSR, Ser. biyal. Navuk *1966* (3): 64-69, 1966. [In Belorus., ab: E.]

4399 - KRAUSE, G.H., BASSHAM, J.A.: Induction of respiratory metabolism in illuminated *Chlorella pyrenoidosa* and isolated spinach chloroplasts by the addition of vitamin K$_5$. - Biochim. biophys. Acta *172*: 553-565, 1969. [Ps.]

4400 - KRAUSS, R.W.: The physiology and biochemistry of algae with special reference to continuous culture techniques for *Chlorella*. - NASA nat. aeron. Space Admin. Sp. *165*: 97-109, 1966. [Ps.]

4401 - KREEB, K.: Eine Feldmethode zur Abschätzung des CO$_2$-Gaswechsels. - Photosynthetica *4*: 158-161, 1970.

4402 - KREIL, W., MATSCHKE, J.: Untersuchungen zur Bestimmung von Weideerträgen mit einem elektronischen Messgerät. - Z. Landeskultur *9*: 75-99, 1968.

4403 - KREĬTSBERG, O.: Metody izucheniya fotosinteza v polevykh usloviyakh. [Methods of studying photosynthesis in field conditions.] - Zernovye maslichnye Kul't. *1970* (11): 38-40, 1970. [In R.]

4404 - KREĬTSBERG, O.E., KRISTKALNE, S.Kh.: Vliyanie urovnya mineral'nogo pitaniya na fotosinteticheskuyu deyatel'nost' rasteniĭ v protsesse formirovaniya urozhaya. [Effect of level of mineral nutrition on photosynthetic activity of plants during yield formation.] - Izv. Akad. Nauk Latv. SSR *1966* (7): 53-64, 1966. [In R.]

4405 - KRENDELEVA, T.E.: O fotosinteticheskom fosforilirovanii v izolirovannykh khloroplastakh vysshikh rasteniĭ. [Photosynthetic phosphorylation in isolated chloroplasts of higher plants.] - Nauch. Dokl. vyssh. Shkoly, biol. Nauki *13* (3): 140, 1970. [In R.]

4406 - **KRENDELEVA, T.E., KORSHUNOVA, V.S., RUBIN, A.B.:** Vliyanie antimytsina *A* na fotosinteticheskoe fosforilirovanie v khloroplastakh gorokha v prisutstvii ekzogennykh
kofaktorov. [Effect of antimycine *A* on photosynthetic phosphorylation in pea chloroplasts in the presence of exogenous cofactors.] - Vestn. mosk. Univ. Ser. 6 -
Biol. Pochvoved. *23* (5): 59-63, 1968. [In R.]

4407 - **KRENDELEVA, T.E., KORSHUNOVA, V.S., RUBIN, A.B.:** Izuchenie kinetiki okislitel'no-
vosstanovitel'nykh prevrashcheniĭ tsitokhromovykh komponentov v elektrontrans-
portnoĭ tsepi fotosinteza. [Study of the kinetics of redox transformations of
cytochrome components in the electron transport chain of photosynthesis.] -
Biofizika *14*: 427-434, 1969. [In R, ab: E.]

4408 - **KRENDELEVA, T.E., KORSHUNOVA, V.S., SHANTORENKO, N.V., RUBIN, B.A.:** O fotofos-
foriliruyushcheĭ aktivnosti fotosinteticheskogo apparata, sformirovannogo v
razlichnykh svetovykh usloviyakh. [Photophosphorylating activity of photosynthetic apparatus formed under various light conditions.] - Nauch. Dokl. vyssh.
Shkoly, biol. Nauki *11* (3): 85-91, 1968. [In R.]

4409 - **KRENZER, E.G. Jr., MOSS, D.N.:** Carbon dioxide compensation in grasses. - Crop
Sci. *9*: 619-621, 1969.

4410 - **KRETCHMAN, D.W., HOWLETT, F.S.:** CO_2 enrichment for vegetable production. - Trans.
ASAE *13*: 252-256, 1970.

4411 - **KRETOVICH, V.L., PERSKAYA, E.B., RACHINSKIĬ, V.V., GEĬKO, N.S.:** Obrazovanie β-
merkaptopirovinogradnoĭ kisloty (ketoanalog tsisteina) v list'yakh *Phaseolus
vulgaris*. [Formation of β-mercaptopyroracemic acid (keto-analogue of cysteine)
in leaves of *Phaseolus vulgaris*).] - Dokl. Akad. Nauk SSSR *194*: 452-454, 1970.
[Ps; in R.]

4412 - **KREUTZ, W.:** The structure of the lamellar system of chloroplasts. - In: GOODWIN,
T.W. (ed.): Biochemistry of Chloroplasts. Vol. I. Pp. 83-88. Academic Press,
London - New York 1966.

4413 - **KREUTZ, W.:** Röntgenographische Strukturuntersuchungen in der Photosyntheseforschung. - Umschau Wiss. Tech. *1966*: 806-813, 1966.

4414 - **KREUTZ, W.:** Über die Tertiärstruktur des Proteins der Chloroplastenlamellen. -
Ber. deut. bot. Ges. *79*: (34)-(43), 1966.

4415 - **KREUTZ, W.:** On the architecture of the photosynthetic apparatus. - Bull. Soc.
franç. Physiol. vég. *14*: 175-193, 1968.

4416 - **KREUTZ, W.:** On the state of chlorophyll *in vivo*. - Z. Naturforsch. *23 b*: 520-527,
1968.

4417 - **KREUTZ, W.:** Neue Untersuchungen zur molekularen Architektur der Thylakoide: Erste
Hinweise für eine "Protonenpumpe". - Ber. deut. bot. Ges. *82*: 459-474, 1969.

4418 - **KREUTZ, W.:** X-ray structure research on the thylakoid. - In: METZNER, H. (ed.):
Progress in Photosynthesis Research. Vol. I. Pp. 91-105. Tübingen 1969.

4419 - **KREUTZ, W.:** On the molecular mechanism of the proton pump in photosynthesis. -
Z. Naturforsch. *25 b*: 88-94, 1970.

4420 - **KREUTZ, W.:** X-ray structure research on the photosynthetic membrane. - Advances
bot. Res. *3*: 53-169, 1970.

4421 - **KREUTZ, W., WEBER, P.:** About the proteinstructure of quantasomes. - Naturwissenschaften *53*: 11-14, 1966.

4422 - **KREUTZER, K.:** Der Einfluss der Mangan-Applikation auf die Pigment- und Nährele-
mentgehalte manganarmer Fichtennadeln. - Z. Pflanzenernähr. Bodenkunde *127*: 84-
91, 1970.

4423 - **KREY, A., GOVINDJEE:** Fluorescence studies on a red alga, *Porphyridium cruentum*.
- Biochim. biophys. Acta *120*: 1-18, 1966.

4424 - **KRIEDEMANN, P.:** The photosynthetic activity of the wheat ear. - Ann. Bot. *30*:
349-363, 1966.

4425 - **KRIEDEMANN, P.E.:** ^{14}C translocation patterns in peach and apricot shoots. -
Aust. J. agr. Res. *19*: 775-780, 1968.

4426 - KRIEDEMANN, P.E.: Observations on gas exchange in the developing sultana berry
       (Vitis vinifera). - Aust. J. biol. Sci. *21*: 907-916, 1968.

4427 - KRIEDEMANN, P.E.: Some photosynthetic characteristics of citrus leaves. - Aust.
       J. biol. Sci. *21*: 895-905, 1968.

4428 - KRIEDEMANN, P.E.: Photosynthesis in vine leaves as a function of light intensi-
       ty, temperature, and leaf age. - Vitis *7*: 213-220, 1968.

4429 - KRIEDEMANN, P.E.: ^{14}C distribution in lemon plants. - J. hort. Sci. *44*: 273-279,
       1969.

4430 - KRIEDEMANN, P.E.: ^{14}C translocation in orange plants. - Aust. J. agr. Res. *20*:
       291-300, 1969.

4431 - KRIEDEMANN, P.E.: The distribution of ^{14}C-labelled assimilates in mature lemon
       trees. - Aust. J. agr. Res. *21*: 623-632, 1970.

4432 - KRIEDEMANN, P.E., CANTERFORD, R.L.: Photosynthesis in pear leaves. - Aust. J.
       biol. Sci. *24*: 197-205, 1970.

4433 - KRIEDEMANN, P.E., KLIEWER, W.M., HARRIS, J.M.: Leaf age and photosynthesis in
       Vitis vinifera L. - Vitis *9*: 97-104, 1970.

4434 - KRINSKY, N.I.: The role of carotenoid pigments as protective agents against
       photosensitized oxidations in chloroplasts. - In: GOODWIN, T.W. (ed.): Biochem-
       istry of Chloroplasts. Vol. I. Pp. 423-430. Academic Press, London-New York 1966.

4435 - KRINSKY, N.I.: The protective function of carotenoid pigments. - In: GIESE, A.C.
       (ed.): Photophysiology. Vol. III. Pp. 123-195. Academic Press, New York-London
       1968.

4436 - KRISHNAMOORTHY, T.M., VISWANATHAN, R.: Primary productivity studies in Bombay
       Harbor Bay using ^{14}C. - Indian J. exp. Biol. *6*: 115-116, 1968.

4437 - KRISHNAMURTHY, K.: Dry-matter production by detached ears of rice. - Indian J. agr.
       Sci. *36*: 229-232, 1966.

4438 - KRISTEN, U.: Untersuchungen über den Zusammenhang zwischen dem CO_2-Gaswechsel
       und der Luftwegigkeit an den CAM-Sukkulenten *Bryophyllum daigremontianum* BERG.
       und *Agave americana* L. - Flora A *160*: 127-138, 1969.

4439 - KRIZEK, D.T., ZIMMERMAN, R.H., KLUETER, H.H., BAILEY, W.A.: Growth and develop-
       ment of crabapple seedlings in controlled environments: Effects of light inten-
       sity and CO_2 concentration. - Plant Physiol. *46* (Suppl.): 7, 1970.

4440 - KROGMANN, D.W., LEE, S.S., YOUNG, M.: Studies on photophosphorylation with *A.
       variabilis*. - Plant Physiol. *43*: S 28, 1968.

4441 - KROGMANN, D.W., MENDELIN, E.A., YOUNG, A.M.: Differential inhibition of the
       Hill reaction to various oxidants. - Fed. Proc. *27*: 827, 1968.

4442 - KRONENBERG, G.H.M.: Some factors influencing the shape of the near-infrared ab-
       sorption spectrum of *Chromatium*, strain *D*. - Meded. Landbouwhogeschool Wagenin-
       gen *69* (10): 1-15, 1969.

4443 - KRUEGER, K.W., RUTH, R.H.: Photosynthesis of red alder, Douglas fir, Sitka
       spruce and Western hemlock seedlings. - In: TRAPPE, J.M., FRANKLIN, J.F.,
       TARRANT, R.F., HANSEN, G.M. (ed.): Biology of Alder. P. 239. US Dept. Agr.,
       Portland, Ore. 1968.

4444 - KRUEGER, K.W., RUTH, R.H.: Comparative photosynthesis of red alder, Douglas-fir,
       Sitka spruce and western hemlock seedlings. - Can. J. Bot. *47*: 519-527, 1969.

4445 - KRUPATKINA (AKININA), D.K.: Vzaimosvyazannye izmeneniya osnovnykh parametrov fo-
       tosinteza u dvukh vidov dinoflagellat. [Interrelated changes of basic parameters
       of photosynthesis in two species of dinoflagellates.] - Fiziol. Rast. *17*: 1091-
       1093, 1970. [In R.]

4446 - KRYLOV, Yu.V.: Rabota s kamerami-prishchepkami pri opredelenii intensivnosti
       fotosinteza. [Application of pincer-chambers for determination of photosynthetic
       rate.] - Fiziol. Rast. *13*: 541-545, 1966. [In R, ab: E.]

4447 - KRYLOV, Yu.V.: Avtomaticheskaya ustanovka dlya ucheta fotosinteza v sisteme malogo ob"ema zamknutogo tsikla. [An automatic arrangement for measurement of photosynthetic rate in a closed system of small volume.] - Fiziol. Rast. *18*: 746-755, 1969. [In R, ab: E.]

4448 - KRYUKOVSKIĬ, F.V.: Opredelenie listovoĭ poverkhnosti u drevesnykh porod. [Determination of leaf area in woody plants.] - Bot. Zh. *51*: 678-681, 1966. [In R.]

4449 - KSENOFONTOVA, T.S., DILUNG, I.I.: O roli molekulyarnykh kompleksov v fotokhimii khlorofilla. [Role of molecular complexes in chlorophyll photochemistry.] - Dokl. Akad. Nauk SSSR *183*: 870-873, 1968. [In R.]

4450 - KUCERA, C.L., DAHLMAN, R.C.: A method for incorporating high activity carbon-14 in prairie grasses for turnover studies. - Iowa State J. Sci. *43*: 13-17, 1968.

4451 - KUDINOV, M.A.: Intensivnost' fotosinteza yasenya oregonskogo, vyrashchennogo iz obluchennykh semyan. [Photosynthetic rate of Oregon ash cultivated from irradiated seeds.] - In: Botanika. Issled. 9. Pp. 134-136. Nauka i Tekhnika, Minsk 1967. [In R.]

4452 - KUDRITSKAYA, S.E., FISHMAN, G.M., SAVINOV, B.G.: O khimicheskoĭ prirode karotinovykh krasyashchikh veshchestv plodov subtropicheskoĭ khurmy. [Chemical nature of carotenoids in fruits of subtropical persimmon.] - Prikl. Biokhim. Mikrobiol. *2*: 340-344, 1966. [In R.]

4453 - KUKHTEVICH, I.L., BEZVERKHIĬ, V.D., BILYAK, A.I.: O strukture khlorofilla i ego metalloproizvodnykh. [Structure of chlorophyll and its metallo-derivatives.] - Zh. neorg. Khim. *13*: 454-456, 1968. [In R.]

4454 - KUKHTEVYCH, I.L., STAROSTIN, A.N.: Doslidzhennya metaloanalogiv khlorofily metodom elektronnogo paramagnitnogo rezonantsu. [Study of metallo-derivatives of chlorophyll with EPR method.]- Ukr. khim. Zh. *35*: 426-427, 1969. [In Ukr.]

4455 - KULAEVA, O.N., DEVYATKO, O.I.: Zaderzhka pozhelteniya list'ev yachmenya na rastenii s pomoshch'yu fitogormonov. [Delay of yellowing of barley leaves on plants by phytochrome treatment.] - Fiziol. Rast. *16*: 288-292, 1969. [In R, ab: E.]

4456 - KULAEVA, O.N., ROMANKO, E.G.: Deĭstvie 6-benzilaminopurina na izolirovannye khloroplasty. [Effect of 6-benzyl amino purine on isolated chloroplasts.] - Dokl. Akad. Nauk SSSR *177*:464-467, 1967. [In R.]

4457 - KULIBABA, Yu.F., SKIPINA, K.P.: O fiziologicheskikh pokazatelyakh vredonosnosti dyrchatoĭ pyatnistosti kostochkovykh. [Physiological symptoms of shot hole disease injuries in stone fruits.] - Sel'skokhoz. Biol. *1*: 784-787, 1966. [Chl; in R., ab: E.]

4458 - KUMAGAI, T., FUJII, Y., TAKAHASHI, H.: Ammonia as an important nitrogen source for the photosynthetic growth of *Rhodospirillum rubrum*. - Plant Cell Physiol. *9*: 13-22, 1968.

4459 - KUMAKOV, V.A.: Svyaz' khozyaĭstvennogo koeffitsienta s rabotoĭ assimilyatsionnogo apparata u yarovoĭ pshenitsy. [The relation between the economic coefficient and the work of the assimilatory apparatus in spring wheat.] - Dokl. Akad. Nauk SSSR *177*: 961-963, 1967. [In R.]

4460 - KUMAKOV, V.A.: Pokazateli fotosinteza kak selektsionnyĭ priznak u pshenitsy. [Photosynthetic characteristics as breeding property in wheat.] - Sel'skokhoz. Biol. *2*: 551-558, 1967. [In R, ab: E.]

4461 - KUMAKOV, V.A.: Napravleniya selektsionnoĭ raboty s tsel'yu uluchsheniya pokazateleĭ fotosinteticheskoĭ deyatel'nosti rasteniĭ. [Direction of selection to improve photosynthetic activity of plants.] - In: Vazhneĭshie Problemy Fotosinteza v Rastenievodstve. Pp. 206-220. Kolos, Moskva 1970. [In R.]

4462 - KUMAKOV, V.A., KUZ'MINA, K.M.: Nekotorye osobennosti fotosinteticheskoĭ deyatel'nosti i struktury urozhaya razlichnykh po skorospelosti rasteniĭ prosa i yarovoĭ pshenitsy. [Some features of photosynthetic activity and crop structure of millet and summer wheat plants differing in earliness.] - Fiziol. Rast. *15*: 41-46, 1968. [In R, ab: E.]

4463 - KUMTA, U.S., SAWANT, P.L., RAMAKRISHNAN, T.V.: Radiation sensitivity studies of plant pigments. I. *In vitro* lability of β-carotene in lipid solvents. - Rad. Bot. *10*: 161-167, 1970.

4464 - KUMURA, A.: [Studies on dry matter production of soybean plant. II. Photosyn-
       thetic rate of soybean plant population as affected by proportion of diffuse
       light.] - Proc. Crop Sci. Soc. Jap. *37*: 570-582, 1968. [In Jap.]

4465 - KUMURA, A.: [Studies on dry matter production of soybean plant. IV. Photosyn-
       thetic properties of leaf as subsequently affected by light conditions.] - Proc.
       Crop Sci. Soc. Jap. *37*: 583-588, 1968. [In Jap.]

4466 - KUMURA, A.: [Studies on dry matter production in soybean plant. V. Photosynthe-
       tic system of soybean plant population.] - Proc. Crop Sci. Soc. Jap. *38*: 74-90,
       1969. [In Jap., ab: E.]

4467 - KUMURA, A.: [Studies on dry matter production of soybean plant. VI. Changes in
       spectral composition of solar radiation penetrating through leaf canopy and
       photosynthetic rate of single leaf as affected by light quality.] - Proc. Crop
       Sci. Soc. Jap. *38*: 408-418, 1969. [In Jap, ab: E.]

4468 - KUNFFY, Z., FARKAS, M.: Examination of the technology of alfalfa dried with hot
       air regarding the conservation and stabilization of carotene. - Acta agron.
       Acad. Sci. hung. *19*: 137-145, 1970.

4469 - KUNG, S.D., WILLIAMS, J.P.: Isolation of chloroplasts free from nuclear DNA con-
       tamination. - Biochim. biophys. Acta *169*: 265-268, 1968.

4470 - KUNTZ, I.D.: Spectroscopic studies of photosynthesis. - Diss. Abstr. *26*: 7062,
       1966.

4471 - KUPERMAN, F.M., ABDEL'-RAKHMAN, K.A.: O parallelizme soderzhaniya khlorofilla
       i intensivnosti vklyucheniya C^{14}-metionina v belok list'ev raznykh yarusov pseh-
       nitsy SaratovskoĬ 38. [Parallelism of chlorophyll content and rate of ^{14}C-me-
       thionine incorporation into albumine of leaves of different insertion levels
       in wheat cv. Saratovskaya 38.] - Dokl. VASKHNIL *1970* (11): 5-7, 1970. [In R.]

4472 - KUPFER, D., MUNSELL, T.: A colorimetric method for the quantitative determina-
       tion of reduced pyridine nucleotides (NADPH and NADH). - Anal. Biochem. *25*: 10-
       16, 1968.

4473 - KUPKA, J., ZAHÁLKOVÁ, V.: Obsah chlorofylů a fotosyntéza. [Chlorophyll content
       and photosynthesis.] - Sbor. VŠZ Praha, Fak. agr., Ser. A/B *1969*: 23-29, 1969.
       [In Czech.]

4474 - KUPREVICH, V.F., REUTSKAYA, L.N.: Fotokhimicheskaya aktivnost' khloroplastov u
       porazhennykh rzhavchinoĬ rasteniĬ. [Photochemical activity of chloroplasts in
       rust infected plants.] - Dokl. Akad. Nauk Belorus. SSR *12*: 373-375, 1968. [In
       R.]

4475 - KURAMOTO, R.T., BLISS, L.C.: Ecology of subalpine meadows in the Olympic Moun-
       tains, Washington. - Ecol. Monogr. *40*: 317-347, 1970. [Chl.]

4476 - KURBANOVA, I.M., KOZLOV, Yu.P., GASANOV, R.A.: O roli lipidov v tsepi perenosa
       elektronov v fotokhimicheskikh reaktsiyakh suspenzii khloroplastov. [Role of
       lipids in electron transport chain of photochemical reactions in a chloroplast
       suspension.] - In: Mineral'nye Elementy i Mekhanizm Fotosinteza. Pp. 177-186.
       Akad. Nauk Mold. SSR, Kishinev 1970. [In R.]

4477 - KURETS, V.K., KURSAKOV, N.F., YAN'KOVA, L.S.: AvtomaticheskiĬ uchet solnechnoĬ
       insolyatsii. [Automatic recording of solar irradiation.] - In: InformatsionnyĬ
       Byulleten' *6*. Sibir. Inst. Fiziol. i Biokhim. Rast. SO Akad. Nauk SSSR. P. 157.
       Irkutsk 1970. [In R.]

4478 - KURGANOVA, L.N.: Vliyanie predposevnogo gamma-oblucheniya semyan grechikhi i
       kukuruzy na fotosintez prorostkov v svyazi s izmeneniem usloviĬ kornevogo pita-
       niya. [Effect of pre-sowing gamma-irradiation of buckwheat and maize seeds on
       seedling photosynthesis in relation with changed conditions of root nutrition.]
       - In: Tezisy Dokladov Konferentsii Molodykh Nauchnykh Rabotnikov. Pp. 9-12. Gor'-
       kov. Univ., Sek. biol. Nauk, Gor'kiĬ 1966. [In R.]

4479 - KURMAEVA, S.A., GALEEVA, S.G., LEMESHKO, L.V.: Vliyanie bikarbonata i CO_2 na
       fotosinteticheskoe fosforilirovanie izolirovannykh khloroplastov. [Effect of bi-
       carbonate and CO_2 on photosynthetic phosphorylation of isolated chloroplasts.]
       - In: Funktsional'nye Osobennosti Khloroplastov. Pp. 60-64. Kazan. Univ., Kazan
       1969. [In R.]

4480 - KUROIWA, S.: Theoretical evaluation of dry-matter production of a crop canopy under isolation- and temperature-climate: a summary. - In: Agroclimatological Methods. Proceedings of the Reading Symposium (1966). Pp. 331-332. UNESCO, Paris 1968.

4481 - KUROIWA, S.: [Theoretical analysis of light factor and photosynthesis in plant communities. (3). Total photosynthesis of a foliage under parallel light in comparison with that under isotropic light condition.] - J. agr. Meteorol. 24: 75-90, 1968. [In Jap., ab: E.]

4482 - KUROIWA, S.: Total photosynthesis of a foliage exposed to a parallel light. - In: Photosynthesis and Utilization of Solar Energy. Level III Experiments. Pp. 29-32. Jap. nat. Subcomm. for PP (JPP), Tokyo 1968.

4483 - KUROIWA, S.: A new calculation method for total photosynthesis of a plant community under illumination consisting of direct and diffused light. - In: ECKARDT, F.E. (ed.): Functioning of Terrestrial Ecosystems at the Primary Production Level. Pp. 391-398. UNESCO, Paris 1968.

4484 - KUROIWA, S.: Total photosynthesis of a foliage in relation to inclination of leaves. - In: Prediction and Measurement of Photosynthetic Productivity. Pp. 79-89. PUDOC, Wageningen 1970.

4485 - KURSANOV, A.L., BROVCHENKO, M.I.: Svobodnoe prostranstvo kak promezhutochnaya zona mezhdu fotosinteziruyushchimi i provodyashchimi kletkami listovoĭ plastinki. [Free space as an intermediate zone between photosynthesizing and conducting cells of leaves.] - Fiziol. Rast. 16: 965-972, 1969. [In R, ab: E.]

4486 - KURSANOV, A.L., SAFONOV, V.I., CHAYANOVA, S.S., SAFONOVA, M.V.: Sravnitel'noe izuchenie belkov khloroplastov metodom elektroforeza v poliakrilamidnom gele. [Comparative study of chloroplast proteins using electrophoresis in polyacrylamide gel.] - In: Funktsional'naya Biokhimiya Kletochnykh Struktur. Pp. 143-153. Nauka, Moskva 1970. [In R.]

4487 - KUR'YANOV, V.I.: Vliyanie fiziologicheski aktivnykh veshchestv na fotosintez u soi. [Effect of physiologically active substances on soybean photosynthesis.] - Zap. voronezh. sel'skokhoz. Inst. 34: 188-192, 1967. [In R.]

4488 - KUSHMAN, L.J.: A rapid method of estimating dry-matter content of sweet-potatoes. - Proc. amer. Soc. hort. Sci. 92: 814-822, 1968.

4489 - KUSHNIRENKO, M.D., MEDVEDEVA, T.N.: Vliyanie zavyadaniya na pigmentnuyu sistemu i razvitie vodouderzhivayushchikh sil list'ev. [Effect of wilting on the pigment system and development of water-retaining forces in leaves.] - Fiziol. Rast. 16: 529-534, 1969. [In R, ab: E.]

4490 - KUSHWAHA, S.C., SUBBARAYAN, C., BEELER, D.A., PORTER, J.W.: The conversion of lycopene-15,15'-^{3}H to cyclic carotenes by soluble extracts of higher plant plastids. - J. biol. Chem. 244: 3635-3642, 1969.

4491 - KUTYURIN, V.M.: Razlozhenie vody pri fotosinteze kak protsess biologicheskogo okisleniya. [Photolysis of water in photosynthesis as a process of biological oxidation.] - In: Mekhanizmy Dykhaniya, Fotosinteza i Fiksatsii Azota. Pp. 248-264. Nauka, Moskva 1967. [In R.]

4492 - KUTYURIN, V.M.: O mekhanizme razlozheniya vody v protsesse fotosinteza. [Mechanism of water splitting in photosynthesis.] - Izv. Akad. Nauk SSSR, Ser. biol. 1970: 569-580, 1970. [In R, ab: E.]

4493 - KUTYURIN, V.M., ARTAMKINA, I.Yu., ANISIMOVA, I.N.: O vzaimodeĭstvii okislennoĭ formy khlorofilla s vodoĭ. [Interaction between the oxidized form of chlorophyll and water.] - Dokl. Akad. Nauk SSSR 180: 1002-1004, 1968. [In R.]

4494 - KUTYURIN, V.M., GRIGOROVICH, V.I.: Opticheskaya aktivnost' khlorofilla i keto-enol'naya tautomeriya karbotsiklicheskogo kol'tsa. [Optical activity of chlorophyll and keto-enol tautomery of carbocyclic ring.] - Khim. prir. Soedin. 3: 51-55, 1967. [In R.]

4495 - KUTYURIN, V.M., MATVEEVA, I.V., ARTAMKINA, I.Yu.: O vodorodnom obmene v karbotsiklicheskom kol'tse khlorofilla. [Hydrogen exchange in the carbocyclic ring of chlorophyll.] - Dokl. Akad. Nauk SSSR 166: 1233-1235, 1966. [In R.]

4496 - KUTYURIN, V.M., NAZAROV, N.M., ANISIMOVA, I.N.: Ob effektivnosti vydeleniya
       kisloroda vodoroslyami pri ikh osveshchenii dlinnovolnovym svetom (λ > 700 mμ).
       [Efficiency of oxygen evolution by algae illuminated by long-wave light
       (λ > 700 nm).] - Dokl. Akad. Nauk SSSR *181*: 1270-1273, 1968. [In R.]

4497 - KUTYURIN, V.M., NAZAROV, N.M., SEMENYUK, K.G.: O kislorodnom obmene mezhdu vodoĭ
       chloroplastov i kompleksom, vydelyayushchim kislorod fotosinteza. [Oxygen ex-
       change between water of chloroplasts and the complex evolving oxygen in photo-
       synthesis.] - Dokl. Akad. Nauk SSSR *171*: 215-217, 1966. [In R.]

4498 - KUTYURIN, V.M., SOLOV'EV, V.P., GRIGOROVICH, V.I.: Issledovanie obratimogo okis-
       leniya khlorofilla i ego proizvodnykh polyarograficheskim metodom. [Reversible
       oxidation of chlorophyll and its derivatives studied by polarographic method.]
       - Dokl. Akad. Nauk SSSR *169*: 479-482, 1966. [In R.]

4499 - KUTYURIN, V.M., ULUBEKOVA, M.V., MATVEEVA, I.V., SHUTILOVA, N.I., ROZONOVA, L.N.:
       Sootnoshenie mezhdu skorost'yu perenosa elektrona i skorost'yu vydeleniya kislo-
       roda khloroplastami v reaktsii Khilla. [Relation between the electron transfer
       rate and Hill reaction oxygen evolution rate in chloroplasts.] - Fiziol. Rast.
       16: 181-186, 1969. [In R, ab: E.]

4500 - KUTYURIN, V.M., ULUBEKOVA, M.V., NAZAROV, N.M.: O sootnoshenii mezhdu intensiv-
       nost'yu vydeleniya kisloroda i reaktsiyami prevrashcheniya ksantofillov u *Elodea
       canadensis* pri razlichnom spektral'nom sostave sveta. [Correlation between rate
       of oxygen evolution and reactions of xanthophyll transformation in *Elodea cana-
       densis*, observed at different spectral composition of radiation.] - Dokl. Akad.
       Nauk SSSR *187*: 470-472, 1969. [In R.]

4501 - KUTYURIN, V.M., ULUBEKOVA, M.V., ROZONOVA, L.N.: Vliyanie metilamina na reaktsiyu
       Khilla pri raznoĭ aktivnosti khloroplastov. [Effect of methylamine on Hill react-
       ion in chloroplasts of different activity.] - Dokl. Akad. Nauk SSSR *188*: 1399-
       1401, 1969. [In R.]

4502 - KUZIN, A.M., KOROLEV, N.P.: Izmenenie aktivnosti polifenoloksidazy, peroksidazy
       i reaktsii Khilla v khloroplastakh list'ev posle γ-oblucheniya. [Change in poly-
       phenoloxidase, peroxidase, and Hill reaction activity in leaf chloroplasts after
       gamma-irradiation.] - Radiobiologiya *6*: 898-901, 1966. [In R.]

4503 - KVĚT, J., ŠETLÍK, I.: An integrating recorder for the measurement of radiation
       in the wavelength region 350 to 750 mμ. - In: BAINBRIDGE, R., EVANS, G.C.,
       RACKHAM, O. (ed.): Light as an Ecological Factor. Pp. 420-423. Blackwell sci.
       Publ., Oxford 1966.

4504 - KVĚT, J., SVOBODA, J.: Development of vertical structure and growth analysis in
       a stand of *Phragmites communis* TRIN. - In: Productivity of Terrestrial Ecosys-
       tems. Production Processes. Czech. nat. Comm. IBP, PT-PP Report No. 1. Pp. 84-
       87. Praha 1970.

4505 - KVĚT, J., SVOBODA, J., FIALA, K.: A simple device for measuring leaf inclinations.
       - Photosynthetica *1*: 127-128, 1967.

4506 - KVĚT, J., SVOBODA, J., FIALA, K.: Canopy development in stands of *Typha latifo-
       lia* L. and *Phragmites communis* TRIN. in South Moravia. - Hidrobiologia (Bucureş-
       ti) *10*: 63-75, 1969. [Growth analysis.]

4507 - KYLIN, A.: The influence of photosynthetic factors and metabolic inhibitors on
       the uptake of phosphate in P-deficient *Scenedesmus*. - Physiol. Plant. *19*: 644-
       649, 1966.

4508 - KYLIN, A., TILLBERG, J.-E.: Action sites of the inhibitor-β complex from potato
       and of phloridzin in light-induced transfer in *Scenedesmus*. - Z. Pflanzenphysiol.
       57: 72-78, 1967. [Ps.]

4509 - KYLIN, A., TILLBERG, J.-E.: The relation between total photophosphorylation,
       level of ATP and oxygen evolution in *Scenedesmus* as studied with DCMU and anti-
       mycin A. - Z. Pflanzenphysiol. *58*: 165-174, 1967.

4510 - KYLIN, A., TILLBERG, J.-E.: Light-induced phosphorylation and formation of ATP
       in *Scenedesmus*. - In: METZNER, H. (ed.): Progress in Photosynthesis Research.
       Vol. III. Pp. 1224-1228. Tübingen 1969.

4511 - **LABAW, L.W., OLSON, R.-A.**: Further electron microscopic observations of bacteriochlorophyll protein crystals. - J. Ultrastruct. Res. *31*: 456-464, 1970.

4512 - **LABER, L.J., BLACK, C.C.**: Site-specific uncoupling of photosynthetic phosphorylation in spinach chloroplasts. - J. biol. Chem. *244*: 3463-3467, 1969.

4513 - **LABORDE, J.A.**: Effect of two fruit color genes in *Capsicum annuum* on chlorophyll,
carotenoids and ultrastructure of the chromoplast. - Diss. Abstr. int. B *31*:
3194-B, 1970.

4514 - **LABOTSKAYA, L.I., BYCHKO, Ya.A., PAL'CHANKA, L.A., BORMATAŬ, U.Ya.**: Ab inten-
siŭnastsi pratsesaŭ fotasintezu i dykhannya raslin tsukrovykh burakoŭ u suvyazi
z poliplaidyTaŭ. [Rates of photosynthesis and respiration of sugar beet plants
in relation to their polyploidy.] - Vestsi Akad. Navuk Belaruss. SSR, Ser.
biyal. Navuk *1968* (2): 66-71, 1968. [In Belorus., ab: R.]

4515 - **LADYGINA, M.E.**: Energeticheskiĭ obmen i transport elektronov v zelenom rastenii
pri virusnoĭ infektsii. [Energy metabolism and electron transport in a virus-infected green plant.] - Sel'skokhoz. Biol. *4*: 848-859, 1969. [In R, ab: E.]

4516 - **LADYGINA, M.E., EDREVA, A.**: Vliyanie virusnoĭ infektsii na fotosinteticheskoe
fosforilirovanie u tabaka. [Effect of virus infection on photosynthetic phosphorylation in tobacco.] - Sel'skokhoz. Biol. *3*: 702-708, 1968. [In R, ab: E.]

4517 - **LADYGINA, M.E., RUBIN, B.A., TIMOFEEV, K.N.**: Vliyanie patogennykh agentov na
protsessy zapasaniya energii i spektry elektronnogo paramagnitnogo rezonansa
rasteniĭ. [Effect of pathogenic agents on processes of energy storage and EPR
spectra of plants.] - Fiziol. Rast. *17*: 416-424, 1970. [Chl; in R, ab: E.]

4518 - **LADYGINA, M.E., RUBIN, B.A., TUKEEVA, M.I.**: Vliyanie virusa tabachnoĭ mozaiki
na energeticheskiĭ obmen raznykh po ustoĭchivosti vidov tabaka. [Effect of tobacco mosaic virus on energy metabolism in tobacco species differing in resistance.] - Fiziol. Rast. *13*: 885-891, 1966. [Chl; in R, ab: E.]

4519 - **LADYGINA, M.E., TUKEEVA, M.I., RUBIN, B.A.**: Deĭstvie gramitsidina *C* na fotofosforilirovanie u tabachnogo rasteniya pri virusnoĭ infektsii. [Action of grami-
cidin *C* on photophosphorylation of virus-infected tobacco.] - Biokhimiya *32*:
1248-1252, 1967. [In R, ab: E.]

4520 - **LAETSCH, W.M.**: Chlorophyll synthesis in cultured tobacco leaf tissue. - Amer. J.
Bot. *53*: 613, 1966.

4521 - **LAETSCH, W.M.**: Chloroplast specialization of dicotyledons possessing the C₄-di-
carboxylic acid pathway of photosynthetic CO_2 fixation. - Amer. J. Bot. *55*:
875-883, 1968.

4522 - **LAETSCH, W.M.**: Chloroplast structure and development. - US Dept. Agr. agr.Res.
Serv. *(74-47)*: 26-31, 1968.

4523 - **LAETSCH, W.M.**: Chloroplast structure with special reference to function. - In:
COOMBS, J. (ed.): Photosynthesis in Sugar Cane. Pp. 41-45. Tate and Lyle Ltd.,
London 1969.

4524 - **LAETSCH, W.M.**: Relationship between chloroplast structure and photosynthetic
carbon-fixation pathways. - Sci. Progr. (Oxford) *57*: 323-351, 1969.

4525 - **LAETSCH, W.M.**: Specialized chloroplast structure of plants exhibiting the dicarboxylic acid pathway of photosynthetic C_2 fixation. - In: METZNER, H. (ed.):
Progress in Photosynthesis Research. Vol. I. Pp. 36-46. Tübingen 1969.

4526 - **LAETSCH, W.M.**: Comparative ultrastructure of chloroplasts in plants with C-4
pathways of carbon fixation. - Plant Physiol. *46* (Suppl.): 22, 1970.

4527 - **LAETSCH, W.M., PRICE, I.**: Development of the dimorphic chloroplasts of sugar
cane. - Amer. J. Bot. *56*: 77-87, 1969.

4528 - **LAETSCH, W.M., STETLER, D.A.**: Regulation of chloroplast development in cultured
plant tissues. - In: SIRONVAL, C. (ed.): Le Chloroplaste, Croissance et Vieillissement. Pp. 291-297. Masson, Paris 1967.

4529 - **LAETSCH, W.M., STETLER, D.A., VLITOS, A.J.**: The ultrastructure of sugar cane
chloroplasts. - Z. Pflanzenphysiol. *54*: 472-474, 1966.

4530 - LAFEBER, A., STEENBERGEN, C.L.M.: Simple device for obtaining synchronous cult-
       ures of algae. - Nature *213*: 527-528, 1967. [Chl.]

4531 - LAFLÈCHE, D., BOVÉ, J.M.: Development of double membrane vesicles in chloroplasts
       from turnip yellow mosaic virus infected cells. - In: METZNER, H. (ed.): Progress
       in Photosynthesis Research. Vol. I. Pp. 74-83. Tübingen 1969.

4532 - LAGERSTEDT, H.B.: A study of kinetin in plants. - Diss. Abstr. *26*: 6281-6282,
       1966. [Chl.]

4533 - LAHUČKÝ, R.: Stabilita chlorofyl-bielkovinového komplexy u listov *Nicotiana ta-
       bacum*. [Stability of the chlorophyll-protein complex in leaves of *Nicotiana ta-
       bacum*.] - Bull. tabák. Priem. *12*: 1-12, 1969. [In Slovak, ab: G.]

4534 - LAÏSK, A.: Perspektivy matematicheskogo modelirovaniya funktsii fotosinteza
       lista. [Prospects of mathematical modelling of leaf photosynthesis function.] -
       In: Fotosintez i Produktivnost' Rastitel'nogo Pokrova. Pp. 5-45. Akad. Nauk Est.
       SSR, Tartu 1968. [In R, ab: E.]

4535 - LAÏSK, A.: Statisticheskiĭ kharakter oslableniya radiatsii v rastitel'nom pokro-
       ve. [Statistical character of light extinction in plant communities.] - In: Rez-
       him Solnechnoĭ Radiatsii v Rastitel'nom Pokrove. Pp. 81-111. Akad. Nauk Est.
       SSR, Tartu 1968. [In R, ab: E.]

4536 - LAÏSK, A.: Fotosintez lista s uchetom adaptatsii ust'its po CO_2. [Leaf photo-
       synthesis, considering stomatal adaptation to CO_2.] - In: Voprosy Effektivnosti
       Fotosinteza. Pp. 64-92. Tartu 1969. [In R, ab: E.]

4537 - LAÏSK, A.: Izmerenie prozrachnosti rastitel'nogo pokrova. [Measurement of plant
       cover transparency.] - In: Fotosinteticheskaya Produktivnost' Rastitel'nogo Po-
       krova. Pp. 174-185. Tartu 1969. [In R, ab: E.]

4538 - LAÏSK, A.: Svetovye krivye fotosinteza dlya opticheski tolstykh list'ev. [Light
       curves of photosynthesis considering light profile of a thick leaf.] - In: Vop-
       rosy Effektivnosti Fotosinteza. Pp. 93-116. Tartu 1969. [In R, ab: E.]

4539 - LAISK, A.: A model of leaf photosynthesis and photorespiration. - In: Prediction
       and Measurement of Photosynthetic Productivity. Pp. 295-306. PUDOC, Wageningen
       1970.

4540 - LAKE, J.V.: Measurement and control of the rate of carbon dioxide assimilation
       by glasshouse crops. - Nature *209*: 97-98, 1966.

4541 - LAKE, J.V.: Respiration of leaves during photosynthesis. I. Estimates from an
       electrical analogue. - Aust. J. biol. Sci. *20*: 487-493, 1967.

4542 - LAKE, J.V.: Respiration of leaves during photosynthesis. II. Effects on the es-
       timation of mesophyll resistance. - Aust. J. biol. Sci. *20*: 495-499, 1967.

4543 - LAKE, J.V.: Glasshouse and leaf canopy enclosures for studying the effects of
       environment on the growth and nature of the photosynaptic apparatus. - In: Pre-
       diction and Measurement of Photosynthetic Productivity. Pp. 405-409. PUDOC,
       Wageningen 1970.

4544 - LAKE, J.V., BROWNE, D.A., BOWMAN, G.E.: A glasshouse as a cuvette. - In: ECKARDT,
       F.E. (ed.): Functioning of Terrestrial Ecosystems at the Primary Production Level.
       Pp. 329-333. UNESCO, Paris 1968. [Ps.]

4545 - LAKE, J.V., SLATYER, R.O.: Respiration of leaves during photosynthesis. III. Res-
       piration rate and mesophyll resistance in turgid cotton leaves, with stomatal
       control eliminated. - Aust. J. biol. Sci. *23*: 529-535, 1970.

4546 - LAMBERT, C., DURANTON, H.: Mise en évidence d'une action de la lumière sur le
       métabolisme de l'arginine dans les tissus du Topinambour cultivés *in vitro*. -
       Physiol. vég. *4*: 263-282, 1966. [Chl.]

4547 - LAMBOLEY, G.: Mesure du rayonnement solaire. - In: Techniques d'Étude des Fac-
       teurs Physiques de la Biosphère. Pp. 45-58. INRA, Paris 1970.

4548 - LAMBOLEY, G.: Utilisation des thermistances. - In: Techniques d'Étude des Fac-
       teurs Physiques de la Biosphère. Pp. 153-157. INRA, Paris 1970.

4549 - LAMMERS, W.T.: Photosynthesis by *Chlorella* after density-gradient centrifuga-
       tion. - Limnol. Oceanogr. *12*: 148-150, 1967.

4550 - **LANDSBERG, J.J., LUDLOW, M.M.**: A technique for determining resistance to mass transfer through the boundary layers of plants with complex structure. - J. appl. Ecol. *7*: 187-192, 1970.

4551 - **LANG, F., VOROB'EVA, L.M., KRASNOVSKIĬ, A.A.**: Obrazovanie a vytsvetanie pigmentov v list'yakh mutantov kukuruzy. [Pigment formation and degradation in mutant maize leaves.] - Dokl. Akad. Nauk SSSR *183*: 711-714, 1968. [In R.]

4552 - **LANG, F., VOROB'EVA, L.M., KRASNOVSKIĬ, A.A.**: Issledovaniya formirovaniya pigmentnoĭ sistemy v normal'nykh i mutantnykh list'yakh kukuruzy. [Study of pigment system formation in normal and mutant maize leaves.] - In: Vsesoyuznyĭ Biokhimicheskiĭ S"ezd. Ser. 19. P. 72. FAN, Tashkent 1969. [In R.]

4553 - **LANG, F., VOROB'EVA, L.M., KRASNOVSKIĬ, A.A.**: Issledovanie zeleneniya etiolirovannykh mutantov kukuruzy. [Study of greening of etiolated maize mutants.] - Biokhimiya *34*: 257-265, 1969. [In R, ab: E.]

4554 - **LANG, F., VOROB'EVA, L.M., KRASNOVSKIĬ, A.A.**: Izmeneniya razlichnykh form pigmentov v list'yakh mutantnykh i normal'nykh rasteniĭ pod deĭstviem ·sveta. [Light-induced changes of different pigment forms in leaves of mutant and normal plants.] - Biofizika *14*: 245-255, 1969. [In R, ab: E.]

4555 - **LANG, F., VOROBYEVA, L.M., KRASNOVSKY, A.A.**: Greening and bleaching processes in mutant maize leaves. - In: METZNER, H. (ed.): Progress in Photosynthesis Research. Vol. II. Pp. 630-634. Tübingen 1969.

4556 - **LANG, N.J.**: The fine structure of blue-green algae. - Annu. Rev. Microbiol. *22*: 15-46, 1968. [Chloroplast.]

4557 - **LANGE, O.L.**: CO_2-Gaswechsel der Flechte *Cladonia alcicornis* nach langfristigem Aufenthalt bei tiefen Temperaturen. - Flora B *156*: 500-502, 1966.

4558 - **LANGE, O.L.**: CO_2-Gaswechsel von Moosen nach Wasserdampfaufnahme aus dem Luftraum. - Planta *89*: 90-94, 1969.

4559 - **LANGE, O.L.**: Die funktionellen Anpassungen der Flechten an die ökologischen Bedingungen arider Gebiete. - Ber. deut. bot. Ges. *82*: 3-22, 1969. [Ps.]

4560 - **LANGE, O.L.**: Experimentell-ökologische Untersuchungen an Flechten der Negev-Wüste. I. CO_2-Gaswechsel von *Ramalina maciformis*(DEL.) BORY unter kontrollierten Bedingungen im Laboratorium. - Flora B *158*: 324-359, 1969.

4561 - **LANGE, O.L., KOCH, W., SCHULZE, E.D.**: CO_2-Gaswechsel und Wasserhaushalt von Pflanzen in der Negev-Wüste am Ende der Trockenzeit. - Ber. deut. bot. Ges. *82*: 39-61, 1969.

4562 - **LANGE, O.L., SCHULZE, E.D., KOCH, W.**: Photosynthese von Wüstenflechten am natürlichen Standort nach Wasserdampfaufnahme aus dem Luftraum. - Naturwissenschaften *55*: 658-659, 1968.

4563 - **LANGE, O.L., SCHULZE, E.-D., KOCH, W.**: Experimentell-ökologische Untersuchungen an Flechten der Negev-Wüste. II. CO_2-Gaswechsel und Wasserhaushalt von *Ramalina maciformis* (DEL.) BORY am natürlichen Standort während der sommerlichen Trockenperiode. - Flora *159*: 38-62, 1970.

4564 - **LANGE, O.L., SCHULZE, E.-D., KOCH, W.**: Experimentell-ökologische Untersuchungen an Flechten der Negev-Wüste. III. CO_2-Gaswechsel und Wasserhaushalt von Krusten- und Blattflechten am natürlichen Standort während der sommerlichen Trockenperiode. - Flora *159*: 525-538, 1970.

4565 - **LANGE, O.L., SCHULZE, E.D., KOCH, W.**: Evaluation of photosynthesis measurements taken in the field. - In: Prediction and Measurement of Photosynthetic Productivity. Pp. 339-352. PUDOC, Wageningen 1970.

4566 - **LANGNER, W.**: Untersuchungen zur Jahresperiodik bei Grünalgen. - Beitr. Biol. Pflanzen *45*: 1-38, 1968. [Ps, Chl.]

4567 - **LAPINA, L.P., BIKMUKHAMETOVA, S.A.**: Vliyanie izoosmoticheskikh kontsentratsiĭ NaCl i Na_2SO_4 na intensivnost' fotosinteza i fotokhimicheskuyu aktivnost' khloroplastov kukuruzy. [Effect of isoosmotic concentrations of NaCl and Na_2SO_4 on photosynthetic rate and photochemical activity of maize chloroplasts.] - Fiziol. Rast. *16*: 638-642, 1969. [In R, ab: E.]

4568 - LAPINA, L.P., POPOV, B.A.: Vliyanie khloristogo natriya na fotosinteticheskiĭ
apparat tomatov. [Effect of NaCl on the photosynthetic apparatus of tomato.] -
Fiziol. Rast. *17*: 580-584, 1970. [In R, ab: E.]

4569 - LAPINA, L.P., POPOV, B.A., STROGONOV, B.P.: Vliyanie izoosmoticheskikh kontsentratsiĭ NaCl, Na$_2$SO$_4$ i dekstrana na strukturu khloroplastov. [Effect of isoosmotic concentration of NaCl, Na$_2$SO$_4$ and dextrane on chloroplast structure.] -
Fiziol. Rast. *15*: 1059-1063, 1968. [In R, ab: E.]

4570 - LAPTEV, V.V., NILOVSKAYA, N.T.: Germeticheskaya ustanovka dlya izucheniya gazoobmena i vodoobmena rasteniĭ. [Hermetic installation for the study of gas and
water exchange in growing plants.] - Sel'.-khoz. Biol. *3*: 892-895, 1968. [In R,
ab: E.]

4571 - LARCHER, W.: Physiological approaches to the measurement of photosynthesis in
relation to dry matter production by trees. - In: 8th National Conference on
Agricultural Meteorology. Pp. 1-14. Amer. meteorol. Soc.,Carleton Univ., Ottawa
1968.

4572 - LARCHER, W.: Physiological approaches to the measurement of photosynthesis in
relation to dry matter production by trees. - Photosynthetica *3*: 150-166, 1969.

4573 - LARCHER, W.: The effect of environmental and physiological variables on the carbon dioxide gas exchange of trees. - Photosynthetica *3*: 167-198, 1969.

4574 - LARCHER, W.: Die Bedeutung des Faktors "Zeit" für die photosynthetische Stoffproduktion. - Ber. deut. bot. Ges. *82*: 71-80, 1969.

4575 - LARIN, A.P.: O fotosinteticheskom ispol'zovanii solnechnoĭ radiatsii sel'skokho-
zyaĭstvennymi kul'turami. [Photosynthetic use of solar radiation by agricultural
plants.] - Uch. Zap. tart. gos. Univ. *185*: 426-434, 1966. [In R.]

4576 - LARIN, A.P.: O faktorakh fotosinteticheskoĭ deyatel'nosti rasteniĭ v posevakh.
[Factors of photosynthetic activity of plants in stands.] - In: Fotosintez i Uroz-
haĭnost' Sel'skokhozyaĭstvennykh Rasteniĭ. Pp. 52-57. Min. sel'. Khoz. SSSR,
Kiev 1970. [In R.]

4577 - LARIN, A.P., LEBEDEV, S.I.: Pogloshchenie i ispol'zovanie solnechnoĭ energii posevami pri razlichnykh usloviyakh proizrastaniya. [Absorption and utilization
of solar energy by stands under various growth conditions.] - In: Aktinometriya
i Optika Atmosfery. Pp. 293-297. Valgus, Tallin 1968. [In R.]

4578 - LARRY, J.R.: Studies of the interaction of chlorophylls and derivatives with tri-
nitrobenzene. - Diss. Abstr. B *27*: 2316-B, 1967.

4579 - LARRY, J.R., van WINKLE, Q.: Charge-transfer interactions of chlorophylls *a* and
b and pheophytins *a* and *b* with *sym*-trinitrobenzene. - J. phys. Chem. *73*: 570-
580, 1969.

4580 - LARSEN, P.: Light as a factor in plant production and growth regulation. - Qual.
Plant. Mat. veg. *13*: 365-367, 1966. [Ps.]

4581 - LARSEN, P.: Light requirements in plant production and growth regulation. -
Acta Agr. scand. *16* (Suppl. 16): 161-172, 1966. [Ps.]

4582 - LARSON, P.R., GORDON, J.C.: Photosynthesis and wood yield. - Agr. Sci. Rev. *7*:
7-14, 1969.

4583 - LARSON, P.R., GORDON, J.C.: Leaf development, photosynthesis, and C^{14} distribution in *Populus deltoides* seedlings. - Amer. J. Bot. *56*: 1058-1066, 1969.

4584 - LASCELLES, J.: The accumulation of bacteriochlorophyll precursors by mutant and
wild-type strains of *Rhodopseudomonas spheroides*. - Biochem. J. *100*: 175-183,
1966.

4585 - LASCELLES, J.: The bacterial photosynthetic apparatus. - Advances microbial
Physiol. *2*: 1-42, 1968.

4586 - LASCELLES, J.: The regulation of haem and chlorophyll synthesis. - In: GOODWIN,
T.W. (ed.): Porphyrins and Related Compounds. Pp. 49-59. Academic Press, London-
New York 1968.

4587 - LASCELLES, J., ALTSHULER, T.: Some properties of mutant strains of *Rhodopseudo-*

monas spheroides which do not form bacteriochlorophyll. - Arch. Mikrobiol. *59*: 204-210, 1967.

4588 - LASCELLES, J., ALTSHULER, T.: Mutant strains of *Rhodopseudomonas spheroides* lacking δ-aminolevulinate synthase: growth, heme, and bacteriochlorophyll synthesis. - J. Bacteriol. *98*: 721-727, 1969.

4589 - LASCELLES, J., HATCH, T.P.: Bacteriochlorophyll and heme synthesis in *Rhodopseudomonas spheroides:* possible role of heme in regulation of the branched biosynthetic pathway. - J. Bacteriol. *98*: 712-720, 1969.

4590 - LASHKHI, A.D., TSITSILASHVILLI, O.K., BERSHTEÏN, B.I.: Vliyanie kaliya na ottok assimilyatov i vinogradnoŤ lozy. [Effect of potassium on photosynthate efflux in vine.] - In: Puti Povysheniya Intensivnosti i Produktivnosti Fotosinteza. Vol. 2. Pp. 198-203. Naukova Dumka, Kiev 1967. [In R.]

4591 - LATZKO, E., v. GARNIER, R., GIBBS, M.: Effect of photosynthesis, photosynthetic inhibitors and oxygen on the activity of ribulose 5-phosphate kinase. - Biochem. biophys. Res. Commun. *39*: 1140-1144, 1970.

4592 - LATZKO, E., GIBBS, M.: Distribution and activity of enzymes of the reductive pentose phosphate cycle in spinach leaves and in chloroplasts isolated by different methods. - Z. Pflanzenphysiol. *59*: 184-194, 1968.

4593 - LATZKO, E., GIBBS, M.: Effect of O_2, arsenite, sulfhydryl compounds and light on the activity of ribulose-5-phosphate kinase. - In: METZNER, H. (ed.): Progress in Photosynthesis Research. Vol. III. Pp. 1624-1630. Tübingen 1969.

4594 - LATZKO, E., GIBBS, M.: Enzyme activities of the carbon reduction cycle in some photosynthetic organisms. - Plant Physiol. *44*: 295-300, 1969.

4595 - LATZKO, E., GIBBS, M.: Level of photosynthetic intermediates in isolated spinach chloroplasts. - Plant Physiol. *44*: 396-402, 1969.

4596 - LATZKO, E., GIBBS, M., LABER, L., O'NEAL, D.: Über den Weg des Kohlenstoffs zwischen Calvin- und Hatch & Slack-Cyclus. - Ber. deut. bot. Ges. *83*: 433-434, 1970.

4597 - LAUDENBACH, B., PIRSON, A.: Über den Kohlenhydratumsatz in *Chlorella* unter dem Einfluss von blauem und rotem Licht. - Arch. Mikrobiol. *67*: 226-242, 1969.

4598 - LAUDI, G.: Ricerche infrastrutturali sui plastidi della piante parassite. II. *Lathraea squamaria*. [Ultrastructural research on plastids of parasitic plants. II. *Lathraea squamaria*.] - Caryologia *19*: 47-54, 1966. [Chloroplasts; in Ital., ab: E.]

4599 - LAUDI, G.: Infrastrutture dei cloroplasti di foglioline di piante eziolate di pisello (*Pisum sativum* L. var. Alaska) poste ad inverdire, alla luce, in presenza di acido indolacetico. [Ultrastructure of chloroplasts of leaves of etiolated seedlings of pea (*Pisum sativum* L. cv. Alaska) treated with indole acetic acid during exposure to light.] - G. bot. ital. *101*: 97-109, 1967. [In Ital., ab: E.]

4600 - LAUDI, G.: Four-layered membrane in the plastids of galls of *Cuscuta australis*. - Experientia *24*: 729, 1968.

4601 - LAUDI, G.: Ultrastructural researches on the plastids of parasitic plants. IV. Galls of *Cuscuta australis*. - G. bot. ital. *102*: 1-19, 1968. [Chl.]

4602 - LAUDI, G., ALBERTINI, A.: Ricerche infrastrutturali sui plastidi delle piante parassite. III. *Orobanche ramosa*. [Infrastructural researches on the plastids of parasitic plants. III. *Orobanche ramosa*.] - Caryologia *20*: 207-216, 1967. [Ps; in Ital., ab: E.]

4603 - LAUDI, G., BONATTI, P., BELLI, G.: Fissazione fotosintetica dell'anidride carbonica in foglie di piante mosaicate di *Nicotiana glutinosa* L. [CO_2 photosynthetic fixation in leaves of *Nicotiana glutinosa* L. infected with cucumber mosaic virus.] - Rev. Patol. veg., Ser. IV, *5*: 135-141, 1969. [In Ital., ab: E.]

4604 - LAUDI, G., BONATTI, P., TROVATELLI, L.D.: Differenze ultrastrutturali di alcune specie di *Trebouxia* poste in condizioni di illuminazione differenti. [Ultrastructural changes in some species of *Trebouxia* under different light conditions.] - G. bot. ital. *103*: 79-107, 1969. [Chl, chloroplast; in Ital., ab: E.]

4605 - LAULHÈRE, J.-P., ALQUIER-BOUFFARD, A.: La disponibilité du fer et la chlorose.
       - Physiol. vég. 7: 277-296, 1969.

4606 - LAUSI, D.: Quantitá di clorofilla negli ecosistemi bentonici del Golfo di Tries-
       te. [Chlorophyll contents of benthic ecosystem of the Gulf of Trieste.] - Nova
       Thalassia 3 (3): 1-29, 1967. [In Ital., ab: E.]

4607 - LAUSI, D., CRISTOFOLINI, G., TARABOCCHIA, M., de CRISTINI, P.: Attività fotosin-
       tetica di alghe marine nella Grotta della Viole (Isola S. Domino-Tremiti). [Pho-
       tosynthetic activity of marine algae in Grotta della Viole (Isola S. Domino-
       Tremiti).] - G. bot. ital. 101: 167-178, 1967. [In Ital., ab: E.]

4608 - LAVAL-MARTIN, D.: Variations de la teneur en chlorophylles au cours de la matu-
       ration et en fonction de la température, chez la poire Passe-Crassane. - Physiol.
       vég. 7: 251-259, 1969.

4609 - LAVOREL, J.: Cinétique de la fluorescence photostimulée de la chlorophylle in
       vivo. - In: THOMAS, J.B., GOEDHEER, J.C. (ed.): Currents in Photosynthesis. Pp.
       39-47. Donker, Rotterdam 1966.

4610 - LAVOREL, J.: Phénomènes primaires d'oxydo-réduction dans la photosynthèse. -
       Bull. Soc. franç. Physiol. vég. 12: 45-68, 1966.

4611 - LAVOREL, J.: The mechanisms of quanta accumulation and photochemical processes
       in photosynthesis. - In: de BROGLIE, L. (ed.): Wave Mechanics and Molecular Bio-
       logy. Pp. 150-169. Addison-Wesley Publ. Co., Inc., Reading, Mass. 1966.

4612 - LAVOREL, J.: On a theory for resonance quenching in photosynthesis. - In: Ab-
       stracts Volume - Seventh International Congress of Biochemistry. Symp. VI - 6,1.
       Tokyo 1967.

4613 - LAVOREL, J.: Resonance quenching as a diffusion-type process. I. Theory. - J.
       chem. Phys. 47: 2235-2240, 1967.

4614 - LAVOREL, J.: Untersuchungen über die Primärreaktionen der Photosynthese. - Um-
       schau Wiss. Tech. 1967: 367, 1967.

4615 - LAVOREL, J.: Sur une relation entre fluorescence et luminescence dans les sys-
       tèmes photosynthétiques. - Biochim. biophys. Acta 153: 727-730, 1968.

4616 - LAVOREL, J.: On a relation between fluorescence and luminescence in photosyn-
       thetic systems. - In: METZNER, H. (ed.): Progress in Photosynthesis Research.
       Vol. II. Pp. 883-898. Tübingen 1969.

4617 - LAVOREL, J., LEVINE, R.P.: Fluorescence properties of wild-type Chlamydomonas
       reinhardi and three mutant strains having impaired photosynthesis. - Plant Phys-
       iol. 43: 1049-1055, 1968.

4618 - LAVRENTOVICH, D.I., MITROFANOV, B.A., MANUIL'SKIŸ, V.D.: Intensivnost' i produk-
       tivnost' fotosinteza kukuruzy v chistykh i smeshannykh posevakh. [Rate and pro-
       ductivity of photosynthesis in pure and mixed stands of maize.] - In: Puti Povys-
       heniya Intensivnosti i Produktivnosti Fotosinteza. Pp. 116-123. Naukova Dumka,
       Kiev 1966. [In R.]

4619 - LAWLER, P.D., ROGERS, L.J.: DDT as an inhibitor of photophosphorylation in cer-
       tain varieties of barley. - Biochem. J. 105: 11 P, 1967.

4620 - LAWLER, P.D., ROGERS, L.J.: Effect of DDT on photosynthesis in certain varieties
       of barley. - Nature 215: 1515-1516, 1967.

4621 - LAWLER, P.D., ROGERS, L.J.: Biochemical investigations into the effect of DDT on
       barley. - Biochem. J. 103: 44 P, 1967. [Ps, chl.]

4622 - LAWLER, P.D., ROGERS, L.J.: Inhibition of photophosphorylation by 1,1,1-trichlo-
       ro-2,2-bis(p-chlorophenyl)ethane (DDT). - Biochem. J. 110: 381-383, 1968.

4623 - LAWLER, P.D., ROGERS, L.J.: Effect of DDT on photosynthesis in certain varieties
       of barley. - In: METZNER, H. (ed.): Progress in Photosynthesis Research. Vol.
       III. Pp. 1761-1768. Tübingen 1969.

4624 - LAWTON, J.R.S.: Translocation in the phloem of Dioscorea spp. I. Distribution
       of radioactive photosynthates as shown by whole-plant autoradiography. - Z.
       Pflanzenphysiol. 58: 1-7, 1967.

4625 - **LAWTON, J.R.S.**: Translocation in the phloem of *Dioscorea* spp. II. Distribution of translocates in the stem. - Z. Pflanzenphysiol. *58*: 8-16, 1967. [Photosynthates.]

4626 - **LEACH, G.J.**: The relation of photosynthesis by phytometers in the profiles of kale crops to leaf area index above them. - J. appl. Ecol. *6*: 499-505, 1969.

4627 - **LEACH, G.J., WATSON, D.J.**: Photosynthesis in crop profiles, measured by phytometers. - J. appl. Ecol. *5*: 381-408, 1968.

4628 - **LEAK, L.V., BURKE, J.F.**: The application of freeze-fracture replication in studying the fine structure of a blue-green alga. - Exp. Cell Res. *48*: 300-306, 1967. [Chloroplasts.]

4629 - **LEBEDEV, O.L., GRYAZNOV, U.M., CHASTOV, A.A.**: O nelineĭnom propuskanii sveta rubinovogo lazera rastvorami khlorofilla. [Non-linear transmission of ruby laser radiation by chlorophyll solutions.] - Opt. Spektrosk. *24*: 622-623, 1968. [In R.]

4630 - **LEBEDEV, S.I.**: Vbyrannya i vykorystannya energiĭ sonyachnoĭ radiatsiĭ i produktyvnist' posiviv sil's'kogospodars'kykh roslyn (pro vzaemozv'yazky mizh strukturoyu i funktsieyu khloroplastiv ta produktyvnist'yu roslyn i posiviv. [Absorption and utilization of solar energy and productivity of agricultural crops (interrelation of chloroplast structure and function, and plant and community productivity).] - In: Pidvyshchennya Vrozhaĭnosti Sil's'kogospodars'kykh Kul'tur. Pp. 179-189. Urozhaĭ, Kiev 1968. [In Ukr.]

4631 - **LEBEDEV, S.I.**: O produktivnosti fotosinteza posevov ozimoĭ pshenitsy v usloviyakh orosheniya. [Photosynthetic productivity of irrigated winter wheat stands.] - Fiziol. Biokhim. kul't. Rast. *1*: 128-133, 1969. [In R, ab: E.]

4632 - **LEBEDEV, S.I.**: O produktivnosti fotosinteza posevov ozimoĭ pshenitsy v usloviyakh orosheniya. [Productivity of photosynthesis of irrigated winter wheat stands.] - In: Fotosintez i Urozhaĭnost' Sel'skokhozyaĭstvennykh Rasteniĭ. Pp. 5-10. Min. sel'.Khoz. SSSR, Kiev 1970. [In R.]

4633 - **LEBEDEV, S.I., KHOSSEÏN, M.M.**: Ob izmeneniyakh adaptirovannoĭ k svetu pigmentnoĭ sistemy fasoli. [Changes of pigment system of light-adapted bean.] - Fiziol. Biokhim. kul't. Rast. *2*: 389-394, 1970. [In R, ab: E.]

4634 - **LEBEDEV, S.I., KIRIENKO, I.M.**: Vliyanie vnekornevoĭ podkormki na aktivnost' fotosinteticheskogo apparata sakharnoĭ svekly. [Effect of extra-root nutrition on the activity of photosynthetic apparatus of sugar beet.] - In: Fotosintez i Urozhaĭnost' Sel'skokhozyaĭstvennykh Rasteniĭ. Pp. 42-47. Min. sel'. Khoz.SSSR, Kiev 1970. [In R.]

4635 - **LEBEDEV, S.I., KIRYATSEVA, O.Kh.**: O roli pigmentov plastid v protsesse rosta rasteniĭ. [Role of plastid pigments in plant growth.] - Fiziol. Rast. *13*: 781-789, 1966. [In R, ab: E.]

4636 - **LEBEDEV, S.I., LITVINENKO, L.G.**: O biosinteze khlorofilla u vysshikh rasteniĭ. [Chlorophyll biosynthesis in higher plants.] - Uch. Zap. tartu. gos. Univ. *185*: 376-387, 1966. [In R.]

4637 - **LEBEDEV, S.I., LITVINENKO, L.G.**: O vzaimosvyazi mezhdu fotokhimicheskoĭ aktivnost'yu, energeticheskimi i strukturnymi osobennostyami khloroplastov. [Interrelationship between photochemical activity and energetic and structural features of chloroplasts.] - Fiziol. Rast. *13*: 411-415, 1966. [In R, ab: E.]

4638 - **LEBEDEV, S.I., LITVINENKO, L.G.**: Sravnitel'noe issledovanie fotokhimicheskoĭ aktivnosti khloroplastov antotsiansoderzhashchikh i zelenykh form rasteniĭ. [Comparison of photosynthetic activity of chloroplasts of green and anthocyanin-containing forms of plants.] - In: Puti Povysheniya Intensivnosti i Produktivnosti Fotosinteza. Pp. 85-91. Naukova Dumka, Kiev 1966. [In R.]

4639 - **LEBEDEV, S.I., LITVINENKO, L.G.**: Fotokhimicheskaya aktivnost' list'ev v svyazi s soderzhaniem khlorofilla. [Photochemical activity of leaves in relation to chlorophyll content.] - Fiziol. Biokhim. kul't. Rast. *2*: 46-51, 1970. [In R, ab: E.]

4640 - **LEBEDEV, S.I., LYSYUK, E.S.**: Opticheskie parametry list'ev i posevov ozimoĭ pshe-

nitsy pri razlichnoĭ vodoobespechennosti rasteniĭ. [Optical parameters of leaves
and crops of winter wheat at different water supply to plants.] - In: Fotosintez
i Urozhaĭnost' Sel'skokhozyaĭstvennykh Rasteniĭ. Pp. 11-15. Min. sel'.Khoz. SSSR,
Kiev 1970. [In R.]

4641 - LEBEDEV, S.I., SAKALO, N.D.: Biokhimicheskie i strukturnye izmeneniya fotosinteticheskogo apparata i produktivnost' grechikhi pri razlichnoĭ vlazhnosti pochvy.
[Biochemical and structural changes of photosynthetic apparatus and productivity
of buckwheat at different soil moisture.] - In: Fotosintez i Urozhaĭnost' Sel'-
skokhozyaĭstvennykh Rasteniĭ. Pp. 21-25. Min. sel'. Khoz. SSSR, Kiev 1970.
[In R.]

4642 - LEBEDEV, S.I., SAKALO, N.D., KIRYATSEVA, O.Kh.: Fiziologicheskoe sostoyanie i
struktura fotosinteticheskogo apparata fasoli pri razlichnoĭ vlazhnosti pochvy.
[Physiological state and structure of photosynthetic apparatus of bean under
different soil moisture.] - Puti Povysheniya Intensivnosti i Produktivnosti Fotosinteza 3: 102-109, 1969. [In R.]

4643 - LEBEDEV, S.I., SAKALO, N.D., KIRYATSEVA, O.Kh.: Izmenenie struktury i funktsii
khloroplastov sel'skokhozyaĭstvennykh rasteniĭ pri razlichnykh usloviyakh proizrastaniya. [Changes in structure and function of chloroplasts of agricultural
plants under different conditions of germination.] - In: Khloroplasty i Mitokhondrii. Pp. 164-172. Nauka, Moskva 1969. [In R.]

4644 - LEBEDEV, S.I., SAKALO, N.D., NAGORNAYA, R.V., SAVCHENKO, N.P.: Fotosinteticheskaya deyatel'nost' rasteniĭ i struktura khloroplastov pri razlichnom sochetanii
elementov mineral'nogo pitaniya. [Photosynthetic activity of leaves and chloroplast structure under various interaction of elements of mineral nutrition.] -
In: Mineral'nye Elementy i Mekhanizm Fotosinteza. Pp. 118-127. Kishinev 1969.
[In R.]

4645 - LEBEDEV, S.I., SAKALO, N.D., NAGORNAYA, R.V., SAVCHENKO, N.P.: O fotosinteticheskom apparate sakharnoĭ svekly vysokoĭ aktivnosti. [Photosynthetic apparatus of
sugar beet of high activity.] - In: Fotosintez i Urozhaĭnost' Sel'skokhozyaĭstvennykh Rasteniĭ. Pp. 36-41. Min. sel'. Khoz. SSSR, Kiev 1970. [In R.]

4646 - LEBEDEV, V.M.: Vliyanie usloviĭ fosfornogo pitaniya na fotosintez yabloni. [Ef-
fect of phosphorus nutrition on apple photosynthesis.] - Nauch. Dokl. vyssh.
Shkoly, biol. Nauki 12 (3): 93-97, 1969. [In R.]

4647 - LEBEDEV, V.N.: K voprosu o deĭstvii ul'trafioletovoĭ radiatsii na reaktsiyu
Khilla. [Effect of ultraviolet radiation on Hill reaction.] - Izv. Akad. Nauk
Tadzh. SSR, Otd. biol. Nauk 1968 (4): 85-88, 1968. [In R.]

4648 - LEBEDEV, V.N., AFANAS'EV, V.P.: Opticheskie svoĭstva list'ev nekotorykh rasteniĭ
Vostochnogo Pamira v oblasti 400-750 nm. [Optical properties at 400-750 nm of
leaves of some plants of Eastern Pamir.] - Izv. Akad. Nauk Tadzh. SSR, Otd. biol.
Nauk 1968 (4): 89-93, 1968. [In R.]

4649 - LEBEDEVA, A.I., MAL'YAN, A.N., MAKAROV, A.D.: III. O nekotorykh osobennostyakh
kinetiki gidroliza ATF na khloroplastakh. [III. Some peculiarities of kinetics
of ATP hydrolysis on chloroplasts.] - Biofizika 14: 1069-1071, 1969. [In R, ab:
E.]

4650 - LEBEDEVA, E.K., ALEKSANDROVA, I.V., VARLAMOV, V.F., GAĬNUTDINOVA, N.A., ILGACH,
G.V., EGOROVA, N.N., KRASOTCHENKO, L.M., UL'YANIN, Yu.N., TSITOVICH, S.I.,
CHERNOVICH, I.L.: Opyt kul'tivirovaniya khlorelly na mineralizovannykh biologi-
cheskim put'em produktakh zhiznedeyatel'nosti cheloveka. [Chlorella cultivation
on biologically mineralized products of human metabolism.] - In: Problemy Sozdaniya Zamknutykh Ekologicheskikh Sistem. Pp. 102-108. Nauka, Moskva 1967.
[Chl, car; in R.]

4651 - LEBEDEVA, K.D., SIVKOV, S.I., FATEEVA, K.A., YANISHEVSKIĬ, Yu.D., YASTREBOVA,
T.K.: O sravnimosti pokazaniĭ piranometrov i balansomerov raznykh tipov. [Com-
parability of measurements made with different types of pyranometers and net
radiometers.] - In: Aktinometriya i Optika Atmosfery. Pp. 230-238. Valgus, Tallin 1968. [In R, ab: E.]

4652 - LEBERMANN, K.W., NELSON, A.I., STEINBERG, M.P.: Post-harvest changes of broccoli
stored in modified atmospheres. 1. Respiration of shoots and color of flower
heads. - Food Technol. 22 (4): 143-146, 1968. [Chl.]

4653 - LEBERMANN, K.W., NELSON, A.I., STEINBERG, M.P.: Post-harvest changes of broccoli stored in modified atmospheres. 2. Acidity and its influence on texture and chlorophyll retention of the stalks. - Food Technol. *22* (4): 146-149, 1968.

4654 - LEBLOVÁ, S., BARTHOVÁ, J., SOFROVÁ, D., KOŠTÍŘ, J.: Příspěvek k řešení vztahu mezi respirací a fotosyntézou. [Relationship of respiration and photosynthesis.] Rostl. Výr. (Praha) *16*: 637-644, 1970. [In Czech, ab: E,G,R.]

4655 - LECLERC, J.-C.: Sur la croissance et la photosynthèse de *Porphyridium*, en fonction de la salinité du milieu de culture. - Photosynthetica *1*: 179-191, 1967.

4656 - LECLERC, J.C.: Pigments et salinité chez *Porphyridium*. - Physiol. Plant. *22*: 1013-1024, 1969.

4657 - LEDIG, F.T.: A growth model for tree seedlings based on the rate of photosynthesis and the distribution of photosynthate. - Photosynthetica *3*: 263-275, 1969.

4658 - LEDIG, F.T., PERRY, T.O.: Variation in photosynthesis and respiration among loblolly pine progenies. - In: Proceedings of the Ninth Southern Conference on Forest Tree Improvement. Pp. 120-128. Knoxville, Tenn. 1967.

4659 - LEDIG, F.T., PERRY, T.O.: Net assimilation rate and growth in loblolly pine seedlings. - Forest Sci. *15*: 431-438, 1969.

4660 - LEDOVSKIĬ, S.Ya.: Nekotorye fiziologo-biokhimicheskie osobennosti rasteniĭ tomatov, vyrashchivaemykh gidroponnym sposobom. [Some physiological and biochemical characteristics of tomato plants grown in hydroponics.] - Dokl. VASKhNIL *1968* (10): 19-21, 1968. [Chl; in R.]

4661 - LEDOVSKIĬ, S.Ya.: Osobennosti rasteniĭ tomatov, vyrashchivaemykh gidroponnym sposobom. [Peculiarities of tomato plants grown in hydroponics.] - Fiziol. Biokhim. kul't. Rast. *2*: 30-33, 1970. [Chl; in R, ab: E.]

4662 - LEE, J.A., WOOLHOUSE, H.W.: A re-appraisal of the electrometric method for the determination of the concentration of carbon dioxide in soil atmospheres. - New Phytol. *65*: 325-330, 1966.

4663 - LEE, J.A., WOOLHOUSE, H.W.: Chlorophyll content of *Deschampsia flexuosa* seedlings grown on a calcareous and a non-calcareous soil. - Nature *209*: 1044-1045, 1966.

4664 - LEE, K.H., YAMAMOTO, H.Y.: Action spectra for light-induced de-epoxidation and epoxidation of xanthophylls in spinach leaf. - Photochem. Photobiol. *7*: 101-107, 1968.

4665 - LEE, R.: Effects of tent type enclosures on the microclimate and vaporization of plant cover. - Oecol. Plant. *1*: 301-326, 1966.

4666 - LEE, S.S., TRAVIS, J., BLACK, C.C. Jr.: Characterization of ferredoxin from nutsedge, *Cyperus rotundus* L., and other species with a high photosynthetic capacity. - Arch. Biochem. Biophys. *141*: 676-689, 1970.

4667 - LEE, S.S., YOUNG, A.M., KROGMANN, D.W.: Site-specific inactivation of the photophosphorylation reactions of *Anabaena variabilis*. - Biochim. biophys. Acta *180*: 130-136, 1969.

4668 - LEE, T.-C.: Studies on the coupling of geranylgeranyl pyrophosphate in the biosynthesis of carotenes. - Diss. Abstr. int. B *31*: 5163-B, 1970.

4669 - LEECH, R.M.: Comparative biochemistry and comparative morphology of chloroplasts isolated by different methods. - In: GOODWIN, T.W. (ed.): Biochemistry of Chloroplasts. Vol. I. Pp. 65-74. Academic Press, London-New York 1966.

4670 - LEECH, R.M.: The chloroplast inside and outside the cell. - In: PRIDHAM, J.B. (ed.): Plant Cell Organelles. Pp. 137-162. Academic Press, London-New York 1968.

4671 - LEECH, R.M., DYER, T.A.: The ribonucleic acids of the chloroplasts of *Vicia faba* L. - In: METZNER, H. (ed.): Progress in Photosynthesis Research. Vol. I. Pp. 359-367. Tübingen 1969.

4672 - LEETE, E., LOUDEN, M.C.L.: Biosynthesis of capsaicin and dihydrocapsaicin in *Capsicum frutescens*. - J. amer. chem. Soc. *90*: 6837-6841, 1968.

4673 - LEFÈBVRE, A., RIBÉREAU-GAYON, P.: La fixation et le dégagement de CO_2 par *V. vi-*

nifera; leurs relations avec les voies du métabolisme. - Compt. rend. Acad. Sci. (Paris), Sér. D *270*: 1727-1729, 1970.

4674 - LEFF, J., KRINSKY, N.I.: A mutagenic effect of visible light mediated by endogenous pigments in *Euglena gracilis*. - Science *158*: 1332-1335, 1967.

4675 - LEFORT-TRAN, M.: Discussion des techniques en vue de la conservation des pigments liposolubles en microscopie électronique. - J. Microscop. (Paris) *9*: 881-890, 1970.

4676 - LEGG, B.J., PARKINSON, K.J.: Calibration of infra-red gas analysers for use with carbon dioxide. - J. sci. Instrum. (J. Phys. E) Ser. 2, *1*: 1003-1006, 1968.

4677 - LEHMANN, J.: Kohlenstoff-14 und die Assimilation des Kohlendioxids. - Chem. unserer Zeit *2* (3): 67-73, 1968.

4678 - LEINA, G.D.: Fotosintez elovogo podrosta pod pologom i na vyrubkakh el'nika chernichnika svezhego v svyazi s davnost'yu rubki. [Photosynthesis of spruce undergrowth in the stand and in bilberry spruce forest fellings in relation to the time of felling.] - In: Svetovoĭ Rezhim, Fotosintez i Produktivnost' Lesa. Pp. 232-236. Nauka, Moskva 1967. [In R.]

4679 - LELĄTKO, Z.: Some aspects of chloroplast movement in leaves of terrestrial plants. - Acta Soc. Bot. Pol. *39*: 453-468, 1970.

4680 - LEMASSON, C.: Contribution à l'étude cinétique de la désactivation en photosynthèse chez *Chlorella pyrenoidosa*. - Compt. rend. Acad. Sci. (Paris), Sér. D *270*: 350-353, 1970.

4681 - LEMOINE, Y.: Évolution du chloroplaste étiolé au cours du verdissement chez le Haricot. - J. Microscop. *7*: 755-770, 1968.

4682 - LEMON, E.: Aerodynamic studies of CO_2 exchange between the atmosphere and the plant. - In: SAN PIETRO, A., GREER, F.A., ARMY, T.J. (ed.): Harvesting the Sun. Pp. 263-290. Academic Press, New York-London 1967.

4683 - LEMON, E.: Gaseous exchange in crop stands. - In: EASTIN, J.D., HASKINS, F.A., SULLIVAN, C.Y., van BAVEL, C.H.M. (ed.): Physiological Aspects of Crop Yield. Pp. 117-142. Amer. Soc. Agron. & Crop Sci. Soc. Amer., Madison, Wisc. 1969.

4684 - LEMON, E., ALLEN, L.H. Jr., MÜLLER, L.: Carbon dioxide exchange of a tropical rain forest. Part II. - BioScience *20*: 1054-1059, 1970.

4685 - LEMON, E.R.: The measurement of height distribution of plant community activity using the energy and momentum balance approaches. - In: ECKARDT, F.E. (ed.): Functioning of Terrestrial Ecosystems at the Primary Production Level. Pp. 381-389. UNESCO, Paris 1968.

4686 - LEMON, E.R., WRIGHT, J.L.: Photosynthesis under field conditions. XA. Assessing sources and sinks of carbon dioxide in a corn (*Zea mays* L.) crop using a momentum balance approach. - Agron. J. *61*: 405-411, 1969.

4687 - LEMON, E.R., WRIGHT, J.L., DRAKE, G.M.: Photosynthesis under field conditions. XB. Origins of short-time CO_2 fluctuations in a cornfield. - Agron. J. *61*: 411-413, 1969.

4688 - LENDZIAN, K., ZIEGLER, H.: Über die Regulation der Glucose-6-phosphat-Dehydrogenase in Spinatchloroplasten durch Licht. - Planta *94*: 27-36, 1970.

4689 - de LEO, P., d'ALESSANDRO, G., ARRIGONI, O.: Effetto dei ritardanti di crescita sulla fosforilazione ossidativa e sulla fotofosforilazione ciclica. II. [Effects of growth retardants in oxidative phosphorylation and cyclic photophosphorylation. II.] - G. bot. ital. *102*: 73-80, 1968. [In Ital., ab: E.]

4690 - LEONARD, D.L., BIDWELL, R.G.S.: The effect of oxygen on carbon dioxide fixation. - Plant Physiol. *44* (Suppl.): 12, 1969.

4691 - LEONARD, D.L.S.: Photosynthetic and post-illumination fixation of carbon dioxide by corn and barley leaves. - Diss. Abstr. int. B *31*: 3182-B-3183-B, 1970.

4692 - LEONARD, O.A., DONALDSON, T.W., BAYER, D.E.: Translocation of labeled assimilates into and out of bean leaves as affected by 2,4-D and benzyl adenine. - Bot. Gaz. *129*: 266-279, 1968.

4693 - LEONARD, O.A., WEAVER, R.J., GLENN, R.K.: Effect of 2,4-D and picloram on trans-location of ^{14}C-assimilates in *Vitis vinifera* L. - Weed Res. *7*: 208-219, 1967.

4694 - LERCH, G., MÜLLER-STOLL, W.R.: Einfluss einer abgestuften Wasserversorgung auf CO_2-Assimilation, Transpiration, Substanzproduktion und Wasserverbrauch von *Avena sativa*. - Kulturpflanze *14*: 381-418, 1966.

4695 - LESHCHENKO, E.V., BATYUK, V.P., OKANENKO, A.S.: Vliyanie narusheniya vodnogo rezhima v assimiliruyushcheï tkani lista na intensivnost' fotosinteza u sakhar-noï svekly. [Effect of disturbance of water relations in leaf assimilatory tis-sue on photosynthetic rate in sugar beet.] - Fiziol. Rast. *13*: 339-342, 1966. [In R, ab: E.]

4696 - LESHCHENKO, E.V., OKANENKO, A.S.: Vliyanie razlichnoï kontsentratsii zol'nykh elementov v list'yakh sakharnoï svekly na intensivnost' fotosinteze. [Effect of different concentrations of ash elements in sugar beet leaves on photosynthetic rate.] - In: Puti Povysheniya Intensivnosti i Produktivnosti Fotosinteza. Vol. 3. Pp. 171-175. Naukova Dumka, Kiev 1969. [In R.]

4697 - LESHEM, Y., THAINE, R.: A note on the measurement of stomatal aperture. - New Phytol. *68*: 1047-1049, 1969.

4698 - LESHINA, A.V.: Vliyanie margantsa na soderzhanie pigmentov i askorbinovoï kis-loty v salate. [Effect of manganese on contents of pigments and ascorbic acid in lettuce.] - Dokl. Akad. Nauk Beloruss. SSR *10*: 279-281, 1966. [In R.]

4699 - LESHINA, A.V.: Vliyanie gibberellovoï kisloty na morfologicheskie i anatomo-fiziologicheskie pokazateli kormovykh bobov. [Effect of gibberellic acid on morphological and anatomo-physiological characteristics of broad bean.] - Dokl. Akad. Nauk Belorus. SSR *10*: 604-606, 1966. [Chl; in R.]

4700 - LETTAU, H.: Note on aerodynamic roughness-parameter estimation on the basis of roughness-element description. - J. appl. Meteorol. *8*: 828-832, 1969.

4701 - LEVENKO, A.A.: Ob otklonenii izmerennykh gradientov temperatury, vlazhnosti i skorosti vetra na vysote 1 m ot rasschitannykh po logarifmicheskomu zakonu pri neravnovesnykh usloviyakh. [On the difference of measured gradients of tempera-ture, humidity and wind speed at a 1 m height from gradients calculated with the logarithmic law at unbalanced conditions.] - Tr. ukr. nauch.-issled. gidro-meteorol. Inst. *62*: 56-64, 1966. [In R.]

4702 - LEVENKO, A.A.: Ob uchete zapazdyvaniya v sutochnom i sezonnom khode temperatury i vlazhnosti vozdukha na vysote 2 m. [On the estimation of lag in diurnal and seasonal course of air temperature and humidity at a 2 m height.] - Tr. ukr. nauch.-issled. gidrometeorol. Inst. *62*: 65-96, 1966. [In R.]

4703 - LEVENKO, A.A.: Utochnennaya metodika rascheta potokov tepla i vlagi po tempera-ture i vlazhnosti vozdukha na vysote 2 m. [An improved method of calculation of heat and humidity fluxes from temperature and humidity of air at a 2 m height.] - Tr. ukr. nauch.-issled. gidrometeorol. Inst. *62*: 97-123, 1966. [In R.]

4704 - LEVENKO, A.A.: Otsenka tochnosti opredeleniya turbulentnykh potokov tepla i vlagi po temperature i vlazhnosti vozdukha na vysote 2 m. [Evaluation of the precision of turbulent fluxes determination of heat and humidity from tempera-ture and humidity of air at a 2 m height.] - Tr. ukr. nauch.-issled. gidrometeo-rol. Inst. *62*: 124-138, 1966. [In R.]

4705 - LEVINE, R.P.: Genetic dissection of photosynthesis. - Science *162*: 768-771, 1968.

4706 - LEVINE, R.P.: A light-induced absorbance change at 564 nm in wild-type and mu-tant strains of *Chlamydomonas reinhardi*. - In: METZNER, H. (ed.): Progress in Photosynthesis Research. Vol. II. Pp. 971-977. Tübingen 1969.

4707 - LEVINE, R.P.: The analysis of photosynthesis using mutant strains of algae and higher plants. - Annu. Rev. Plant Physiol. *20*: 523-540, 1969.

4708 - LEVINE, R.P.: The mechanism of photosynthesis. - Sci. Amer. *221* (6): 58-64, 69, 70, 152, 1969.

4709 - LEVINE, R.P., GOODENOUGH, U.W.: The genetics of photosynthesis and of the chloro-plast in *Chlamydomonas reinhardi*. - Annu. Rev. Genet. *4*: 397-408, 1970.

4710 - LEVINE, R.P., GORMAN, D.S.: Photosynthetic electron transport chain of *Chlamydomonas reinhardi*. III. Light-induced absorbance changes in chloroplast fragments of the wild type and mutant strains. - Plant Physiol. *41*: 1293-1300, 1966.

4711 - LEVINE, R.P., GORMAN, D.S., AVRON, M., BUTLER, W.L.: Light-induced absorbance changes in wild-type and mutant strains of *Chlamydomonas reinhardi*. - In: Energy Conversion by the Photosynthetic Apparatus. Brookhaven Symp. Biol. *19*: 143-148, 1967.

4712 - LEVINE, R.P., PASZEWSKI, A.: Chloroplast structure and function in ac-20, a mutant strain of *Chlamydomonas reinhardi*. II. Photosynthetic electron transport. - J. Cell Biol. *44*: 540-546, 1970.

4713 - LEVRING, T.: Submarine light and algal shore zonation. - In: BAINBRIDGE, R., EVANS, G.C., RACKHAM, O. (ed.): Light as an Ecological Factor. Pp. 305-318. Blackwell sci. Publ., Oxford 1966. [Ps.]

4714 - LEVRING, T.: Light conditions, photosynthesis and growth of marine algae in coastal and clear oceanic water. - In: MARGALEF, R. (ed.): Proceedings of the Sixth International Seaweed Symposium. Pp. 235-244. Dirección gen. Pesca mar., Madrid 1969.

4715 - LEWINGTON, R.J., SIMON, E.W.: The effect of light on the senescence of detached cucumber cotyledons. - J. exp. Bot. *20*: 138-144, 1969. [Chl.]

4716 - LEWINGTON, R.J., TALBOT, M., SIMON, E.W.: The yellowing of attached and detached cucumber cotyledons. - J. exp. Bot. *18*: 526-534, 1967. [Chl.]

4717 - LHOSTE, J.: Les régulateurs de croissance. Emploi du C.C.C. en culture céréalière. - Nucleus *8*: 207-212, 1967. [Chl.]

4718 - LHOSTE, J.-M.: Résonance paramagnétique électronique du premier état triplet photo-excité de la chlorophylle *b*. - Compt. rend. Acad. Sci. (Paris), Sér. D *266*: 1059-1062, 1968.

4719 - LHOSTE, J.-M.: Les états électroniquement excités des chlorophylles: états singulets et états triplets. - Bull. Soc. franç. Physiol. vég. *14*: 379-408, 1968.

4720 - LHOSTE, J.M.: The electronic structure of some triplet states of photo-biological interest as studied by ESR spectroscopy. - Stud. biophys. *12*: 135-144, 1968. [Chl.]

4721 - LIAAEN JENSEN, S.: Recent studies on the structure and distribution of carotenoids in photosynthetic bacteria. - In: GOODWIN, T.W. (ed.): Biochemistry of Chloroplasts. Vol. I. Pp. 437-441. Academic Press, London-New York 1966.

4722 - LIAAEN JENSEN, S.: Recent advances in the chemistry of naturally occurring carotenoids. - Tidsskr. Kjemi, Berg., Metallurgi *26*: 128, 1966.

4723 - LIAAEN JENSEN, S.: Bacterial carotenoids XXII. The carotenoids of *Thiorhodaceae*. 5. Structural elucidation of okenone. - Acta chem. scand. *21*: 961-969, 1967.

4724 - LIAAEN JENSEN, S.: Recent advances in the chemistry of natural carotenoids. - Pure appl. Chem. *14*: 227-244, 1967.

4725 - LIAAEN JENSEN, S.: Naturlige carotenoider - nyere variasjoner over et klassisk tema. [Natural carotenoids. Newer variations on a classical theme.] - Tidsskr. Kjemi, Bergv. Metallurgi *29* (3): 62-73, 1969. [In Norweg., ab: E.]

4726 - LIAAEN JENSEN, S.: Selected examples of structure determination of natural carotenoids. - Pure appl. Chem. *20*: 421-447, 1969.

4727 - LIAAEN-JENSEN, S.: Developments in the carotenoid field. - Experientia *26*: 697-710, 1970.

4728 - LIAAEN-JENSEN, S., HERTZBERG, S.: Selective preparation of the lutein monomethyl ethers. - Acta chem. scand. *20*: 1703-1709, 1966.

4729 - LIAN, S., TANAKA, A.: Behaviour of photosynthetic products associated with growth and grain production in the rice plant. - Plant Soil *26*: 333-347, 1967.

4730 - LIBERMAN, E.A., TSOFINA, L.M.: Aktivnyĭ transport pronikayushchikh anionov fragmentami mitokhondriĭ i fotofosforiliruyushchikh bakteriĭ. [Active transport of penetrating anions by fragments of mitochondria and photophosphorylating bacteria.] - Biofizika *14*: 1017-1022, 1969. [In R, ab: E.]

4731 - LICHTENTHALER, H.K.: Plastoglobuli und Plastidenstruktur. - Ber. deut. bot. Ges. *79*: 82-88, 1966.

4732 - LICHTENTHALER, H.K.: Beziehungen zwischen Zusammensetzung und Struktur der Plastiden in grünen und etiolierten Keimlingen von *Hordeum vulgare* L. - Z. Pflanzenphysiol. *56*: 273-281, 1967.

4733 - LICHTENTHALER, H.K.: Die Verbreitung der lipophilen Plastidenchinone in nichtgrünen Pflanzengeweben. - Z. Pflanzenphysiol. *59*: 195-210, 1968.

4734 - LICHTENTHALER, H.K.: Plastoglobuli and the fine structure of plastids. - Endeavour *27*: 144-149, 1968.

4735 - LICHTENTHALER, H.K.: Verbreitung und relative Konzentration der lipophilen Plastidenchinone in grünen Pflanzen. - Planta *81*: 140-152, 1968. [Chl, car.]

4736 - LICHTENTHALER, H.K.: Die Bildung überschüssiger Plastidenchinone in den Blättern von *Ficus elastica* ROXB. - Z. Naturforsch. *24 b*: 1461-1466, 1969.

4737 - LICHTENTHALER, H.K.: Die Plastoglobuli von Spinat, ihre Grösse, Isolierung und Lipochinonzusammensetzung. - Protoplasma *68*: 65-77, 1969.

4738 - LICHTENTHALER, H.K.:Die Plastoglobuli von Spinat, ihre Grösse und Zusammensetzung während der Chloroplastendegeneration. - Protoplasma *68*: 315-326, 1969.

4739 - LICHTENTHALER, H.K.: Light-stimulated synthesis of plastid quinones and pigments in etiolated barley seedlings. - Biochim. biophys. Acta *184*: 164-172, 1969.

4740 - LICHTENTHALER, H.K.: Localization and functional concentrations of lipoquinones in chloroplasts. - In: METZNER, H. (ed.): Progress in Photosynthesis Research. Vol. I. Pp. 304-314. Tübingen 1969.

4741 - LICHTENTHALER, H.K.: Plastoglobuli und Lipochinongehalt der Chloroplasten von *Cereus peruvianus* (L.) MILL. - Planta *87*: 304-310, 1969.

4742 - LICHTENTHALER, H.K.: Zur Synthese der lipophilen Plastidenchinone und Sekundärcarotinoide während der Chromoplastenentwicklung. - Ber. deut. bot. Ges. *82*: 483-497, 1969.

4743 - LICHTENTHALER, H.K.: Die Lokalisation der Plastidenchinone und Carotinoide in den Chromoplasten der Petalen von *Sarothamnus scoparius* (L.) WIMM ex KOCH. - Planta *90*: 142-152, 1970.

4744 - LICHTENTHALER, H.K.: Formation and function of plastoglobuli in plastids. - In: 7th International Congress on Electron Microscopy. Vol. III. Pp. 205-206. Grenoble 1970.

4745 - LICHTENTHALER, H.K., BECKER, K.: Inhibition of the light-induced vitamin K_1 and pigment synthesis by abscisic acid. - Phytochemistry *9*: 2109-2113, 1970.

4746 - LICHTENTHALER, H.K., PEVELING, E.: Osmiophile Lipideinschlüsse in den Chloroplasten und im Cytoplasma von *Hoya carnosa* R. BR. - Naturwissenschaften *53*: 534, 1966.

4747 - LICHTENTHALER, H.K., PEVELING, E.: Plastoglobuli in verschiedenen Differenzierungsstadien der Plastiden bei *Allium cepa* L. - Planta *72*: 1-13, 1967.

4748 - LICHTENTHALER, H.K., PEVELING, E.: Plastoglobuli und osmiophile cytoplasmatische Lipideinschlüsse in grünen Blättern von *Hoya carnosa* R. Br. - Z. Pflanzenphysiol. *56*: 153-165, 1967.

4749 - LICHTENTHALER, H.K., SPREY, B.: Über die osmiophilen globulären Lipideinschlüsse der Chloroplasten. - Z. Naturforsch. *21 b*: 690-697, 1966.

4750 - LICHTENTHALER, H.K., TEVINI, M.: Die Wirkung von UV-Strahlen auf die Lipochinon-Pigment-Zusammensetzung isolierter Spinatchloroplasten. - Z. Naturforsch. *24 b*: 764-769, 1969.

4751 - LICHTENTHALER, H.K., TEVINI, M.: Messung der Hill-Reaktion mit der Sauerstoffelektrode. - Naturwissenschaften *56*: 284-285, 1969.

4752 - LICHTENTHALER, H.K., TEVINI, M.: Die Verteilung der Pigmente, Plastidenchinone und Plastoglobuli auf verschiedene aus beschallten Spinatchloroplasten gewonnene Partikelfraktionen. - Z. Pflanzenphysiol. *62*: 33-50, 1970.

4753 - LICHTENTHALER, H.K., VERBEEK, L., BECKER, K.: Promotion of vitamin K_1-synthesis by naphthoquinones. - Phytochemistry *10*: 79-84, 1970. [Car.]

4754 - LICHTENTHALER, H.K., WEINERT, H.: Die Beziehungen zwischen Lipochinonsynthese und Plastoglobulibildung in den Chloroplasten von *Ficus elastica* ROXB. - Z. Naturforsch. *25 b*: 619-623, 1970.

4755 - LICHTLÉ, C., GIRAUD, G.: Étude ultrastructurale de la zone apicale du thalle du *Polysiphonia elongata* (HARV.) Rhodophycée, Floridée. Évolution des plastes. - J. Microscop. *8*: 867-874, 1969. [Chloroplasts.]

4756 - LICHTLÉ, C., GIRAUD, G.: Aspects ultrastructuraux particuliers au plaste du *Batrachospermum virgatum* (SIRDT): Rhodophycée: Nemalionale. - J. Phycol. *6*: 281-289, 1970.

4757 - LIDWELL, O.M., WYON, D.P.: A rapid response radiometer for the estimation of mean radiant temperature in environmental studies. - J. sci. Instrum. (J. Phys. E) Ser. 2, *1*: 534-538, 1968.

4758 - LIETH, H.: The determination of plant dry-matter production with special emphasis on the underground parts. - In: ECKARDT, F.E. (ed.): Functioning of Terrestrial Ecosystems at the Primary Production Level. Pp. 179-186. UNESCO, Paris 1968.

4759 - LIETH, H.: The measurement of calorific values of biological material and the determination of ecological efficiency. - In: ECKARDT, F.E. (ed.): Functioning of Terrestrial Ecosystems at the Primary Production Level. Pp. 233-242. UNESCO, Paris 1968.

4760 - LIGHTBODY, J.L., KROGMANN, D.W.: The role of plastoquinone in the photosynthetic reactions of *Anabaena variabilis*. - Biochim. biophys. Acta *120*: 57-64, 1966.

4761 - LIGHTBODY, J.J., KROGMANN, D.W.: Isolation and properties of plastocyanin from *Anabaena variabilis*. - Biochim. biophys. Acta *131*: 508-515, 1967.

4762 - LILJENBERG, C.: The effect of light on the phytolization of chlorophyllide *a* and the spectral dependence of the process. - Physiol. Plant. *19*: 848-853, 1966.

4763 - LILJENBERG, C., ODHAM, G.: Gas chromatographic determination of phytol in plant material. - Physiol. Plant. *22*: 686-693, 1969.

4764 - LILOV, D.: Effect of NGS on the content of some plastid pigments in vine leaves. - Dokl. bolg. Akad. Nauk *23*: 105-108, 1970.

4765 - LIMAR', R.S.: Izuchenie intensivnosti fotosinteza "Khoranskoĭ" gruppy tverdykh pshenits. [Photosynthetic rate in the "Khoranka" cultivars of durum wheat.] - Dokl. vsesoyuz. Akad. sel'.-khoz. Nauk *1968* (12): 4-6, 1968. [In R.]

4766 - LIN,T.-H., POHLIT, H.: An improved method for the liquid scintillation counting of $^{14}CO_2$. - Anal. Biochem. *28*: 150-155, 1969.

4767 - LIN, Y.-H., SU, J.-C.: Chromoproteins of marine red alga *Porphyra crispata*. - Bot. Bull. Acad. sinica (Taiwan) *11*: 27-35, 1970.

4768 - LINACRE, E.T.: Further studies of the heat transfer from a leaf. - Plant Physiol. *42*: 651-658, 1967.

4769 - LINACRE, E.T.: Estimating the net-radiation flux. - Agr. Meteorol. *5*: 49-63, 1968.

4770 - LINACRE, E.T., HARRIS, W.J.: A thermistor leaf thermometer. - Plant Physiol. *46*: 190-193, 1970.

4771 - LINDAHL, P.E.B.: On the mechanisms of inhibition of growth and photosynthesis in aquatic plants by *N*-disubstituted dithiocarbamate derivatives. - Acta Univ. upsal. Abst. Uppsala Diss. Sci. *69*: 1-17, 1966.

4772 - LINDAHL, P.E.B.: On the reversal of the inhibition of photosynthesis induced by sodium dimethyldithiocarbamate and tetramethylthiuram disulphide. - Physiol. Plant. *19*: 87-98, 1966.

4773 - LINDAUEROVÁ, T., STRYCH, N.A.: Chloroplasts and botanical taxonomy. - Acta agron. Acad. Sci. hung. *18*: 412-415, 1968.

4774 - LINGOVA, S., STANEV, V.: Fotosinteticheski aktivna radiatsiya i razvitie na tsa-

revichniya posev v zavisimost ot g'stotata mu. [Photosynthetically active radiation and the dependence of maize stand development on its density.] – Khidrol.
Meteorol. (Sofia) *18* (4): 47–56, 1969. [In Bulg., ab: E, R.]

4775 – LINNANE, A.W., STEWART, P.R.: The inhibition of chlorophyll formation in *Euglena* by antibiotics which inhibit bacterial and mitochondrial protein synthesis.
– Biochem. biophys. Res. Commun. *27*: 511–516, 1967.

4776 – LINTILHAC, P.M., PARK, R.B.: Localization of chlorophyll in spinach chloroplast
lamellae by fluorescence microscopy. – J. Cell Biol. *28*: 582–585, 1966.

4777 – LIPPERT, K.-D., KLEMME, J.-H.: Untersuchungen zum Mechanismus der Photoreduktion
von Pyridinnucleotid durch Chromatophoren aus *Rhodospirillum rubrum*. – Arch. Mikrobiol. *62*: 307–321, 1968.

4778 – LIPPERT, K.-D., PFENNIG, N.: Die Verwertung von molekularem Wasserstoff durch
Chlorobium thiosulfatophilum. Wachstum und CO_2-Fixierung. – Arch. Mikrobiol. *65*:
29–47, 1969.

4779 – LIPS, S.H., ROTH-BEJERANO, N., BEN-ZIONI, A.: Kinetin, leaf age and photosynthesis. – Israel J. Bot. *17*: 130, 1968.

4780 – LIPSKAYA, G.A.: Sovremennoe sostoyanie voprosa ob ul'trastrukture khloroplastov
i ee izmenenii pod vliyaniem faktorov vneshneĭ sredy. [Recent state of knowledge
of chloroplast ultrastructure and its changes with external factors.] – Uspekhi
sovrem. Biol. *65*: 362–383, 1968. [In R, ab: E.]

4781 – LIPSKAYA, G.A.: Nakoplenie khlorofilla v khloroplastakh sakharnoĭ svekly pod deĭstviem kobal'ta, vnesennogo razdel'no i sovmestno s B, Mn, Cu, Zn, Mo. [Accumu-
lation of chlorophyll in the chloroplasts of sugar beet to which cobalt is ap-
plied separately and in combination with boron, manganese, copper, zinc and molybdenum.] – Agrokhimiya *1970* (2): 105–110, 1970. [In R.]

4782 – LIPSKAYA, G.A., ANTIPOVA, A.I.: Fotosinteticheskiĭ apparat sakharnoĭ svekly pri
razlichnom sootnoshenii mikroelementov v pitatel'noĭ srede. [Photosynthetic apparatus of sugar beet plants under different content of microelements in the nutrient medium.] – Fiziol. Rast. *16*: 734–737, 1969. [In R.]

4783 – LIPSKAYA, G.A., FARTOTSKAYA, I.K.: Uzdzeyanne parnykh spaluchennyaŭ kobal'tu z
boram, margantsam, meddzyu, tsynkam i malibdenam na pigmentnuyu sistemu listsyaŭ
agurkoŭ. [Action of new combinations of cobalt with boron, manganese, copper,
zinc and molybdenum on the pigment system of cucumber leaves.] – Vestsi Akad.
Navuk Belarus. SSR, Ser. biyal. Navuk *1970* (5): 57–61, 1970. [In Belorus., ab: R.]

4784 – LIPSKAYA, G.A., KALISHEVICH, S.V.: Uplyŭ malibdenu na nakaplenne pigmentaŭ u
listsyakh tsukrovykh burakoŭ. [Effect of molybdenum on pigment accumulation in
sugar beet leaves.] – Vestsi Akad. Navuk Belarus. SSR, Ser. biyal. Navuk *1969*
(5): 26–29, 1969. [In Belorus., ab: R.]

4785 – LIPSKAYA, G.A., ZHAŬNYAROVICH, N.I.: Nakaplenne khlarafilu *a*, *b* i karatsinoidaŭ
pad dzeyannem Co i Mo u listsi tsukrovykh burakoŭ. [Accumulation of chlorophyll
a and *b* and carotenoids affected by Co and Mo in sugar beet leaves.] – Vestsi
Akad. Navuk Belarus. SSR, Ser. biyal. Navuk *1966* (2): 31–34, 1966. [In Belorus.,
ab: R.]

4786 – LISOVSKIĬ, G.M., YAN, N.A., SYPNEVSKAYA, E.K., NIKOLAĬCHUK, L.P., SAKASH, V.G.,
SHILENKO, M.P.: Produktivnost' razlichnykh form vodorosleĭ pri ikh nepreryvnom
kul'tivirovanii. [Productivity of various forms of algae during their continuous
cultivation.] – In: Upravlyaemyĭ Biosintez. Pp. 276–282. Nauka, Moskva 1966.
[Chl; in R.]

4787 – LISTER, G.R., SLANKIS, V., KROTKOV, G., NELSON, C.D.: Physiology of *Pinus strobus* L. seedlings grown under high or low soil moisture conditions. – Ann. Bot.
31: 121–132, 1967. [Ps.]

4788 – LISTER, G.R., SLANKIS, V., KROTKOV, G., NELSON, C.D.: The growth and physiology
of *Pinus strobus* L. seedlings as affected by various nutritional levels of nitrogen and phosphorus. – Ann. Bot. *32*: 33–43, 1968. [Ps.]

4789 – LITVIN, F.F.: Sistemy nativnykh form fotosinteticheskikh pigmentov i ikh uchastie v protsessakh fotosinteza. [Systems of native forms of photosynthetic pig-

ments and their participation in processes of photosynthesis.] - Nauch. Dokl. vyssh. Shkoly, biol. Nauki *1967* (1): 144-145, 1967. [In R.]

4790 - LITVIN, F.F.: Kompleksnye metody spektral'nogo issledovaniya fotosinteza i pigmentnoĭ sistemy fotosinteziruyushchikh organizmov. [Complex spectral methods of research of photosynthesis and pigment system of photosynthesizing organisms.] - Nauch. Dokl. vyssh. Shkoly, biol. Nauki *1968* (4): 134-146, 1968. [In R.]

4791 - LITVIN, F.F., BELYAEVA, O.B.: Issledovanie fotokhimicheskikh reaktsiĭ biosinteza khlorofilla. [Photochemical reactions of chlorophyll biosynthesis.] - Biokhimiya *33*: 928-936, 1968. [In R, ab: E.]

4792 - LITVIN, F.F., EFIMTSEV, E.I.: Vnutrikletochnoe issledovanie mikroelektrodnym metodom fotosinteza i sopryazhennykh s nim protsessov. [Intracellular studies of photosynthesis and related processes by the microelectrode method.] - Tr. mosk. Obshch. Ispyt. Prirody, Otd. Biol. *28*: 196-199, 1968. [In R, ab: E.]

4793 - LITVIN, F.F., GULYAEV, B.A.: "Krasnoe smeshchenie", i uslozhnenie struktury spektra pogloshcheniya fotosinteziruyushchikh organizmov kak sledstvie sushchestvovaniya sistemy agregirovannykh form pigmentov. [" Red shift" and complication of structure of absorption spectrum in photosynthesizing organisms, as a result of existence of a system of aggregated pigment forms.] - Dokl. Akad. Nauk SSSR *169*: 1187-1190, 1966. [In R.]

4794 - LITVIN, F.F., GULYAEV, B.A.: Proizvodnaya spektrofotometriya i matematicheskiĭ analiz spektrov pogloshcheniya pigmentov v rastitel'noĭ kletke. 1. Eksperimental'nye metody i rezul'taty issledovaniya nativnykh form pigmentov. [Derivative spectrophotometry and mathematical analysis of absorption spectra of pigments in a plant cell. I. Experimental methods and results of studying native forms of pigments.] - Nauch. Dokl. vyssh. Shkoly, biol. Nauki *12* (2): 118-135, 1969. [In R.]

4795 - LITVIN, F.F., GULYAEV, B.A.: Proizvodnaya spektrofotometriya i matematicheskiĭ analiz spektrov pogloshcheniya pigmentov v rastitel'noĭ kletke. II. Matematicheskiĭ analiz spektrov pogloshcheniya, proizvodnykh spektrov i interpretatsiya struktury nativnykh agregirovannykh form khlorofilla. [Derivative spectrophotometry and mathematical analysis of absorption spectra of pigments in a plant cell. II. Mathematical analysis of absorption spectra, derivative spectra and the interpretation of structure of native aggregated chlorophyll forms.] - Nauch. Dokl. vyssh. Shkoly, biol. Nauki *12* (5): 130-139, 1969. [In R.]

4796 - LITVIN, F.F., GULYAEV, B.A.: Razreshenie struktury spektra pogloshcheniya khlorofilla *a* i ego bakterial'nykh analogov v kletke metodom izmereniya vtoroĭ proizvodnoĭ pri 20 i -196°. [Resolution of absorption spectrum structure of chlorophyll *a* and its bacterial analogues in the cell by measuring their second derivative at 20 and -196 °C.] - Dokl. Akad. Nauk SSSR *189*: 1385-1388, 1969. [In R.]

4797 - LITVIN, F.F., GULYAEV, B.A.: Sistema agregirovaniya form bakterial'nykh pigmentov. [System of aggregated forms of bacterial pigments.] - Izv. Akad. Nauk SSSR, Ser. biol. *1970*: 43-52, 1970. [In R, ab: E.]

4798 - LITVIN, F.F., GULYAEV, B.A., KARNEEVA, N.V.: Issledovanie nativnykh form khlorofilla po nizkotemperaturnym proizvodnym spektram pogloshcheniya. [Native forms of chlorophyll studied from low-temperature derivative absorption spectra.] - Nauch. Dokl. vyssh. Shkoly, biol. Nauki *13* (4): 95-105, 1970. [In R.]

4799 - LITVIN, F.F., I-TAN', Kh.: Issledovanie spektrov deĭstviya fotosinteza i effekta Emersona u vysshikh rasteniĭ. [Action spectra of photosynthesis and Emerson effect in higher plants.] - Dokl. Akad. Nauk SSSR *167*: 1187-1190, 1966. [In R.]

4800 - LITVIN, F.F., I-TAN', Kh.: Spektry deĭstviya fotosinteza, effekt Emersona i induktsionnye yavleniya u vysshikh rasteniĭ. [Action spectra of photosynthesis, Emerson effect and induction phenomena in higher plants.] - Fiziol. Rast. *14*: 219-231, 1967. [In R, ab: E.]

4801 - LITVIN, F.F., KRASNOVSKIĬ, A.A. Jr.: Fotokhemilyuminestsentsiya khlorofilla i ego analogov v rastvorakh. [Photochemiluminescence of chlorophyll and its analogues in solutions.] - Dokl. Akad. Nauk SSSR *173*: 451-454, 1967. [In R.]

4802 - LITVIN, F.F., KRASNOVSKIĬ, A.A. Jr., SHUVALOV, V.A.: Izuchenie dlitel'nogo poslesvecheniya list'ev rasteniĭ. [Study of delayed light emission of plant leaves.] – Tr. mosk. Obshch. Ispyt. Prirody, Otd. Biol. *16*: 261-271, 1966. [In R, ab: E.]

4803 – LITVIN, F.F., PERSONOV, R.I., KOROTAEV, O.N.: Tonkaya struktura elektronnykh spektrov khlorofilla v kristallicheskikh rastvorakh pri 4°K. [Fine structure of electron chlorophyll spectra in crystalline solutions at 4 K.] – Dokl. Akad. Nauk SSSR *188*: 1169-1171, 1969. [In R.]

4804 – LITVIN, F.F., SHUVALOV, V.A.: Izuchenie fotosinteticheskikh pigmentnykh sistem po spektram izlucheniya i spektram vozbuzhdeniya khemilyuminestsentsii khlorofilla v vysshikh rasteniyakh. [Photosynthetic pigment systems studied by emission spectra and chemiluminescence action spectra of chlorophyll in higher plants.] – Biokhimiya *31*: 1264-1275, 1966. [In R, ab: E.]

4805 – LITVIN, F.F., SHUVALOV, V.A.: Dlitel'noe poslesvechenie khlorofilla v rasteniyakh i lovushki energii pri fotosinteze. [Long-term afterglow of chlorophyll in plants and energy traps during photosynthesis.] – Dokl. Akad. Nauk SSSR *181*: 733-736, 1968. [In R.]

4806 – LITVIN, F.F., SHUVALOV, V.A., KRASNOVSKIĬ, A.A. Jr.: Uchastie pigmentnykh sistem vysshikh rasteniĭ v protsessakh dlitel'nogo poslesvecheniya. Spektry deĭstviya svecheniya etiolirovannykh i zeleneyushchikh list'ev, effekty usileniya i ingibirovaniya. [Participation of pigment systems of higher plants in processes of long afterglow. Action spectra of luminescence of etiolated and greening leaves, enhancement and inhibition of luminescence.] – Dokl. Akad. Nauk SSSR *168*: 1195-1198, 1966. [In R.]

4807 – LITVIN, F.F., SINESHCHEKOV, V.A.: Dokazatel'stva sushchestvovaniya sistemy agregirovannykh form khlorofilla v rastvore i plenkakh metodom issledovaniya raznostnykh spektrov vozbuzhdeniya i lyuminestsentsii. [Existence of system of aggregated chlorophyll forms in solution and films, proved by investigating differential spectra of excitation and luminescence.] – Dokl. Akad. Nauk SSSR *175*: 1175-1178, 1967. [In R.]

4808 – LITVIN, F.F., SINESHCHEKOV, V.A.: Issledovanie vzaimodeĭstviya form khlorofilla v protsessakh lyuminestsentsii, migratsii energii i fotosinteza metodom spektrov deĭstviya. [Interaction of chlorophyll forms in processes of luminescence, energy migration and photosynthesis studied by the method of action spectra.] – Biofizika *12*: 433-444, 1967. [In R.]

4809 – LITVIN, F.F., SINESHCHEKOV, V.A., SHUVALOV, V.A.: Rol' pigmentnykh sistem i svetovogo faktora v regulirovanii fotosinteza vysshikh rasteniĭ. [Role of pigment systems and light factor in regulation of higher plants photosynthesis.] – Tr. mosk. Obshch. Ispyt. Prirody, Otd. Biol. *28*: 43-62, 1968. [In R, ab: E.]

4810 – LITVIN, F.F., ZVALINSKIĬ, V.I.: Fotoelektricheskie svoĭstva khloroplastov i khromatoforov. [Photoelectric properties of chloroplasts and chromatophores.] – Nauch. Dokl. vyssh. Shkoly, biol. Nauki *10* (1): 146-147, 1967. [In R.]

4811 – LITVIN, F.F., ZVALINSKIĬ, V.I.: Poluprovodnikovye svoĭstva khloroplastov rasteniĭ i khromatoforov fotosinteziruyushchikh bakteriĭ. [Semi-conductor properties of plant chloroplasts and photosynthesizing bacteria chromatophores.] – Biofizika *13*: 241-254, 1968. [In R.]

4812 – LITVINENKO, L.G.: Diya blyz'koĭ infrachervonoĭ radiatsiĭ na zaklyuchnyĭ etap biosyntezu khlorofilu. [Effect of near infra-red radiation in the final period of chlorophyll biosynthesis.] – Dopovidi Akad. Nauk URSR *1966*: 1078-1080, 1966. [In Ukr., ab: E.]

4813 – LITVINENKO, L.G.: Biosyntez khlorofilu u riznykh chastynakh spektru. [Chlorophyll biosynthesis in different parts of the spectrum.] – In: Fotosintez, yak Faktor Pidvyshchennya Urozhayu Sil'skogospodars'kykh Roslyn. Vol. 4. Pp. 18-23. Ukr. Sel'skokhoz. Akad., Kiev 1966. [In Ukr.]

4814 – LITVINENKO, L.G.: Opyt ispol'zovaniya reaktsii Khilla pri izuchenii fotosinteza sel'skokhozyaĭstvennykh rasteniĭ. [Utilization of Hill reaction in studying photosynthesis of agricultural plants.] – In: Fotosintez i Urozhaĭnost' Sel'skokhozyaĭstvennykh Rasteniĭ. Pp. 85-95. Min.sel'. Khoz. SSSR, Kiev 1970. [In R.]

4815 – LITVINENKO, L.G., BIDZILYA, N.I.: Reaktsiya Khilla i obrazovanie paramagnitnykh tsentrov v khloroplastakh razlichnykh kul'tur. [Hill reaction and formation of paramagnetic centres in chloroplasts of different cultures.] – Fiziol. Biokhim. kul't. Rast. 2: 540-543, 1970. [In R, ab: E.]

4816 – LIVINGSTON, A.L., KNOWLES, R.E.: The occurrence of zeinoxanthin in alfalfa. – Phytochemistry 8: 1311-1312, 1969.

4817 – LIVINGSTON, A.L., KNOWLES, R.E., ISRAELSEN, M., NELSON, J.W., MOTTOLA, A.C., KOHLER, G.O.: Xanthophyll and carotene stability during alfalfa dehydration. – J. agr. Food Chem. 14: 643-644, 1966.

4818 – LIVINGSTON, A.L., KNOWLES, R.E., NELSON, J.W., KOHLER, G.O.: Xanthophyll and carotene loss during pilot and industrial scale alfalfa processing. – J. agr. Food Chem. 16: 84-87, 1968.

4819 – LIVINGSTON, A.L., SMITH, D., CARNAHAN, H.L., KNOWLES, R.E., NELSON, J.W., KOHLER, G.O.: Variation in the xanthophyll and carotene content of lucerne, clovers and grasses. – J. Sci. Food Agr. 19: 632-636, 1968.

4820 – LIVINGSTON, R., CONRAD, J.R.: A reversible photochemical reaction of chlorophyll a with I_2. – Photochem. Photobiol. 9: 151-163, 1969.

4821 – LIVINGSTON, R., WONG, H.K.: Quenching of the fluorescent and triplet states of chlorophyll–a safranine. – Photochem. Photobiol. 5: 271-272, 1966.

4822 – LIVNE, A., RACKER, E.: A new coupling factor for photophosphorylation. – Biochem. biophys. Res. Commun. 32: 1045-1049, 1968.

4823 – LIVNE, A., RACKER, E.: Partial resolution of the enzymes catalyzing photophosphorylation. V. Interaction of coupling factor 1 from chloroplasts with ribonucleic acid and lipids. – J. biol. Chem. 244: 1332-1338, 1969.

4824 – LIVNE, A., RACKER, E.: Partial resolution of the enzymes catalyzing photophosphorylation. VI. Interaction of coupling factor 1 from chloroplasts with a new coupling factor for photophosphorylation. – J. biol. Chem. 244: 1339-1344, 1969.

4825 – LJUBEŠIĆ, N.: Ultrastructural changes of chloroplasts during the processes of yellowing (senescence) and regreening (rejuvenation) of leaves. – In: Photochemistry and Photobiology in Plant Physiology. European Photobiol. Symp. Hvar, Yugoslavia, 19th-22nd September 1967. Book of Abstracts. Pp. 51-54. Hvar 1967.

4826 – LJUBEŠIĆ, N.: Feinbau der Chloroplasten während der Vergilbung und Wiederergrünung der Blätter. – Protoplasma 66: 369-379, 1968.

4827 – LJUBEŠIĆ, N.: Fine structure of developing chromoplasts in outer yellow fruit parts of Cucurbita pepo cv. pyriformis. – Acta bot. croat. 29: 51-56, 1970.

4828 – LLOYD, D.: The development of organelles concerned with energy production. – Symp. Soc. gen. Microbiol. 19: 299-332, 1969. [Chloroplasts.]

4829 – LLOYD-JONES, C.P.: Determination of carbon-14 in biological material by flask combustion and suspension scintillation counting of carbon-14 labelled calcium carbonate. – Analyst 95: 366-369, 1970.

4830 – LOACH, K.: Shade tolerance in tree seedlings. I. Leaf photosynthesis and respiration in plants raised under artificial shade. – New Phytol. 66: 607-621, 1967.

4831 – LOACH, K.: Analysis of differences in yield between six sugar-beet varieties. – Ann. appl. Biol. 66: 217-223, 1970. [Growth analysis.]

4832 – LOACH, P.A.: Primary oxidation-reduction changes during photosynthesis in Rhodospirillum rubrum. – Biochemistry 5: 592-600, 1966.

4833 – LOACH, P.A., HADSELL, R.M., SEKURA, D.L., STEMER, A.: Quantitative dissolution of the membrane and preparation of photoreceptor subunits from Rhodospirillum rubrum. – Biochemistry 9: 3127-3135, 1970.

4834 – LOACH, P.A., SEKURA, D.L.: A comparison of decay kinetics of photoproduced absorbance, EPR, and luminescence changes in chromatophores of Rhodospirillum rubrum. – Photochem. Photobiol. 6: 381-393, 1967.

4835 – LOACH, P.A., SEKURA, D.L.: Primary photochemistry and electron transport in Rhodospirillum rubrum. – Biochemistry 7: 2642-2649, 1968.

4836 - LOACH, P.A., SEKURA, D.L., HADSELL, R.M., STEMER, A.: Quantitative dissolution
of the membrane and preparation of photoreceptor subunits from *Rhodopseudomonas
spheroides*. - Biochemistry *9*: 724-733, 1970.

4837 - LOACH, P.A., WALSH, K.: Quantum yield for the photoproduced electron paramagnetic resonance signal in chromatophores from *Rhodospirillum rubrum*. - Biochemistry
8: 1908-1913, 1969.

4838 - LOBOTSKAYA, L.I., PROTASEVICH, R.T., SHCHERBAKOVA, A.M., BORMOTOV, V.E.: K izucheniyu khlorofillonosnoĭ sistemy poliploidnoĭ sakharnoĭ svekly. [Chlorophyll-
bearing system of polyploid sugar beet.] - Dokl. Akad. Nauk Belorus. SSR *12*:
174-176, 1968. [In R.]

4839 - LOEBLICH, A.R. III.: Aspects of the physiology and biochemistry of the *Pyrrho-
phyta*. - Phykos *5*: 216-255, 1966. [Ps, chl, car.]

4840 - LOEBLICH, A.R. III., SMITH, V.E.: Chloroplast pigments of the marine dinoflagellate *Gyrodinium resplendens*. - Lipids *3*: 5-13, 1968.

4841 - LOGAN, K.T.: Adaptations of the photosynthetic apparatus of sun- and shade-
grown yellow birch (*Betula alleghaniensis* BRITT.). - Can. J. Bot. *48*: 1681-1688,
1970.

4842 - LOGAN, K.T., KROTKOV, G.: Adaptations of the photosynthetic mechanism of sugar
maple *(Acer saccharum)* seedlings grown in various light intensities. - Physiol.
Plant. *22*: 104-116, 1969.

4843 - LOGINOV, M.A.: O fotosinteze list'ev razlichnykh yarusov posevov sorgo. [Photo-
synthesis of leaves of different insertion in sorghum crops.] - In: Fotosintez
i Ispol'zovanie Energii Solnechnoĭ Radiatsii. Pp. 19-20. Dushanbe 1967. [In R.]

4844 - LOGINOV, M.A.: O svetovykh krivykh fotosinteza - *Phargmites communis* TRIN.
[Light curves of photosynthesis of *Phragmites communis* TRIN.] - Dokl. Akad. Nauk
Tadzh. SSR *11*: 52-54, 1968. [In R.]

4845 - LOGINOV, M.A., KICHITOV, V.K.: Osobennosti dnevnykh i sezonnykh izmeneniĭ intensivnosti fotosinteza sorgo v zavisimosti ot rezhima orosheniya. [Peculiari-
ties of daily and seasonal changes in photosynthetic rate of *Sorghum* in relation
to irrigation regime.] - Izv. Akad. Nauk Tadzh. SSR, Otd. biol. Nauk *1966* (2 (23))
39-47, 1966. [In R, ab: Tadzh.]

4846 - LOGINOV, M.A., NASYROV, Yu.S.: Zavisimost' svetovykh krivykh fotosinteza khlopchatnika ot temperatury. [Temperature dependence of light curves of cotton photosynthesis.] - In: Issledovaniya po Fotosintezu. Pp. 46-53. Akad. Nauk Tadzh. SSR,
Dushanbe 1967. [In R, ab: Tadzh.]

4847 - LOGINOV, M.A., NASYROV, Yu.S.: Ekologo-fiziologicheskiĭ analiz fotosinteza vy-
sotnozameshchayushchikh vidov osoki Zapadnogo Pamiro-Altaya. [Eco-physiological
analysis of photosynthesis of sedge species growing at high altitude of Western
Pamiro-Altai.] - Izv. Akad. Nauk Tadzh. SSR, Otd. biol. Nauk *1969* (2): 15-20,
1969. [In R.]

4848 - LOGINOV, M.A., NASYROV, Yu.S., KICHITOV, V.K.: O fotosinteze i produktivnosti
nekotorykh introdutsiruemykh kormovykh kul'tur Tadzhikistana. [Photosynthesis
and productivity of some introduced forage crops of Tadzhikistan.] - In: Obshchie
Teoreticheskie Problemy Biologicheskoĭ Produktivnosti. Pp. 103-110. Nauka, Leningrad 1969. [In R.]

4849 - LÖHR, E.: Trehalose, Respiration und Photosynthese in dem Farn *Ophioglossum*.
Ein neuer Schattenblattypus. - Physiol. Plant. *21*: 668-672, 1968.

4850 - LOMAGIN, A.G.: Vliyanie sveta na ustoĭchivost' rastitel'nykh kletok k provrezh-
deniyu. [Influence of light on the resistance of plant cells to damage.] - Uspekhi sovrem. Biol. *67*: 147-163, 1969. [Chl; in R.]

4851 - LOMAGIN, A.G., ANTROPOVA, T.A.: Photodynamic injury to heated leaves. - Planta
68: 297-309, 1966. [Ps, chl.]

4852 - LOMAGIN, A.G., ANTROPOVA, T.A.: Vliyanie kinetina na ustoĭchivost' list'ev tradeskantsii k ul'trafioletovym lucham. [Effect of kinetin on the resistance of
Tradescantia leaves to ultraviolet rays.] - Dokl. Akad. Nauk SSSR *172*: 964-965,
1967. [Chl; in R.]

4853 - LOMNICKI, A., BANDOLA, E., JANKOWSKA, K.: Modification of the Wiegert-Evans
       method for estimation of net primary production. - Ecology *49*: 147-149, 1968.

4854 - LONA, F.: Fitoregolatori abscisino-simili ottenuri per degradazione *in vitro* di
       carotenoidi e retineni. [Abscisin-like phytoregulators obtained by *in vitro* de-
       gradation of carotenoids and retinene.] - Atti Acad. naz. Lincei, Cl. Sci. Fis.,
       Mat., Nat., Rendiconti *46*: 753-758, 1969. [In Ital.]

4855 - LONG, I.F.: Instruments and techniques for measuring the microclimate of crops.
       - In: WADSWORTH, R.M., CHAPAS, L.C., RUTTER, A.L., SOLOMON, M.E., WARREN WILSON,
       J. (ed.): The measurement of Environmental Factors in Terrestrial Ecology. Pp.
       1-32. Blackwell sci. Publ., Oxford-Edinburgh 1968.

4856 - LOOMIS, R.S., WILLIAMS, W.A.: Productivity and the morphology of crop stands:
       patterns with leaves. - In: EASTIN, J.D., HASKINS, F.A., SULLIVAN, C.Y., van
       BAVEL, C.H.M. (ed.): Physiological Aspects of Crop Yield. Pp. 27-51. Amer. Soc.
       Agron. & Crop Sci. Soc. Amer., Madison, Wisc. 1969.

4857 - LOOMIS, R.S., WILLIAMS, W.A., DUNCAN, W.G.: Community architecture and the pro-
       ductivity of terrestrial plant communities. - In: SAN PIETRO, A., GREER, F.A.,
       ARMY, T.J. (ed.): Harvesting the Sun. Pp. 291-308. Academic Press, New York-
       London 1967.

4858 - LOONEY, N.E.: Effect of N-dimethylaminosuccinamic acid on ripening and respira-
       tion of apple fruits. - Can. J. Plant Sci. *47*: 549-553, 1967. [Chl.]

4859 - LOONEY, N.E.: Comparison of photosynthetic efficiency of two apple cultivars
       with their compact mutants. - Proc. Amer. Soc. hort. Sci. *92*: 34-36, 1968.

4860 - LOONEY, N.E.: Light regimes within standard size apple trees as determined
       spectrophotometrically. - Proc. amer. Soc. hort. Sci. *93*: 1-6, 1968.

4861 - LOONEY, N.E., PATTERSON, M.E.: Chlorophyllase activity in apples and bananas
       during the climacteric phase. - Nature *214*: 1245-1246, 1967.

4862 - LOOS, E.: Emerson enhancement at different intensities and ratios of two light
       beams. - Carnegie Inst. Year Book *67*: 520-525, 1969.

4863 - LOOS, E.: An action spectrum for methyl viologen reduction by fractionated
       spinach chloroplasts. - Carnegie Inst. Year Book *68*: 574-578, 1970.

4864 - LOPONITSINA, V.V., NOVOZHILOVA, M.I., KONDRAT'EVA, E.N.: Fotosinteziruyushchie
       bakterii, vydelennye iz Aral'skogo morya. [Photosynthetic bacteria found in the
       Aral sea.] - Mikrobiologiya *38*: 358-363, 1969. [In R, ab: E.]

4865 - LORD, J.M., MERRETT, M.J.: The pathway of glycollate utilization in *Chlorella
       pyrenoidosa*. - Biochim. J. *117*: 929-937, 1970.

4866 - LORD, J.M., MERRETT, M.J.: The regulation of glycollate oxidoreductase with
       photosynthetic capacity in *Chlamydomonas mundana*. - Biochem. J. *119*: 125-127,
       1970.

4867 - LORD, M.J., MERRETT, M.J.: The effect of hydroxymethanesulphonate on photosyn-
       thesis in *Chlorella pyrenoidosa*. - J. exp. Bot. *20*: 743-750, 1969.

4868 - LORENZEN, C.J.: A method for the continuous measurement of *in vivo* chlorophyll
       concentration. - Deep-Sea Res. *13*: 223-227, 1966.

4869 - LORENZEN, C.J.: Vertical distribution of chlorophyll and phaeo-pigments: Baja
       California. - Deep-Sea Res. *14*: 735-745, 1967.

4870 - LORENZEN, C.J.: Carbon chlorophyll relationships in an upwelling area. - Limnol.
       Oceanogr. *13*: 202-204, 1968.

4871 - LORENZEN, H.: Aspects of synchronous culturing of *Chlorella*. - Phykos *7*: 50-57,
       1968. [Ps production.]

4872 - LORENZEN, H.: Synchronous cultures. - In: HALLDAL, P. (ed.): Photobiology of
       Microorganisms. Pp. 187-212. Wiley-Interscience, London-New York-Sydney-Toronto
       1970. [Chl, Ps.]

4873 - LORENZEN, H., HESSE, M.: Nachweis endogen bedingter Leistungsschwankungen bei
       Chlorella. - Z. Pflanzenphysiol. *58*: 454-456, 1968. [Ps.]

4874 - LOSADA, M., APARICIO, P.J., PANEQUE, A.: Separation of two enzyme activities in
       the reduction of nitrate with NADH. - In: METZNER, H. (ed.): Progress in Photo-

synthesis Research. Vol. III. Pp. 1504-1509. Tübingen 1969.

4875 - LOSADA, M., PANEQUE, A.: Light reduction of nitrate by chloroplasts depending on ferredoxin and NAD^+. - Biochim. biophys. Acta 126: 578-580, 1966.

4876 - LOSASA, M.: Nitrate assimilation in a reconstituted chloroplast system. - In: THOMAS, J.B., GOEDHEER, J.C. (ed.): Currents in Photosynthesis. Pp. 431-439. Donker, Rotterdam 1966.

4877 - LOSEV, A.P., GURINOVICH, G.P.: O prirode form vosstanovlennogo khlorofilla. [Nature of forms of reduced chlorophyll.] - Biokhimiya 32: 409-415, 1967. [In R, ab: E.]

4878 - LOSEV, A.P., GURINOVICH, G.P.: Izuchenie migratsii energii i sostoyaniya pigmentov v gomogenatakh zeleneyushchikh etiolirovannykh rasteniĭ. [Energy migration and pigment state in homogenates of greening etiolated leaves.] - Biofizika 14: 110-118, 1969. [In R, ab: E.]

4879 - LOSEV, A.P., SHLYK, A.A.: Sopryazhennost' metabolizma i geterogennosti karotinoidov i khlorofilla. [Coupling of metabolism and heterogeneity of chlorophyll and carotenoids.] - Dokl. Akad. Nauk SSSR 186: 971-974, 1969. [In R.]

4880 - LOSEV, A.P., ZENKEVICH, E.I.: Izuchenie migratsii energii v rastvorakh khlorofillov. [Migration of energy in chlorophyll solutions.] - Zh. prikl. Spektroskop. 9: 144-151, 1968. [In R.]

4881 - LÜTSCH, B.: Lichtatmung und photosynthetische Primärreaktionen. Ihr Verständnis aus der evolutionären Entstehung. - Ber. deut. bot. Ges. 83: 41-54, 1970.

4882 - LOTT, J.N.A.: Changes in the cotyledons of Cucurbita maxima during germination. III. Plastids and chlorophylls. - Can. J. Bot. 48: 2259-2265, 1970.

4883 - LOUWERSE, W., van OORSCHOT, J.L.P.: An assembly for routine measurements of photosynthesis, respiration and transpiration of intact plants under controlled conditions. - Photosynthetica 3: 305-315, 1969.

4884 - LOVELL, P.H., LOVELL, P.J.: Fixation of CO_2 and export of photosynthate by the carpel in Pisum sativum. - Physiol. Plant. 23: 316-322, 1970.

4885 - LOVELL, P.H., MOORE, K.G.: A comparative study of cotyledons as assimilatory organs. - J. exp. Bot. 21: 1017-1030, 1970.

4886 - LØVLIE, A.: increase in function during growth of a single cell organelle. - J. Cell Biol. 43: 389-395, 1969. [Ps.]

4887 - LOWRY, L.K., CHICHESTER, C.O.: ß-ionone and its effect on the incorporation of ^{14}C mevalonate into carotenes and high counting fractions in carrot root. - Phytochemistry 6: 367-370, 1967.

4888 - LOZOVA, G.I.: Rol' endospermu v proyavi vlastyvosteĭ pigmentnoĭ systemy plastyd prorostkiv kukuruzy. [Role of endosperm in manifestation of properties of the pigment system of plastids of maize seedlings.] - Ukr. bot. Zh. 23 (1): 17-22, 1966. [In Ukr., ab: E, R.]

4889 - LOZOVA, G.I., SEREBROVS'KA, K.B.: Vyvchennya fotosensybilizuyuchoĭ aktyvnosti khlorofil-bilkovo-lipoĭdnykh kompleksiv (koatservativ). [Photosensibilizing activity of chlorophyll-protein-lipoid complexes (coacervates).] - Ukr. biokhim. Zh. 39: 78-82, 1967. [In Ukr., ab: E, R.]

4890 - LOZOVA, G.I., SUD'INA, O.G., SAMSONOVA, C.M.: Fotosensibilizatsiĭni vlastyvosti protokhlorofilu v shtuchnykh kompleksakh. [Photosensibilizing properties of protochlorophyll in artificial complexes.] - Ukr. bot. Zh. 24 (4): 3-5, 1967. [In Ukr., ab: E, R.]

4891 - LOZOVA, G.I., VOĬTSEKHOVS'KA, K.Yu.: Spektral'ni ta fotosensybilizatsiĭni vlastyvosti khlorofilidu v shtuchnykh kompleksakh (koatservatakh). [Spectral and photosensibilization properties of chlorophyllide in artificial complexes (coacervates).] - Ukr. bot. Zh. 26 (4): 97-102, 1969. [In Ukr., ab: E, R.]

4892 - LOZOVAYA, G.I.: Rol' endosperma v proyavlenii svoĭstv pigmentnoĭ sistemy plastid prorostkov kukuruzy. [Role of endosperm in manifestation of properties of the plastid pigment system of maize seedlings.] - In: Materialy IV Konferentsii Molodykh Uchenykh Moldavii 1964, Sekts. Fiziol., Biokhim. i Genet. Rast. Pp. 98-99. Kishinev 1966. [In R.]

4893 - LOZOVAYA, G.I.: Modelirovanie fotoprotsessov v koatservatnykh sistemakh. [Model-ling of photoprocesses in coacervate systems.] - In: Tezisy Dokladov II Vsesoyuznogo Biokhimicheskogo S"ezda. Pp. 13-15. FAN, Tashkent 1969. [In R.]

4894 - LOZOVAYA, G.I.: Vydelenie pigment-belkovykh kompleksov iz khloroplastov. [Isola-tion of pigment-protein complexes from chloroplasts.] - In: KIRICHENKO, E.B. (ed.): Metody Vydeleniya Khloroplastov. Pp. 64-76. Pushchino-na-Oke 1970. [In R, ab: E.]

4895 - LOZOVAYA, G.I., SEMICHAEVSKIĬ, V.D.: Zavisimost' funktsional'noĬ aktivnosti khlorofilla ot komponentov iskusstvennykh pigmentsoderzhashchikh kompleksov. [Dependence of functional activity of chlorophyll on the components of artifi-cial complexes containing pigments.] - In: Tezisy Dokladov II Vsesoyuznogo Bio-khimicheskogo S"ezda. 19. Sektsiya: Problemy Fotosinteza. Pp. 76-77. FAN, Tash-kent 1969. [In R.]

4896 - LUCENA CONDE, F., SANCHEZ DE LA PUNTE, L.: Determinacion del equilibrio mineral de la planta mediante la espectrofotometria de sus pigmentos organicos. [Spec-trophotometry of organic pigments for the determination of mineral equilibrium in plants.] - In: Atti VI Simp. int. agrochimica. Pp. 371-382. Varenna 1966. [Chl; in Span., ab: E, F, G, Ital.]

4897 - LUCERO, D.P.: Performance characteristics of membrane-covered polarographic gas detectors. - Anal. Chem. 40: 707-711, 1968.

4898 - LUCKWILL, L.C., CASELEY, J.C.: The effect of herbicides on fruit plants. - In: FRYER, J.D. (ed.): Herbicides in British Fruit Growing. Pp. 81-100. Blackwell sci. Publ., Oxford 1966. [Ps.]

4899 - LUDLOW, C.J.: Photosynthesis in shade and sun ferns. - Diss. Abstr. B 28: 4022-B-4023-B, 1968.

4900 - LUDLOW, C.J., PARK, R.B.: Action spectra for photosystems I and II in formalde-hyde fixed Anacystis nidulans. - Plant Physiol. 43: S-29, 1968.

4901 - LUDLOW, C.J., PARK, R.B.: Action spectra for photosystems I and II in formalde-hyde fixed algae. - Plant Physiol. 44: 540-543, 1969.

4902 - LUDLOW, C.J., WOLF, F.T.: Photosynthesis in relation to chlorophyll content, stomatal distribution and Hill reaction activity in shade and sun ferns. - ASB Bull. 14 (2): 32, 1967.

4903 - LUDLOW, C.J., WOLF, F.T.: Photosynthesis in shade and sun ferns in relation to chlorophyll content, stomatal distribution and Hill reaction activity. - Plant Physiol. 42: S-46, 1967.

4904 - LUDLOW, M.M.: Effect of oxygen concentration on leaf photosynthesis and resis-tances to carbon dioxide diffusion. - Planta 91: 285-290, 1970.

4905 - LUDLOW, M.M., WILSON, G.L.: Studies on the productivity of tropical pasture plants. I. Growth analysis, photosynthesis, and respiration of Hamil grass and Siratro in a controlled environment. - Aust. J. agr. Res. 19: 35-45, 1968.

4906 - LUDLOW, M.M., WILSON, G.L.: Studies on the productivity of tropical pasture plants. II. Growth analysis, photosynthesis, and respiration of 20 species of grasses and legumes in a controlled environment. - Aust. J. agr. Res. 21: 183-194, 1970.

4907 - LUDWIG, J., KROTKOV, G., CANVIN, D.T.: The relationship of the products of pho-tosynthesis to the substrates for CO_2-evolution in light and in darkness. - In: METZNER, H. (ed.): Progress in Photosynthesis Research. Vol. I. Pp. 494-502. Tübingen 1969.

4908 - LUGANSKAYA, A.N., KRASNOVSKIĬ, A.A.: Issledovanie mekhanizma fotosensibiliziro-vannykh khlorofillom okislitel'no-vosstanovitel'nykh reaktsiĬ v prisutstvii kisloroda. [Mechanism of chlorophyll-photosensitized oxido-reduction reactions in the presence of oxygen.] - Mol. Biol. (Moskva) 4: 848-859, 1970. [In R, ab: E.]

4909 - LUKINA, G.A.: DeĬstvie fenola na fotosintez i dykhanie khlorelly. [Action of phenol on Chlorella photosynthesis and respiration.] - Tr. Inst. biol. vnutr. vod Akad. Nauk SSSR 1969 (19 (22)): 87-89, 1969. [In R.]

4910 - LUKINA, G.A.: DeТstvie fenola na fotosintez i dykhanie khlorelly. [Action of
phenol on *Chlorella* photosynthesis and respiration.] - In: Voprosy VodnoТ Tok-
sikologii. Pp. 183-185. Nauka, Moskva 1970. [In R.]

4911 - LUKINA, G.A., SINEL'NIKOV, V.E.: Vliyanie fotosinteticheskoТ deyatel'nosti vo-
dorosleТ na ingibitory svobodnoradikal'nykh reaktsiТ. [Effect of photosynthetic
activity of algae on inhibitors of reactions of free radicals.] - Gidrobiol.
Zh. *5* (2): 44-50, 1969. [In R.]

4912 - LUKMANOV, F.G., MAZIL'NIKOV, G.V., SAKHIPOV, R.T.: Metabolizm ugleroda v protsesse fotosinteza u gorokha v zavisimosti ot usloviТ vneseniya kaliТnykh udo-
breniТ. [Carbon metabolism in pea photosynthesis in relation to application of
potassium nutrients.] - In: Mineral'nye Elementy i Mekhanizm Fotosinteza. Pp.
49-53. Kishinev 1969. [In R.]

4913 - LUKOVNIKOVA, G.A., GLUSHCHENKO, E.Ya.: Karotinoidnye pigmenty tomatov. [Carote-
noid pigments of tomato.] - Dokl. VASKHNIL *1966* (8): 19-21, 1966. [In R.]

4914 - LUK'YANOV, V.: Solnechnaya radiatsiya v kronakh yablon'. [Solar radiation in
apple crowns.] - In: Fotosinteticheskaya Produktivnost' Rastitel'nogo Pokrova.
Pp. 186-197. Tartu 1969. [In R.]

4915 - LUK'YANOV, V.M., DENISOV, A.M.: Metodika opredeleniya svetovogo rezhima v kronakh plodovykh derev'ev. [Methods of light regime evaluation in crowns of fruit
trees.] - Sel'skokhoz. Biol. *3*: 582-584, 1968. [In R.]

4916 - LULL, H.W., REIGNER, I.C.: Radiation measurements by various instruments in the
open and in the forest. - U.S. Forest Serv. Res. Pap. NE-84: 1-21, 1967.

4917 - LUNDEGÅRDH, H.: Action spectra of photosynthetic activities of *Chlorella ellip-
soidea.* - Physiol. Plant. *19*: 541-553, 1966.

4918 - LUNDEGÅRDH, H.: Action spectra and the role of carotenoids in photosynthesis.
- Physiol. Plant. *19*: 754-769, 1966.

4919 - LUNDEGÅRDH, H.: β-carotene as a photoreductone for ferredoxin and triphosphopyridine nucleotide *in vitro.* - Nature *212*: 606-608,1966.

4920 - LUNDEGÅRDH, H.: The role of carotenoids in the photosynthesis of green plants.
- Proc. nat. Acad. Sci. U.S.A. *55*: 1062-1065, 1966.

4921 - LUNDEGÅRDH, H.: Role of carotenoids in photosynthesis of green plants. - Nature
216: 981-985, 1967.

4922 - LUNDEGÅRDH, H.: The systems I, II and III in the photosynthetic cycle of electron transfer. - Physiol. Plant. *21*: 148-167, 1968.

4923 - LUNDEGÅRDH, H.: Relative quantum efficiency of photosynthetic oxygen production
in chloroplasts of spinach. - Nature *221*: 280-281, 1969.

4924 - LUNDEGÅRDH, H.G.: Relative quantum efficiency of visible and infrared light on
photosynthetic O_2 production. - Proc. nat. Acad. Sci. U.S.A. *59*: 293-295, 1968.

4925 - LUND-HÖIE, K.: The effect of simazine on the photofixation of CO_2 and on translocation of assimilates in Norway spruce *(Picea abies).* - Weed Res. *9*: 185-191,
1969.

4926 - LUNTS, A.M.: O periodichnosti deleniТ u *Eudorina elegans* EHRBG. [Periodicity of
cell division in *Eudorina elegans* EHRBG.] - Zh. obshch. Biol. *29*: 250-251, 1968.
[Ps; in R.]

4927 - LUPTON, F.G.H.: Translocation of photosynthetic assimilates in wheat. - Ann.
appl. Biol. *57*: 355-364, 1966.

4928 - LUPTON, F.G.H.: Physiological parameters of yield in wheat. - Eucarpia, 5th
Congr. europe. Ass. Res. Plant Breeding *5*: 457-464, 1968. [Ps.]

4929 - LUPTON, F.G.H.: The analysis of grain yield of wheat in terms of photosynthetic
ability and efficiency of translocation. - Ann. appl. Biol. *61*: 109-119, 1968.

4930 - LUPTON, F.G.H.: Estimation of yield in wheat from measurements of photosynthesis
and translocation in the field. - Ann. appl. Biol. *64*: 363-374, 1969.

4931 - LUPTON, F.G.H., ALI, M.A.M.: Studies on photosynthesis in the ear of wheat. -
Ann. appl. Bot. *57*: 281-286, 1966.

4932 - LUPTON, F.G.H., ALI, M.A.M., SUBRAMANIAM, S.: Varietal differences in growth pa-
       rameters of wheat and their importance in determining yield. - J. agr. Sci. *69*:
       111-123, 1967.

4933 - LÜRSSEN, K.: Volumen, Trockengewicht und Zusammensetzung von Etioplasten und
       ihren Ergrünungsstadien. - Z. Naturforsch. *25 b*: 1113-1119, 1970.

4934 - LUTSISHINA, E.G., GRODZINSKIĬ, D.M.: Nativnye plastokhinony v fotosinteticheskom
       fosforilirovanii izolirovannykh khloroplastov. [Natural plastoquinones in photo-
       synthetic phosphorylation in isolated chloroplasts.] - In: Puti Povysheniya In-
       tensivnosti i Produktivnosti Fotosinteza. Vol. 3. Pp. 151-156. Naukova Dumka,
       Kiev 1969. [In R.]

4935 - LUTSISHINA, E.G., GRODZINSKIĬ, D.M.: Primenenie teorii misheni pri izuchenii
       fotosinteticheski aktivnykh edinits khloroplastov. II. Reaktsiya fotovosstanov-
       leniya NADF. [Use of the target theory in studying photosynthetically active
       chloroplast units. II. Reaction of NADP photoreduction.] - Tsitol. Genet. *4*:60-
       63, 1970. [In R.]

4936 - LÜTGGE, U.: Die Photosynthese-abhängige Ionenaufnahme durch die grünen Zellen
       von Luft-Blättern höherer Pflanzen. - Ber. deut. bot. Ges. *83*: 473-479, 1970.

4937 - LÜTGGE, U., PALLAGHY, C.K.: Light triggered transient changes of membrane poten-
       tials in green cells in relation to photosynthetic electron transport. - Z.
       Pflanzenphysiol. *61*: 58-67, 1969.

4938 - LÜTTGE, U., PALLAGHY, C.K., OSMOND, C.B.: Coupling of ion transport in green
       cells of *Atriplex spongiosa* leaves to energy sources in the light and in the
       dark. - J. Membrane Biol. *2*: 17-30, 1970.[Ps.]

4939 - LWOWSKI, W.: The synthesis of chlorophyll *a*. - In: VERNON, L.P., SEELY, G.R.
       (ed.): The Chlorophylls. Pp. 119-153. Academic Press, New York-London 1966.

4940 - LYAKHNOVICH, Ya.P.: Osobennosti pigmentnoĭ sistemy nekotorykh burykh mutantov
       khlorelly. [Properties of pigment system of some brown mutants of *Chlorella*.]
       - In: Metabolizm i Stroenie Fotosinteticheskogo Apparata. Pp. 164-172. Nauka
       i Tekhnika, Minsk 1970. [In R.]

4941 - LYAKHNOVICH, Ya.P., GODNEV, T.N.: O roste, razmnozhenii i nakoplenii pigmentov
       v suspenzii khlorelly, obrabotannoĭ mnogokratno korotkimi vysokotemperaturnymi
       impul'sami pri vyrashchivanii v razlichnykh usloviyakh osveshcheniya i aeratsii.
       [Growth, propagation and pigment accumulation in *Chlorella* suspension many times
       treated with short high-temperature impulses during cultivation under various
       conditions of light and aeration.] - In: Issledovaniya po Fiziologii i Biokhimii
       RasteniĬ. Pp. 21-24. Nauka i Tekhnika, Minsk 1966. [In R.]

4942 - LYAKHNOVICH, Ya.P., KOLESHKO, O.I.: K voprosu ispol'zovaniya sokovykh vod v
       smeshannykh kul'turakh bakteriĬ i khlorelly. [Utilization of juice waters in
       mixed cultures of bacteria and *Chlorella*.] - In: Fiziologo-biokhimicheskie Iss-
       ledovaniya RasteniĬ. Pp. 21-26. Nauka i Tekhnika, Minsk 1967. [Chl; in R.]

4943 - LYAKHNOVICH, Ya.P., VECHER, A.S., GODNEV, T.N., ORLOVSKAYA, K.I.: Rost i nako-
       plenie pigmentov u khlorelly na srede Tamiya s dobavleniem kartofel'nogo soka.
       [Growth and pigment accumulation in *Chlorella* on the Tamiya medium with added
       potato juice.] - In: Botanika. Issledovaniya *9*. Pp. 3-9. Nauka i Tekhnika,
       Minsk 1967. [In R.]

4944 - LYALIN, O.O., PASICHNYĬ, A.P.: Sravnitel'noe izuchenie bioelektricheskoĭ reakt-
       sii lista rasteniya na deĭstvie CO_2 i sveta. [Comparative study of bioelectric
       reaction of a plant leaf on the action of CO_2 and light.] - Dokl. Akad. Nauk
       SSSR *188*: 1402-1404, 1969. [In R.]

4945 - LYAPSHINA, Z.F.: Vliyanie temperatury v period formirovaniya i naliva zerna na
       produktivnost' yarovoĭ pshenitsy. [Effect of temperature during grain formation
       and filling on productivity of spring wheat.] - Fiziol. Rast. *13*: 327-331, 1966.
       [In R, ab: E.]

4946 - LYAPSHINA, Z.F.: Zavisimost' velichiny urozhaya zerna ot razmerov listovoĭ pov-
       erkhnosti i nakopleniya sukhogo veshchestva v ontogeneze myagkoĭ yarovoĭ pshe-
       nitsy. [Dependence of grain yield on leaf area dimensions and dry matter accumu-
       lation during ontogenesis of soft spring wheat.] - Fiziol. Rast. *14*: 70-74,
       1967. [In R, ab: E.]

4947 - **LYAPSHINA, Z.F.**: Zavisimost' urozhaya zerna ot produktivnosti fotosinteza yaro-voĭ pshenitsy. [Dependence of grain yield on productivity of photosynthesis in spring wheat.] - In: Vazhneĭshie Problemy Fotosinteza v Rastenievodstve. Pp. 161-170. Kolos, Moskva 1970. [In R.]

4948 - **LYASHCHENKO, I.F.**: O kharaktere spontannogo mutirovaniya khlorofil'nykh mutatsiĭ podsolnechnika. [Character of spontaneous mutability of sunflower chlorophyll mutants.] - Genetika *1966* (5): 85-88, 1966. [In R.]

4949 - **LYASHCHENKO, I.F., VILOR, T.A.**: Gibridizatsiya razlichnykh tipov khlorofil'nykh mutatsiĭ podsolnechnika mezhdu soboĭ i zelenymi rasteniyami. [Hybridization be-tween different types of sunflower chlorophyll mutants *inter se* and with green plants.] - Genetika *5* (5): 174-176, 1969. [In R, ab: E.]

4950 - **LYASHCHENKO, I.I.**: Vliyanie privivok na pigmentnyĭ sostav al'binosnykh rasteniĭ, privitykh na zelenye rasteniya podsolnechnika. [Effect of grafts on pigment com-position of albino plants grafted on green sunflower plants.] - Genetika *1966* (8): 25-29, 1966. [In R., ab: E.]

4951 - **LYFORD, J.H. Jr., PHINNEY, H.K.**: Primary productivity and community structure of an estuarine impoundment. - Ecology *49*: 854-866, 1968.

4952 - **LYNN, R., BROCK, T.D.**: Notes on the ecology of a species of *Zygogonium* (KÜTZ.) in Yellowstone National Park. - J. Phycol. *5*: 181-185, 1969. [Ps.]

4953 - **LYNN, W.S.**: Inhibition of photophosphorylation by phenazine methosulfate. - J. biol. Chem. *242*: 2186-2191, 1967.

4954 - **LYNN, W.S.**: Changes in internal hydrogen ion concentration associated with photophosphorylation in intact and sonically treated chloroplasts. - J. biol. Chem. *243*: 1060-1064, 1968.

4955 - **LYNN, W.S.**: H^+ and electron poising and photophosphorylation in chloroplasts. - Biochemistry *7*: 3811-3820, 1968.

4956 - **LYNN, W.S., BROWN, R.H.**: $P/2e^-$ ratios approaching 4 in isolated chloroplasts. - J. biol. Chem. *242*: 412-417, 1967.

4957 - **LYNN, W.S., BROWN, R.H.**: P : O and ADP : O ratios and quantum yields in chloro-plasts, with the use of chloranil as electron acceptor. - J. biol. Chem. *242*: 418-425, 1967.

4958 - **LYNN, W.S., BROWN, R.H.**: Competition between phosphate and protons on phosphor-ylation and cation exchange in chloroplasts. - J. biol. Chem. *242*: 426-432, 1967.

4959 - **LYNN, W.S., STRAUB, K.D.**: ADP kinase and ATPase in chloroplasts. - Proc. nat. Acad. Sci. U.S.A. *63*: 540-547, 1969.

4960 - **LYNN, W.S., STRAUB, K.D.**: Isolation and properties of a protein from chloroplasts required for phosphorylation and H^+ uptake. - Biochemistry *8*: 4789-4793,1969.

4961 - **LYNN CO, D.Y.-C.**: Detection, isolation and characterization of chlorophylls and related pigments during ripening of fruits and vegetable. - Diss. Abstr. B *28*: 227-B, 1967.

4962 - **LYNN CO, D.Y.C., SCHANDERL, S.H.**: Separation of chlorophylls and related plant pigments by two-dimensional thin layer chromatography. - J. Chromatogr. *26*: 442-448, 1967.

4963 - **LYNN CO, D.Y.C., SCHANDERL, S.H.**: The occurrence of 418 and 444 nm chlorophyll-type compounds in some green plant tissues. - Phytochemistry *6*: 145-148, 1967.

4964 - **LYSEK, G.**: Effect of photophosphorylation and substrate uptake on the metabolism of polyphosphates in *Ankistrodesmus braunii*. - In: METZNER, H. (ed.): Progress in Photosynthesis Research. Vol. III. Pp. 1149-1154. Tübingen 1969.

4965 - **LYSEK, G., SIMONIS, W.**: Substrataufnahme und Phosphatstoffwechsel bei *Ankistro-demus braunii*. I. Beteiligung der Polyphosphate an der Aufnahme von Glucose und 2-Desoxy-glucose im Dunkeln und im Licht. - Planta *79*: 133-145, 1968.

4966 - **LYSEK, G., SIMONIS, W.**: Substrataufnahme und Phosphatstoffwechsel bei *Ankistro-desmus braunii*. II. Einfluss der Glykolsäure auf den Einbau ^{32}P-markierter Or-thophosphorsäure im Dunkeln und im Licht. - Planta *79*: 319-323, 1968.

4967 - LYTTLETON, J.W.: Use of colloidal silica in density gradients to separate intact chloroplasts. - Anal. Biochem. *38*: 277-281, 1970.

4968 - LYUBIMOVA, E.E., MIKHAÏLICHENKO, B.P., SHATILOV, I.S.: Vliyanie mikroelementov na fotosinteticheskiÏ apparat travostoya kul'turnogo pastbischa. [Effect of microelements on photosynthetic apparatus of grasses of cultivated pasture.] - Izv. timiryaz. sel'skokhoz. Akad. *1969* (3): 77-83, 1969. [In R.]

4969 - LYUBIMOVA, E.E., SHATILOV, I.S.: Vzaimosvyaz' mineral'nogo pitaniya i fotosinte-ticheskoÏ deyatel'nosti pastbishchnykh trav. [Interrelation of mineral nutrition and photosynthetic activity of pasture grasses.]- Izv. timiryaz. sel'skokhoz. Akad. *1969* (2): 43-50, 1969. [In R.]

4970 - LYUTOVA, M.I., KISLYUK, I.M., AGEEVA, O.G.: Voprosy interpretatsii rezul'tatov issledovaniya deÏstviya temperatur na fotokhimicheskuyu aktivnost' khloroplastov. [Interpretation of results of studying temperature action on photochemical ac-tivity of chloroplasts.] - In: KIRICHENKO, E.B. (ed.): Metody Issledovaniya Fo-tofosforilirovaniya. Pp. 167-181. Pushchino-na-Oke 1970. [In R, ab: E.]

4971 - MacCOLL, D., COOPER, J.P.: Climatic variation in forage grasses III. Seasonal changes in growth and assimilation in climatic races of *Lolium, Dactylis* and *Festuca*. - J. appl. Ecol. *4*: 113-127, 1967.

4972 - MACDOWALL, F., WALKER, M.: Fluorescence lifetime of phycoerythrin. - Photochem. Photobiol. *7*: 109-111, 1968.

4973 - MACDOWALL, F.D.H., BEDNAR, T., ROSENBERG, A.: Conformation dependence of intra-molecular energy transfer in phycoerythrin. - Proc. nat. Acad. Sci. U.S.A. *59*: 1356-1363, 1968.

4974 - MACFADYEN, A.: Simple methods for measuring and maintaining the proportion of carbon dioxide in air, for use in ecological studies of soil respiration. - Soil Biol. Biochem. *2*: 9-18, 1970.

4975 - MACHOLD, O.: Untersuchungen an stoffwechseldefekten Mutanten der Kulturtomate. II. Einfluss des Eisenstoffwechsels auf die Ausbildung des Chlorophylldefekts. - Flora A *157*: 183-199, 1966.

4976 - MACHOLD, O.: Untersuchungen an stoffwechseldefekten Mutanten der Kulturtomate. III. Die Wirkung von Ammonium- und Nitratstickstoff auf den Chlorophyllgehalt. - Flora A *157*: 536-551, 1967.

4977 - MACHOLD, O.: Untersuchungen an stoffwechseldefekten Mutanten der Kulturtomate. IV. Einfluss des Redoxpotentials der Nährlösung auf den Chlorophyllgehalt ver-schiedener Mutanten. - Kulturpflanze *15*: 75-83, 1967.

4978 - MACHOLD, O.: Einfluss der Ernährungsbedingungen auf den Zustand des Eisens in den Blättern, den Chlorophyllgehalt und die Katalase- sowie Peroxydaseaktivität. - Flora A *159*: 1-25, 1968.

4979 - MACHOLD, O.: Protein- und Chlorophyll-Synthese bei verändertem Eisenstoffwech-sel. - Agrochimica *13*: 64-74, 1969.

4980 - MACHOLD, O., GRÜBER, K.: Untersuchungen an stoffwechseldefekten Mutanten der Kulturtomate. I. Beziehungen zwischen Kalium: Calzium-Verhältnis und Chloro-phyllgehalt. - Flora A *157*: 170-182, 1966.

4981 - MACHOLD, O., SCHOLZ, G.: Eisenhaushalt und Chlorophyllbildung bei höheren Pflan-zen. - Naturwissenschaften *56*: 447-452, 1969.

4982 - MACHOLD, O., STEPHAN, U.W.: The function of iron in porphyrin and chlorophyll biosynthesis. - Phytochemistry *8*: 2189-2192, 1969.

4983 - MACKENDER, R.O., LEECH, R.M.: Isolation of chloroplast envelope membranes. - Nature *228*: 1347-1349, 1970.

4984 - MACKINNEY, G.: Carotenoid patterns in fruits. - Qual. Plant. Mater. veg. *13*: 228-235, 1966.

4985 - MACKINNEY, G.: Recherches sur les pigments des chloroplastes. - Compt. rend. Acad. agr. France *53*: 1209-1216, 1967.

4986 - MACKINNEY, G.: Sur les carotenoïdes des plantes. - Fruits 23: 569-571, 1968.

4987 - MACLEAN, F.I., FUJITA, Y., FORREST, H.S., MYERS, J.: Stimulation of photophosphorylation and cytochrome c photooxidation by pteridines. - Plant Physiol.
41: 774-779, 1966.

4988 - MADGWICK, H.A.I.: Seasonal changes in biomass and annual production of an old-
field Pinus virginiana stand. - Ecology 49: 149-152, 1968.

4989 - MADGWICK, H.A.I.: Caloric values of Pinus virginiana as affected by time of
sampling, tree age, and position in stand. - Ecology 51: 1094-1097, 1970.

4990 - MADGWICK, J.C.: Chromatographic determination of chlorophylls in algal structures and phytoplankton. - Deep-Sea Res. 13: 459-466, 1966.

4991 - MADSEN, E.: Effect of CO_2-concentration on the accumulation of starch and sugar
in tomato leaves. - Physiol. Plant. 21: 168-175, 1968. [Ps.]

4992 - MADSEN, E.: CO_2-koncentrationens indflydelse på udvikling og tørstofproduktion
hos unge tomatplanter. [Effect of CO_2 concentrations on development and dry
matter production of tomato plants.] - Horticultura 24 (1): 1-4, 1970. [In Danish,
ab: E.]

4993 - MADSEN, E.: Morfologiske og anatomiske aedringer hos tomatplanter opvokset ved
forskellige CO_2-koncentrationer. [Morphological and anatomical changes of tomato
plants grown under increased CO_2 concentrations.] - Horticultura 24 (1): 4-7,
1970. [In Danish, ab: E.]

4994 - MADSEN, E.: CO_2-koncentrationens indflydelse på fotosyntesen hos tomat. [CO_2
concentrations affecting tomato photosynthesis.]- Horticultura 24 (1): 8-10,
1970. [In Danish, ab: E.]

4995 - MAEDA, O.: On the dry matter productivity of two ferns, Osmunda cinnamomea and
Dryopteris crassirhizoma, in relation to their geographical distribution in
Japan. - Jap. J. Bot. 20: 237-267, 1970.[Ps.]

4996 - MAESTRINI, S.: Construction d'un appareil permettant l'étude de la productivité
d'une culture abactérienne (*) d'algues planctoniques, soumises à des éclairements d'énergie constante. - Recueil Trav. Sta. mar. Endoume 1966(57): 25-32,
1966.

4997 - MAFFLY, R.H.: A conductometric method for measuring micromolar quantities of
carbon dioxide. - Anal. Biochim. 23: 252-262, 1968.

4998 - MAGGIONI, A., PASSERA, C., RENOSTO, F.: The effect of chloramphenicol on the
photosynthesis in Chlorella vulgaris. - Ricerca sci. 39: 696-700, 1969.

4999 - MAGGIORA, G.M., INGRAHAM, L.L.: Chlorophyll triplet states. Some theoretical
considerations on triplet formation. - Structure Bonding 2: 126-159, 1967.

5000 - MAGGS, D.H., ALEXANDER, D.McE.: Tests of a uranyl oxalate light integrator for
use in fruit tree canopies. - J. appl. Ecol. 7: 639-646, 1970.

5001 - MAGREE, L., HENNINGER, M.D., CRANE, F.L.: Electron transport in chloroplasts.
II. Effect of hydrocarbon solvent extraction on chloroplast membrane structure.
- J. biol. Chem. 241: 5197-5200, 1966.

5002 - MAHER, F.J.: A sensitive thermometer. - J. sci. Instrum. (J. Phys. E) Ser. 2,
1: 584-585, 1968.

5003 - MAHLBERG, P.G., VENKETESWARAN, S.: Pigment analysis of normal and proliferated
genetical strains of Nicotiana under cultural conditions. - Bot. Gaz. 127: 114-
119, 1966.

5004 - MAÏSEENKA, I.F.: Uplyŭ vodnaga rezhymu tarfyana-balotnaĬ gleby na kol'kasts'
asimilyuyuchykh pigmentaŭ u sasny. [Effect of water regime of peat-bog soil on
the amount of assimilatory pigments in pine.] - Vestsi Akad. Navuk Belarus. SSR,
Ser. biyal.Navuk 1969 (4): 22-25, 123, 1969. [In Belorus., ab: R.]

5005 - MAJAK, W., CRAIGIE, J.S., McLACHLAN, J.: Photosynthesis in algae. I. Accumulation
products in the Rhodophyceae. - Can. J. Bot. 44: 541-549, 1966.

5006 - MAJUMDER, S.K.: Heat stability of chlorophyll as an index of adaptation for over-
wintering. - Biochem. Physiol. Pflanzen 161: 174-177, 1970.

5007 - MAKARENKO, K.I., GRODZINSKIĬ, D.M., OKANENKO, A.S.: K voprosu o perekhodnykh protsessakh fotosinteza. [Transitional processes of photosynthesis.] - Biofizika 12:
730-732, 1967. [In R.]

5008 - MAKAROV, A.D.: II. Ob osobennostyakh kinetiki fotovosstanovleniya NADF khloroplastami v prisutstvii metilviologena. [II. Peculiarities of kinetics of NADP
photoreduction by chloroplasts in the presence of methylviologen.] - Biofizika
14: 1121-1122, 1969. [In R, ab: E.]

5009 - MAKAROV, A.D., MAL'YAN, A.N., LEBEDEVA, A.I., KUZNETSOV, V.P.: O svyazi strukturnykh izmeneniĭ khloroplastov s mekhanizmom fotofosforilirovaniya. [Relation
between structural changes of chloroplasts and the mechanism of photophosphorylation.] - Dokl. Akad. Nauk SSSR 193: 1201-1203, 1970. [In R.]

5010 - MAKAROV, A.D., MAL'YAN, A.N., OPANASENKO, V.K., KARTASHOV, I.M.: O nekotorykh
aspektakh pervichnogo vzaimodeĭstviya komponentov protsessa fotofosforilirovaniya
s khloroplastami. [Some aspects of the primary interaction of photophosphorylation components with chloroplasts.] - In: KIRICHENKO, E.B. (ed.): Metody Issledovaniya Fotofosforilirovaniya. Pp. 151-166. Pushchino-na-Oke 1970. [In R, ab: E.]

5011 - MAKAROV, A.D., STAKHOV, L.F.: O khimicheskoĭ prirode kofaktora fotofosforilirovaniya - fosfodoksina. [Chemical nature of phosphodoxin, a photophosphorylation
co-factor.] - Dokl. Akad. Nauk SSSR 191: 237-239, 1970. [In R.]

5012 - MAKAROV, A.D., SUROVTSEV, V.I.: O nekotorykh osobennostyakh kinetiki fotovosstanovleniya NADF khloroplastami. [Some properties of the kinetics of NADP photoreduction in chloroplasts.] - Biofizika 14: 653-657, 1969. [In R, ab: E.]

5013 - MAKEDONSKA, Ts.: Vliyanie na zasushavaneto v'rkhu nyakoi produkti na fotosintezata v lista na *Pinus silvestris* L. [Effect of dry period on photosynthates in
Pinus sylvestris L. needles.] - Izv. bot. Inst. b"lg. Akad. Nauk 18: 79-87,
1968. [In Bulg.]

5014 - MAKEDONSKA, Ts.: Photosynthesis activity and synthesis of starch and proteins
in isolated white pine needles in the course of wilting. - Dokl. bolg. Akad.
Nauk 21: 929-932, 1968.

5015 - MAKEDONSKA, Ts., YORDANOV, I.: Seasonal changes in the intensity and products
of photosynthesis in some poplars. - Dokl. bolg. Akad. Nauk 22: 1075-1078, 1969.

5016 - MAKEEVA-GUR'YANOVA, L.T., CHKANIKOV, D.I.: Nekotorye dopolnitel'nye dannye o
vozdeĭstvii gerbitsidov-proizvodnykh mocheviny na fotosinteticheskie protsessy.
[The effect of urea herbicides on photosynthetic processes.] - Agrokhimiya 1968
(3): 93-98, 1968. [In R.]

5017 - MAKHARINETS, S.N.: Sostoyanie pigmentnoĭ sistemy i fotokhimicheskaya aktivnost'
khloroplastov ozimoĭ pshenitsy. [State of pigment system and photochemical activity of chloroplasts of winter wheat.] - In: Izuchenie Fotosinteza Odnoletnikh
Rasteniĭ. Pp. 93-104. Akad. Nauk Mold. SSR, Kishinev 1970. [In R.]

5018 - MAKHARINETS, S.N.: Fotokhimicheskaya aktivnost' khloroplastov razlichnykh organov ozimoĭ pshenitsy v reproduktivnyĭ period ee razvitiya. [Photochemical
activity of chloroplasts of different organs of winter wheat in the reproduction
period of its development.] - In: Fotosinteticheskaya Deyatel'nost' Rasteniĭ i
Vliyanie na nee Mineral'nogo Pitaniya. Pp. 126-132. Kishinev 1970. [In R.]

5019 - MAKHMADBEKOVA, L.M.: PosledeĬstvie ponizhennoĭ temperatury na vklyuchenie ugleroda (C^{14}) v polisakharidy list'ev khlopchatnika. [Aftereffect of lowered temperature on ^{14}C incorporation into polysaccharides of cotton leaves.] - In: Issledovaniya po Fotosintezu. Pp. 54-58. Akad. Nauk Tadzh. SSR, Dushanbe 1967. [In
R, ab: Tadzh.]

5020 - MAKHMADBEKOVA, L.M., BATALOVA, A.G., NASYROV, Yu.S.: Metabolizm i transport produktov fotosinteza u khlopchatnika v ontogeneze. [Metabolism and transport of
photosynthates in cotton plants during ontogenesis.] - In: Tezisy Dokladov Vtorogo Vsesoyuznogo Biokhimicheskogo S"ezda, Sektsiya "Problemy Fotosinteza". Pp.
83-84. FAN, Tashkent 1969. [In R.]

5021 - MAKSIMOV, V.A.: Fotosintez i produktivnost' zelenoĭ vodorosli *Scenedesmus acuminatus* (LAGERH.) CHOD. pri razlichnykh usloviyakh fosfornogo pitaniya. [Photo-
synthesis and productivity of the green alga *Scenedesmus acuminatus* (LAGERH.)

CHOD, at different phosphorus nutrition.]- Fiziol. Biokhim. kul't. Rast. 2: 548-552, 1970. [In R, ab: E.]

5022 - **MAKSYMOV, V.O.**: Vplyv umov fosfornogo zhyvlennya na rist, prodyktyvnist' ta de-yaki protsesy obminu v *Ankistrodesmus braunii* BRUNNTH. [Effect of phosphorus nutrition on growth, productivity and some metabolic processes in *Ankistrodesmus braunii* BRUNNTH.] - Ukr. bot. Zh. *25* (6): 15-21, 1968. [Ps; in Ukr., ab: E, R.]

5023 - **MALASHEVICH, A.V.**: Vliyanie tripsina na sostoyanie i metabolizm khlorofillov *a* i *b*. [Effect of trypsin on the state and metabolism of chlorophylls *a* and *b*.] - In: Tezisy IV Nauchnoĭ Konferentsii Molodykh Uchenykh po Sovremennym Problemam Biologii. Pp. 60-62. Minsk 1970. [In R.]

5024 - **MALESZEWSKI, S., NELSON, C.D.**: The effect of oxygen concentration on the rate of formation of the ^{14}C-products of photosynthesis in balsam fir. - Plant Physiol. *43*: S-20, 1968.

5025 - **MALHOTRA, H.C., BRITTON, G., GOODWIN, T.W.**: The identification of spheroidene and hydroxyspheroidene in diphenylamine-inhibited cultures of *Rhodospirillum rubrum*. - Phytochemistry *8*: 1047-1049, 1969.

5026 - **MALHOTRA, H.C., BRITTON, G., GOODWIN, T.W.**: The mono- and dimethoxy-carotenoids of diphenylamine-inhibited cultures of *Rhodospirillum rubrum*. - Phytochemistry *9*: 2369-2375, 1970.

5027 - **MALKIN, R., RABINOWITZ, J.C.**: Nonheme iron electron-transfer proteins. - Annu. Rev. Biochem. *36*: 113-148, 1967.

5028 - **MALKIN, S.**: Fluorescence induction studies in isolated chloroplasts. II. Kinetic analysis of the fluorescence intensity dependence on time. - Biochim. biophys. Acta *126*: 433-442, 1966.

5029 - **MALKIN, S.**: Theoretical analysis of the enhancement effect in photosynthesis. Evidence for the "spill-over" model. - Biophys. J. *7*: 629-649, 1967.

5030 - **MALKIN, S.**: Analysis of the "weak light effect" on the fluorescence yield in isolated chloroplasts. - Biochim. biophys. Acta *153*: 188-196, 1968.

5031 - **MALKIN, S.**: Kinetic studies on electron-transport components in isolated chloroplasts. I. The effect of the pool electron carriers between the two photosystems on P_{700} changes. - Biochim. biophys. Acta *162*: 392-401, 1968.

5032 - **MALKIN, S.**: On the equilibrium between the reaction centers of the two photosystems in photosynthesis. The effect of independent electron-transport chains. - Biophys. J. *9*: 489-499, 1969.

5033 - **MALKIN, S.**: The topological structure of the photosynthetic unit; kinetic experiments and considerations. - In: METZNER, H. (ed.): Progress in Photosynthesis Research. Vol. II. Pp. 845-856. Tübingen 1969.

5034 - **MALKIN, S., JONES, L.W.**: Photoinhibition and excitation quenching in photosystem II of photosynthesis, from fluorescence induction measurements. - Biochim. biophys. Acta *162*: 297-299, 1968.

5035 - **MALKIN, S., KOK, B.**: Fluorescence induction studies in isolated chloroplast. I. Number of components involved in the reaction and quantum yields. - Biochim. biophys. Acta *126*: 413-432, 1966.

5036 - **MALKINA, I.S.**: Ob izmenchivosti svetovykh krivykh fotosinteza *Carex pilosa* L. [Variability of light curves of *Carex pilosa* L. photosynthesis.] - Bot. Zh. *51*: 1516-1519, 1966. [In R.]

5037 - **MALKINA, I.S.**: Svetovye krivye fotosinteza podrosta listvennykh porod. [Light curves of photosynthesis of undergrowth of broad-leaved trees.] - In: Svetovoĭ Rezhim, Fotosintez i Produktivnost' Lesa. Pp. 220-231. Nauka, Moskva 1967. [In R.]

5038 - **MALKINA, I.S.**: Photosynthese-Lichtkurven von Jungholz unter Bestandesschirm. - Tagungsber. Deut. Akad. Landwirtschaftswiss. Berlin *100*: 197-210, 1968.

5039 - **MALKINA, I.S.**: Prikhodnaya chast' balansa organicheskogo veshchestva drevesnogo podrosta v zavisimosti ot svetovykh uslovĭ vyrashchivaniya. [Received part of organic matter balance of forest woody undergrowth depending on irradiance during growing.] - In: Aktinometriya i Optika Atmosfery. Pp. 348-352. Valgus, Tallin 1968. [Ps; in R.]

5040 - MALOFEEV, V.M.: Ispol'zovanie metodov fotointegriruyushcheĭ sfery i blizkogo detektora pri izuchenii dinamicheskoĭ optiki otchlenennogo lista. [Use of photointegrating sphere and attached detector in studying the dynamic optics of a detached leaf.] - Izv. timiryaz. sel'.-khoz. Akad. *1969* (3): 20-29, 1969. [In R, ab: E.]

5041 - MALOFEEV, V.M.: Zavisimost' skorosti fotosinteza ot intensivnosti sveta v usloviyakh vnezapnogo narusheniya vodoobmena lista. [Dependence of photosynthetic rate on irradiance under the immediate disturbance of leaf water metabolism.] - Dokl. TSKhA (Moskva) *154*: 201-206, 1969. [In R.]

5042 - MALOFEEV, V.M., AVAKIMOVA, L.G.: Fotosintez, transpiratsiya i nekotorye opticheskie svoĭstva otrezannogo lista fasoli. [Photosynthesis, transpiration and some optical properties of a detached bean leaf.] - Izv. timiryaz. sel'skokhoz. Akad. *1969* (2): 13-21, 1969. [In R, ab: E.]

5043 - MALOFEEV, V.M., BELIKOV, P.S.: Ustanovka dlya izucheniya vremennogo khoda intensivnosti fluorestsentsii i skorosti fotosinteza na tselom rastenii. [Device for studying time course of fluorescence intensity and photosynthetic rate of a whole plant.] - Izv. timiryaz. sel'.-khoz. Akad. *1967* (1): 212-215, 1967. [In R.]

5044 - MALOFEEV, V.M., BELIKOV, P.S.: Intensivnost' fotosinteza v usloviyakh narastayushchego obezvozhivaniya lista. [Photosynthetic rate under increasing water deficit of a leaf.] - Sel'.-khoz. Biol. *5*: 869-873, 1970. [In R, ab: E.]

5045 - MALOFEEV, V.M., SHATILOV, I.S., ABISALOV, R.S., VAULIN, A.V.: Avtomaticheskaya ustanovka dlya nepreryvnoĭ registratsii fotosinteza i dykhaniya rasteniĭ v polevykh usloviyakh. [Automatic equipment for continuous recording of photosynthesis and respiration of plants in field conditions.] - Izv. timiryaz. sel'.-khoz. Akad. *1970* (2): 33-40, 1970. [In R, ab: E.]

5046 - MAMAENKO, G.E., STEFANOV, N.G., ANDREEV, V.D.: O rezul'tatakh izmereniya temperatury vnutrennikh poverkhnosteĭ dvustoronnego pirgeometra. [Measurement of temperature of internal surfaces of double net radiometer.] - In: Aktinometriya i Optiķa Atmosfery. Pp. 239-243. Valgus, Tallin 1968. [In R, ab: E.]

5047 - MAMUSHINA, N.S., DALETSKAYA, I.A.: Primenenie radioaktivnogo metoda dlya izucheniya fotosinteza khlorelly. [Use of the radioactive method for studying photosynthesis in *Chlorella*.] - Bot. Zh. *53*: 960-966, 1968. [In R, ab: E.]

5048 - MAMUSHINA, N.S., ZALENSKIĬ, O.V.: Vliyanie temperatury na metabolizm mechenogo ugleroda, pogloshchennogo pri fotosinteze, u raznykh shtammov khlorelly. [Effect of temperature on the metabolism of labelled carbon absorbed during photosynthesis in different *Chlorella* strains.] - Bot. Zh. *53*: 1274-1285, 1968. [In R, ab: E.]

5049 - MANDELLI, E.F.: Enhanced photosynthetic assimilation ratios in antarctic polar front (convergence) diatoms. - Limnol. Oceanogr. *12*: 484-491, 1967.

5050 - MANDELLI, E.F.: Carotenoid pigments of the dinoflagellate *Glenodinium foliaceum* STEIN. - J. Phycol. *4*: 347-348, 1968.

5051 - MANDELLI, E.F.: Carotenoid interconversion in light-dark cultures of the dinoflagellate *Amphidinium klebsii*. - J. Phycol. *5*: 382-384, 1969.

5052 - MANN, J.D., COTA-ROBLES, E., YUNG, K.-H., PU, M., HAID, H.: Phenylurethane herbicides: inhibitors of changes in metabolic state. I. Botanical aspects. - Biochim. biophys. Acta *138*: 133-139, 1967. [Chl.]

5053 - MANN, J.D., YUNG, K.-H., STOREY, W.B., PU, M., CONLEY, J.: Similarity between phytokinins and herbicidal phenylurethanes. - Plant Cell Physiol. *8*: 613-622, 1967. [Chl.]

5054 - MANN, J.E., MYERS, J.: Enhancement in the marine diatom, *Phaeodactylum tricornutum*. - Plant Physiol. *42*: S-34, 1967. [Ps.]

5055 - MANN, J.E., MYERS, J.: On pigments, growth, and photosynthesis of *Phaeodactylum tricornutum*. - J. Phycol. *4*: 349-355, 1968.

5056 - MANN, J.E., MYERS, J.: Photosynthetic enhancement in the diatom *Phaeodactylum tricornutum*. - Plant Physiol. *43*: 1991-1995, 1968.

5057 - MANNY, B.A.: The relationship between organic nitrogen and the carotenoid to chlorophyll *a* ratio in five freshwater phytoplankton species. - Limnol. Oceanogr. *14*: 69-79, 1969.

5058 - MANORYK, A.V., STARCHENKOV, Yu.P., DATSENKO, V.K.: Znachennya produktiv fotosyntezu dlya fiksatsiï atmosfernogo azotu roslynamy lyupynu. [Role of photosynthates in fixation of air nitrogen by *Lupinus* plants.] - Ukr. bot. Zh. *26* (6): 104-108, 1969. [In Ukr., ab: R, E.]

5059 - MANSFIELD, T.A.: Carbon dioxide compensation points in maize and *Pelargonium*. - Physiol. Plant. *21*: 1159-1162, 1968.

5060 - MANTAI, K.E.: Carbonylcyanide *m*-chlorophenylhydrazone as an inhibitor of coupled electron transport in trypsin treated spinach chloroplasts. - Biochim. biophys. Acta *189*: 449-451, 1969.

5061 - MANTAI, K.E.: Electron transport and degradation of chloroplasts by hydrolytic enzymes and ultraviolet irradiation. - Carnegie Inst. Year Book *68*: 598-603, 1970.

5062 - MANTAI, K.E.: Some effects of hydrolytic enzymes on coupled and uncoupled electron flow in chloroplasts. - Plant Physiol. *45*: 563-566, 1970.

5063 - MANTAI, K.E., BISHOP, N.I.: Studies on the effects of ultraviole' irradiation on photosynthesis and on the 520 nm light-dark difference spectra in green algae and isolated chloroplasts. - Biochim. biophys. Acta *131*: 350-356, 1967.

5064 - MANTAI, K.E., BISHOP, N.I.: The effects of ultraviolet (UV) irradiation on the photosynthetic apparatus. - Plant Physiol. *43*: S-21, 1968.

5065 - MANTAI, K.E., WONG, J., BISHOP, N.I.: Comparison studies on the effects of ultraviolet irradiation on photosynthesis. - Biochim.biophys. Acta *197*: 257-266, 1970.

5066 - MANTON, I.: Some possible significant structural relations between chloroplasts and other cell components. - In: GOODWIN, T.W. (ed.): Biochemistry of Chloroplasts. Vol. I. Pp. 23-47. Academic Press, London-New York 1966.

5067 - MANUELYAN, Kh.: Podobren nachin na khromatografsko otdelyane na karotina pri domatite. [Improved method of chromatographic separation of carotene in tomatoes.] - Gradin. lozar. Nauka *4* (2): 53-58, 1967. [In Bulg., ab: R, E.]

5068 - MANZ, U.: Die Anwendung und Bedeutung von synthetischen Carotinoiden in der Lebens- und Futtermittel- sowie in der pharmazeutischen Industrie. - Chimia *21*: 329-335, 1967.

5069 - MARANVILLE, J.W., PAULSEN, G.M.: Alteration of carbohydrate composition of corn (*Zea mays* L.) seedlings during moisture stress. - Agron. J. *62*: 605-608, 1970. [Chl.]

5070 - MARCELLIN, P., PHAN PHUC, A.: Mesure de la surface spécifique des pommes. - Physiol. vég. *8*: 173-187, 1970.

5071 - MARCHAND, C.: Actions comparées des pressions hydrostatiques élevées sur la bioluminescence, la photosynthèse et la fermentation alcoolique. - Compt. rend. Acad. Sci. Paris, Sér. D *267*: 2376-2378, 1968.

5072 - MARCHANT, R.H.: Light-induced hydrolysis of ATP by chloroplasts. - In: METZNER, H. (ed.): Progress in Photosynthesis Research. Vol. III. Pp. 1176-1182. Tübingen 1969.

5073 - MARCHANT, R.H.: Indirect evidence that light-induction of adenosine triphosphate hydrolysis by chloroplasts depends on electron-transfer reactions. - Biochem. J. *118*: 35 P, 1970.

5074 - MARENČÍK, A.: Príspevok k štúdiu denného chodu fotosyntézy kukurice (*Zea mays*). [Diurnal course of photosynthetic rate in maize (*Zea mays*).] - Sborník pedagogickej Fakulty v Nitre, prírodné Vedy *12*: 161-171, 1967. [In Slovak, ab: G, R.]

5075 - MARFINA, K.G., IBRAGIMOV, A.P.: Vliyanie gamma-luchei Co^{60} na intensivnost' fotosinteza u nekotorykh vysshikh rasteniĭ. [Effect of ^{60}Co gamma rays on photo-synthetic rate in some higher plants.] - Uzb. biol. Zh. *11* (4): 7-9, 1967. [In R, ab: E, Uzb.]

5076 - MARFINA, K.G., IBRAGIMOV, A.P.: Vliyanie radiatsii na raspredelenie C^{14} poglosh-chennogo pri fotosinteze, sredi razlichnykh grupp organicheskikh veshchestv. [Effect of radiation on the distribution of ^{14}C absorbed during photosynthesis among various groups of organic substances.] - Uzb. biol. Zh. *12* (6): 51, 1968. [In R.]

5077 - MARGAĬLIK, G.I., TRUKHANOŬSKI, D.S.: Dynamika nakaplennya khlarafilu listsyami drevavykh raslin u zalezhnastsi ad faz ikh razvitstsya. [Dynamics of chlorophyll accumulation in leaves of woody plants in relation to their developmental pha-ses.] - Vestsi Akad. Navuk BSSR, Ser. biyal. Navuk *1968* (5): 100-106, 142, 1968. [In Belorus., ab: R.]

5078 - MARGALEF, R., BALLESTER, A.: Fitoplancton y producción primaria de la costa ca-talana, de junio de 1965 a junio de 1966. [Phytoplankton and primary production of Catalanian coast from June, 1965, to June, 1966.] - Invest. Pesquera *31*: 165-182, 1967. [In Span.]

5079 - MARGOZZI, A.P., HENDERSON, M., WEAVER, E.C.: A dual-beam source of actinic light for photosynthesis research. - Photochem. Photobiol. *9*: 549-553, 1969.

5080 - MARGULIES, M.M.: Effect of chloramphenicol on formation of chloroplast struc-ture and protein during greening of etiolated leaves of *Phaseolus vulgaris*. - Plant Physiol. *41*: 992-1003, 1966.

5081 - MARGULIES, M.M.: Concerning the preparation of chloroplasts active in Hill and photosynthetic phosphorylation activities from leaves of *Phaseolus vulgaris*. - Plant Physiol. *41*: 1320-1322, 1966.

5082 - MARGULIES, M.M.: Effect of chloramphenicol on chlorophyll synthesis of bean leaves. - Plant Physiol. *42*: 218-220, 1967.

5083 - MARGULIES, M.M.: *In vitro* protein synthesis by plastids of *Phaseolus vulgaris*. V. Incorporation of ^{14}C-leucine into a protein fraction containing ribulose 1,5-diphosphate carboxylase. - Plant Physiol. *46*: 136-141, 1970.

5084 - MARGULIES, M.M.: Changes in absorbance spectrum of the diatom *Phaeodactylum tricornutum* upon modification of protein structure. - J. Phycol. *6*: 160-164, 1970. [Chl.]

5085 - MARGULIES, M.M., BRUBAKER, C.: Effect of chloramphenicol on amino acid incorpo-ration by chloroplasts and comparison with the effect of chloramphenicol on chloroplasts development *in vivo*. - Plant Physiol. *45*: 632-633, 1970.

5086 - MARGULIES, M.M., GANTT, E., PARENTI, F.: *In vitro* protein synthesis by plastids of *Phaseolus vulgaris*. II. The probable relation between ribonuclease insensi-tive amino acid incorporation and the presence of intact chloroplasts. - Plant Physiol. *43*: 495-503, 1968.

5087 - MARGULIES, M.M., PARENTI, F.: *In vitro* protein synthesis by plastids of *Phaseo-lus vulgaris*. III. Formation of lamellar and soluble chloroplast protein. - Plant Physiol. *43*: 504-514, 1968.

5088 - MARINCHIK, A.F., DAVYDOVA, N.A., KURINNYĬ, F.I.: Vliyanie mineral'nogo pitaniya na soderzhanie pigmentov v list'yakh sakharnoĭ svekly v zavisimosti ot vozrasta ikh i perioda vegetatsii. [Effect of mineral nutrition on pigment content in leaves of sugar beet in dependence on their growth and vegetation period.] - In: Osnovnye Vyvody Nauchno-issledovatel'skikh Rabot po Sakharnoĭ Svekle za 1966 g. Vol. I. Pp. 21-24. Kiev 1968. [In R.]

5089 - MARKER, A.F.H., WHITTINGHAM, C.P.: The photoassimilation of glucose in *Chlorella* with reference to the role of glycollic acid. - Proc. roy. Soc. B *165*: 473-485, 1966.

5090 - MARKER, A.F.H., WHITTINGHAM, C.P.: The site of synthesis of sucrose in green plant cells. - J. exp. Bot. *18*: 732-739, 1967.

5091 - MARKHAM, M.C., LIAAEN-JENSEN, S.: Carotenoids of higher plants - I. The struc-tures of lycoxanthin and lycophyll. - Phytochemistry *7*: 839-844, 1968.

5092 - MARKOSYAN, L.S.: Ob izmenenii funktsional'noT aktivnosti list'ev klena amerikans-
kogo pri narushenii floemnoT svyazi. [Changes in functional activity of leaves
of ash-leaved maple induced by impairing the phloem communication.] - Biol. Zh.
Armenii 22 (7): 94-97, 1969. [Ps; in R.]

5093 - MARKS, G.S.: The biosynthesis of heme and chlorophyll. - Bot. Rev. 32: 56-94,
1966.

5094 - MARLBOROUGH, D.I., HALL, D.O., CAMMACK, R.: Magneto-optiçal rotatory dispersion
(MORD) studies on spinach ferredoxin. - Biochem. biophys. Res. Commun. 35: 410-
413, 1969.

5095 - MAROC, J., de KLERK, H., KAMEN, M.D.: Quinones of Athiorhodaceae. - Biochim.
biophys. Acta 162: 621-623, 1968.

5096 - MARRÈ, E., SCHWENDIMANN, M., LADO, P.: Ricerche sull' invecchiamento in foglie
recise. I. Effetto protettivo degli zuccheri e della luce sui sitemi fotosin-
tetico e respiratorio di fogli recise de Solanum tuberosum. [Ageing of detached leaves. I. Protective effects of sugars and radiation on photosynthetic and
respiratory systems in detached leaves of Solanum tuberosum.] - Atti Acad. naz.
Lincei, Rend. Cl. Sci. fis., mat. nat., Ser. 8, 40: 1089-1094, 1966. [In Ital.]

5097 - MARRIOTT, J., NEUBERGER, A., TAIT, G.H.: Control of δ-aminolaevulate synthetase
activity in Rhodopseudomonas spheroides. - Biochem. J. 111: 385-394, 1969. [Chl.]

5098 - MARŠÁLEK, L., VOŘÍŠEK, V.: Analyse einiger physiologisch aktiver Stoffe bei Mais-
Inzuchtlinien und Einfachhybriden in Beziehung zum Heterosiseffekt. - Z. Pflanzenzücht. 59: 259-272, 1968. [Chl, car.]

5099 - MARSCHNER, H., GÜNTHER, I.: Veränderungen der Feinstruktur der Chloroplasten in
Gerstensprossen unter dem Einfluss von Cäsium. - Flora A 156: 684-696, 1966.

5100 - MARSH, J.A. Jr.: Primary productivity of reef-building calcareous red algae. -
Ecology 51: 255-263, 1970.

5101 - MARSHALL, C., SAGAR, G.R.: The distribution of assimilates in Lolium multiflorum
LAM. following differential defoliation. - Ann. Bot. 32: 715-719, 1968.

5102 - MARSHALL, J.K.: The photographic-photoelectric planimeter combination method for
leaf area measurement. - Photosynthetica 2: 1-9, 1968.

5103 - MARSHALL, J.K.: Methods for leaf area measurement of large and small leaf samples. - Photosynthetica 2: 41-47, 1968.

5104 - MARSHO, T.V., KOK, B.: Interaction between electron transport components in chloroplasts. - Biochim. biophys. Acta 223: 240-250, 1970.

5105 - MARSTOLF, J.D., DECKER, W.L.: Microclimate modification by manipulation of net
radiation. - Agr. Meteorol. 7: 197-216, 1970.

5106 - MARTINEZ GARRIDO, J.: El color del tomate. [Colour of tomato.] - Ion (Madrid)
28: 484-489, 1968. [Chl, car; in Span.]

5107 - MARTY, D.: Variations infrastructurales dans les chloroplastes des feuilles pa-
nachées de Coleus hybrida. - Ann. Sci. nat., Bot. Biol. vég. 9: 575-608, 1968.

5108 - MASHTAKOV, S.M., PAROMCHIK, I.I.: Izmenenie fotosinteza i reaktsii Khilla u raznykh form kukuruzy pod deTstviem 2,4-D i 2 M-4 X. [Photosynthesis and Hill reaction changes induced by 2,4-D and 2 M-4 X in various maize cultivars.] - Dokl.
Akad. Nauk Belorus. SSR 10: 691-694, 1966. [In R.]

5109 - MASHTAKOV, S.M., PAROMCHIK, I.I.: Izmenenie prochnosti khlorofillbelkovolipoid-
nogo kompleksa rasteniT, obrabotannykh natrievymi solyami khlorfenoksiuksusnykh
kislot. [Change of stability of the chlorophyll-protein-lipoid complex of plants
treated with sodium salts of chlorphenoxyacetic acids.] - Dokl. Akad. Nauk Beloruss. SSR 10: 792-795, 1966. [In R.]

5110 - MASHTAKOV, S.M., SHCHERBAKOV, V.A.: DeTstvie otechestvennogo khlorkholinkhlorida
(CCC) na nachal'nyT rost i soderzhanie khlorofilla u rasteniT pshenitsy. [In-
fluence of native CCC on primary growth and chlorophyll content in wheat plants.]
- Dokl. Akad. Nauk Beloruss. SSR 11: 169-173, 1967. [In R.]

5111 - MASLOVA, T.G., MEISTER, A.: Einfluss einiger Faktoren auf die lichtinduzierten

Absorptionsänderungen des Blattes im blauen Spektralbereich. - Z. Pflanzenphysiol. *60*: 114-122, 1969.

5112 - MASLOVA, T.G., POPOVA, I.A.: Prevrashchenie ksantofillov v list'yakh, osveshchaemykh krasnym svetom. [Xanthophyll transformation in leaves irradiated with red wavelengths.] - Fiziol. Rast. *16*: 106-110, 1969. [In R, ab: E.]

5113 - MASLOWA, T., POPOWA, I.: Über die spektrale Abhängigkeit der Umwandlungsreaktion der Xanthophylle. - Stud. biophys. *5*: 217-224, 1967.

5114 - MASSINI, P., VOORN, G.: The effect of ferredoxin and ferrous ion on the chlorophyll sensitized photoreduction of dinitrophenol. - Photochem. Photobiol. *6*: 851-856, 1967.

5115 - MASSINI, P., VOORN, G.: Optical and photochemical properties of chlorophyll *a* solubilized in aqueous solutions of surfactants. - Biochim. biophys. Acta *153*: 589-601, 1968.

5116 - MASYUK, N.P.: Masova kul'tura karotynonosnoï vodorosti *Dunaliella salina* TEOD. [Mass culture of the carotene containing alga *Dunaliella salina* TEOD.] - Ukr. bot. Zh. *23* (2): 12-19, 1966. [In Ukr., ab: E, R.]

5117 - MASYUK, N.P., ABDULA, E.G.: Pershyï dosvid vyroshchuvannya karotynonosnykh vodorosteï v napivpromyslovykh umovakh. [First experiment of growing carotene-containing algae under semi-industrial conditions.] - Ukr. bot. Zh. *26* (3): 21-27, 1969. [In Ukr., ab: E, R.]

5118 - MASYUK, N.P., BERENSHTEÏN, O.P., YURKOVA, G.N.: Vplyv vidkhodiv kharchovikh ta brodyl'nykh vyrobnytstv na rist ta nagromadzhennya karotynu v kul'turi *Dunaliella salina* TEOD. [Effect of by-products of food and brewing industries on growth and carotene accumulation in the culture of *Dunaliella salina* TEOD.] - Ukr. bot. Zh. *23*: 35-43, 1966. [In Ukr., ab: E, R.]

5119 - MASYUK, N.P., RADCHENKO, M.I.: Izvlechenie pigmentov iz vodorosteï s plazmaticheskoï obolochkoï. [Extraction of pigments from algae with a plasmatic membrane.] - Gidrobiol. Zh. *3* (6): 77-78, 1967. [In R.]

5120 - MASYUK, N.P., RADCHENKO, M.Ï.: Do metodyky vyluchennya pigmentiv z poliblefarydovykh vodorosteï. [Method of pigment isolation from *Polyblepharidaceae*.] - Ukr. bot. Zh. *25* (5): 91-98, 1968. [In Ukr., ab: E, R.]

5121 - MASYUK, N.P., RADCHENKO, M.I.: Sravnitel'noe khromatograficheskoe izuchenie pigmentov nekotorykh vidov i shtammov *Dunaliella* TEOD. [Comparative chromatographic study of pigments in some species and strains of *Dunaliella* TEOD.] - Gidrobiol. Zh. *6* (3): 51-58, 1970. [PC, TCL, spectra; in R, ab: E.]

5122 - MATHIEU, Y.: Sur les activités glutamate-pyruvate et glutamate-oxaloacétate transaminases dans les préparations chloroplastiques de feuilles de *Bryophyllum daigremontianum* BERGER. - Physiol. vég. *4*: 299-316, 1966.

5123 - MATHIEU, Y.: Sur l'isolement, en milieu aqueux, de chloroplastes "intacts" à partir de feuilles de plantules d'Orge. - Photosynthetica *1*: 57-63, 1967.

5124 - MATHIEU, Y.: Influence de l'oxygène sur les transferts d'électrons de la photosynthèse. I. Influence de diverses concentrations d'oxygène sur quelques réactions de Hill. - Biochim. biophys. Acta *189*: 411-421, 1969.

5125 - MATHIEU, Y.: Influence de l'oxygène sur les transferts d'électrons de la photosynthèse. II. Influence de très faibles concentrations en oxygène sur la réduction du NADP+ par les chloroplastes isolés. - Biochim. biophys. Acta *189*: 422-428, 1969.

5126 - MATHIEU, Y., MIGINIAC-MASLOW, M., REMY, R.: Étude du rôle de la ferrédoxine dans les transferts d'électrons et la photophosphorylation des chloroplastes d'épinard et de fève en aérobiose. - Biochim. biophys. Acta *205*: 95-101, 1970.

5127 - MATHIS, B.J., ROOT, R.J., WEIMAN, G.W.: Chlorophyll *a* in the Illinois river. - Trans. Ill. State Acad. Sci. *61*: 416-420, 1968.

5128 - MATHIS, P.: Variation d'absorption de courte durée, induite dans une suspension de chloroplastes par un éclair laser. - Compt. rend. Acad. Sci. Paris, Sér. D *263*: 1770-1772, 1966.

5129 - MATHIS, P.: Étude par spectroscopie d'éclairs du transfert d'énergie chlorophyl-
le-carotenoide. - Photochem. Photobiol. 9: 55-63, 1969.

5130 - MATHIS, P.: Triplet-triplet energy transfer from chlorophyll a to carotenoids
in solution and in chloroplasts. - In: METZNER, H. (ed.): Progress in Photosynthesis Research. Vol. II. Pp. 818-822. Tübingen 1969.

5131 - MATHIS, P., GALMICHE, J.M.: Action des gaz paramagnétiques sur un état transi-
toire unduit par un éclair laser dans une suspension de chloroplastes. - Compt.
rend. Acad. Sci. Paris, Sér. D 264: 1903-1906, 1967.

5132 - MATHRE, D.E.: Photosynthetic activities of cotton plants infected with Verticillium albo-atrum. - Phytopathology 58: 137-141, 1968.

5133 - MATIENKO, B.T., SALINSKIĬ, S.M., SOLOVEĬ, V.K.: O vozmozhnosti prevrashcheniya
karotinoidoplastov (khromoplastov) v khloroplasty v subepidermise ploda arbuza.
[Possibility of transformation of carotenoid-plasts (chromoplasts) into chloroplasts in subepidermis of watermelon fruit.] - In: Khloroplasty i Mitokhondrii.
Pp. 190-198. Nauka, Moskva 1969. [In R.]

5134 - MATSUBARA, H.: Purification and some properties of Scenedesmus ferredoxin. - J.
biol. Chem. 243: 370-375, 1968.

5135 - MATSUBARA, H., SASAKI, R.M., CHAIN, R.K.: The amino acid sequence of spinach
ferredoxin. - Proc. nat. Acad. Sci. U.S.A. 57: 439-445, 1967.

5136 - MATSUIKE, K.: The optical characteristics of the water in the three oceans. Part
III. The distribution of solar energy reached to and penetrated in the water of the
Antarctic ocean in the summer and its comparison to other oceans. - J. oceanogr.
Soc. Jap. 25: 81-90, 1969.

5137 - MATSUKA, M., HASE, E.: The role of respiration and photosynthesis in the chloroplast regeneration in the "glucose-bleached" cells of Chlorella protothecoides.
- Plant Cell Physiol. 7: 149-162, 1966.

5138 - MATSUKA, M., HASE, E.: Effect of cycloheximide on the process of "glucose-
bleaching" in Chlorella protothecoides. - Plant Cell Physiol. 10: 277-282, 1969.
[Chl.]

5139 - MATSUKA, M., MIYACHI, S., HASE, E.: Further studies on the metabolism of glucose
in the process of "glucose-bleaching"of Chlorella protothecoides. - Plant Cell
Physiol. 10: 513-526, 1969. [Chl.]

5140 - MATSUKA, M., MIYACHI, S., HASE, E.: Acetate metabolism in the process of "acetate-
bleaching" of Chlorella protothecoides. - Plant Cell Physiol. 10: 527-538, 1969.
[Chl.]

5141 - MATSUKA, M., OTSUKA, H., HASE, E.: Changes in contents of carbohydrate and fatty
acid in the cells of Chlorella protothecoides during the process of de- and regeneration of chloroplasts. - Plant Cell Physiol. 7: 651-652, 1966.

5142 - MATSUMOTO, C., SUGIYAMA, T., AKAZAWA, T., MIYACHI, S.: Structure and function
of chloroplast proteins. IX. Further comparative studies on Chlorella and spinach leaf ribulose-1,5-diphosphate carboxylase. - Arch. Biochem. Biophys. 135:
282-287, 1969.

5143 - MATSUSHITA, K., TSURUDA, K.: Infectious entity in chloroplasts of tobacco
leaves infected with tobacco mosaic virus. - Mem. Fac. Sci., Kyushu Univ. Ser.
E (Jap.) 5: 57-64, 1969.

5144 - MATSUURA, K., IWATA, T., HASEGAWA, T.: [Studies on the effect of deep-layer application of fertilizers in rice plant. I. On the analysis of yield.] - Proc.
Crop Sci. Soc. Jap. 38: 215-221, 1969. [Growth analysis; in Jap., ab: E.]

5145 - MATSYUK, L.S., GRINBERG, I.P.: Povyshenie urozhaĭnosti silosnogo sorgo putem
predposevnoĭ obrabotki semyan stimulyatorami rosta. [Increase in yield of silage
sorghum induced by pre-sowing treatment of seeds by growth stimulators.] - Tr.
kishinev. sel'skokhoz. Inst. 51: 177-178, 1968. [Chl; in R.]

5146 - MATTERNE, M.: Vergleiche zwischen Primärproduktion und Syntheseraten organischer
Zellbestandteile mariner Phytoplankter. - Kiel. Meeresforsch. 25: 290-313, 1969.

5147 - MATTHERN, R.O., KOSTICK, J.A., OKADA, I.: Effect of total illumination upon continuous *Chlorella* production in a high intensity light system.- Biotechnol. Bioeng. *11*: 863-874, 1969. [Chl.]

5148 - MATTOO, A.K., MODI, V.V., REDDY, V.V.R.: Oxidation & carotenogenesis regulating factors in mangoes. - Indian J. Biochem. *5*: 111-114, 1968.

5149 - MAXWELL, M.A.B., WILLIAMS, J.P.: The purification of lipid extracts using Sephadex LH-20. - J. Chromatogr. *31*: 62-68, 1967. [Chl.]

5150 - MAYER, F.: Lichtinduzierte Chloroplasten-Verlagerungen bei *Selaginella martensii*. - Z. Pflanzenphysiol. *55*: 65-70, 1966.

5151 - MAYER, F., CZYGAN, F.-C.: Änderungen der Ultrastrukturen in den Grünalgen *Ankistrodesmus braunii* und *Chlorella fusca* var. *rubescens* bei Stickstoffmangel. - Planta *86*: 175-185, 1969. [Chl, car.]

5152 - MAYEUX, J.V., JOHNSON, E.J.: Effect of adenosine monophosphate, adenosine diphosphate, and reduced nicotinamide adenine dinucleotide on adenosine triphosphate-dependent carbon dioxide fixation in the autotroph *Thiobacillus neapolitanus*. - J. Bacteriol. *94*: 409-414, 1967.

5153 - MAYHEW, S.G., PETERING, D., PALMER, G., FOUST, G.P.: Spectrophotometric titration of ferredoxins and *Chromatium* high potential iron protein with sodium dithionite. - J. biol. Chem. *244*: 2830-2834, 1969.

5154 - MAYNARD, D.N.: The effects of nutrient stress on the growth and composition of spinach. - J. amer. Soc. hort. Sci. *95*: 598-600, 1970. [Chl.]

5155 - MAYNE, B.C.: Chemiluminescence of chloroplasts. - In: Energy Conversion by the Photosynthetic Apparatus. Brookhaven Symp. Biol. *19*: 460-466, 1967.

5156 - MAYNE, B.C.: The effect of inhibitors and uncouplers of photosynthetic phosphorylation on the delayed light emission of chloroplasts. - Photochem. Photobiol. *6*: 189-197, 1967.

5157 - MAYNE, B.C.: The light requirement of acid-base transition induced luminescence of chloroplasts. - Photochem. Photobiol. *8*: 107-113, 1968. [Chl.]

5158 - MAYNE, B.C.: The light requirement for the chemiluminescence of chloroplasts. - In: METZNER, H. (ed.): Progress in Photosynthesis Research. Vol. II. Pp. 947-951. Tübingen 1969.

5159 - MAYNE, B.C., CLAYTON, R.K.: Luminiscence of chlorophyll in spinach chloroplasts induced by acid-base transition. - Proc. nat. Acad. Sci. U.S.A. *55*: 494-497, 1966.

5160 - MAYNE, B.C., RUBINSTEIN, D.: Absorption changes in the blue-green algae at the temperature of liquid nitrogen. - Nature *210*: 734-735, 1966.

5161 - MAZIL'NIKOV, G.V.: Vliyanie kaliTnoT vnekornevoT podkormki na khimizm fotosinteza gorokha v usloviyakh zasukhi. [Effect of potassium outside root application on chemistry of pea photosynthesis under conditions of drought.] - In: Funktsional'nye Osobennosti Khloroplastov. Pp. 101-104. Kazan. Univ., Kazan 1969. [In R.]

5162 - McBRIEN, D.C.H., HASSAL, K.A.: The effect of toxic doses of copper upon respiration, photosynthesis and growth of *Chlorella vulgaris*. - Physiol. Plant. *20*: 113-117, 1967.

5163 - McCALLA, D.R.: Action of some analogs of nitrosoguanidine on the chloroplast of *Euglena gracilis*. - J. Protozool. *13*: 472-474, 1966.

5164 - McCALLA, D.R.: Mutation of the *Euglena* chloroplast system: the mechanism of bleaching by nitrosoguanidine and related compounds. - J. Protozool. *14*: 480-482, 1967. [Chl.]

5165 - McCALLA, D.R., BAERG, W.: Action of myxin on the chloroplast system of *Euglena gracilis*. - J. Protozool. *16*: 425-428, 1969.

5166 - McCARTY, R.E.: Relation of photophosphorylation to hydrogen ion transport. - Biochem. biophys. Res. Commun. *32*: 37-43, 1968.

5167 - McCARTY, R.E.: The uncoupling of photophosphorylation by valinomycin and ammonium chloride. - J. biol. Chem. *244*: 4292-4298, 1969.

5168 - McCARTY, R.E., COLEMAN, C.H.: Effect of hydrocarbon chain length on the uncoup-
         ling of photophosphorylation by amines. - Arch. Biochem. Biophys. *141*: 198-206,
         1970.

5169 - McCARTY, R.E., RACKER, E.: Effects of an antiserum to the chloroplast coupling
         factor on photophosphorylation and related processes. - Fed. Proc. *25*: 226, 1966.

5170 - McCARTY, R.E., RACKER, E.: Effect of a coupling factor and its antiserum on
         photophosphorylation and hydrogen ion transport. - In: Energy Conversion by the
         Photosynthetic Apparatus. Brookhaven Symp. Biol. *19*: 202-214, 1967.

5171 - McCARTY, R.E., RACKER, E.: Partial resolution of the enzymes catalyzing photo-
         phosphorylation. II. The inhibition and stimulation of photophosphorylation by
         N,N'-dicyclohexylcarbodiimide. - J. biol. Chem. *242*: 3435-3439, 1967.

5172 - McCARTY, R.E., RACKER, E.: Partial resolution of the enzymes catalyzing photo-
         phosphorylation. III. Activation of adenosine triphosphatase and ^{32}P-labeled
         orthophosphate-adenosine triphosphate exchange in chloroplasts. - J. biol. Chem.
         243: 129-137, 1968.

5173 - McCLURE, W.F.: Spectral characteristics of tobacco in the near-infrared region
         from 0.6 to 2.6 microns. - Tobacco *167* (24): 38-41, 1968. [Chl.]

5174 - McCLURE, W.F.: An instrument for nondestructive measurement of total chlorophyll
         in tobacco. - Tobacco *168* (7): 24-26, 1969; Tobacco Sci. *13*: 22-24, 1969.

5175 - McCLURE, W.F.: Fiber-optic spectrophotometer for *in vivo* analysis of biological
         materials: chlorophyll measurements. - Trans. ASAE *12*: 319-321, 1969.

5176 - McCLURE, W.F.: Fiber optic biophotometers. - Amer. Lab. *1970* (Oct.): 35-38, 40,
         1970. [Chl.]

5177 - McCLURE, W.F., JOHNSON, W.H., HASSLER, F.J.: An instrument for determining the
         spectrofluorometric properties of biological materials. - Trans. ASAE *11*: 112-
         115, 1968.

5178 - McCONNELL, W.B., MAZUREK, M.: The transport of carbon in wheat plants. - Can. J.
         Biochem. *45*: 1853-1861, 1967. [Ps.]

5179 - McCORMICK, A., LIAAEN JENSEN, S.: Silylation as a method for establishment of
         tertiary hydroxyl groups in carotenoids. - Acta chem. scand. *20*: 1989-1991, 1966.

5180 - McCREE, K.J.: A solarimeter for measuring photosynthetically active radiation.
         - Agr. Meteorol. *3*: 353-366, 1966.

5181 - McCREE, K.J.: Light and growth of plants. - New Zealand sci. Rev. *25* (3): 31-33,
         1967. [Ps.]

5182 - McCREE, K.J.: A solar radiation recorder for plant growth studies. - In: ECKARDT,
         F.F. (ed.): Functioning of Terrestrial Ecosystems at the Primary Production
         Level. Pp. 463-466. UNESCO, Paris 1968.

5183 - McCREE, K.J.: Infrared-sensitive colour film for spectral measurements under
         plant canopies. - Agr. Meteorol. *5*: 203-208, 1968.

5184 - McCREE, K.J.: Towards a definition of photosynthetically active radiation. -
         Biometeorology *4* (11): 59, 1970.

5185 - McCREE, K.J.: An equation for the rate of respiration of white clover plants
         grown under controlled conditions. - In: Prediction and Measurement of Photo-
         synthetic Productivity. Pp. 221-229. PUDOC, Wageningen 1970. [Ps.]

5186 - McCREE, K.J., LOOMIS, R.S.: Photosynthesis in fluctuating light. - Ecology *50*:
         422-428, 1969.

5187 - McCREE, K.J., MORRIS, R.A.: A transmission meter for photosynthetically active
         radiation. - J. agr. eng. Res. *12*: 246-248, 1967.

5188 - McCREE, K.J., TROUGHTON, J.H.: Prediction of growth rate at different light
         levels from measured photosynthesis and respiration rates. - Plant Physiol. *41*:
         559-566, 1966.

5189 - McCREE, K.J., TROUGHTON, J.H.: Non-existence of an optimum leaf area index for
         the production rate of white clover grown under constant conditions. - Plant
         Physiol. *41*: 1615-1622, 1966. [Ps.]

5190 - McCREE, K.J., TROUGHTON, J.H.: The prediction of growth rate from incident light or carbon dioxide uptake: a laboratory experiment with white clover. - In: ECKARDT, F.E. (ed.): Functioning of Terrestrial Ecosystems at the Primary Production Level. Pp. 409-414. UNESCO, Paris 1968. [Ps.]

5191 - McDOUGALL, B.M.: Movement of ^{14}C-photosynthate into the roots of wheat seedlings and exudation of ^{14}C from intact roots. - New Phytol. 69: 37-46, 1970.

5192 - McELROY, J.D., FEHER, G., MAUZERALL, D.C.: On the nature of the free radical formed during the primary process of bacterial photosynthesis. - Biochim. biophys. Acta 172: 180-183, 1969.

5193 - McEVOY, F.A., LYNN, W.S.: Proton uptake and phosphorylation in digitonin-treated chloroplasts particles. - FEBS Letters 10: 299-300, 1970.

5194 - McFEETERS, R.F., SCHANDERL, S.H.: Biological degradation of chlorophyll in a system using bell peppers (Capsicum frutescens). - J. Food Sci. 33: 547-553, 1968.

5195 - McGEER, E.G.: Gelatin capsules as disposable wells for $^{14}CO_2$ absorption. - Anal. Biochem. 35: 300-301, 1970.

5196 - McGINNIS, R.C.,DYCK, P.L., HILDEBRANDT, S.G., LIN, C.C.: The association of a third chromosome with chlorophyll production in Avena sativa. - Can. J. Genet. Cytol. 10: 228-231, 1968.

5197 - McHALE, J.S., DOVE, L.D.: Mobilization-independent effects of a cytokinin on senescing tomato leaves. - Naturwissenschaften 55: 141, 1968. [Chl.]

5198 - McHALE, J.S., DOVE, L.D.: Ribonuclease activity in tomato leaves as related to development and senescence. - New Phytol. 67: 505-515, 1968. [Chl.]

5199 - McKEE, G.W., YOCUM, J.O.: Coefficients for computing leaf area in Type 41, Pennsylvania Broadleaf, tobacco. - Agron. J. 62: 433-434, 1970.

5200 - McKENNA, J.M., BISHOP, N.I.: Studies on the photooxidation of manganese by isolated chloroplasts. - Biochim. biophys. Acta 131: 339-349, 1967.

5201 - McLAREN, I., THOMAS, D.R.: CO_2 fixation, organic acids and some enzymes in green and colourless tissue cultures of Kalanchoe crenata. - New Phytol. 66: 683-695, 1967.

5202 - McLAUGHLIN, S.B., MADGWICK, H.A.I.: The effects of position in crown on the morphology of needles of loblolly pine (Pinus taeda L.). - Amer. Midland Natur. 80: 547-550, 1968. [Ps.]

5203 - McLEAN, R.J.: Desiccation and heat resistance of the green alga Spongiochloris typica. - Can. J. Bot. 45: 1933-1938, 1967. [Chl.]

5204 - McLEAN, R.: Primary and secondary carotenoids of Spongiochloris typica. - Physiol. Plant. 20: 41-47, 1967.

5205 - McLEAN, R.J.: Physiological changes of Spongiochloris typica in culture. - J. Phycol. 4: 73-75, 1968. [Ps, Chl.]

5206 - McLEAN, R.J.: Ultrastructure of Spongiochloris typica during senescence. - J. Phycol. 4: 277-283, 1968. [Ps.]

5207 - McLEAN, R.J.: New taxonomic criteria in the classification of Chlorococcum species. I. Pigmentation. - J. Phycol. 4: 328-332, 1968.

5208 - McLEAN, R.J.: Physiological changes in aging cultures of the green alga Spongiochloris. - Diss. Abstr. B 28: 3179-B, 1968. [Chl, car.]

5209 - McLEAN, R.J., PESSONEY, G.F.: A large scale quasi-crystalline lamellar lattice in chloroplasts of the green alga Zygnema. - J. Cell Biol. 45: 522-531, 1970.

5210 - McLURE, W.F.: Fiber-optic spectrophotometer for in vivo analysis of biological materials: chlorophyll measurements. - Trans. A.S.A.E. 12: 319-321, 1969.

5211 - McMAHON, D., BOGORAD, L.: Some kinetic studies of ribulose-1,5-diphosphate carboxylase (carboxydismutase) from races of Mimulus cardinalis. - Carnegie Inst. Year Book 65: 459-461, 1967.

5212 - McMAHON, D., BOGORAD, L.: Inhibition of the formation of photosynthetic enzymes by inhibitors of photosynthesis. - Plant Physiol. 43: 188-192, 1968.

5213 - McNAUGHTON, F.: Net primary production of sycamore (*Acer pseudoplatanus*) in Western Scotland. - J. appl. Ecol. *7*: 577-590, 1970.

5214 - McNAUGHTON, S.J.: Photosynthetic system II: Racial differentiation in *Typha latifolia*. - Science *156*: 1363, 1967.

5215 - McNAUGHTON, S.J., FULLEM, L.W.: Photosynthesis and photorespiration in *Typha latifolia*. - Plant Physiol. *45*: 703-707, 1970.

5216 - McPHERSON, H.G.: Photocell-filter combinations for measuring photosynthetically active radiation. - Agr. Meteorol. *6*: 347-356, 1969.

5217 - McSWAIN, B.D.: Quantum requirements and enhancement effects of photosynthetic electron transport in chloroplasts. - Diss. Abstr. B *30*: 993-B-994-B, 1969.

5218 - McSWAIN, B.D., ARNON, D.I.: Enhancement effects and the identity of the two photochemical reactions of photosynthesis. - Proc. nat. Acad. Sci. U.S.A. *61*: 989-996, 1968.

5219 - McWEENY, D.J.: Deterioration of β-carotene in certain hydrogenated fats. I. Incidence of green discoloration during storage. - J. Sci. Food Agr. *19*: 250-253, 1968.

5220 - McWEENY, D.J.: Deterioration of β-carotene in certain hydrogenated fats. II. Products of β-carotene deterioration and nature of the green pigment. - J. Sci. Food Agr. *19*: 254-258, 1968.

5221 - McWEENY, D.J.: Deterioration of β-carotene in certain hydrogenated fats. III. Factors affecting the rate at which green discoloration occurs. - J. Sci. Food Agr. *19*: 259-265, 1968.

5222 - McWILLIAM, J.R., NAYLOR, A.W.: Temperature and plant adaptation. I. Interaction of temperature and light in the synthesis of chlorophyll in corn. - Plant Physiol. *42*: 1711-1715, 1967.

5223 - MEDINA, E.: Intercambio gaseoso de arboles de las sabanas de *Trachypogon* en Venezuela. [Gas exchange in trees of *Trachypogon* savanas in Venezuela.] - Bol. Soc. Venez. Cienc. nat. *27*: 56-69, 1967. [In Span.]

5224 - MEDINETS, V.D.: O povyshenii koeffitsienta khozyaĭstvennoĭ polnotsennosti fotosinteza. [Increase of coefficient of the agricultural full-value of photosynthesis.] - In: Fotosinteziruyushchie Sistemy Vysokoĭ Produktivnosti. Pp. 162-168. Nauka, Moskva 1966. [In R.]

5225 - MEDVEDEV, Y.V., FURSOVA, L.E.: Vliyanie razlichnoĭ stepeni osveshchennosti na adenilovuyu sistemu list'ev kukuruzy. [Effect of various illuminance on the adenylate system of maize leaves.] - Sb. Rab. Aspir. tadzh. Univ., Ser. biol.Nauk *3*: 33-39, 1969. [Ps; in R.]

5226 - MEDVEDEVA, T.N.: Izmenenie komponentov pigmentnoĭ sistemy list'ev plodovykh rasteniĭ v techenie vegetatsii. [Change in the components of the pigment system of the leaves of fruit-bearing plants during vegetation.] - In: Fiziologiya Zimostoĭkosti i Zasukhoustoĭchivosti Plodovykh i Vinograde. Pp. 100-104. Kishinev 1970. [In R.]

5227 - MEFFERT, M.-E., OVERBECK, J.: Wachstum von *Scenedesmus obliquus* in Abhängigkeit von der Eisenversorgung. - Planta *78*: 39-48, 1968. [Chl.]

5228 - MEGO, J.L.: Inhibitors of the chloroplast system in *Euglena*. - In: BUETOW, D.E. (ed.): The Biology of *Euglena*. Vol. 2. Pp. 351-381. Academic Press, New York-London 1968.

5229 - MEGO, J.L., BUETOW, D.E.: Studies on chloroplast development in heat-bleached *Euglena*. - In: SIRONVAL, C. (ed.): Le Chloroplaste, Croissance et Vieillisement. Pp. 274-290. Masson, Paris 1967. [Chl.]

5230 - MEGURO, H., ITO, K., FUKUSHIMA, H.: Ice flora (bottom type): a mechanism of primary production in polar seas and the growth of diatoms in sea ice. - Arctic *20*: 114-133, 1967. [Also Chl.]

5231 - MEIDNER, H.: Further observations on the minimum intercellular space carbon-dioxide concentration (Γ) of maize leaves and the postulated roles of "photo-respiration" and glycollate metabolism. - J. exp. Bot. *18*: 177-185, 1967.

5232 - MEIDNER, H.: The effect of kinetin on stomatal opening and the rate of intake of carbon dioxide in mature primary leaves of barley. - J. exp. Bot. *18*: 556-561, 1967.

5233 - MEIDNER, H.: Infra-red gas analysis in the study of photosynthesis. - Hilger J. *11*: 3-6, 24, 1967/68.

5234 - MEIDNER, H.: "Rate limiting" resistances and photosynthesis. - Nature *222*: 876-877, 1969.

5235 - MEIDNER, H.: Effects of photoperiodic induction and debudding in *Xanthium pennsylvanicum* and of partial defoliation in *Phaseolus vulgaris* on rates of net photosynthesis and stomatal conductances. - J. exp. Bot. *21*: 164-169, 1970.

5236 - MEIDNER, H.: A critical study of sensor element diffusion porometers. - J. exp. Bot. *21*: 1060-1066, 1970.

5237 - MEIDNER, H.: Precise measurements of carbon dioxide exchange by illuminated leaves near the compensation point. - J. exp. Bot. *21*: 1067-1075, 1970.

5238 - MEIDNER, H.: Light compensation points and photorespiration. - Nature *228*: 1349, 1970.

5239 - MEIDNER, H., MANSFIELD, T.A.: Rates of photosynthesis and respiration in relation to stomatal movements in leaves treated with α-hydroxysulphonate and glycollate. - J. exp. Bot. *17*: 502-509, 1966.

5240 - MEIKLE, R.W.: Inhibition of photosynthesis by Pyriclor. - Weed Sci. *18*: 475-478, 1970.

5241 - MEÏLANOV, I.S., BENDERSKIÏ, V.A., BLYUMENFEL'D, L.A.:Fotoelektricheskie svoĭstva sloev khlorofillov *a* i *b*. I. Fototoki pri postoyannum osveshchenii. [Photoelectric properties of chlorophyll *a* and *b* layers. I. Photofluxes at continuous illumination.] - Biofizika *15*: 822-827, 1970. [In R, ab: E.]

5242 - MEÏLANOV, I.S., BENDERSKIÏ, V.A., BLYUMENFEL'D, L.A.: Fotoelektricheskie svoĭstva sloev khlorofillov *a* i *b*. II. Fototoki pri impul'snom osveshchenii. [Photoelectric properties of chlorophyll *a* and *b* layers. II. Photofluxes at impulse illumination.] - Biofizika *15*: 959-964, 1970. [In R, ab: E.]

5243 - MEINL, G.: Assimilationsvermögen als Sortenmerkmal. I. Vergleich der apparenten Assimilation von Kartoffelsorten verschiedener Reifezeit. - Photosynthetica *1*: 51-56, 1967.

5244 - MEINL, G.: Assimilationsvermögen als Sortenmerkmal. II. Trockenmassenproduktion, apparente Assimilation, Respiration und Transpiration von Kartoffelklonen unterschiedlichen Valenzstufen und Sorten bei unterschiedlich hoher NPK-Versorgung. - Photosynthetica *3*: 9-19, 1969.

5245 - MEISTER, A.: Ein registrierendes Spektrophotometer zur Aufzeichnung der Extinktion, ihrer ersten und zweiten Ableitung nach der Wellenlänge. - Exp. Tech. Phys. *14*: 168-173, 1966. [Chl.]

5246 - MEISTER, A.: Zur Untersuchung der verschiedenen Formen von Chlorophyll in der lebenden Pflanze durch Anwendung der Derivativ-Spektrophotometrie. - Kulturpflanze *14*: 235-255, 1966.

5247 - MEISTER, A.: Das Verhalten der Formen von Chlorophyll *a* und *b* im ergrünenden Blatt während Dunkelheit. - Flora A *158*: 512-518, 1967.

5248 - MEISTER, A.: Untersuchung der Chlorophyllbildung kurzzeitig belichteter etiolierter Blätter von *Phaseolus vulgaris* mit Hilfe der Derivativ-Spectrophotometrie. - Photosynthetica *1*: 149-156, 1967.

5249 - MEISTER, A.: Untersuchungen zur Chlorophyllbildung in etiolierten Pflanzen mit Hilfe der Derivativ-Spektrophotometeie. - Stud. biophys. *5*: 59-66, 1967.

5250 - MEISTER, A.: Zur Chlorophyllbildung in kurzzeitig beleuchteten etiolierten Blättern von *Phaseolus vulgaris*. - Kulturpflanze *16*: 91-96, 1968.

5251 - MEISTER, A., MASLOVA, T.G.: Die Bestimmung der lichtinduzierten Absorptionsänderungen durch Messung der 2. Ableitung der Extinktion. - Photosynthetica *2*: 261-267, 1968.

5252 - MEISTER, A., MASLOVA, T.G.: Spektroskopische Untersuchungen der Formen von Chlorophyll im Blatt unter dem Einfluss einiger schädigender Faktoren. - Photosynthetica 3: 63-68, 1969.

5253 - MEISTER, A., SAGROMSKY, H.: Studies of chlorophyll formation by derivative spectrophotometry. - In: Photochemistry and Photobiology in Plant Physiology. Europe. Photobiol. Symp. Hvar, Yugoslavia, 19th-22nd September 1967. Book of Abstracts. Pp. 55-58. Hvar 1967.

5254 - MEISTER, A., STRUS, F.: Experimentelle Untersuchungen des Auflösungsvermögens in der Derivativ-Spektrophotometrie. - Stud. biophys. 8: 135-142, 1968. [Chl.]

5255 - MEKHTIZADE, R.M., FATALIEV, A.T.: Izmenenie soderzhaniya pigmentov v list'yakh yabloni v zavisimosti ot vertikal'noT zonal'nosti. [Changes in pigment content in apple leaves as related to vertical profile.] - Vestn. sel'.-khoz. Nauki 1969 (6): 59-62, 1969. [In R.]

5256 - MEKLER, L.B., BYCHOVSKY, A.F., KRIKUN, B.L.: Electron microscope study of the viricidal properties of sodium magnesium-chlorophyllin. - Nature 222: 574-575, 1969.

5257 - MELANDRI, B.A., BACCARINI, A., FORTI, G.: Light-induced formation of ferredoxin in bean leaves. - Physiol. vég. 5: 337-339, 1967.

5258 - MELANDRI, B.A., BACCARINI, A., FORTI, G.: Selective inhibition by actinomycin D of the synthesis in photosynthetic and non-photosynthetic enzymes during the greening of etiolated bean leaves. - Plant Physiol. 44: 95-100, 1969.

5259 - MELANDRI, B.A., BACCARINI, A., PUPILLO, P.: Glyceraldehyde-3-phosphate dehydrogenase in photosynthetic tissues: kinetic evidence for competitivity between NADP and NAD. - Biochem. Biophys. Res. Commun. 33: 160-164, 1968.

5260 - MELANDRI, B.A., BACCARINI-MELANDRI, A., GEST, H., SAN PIETRO, A.: A bacterial photophosphorylation coupling factor; relevance to light-induced H⁺ uptake. - Plant Physiol. 46 (Suppl.): 40,1970.

5261 - MELANDRI, B.A., BACCARINI-MELANDRI, A., SAN PIETRO, A.: The photosynthetic apparatus of Euglena gracilis. I. Adaptation to population density change. - Arch. Biochem. Biophys. 138: 598-605, 1970.

5262 - MELANDRI, B.A., BACCARINI-MELANDRI, A., SAN PIETRO, A., GEST, H.: Role of phosphorylation coupling factor in light-dependent proton translocation by Rhodopseudomonas membrane preparations. - Proc. nat. Acad. Sci. U.S.A. 67: 477-484, 1970.

5263 - MELANDRI, B.A., PUPILLO, P., BACCARINI-MELANDRI, A.: D-glyceraldehyde-3-phosphate dehydrogenase in photosynthetic cells. I. The reversible light-induced activation in vivo of NADP-dependent enzyme and its relationship to NAD-dependent activities. - Biochim. biophys. Acta 220: 178-189, 1970.

5264 - MELESHCHENKO, S.N.: Dinamika perekhodnykh protsessov zelenogo lista pri izmenenii svetovogo rezhima. [Dynamics of transient processes in a green leaf after a change in light regime.] - Biofizika 11: 933-935, 1966. [In R.]

5265 - MELESHKO, G.I.: Nekotorye kharakteristiki populyatsii khlorelly kak zvena zamknutoT ekologicheskoT sistemy. [Some characteristics of Chlorella population as a link of a closed ecological system.] - In: Problemy Sozdaniya Zamknutykh Ekologicheskikh Sistem. Pp. 73-78. Nauka, Moskva 1967. [Ps; in R.]

5266 - MELHUISH, F.M.: A precise technique for measurement of roots and root distribution in soils. - Ann. Bot. 32: 15-22, 1968.

5267 - MEL'NIKOV, E.S., RODICHEVA, E.K.: Fotoelektricheskoe izmerenie intensivnosti agglyutinatsii. [Photoelectric measurement of agglutination intensity.] - In: Nepreryvnoe Upravlyaemoe Kul'tivirovanie Mikroorganizmov. Pp. 33-37. Nauka, Moskva 1967. [In R.]

5268 - MEL'NIKOV, V.N.: Kharakteristika nekotorykh sposobov opredeleniya estestvennoT osveshchennosti v vodoemakh. [The characteristic of some methods of determination of natural illuminance in water reservoirs.] - Nauch. Dokl. vyssh. Shkoly, biol. Nauki 11 (12): 141-146, 1968. [In R.]

5269 – MEL'NIKOVA, L.M.: O khlorofillaznoĭ aktivnosti khvoi sosny obyknovennoĭ v svyazi s degradatsieĭ pigmentov v osenne-zimniĭ period. [Chlorophyllase activity of pine needles in relation to pigment degradation during autumn and winter.] – In: Tezisy IV NauchnoĭKonferentsii Molodykh Uchenykh po Sovremennym Problemam Biologii. Pp. 3-4. Minsk 1970. [In R.]

5270 – MENCÁKOVÁ, A.: Genetic analysis of chlorophyll content in maize and tobacco. – Photosynthetica 1: 77-88, 1967.

5271 – MENDIOLA, L.R., KOVACS, A.E., PRICE, C.A.: Separation of chloroplast from cytoplasmic ribosomes in Euglena by zonal centrigugation. Plant Physiol. 43: S 6, 1968.

5272 – MENKE, W.: The structure of the chloroplasts. – In: GOODWIN, T.W. (ed.): Biochemistry of Chloroplasts. Vol. I. Pp. 3-18. Academic Press, London-New York 1966.

5273 – MENKE, W.: The molecular structure of photosynthetic lamellar systems. – In: Energy Conversion by the Photosynthetic Apparatus. Brookhaven Symp. Biol. 19: 328-340, 1967.

5274 – MENKE, W.: Far ultraviolet circular dichroism and infrared absorption of thylakoids. – Z. Naturforsch. 25b: 849-855, 1970.

5275 – MENKE, W., WOLFERSDORF, B.: Über die Plastiden von Neottia nidus-avis. – Planta 78: 134-143, 1968.

5276 – MENZ, K.M., MOSS, D.N., CANNELL, R.Q., BRUN, W.A.: Screening for photosynthetic efficiency. – Crop Sci. 9: 692-694, 1969.

5277 – MENZEL, D.W., ANDERSON, J., RANDTKE, A.: Marine phytoplankton vary in their response to chlorinated hydrocarbons. – Science 167: 1724-1726, 1970. [Ps.]

5278 – MENZHULIN, G.V.: Metodika rascheta fotosinteza rastitel'nykh soobshchestv pri dostatochnom uvlazhnenii. [Method of calculating photosynthesis of well irrigated plant communities.] – Tr. glav. geofiz. Observ. 229: 81-103, 1968. [In R.]

5279 – MERAKCHYSKA, M.: Photosynthetic and photochemical activity under streptomycin and iron chlorosis. – Dokl. bolg. Akad. Nauk 23: 1299-1302, 1970.

5280 – MERCER, E.I., PUGHE, J.E.: The effects of abscisic acid on the biosynthesis of isoprenoid compounds in maize. – Phytochemistry 8: 115-122, 1969. [Chl, car.]

5281 – MERCER, E.I., THOMAS, G.: The occurrence of ATP-adenylsulphate 3'-phosphotransferase in the chloroplasts of higher plants. – Phytochemistry 8: 2281-2285, 1969.

5282 – MERCER, E.I., TREHARNE, K.J.: Occurrence of sterols in chloroplasts. – In: GOODWIN, T.W. (ed.): Biochemistry of Chloroplasts. Vol. I. Pp. 181-185. Academic Press, London-New York 1966.

5283 – MEREDITH, F.I., PURCELL, A.E.: Changes in the concentration of carotenes of ripening Homestead tomatoes. – Proc. amer. Soc. hort. Sci. 89: 544-548, 1966.

5284 – MEREDITH, F.I., YOUNG, R.H.: Effect of temperature on pigment development in Red Blush grapefruit and Ruby blood oranges. – Proc. first int. Citrus Symp. 1: 271-276, 1969. [Chl, Car.]

5285 – MEREZHINSKIĬ, Yu.G., LAPINA, T.V., IVANISHCHEV, V.N., BELOUS, I.I.: Vliyanie gerbitsidov na nekotorye storony fotosinteza. [Herbicide effect on some aspects of photosynthesis.] – Fiziol. Biokhim. kul't. Rast. 1: 255-260, 1969. [Ps, Chl; in R, ab: E.]

5286 – MEREZHKO, A.I.: Ob istochnikakh ugleroda pri avtotrofnom pitanii sinezelenykh vorodoslĕĭ. [Carbon sources in autotrophic nutrition of blue-green algae.] – In: Tsvetenie Vody. Pp. 187-196. Naukova Dumka, Kiev 1968. [Ps; in R.]

5287 – MERKELO, H., HARTMAN, S.R., MAR, T., SINGHAL, G.S., GOVINDJEE: Mode-locked lasers: Measurements of very fast radiative decay in fluorescent systems. – Science 164: 301-302, 1969.

5288 – MERMIER, M., MÉTHY, M.: Mesure du rayonnement net. – In: Techniques d'Études des Facteurs Physiques de la Biosphère. Pp. 71-77. INRA, Paris. 1970.

5289 – MERRETT, M.: Observations on the fine structure of Chlamydobotrys stellata, with particular reference to its unusual chloroplast structure. – Arch. Mikrobiol. 65: 1-11, 1969.

5290 - MERRETT, M.J.: Carbon dioxide fixation in *Pyrobotrys stellata*. - Brit. phycol. Bull. *3*: 219-223, 1967.

5291 - MESSER, Y., BEN-SHAUL, Y., NEUMANN, J.: Effects of sonication on chloroplast ultrastructure. - Israel J. Chem. *6*: 131 p, 1968.

B5292 - Metabolizm i Stroenie Fotosinteticheskogo Apparata. [Metabolism and Structure of Photosynthetic Apparatus.] - Nauka i Tekhnika, Minsk 1970. [In R.]

5293 - MÉTHY, M.: Les méthodes de mesure de la composition spectrale du rayonnement global. - In: Techniques d'Étude des Facteurs Physiques de la Biosphère. Pp. 79-87. INRA, Paris 1970.

B5294 - Metody Kompleksnogo Izucheniya Fotosinteza. (Metodicheskiĭ Sbornik). [Methods of Complex Study of Photosynthesis. (Methodological Collection).] - Akad. sel'.-khoz. Nauk, Leningrad 1969. [In R.]

5295 - METSNER, Kh., FISHER, K., SHTRUSS, Z.: Primenenie metoda protivotochnogo raspredeleniya v tsitokhimicheskikh issledovaniyakh. [Use of counter-current distribution in cytochemical studies.] - In: Funktsional'naya Biokhimiya Kletochnykh Struktur. Pp. 110-113. Nauka, Moskva 1970. [Chl; in R.]

5296 - METZNER, H.: Photosynthese und Bedeutung wichtiger Kohlenhydrate in Pflanzen. - In: *D*-Glucose und verwandte Verbindungen in Medizin und Biologie. Pp. 192-228. Ferdinand Enke Verlag, Stuttgart 1966.

5297 - METZNER, H.: Photochemische Aktivität isolierter Chloroplasten. - Naturwissenschaften *53*: 141-150, 1966.

5298 - METZNER, H.: B. Physiologie. I. Photosynthese. - In: Fortschritte der Botanik. Vol. 28. Pp. 45-59. Springer-Verlag, Berlin-Heidelberg-New York 1966.

5299 - METZNER, H.: Spektroskopische Messungen an lebenden Algenzellen. - In: THOMAS, J.B., GOEDHEER, J.C. (ed.): Currents in Photosynthesis. Pp. 17-30. Donker, Rotterdam 1966. [Chl.]

5300 - METZNER, H.: Untersuchungen zur Chlorophyll-Absorption *in vivo*. - Z. Pflanzenphysiol. *54*: 183-194, 1966.

5301 - METZNER, H.: IV. Photosynthese. - In: Fortschritte der Botanik. Vol. 29. Pp. 97-118. Springer-Verlag, Berlin-Heidelberg-New York 1967.

5302 - METZNER, H.: Über eine Eingefäss-Methode zur kontinuierlichen Messung von Photosynthese- und Atmungsquotienten. - Photosynthetica *1*: 249-252, 1967.

5303 - METZNER, H.: Untersuchungen über das Pigmentsystem der Cyanophyceen. - Stud. biophys. *5*: 7-16, 1967.

5304 - METZNER, H.: Lichtinduzierte Wasserspaltung im Photosynthese-Modell. - Hoppe-Seyler's Z. physiol. Chem. *349*: 1586-1588, 1968.

5305 - METZNER, H.: Zur Messung der $^{14}CO_2$-Aufnahme in pflanzliche Zellen. - Rev. roum. Biol.,Sér. Bot. *13*: 69-72, 1968.

5306 - METZNER, H.: Comparative biochemistry of photosynthesis. A report of the seminar held in Gwatt, Switzerland, 21-26 July 1969. - FEBS Letters *5*: 93-95, 1969.

5307 - METZNER, H.: Die Photolyse des Wassers im Reagenzglass. - Umschau Wiss. Tech. *1969* (5): 147, 1969.

B5308 - METZNER, H. (ed.): Progress in Photosynthesis Research. Vol. I-III. - Proceedings of the International Congress of Photosynthesis Research, Freudenstadt, Germany, June 4-8, 1968. Tübingen 1969.

5309 - METZNER, H.: Optical properties of algal suspensions. - In: Prediction and Measurement of Photosynthetic Productivity. Pp. 503-505. PUDOC, Wageningen 1970. [Chl.]

5310 - METZNER, H.: Photosynthese. - Fortschr. Bot. *32*: 55-68, 1970.

5311 - METZNER, H.: Photosynthese - *in vivo* und *in vitro*. - Fortschr. Med. *88*: 209-210, 1970.

5312 - METZNER, H., FISCHER, K.: Becquerel effect and Hill reaction in model systems. - In: METZNER, H. (ed.): Progress in Photosynthesis Research. Vol. II. Pp. 1027-1031. Tübingen 1969.

5313 - METZNER, H., FISCHER, K., SCHREIBER, U., HOFFMANN, N.: Fluoreszenzinduktion an Chlorophyll-beschichteten Halbleitern. - Naturwissenschaften *57*: 494, 1970.

5314 - MEYER, T.E.: Comparative studies on soluble iron-containing proteins in photosynthetic bacteria and some algae. - Diss. Abstr. B *31*: 3233, 1970.

5315 - MEYER, T.E., BARTSCH, R.G., CUSANOVICH, M.A., MATHEWSON, J.H.: The cytochromes of *Chlorobium thiosulfatophilum*. - Biochim. biophys. Acta *153*: 854-861, 1968.

5316 - MICHAEL, G.: Untersuchungen über die winterliche Dürreresistenz einiger immergrüner Gehölze im Hinblick auf eine Frosttrocknisgefahr. - Flora B *155*: 350-372, 1966. [Ps.]

5317 - MICHAEL, G.: Prüfung der Lebensfähigkeit geschädigter Pflanzen mit Hilfe der kolorimetrischen Methode nach KAUKO/ÅLVIK. - Tagungsber. deut. Akad. Landwirtschaftswiss. Berlin, *100*: 107-112, 1968. [Ps.]

5318 - MICHAEL, G.: Eine Methode zur Bestimmung der Spaltöffnungsweite von Koniferen. - Flora A *159*: 559-561, 1969.

5319 - MICHAEL, G., KOUHSIAHI-TORK, K., WILBERG, E.: Einfluss unterschiedlicher Luftfeuchtigkeit auf Chlorophyll- und Eiweissabbau in Blättern von Tabakpflanzen. - Flora A *160*: 186-195, 1969.

5320 - MICHAELIS, P., FRITZ, H.G.: Beiträge zum Problem der Plastiden-Abänderung II. Chlorophyllbestimmungen an Pflanzen mit Plastiden-"Rückmutationen". - Z. Naturforsch. *21 b*: 66-71, 1966.

5321 - MICHEL, J.M.: Electrophoretic study of the chlorophyll-lipoprotein complexes of *Euglena*. - Carnegie Inst. Year Book *66*: 186-189, 1968.

5322 - MICHEL, J.-M., MICHEL-WOLWERTZ, M.-R.: Fractionation of the photosynthetic apparatus from broken spinach chloroplasts by sucrose density-gradient centrifugation. - Carnegie Inst . Year Book *67*: 508-514, 1969.

5323 - MICHEL, J.M., MICHEL-WOLWERTZ, M.R.: On the fractionation of the photosynthetic apparatus of spinach chloroplasts. - In: METZNER, H. (ed.): Progress in Photosynthesis Research. Vol. I. Pp. 115-121. Tübingen 1969.

5324 - MICHEL, J.-M., MICHEL-WOLWERTZ, M.-R.: Fractionation and photochemical activities of photosystems isolated from broken spinach chloroplasts by sucrose-density gradient centrifugation. - Photosynthetica *4*: 146-155, 1970.

5325 - MICHEL, J.-P., GUÉRIN de MONTGAREUIL, P.: Phosphorylations associées à l'activation par le rouge lointain d'une capacité stable en oxygène photosynthétique. - Compt. rend. Acad. Sci. (Paris), Sér. D *270*: 2655-2658, 1970.

5326 - MICHEL, J.-P., THIBAULT, P.: Effet antagoniste rouge-rouge lointain au niveau des photophosphorylations associées à l'activation du dégagement d'oxygène photosynthétique. - Compt. rend. Acad. Sci. (Paris), Sér. D *271*: 976-979, 1970.

5327 - MICHEL-WOLWERTZ, M.-R.: The chlorophylls extracted from plants by organic solvents. - Carnegie Inst. Year Book *66*: 189-192, 1968.

5328 - MICHEL-WOLWERTZ, M.-R.: Effect of enzymatic digestion of chloroplast lamellae on chlorophyll absorption. - Carnegie Inst. Year Book *67*: 505-508, 1969.

5329 - MICHEL-WOLWERTZ, M.-R., MICHEL, J.-M.: Absorption spectra of fractions obtained by sucrose gradient centrifugation from *Chlorella pyrenoidosa*. - Carnegie Inst. Year Book *67*: 514-516, 1969.

5330 - MICHNA, M.: Kształtowanie się niektórych składników chemicznych w sależności od stopnia dojrzałości owoców u kilku odmian papryki pochodzienia zagranicznego. [Formation of some chemical components in several red pepper cultivars of foreign origin depending on the degree of fruit ripeness.] - Roczn. Nauk roln. A *91*: 421-428, 1966. [Car; in Pol., ab: E, R.]

5331 - MICHNIEWICZ, M., CHROMIŃSKI, A., BELT, H.: Retardation of chlorophyll degradation as possible effect of (2-chloroethyl)trimethylammonium chloride on increase of winter grain yield. - Bull. Acad. pol. Sci., Sér. Sci. biol. *16*: 451-453, 1968.

5332 - MICKIEWICZ, E.: Współczesne poglądy na przemiany energii świetlnej w fotosyntezie. [Recent views on transformation of radiant energy in photosynthesis.] - Wiadom. bot. *11*: 89-111, 1967. [In Pol.]

5333 - MIEDZIEJKO, E.: Optical properties of chlorophyllide-cytochrome solutions. -
       Bull. Acad. pol. Sci., Sér. Sci. math., astr., phys. *18*: 701-706, 1970.

5334 - MIEDZIEJKO, E., FRACKOWIAK, D.: Excitation energy transfer in bilirubin-chloro-
       phyll aggregates. - Photochem. Photobiol. *10*: 97-108, 1969.

5335 - MIEN, Ch.Kh.: Vliyanie vlazhnosti pochvy i urovnya mineral'nogo pitaniya na
       produktivnost' fotosinteza kukuruzy. [Effect of soil moisture and mineral nut-
       rition on photosynthetic productivity of maize.] - In: Fotosintez i Urozhaĭnost'
       Sel'skokhozyaĭstvennykh Rasteniĭ. Pp. 16-20. Min. sel'. Khoz. SSSR, Kiev 1970.
       [In R.]

5336 - MIFLIN, B.J., HAGEMAN, R.H.: Activity of chloroplasts isolated from maize inbreds
       and their F_1 hybrids. - Crop Sci. *6*: 185-187, 1966.

5337 - MIFLIN, B.J., MARKER, A.F.H., WHITTINGHAM, C.P.: The metabolism of glycine and
       glycollate by pea leaves in relation to photosynthesis. - Biochim. biophys. Acta
       120: 266-273, 1966.

5338 - MIFLIN, B.J., WHITTINGHAM, C.P.: The effect of inhibitors on the path of carbon
       in photosynthesis by *Chlorella* at low partial pressures of CO_2. I. Methylamine.
       - Ann. Bot. *30*: 329-337, 1966.

5339 - MIFLIN, B.J., WHITTINGHAM, C.P.: The effect of inhibitors on the path of carbon
       in photosynthesis by *Chlorella* at low partial pressures of CO_2. II. The effect
       of inhibitors on oxygen evolution. - Ann. Bot. *30*: 339-347, 1966.

5340 - MIGINIAC-MASLOW, M.: Effet de quelques glucides sur l'intensité de la photophos-
       phorylation. - Bull. Soc. franç. Physiol. vég. *13*: 279-283, 1967.

5341 - MIGINIAC-MASLOW, M.: Action comparée de l'atractyloside sur les phosphorylations
       réalisées par les mitochondries et les chloroplastes isolées des végétaux. -
       Compt. rend. Séances Soc. Biol. *162*: 12-16, 1968.

5342 - MIGINIAC-MASLOW, M., MOYSE, A.: Some features of the photophosphorylation of
       isolated intact chloroplasts. - In: METZNER, H. (ed.): Progress in Photosynthe-
       tic Research. Vol. III. Pp. 1203-1212. Tübingen 1969.

5343 - MIGINIAC-MASLOW, M., MOYSE, A.: Structure et activité de photophosphorylation
       de chloroplastes isolés en présence de ClNa ou de saccharose. - Compt. rend.
       Séances Soc. Biol. *163*: 2491-2496, 1969 (1970).

5344 - MIHARA, S., KIMURA, K., HASE, E.: Studies on ribonucleic acids from *Chlorella*
       protothecoides with special reference to the degradation of chloroplast RNA
       during the process of "glucose-bleaching". - Plant Cell Physiol. *9*: 87-102, 1968.

5345 - MIHNEA, P.: Despre productivitatea primara a bazinelor acvatice. [On the primary
       productivity of aquatic basins.] - Natura, Ser. Biol. *19* (5): 38-45, 1967. [In
       Rum.]

5346 - MIKA, A.: Effects of shoot bending of apple trees on accumulation and transloca-
       tion of ^{14}C-labelled assimilates. - Biol. Plant. *11*: 175-182, 1969.

5347 - MIKABERIDZE, V.E.: Vliyanie gustoty razmeshcheniya limonnykh derev'ev na intensiv-
       nost' fotosinteza v svyazi s ikh morozoustoĭchivost'yu. [Effect of stand density
       of lemon trees on photosynthetic rate in relation to their frost resistance.] -
       Subtrop. Kul't. *1966* (4): 32-36, 1966. [In R, ab: E.]

5348 - MIKABERIDZE, V.E.: Vliyanie gustoty razmeshcheniya derev'ev na sezonnye izmene-
       nyya pigmentov limonnogo rasteniya i na ego morozoustoĭchivost'. [Effect of tree
       stand density on seasonal changes in pigments of lemon trees and on their frost
       resistance.] - Subtrop. Kul't. *1967* (1): 80-84, 1967. [In R, ab: E.]

5349 - MIKHAÏLOVA, L.P.: O deĭstvii nematotsidov na nekotorye fiziologicheskie protses-
       sy ogurtsev i pomidorov. [Effect of nematocides on some physiological properties
       in cucumber and tomato.] - In: Nematodnye Bolezni Sel'skokhozyaĭstvennykh Raste-
       niĭ. Pp. 90-100. Kolos, Moskva 1967. [In R.]

5350 - MIKHAÏLOVA, S.: Vliyanie na mikroelementa bor v"rkhu nyakoi fiziologichni prot-
       sesi i dobiva pri sort Bolgar. [Effect of boron on some physiological processes
       and yield of the variety Bolgar.] - Grad. lozar. Nauka *4* (8): 75-83, 1967. [Ps
       and Chl in vine; in Bulg.]

5351 - MIKHAÏLOVA, S.A.: O kharaktere izvlecheniya khlorofilla iz kletok khlorelly pos-
       ledovatel'no primenyaemymi rastvoritelyami raznoĭ prirody. [Character of chloro-
       phyll extraction from *Chlorella* cells by subsequent use of different solvents.]
       - In: Tezisy IV Nauchnoĭ Konferentsii Molodykh Uchenykh po Sovremennym Problemam
       Biologii. Pp. 51-52. Minsk 1970. [In R.]

5352 - MIKHAÏLOVA, S.A.: K voprosu o prostranstvennoĭ blizosti metabolicheski geterogen-
       nykh form khlorofilla u khlorelly. [Spatial proximity of metabolically heteroge-
       nous chlorophyll forms in *Chlorella*.] - In: Tezisy IV Nauchnoĭ Konferenstii Molo-
       dykh Uchenykh po Sovremennym Problemam Biologii. Pp. 52-54. Minsk 1970. [In R.]

5353 - MIKHAÏLOVA, T.L.: Ingibirovanie fotosinteza ftoridom natriya. [Inhibition of
       photosynthesis by sodium fluoride.] - Uch. Zap. ural'. gos. Univ., Ser. biol.
       58: 58-66, 1967. [In R.]

5354 - MIKHALEVA, E.N., KONOVALOV, I.N.: O fiziologicheskoĭ raznokachestvennosti pobe-
       gov gretskogo orekha. [Physiological heterogeneity of walnut shoots.] - Tr.
       bot. Inst. V.L. Komarova Akad. Nauk SSSR, Ser. 4 - eksp. Bot. *19*: 206-215, 1967.
       [Ps; in R, ab: E.]

5355 - MILAEV, Ya.I., BURDIN, A.G.: Metodika izmereniya intensivnosti fotosinteza u ku-
       kuruzy v polevykh usloviyakh. [Method of measuring photosynthetic rate in maize
       in field conditions.] - Selektsiya Semenovodstvo (Kiev) *12*: 60-68, 1969. [In R.]

5356 - MILAEV, Ya.I., PRIMAK, N.N.: Sravnitel'noe opredelenie kolichestva pigmentov v
       list'yakh kukuruzy i tabaka uskorennym metodom. [Comparative determination of
       pigment amount in leaves of maize and tobacco by a rapid method.] - Selektsiya
       Semenovodstvo (Kiev) *12*: 69-72, 1969. [In R.]

5357 - MILES, C.D., JAGENDORF, A.T.: Ionic and pH transitions triggering chloroplast
       post-illumination luminescence. - Arch. Biochem. Biophys. *129*: 711-719, 1969.

5358 - MILES, C.D., JAGENDORF, A.T.: Evaluation of electron transport as the basis of
       adenosine triphosphate synthesis after acid-base transition by spinach chloro-
       plasts. - Biochemistry *9*: 429-434, 1970.

5359 - MILES, D., JAGENDORF, A.: Induction of chloroplast luminescence by a salt gra-
       dient. - Plant Physiol. *43*: S 28, 1968.

5360 - MILLER, D.H., MACHLIS, L.: Phytochrome-mediated chlorophyll synthesis in the
       liverwort *Sphaerocarpos donnellii*. - Plant Physiol. *42*: S-10, 1967.

5361 - MILLER, D.H., MACHLIS, L.: Effects of light on the growth and development of the
       liverwort, *Sphaerocarpos donnellii* AUST. - Plant Physiol. *43*: 714-722, 1968.
       [Ps, Chl.]

5362 - MILLER, D.H., MACHLIS, L.: Light mediated changes in the chloroplasts of the
       liverwort, *Sphaerocarpos donnellii* AUST. - Plant Physiol. *43*: 723-729, 1968.

5363 - MILLER, P.C.: Tests of solar radiation models in three forest canopies. - Ecolo-
       gy *50*: 878-885, 1969. [Growth analysis.]

5364 - MILLER, P.R.: The relationship of ozone to suppression of photosynthesis and to
       the cause of the chlorotic decline of ponderosa pine. - Diss. Abstr. *26*: 3574-
       3575, 1966.

5365 - MILLER, W.F.: Seasonal discoloration of loblolly pine foliage. - Forest Sci. *12*:
       296-297, 1966. [Chl.]

5366 - MILLERD, A., GOODCHILD, D.J., SPENCER, D.: Studies on a maize mutant sensitive
       to low temperature II. Chloroplast structure, development, and physiology. -
       Plant Physiol. *44*: 567-583, 1969.

5367 - MILLERD, A., McWILLIAM, J.R.: Studies on a maize mutant sensitive to low temper-
       ature. I. Influence of temperature and light on the production of chloroplast
       pigments. - Plant Physiol. *43*: 1967-1972, 1968.

5368 - MILLETTI, G., de CAPITE, L.: Fotosintesi e respirazione in *Hydrangea macrophylla*
       SER. trattata con CCC e B-995. [Photosynthesis and respiration in *Hydrangea ma-
       crophylla* SER. treated with CCC and B-995.] - Ann. Fac. Agr. Univ. Studi Perugia
       23: 183-193, 1968. [In Ital.]

5369 - MILLINGTON, R.J., PETERS, D.B.: Exchange (mass transfer) coefficients in crop
       canopies. - Agron. J. *61*: 815-819, 1969. [Ps.]

5370 - MILNE, P.R., WELLS, J.R.E.: Structural and molecular weight studies on the small copper protein, plastocyanin. - J. biol. Chem. 245: 1566-1574, 1970.

5371 - MILOGRADOVA, E.I., BERDYKULOV, Kh.A.: Vliyanie temperatury i intensivnosti osveshcheniya na fotosintez *Chlorella pyrenoidosa* CHICK. v ustanovkakh pod otkrytym nebom. [Effect of temperature and irradiance on photosynthesis of *Chlorella pyrenoidosa* CHICK. in open-air cultivation units.] - In: Dikorastushchie i Vvodimye v Kul'turu Rasteniya Uzbekistana. Pp. 5-11. FAN, Tashkent 1966. [In R.]

5372 - MILOSAVLJEVIC, M.: Einfluss der Unterlage auf die Intensität der Photosynthese und den Transport der Assimilate bei jungen Pfropfreben. - Vitis 7: 6-9, 1968.

5373 - MILOSAVLJEVIĆ, M., POPOVIĆ, R.: Uticaj bora i mangana, na intenzitet fotosinteze vinove loze. [Effect of boron and manganese on photosynthetic rate of grape vines.] - Arh. poljopr. Nauke 23 (83): 15-24, 1970. [In Croat., ab:E.]

5374 - MILTHORPE, F.L.: Some physiological principles determining the yield of root crops. - In: Proceedings of the International Symposium on Root Crops. Vol. I. Pp. II-1-II-18. Univ. West Indies, St. Augustine 1969. [Growth analysis.]

5375 - MILTHORPE, F.L., PENMAN, H.L.: The diffusive conductivity of the stomata of wheat leaves. - J. exp. Bot. 18: 422-457, 1967.

B5376 - Mineral'nye Elementy i Mekhanizm Fotosinteza. [Mineral elements and photosynthesis Mechanism.] - Red.-izdat. Otdel Akad. Nauk MSSR, Kishinev 1969. [In R.]

5377 - MINOHARA, Y., TAKANASHI, S., MATUO, E.: Growth and development of horticultural plants in artificial lights: 2. Effect of the light quality on chlorophyll formation in the cucumber seedlings. - J. agr. Lab. (Chiba) 9: 65-70, 1967.

5378 - MIRONYUK,V.I., EĬNOR, L.O.: Kislorodnyĭ obmen i soderzhanie pigmentov u raznykh form *Dunaliella salina* TEOD. v usloviyakh povysheniya soderzhaniya khloristogo natriya. [Oxygen metabolism and pigment content in different forms of *Dunaliella salina* TEOD. under increased sodium chloride content.] - Gidrobiol. Zh. 4: 23-29, 1968. [In R.]

5379 - MIRONYUK,V.I., SEMICHAEVSKIĬ, V.D.: Vliyanie karotina na fotosensibilizatsionnuyu aktivnost' khlorofilla *a*. [Effect of carotene on photosensibilization activity of chlorophyll *a*.] - In: Tezisy Dokladov II Vsesoyuznogo Biokhimicheskogo S"ezda. 19. Sektsiya: Problemy Fotosinteza. Pp. 84-85. FAN, Tashkent 1969. [In R.]

5380 - MIROSLAVOV, E.A.: O svoeobraznykh strukturakh v plastidakh zamykayushchikh kletok ust'its lista *Vicia faba* L. [Specific structures in plastids of stomatal guard cells in a *Vicia faba* L. leaf.] - Bot. Zh. 51: 982-983, 1966. [In R.]

5381 - MIROSLAVOVA, S.A.: Pererabotka organicheskikh kislot na svetu diskami lista *Bryophyllum daigremontianum*. [Transformation of organic acids in light by leaf discs of *Bryophyllum daigremontianum*.] - Vestn. leningrad. Univ. 1968 (15): 116-121, 1968. [Ps; in R, ab: E.]

5382 - MIRSAGATOV, M.M.: Transpiratsiya i fotosintez kostochkovykh porod v gornykh raĭonakh. [Transpiration and photosynthesis of stone-fruit crops in mountain areas.] - Uzb. biol. Zh. 13 (2): 25-26, 1969. [In R, ab: E, Uzb.]

5383 - MISHRA, A.K., JHA, A.: Paper chromatography of leaf juice from healthy and mosaic infected chili plants. - Indian Phytopathol. 20: 387-388, 1967. [Chl, Car.]

5384 - MISHRA, D.: Interaction of benzimidazole and allantoin on the chlorophyll content of detached wheat leaves. - Naturwissenschaften 53: 483-484, 1966.

5385 - MISHRA, D., MISRA, B.: Effect of growth regulating chemicals on degradation of chlorophyll and starch in detached leaves of crop plants. - Z. Pflanzenphysiol. 58: 207-211, 1968.

5386 - MISHRA, D., WAYGOOD, E.R.: Effect of benzimidazole and kinetin on the nicotinamide nucleotide content of senescing wheat leaves. - Can. J. Biochem. 46: 167-178, 1968.

5387 - MISRA, R.: Form, function and factors in ecology, - J. indian bot. Soc. 46: 144-153, 1967. [Primary production.]

5388 - MISRA, R., SINGH, J.S., SINGH, K.P.: Preliminary observations on the production of dry matter by sal (*Shorea robusta* GAERTN. F.). - Trop. Ecol. 8: 94-104, 1967.

5389 - MISRA, R., SINGH, J.S., SINGH, K.P.: Dry matter production in sun and shade
       leaves and a simple method for the measurement of primary production. - Curr.
       Sci. *37*: 306-307, 1968.

5390 - MISRA, R., SINGH, J.S., SINGH, K.P.: A new hypothesis to account for the opposite
       trophic-biomass structure on land and in water. - Curr. Sci. *37*: 382-383, 1968.

5391 - MITCHELL, P.: Chemiosmotic coupling in oxidative and photosynthetic phosphoryla-
       tion. - Biol. Rev. Cambridge phil. Soc. *41*: 445-502, 1966.

5392 - MITCHELL, P.: Proton-translocation phosphorylation in mitochondria, chloroplasts
       and bacteria: Natural fuel cells and solar cells. - Fed. Proc. *26*: 1370-1379,
       1967.

5393 - MITCHELL, R.L., ANDERSON, I.C.: Effect of gibberellic acid in reducing Fe chloro-
       sis in soybeans. - Crop Sci. *6*: 111-112, 1966.

5394 - MITRANI, L., BALABANOV, N., MITRIKOV, M.: Izsledovane kinetikata na fotosintetich-
       nata biokhemiluminestsentsiya po metoda na neprek"snatiya potok. [Study on the
       kinetics of the photosynthetic bioluminescence by the constant stream method.] -
       Nauch. Tr. vissh. ped. inst. Plovdiv *4* (2): 83-92, 1966. [In Bulg., ab: R, E.]

5395 - MITROFANOV, B.A., GULYAEV, B.I., MAKHOVSKAYA, M.A., LAVRENTOVICH, D.I., POCHINOK,
       Kh.N., OKANENKO, A.S.: Rol' list'ev, stebleĭ i kolos'ev ozimoĭ pshenitsy v foto-
       sinteze poseva. [Role of leaves, stems and ears in photosynthesis of a winter
       wheat stand.] - In: Puti Povysheniya Intensivnosti i Produktivnosti Fotosinteza.
       Vol. 3. Pp. 69-86. Naukova Dumka, Kiev 1969. [In R.]

5396 - MITSCHERLICH, G., MOLL, W., KÜNSTLE, E., MAURER, P.: Ertragskundlich-ökologische
       Untersuchungen im Rein- und Mischbestand. V. Relative Luftfeuchte, Evaporation
       und CO_2-Gehalt der Luft. - Allg. Forst - Jagdzeitung *137* (2): 25-33, 1966. [CO_2
       content in forests.]

5397 - MITSUDA, H., NAKAMURA, H., YASUMOTO, K.: Properties of *Chlorella* cells grown un-
       der various photo-heterotrophic conditions. - Plant Cell Physiol. *11*: 281-292,
       1970. [Chl, Car.]

5398 - MIYACHI, S.: Labile $C^{14}O_2$ fixation products by chloroplasts and their possible
       relation to the mechanism of photosynthesis. - In: Energy Conversion by the Pho-
       tosynthetic Apparatus. Brookhaven Symp. Biol. *19*: 491-498, 1967.

5399 - MIYACHI, S.: Carbon dioxide fixation in green plants. - In: METZNER, H. (ed.):
       Progress in Photosynthesis Research. Vol. III. Pp. 1678-1681. Tübingen 1969.

5400 - MIYACHI, S.: [Carbon dioxide fixation in green plants.] - Chem. Life [Kagaku to
       Seibutsu] *8*: 129-138, 1970. [Ps; in Jap.]

5401 - MIYACHI, S., HOGETSU, D.: Light-enhanced carbon dioxide fixation in isolated
       chloroplasts. - Plant Cell Physiol. *11*: 927-936, 1970.

5402 - MIYACHI, S., HOGETSU, D.: Effect of preillumination with light of different
       wavelengths on subsequent dark CO_2-fixation in *Chlorella* cells. - Can. J. Bot.
       48: 1203-1207, 1970.

5403 - MIYACHI, S., KANAI, R., BENSON, A.A.: Aerobically bound CO_2 in *Chlorella* cells.
       - In: SHIBATA, K., TAKAMIYA, A., JAGENDORF, A.T., FULLER, R.C. (ed.): Compara-
       tive Biochemistry and Biophysics of Photosynthesis. Pp. 246-252. Univ. Tokyo
       Press, Tokyo; Univ. Park Press, State College, Pa. 1968.

5404 - MIYACHI, S., MIYACHI, S.: Sulfolipid metabolism in *Chlorella*. - Plant Physiol.
       41: 479-486, 1966. [Ps, Chl.]

5405 - MIYASAKA, A., IIO, S.: [On the measurements of the distribution of reflected
       and transmitted light energy of the leaf by the spectroradiometer with the fiber
       photo-guide head.] - Proc. Crop Sci. Soc. Jap. *39*: 533-534, 1970. [In Jap.]

5406 - MIYAZAKI, T., TATEMICHI, Y.: [The change of photosynthetic activity and the va-
       rietal difference in the ripening of tobacco.] - Proc. Crop Sci. Soc. Jap. *37*:
       135-139, 1968. [In Jap., ab: E.]

5407 - MIYOSHI, Y., TSUBO, Y.: Permanent bleaching of *Euglena* by chloramphenicol. -
       Plant Cell Physiol. *10*: 221-225, 1969.

5408 - MŁODZIANOWSKI, F., SZWEYKOWSKA, A., SCHNEIDER, J.: The effect of chloramphenicol on the ultrastructure of chloroplasts in the protonema of *Funaria hygrometrica*. - Acta Soc. Bot. Pol. *39*: 37-43, 1970.

5409 - MOCHALKIN, A.I., SAMOLADAS, T.Kh.: Izuchenie reemissii sveta list'yami rasteniĭ tsitrusovykh i drugikh subtropicheskikh kul'tur, obrabotannykh gerbitsidami. [Light re-emission of leaves of *Citrus* and other subtropical plants treated with herbicides.] - Subtrop. Kul't. *1970* (5): 145-147, 1970. [In R.]

5410 - MOELKER, W.H.: Photoelectric equipment for a quick determination of leaf areas. - Plant Soil *25*: 305-308, 1966.

5411 - MOGILEVA, G.A.: K izucheniyu fotosinteza nekotorykh sortov kapusty, tomatov i ogurtsov. [Photosynthesis in some cultivars of cabbage, tomato and cucumber.] - Byul. vsesoyuz. Ord. Lenina Inst. Rastenievod. I.I. Vavilova *14*: 43-45, 1969. [In R.]

5412 - MOHANTY, P., MUNDAY, J.C. Jr., GOVINDJEE: Time-dependent quenching of chlorophyll *a* fluorescence from (pigment) system II by (pigment) system I of photosynthesis in *Chlorella*. - Biochim. biophys. Acta *223*: 198-200, 1970.

5413 - MOHSIN, M.A.: A field porometer for use on paddy plants. - Riso *19*: 169-173, 1970.

5414 - MOIZ, A.: Assimilyatsiya CO_2 preparatami izolirovannykh khloroplastov (faktory, limitiruyushchie assimilyatsiyu). [CO_2 assimilation in isolated chloroplasts (factors limiting assimilation).] - In: Funktsional'naya Biokhimiya Kletochnykh Struktur. Pp. 87-96. Nauka, Moskva 1970. [In R.]

5415 - MOKIEVSKIĬ, K.A., RYCHKOVA, M.A.: Zavisimost' fotosinteza pogruzhennoĭ vodnoĭ rastitel'nosti ot intensivnosti pronikayushcheĭ radiatsii. [Dependence of photosynthesis in submersed plants on density of penetrating radiant flux.] - In: Aktinometriya i Optika Atmosfery. Pp. 331-335. Valgus, Tallin 1968. [In R.]

5416 - MOKRONOSOV, A.T.: Ispol'zovanie produktov fotosinteza v rostovykh protsessakh. [Utilization of photosynthates in growth processes.] - In: Fotosinteziruyushchie Sistemy Vysokoĭ Produktivnosti. Pp. 157-161. Nauka, Moskva 1966. [In R.]

5417 - MOKRONOSOV, A.T.: Nekotorye voprosy metodiki primeneniya izotopa ugleroda-14 dlya izucheniya fotosinteza. [Some methodological questions of the use of carbon-14 for studying photosynthesis.] - Zap. sverdl. Otd. vsesoyuz. bot. Obshch. *1966* (4): 3-13, 1966. [In R.]

5418 - MOKRONOSOV, A.T.: Regulyatsiya fotosinteza. [Regulation of photosynthesis.] - Uch. Zap. ural'. gos. Univ., Ser. biol. *58*: 3-16, 1967. [In R.]

5419 - MOKRONOSOV, A.T.: Endogennaya determinatsiya fotosinteza v sisteme rasteniya. [Endogenous control of photosynthesis in a plant.] - Uch. Zap. ural'. gos. Univ. *113*: 3-19, 1970. [In R.]

5420 - MOKRONOSOV, A.T., BAGAUTDINOVA, R.I.: Kompensatornye yavleniya v regulirovanii fotosinteza. [Compensation phenomena in control of photosynthesis.] - Zap. sverdl. Otd. vsesoyuz. bot. Obshch. *1970* (5): 68-76, 1970. [In R.]

5421 - MOKRONOSOV, A.T., IVANOVA, N.A.: Fotosinteticheskaya funktsiya lista kartofelya v avtonomnom i sistemnom rezhimakh. [Photosynthetic function of a potato leaf in autonomous and system conditions.] - Fiziol. Rast. *17*: 265-273, 1970. [In R, ab: E.]

5422 - MOKRONOSOV, A.T., NEKRASOVA, G.F.: Kataliticheskaya rol' CO_2 v assimilyatsii ugleroda. [Catalytic function of CO_2 in carbon assimilation.] - Dokl. Akad. Nauk SSSR *165*: 1200-1203, 1966. [In R.]

5423 - MOKRONOSOV, A.T., NEKRASOVA, G.F.: Metabolicheskie aspekty perekhodnykh sostoyaniĭ fotosinteza v induktsionnom periode. [Metabolic aspects of the transition states of photosynthesis during the induction period.] - Fiziol. Rast. *13*: 385-397, 1966. [In R, ab: E.]

5424 - MOKRONOSOV, A.T., NEKRASOVA, G.F.: Fotosinteticheskiĭ metabolizm ugleroda pri defitsite CO_2. [Photosynthetic carbon metabolism accompanying CO_2 deficiency.] - Dokl. Akad. Nauk SSSR *173*: 1463-1465, 1967. [In R.]

5425 - MOKRONOSOV, A.T., NEKRASOVA, G.F., ZINOV'EVA, S.D.: Izmenenie fotosinteliches-
kogo metabolizma khlorelly pri adaptatsii k raznym kontsentratsiyam CO_2 i k raz-
nyn istochnikam azota. [Change of photosynthetic metabolism in *Chlorella* during
adaptation to various CO_2 concentrations and nitrogen sources.] - In: Materialy
II. Vsesoyuznogo Soveshchaniya "UpravlyaemyT Fotosintez i Biofizika PopulyatsiT."
P. 22. Krasnoyarsk 1969. [In R.]

5426 - MOLCHANOV, M.I., BALAUR, N.S., BEZINGER, E.N.: Sinter lamellyarnogo belka *in
vivo* pri formatsii struktury khloroplastov kukuruzy pod vozdeTstviem sveta. [Syn-
thesis of lamellar protein *in vivo* during formation of the structure of maize
chloroplasts under influence of light.] - Dokl. Akad. Nauk SSSR *187*: 935-936,
1969. [In R.]

5427 - MOLDAU, Kh.: Opticheskaya model' lista rasteniya. [Optical model of plant leaf.]
- In: Fitoaktinometricheskie Issledovaniya Rastitel'nogo Pokrova. Pp. 89-109.
Valgus, Tallin 1967. [In R, ab: E.]

5428 - MOLDAU, Kh.: O soprotivleniyakh transpiratsii v gazovoT faze. [Transpiration
resistances in the vapour phase.] - In: Voprosy Effektivnosti Fotosinteza. Pp.
117-131. Tarty 1969. [Comparison of methods; in R, ab: E.]

5429 - MOLDAU, Kh.: Soprotivleniya ust'its v estestvennykh usloviyakh. [Stomatal resistances *in situ*.] - In: Voprosy Effektivnosti Fotosinteza. Pp. 132-142. Tartu
1969. [In R, ab: E.]

5430 - MOLDAU, Kh., KERES, S.: NelineTnye opticheskie yavleniya v list'yakh nekotorykh
rasteniT. [Non-linear optical phenomena in leaves of some plants.] - Izv. Akad.
Nauk Est. SSR, Ser. fiz.-mat. tech. Nauk *15*: 511-518, 1966. [In R, ab: E, Estonian.]

5431 - MOLL, B., LEVINE, R.P.: Characterization of a photosynthetic mutant strain of
Chlamydomonas reinhardi deficient in phosphoribulokinase activity. - Plant Physiol. *46*: 576-580, 1970.

5432 - MOLOTKOVSKIÏ, Yu.G.: Gidroliz fosfolipidov i obrazovanie svobodnykh zhirnykh kislot v izolirovannykh khloroplastakh. [Hydrolysis of phospholipids and formation
of free acids in isolated chloroplasts.] - Biokhimiya *33*: 961-968, 1968. [In R,
ab: E.]

5433 - MOLOTKOVSKIÏ, Yu.G.: PotentsiometricheskiT metod opredeleniya aktivnosti fotofosforilirovaniya. [Potentiometric determination of photophosphorylaction activity.] - In: KIRICHENKO, E.B. (ed.): Metody Issledovaniya Fotofosforilirovaniya.
Pp. 140-141. Pushchino-na-Oke 1970. [In R, ab: E.]

5434 - MOLOTKOVSKIÏ, Yu.G., DZYUBENKO, V.S.: Sveto-aktiviruemoe nabukhanie khloroplastov i mekhanizm ego regulyatsii. [Light-stimulated swelling of chloroplasts and
the mechanism of its regulation.] - Fiziol. Rast. *16*: 78-88, 1969. [In R, ab: E.]

5435 - MOLOTKOVSKIÏ, Yu.G., DZYUBENKO, V.S.: Indutsiruemoe svetom pogloshchenie H^+ izolirovannymi khloroplastami. [Light-induced uptake of H^+ by isolated chloroplasts.]
- Mol. Biol. (Moskva) *4*: 383-394, 1970. [In R, ab: E.]

5436 - MOLOTKOVSKIÏ, Yu.G., DZYUBENKO, V.S.: Stimuliruyushchee vliyanie K^+ i Na^+ na
svetozavisimye reaktsii izilirovannykh khloroplastov. [Stimulating effect of K^+
and Na^+ on light-dependent reactions of isolated chloroplasts.] - Fiziol. Rast.
17: 280-289, 1970. [In R, ab: E.]

5437 - MOLOTKOVSKIÏ, Yu.G., DZYUBENKO, V.S., TIMONINA, V.N.: Energozavisimye konformat-
sionnye izmeneniya izolirovannykh khloroplastov. [Energy-dependent conformational
changes of isolated chloroplasts.] - In: KIRICHENKO, E.B. (ed.): Metody Vydeleniya Khloroplastov. Pp. 77-93. Pushchino-na-Oke 1970. [In R, ab: E.]

5438 - MOLOTKOVSKIÏ, Yu.G., MORYAKOVA, V.F.: Primery indutsirovannogo sinteza belka v
rasteniyakh. [Examples of induced protein synthesis in plants.] - Uch. Zap. tart.
gos. Univ. *185*: 401-408, 1966. [Chl; in R.]

5439 - MOLOTKOVSKIÏ, Yu.G., ZHESTKOVA, I.M.: Morfologicheskie i funktsional'nye izmeneniya izolirovannykh khloroplastov pri progreve. [Morphological and functional
changes of isolated chloroplasts during heating.] - Dokl. Akad. Nauk SSSR *166*:
488-491, 1966. [In R.]

5440 - MOLOTKOVSKIĬ, Yu.G., ZHESTKOVA, I.M.: Vliyanie sakharozy na ustoĭchivost' izolirovannykh khloroplastov. [Effect of sucrose on the stability of isolated chloroplasts.] - Fiziol. Rast. *14*: 367-371, 1967. [In R.]

5441 - MOLOTKOVSKIĬ, Yu.G., ZHESTKOVA, I.M., KASHURO, V.F.: Mekhanizm regulyatsii struk-
turnogo sostoyaniya i fotokhimicheskoĭ aktivnosti izolirovannykh khloroplastov.
[Mechanism of regulation of structural state and photochemical activity of isolated chloroplasts.] - Zh. evolyuts. Biokhim. Fiziol. *2*: 159-165, 1966. [In R, ab: E.]

5442 - MOLOTKOVSKY, Y.G., DZYUBENKO, V.S.: Regulation of light-induced swelling of isolated chloroplasts. - Biochem. biophys. Res. Commun. *29*: 298-302, 1967.

5443 - MOLOTKOVSKY, Y.G., DZYUBENKO, V.S.: Light-induced pH gradient through chloroplast
membrane and translocation of cations. - Nature *219*: 496-498, 1968.

5444 - MOLOTKOVSKY, Y.G., ZHESTKOVA, I.M.: Morphological and functional changes in isolated chloroplast under the influence of oleate. - Biochim. biophys. Acta *112*:
170-172, 1966.

5445 - MOLOTKOVS'KYĬ, G.Kh., DEREVENKO, O.S.: Polyarnist', asymetriya ta geterozys u
roslyn. [Polarity, asymmetry and heterosis in plants.] - Ukr. bot. Zh. *26* (2):
58-64, 1969. [Chl, Car; In Ukr., ab: R, E.]

5446 - MONÉGER, R.: Incorporation de carbone radioactif dans les caroténoïdes de frondes de *Spirodela polyrrhiza* SCHLEID. exposées à la lumière sur milieu minéral
contenant du bicarbonate de sodium ^{14}C, après un séjour prolongé sur milieu su-
cré. - Compt. rend. Acad. Sci. (Paris), Sér. D *263*: 52-54, 1966.

5447 - MONÉGER, R.: Contribution à l'étude de l'influence exercée par la lumière sur
la biosynthèse des caroténoïdes chez la *Spirodela polyrrhiza* (L.) SCHLEIDEN. -
Physiol. vég. *6*: 165-202, 1968.

5448 - MONÉGER, R.: Mise au point d'une méthode de micro-analyse pour l'étude des carotenoïdes de petits échantillons de végétaux étiolés. - Physiol. vég. *6*: 367-402,
1968.

5449 - MONÉGER, R.: Sur la réalisation d'un dispositif ultrasensible permettant de sé-
parer et de doser les caroténoïdes de petits échantillons de frondes étiolées
de *Spirodela polyrrhiza* SCHLEID. - Compt. rend. Acad. Sci. (Paris), Sér. D *266*:
672-675, 1968.

5450 - MONÉGER, R.: Action de radiations oligochromatiques sur les teneurs en caroténoïdes des frondes étiolées de *Spirodela polyrrhiza* SCHLEID. - Compt. rend. Acad.
Sci. (Paris), Sér. D *267*: 313-316, 1968.

5451 - MONÉGER, R.: Incorporations de radio-activité, à partir de bicarbonate ^{14}C et
d'acétate 2-^{14}C de sodium, dans les caroténoïdes de frondes étiolées de *Spirodela polyrrhiza* SCHLEID. exposées à des radiations oligochromatiques. - Compt.
rend. Acad. Sci. (Paris), Sér. D *267*: 605-608, 1968.

5452 - MONÉGER, R.: Influence de la durée d'éclairement préalable de frondes étiolées
de *Spirodela polyrrhiza* SCHLEID. sur les radio-activités incorporées à l'obscuri-
té dans leurs caroténoïdes, à partir d'acétate 2-^{14}C de sodium. - Compt. rend.
Acad. Sci. (Paris), Sér. D *267*: 733-736, 1968.

5453 - MONÉGER, R.: Un example d'application des techniques du froid en physiologie végétale. Mise au point d'une méthode pour analyser les effets de la lumière sur
la synthèse des caroténoïdes chez les végétaux étiolés. - Rev. gen. Froid *1968*:
425-451, 1968.

5454 - MONÉGER, R.: Un mécanisme possible d'action de la lumière sur les biosynthèses
de caroténoïdes chez les végétaux supérieurs chlorophylliens. - Bull. Soc. franç.
Physiol. vég. *14*: 473-497, 1968.

5455 - MONÉGER, R., JACQUES,R.: Action de radiations oligochromatiques sur les teneurs
en caroténoïdes des frondes étiolées de *Spirodela polyrrhiza* SCHLEID. - Compt.
rend. Acad. Sci. (Paris), Sér. D *267*: 313-316, 1968.

5456 - MONSI, M.: Mathematical models of plant communities. - In: ECKARDT, F.E. (ed.):
Functioning of Terrestrial Ecosystems at the Primary Production Level. Pp. 131-
149. UNESCO, Paris 1968. [Ps, Chl.]

5457 - MONSI, M., MURATA, Y.: Development of photosynthetic systems as influenced by
       distribution of matter. - In: Prediction and Measurement of Photosynthetic Produc-
       tivity. Pp. 115-129. PUDOC, Wageningen 1970.

5458 - MONTAL, M., NISHIMURA, M., CHANCE, B.: Uncoupling and charge transfer in bacterial
       chromatophores. - Biochim. biophys. Acta 223: 183-188, 1970.

5459 - MONTEITH, J.L.: Physical limitations to crop growth. - Agr. Progress 41: 9-23,
       1966.

5460 - MONTEITH, J.L.: The photosynthesis and transpiration of crops. - Exp. Agr. 2:
       1-14, 1966.

5461 - MONTEITH, J.L.: Climatological measurements. - Photosynthetica 1: 129-132, 1967.
       [For Ps studies.]

5462 - MONTEITH, J.L.: Analysis of the photosynthesis and respiration of field crops
       from vertical fluxes of carbon dioxide. - In: ECKARDT, F.E. (ed.): Functioning of
       Terrestrial Ecosystems at the Primary Production Level. Pp. 349-358. UNESCO,
       Paris 1968.

5463 - MONTEITH, J.L.: Light interception and radiative exchange in crop stands. - In:
       EASTIN, J.D., HASKINS, F.A., SULLIVAN, C.Y., van BAVEL, C.H.M. (ed.): Biological
       Aspects of Crop Yield. Pp. 89-115. Amer. Soc. Agron. & Crop Sci. Soc. Amer.,
       Madison, Wisc. 1969.

5464 - MONTEITH, J.L.: Prospects for photosynthesis from A.D. 1970 to A.D. 2000. -
       Weather 25: 456-462, 1970.

5465 - MONTEITH, J.L., BULL, T.A.: A diffusive resistance porometer for field use. II.
       Theory, calibration and performance. - J. appl. Ecol. 7: 623-638, 1970.

5466 - MONTIES, B.: Propriétés chimiques des chlorophylles: réaction de la chlorophylle
       a avec $AlCl_3$ et quelques acides de Lewis. - Compt. rend. Acad. Sci. (Paris),
       Sér. D 267: 921-924, 1968.

5467 - MONTIES, B., COSTES, C.: Propriétés des caroténoïdes chloroplastiques: réaction
       de la violaxanthine et du chlorure d'aluminium. - Compt. rend. Acad. Sci. (Paris)
       Sér. D 266: 481-484, 1968.

5468 - MONTROLL, E.W.: Random walks on lattices. III. Calculation of first-passage
       times with application to exciton trapping on photosynthetic units. - J. math.
       Phys. 10: 753-765, 1969.

5469 - MONTROLL, E.W.: Random walks on lattices containing traps. - J. phys. Soc. Jap.
       26 (Suppl.): 6-10, 1969. [Ps, Chl.]

5470 - MOONEY, H.A., DUNN, E.L.: Photosynthetic systems of mediterranean-climate shrubs
       and trees of California and Chile. - Amer. Naturalist 104: 447-453, 1970.

5471 - MOONEY, H.A., HARRISON, A.T.: The influence of conditioning temperature on sub-
       sequent temperature-related photosynthetic capacity in higher plants. - In: Pre-
       diction and Measurement of Photosynthetic Productivity. Pp. 411-417. PUDOC,
       Wageningen 1970.

5472 - MOONEY, H.A., STRAIN, B.R., WEST, M.: Photosynthetic efficiency at reduced carbon
       dioxide tensions. - Ecology 47: 490-491, 1966.

5473 - MOONEY, H.A., WEST, M., BRAYTON, R.: Field measurements of the metabolic respon-
       ses of bristlecone pine and big sagebrush in the White Mountains of California.
       - Bot. Gaz. 127: 105-113, 1966. [Ps.]

5474 - MOORBY, J.: The production, storage and translocation of carbohydrates in deve-
       loping potato plants. - Ann. Bot. 34: 297-308, 1970.

5475 - MOORE, K., LOVELL, P.: Chlorophyll content and the pattern of yellowing in senes-
       cent leaves. - Ann. Bot. 34: 1097-1100, 1970.

5476 - MOORE, K., LOVELL, P.: Differential effects of the embryonic axis on chlorophyll
       production and photosynthesis of mustard cotyledons. - Planta 93: 289-294, 1970.

5477 - MOORE, K.G.: Senescence in leaves of Acer pseudoplatanus L. and Parthenocissus
       tricuspidata PLANCH. - Ann. Bot. 30: 683-699, 1966. [Chl.]

5478 - MOORE, R.B., CRANE, C.A., FRANTZ, I.D. Jr.: An apparatus and a method using phenethylamine for liquid scintillation counting of $C^{14}O_2$ obtained by wet oxidation of biological materials. - Anal. Biochem. *24*: 545-554, 1968.

5479 - MOORE, R.E., SPRINGER-LEDERER, H., OTTENHEYM, H.C.J., BASSHAM, J.A.: Photosynthesis by isolated chloroplasts. IV. Regulation by factors from leaf cells. - Biochim. biophys. Acta *180*: 368-376, 1969.

5480 - MORALES, M.T., THOMPSON, N.R.: Leaf area in relation to economic photosynthetic efficiency in the potato. - Amer. Potato J. *46*: 435, 1969.

5481 - MORALES, M.T., THOMPSON, N.R.: Genetics of economic photosynthetic efficiency in the potato.-Amer. Potato J. *46*: 436, 1969.

5482 - MORELAND, D.E.: Mechanisms of action of herbicides. - Annu. Rev. Plant Physiol. *18*: 365-386, 1967. [Ps.]

5483 - MORELAND, D.E.: Inhibitors of chloroplast electron transport: structure-activity relations. - In: METZNER, H.:(ed.): Progress in Photosynthesis Research. Vol. III. Pp. 1693-1711. Tübingen 1969.

5484 - MORELAND, D.E., BLACKMON, W.J.: Effects of 3,5-dibromo-4-hydroxybenzaldehyde O-(2,4-dinitrophenyl)oxime on reactions of mitochondria and chloroplasts. - Weed Sci. *18*: 419-426, 1970.

5485 - MORESHET, S.: Effect of environmental factors on cuticular transpiration resistance. - Plant Physiol. *46*: 815-818, 1970. [Double diffusion porometer.]

5486 - MORESHET, S., KOLLER, D., STANHILL, G.: The partitioning of resistances to gaseous diffusion in the leaf epidermis and the boundary layer. - Ann. Bot. *32*: 695-701, 1968.

5487 - MORESHET, S., STANHILL, G., KOLLER, D.: A new method of measuring stomatal diffusion resistance. - Israel J. Bot. *16*: 50, 1967.

5488 - MORESHET, S., STANHILL, G., KOLLER, D.: A radioactive tracer technique for the direct measurement of the diffusion resistance of stomata. - J. exp. Bot. *19*: 460-467, 1968.

5489 - MORETH, C.M., YENTSCH, C.S.: The role of chlorophyllase and light in the decomposition of chlorophyll from marine phytoplankton. - J. exp. mar. Biol. Ecol. *4*: 238-249, 1970.

5490 - MORGAN, D.G.: A quantitative study of the effects of gibberellic acid on the growth of *Festuca arundinacea*. - Aust. J. agr. Res. *19*: 221-225, 1968. [Ps.]

5491 - MORGAN, R.C.: Chemical studies on concentrated pineapple juice. I. Carotenoid composition of fresh pineapples. - J. Food Sci.*31*: 213-217, 1966.

5492 - MORI, M., SANO, S.: Protoporphyrin formation from coproporphyrinogen III by *Chromatium* cell extracts. - Biochem. biophys. Res. Commun. *32*: 610-615, 1968.

5493 - MORITA, S.: Purification and some physico-chemical properties of *Rhodopseudomonas spheroides* cytochrome 550. - Bot. Mag. (Tokyo) *79*: 630-633, 1966.

5494 - MORITA, S.: Evidence for three photochemical systems in *Chromatium* D. - Biochim. biophys. Acta *153*: 241-247, 1968.

5495 - MORITA, S.: Possibility of differentiation of three photoreaction systems in *Chromatium* D. - In: SHIBATA, K., TAKAMIYA, A., JAGENDORF, A.T., FULLER, R.C. (ed.): Comparative Biochemistry and Biophysics of Photosynthesis. Pp. 133-139. Univ. of Tokyo Press, Tokyo; Univ. Park Press, State College, Pa. 1968.

5496 - MORRIS, I.: Inhibition of spinach chloroplast fructose-1,6-diphosphatase by $MgATP^{2-}$, $MgADP^-$, and magnesium pyrophosphate ($MgP_2O_7^{2-}$). - Biochim. biophys. Acta *162*: 462-464, 1968.

5497 - MORRIS, I.: The effect of methyl glyoxal on growth and cell division of *Chlamydomonas reinhardii*. - Physiol. Plant. *22*: 1059-1068, 1969. [Ps.]

5498 - MORRIS, J.Y., TRANQUILLINI, W.: Über den Einfluss des osmotischen Potentiales des Wurzelsubstrates auf die Photosynthese von *Pinus contorta*-Sämlingen im Wechsel der Jahreszeiten. - Flora B *158*: 277-287, 1969.

5499 - MORRISON, R.G., YARRANTON, G.A.: An instrument for rapid and precise point samp-
       ling of vegetation. - Can. J. Bot. 48: 293-297, 1970.

5500 - MÖRTEL, G.: Chloroplastenkontraktion als Lichtreaktion bei *Mougeotia*. - Ber.
       deut. Bot. Ges. 83: 199, 1970.

5501 - MOSER, W.: Neues von der botanischen Forschungsstation "Hoher Nebelkogel"/Tirol.
       - Jahrb. Vereins Schutz.Alpenpfl. - Tiere e.V. 33: 1-9, 1968. [Ps.]

5502 - MOSER, W.: Die Photosyntheseleistung von Nivalpflanzen. - Ber. deut. bot. Ges.
       82:63-64, 1969.

5503 - MOSHKOV, B.S.: Znachenie svetofiziologicheskikh issledovaniĭ dlya poznaniya po-
       tentsial'nykh vozmozhnosteĭ rasteniĭ. [Role of light-physiological studies for
       discovering potential posibilities of plants.] - Sel'.-khoz. Biol. 1: 243-247,
       1966. [In R, ab: E.]

5504 - MOSHKOV, B.S., ODUMANOVA-DUNAEVA, G.A., KHOVANSKAYA, N.V.:Izuchenie chuvstvitel'-
       nosti vegetativnogo razmnozheniya *Kalanchoe daigremontiana* (R. HAMMET et H. PERR
       de la BÂTH.) JACOB. k isklyucheniyu fotosinteza. [Sensitivity of vegetative pro-
       pagation of *Kalanchoe daigremontiana* (R. HAMMET et H. PERR de la BÂTH.) JACOB.
       to elimination of photosynthesis.] - Sb. Tr. agron. Fiz. (Leningrad) 21: 94-102,
       1970. [In R.]

5505 - MOSS, B.: A spectrophotometric method for the estimation of percentage degrada-
       tion of chlorophylls to pheo-pigments in extracts of algae. - Limnol. Oceanogr.
       12: 335-340, 1967.

5506 - MOSS, B.: Studies on the degradation of chlorophyll *a* and carotenoids in fresh-
       waters. - New Phytol. 67: 49-59, 1968.

5507 - MOSS, B.: Algae of two Somersetshire pools: Standing crops of phytoplankton and
       epipelic algae as measured by cell numbers and chlorophyll a^1. - J. Phycol. 5:
       158-168, 1969.

5508 - MOSS, D.N.: Respiration of leaves in light and darkness. - Crop Sci. 6: 351-354,
       1966.

5509 - MOSS, D.N.: High activity of the glycolic acid oxidase system in tobacco leaves.
       - Plant Physiol. 42: 1463-1464, 1967.

5510 - MOSS, D.N.: Nutrient deficiencies reduce photosynthesis. - Plant Food Rev. 13:
       15-16, 1967.

5511 - MOSS, D.N.: Photorespiration and glycolate metabolism in tobacco leaves. - Crop
       Sci. 8: 71-76, 1968.

5512 - MOSS, D.N.:Relation in grasses of high photosynthetic capacity and tolerance to
       atrazine. - Crop Sci. 8: 774, 1968.

5513 - MOSS, D.N.: Photosynthetic and respiratory uniqueness in certain tropical plants.
       - Proc. Soil Crop Sci. Soc. Florida 29: 268-272, 1969.

5514 - MOSS, D.N.: Laboratory measurements and breeding for photosynthetic efficiency.
       - In: Prediction and Measurement of Photosynthetic Productivity. Pp. 323-330.
       PUDOC, Wageningen 1970.

5515 - MOSS, D.N., KRENZER, E.G. Jr., BRUN, W.A.: Carbon dioxide compensation points in
       related plant species. - Science 164: 187-188, 1969.

5516 - MOSS, D.N., RASMUSSEN, H.P.: Cellular localization of CO_2 fixation and translo-
       cation of metabolites. - Plant Physiol. 44: 1063-1068, 1969.

5517 - MOTODA, S.: An assessment of primary productivity of a coral reef lagoon in Palau,
       Western Caroline Islands, based on the data obtained during 1935-37. - Rec. ocea-
       nogr. Works Jap. 10: 65-74, 1969. [Ps, Chl.]

5518 - MOTOTANI, I.: Horizontal distribution of light intensity in plant communities. -
       In: Photosynthesis and Utilization of Solar Energy. Level III Experiments. Pp.
       25-28. Jap. nat. Subcomm. for PP (JPP), Tokyo 1968.

5519 - MOTOYAMA, E., FUKUI, H., KUBOTA, S.: [Studies on the rational application of fer-
       tilizer on citrus trees in Seto Inland sea area: V. Translocation and fruit

productivity of photosynthate at certain times of the year: 1. Preliminary test.]
- Bull. Shikoku agr. exp. Sta. *14*: 91-99, 1966. [In Jap.]

5520 - **MOTOYOSHI, F.**: Chlorophyll formation in several chlorophyll mutants of sand oats
and barley cultured in nutrient media. - Jap. J. Genet. *42*: 291-297, 1967.

5521 - **MOUDRIANAKIS, E.N.**: Structural and functional aspects of photosynthetic lamellae.
- Fed. Proc. *27*: 1180-1185, 1968.

5522 - **MOUDRIANAKIS, E.N., HOWELL, S.H., KARU, A.E.**: Characterization of the "quantasome"
and its role in photosynthesis. - In: SHIBATA, K., TAKAMIYA, A., JAGENDORF, A.T.,
FULLER, R.C. (ed.): Comparative Biochemistry and Biophysics of Photosynthesis,
Pp. 67-81. Univ. Tokyo Press, Tokyo; Univ. Park Press, State College, Pa. 1968.

5523 - **MOURAVIEFF, I.**: Sur les propriétés optiques des cellules stomatiques. - I. Transmission de la lumière dans les longuers d'ondes de 440, 530 et 662 nm par les
cellules stomatiques et épidermiques. - Bull. Soc. bot. France *113*: 220-223,
1966.

5524 - **MOURAVIEFF, I.**: Mise en évidence de la réduction du bleu de tétrazolium par les
plastes épidermiques en présence et en absence de gaz carbonique, en lumière
rouge claire (662 nm) ou rouge sombre (720 nm). - Compt. rend. Acad. Sci. (Paris),
Sér. D *264*: 828-829, 1967. [Ps.]

5525 - **MOURAVIEFF, I.**: Photoréduction du bleu et du nitrobleu de tétrazolium par les
plastes épidermiques et stomatiques *in situ* après éclairage par les radiations
de 400, 530 et 662 nm à diverses énergies. - Compt. rend. Acad. Sci. (Paris),
Sér. D *264*: 58-60, 1967.

5526 - **MOURAVIEFF, I.**: Sur les propriétés optiques de l'appareil stomatiques. Absorption des rayonnements de 662, 530 et 440 nm par les cellules stomatiques et
leurs plastes chez quelques espèces du midi méditerranéen au cours de la saison
sèche. - Bull. mens. Soc. linnéene Lyon *36*: 169-173, 1967. [Chl.]

5527 - **MOURAVIEFF, I.**: Microphotométrie comparative de la fluorescence et de l'absorption de radiations rouge et verte par les plastes stomatiques et les plastes
du mésophylle. - Compt. rend. Acad. Sci. (Paris), Sér. D *270*: 796-798, 1970.
[Chl.]

5528 - **MOUSSEAU, M.**: Influence de l'éclairement sur l'assimilation journalière et annuel-
le du *Teucrium scorodonia* L. en conditions naturelles. - Oecol. Plant. *1*: 103-116,
1966.

5529 - **MOUSSEAU, M.**: Les phénomènes de régulation structurale et fonctionelle de l'appareil photosynthétique de *Teucrium scorodonia*: un mécanisme d'adaptation aux
conditions d'éclairement. - Oecol. Plant. *2*: 15-26, 1967.

5530 - **MOUSSEAU, M.**: Action comparée de la lumière sur l'intensité photosynthétique de
feuilles entières coupées ou sur pied, selon les conditions d'éclairement pendant
la croissance. - Compt. rend. Acad. Sci. (Paris), Sér. D *266*: 1391-1393, 1968.

5531 - **MOUSSEAU, M., BOURDU, R.**: Influence des conditions écologiques d'éclairement
pendant la croissance sur la structure et l'activité des chloroplastes de *Teucrium scorodonia* L. - Bull. Soc. franç. Physiol. vég. *14*: 307-315, 1968.

5532 - **MOUSSEAU, M., COSTE, F., de KOUCHKOVSKY, Y.**: Influence des conditions d'éclairement pendant la croissance sur l'activité photosynthétique de feuilles entières
et de chloroplastes isolés. - Compt. rend. Acad. Sci. (Paris), Sér. D *264*: 1158-
1161, 1967.

5533 - **MOUSSERON-CANET, M., DALLE, J.-P., MANI, J.-C.**: Photooxydation sensibilisée de
composés apparentes aux caroténoïdes. - Photochem. Photobiol. *9*: 91-94, 1969.

5534 - **MOYSE, A.**: Les chloroplastes: activité photosynthétique et métabolisme des glucides. - In: GOODWIN, T.W. (ed.): Biochemistry of Chloroplasts. Vol. II. Pp. 91-
129. Academic Press, London-New York 1967.

5535 - **MOYSE, A.**: Les bactérie phototrophes. - Atomes *243*: 300-306, 1967. [Ps.]

5536 - **MOYSE, A.**: L'assimilation de CO_2 par les préparations de chloroplastes isolés.
Examen critique de ses facteurs limitants. - Physiol. vég. *7*: 43-56, 1969.

5537 - **MOYSE, A.**: Effect of nitrogen starvation and refeeding on the structure of the photosynthetic apparatus in *Chlorella*. - In: Prediction and Measurement of Photosynthetic Productivity. Pp. 551-553. PUDOC, Wageningen 1970.

5538 - MÜHLETHALER, K.: Der Feinbau des Photosynthese-Apparates. - Umschau Wiss. Tech. *66*: 659-662, 1966.

5539 - MÜHLETHALER, K.: The ultrastructure of the plastid lamellae. - In: GOODWIN. T.W. (ed.): Biochemistry of Chloroplasts. Vol. I. Pp. 49-64. Academic Press, London-New York 1966.

5540 - MÜHLETHALER, K.: L'ultrastructure de la lamelle des chloroplastes. - In: SIRON-VAL, C. (ed.): Le Chloroplaste, Croissance et Vieillissement. Pp. 42-47. Masson, Paris 1967.

5541 - MÜHLETHALER, K., WEHRLI, E.: Freeze-etch studies on photosynthetic lamellae. - In: METZNER, H. (ed.): Progress in Photosynthesis Research. Vol. I. Pp. 87-90. Tübingen 1969.

5542 - MUKERJI, S.K., TING, I.P.: Isolation and characterization of phosphoenolpyruvate carboxylase isoenzymes from cotton leaf tissue. - Plant Physiol. *44* (Suppl.): 35, 1969.

5543 - MUKHERJEE, D.C., CHO, D.H., TOLLIN, G.: ESR studies of chlorophyll one-electron photochemistry in solution: kinetics of reversible quinone reduction and general redox mechanisms. - Photochem. Photobiol. *9*: 273-289, 1969.

5544 - MUKHIN, E.N.: Rol' ferredoksina v fotosinteze. [Role of ferredoxin in photosynthesis.] - Uspekhi sovrem. Biol. *67*: 201-221, 1969. [In R.]

5545 - MUKHIN, E.N., AKULOVA, E.A.: O svetovoĭ adaptatsii mekhanizma fotovosstanovleniya NADF khloroplastami. [Light adaptation mechanism of NADP photoreduction by chloroplasts.] - Dokl. Akad. Nauk SSSR *169*: 699-702, 1966. [In R.]

5546 - MUKHIN, E.N., AKULOVA, E.A.: Ob uchastii ferredoksina iz prorostkov gorokha v fotovosstanovlenii NADF. [Participation of ferredoxin from pea seedlings in NADP photoreduction.] - Dokl. Akad. Nauk SSSR *167*: 1177-1180, 1966. [In R.]

5547 - MUKHIN, E.N., AKULOVA, E.A.: Rol' ferredoksina v pervichnykh protsessakh zapasaniya energii sveta pri fotosinteze. [Role of ferredoxin in primary processes of storage of radiant energy in photosynthesis.] - In: Bioenergetika i Biologicheskaya Spektrofotometriya. Pp. 157-162. Nauka, Moskva 1967. [In R.]

5548 - MUKHIN, E.N., AKULOVA, E.A.: Nekotorye usloviya funktsionirovaniya ferredoksina v tsepi fotosinteticheskogo perenosa elektrona. [Some conditions of functioning of ferredoxin in the chain of photosynthetic electron transfer.] - In: Mekhanizmy Dykhaniya, Fotosinteza i Fiksatsii Azota. Pp. 292-298. Nauka, Moskva 1967. [In R.]

5549 - MUKHIN, E.N., AKULOVA, E.A., POTAPOVA, V.M.: Nizkomolekulyarnaya vodorastvorimaya fraktsiya iz list'ev gorokha - vozmozhnyĭ uchastnik transporta elektronov pri fotosinteze. [Low-molecular water-soluble fraction from pea leaves as a possible participant in the photosynthetic electron transfer.] - Dokl. Akad. Nauk SSSR *174*: 1215-1218, 1967. [In R.]

5550 - MUKHIN, E.N., KHRUSLOVA, S.G., EGOROVA, E.F., SHMELEVA, V.L.: Vliyanie intensivnosti osveshcheniya pri vyrashchivanii rasteniĭ na skorost' zavisimogo ot askorbata fotovosstanovleniya NADF khloroplastami. [Effect of irradiance during cultivation of plants on the rate of ascorbate-dependent NADP photoreduction by chloroplasts.] - Dokl. Akad. Nauk SSSR *193*: 940-943, 1970. [In R.]

5551 - MUKHIN, E.N., KHRUSLOVA, S.G., GINS, V.K.: Izmenenie svoĭstv i kolichestva ferredoksina u fotosinteziruyushchego organizma pod vliyaniem usloviĭ osveshcheniya. [Changes in properties and amount of ferredoxin in a photosynthesizing organism under influence of irradiance conditions.] - Fiziol. Rast. *17*: 1193-1197, 1970. [In R, ab: E.]

5552 - MUKHIN, E.N., POPOVA, N.B.: Fosfodoksin iz list'ev *Pisum sativum* L. [Phosphodoxin from leaves of *Pisum sativum* L.] - Izv. Akad. Nauk SSSR, Ser. biol. *1968*: 133-136, 1968. [In R, ab: E.]

5553 - MUKHIN, E.N., POPOVA, N.B., KALASHNIKOV, Yu.E.: O dvukh osnovnykh komponentakh fosfodoksina. [Two basic components of phosphodoxin.] - Mol. Biol. (Moskva) *2*: 588-596, 1968. [In R, ab: E.]

5554 - **MUKOHATA, Y.**: On the light scattering pattern of spinach chloroplast suspension. - In: SHIBATA, K., TAKAMIYA, A., JAGENDORF, A.T., FULLER, R.C. (ed.): Comparative Biochemistry and Biophysics of Photosynthesis. Pp. 89-96. Univ. Tokyo Press, Tokyo; Univ. Park Press, State College, Pa. 1968.

5555 - **MULCHI, C.L., VOLK, R.J., JACKSON, W.A.**: Oxygen exchange by illuminated soybean leaves at carbon dioxide concentrations limiting photosynthesis. - Plant Physiol. *43*: S 13, 1968.

5556 - **MÜLLER, A., LUMRY, R., WALKER, M.S.**: Light-intensity dependence of the *in vivo* fluorescence lifetime of chlorophyll. - Photochem. Photobiol. *9*: 113-126, 1969.

5557 - **MÜLLER, B.**: Trennung ganzer von zerstörten Chloroplasten. - Naturwissenschaften *54*: 520-521, 1967.

5558 - **MÜLLER, B.**: On the mechanism of the light-induced activation of the NADP-dependent glyceraldehyde phosphate dehydrogenase. - Biochim. biophys. Acta *205*: 102-109, 1970.

5559 - **MÜLLER, B., ZIEGLER, H.**: Die lichtinduzierte Aktivitätssteigerung der NADP$^+$-abhängigen Glycerinaldehyd-3-phosphat-Dehydrogenase. IX. Die Reaktion in isolierten Chloroplasten. - Planta *85*: 96-104, 1969.

5560 - **MÜLLER, B., ZIEGLER, I., ZIEGLER, H.**: Lichtinduzierte, reversible Aktivitätssteigerung der NADP-abhängigen Glycerinaldehyd-3-phosphat-Dehydrogenase in Chloroplasten. Zum Mechanismus der Reaktion. - Europe. J. Biochem. *9*: 101-106, 1969.

5561 - **MÜLLERSTAËL, H.**: Untersuchungen über den Gaswechsel zweijähriger Holzpflanzen bei fortschreitender Bodenaustrocknung. - Beitr. Biol. Pflanzen *44*: 319-341, 1968.

5562 - **MULLINS, J., JONES, L.**: Environmental effects upon the photosynthetic capacity of a blue-green alga, *Anacystis nidulans*. - Plant Physiol. *44* (Suppl.): 12, 1969.

5563 - **MULLINS, M.G.**: Transport of ^{14}C-assimilates in seedlings of *Phaseolus vulgaris* L. in relation to vascular anatomy. - Ann. Bot. *34*: 889-896, 1970.

5564 - **MULLINS, M.G.**: Hormone-directed transport of assimilates in decapitated internodes of *Phaseolus vulgaris* L. - Ann. Bot. *34*: 897-909, 1970.

5565 - **MUMINOV, F.A.**: Radiatsionnyĭ rezhim i teplovoĭ balans khlopkovogo polya i urozhaĭ khlopchatnika. [Radiation regime and heat balance of a cotton field and its yield.] - In: Aktinometriya i Optika Atmosfery. Pp. 309-314. Valgus, Tallin 1968. [In R, ab: E.]

5566 - **MUNDAY, J.C. Jr., GOVINDJEE**: Fluorescence transients in *Chlorella*: effects of supplementary light, anaerobiosis, and methyl viologen. - In: METZNER, H. (ed.): Progress in Photosynthesis Research. Vol. II. Pp. 913-922. Tübingen 1969.

5567 - **MUNDAY, J.C. Jr., GOVINDJEE**: Light-induced changes in the fluorescence yield of chlorophyll *a in vivo*. III. The dip and the peak in the fluorescence transient of *Chlorella pyrenoidosa*. - Biophys. J. *9*: 1-21, 1969.

5568 - **MUNDAY, J.C. Jr., GOVINDJEE**: Light-induced changes in the fluorescence yield of chlorophyll *a in vivo*. IV. The effect of preillumination on the fluorescence transient of *Chlorella pyrenoidosa*. - Biophys. J. *9*: 22-35, 1969.

5569 - **MUNSCHE, D., ENGELBRECHT, L., GÖHLER, K.-D., CONRAD, K.**: Beitrag zur Frage der Beziehungen zwischen Struktur und Cytokininwirksamkeit. - Flora A *159*: 268-273, 1968. [Chl.]

5570 - **MÜNTZ, K.**: Über Wachstum und Stoffwechsel von *Chlorella pyrenoidosa* während Kälteeinwirkung im Dunkeln. - Flora A *160*: 139-157, 1969. [Chl.]

5571 - **MÜNZEL, E.**: Photosynthese und Hydrolasenaktivität in Blattstücken von *Begonia rex* PUTZEY unter Einfluss von Gibberellinsäure. - Beitr. Biol. Pflanzen *47*: 117-125, 1970.

5572 - **MURAKAMI, S.**: On the nature of the particles attached to the thylakoid membrane of spinach chloroplasts. - In: SHIBATA, K., TAKAMIYA, A., JAGENDORF, A.T., FULLER, R.C. (ed.): Comparative Biochemistry and Biophysics of Photosynthesis. Pp. 82-88. Univ. Tokyo Press, Tokyo; Univ. Park Press, State College, Pa. 1968.

5573 - **MURAKAMI, S., NOBEL, P.S.**: Lipids and light-dependent swelling of isolated spinach chloroplasts. - Plant Cell Physiol. *8*: 657-671, 1967.

5574 - MURAKAMI, S., PACKER, L.: Reversible changes in the conformation of thylakoid membranes accompanying chloroplast contraction or expansion. - Biochim. biophys. Acta *180*: 420-423, 1969.

5575 - MURAKAMI, S., PACKER, L.: Light-induced changes in the conformation and configuration of the thylakoid membrane of *Ulva* and *Porphyra* chloroplasts *in vivo*. - Plant Physiol. *45*: 289-299, 1970.

5576 - MURAKAMI, S., PACKER, L.: Protonation and chloroplast membrane structure. - J. Cell Biol. *47*: 332-351, 1970.

5577 - MURANO, F., FUJITA, Y.: Comparative studies of photochemical oxidation-reduction reactions in lamellar fragments of various algae and spinach. - Plant Cell Physiol. *8*: 673-682, 1967.

5578 - MURATA, N.: [Fluorescence of photosynthetic bacteria.] - Protein, nucleic Acid, Enzyme [Tampakushitsu, Kakusan, Koso] *13*: 402-419, 1968. [In Jap.]

5579 - MURATA, N.: Fluorescence of chlorophyll in photosynthetic systems. IV. Induction of various emissions at low temperatures. - Biochim. biophys. Acta *162*: 106-121, 1968.

5580 - MURATA, N.: Control of excitation transfer in photosynthesis. I. Light-induced change of chlorophyll *a* fluorescence in *Porphyridium cruentum*. - Biochim. biophys. Acta *172*: 242-251, 1969.

5581 - MURATA, N.: Control of excitation transfer in photosynthesis. II. Magnesium ion-dependent distribution of excitation energy between two pigment systems in spinach chloroplasts. - Biochim. biophys. Acta *189*: 171-181, 1969.

5582 - MURATA, N.: Control of excitation transfer in photosynthesis. IV. Kinetics of chlorophyll *a* fluorescence in *Porphyra yezoensis*. - Biochim. biophys. Acta *205*: 379-389, 1970.

5583 - MURATA, N., BROWN, J.: Photochemical activities of spinach chloroplast particles fractionated after French press treatment. - Plant Physiol. *45*: 360-361, 1970.

5584 - MURATA, N., NISHIMURA, M., TAKAMIYA, A.: Fluorescence of chlorophyll in photosynthetic systems. I. Analysis of "weak light effect" in isolated chloroplasts. - Biochim. biophys. Acta *112*: 213-222, 1966.

5585 - MURATA, N., NISHIMURA, M., TAKAMIYA, A.: Fluorescence of chlorophyll in photosynthetic systems. II. Induction of fluorescence in isolated spinach chloroplasts. - Biochim. biophys. Acta *120*: 23-33, 1966.

5586 - MURATA, N., NISHIMURA, M., TAKAMIYA, A.: Fluorescence of chlorophyll in photosynthetic systems. III. Emission and action spectra of fluorescence-three emission bands of chlorophyll *a* and the energy transfer between two pigment systems. - Biochem. biophys. Acta *126*: 234-243, 1966.

5587 - MURATA, N., SUGAHARA, K.: Control of excitation transfer in photosynthesis. III. Light-induced decrease of chlorophyll *a* fluorescence related to photophosphorylation systems in spinach chloroplasts. - Biochim. biophys. Acta *189*: 182-192, 1969.

5588 - MURATA, N., TAKAMIYA, A.: Changes in emission spectra of photosynthetic pigments *in vivo*. - Plant Cell Physiol. *8*: 683-394, 1967.

5589 - MURATA, N., TAKAMIYA, A.:Nature of light-induced absorbance changes at 682 mµ and 702 mµ in photosynthesis of *Anacystis nidulans*. - Plant Cell Physiol. *10*: 193-202, 1969.

5590 - MURATA, N., TASHIRO, H., TAKAMIYA, A.: Effects of divalent metal ions on chlorophyll *a* fluorescence in isolated spinach chloroplasts. - Biochim. biophys. Acta *197*: 250-256, 1970.

5591 - MURATA, T., ODAKA, Y., UCHINO, K., YAKUSHIJI, E.: Reconstitution of the photosensitive form of *Chenopodium* chlorophyll protein from its apoprotein. - In: SHIBATA, K., TAKAMIYA, A., JAGENDORF, A.T., FULLER, R.C. (ed.): Comparative Biochemistry and Biophysics of Photosynthesis. Pp. 222-228. Univ. Tokyo Press, Tokyo; Univ. Park Press, State College, Pa. 1968.

5592 - MURATA, Y.: On the influence of solar radiation and air temperature upon the local differences in the productivity of paddy rice in Japan. - Int. Rice Comm. Newsletter *15*: 20-30, 1966.[Ps.]

5593 - MURATA, Y.: Features and problems in rice culture in Japan. - Rep. Inst. Agr. Res. Tohoku Univ. *18*: 23-43, 1967. [Ps.]

5594 - MURATA, Y.: On a new automatic leaf-area meter. - Jap. agr. Rev. quart. (Tokyo) *2*: 35-37, 1967.

5595 - MURATA, Y.: Physiological responses to nitrogen in plants. - In: EASTIN, J.D., HASKINS, F.A., SULLIVAN, C.Y., van BAVEL, C.H.M. (ed.): Physiological Aspects of Crop Yield. Pp. 235-259, 260-263 (disc.). Amer. Soc. Agron. & Crop Sci. Soc. Amer., Madison, Wisc. 1969. [Ps.]

5596 - MURATA, Y., AKITA, S., HONMA, T.: Studies on the photosynthesis of rice plants. XIV. The effect of white mulch on dry matter production and yield of rice. - Proc. Crop Sci. Soc. Jap. *36*: 447, 1967.

5597 - MURATA, Y., HAYASHI, K.: On a new, automatic device for leaf-area-measurement. - Proc. Crop Sci. Soc. Jap. *36*: 463-467, 1967.

5598 - MURATA, Y., HAYASHI, K.: On a new, automatic leaf-area meter. - In: Photosynthesis and Utilization of Solar Energy. Level III. Experiments. Pp. 33-34. Jap. nat. Subcomm. for PP (JPP), Tokyo 1968.

5599 - MURATA, Y., IYAMA, J., HIMEDA, M., IZUMI, S., KAWABE, A., KANZAKI, Y.: Studies on the deep-plowing, dense-planting cultivation of rice plants from the point of view of photosynthesis and production of dry matter. - Bull. nat. Inst. agr. Sci. Ser. D *15*: 1-53, 1966.

5600 - MURATA, Y., IYAMA, J., HONMA, T.: Studies on the photosynthesis of forage crops. V. The influence of soil moisture content on the photosynthesis and respiration of seedlings in various forage crops. - Proc. Crop Sci. Soc. Jap. *34*: 385-390, 1966.

5601 - MURATA, Y., MIYASAKA, A., AKITA, S., MUNAKATA, K.: Estimation of dry matter increase of a rice stand by continuous measurement using a large assimilation chamber. - In: Photosynthesis and Utilization of Solar Energy. Level III Experiments. Pp. 35-38. Jap. nat. Subcomm. for PP (JPP), Tokyo 1968.

5602 - MURESAN, T., HURDUC, N., NASTASIA, I.: Relaţile dintre fotosinteză şi heterozis la porumb. [Correlations between photosynthesis and heterosis in corn.] - An. Inst. Cercet. Pentru Cereale Plante teh-fundulea, Ser. C *35*: 339-356, 1967. [In Rum.]

5603 - MURPHY, R.F., O'CARRA, P.: Reversible denaturation of *C*-phycocyanin. - Biochim. biophys. Acta *214*: 371-373, 1970.

5604 - MURRAY, A.E.: Convertible and non-convertible protochlorophyllide in bean leaves. - Diss. Abstr. int. B *31*: 3678, 1970.

5605 - MURRAY, A.E., KLEIN, A.O.: Controls of protochlorophyllide synthesis. - Plant Physiol. *44* (Suppl.): 12-13, 1969.

5606 - MURRAY, C.N., RILEY, J.P.: The solubility of gases in distilled water and sea water. - II. Oxygen. - Deep-Sea Res. *16*: 311-320, 1969.

5607 - MURRAY, C.N., RILEY, J.P., WILSON, T.R.S.: The solubility of oxygen in Winkler reagents used for the determination of dissolved oxygen. - Deep-Sea Res. *15*: 237-238, 1968.

5608 - MURRAY, S.A.: Shock effects on plants: oxygen evolution of *Elodea*. - Experientia *26*: 710-711, 1970. [Ps.]

5609 - MUSHAK, P.O.: Vplyv umov osvitlennya na kul'turu *Chlorella vulgaris* BEYER. [Effect of light conditions on the culture of *Chlorella vulgaris* BEYER.] - Ukr. bot. Zh. *24*: 18-21, 1967. [Chl; in Ukr., ab: E, R.]

5610 - MUSHAK, P.O., SUD'INA, O.G.: Biokhimichna kharakterystyka riznykh fraktsiĭ klityn *Chlorella vulgaris* BEIJER. [Biochemical characteristics of different fractions of *Chlorella vulgaris* BEIJER. cells.] - Ukr. bot. Zh. *27*: 306-309, 1970. [Ps; in Ukr., ab: E, R.]

5611 - MUSSELMAN, R.C., GATHERUM, G.E.: Effects of light and moisture on red oak seed-
       lings. - Iowa State J. Sci. 43: 273-284, 1969. [Ps.]

5612 - MUTOH, N., YOSHIDA, K.H., YOKOI, Y., KIMURA, M., HOGETSU, K.: Studies on the
       production processes and net production of Miscanthus sacchariflorus community.
       - Jap. J. Bot. 20: 67-92, 1968.

5613 - MUTOH, N., YOSHIDA, K.H., YOKOI, Y., KIMURA, M., MIDORIKAWA, B.: Studies on
       the production processes and net production of the Miscanthus sacchariflorus
       community. - In: Photosynthesis and Utilization of Solar Energy. Level III Ex-
       periments. Pp. 47-54. Jap. nat. Subcomm. for PP (JPP), Tokyo 1968.

5614 - MUTUSKIN, A.A., PSHENOVA, K.V., KAVERINA, G.P., KOLESNIKOV, P.A.: Ferredoksin
       kladofory. [Cladophora ferredoxin.] - Dokl. Akad. Nauk SSSR 183: 715-718,
       1968. [In R.]

5614 - MUTUSKIN, A.A., PSHENOVA, K.V., KOLESNIKOV, P.A.: Ferredoksin iz list'ev pshe-
       nitsy. [Ferredoxin from wheat leaves.] - Biokhimiya 31: 924-926, 1966. [In R,
       ab: E.].

5616 - MUZAFAROV, A.M., MILOGRADOVA, E.I., BERDYKULOV, Kh.A.: Vliyanie dlitel'nosti
       osveshcheniya na fotosintez i vykhod autospor Chlorella pyrenoidosa CHICK.
       [Effect of illumination duration on photosynthesis and exit of autospores of
       Chlorella pyrenoidosa CHICK.] - Uzb. biol. Zh. 12 (5): 27-29, 1968. [In R, ab: E,
       Uzb.]

5617 - MYAGI, Kh., ROSS, Yu.: Fitometricheskie kharakteristiki i fotosinteticheskaya
       produktivnost' poseva yachmenya. I. Agrometeorologicheskaya kharakteristika i
       geometricheskaya struktura poseva. [Phytometric characteristics and photosyn-
       thetic productivity of the barley stand. I. Agrometeorological conditions and
       geometrical structure of the stand.] - In: Fotosinteticheskaya Produktivnost'
       Rastitel'nogo Pokrova. Pp. 102-143. Tartu 1969. [In R, ab: E.]

5618 - MYAGI, Kh., ROSS, Yu.: Fitometricheskie kharakteristiki i fotosinteticheskaya
       produktivnost' poseva yachmenya. II. Dinamika rosta assimiliruyushcheT plosh-
       chadi i nakopleniya fitomassy poseva. [Phytometric characteristics and photo-
       synthetic productivity of the barley stand. II. Growth dynamics of the assimi-
       lation surface and increase in dry matter of the stand.] - In: Fotosinteliches-
       kaya Produktivnost' Rastitel'nogo Pokrova. Pp. 144-173. Tartu 1969. [In R,
       ab: E.]

5619 - MYERS, J.: Genetic and adaptive physiological characteristics observed in the
       chlorellas. - In: Prediction and Measurement of Photosynthetic Productivity.
       Pp. 447-454. PUDOC, Wageningen 1970.

5620 - MYRONYUK, V.I.: Katalaza ta peroksydaza Dunaliella salina TEOD. [Catalase and
       peroxidase of Dunaliella salina TEOD.] - Ukr. bot. Zh. 26 (2): 92-95, 1969.
       [Car; in Ukr., ab: R, E.]